Florida's Climate

CHANGES, VARIATIONS, & IMPACTS

Edited by
Eric P. Chassignet, James W. Jones,
Vasubandhu Misra, & Jayantha Obeysekera

Florida Climate Institute
GAINESVILLE, FLORIDA

Copyright © 2017 by Florida Climate Institute.

All rights reserved. No part of this publication may be reproduced, distributed or transmitted in any form or by any means, including photocopying, recording, or other electronic or mechanical methods, without the prior written permission of the publisher, except in the case of brief quotations embodied in critical reviews and certain other noncommercial uses permitted by copyright law. For permission requests, write to the publisher, addressed "Attention: Permissions Coordinator," at the address below.

Florida Climate Institute
P.O. Box 110570
Gainesville, FL 32611
www.floridaclimateinstitute.org

Florida's Climate: Changes, Variations, & Impacts / [Edited by] Eric P. Chassignet, James W. Jones, Vasubandhu Misra, & Jayantha Obeysekera. —1st ed.
ISBN-13: 978-1979091046
ISBN-10: 1979091048

Cover photo credits:
Front: Passing Storms Churn Gulf Coast Waters, J. Allen, based on data from the MODIS Rapid Response Team at NASA GSFC,
www.earthobservatory.nasa.gov/IOTD/view.php?id=3985.
Back (top): Stained Glass, Everglades National Park Service, G. Gardner,
www.flickr.com/photos/evergladesnps/9100071511.
Back (bottom): Miami-Dade Skyline, Spcc111 (CC BY-SA 4.0),
www.commons.wikimedia.org/wiki/File:Miami-Dade_Skyline.jpeg.

Contents

Executive Summary   iii
Eric P. Chassignet, James W. Jones, Vasubandhu Misra, and Jayantha Obeysekera

Societal Challenges Associated with Climate and Climate Change in Florida

1. Human Dimensions and Communication of Florida's Climate   1
Peter J. Jacques, Kenneth Broad, William Butler, Christopher Emrich, Sebastian Galindo, Claire Knox, Keith W. Rizzardi, and Kathryn Ziewitz

2. Florida Land Use and Land Cover Change in the Past 100 Years   51
Michael I. Volk, Thomas S. Hoctor, Belinda B. Nettles, Richard Hilsenbeck, Francis E. Putz, and Jon Oetting

3. Implications of Climate Change on Florida's Water Resources   83
Jayantha Obeysekera, Wendy Graham, Michael C. Sukop, Tirusew Asefa, Dingbao Wang, Kebreab Ghebremichael, and Benjamin Mwashote

4. Climate Change Impacts on Human Health   125
Song Liang, Kristina Kintziger, Phyllis Reaves, and Sadie J. Ryan

5. Climate Change Impacts on Florida's Energy Supply and Demand   153
Wendell A. Porter and Hal Knowles III

6. Climate Change Impacts on Insurance in Florida   179
Lorilee Medders

7. Climate Change Impacts on Law and Policy in Florida   209
Thomas Ruppert and Erin L. Deady

Economic and Environmental Challenges Associated with Climate Change in Florida

8. Climate Change Impacts and Adaptation in Florida's Agriculture   235
Young Gu Her, Kenneth J. Boote, Kati W. Migliaccio, Clyde Fraisse, David Letson, Odemari Mbuya, Aavudai Anandhi, Hongmei Chi, Lucy Ngatia, and Senthold Asseng

9. Managing Florida's Plantation Forests in a Changing World   269
Timothy A. Martin, Damian C. Adams, Matthew J. Cohen, Raelene M. Crandall, Carlos A. Gonzalez-Benecke, Jason A. Smith, and Jason G. Vogel

10. Florida Tourism   297
Julie Harrington, Hongmei Chi, and Lori Pennington Gray

11. Adaptation to Florida's Urban Infrastructure to Climate Change   311
Frederick Bloetscher, Serena Hoermann, and Leonard Berry

12. Climate Change Impacts on Florida's Biodiversity and Ecology   339
Beth Stys, Tammy Foster, Mariana M.P.B. Fuentes, Bob Glazer, Kimberly Karish, Natalie Montero, and Joshua S. Reece

13. Florida's Oceans and Marine Habitats in a Changing Climate   391
Steven Morey, Marguerite Koch, Yanyun Liu, and Sang-Ki Lee

14. Climate Change Impacts on Florida's Fisheries and Aquaculture Sectors and Options for Mitigation   427
Kai Lorenzen, Cameron Ainsworth, Shirley Baker, Luiz Barbieri, Edward Camp, Jason Dotson, and Sarah Lester

Florida's Physical Climate: Past, Present, and Future

15. Paleoclimate of Florida   457
Albert C. Hine, Ellen E. Martin, John M. Jaeger, and Mark Brenner

16. Terrestrial and Ocean Climate of the 20th Century   485
Vasubandhu Misra, Christopher Selman, Amanda J. Waite, Satish Bastola, and Akhilesh Mishra

17. Florida Climate Variability and Prediction   511
Ben P. Kirtman, Vasubandhu Misra, Robert J. Burgman, Johnna Infanti, and Jayantha Obeysekera

18. Future Climate Change Scenarios for Florida   533
Ben P. Kirtman, Vasubandhu Misra, Aavudai Anandhi, Diane Palko, and Johnna Infanti

19. Sea Level Rise   557
Gary Mitchum Andrea Dutton, Don P. Chambers, and Shimon Wdowinski

20. Climate and Weather Extremes   579
Jennifer M. Collins, Charles H Paxton, Thomas Wahl, and Christopher T. Emrich

Index   617

Executive Summary

Florida's climate has been and continues to be one of its most important assets. It has enabled the growth of many major industries, including tourism and agriculture, which both rank at the top of Florida's diverse economic activities. Florida's weather and the natural beauty of its native ecosystems—including more than 11,000 miles of rivers, streams, and waterways and 663 miles of beaches—attract visitors and new residents from other states and around the world. The Sunshine State's dependency on climate is widely recognized and generally taken for granted. However, as we observe climate around the world changing, questions arise about whether or not Florida's climate is changing as well, how rapidly these changes might occur, how Florida might adapt to anticipated changes, and how Floridians might support efforts to reduce the rates of change. This includes questions about how the state's major economic activities might be affected by climate change, how they might adapt, and how they might contribute to reducing human activities' influence.

Although many scientific papers and books have been written on climate trends and changes at global, national, or regional scales, this book focuses on Florida—its climate, changing sea levels, the impacts of these changes, and how our societal and natural systems may adapt to anticipated changes. It addresses the unique conditions in our state and provides a thorough review of the current state of research on Florida's climate, including physical climate benchmarks; climate prediction, projection, and attribution; and the impacts of climate and climate change on the people and natural resources in the state. Over ninety researchers from universities across the state and beyond have contributed to this volume, summarizing important topics such as sea level rise, water resources, and how climate affects various sectors, including energy, agriculture, forestry, tourism, and insurance. Authors of each chapter summarize key messages to help readers understand important risks associated with climate change and societal responses that would be beneficial to Florida's economy, its natural resources, and the well-being of its citizens.

The purpose of this book is to provide accessible, accurate information to a broad audience, to serve as a reference on Florida's climate and its influences on important sectors, and to introduce the various approaches that Floridians are considering for adapting to pending changes in climate and sea levels. It is intended for use by a broad audience that includes: students and educators, both at the secondary school level and at universities; policymakers; business and industry planners; and the general public. It focuses on the unique characteristics of Florida's natural and built environments, including its peninsular features and highly active economic activities, how these characteristics are influenced by our climate systems, how changes have influenced these current and past features, and how they are likely to affect these features in the future. Furthermore, it provides information on how various components of our state's natural and built systems may be able to adapt to future changes.

Climate change poses significant challenges to a wide range of economic activities in Florida, some of which are more critical to our state's economy than others; for example, tourism, agriculture, forestry, fisheries and the infrastructures that support those activities. The science that addresses these changes in climate and oceans is offered, not as an afterthought, but rather as key evidence that supports the societal impacts and responses to the anticipated changes presented in earlier chapters. The book is therefore divided in three sections: societal challenges associated with climate and climate change, economic activity challenges associated with climate change, and past, current and future physical climate.

Societal Challenges Associated with Climate and Climate Change in Florida

There are many societal challenges associated with climate variability and climate change in Florida. These challenges include human health impacts, the availability of and access to essential resources such as water and energy, changes in land use, and the need for new laws and policies associated with all of these issues. Although many direct effects of climate change on Florida are of major concern, it is clear that sea level rise, and its impacts on coastal communities, infrastructure, ecosystems, and businesses, is the state's most immediate climate-related threat.

The interaction between Florida's climate and societal systems is bidirectional: climate change affects social systems *and* social systems influence climate change. Furthermore, communicating this can instigate social action and facilitate (or obstruct) adaptive responses. But unfortunately, while scientists develop skills for communicating with other scientists, many lack the necessary skills to effectively communicate with the broader public. This is particularly true and problematic in scientists' discussions of climate change, in part due to the complexity and potential impacts of the issue. In particular, there is a nationwide organized effort embraced by some Florida politicians to reject climate change science, which makes it difficult to consider practical and rational policies at the state level. However, as this book points out, Florida has a wide array of response capabilities for imminent threats, and they must be improved if we are to address longer-term challenges such as climate change and sea level rise.

Although there are many specific areas where climate change will impact the daily lives of Floridians across different sectors, several are of concern to all residents of the state. For instance, the chapters in this section address connections between climate change and human health. Research has demonstrated that increased risks to human health are likely to occur due to heat waves and temperature-related illnesses, waterborne and vector-borne diseases, and direct human exposure to stronger hurricanes, storm surge, intensive rainfall and flooding, and other extreme events. There are also societal challenges associated with land use, particularly those related to population growth and the possible inward migration from coastal areas due to climate change and sea level rise.

The availability and accessibility of essential resources are likely to become more limited as our climate continues to change. For example, competition for water and energy resources will continue to increase, intensified by greater demands under expected temperature increase. Policies and investment decisions will be essential to address these growing needs. Another key societal issue affecting property, infrastructure, and businesses is the changing risk levels, particularly along the coasts where vulnerabilities to climate change and sea level rise are highest. Climate change and sea level rise increase the risks of losses, thus increasing the cost of insurance and making it very likely that consumers will see substantial changes in the insurance products available to them; some of these modifications are already being made. Finally, the authors in this section point out that some policies and laws, which were developed under more stable climate and sea level conditions, have now become obsolete. New policies and laws are needed to ensure that our social support systems adapt to climate change and sea level rise and to address these societal challenges in order to make communities and infrastructure more resilient to anticipated changes.

Economic and Environmental Challenges Associated with Climate Change in Florida

Florida has one of the most vibrant state economies in the US, and all its economic sectors benefit from the pleasant climate and abundant natural resources in the state. Economic growth in Florida is predicted to outpace national trends. However, Florida's economic activities may be severely impacted by higher sea levels and projected climate conditions, including the potential for more powerful hurricanes. To achieve the predicted economic growth, business leaders must address the serious threats associated with climate change and sea level rise and develop technological and policy solutions that ensure that the state's potential is achieved. Fortunately, Florida has highly educated scientists backed by nationally acclaimed research universities as well as political and business leaders, and we have the human, physical, and natural resources necessary to address these threats.

Currently, the state's two largest economic sectors are tourism and agriculture. Both are highly dependent on climate conditions and both are vulnerable to the projected changes. The tourism industry in Florida accounts for about 2.5 million jobs, either directly or indirectly, and serves more than 106 million tourists per year, generating nearly $90 billion in economic impact annually. Because much of the state's economic activity and projected growth are in coastal areas, solutions to overcome risks associated with sea level rise, storm surge, wind damage, and other detrimental events must be developed. And, if projected increases in temperature and the frequency of intense rainstorms disrupt many tourism activities, these risks must also be addressed.

Similarly, Florida's agricultural and forestry industries are highly vulnerable to climate, most notably to the availability of water and energy resources that are likely to be negatively impacted by climate change. Together, these two industries contribute more than $120 billion to the state's economic revenues and support more than 2 million jobs. Fortunately, Florida agriculture has a successful history of adapting to the vagaries of weather. But climate change is occurring at an unprecedented rate, thus increasing the challenges of continuing or increasing the economic contributions of the state's agriculture sector. In the case of forestry, projected increases in temperature and atmospheric CO_2 concentration in Florida may actually result in increased plantation production *if*, as projected, rainfall remains constant or increases slightly.. Forestry provides a number of positive ecosystem services and withdraws more carbon dioxide from the atmosphere than it requires for production. Thus, this "carbon sequestration" could contribute even more to a reduction in greenhouse gases in the atmosphere. That said, science-based management approaches are essential for society to fully benefit from forest ecosystem services in the future.

Urban infrastructure is also crucial to our economic activity, supporting many businesses and contributing to the state's economic activity through construction, revitalization, and maintenance. Similar to agriculture and tourism, Florida's urban development and populations are concentrated in the coastal areas. This means that infrastructure is also highly vulnerable to sea level rise and climate change, particularly near the coasts, where we continue to see increasing coastal development and population densities. Thus, a large portion of Florida's urban infrastructure is susceptible to damage due to storm surge, flooding, and wind. New technologies, policies, and strategies are needed at the individual and community-level that will lead to the development of a more resilient and sustainable urban infrastructure.

Florida's biodiversity and ecology are central to most of the state's economic activities and valued greatly by its residents. The rich biodiversity of Florida is the product of climate conditions, geographic position, and geology, all of which contribute to the unique ecosystems that exist. This biodiversity and the resulting ecosystems are highly vulnerable to changes in climate and local hydrology; they are dependent upon society's willingness and ability to protect them. They may disappear if coastal populations continue to grow, causing development that displaces coastal habitats, and if projected increases in sea level lead to coastal habitat inundation. Some of this is already happening. Migration of wildlife from affected areas may also be threatened due to loss of habitats, making the biodiversity of coastal ecosystems extremely vulnerable. Adaptive capacity to climate change must be improved to avoid unwanted losses in these critical areas.

Florida's extensive shorelines support a diverse marine life that contributes significantly to the economy of the state in a variety of ways. This section also deals with potential impacts of climate change on the vast marine habitat of Florida, such as coral bleaching, warming of the Gulf of Mexico, and ocean acidification, which can affect larval and nutrient transport and fisheries production and lead to harmful algal blooms.

Finally, marine and freshwater fisheries and other large aquaculture industries contribute about $15 billion annually to Florida's revenue. These types of industries are also highly vulnerable to higher temperatures, more frequent severe storms, and sea level rise. The interactions between climate change and fish production/catches are complex, making it difficult to accurately assess the impacts. However, it is likely that there will be negative effects from climate change in Florida that will need to be addressed to help sustain the viability of these industries.

Florida's Physical Climate: Past, Current, and Future

The peninsular geography of Florida is an outcome of tectonic movements that took place some ~200 million years ago with the simultaneous creation of the Gulf of Mexico and the Caribbean Sea. Consequently, Florida, surrounded by water, has a unique climate characterized by sea breezes that bring moist air to both the east and west coasts and by a strong seasonality in rainfall. The cyclic growth and retreat of huge ice sheets in the Northern Hemisphere lead to strong fluctuations in sea level, alternatively exposing or covering the Florida Peninsula. A paleoclimate analysis of Florida reveals clearly the interglacial cycles punctuated by the glacial periods, which make detection of regional climate change trends non-trivial. To compound this intricate entanglement, Florida's climate is dependent upon phases of the El Niño and the Southern Oscillation (ENSO), the Atlantic Multi-decadal Oscillation (AMO), and the Pacific Decadal Oscillation (PDO).

Despite this complexity of regional climate variations, significant progress has been achieved in the seasonal prediction of winter climate over Florida on the basis of the persistence of large-scale variations of ENSO. However, regional climate predictions are daunting and challenge our current understanding and state-of-the-art climate models to account for the correct balance of influences including remote large-scale climate drivers (e.g., ENSO), local feedback (e.g., land-biosphere-atmosphere interactions), and climate change.

There is undeniable evidence linking increased greenhouse gas emissions to sea level rise. The discussion of sea level rise presents this evidence and shares an in-depth discussion on the challenges of arriving at quantitative sea level rise estimates for Florida. There is also the issue of climate and weather extremes affecting Florida that increases the vulnerability profile of the state and makes it a focus of the risk insurance market. This section gives a thorough review of events such as droughts, land-falling tropical cyclones, tornadoes, severe thunderstorms, and lightning, which cause collateral damage and human fatality periodically across Florida. Finally, the variations and triggers of such extreme events are discussed in this book, providing a complete picture of the physical (earth, ocean, and atmosphere) climate system of Florida.

Conclusion

There is no doubt that development, climate change, vulnerability, and risk go hand in hand, particularly in Florida. These vulnerabilities and risks exist in all segments of Florida, but they are particularly severe along our coastline where negative impacts are already being felt. Understanding these linkages is important and will help in the development of effective mitigation and adaptation strategies to address the impacts of climate change. The purpose of this book is to share knowledge about this linkage that has been gathered over years of research from experts around the state, and to provide candid views on the known, unknown, and the unknowable. There is no silver bullet to eliminate the risks imposed on Florida by climate change. But making informed decisions in light of potential future climate evolution would go a long way toward mitigating at least some of future vulnerabilities and risks. It is our sincere hope that the contents of this book will help move us in this direction.

Acknowledgements

The authors and editors of this book are extremely grateful to Meredith Field, coordinator for the Florida Climate Institute (FCI) at Florida State University, for her tireless dedication to its assembly and publication. We would also like to express our thanks to Tracy Ippolito for editing each of the chapters and Carolyn Cox, coordinator for the FCI at the University of Florida, for her ongoing contributions to and support of the FCI and its publication of this book.

Eric P. Chassignet, Florida State University
James W. Jones, University of Florida
Vasubandhu Misra, Florida State University
Jayantha "Obey" Obeysekera, South Florida Water Management District

December 2017

Note: Access to eBook and Color Figures

For access to the electronic, full-color version of this book, please visit:
http://floridaclimateinstitute.org.

CHAPTER 1

Human Dimensions and Communication of Florida's Climate

Peter J. Jacques[1], Kenneth Broad[2, 3], William Butler[4], Christopher Emrich[5], Sebastian Galindo[6], Claire Knox[5], Keith W. Rizzardi[7], and Kathryn Ziewitz[8]

[1]*Department of Political Science, University of Central Florida, Orlando, FL;* [2]*Abess Center for Ecosystem Science & Policy, University of Miami, Coral Gables, FL;* [3]*Department of Marine Ecosystems and Society, Rosenstiel School of Marine and Atmospheric Science, University of Miami, Miami, FL;* [4]*Department of Urban and Regional Planning, Florida State University, Tallahassee, FL;* [5]*School of Public Administration, University of Central Florida, Orlando, FL;* [6]*Department of Agricultural Education and Communication, University of Florida, Gainesville, FL;* [7]*School of Law, St. Thomas University, Miami Gardens, FL;* [8]*Sustainability Institute, Florida A&M University, Tallahassee, FL*

Florida's climate system, which is nested within regional and global climate systems, cannot be fully understood without including human dimensions that interact with the climate systems in two principal ways: 1) where social systems facilitate or dominate causes of climate change, and 2) where climate change affects social systems. These aspects include complex social interactions and feedbacks, but can be broken down into the impacts, risks, and causes of climate change specific to Florida. Further, communication of these elements can interact with social in/action and facilitate or obstruct adaptive responses. It is important to view the organization of these interactions through social structure, where essential drivers of social forces include the political-economy, demographic, and attitudinal architecture of Florida social systems. In this chapter, we review key social drivers of specific impacts, risks, and causes of climate change within Florida.

Key Messages

Mitigation
- Florida faces a series of threats from climate change that will affect social groups and geographic areas differently. Florida's future depends critically on global reductions of greenhouse gases (GHGs). Florida itself is the 27th largest GHG emitter across all states and other countries. This makes it essential that Florida contribute to global reductions.
- To understand how to reduce GHGs, one needs to be familiar with the development of land and energy in the state that determines sources of power for buildings, transportation infrastructure, and the institutions (rules and laws) that ultimately guide the consumption of hydrocarbon-based energy. Florida has been guided heavily by land and highway development, with almost all the energy consumed in the state coming from sources that directly emit GHGs, with the exception of nuclear power plants.
- There are important macro and micro obstacles to change that must be understood. At the macro level, social structures guide the behavior of large groups, and individuals acting alone are less effective in reducing GHGs than changing these social structures, including institutions. A significant obstacle has been a national organized effort to reject climate change science that some Florida politicians have reproduced at the state level, making policy efforts in this area difficult to even discuss.
- At the micro level (the individual), communication does not necessarily consist of what people say, but instead what is heard. Improvements in reaching out to different target audiences will require engaging creative approaches to communication.

- And providing information is only one step in the process. Sustaining motivation for change remains one of the biggest challenges and will require collaboration among academics, practitioners, and community leaders to ensure that we continue to move forward.

Impacts and Adaptation
- Effective change strategies will require the coordinated collaboration of multiple sectors of society.
- GHG emissions have already exceeded the mid-scenario emission posed by the IPCC 2005 reports and are expected a continued increase in carbon emissions associated with SLR.
- The most realistic SLR estimates place Florida's shoreline 4 ft higher in 2100 than it was in 1990. However, potential impacts from SLR go beyond inundation in coastal and near-coastal areas, and include decreases in potable water for consumption and fresh water for crop growth.
- Sea level hazards have far-reaching implications beyond near-shore areas including loss of tax bases, changes in vector-borne illnesses associated with standing water, and water system failures leading to polluted waters entering the water supply.
- We can identify those areas where populations are least able to adequately prepare for, respond to, and rebound from disasters. When overlaid with potential impacts, we can create a clear path forward for adaptation, mitigation, and resilience building.
- Focusing efforts on those areas designated highly vulnerable will ensure that their respective populations will have more opportunities to increase their resilience to disasters.
- Social vulnerability results from the dynamic interaction of many socio-demographic characteristics and is specific to distinct places, beyond any one factor such as race or economic class. This means that the drivers of vulnerability are different across the landscape and policies for improving resilience need to be aware of what issues and specific drivers exist at the local level.
- Florida has a bipartisan history of funding conservation, a critically important component of both climate mitigation and adaptation efforts so long as funding is used for acquiring land and setting it aside for conservation—which does not always occur in the state.
- Florida has an array of planning tools at its disposal to mitigate imminent future threats but there is significant room for improvement in these plans, which are often incremental and disconnected from other planning documents. Efforts such as the Southeast Florida Climate Change Compact are among the most promising models for regional collaboration geared toward adapting to climate risks.

Keywords

Vulnerability; Adaptation; Governance; Political economy; Institutions; Barriers to communication

Introduction

Florida's population faces some of the most serious threats from climate change in the United States through of a series of interconnected changes including but not limited to sea level rise, intensifying tropical storms, severe erosion of the coastline, inland flooding, saltwater intrusion, extreme heat, and changing hydrological and tidal patterns. Direct effects can then give rise to cascading secondary effects such as falling real estate prices and other economic problems, increased costs to government, and social disruption and disorganization. These are but a few examples of how climate change will be felt by society in Florida.

We can think of the human dimensions of climate change as the way humans cause global climate change within a nested hierarchical set of scales, from the local to global, as well as the risks and adaptations to climate-related impacts to human society. Each element is incredibly complex. This chapter can only provide an overview of these issues, treating each element in turn, starting with human causes of climate change specific to Florida and then laying out the social impacts, risks, and adaption to climate change.

Structural Causes of Greenhouse Gas Emissions: Florida's Development, Demographic Change, and Urbanization

Human greenhouse gas emissions are the dominant reason for contemporary global warming and climate change (Pachauri et al. 2014). We must understand the structural social conditions in which emissions are produced to reduce them. Social structures are the political, economic, demographic, and attitudinal forces that contextualize group behavior (Dunlap and Brulle 2015). Individuals may want to reduce their personal emissions, but their behavior is systematically constrained by institutions, social norms, and even available infrastructure. For example, a person may want to reduce their transportation emissions but will have little opportunity to do so without the availability of a robust public transit system; and, public transit is an artifact of political and economic commitments.

Political-Economic Structures and Change

Florida emissions are deeply tied to the political economy of Florida. Political economy refers to the structural components of the economy, such as its composition and the rules that are in place. The political economy of Florida has radically changed over the last 100 years: "Florida went from a small, poor, rural state early in the 20th century to a large state with dense urban areas and average incomes late in the 20th century" (Dewey and Denslow 2014). In that time, the composition of Florida's economy has focused heavily on tourists and retirees. For example, 106 million people visited Florida in 2015; tourism has continued to grow each decade since 1980 and is a major driver of Florida's economy (Bureau of Economic and Business Research 2016), generating an estimated $65-90 billion. Comparatively, in 2013 there were nearly 50,000 farms in Florida producing $8.45 billion in receipts (Florida Department of Agriculture and Consumer Services 2016). Unfortunately, jobs tied to tourism and serving retirees tend to be low-wage, and retirees are often less willing to pay taxes for public services such as education and infrastructure; consequently, the 2012 Florida middle-class had a per capita income equal to that seen in 1978 (95% of the US average) (Dewey and Denslow 2014). Since income/poverty is an indicator for vulnerability to hazards (as detailed below), the structure of Florida's economy is partly responsible for this pattern of vulnerability alongside government policies, which have focused on bringing more tourists and retirees, resulting in increased development and subsequent carbon emissions. These dynamics partially guide the rationale and benefits of increased urban and suburban development, as well as the structure of energy production and consumption.

Institutional and Behavioral Change

Effective responses to climate change will require human behavior modifications. However, individual behavior changes are, "... not autonomous: they are constrained by institutional processes such as regulatory structures, property rights and social norms associated with rules in use" (Adger et al. 2005). These larger forces tend to change incrementally or slowly until a breakpoint opens up a cascade of change (Baumgartner et al. 2009). This means that changes *within* social structures are important for climate mitigation and adaptation, and this requires an astute and powerful civil society that is not confused about climate-related public interests -- but unfortunately, civil society is often easily misled by powerful interests (Jacques 2014, 2016; Gramsci 2011).

Often, thoughts about social change are overly focused on getting individuals to "plant a tree, ride a bike, save the world" (Maniates 2001), ignoring social structures that may impede meaningful change. Certainly, individual behavior is important; but institutional, organizational, or governmental change is also required for any real and sustainable environmental outcomes. For example, "adaptation pathways" visualize potential actions in response to changes in climate (Wise 2014), but the Florida Center for Environmental Studies at Florida Atlantic University (Polski 2016) discovered these decision-making tools need to take into account the interaction social sectors to be most effective (Wise 2014). Meanwhile, social and institutional obstacles are the most commonly reported barriers to climate change responses (Biesbroek 2013).

Culture is also important. According to Adger et al. (2013), "society's response to every dimension of global climate change is mediated by culture," which they define as "symbols that express meaning, including beliefs, rituals, art and stories that create collective outlooks and behaviors, and from which strategies to respond to problems are devised and implemented" and culture is "...embedded in the dominant modes of production, consumption, lifestyles and social organization that give rise to emissions of greenhouse gases," (Adger et al. 2013, 112). Thus, the stories we tell ourselves as a culture influence how people identify risk and response (Adger et al. 2013).

Further, information alone does not inspire change. Often, individuals with adequate knowledge and who want to change still fail to actually modify their behavior or practice (Rogers 2003). At the policy level, stakeholders in governments, planners, communities, individuals, industry, and interest groups frequently disagree about the relevance and effectiveness of mitigation and adaptation strategies due to differences in culture and values (Adger et al. 2013).

Thus, participatory methods that bring attention to a plurality of vulnerabilities and solutions can more effectively plan for collective risk (Smit and Wandel 2006; van Aalst et al. 2008; Oppenheimer 2013; Susskind et al. 2015).

Many different models of individual and organizational change are available to assist in the design of social interventions aimed at modifying behavior (Rogers 2003; Burke and Litwin, 1992; Burke 2014; Warner et al. 2014). For example, Rogers (2003) developed the Innovation-Decision Process model to explain the process followed by individuals to adopt or reject a technology or idea. The model, consisting of five progressive stages — knowledge, persuasion, decision, implementation, and confirmation — accounts for characteristics of the context, the decision-making unit, and the innovation. Using this model as a guide for the design and implementation of mitigation and adaptation strategies in response to climate change can help us identify the feasibility of the change, the types of information that should be communicated at different stages of the process, the best ways to communicate the required information, the effectiveness of the change process, and the sustainability of the proposed change.

It is imperative that current and future mitigation and adaptation efforts are framed using sound models and methods for planned change. To be effective, adaptation and mitigation responses must connect with what is important to the target population (Adger et al. 2013). Barriers will always be present in any change initiative, but what matters is to ask why and how those barriers emerged (Biesbroek et al. 2013), and what are the best approaches to overcome them. However, the barriers to effective climate change communication are formidable.

Communication of Climate Change in Florida

Another set of challenges to institutional and behavioral change are cognitive barriers to communicating climate change. We like to fancy ourselves a rational species, evolved to a state of consciousness that Descartes described as *cogito ergo sum*: "I think, therefore I am." However, the basis for decisions, from the personal to financial to environmental realms, is more accurately characterized as "I feel, therefore I am" (Damasio 1994; Eakin 2003). We respond to information instinctually, at an unconscious level in fractions of a second and driven by emotions. We often rationalize these snap judgments post hoc, with convenient explanations or selective use of evidence that supports our decisions (i.e., "confirmation bias"). The more uncertain or emotionally charged the situation is, the more we tend to rely on these unconscious processes (Kahneman 1982). Honed by trial and error learning, instincts served us well for survival in the wild and for undertaking tasks for which we have repeated experience or formalized training. We experience "risk as feelings" (Loewenstein 2001). However, as our technological and sociopolitical systems have become more interdependent and complex, we have difficulty comprehending complex multi-causal events or concretely envisioning and accurately planning for the future, resulting in what behavioral economists refer to as "hyperbolic discounting" and social psychologists have described in "construal theory," where distance to an event/value makes it too abstract for prudent decision-making (Trope and Liberman 2010).

In many cases, there are plausible explanations for why these mental shortcuts, or heuristics, may have pro-social benefits regarding fairness, justice, and cooperation (Atran and Henrich 2010; Lind 2001). However, for climate change-related decisions and other environmental risks, these biases lead to suboptimal outcomes for the larger population. Dozens of cognitive biases are known (Figure 1.1) and are increasingly implicated in human decision-making (for an overview of the research history and recent advances, see Kahneman 2011).

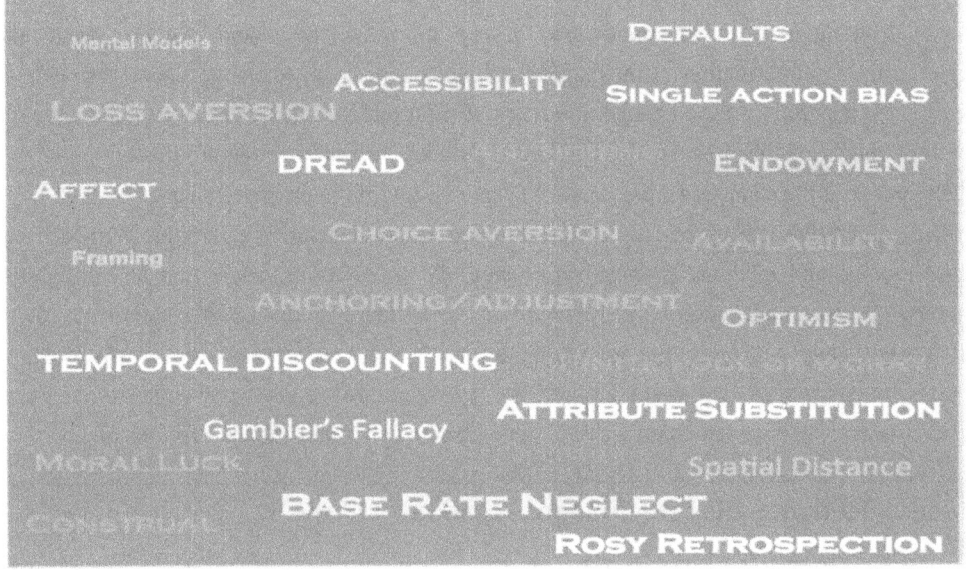

Figure 1.1. Known cognitive biases.

The decision sciences have increasingly influenced climate change communication efforts as well (Center for Research on Environmental Decisions 2009). Warnings and subsequent outreach campaigns regarding climate change originated from the scientific community, whose currency is data. Analytical information is presented to the public in hopes that we would act rationally. Such an assumption is naïve and recent work has shown that increased scientific literacy does not increase belief in climate change (Kahan et al. 2012). A cursory look at other realms where there is much less scientific uncertainty (e.g., smoking, obesity, texting while driving) reminds us of the challenges to shifting behavior.

Further, climate change has many of the characteristics that make concerted, coordinated action at the individual and group level difficult. We have not had multiple experiences with global phenomena that have effectively informed learning; in fact, members of the general public often repeat the same suboptimal choices in terms of preparing for hurricanes, which are relatively frequent compared to climate change! (Broad et al. 2007). Unfortunately, climate change is "a slow, creeping and invisible phenomenon" that does not evoke the emotional triggers that drive individual or policy action and that we've seen in response to other threats such as nuclear power or oil tanker spills (Slovic 1987; Beamish 2002). Since climate change is a collective action problem, individual responses are often seen as a drop in the bucket, and responsibility can be deferred. Perhaps most significantly, the impacts are psychologically distant; considered to manifest in the distant future and in far away, remote places (Weber 2006).

Some of these biases are not solely a result of our evolutionary heritage, but are initiated and exacerbated by external forces such as misinformation campaigns (Jacques et al. 2008) put forth by so-called skeptics and those with large financial stakes in the oil and gas industry, which serve to exacerbate an ideological divide in the U.S. (Jacques et al. 2008; McCright and Dunlap 2000; Oreskes and Conway 2010). All of these factors combine to make effective climate change communication extremely challenging (Moser 2010).

While Florida is among the most economically-vulnerable states to climate change, Floridians are also particularly vulnerable to some of these cognitive biases. Our population, especially around the coasts, is highly transient with many recent immigrants from other parts of the U.S. or internationally who may lack a strong sense of place or connective history to Florida (LiPuma and Koelble 2012) or the development of multigenerational knowledge and close interactions with nature that would allow for the perception of subtle changes (Marlen et al. in review). This might relieve one obstacle to inland migration (as discussed later in this chapter). That said, humans cannot perceive subtle, slow changes in a statistically-accurate manner. But just because we are not skilled at picking up these subtle changes or comprehending complex causal relationships does not mean that we do not do it heuristically, as noted above. For example, Zaval et al. (2014) found a strong correlation between the local daily temperature and peoples' belief in climate change.

Floridians associate impacts of climate change with increased heat and storms, two threats that they are relatively well-adapted to (Leiserowitz and Broad 2008). The largest threat by most accounts, sea level rise (SLR), however, is imperceptible to the eye. A critical impact of SLR is on our groundwater supply, and most residents are not even aware of where our potable water comes from or the mechanisms through which saltwater can make its way underground. Relatedly, our limestone geology does not lend itself to the barrier adaptations that have been implemented in other SLR-prone areas such as The Netherlands. These "out of sight, out of mind" interlinked impacts are not unique to Florida's land. For example, ocean acidification can negatively impact reefs, which can hurt fisheries and related tourism as well as the shoreline buffer that the reefs provide, increasing coastal erosion (Ferrario et al. 2014). Yet, most residents are not aware of these coastal processes.

That said, the information on cognitive bias is being incorporated into creative initiatives in the arts to produce unique creative approaches to gain attention and motivate behavioral change (Levin 2015). But there is no silver bullet for information presentation, and the grand challenge is shifting behavior from being reactive to proactive. Alternate framings of the climate issue are well underway and what was once pitted as environmental versus economic choices is increasingly framed as a moral imperative (e.g., the "What would Jesus drive?" campaign). Creative uses of shaming have also been employed (Jacquet 2015) and advances in computer technology hold promise for "information acceleration" (Meyer et al. 2013). For example, researchers are now using computers and augmented/virtual reality to observe people's decision-making in realistic scenarios, which has the potential to advance decision research in ways that will facilitate a shift from simple information provision to behavioral motivation. These problems not only inform individual behavior, but the context for civil society preferences and the legitimacy of social forces such as social stratification and institutions that will be crucial for mitigating and adapting to climate change (see Adger 2010 and Agrawal 2010).

Demographic Change and Urbanization

Demography is another structural force that provides opportunities and barriers to change. At the start of the 20th century, Florida's population was 528,000 with almost 80% of the population residing in rural areas (US Bureau of the Census 1995). After World War II, Florida's population growth escalated, especially in the central and southern portions of the state. The advent of air conditioning and easy highway travel were crucial in making Florida an appealing destination, attracting full-time and seasonal residents, as well as tourists (Mormino 2005). In the decades from 1950 to 2000, Florida was the fastest-growing state in the nation (Smith 2005), rising from a population of 2.8 million in 1950 to almost 16 million in 2000, and with decadal growth rates ranging from 20-80% (Smith 2005). In the latter half of the 20th century, the majority of Florida's population shifted from residing within city boundaries to living in more dispersed, unincorporated suburban locations (Mormino 2005).

More and more people choose to make Florida their home (Smith 2005). Immigrants to the state have long included retirees; retiring baby boomers continue to move to Florida (Smith and House 2006). In addition, younger newcomers are migrating to take jobs in Florida as the economies and cultures diversify (Florida Trend 2015). The housing crisis and recession caused a brief decline in population from 2008 to 2009 (University of Florida Bureau of Economic and Business Research 2010); however by 2010, the state's population was growing again.

More recently, Florida's population growth has continued to be among the strongest nationwide, and real estate development is still a major driver of the state's economy (US Bureau of Labor Statistics 2015). In 2015, the Miami-Ft. Lauderdale-West Palm Beach Metropolitan Statistical Area (MSA) ranked as the 8th most populous in the nation, followed by the Tampa-St. Petersburg-Clearwater MSA, which was 18th, and the Orlando-Kissimmee-Sanford MSA, which was 26th (Bureau, U.C. 2015).

Residential growth has included the building of many subdivisions spanning large acreages. Housing development consists of primary residences as well as second homes for seasonal residents and domestic or international affluent buyers (Smith and House 2006). Citrus, cattle ranching, forestry, and crop land uses continue to diminish, giving way to development of urban, suburban, and exurban developments (Mulkey 2006). While Florida's earliest towns and cities were founded along coasts and rivers accessible by water transportation, wetland drainage and roadway construction increased the areas easily accessible and suitable for development, allowing urbanization to proceed from the coastal to interior lands (Mormino 2005; Derr 1989). This shifting land use sets the preconditions for transportation and other sources of emissions.

Historic Development in Florida

Development does not occur in a vacuum, but rather is somewhat path-dependent —what came before affects what can happen in the future. For example, roads are often precursors to future development, providing easier access to previously unavailable interior areas.

Evidence of ancient human habitation has revealed that in the distant past, Florida's first peoples adapted to a changing climate: prehistoric residents exploited the larger land mass that was available during the "ice age" of the late Pleistocene and, over time approximately 5,000 to 10,000 years ago (Faught 2004). *[Cf: Ch. 2]*, retreated when shorelines assumed configurations close to the present ones Since those prehistoric times, subsequent Florida settlements have reshaped large portions of the state's landscape (see Chapter 2). As the built environment has expanded, the proportion of land in forests, wetlands, prairies, and other natural areas has diminished; urbanization in the last century also reduced the amount of land in pastures and farmlands. These changes in land cover affect how the chief greenhouse gas, carbon dioxide, cycles through various reservoirs including soil, vegetation, and the atmosphere. Reductions in vegetated landscapes serve to diminish the rate of photosynthesis taking place and thus the land's overall capacity to absorb carbon dioxide (DeFries et al. 1999). As a result, greenhouse gas emissions from human activities have increased. In Florida, as globally, humans have become important driving agents of geologic change (Steffen et al. 2007) including climate change (Pachauri et al. 2014), emitting 218 million metric tons of CO_2 equivalent in 2013, ranking sixth among the US states (US Energy Information Administration 2014).

Land-use change in Florida is strongly tied to residential real estate development, which has transformed vast areas of the state from natural and agricultural areas into housing subdivisions of varying densities (Derr 1989; Zwick and Carr 2006).

Energy Use and Greenhouse Gas Emissions in Florida

The state's energy use and policies bear heavily upon how Florida addresses its role in climate change. Forging effective and complementary climate and energy policies is complicated by conflicts between economic growth and development goals, missions of government bodies and utilities, political will, and varied perspectives on climate change.

Use of fossil fuels has enabled growth and development in Florida, supplying the vast majority of fuel used to generate electricity to heat, light, and, very importantly, to cool buildings.

When ranked globally with all US states and nations in a 2004 study, Florida was the 27th largest global emitter of greenhouse gases, outstripping emissions of entire nations including Turkey, Taiwan, and the Netherlands (Peterson and Rose 2006).

In Florida, electric power generation is the chief source of carbon dioxide emissions, responsible for more than half of the emissions in 2015. Residences use more than 50% of this electricity. Transportation was the next largest source of carbon dioxide emissions, contributing almost nine times more than Florida's relatively small industrial sector (US Energy Information Administration 2016).

Florida's energy profile is characterized by a high proportion of residential and commercial customers and a low proportion of industrial ones; the state also experiences the highest number of cooling degree days of any state and the lowest number of heating degree days of any continental state (Florida Public Service Commission 2014). Supplying energy to meet peak demand is driven by the weather-imposed demand for cooling.

Development and Climate Impacts: Transportation and Vegetated Areas

The layout of urban structures affect climate in multiple and complex ways. Scientists are seeking to better understand these impacts separately and in concert with one another. One important consideration is the extent to which urban forms reduce or increase dependence upon private transit (e.g., passenger vehicles, motorbikes, etc.). Urbanization also affects the distribution and extent of vegetative land cover, influencing the degree to which green areas mediate urban "heat island" impacts through the cooling effects of evapotranspiration (Stone and Rodgers 2001). In addition, the presence or absence of vegetation determines how much carbon is sequestered (i.e., taken up in plant growth), which affects the overall carbon balance for a chosen geographic area (Imhoff et al. 2004).

The relationship between transportation and greenhouse gas emissions is well known. Each gallon of gasoline burned emits just under 20 pounds of carbon dioxide when combusted in the atmosphere (US Energy Information Administration 2014) and the vast majority of greenhouse gas emissions in the US result from fossil fuel combustion from passenger vehicles, light trucks, sports utility vehicles, and minivans, which contribute more than half of transportation emissions nationwide (U.S. Environmental Protection Agency 2014).

Globally, the shape into which urban areas evolve is dependent upon the interaction of transportation pathways with the locations of residences, workplaces, and other typical destinations (Hillier 2008). Roadway placement influences the location of residential developments and workplaces, and vice versa, thus influencing the creation of a growing urban structure.

Since the majority of Florida was developed after the automobile age, most of the built environment has been laid with car travel in mind. This has fostered a sprawling land use pattern. Private vehicular travel is the default mode of travel in Florida, exacerbated by conventional zoning that separates land uses — dividing residential from commercial, industrial, and civic activities. Low-density sprawling land-use configurations with separately zoned land uses, typical in Florida, are associated with heavy dependence on fossil fuel-powered personal automobiles and high rates of vehicle miles traveled (VMT), ultimately contributing a greater share of greenhouse gas emissions from transportation than what occurs in compact cities with mixed land uses (Ewing et al. 2007; Steiner et al. 2010). In 2014, approximately 80% of Florida workers drove alone, compared to 76% nationally, and just over 2% used public transportation compared to a national rate of 5% (U.S. Bureau of the Census 2016). Florida ranks third among states for the VMT, with a collective 192,702 million miles traveled in 2013 (US Energy Information Administration 2016).

Transportation is the state's largest "end use" of energy at 36% (US Energy Information Administration 2016). Use of motor fuel by residents and jet fuel to move visitors and cargo via air travel makes Florida one of the nation's leading consumers of motor gasoline and jet fuel (US Energy Information Administration 2016). Consequently, the transportation sector accounts for the state's second highest source of carbon dioxide emissions following electric power generation (US Energy Information Administration 2016), mirroring national trends.

While options for mass and alternative transportation are improving in many communities, lack of comprehensive mass transit services and poor infrastructure for non-motorized travel discourages alternative modes of transit in many Florida cities and counties. Lack of adequate mass transit options has social as well as environmental impacts, as it disproportionately impacts employment and life choices for lower-income residents (Bullard 2003).

Climate is affected by how the expanding built environment changes vegetated areas. Loss of green space in urban regions drives up temperatures, increasing the demand for cooling and negatively impacting human health through the effects of heat stress and impaired air quality (Stone and Rodgers 2001). Addressing the "urban warming" effect is a pressing need as the process of urbanization continues to expand. A study of Atlanta's urban area found that compact-to high-density development contributed less radiant heat energy to surface "heat island" formation than lower-density patterns, prompting policy suggestions to favor compact development in combination with urban tree planting (Church et al. 2013).

As for the relationship between urban form and carbon sequestration, relative to its pre-development status, the overall loss of forests, inland and coastal wetlands, and agricultural landscapes has indisputably reduced the rate of primary productivity and thus carbon uptake in Florida. At a finer scale, changes from human impacts are more complex. Sequestration rates vary according to the types of developed areas and the previous land use of an area that was converted (Zhao et al. 2012). One analysis of Florida's carbon balance showed that residential carbon emissions from energy and transportation fuel consumption were compensated by carbon sequestration in exurban and rural areas, with the state still functioning as a net carbon sink. However, the study did not consider the commercial sector or embodied carbon associated with products created from emissions in distant locations (Zhao et al. 2011).

Understanding of the relationships between population density, land use and cover, and carbon impacts is incomplete; some authors find countervailing positive and negative effects for urban densification when factors extending to distant consumption are considered (Elliott and Clement 2014). It appears that striving for lower-carbon cities will require further attention to reconciling goals for density to support more fuel-efficient transportation with goals for configuration of urban areas that include adequate green and treed spaces to sequester carbon and moderate local and regional climate.

Into the future, human development will not only affect the state's carbon balance, but also its ability to provide other ecosystem services needed to sustain Florida's unique natural and human ecosystems. For example, Florida lost an estimated 84,000 acres of wetlands between 1990-2003, despite the US Clean Water Act requirement for "no net loss" of wetlands to development — this represents a clear policy failure, and perhaps even a corruption (Pittman and Waite 2009). Future "build out" scenarios show vastly different possibilities, with varying portions given to development, conservation, and agriculture depending on the efficacy of growth management efforts to alleviate sprawl (Zwick and Carr 2006; University of Central Florida 2007). Historically, market forces have been strongly influential in driving development. Growth management efforts are tempered by strong "home rule" tendencies and property rights legislation, and Florida's approach to growth management in the past had little effect on the amount, more so on the location and timing of development, ironically facilitating rapid housing construction and increasing population in suburban areas (Boarnet et al. 2011).

Florida Energy Sources

Florida's energy sources are mainly carbon-based fuels, natural gas (mostly used by power plants), and coal. Natural gas has displaced coal as the primary fuel for electricity generation (US Energy Information Administration 2016). In addition to these carbon-based energy sources, two nuclear power plants on the Atlantic Coast supplied approximately 12% of the state's needs for electricity generation in 2014 (Florida Public Service Commission 2016).

Renewable resources accounted for just over 2% of Florida's net electricity generation in 2015, mostly supplied by biomass, including municipal, forestry, and agricultural waste. Solar energy provided less than 10% of the state's renewable net generation as of 2016 (meaning solar energy provided only 0.002 percent of generation). The potential for continued expansion of solar power is strong because of Florida's bountiful solar thermal and photovoltaic resources (US Energy Information Administration 2016). A small fraction of the state's energy comes from hydroelectric power generated in the Florida Panhandle, and the state's wind resources are considered to be limited with no commercial wind facilities currently in existence (Chamlee-Wright and Storr 2009).

As prices of solar technology have dropped, Florida's capacity in this arena has expanded; in 2015, Florida installed 41 megawatts of solar electric capacity, nearly doubling its prior capacity

(Solar Energy Industries Association 2016). Large utility projects include Florida Power and Light's 75-megawatt solar power plant in Martin County and the 12-megawatt Jacksonville Solar photovoltaic facility. Businesses and homeowners are increasing solar use, but by 2016 that amounted to just one-tenth of 1% of Florida utility customers (Klas 2016); and, while Florida has the nation's third highest potential for rooftop solar (Solar Energy Industries Association 2016), it ranked 14th in installed solar capacity in the US in 2016.

Florida Energy Policy

As of 2016, Florida lacked overarching climate change and greenhouse gas reduction (climate mitigation) policies, as well as a comprehensive energy policy at the state level. Florida has standards for energy efficiency in buildings and a net metering policy that allows onsite electricity generators to offset energy they consume, but it lacks other more assertive policy tools that other states use to reduce greenhouse gases and promote renewable energy (e.g., performance-based programs to reward mitigation of carbon emissions or a renewable energy portfolio mandating a minimum level of supply by renewable energy sources). The state supports conservation programs through utilities, as mandated by a 1980 plan, however these were dramatically scaled back in 2014.

In Florida, lawmakers discussed a state energy and climate policy in 2007 and 2008, including a reduction of greenhouse gas emissions from state utilities, but the proposals failed. A 2007 executive order (Florida Public Service Commission 2014) mandated state agencies track and reduce greenhouse gas emissions. Other initiatives included on that prohibited idling heavy-duty vehicles in 2008 and a 2009 law that adopted California's motor vehicle emission standards, but both were repealed in 2012 (FS Chapter 62-285). A 2009 clean diesel rebate program remains in effect, and a 2011 provision in the Community Planning Act that allows for regional-scale planning for Adaptive Action Areas (targeted areas for investment specifically aimed at SLR adaptation) allows the Southeast Regional Climate Compact to coordinate its climate action plan (Bolstad 2016). Meanwhile, energy efficiency in buildings substantially improved as a result of a state energy code enacted in 2007 and updated in 2009, but home sizes increased, which reducing savings (Florida Solar Energy Center 2009).

The Florida Energy Efficiency and Conservation Act of 1980 directs the state's utility regulatory body, the Public Service Commission (PSC), to conserve and reduce peak energy demand. It required investor-owned utilities to submit demand-side management (DSM) proposals for approval by the PSC every five years. PSC approval considers efficiency measures, including energy audits and incentive programs, and a popular solar rebate program. Over the years, the PSC estimates that the utility DSM programs eliminated the need for 45 150-megawatt power plant equivalents (Florida Public Service Commission 2014).

However in 2014, the PSC approved a major reduction in utility conservation programs, and in 2015 the PSC ended solar rebates. Utilities argued that availability of affordable natural gas countered the need for conservation programs (Klas 2016). Although natural gas produces fewer carbon emissions per units of energy than coal, the rising use of natural gas consumption has resulted in natural gas-related CO_2 emissions surpassing those from coal nationally (US Energy Information Administration 2014).

Mitigation policies in other states include renewable energy portfolios requiring providers to use a minimum percent of renewable sources by target dates and power purchase agreements that allow third-party businesses to install solar systems and sell the power to customers independent of the utility (Solar Energy Industries Association 2012). Other states have also used market-based incentives and performance-based programs to mitigate carbon use (U.S. Environmental Protection Agency 2014) because states are likely to shoulder the costs of climate-related impacts (Peterson and Rose 2006). Successful policies are based on input from key stakeholders to tailor effective measures suited to the state's unique situation (U.S. Environmental Protection Agency 2014). Minimization of conflicts can be achieved by working in "an open and self-determined policy process" to reduce mitigation costs and promote equity across regions, socioeconomic groups, and generations (Peterson and Rose 2006).

Florida Conservation Policy

Conservation of ecosystems is among the most important strategies for both climate change mitigation and adaptation because, in addition to their intrinsic value, natural resources support the major tourist industries of Florida and conserved ecosystems preserve options and ecosystem services critical to the lives of Floridians. Conservation is mainly achieved through governmental institutions, including Adaptive Action Areas (above). Institutions are rules and decision-making procedures that guide large-scale behaviors and action. Conservation protects areas such as forests, mangrove swamps, and salt marshes that absorb carbon (Chmura et al. 2003). Biodiverse ecosystems promote ecosystem stability (Ives and Carpenter 2007), and coastal ecosystems reduce vulnerability where intact coral reefs, dunes, and wetlands all absorb water and energy that reduce coastal vulnerability (e.g., wetlands provide $23 billion/year in storm protection alone, and every lost hectare adds about $33,000 in storm damage (Costanza et al. 2008)). A thorough analysis of institutions focused on conservation should start at the global level and move incrementally to the local level, but space in this chapter forces us to focus only state institutions.

Relatively open rules for development in the earlier part of the 20th century led to serious environmental problems. In 1972, Florida passed a raft of environmental legislation (Carter 2013) to help Floridians handle climate-related threats such as changes to the water cycle. Among the laws passed were the Florida Water Resources Act, the Land Conservation Act, the Environmental Land and Water Management Act, and the Comprehensive Planning Act. Each of these acts worked at different levels to ensure better science, flexibility, and enforcement of land and water conservation, including the establishment of funds to purchase sensitive and valuable lands. Funding the purchase of select lands started in 1964 with the Land Acquisition Trust Fund. Since then, Florida voters and legislatures have passed multiple conservation programs, including the 1989 "Preservation 2000," which spent $3 billion and purchased the most land of the conservation programs. In 2014, a voter-approved constitutional amendment passed that allocated 33% of taxes on real estate sales to raise Florida's conservation funding to ~$10 billion over 20 years. But, the Florida Legislature and Administration refused to use the money to purchase lands as intended, instead using the funding for operating costs normally paid by the state's general fund, probably because it interfered with the pro-development political agenda that has long been a major driver of Florida politics (Staletovich 2016). Clearly, however, Florida voters have a bipartisan history of supporting conservation, and civil society support is critical to legitimizing government conservation policy.

The 1972 Water Resources Act created the water management districts organized around watersheds, designating regulatory authority at the regional level. This regional approach was intended to empower water management districts to make fast-moving decisions as conditions change and provide the expertise to understand how water systems work, since the districts have researchers on staff and collaborate with the various research universities across the state. The promise of this design was that it was a science-based, watershed-specific approach. However, in practice, the districts have heavily favored growth and development to the point that within the first decade of the 21st century the St. John's Water Management District had reached its "sustainable maximum," exhausting the amount of water available for growth and leaving very little for conservation of critical seeps and springs.

Finally, growth management is critical to conservation efforts to avoid low-density development (sprawl) that fills in an area's available green space and habitats. The 1985 Local Government Comprehensive Planning and Land Development Regulation Act dictated coordinated regional planning through the Florida Department of Community Affairs. However, in 2011 the act was replaced with the Community Planning Act, which removed strict state oversight for "expedited" reviews handled through the state's Department of Economic Opportunity (DEO). This change was meant to open development and remove government obstacles.

Overall, there is a substantial history of institutions for conservation in place in Florida that can be effective, but often these institutions end up favoring development for political reasons.

Paths Forward in Climate Mitigation in Florida

As Florida continues to grow, opportunities exist to improve land-use planning for climate mitigation and improve land conservation. This will increase the resilience of ecological conditions that provide critical life support for Floridians (e.g., more sustainable hydrological conditions) and keep options on the table for the future.

As the "Sunshine State," Florida's energy policy certainly has enormous room for growth in the area of solar energy. For example, solar panels could be a requirement for all new home and business construction. This would provide increased energy independence, lower energy bills for homeowners, and added employment for solar panel installation and maintenance.

Regardless of the specific strategies adopted, attention to disparate impacts on disadvantaged populations should always remain a priority, and efforts toward creating "green" cities (e.g. better mass transit, that mitigate carbon emissions) have the added benefit of resulting in many other sustainability benefits to residents. As with past efforts to manage growth, large challenges exist to obtaining funding, affecting intergovernmental cooperation, and achieving equity in infrastructure spending and policies. Innovative collaborative community involvement and effective communication will be invaluable in addressing collective risk and opportunity (Susskind et al. 2015).

Impacts of Climate Change on Florida's Human Population

We now turn to the impacts of climate change that Florida, specifically, will need to adapt to including SLR, social vulnerability, economic impacts, and the loss/damage for which governments, businesses, and residents will be unable to prepare for. Then, we discuss the planning obstacles and initiatives, including emergency management, to climate-related impacts in Florida.

Sea Level Rise (SLR)

Developing a spatial understanding of SLR is not a new scientific endeavor. Many have modeled potential rising water impact areas from the scientific perspective (Allison et al. 2011; Camber 1992; Hoffman et al. 1983; Rahmstorf 2007) as well as hypotheses and theories on global (Awosika et al. 1992, Stocher et al. 2010), national (Dunbar et al. 1992; FEMA 1991; Smith and Tirpak 1989; Titus 1986; Yohe 1990; Yohe 1996), and more localized (Kana et al. 1984, 1986, 1988) levels (Diaz and Murnane 2008). Fortunately, the geospatial processes required to understand the spatial relationship between estimated water height and potential areas of inundation are very sound (Engelen et al. 1995) and have been used widely over the past two decades to describe the physical impacts of SLR (Dasgupta et al. 2009; Li et al. 2009; Neumann et al. 2010). Titus and Narayanan (1995) described probabilities (by 2010) associated with non-anthropogenic climate change SLR ranging from 55cm to 120 cm. More recent projections range from a .5- to a 1.4-meter rise from 1990 levels by 2100 (Rahmstorf 2007). Planning for the possible effects of a changing climate first requires an understanding of the spatial "footprint" of adverse impacts. One way to understand this is through a geospatial assessment of areas at or below suggested sea level rise estimates.

Florida's vulnerability to sea level rise hazards is represented through a combination of quality digital elevation models (DEM) derived for the entire state using Light Detection and Ranging (LIDAR) systems. This DEM product, available from the Florida Geographic Data Library (FGDL), represents the best available statewide elevation data. FGDL cites four sources for this mosaic dataset of elevation:

1. Northwest Florida Water Management District (NWFWMD) DEM. Reported vertical accuracy ranges from 13 to 30 centimeters.
2. NOAA LIDAR Coastal DEM. Produced using Federal Emergency Management Agency (FEMA) accuracy standards from the Guidelines and Specifications for Flood Hazard Mapping Partners (Federal Emergency Management Agency 2013).
3. Florida Fish and Wildlife Conservation Commission (FWC) Florida Statewide 5-Meter DEM. Produced using U.S. National Map accuracy standards (U.S. National Map 2013).
4. Contour Derived DEM. Based on 2 ft contours from the coastal LiDAR project. The biggest portion of this source data is for the area around Lake Okeechobee, where LIDAR data was from provided by Merrick & Company.

Potential inundation zones were identified spatially through a standard "bathtub" fill flood modeling approach similar to those used in other studies (Rowley et al. 2007; Poulter and Halpin 2008; Mazria and Kershner 2007). Specifically, the DEM was classified as flooded/not flooded based on the value of each grid cell in relation to a given sea level rise scenario. Here, we present the "intermediate-high" scenario from the National Climate Assessment in Table 1.1 and Fig. 1.2 (Parris et al. 2012). We chose to highlight this scenario because lower scenarios primarily take into account the ocean warming, but not ice sheet loss in Antarctica, Greenland, and glaciers; thus, these lower estimates are not realistic (DeConto and Pollard 2016). Florida should plan for at least the SLR presented here, which are based on predictions by Rahmstorf (2007) and imply 126.3 cm SLR by 2100 compared to 1990 levels.

The resulting grid representations show all areas in the state with elevations at or below each scenario threshold, regardless of their situation to the coast. A spatial cost distance algorithm (McCoy et al. 2001) was used to remove those grid cells that met the elevation criteria but were "disconnected" from the coastline.

Caveats: Sea Level Rise Measurement Complexity

Hypothesizing about the potential impacts of possible sea level rise across the coast of Florida is not an exact science. Not only do projections of sea levels in 10, 20, 50, or 100 years continue to be moving targets, but methods, tools, data, and processes for measuring such changes are continuously evolving. Changes in any one of these can have a dramatic effect on the resulting "knowledge" about sea level rise inundation, especially when looking at very small scales. We can, however, with some regional certainty, begin to identify by census tracts those areas where environmental threats such as SLR will interfere with the current human use system. Census tracts are subdivisions of a county that are fairly permanent over time, monitored and delineated by the U.S. Census Bureau. We can also overlay discrete entities on the ground, such as critical facilities, with representations of SLR inundation areas to map specific possible impacts. However, spatial differences between elevation and potential SLR could produce spatial inaccuracies at the local level and these generalized results should not be employed beyond simple visual display.

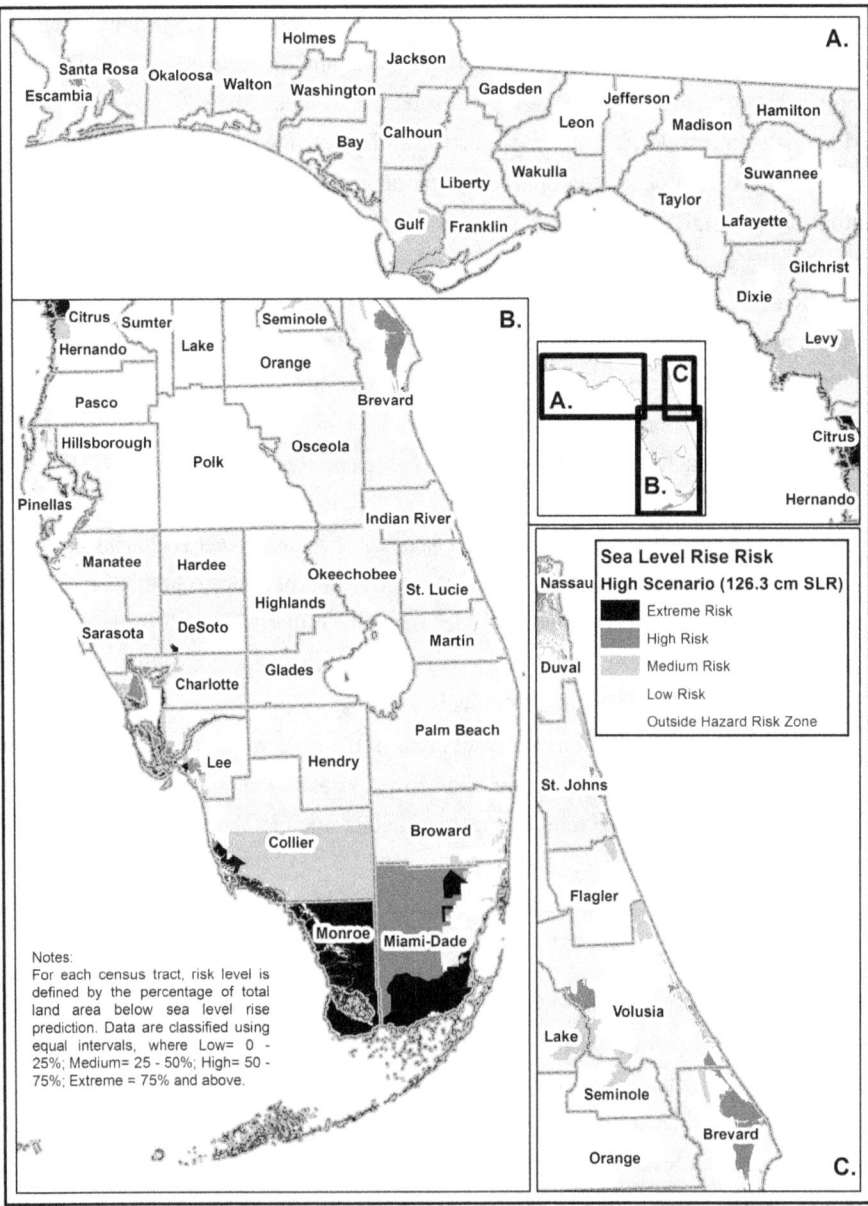

Figure 1.2. Sea level rise risk in Florida (126.3cm by 2100). Areas included are contiguous from the shore.

Table 1.1. Sea level rise (SLR) estimated risk by census tract within counties.

County Name	SLR - High Estimate (Connected Area Under 126.3 cm) Hazard Risk				
	Extreme (75%)	High (50%-75%)	Medium (25%-50%)	Low (<25%)	Out
Alachua	0	0	0	16,164	231,172
Baker	0	0	0	0	27,115
Bay	0	0	6,946	133,878	28,028
Bradford	0	0	0	0	28,520
Brevard	3,300	23,025	25,929	296,824	194,291
Broward	8,638	26,566	147,664	940,949	624,249
Calhoun	0	0	0	14,625	0
Charlotte	0	18,010	24,122	115,936	1,910
Citrus	9,092	0	0	21,077	111,067
Clay	0	0	13,596	154,992	22,277
Collier	11,601	11,861	23,527	159,380	115,151
Columbia	0	0	0	24,177	43,354
DeSoto	1,218	0	0	22,672	10,972
Dixie	0	0	0	11,432	4,990
Duval	0	6,261	70,385	413,209	374,408
Escambia	0	0	3,978	136,281	157,360
Flagler	0	3,217	3,986	35,001	53,492
Franklin	0	1,690	2,804	7,055	0
Gadsden	0	0	0	26,582	19,807
Gilchrist	0	0	0	10,510	6,429
Glades	0	0	0	12,884	0
Gulf	0	0	4,450	11,413	0
Hamilton	0	0	0	14,799	0
Hardee	0	0	0	26,772	959
Hendry	0	0	0	39,140	0
Hernando	0	3,027	5,516	3,686	160,549
Highlands	0	0	0	26,792	71,994
Hillsborough	15	4,547	16,947	377,145	830,572
Holmes	0	0	0	5,544	14,383
Indian River	0	3,212	19,765	88,621	26,430
Jackson	0	0	0	25,398	24,348
Jefferson	0	0	0	4,380	10,381
Lafayette	0	0	0	8,870	0
Lake	0	0	1,634	21,594	273,824

	SLR - High Estimate (Connected Area Under 126.3 cm) Hazard Risk				
County Name	Extreme (75%)	High (50%-75%)	Medium (25%-50%)	Low (<25%)	Out
Lee	8,607	39,046	72,318	320,537	178,246
Leon	0	0	0	18,183	257,304
Levy	0	0	3,289	10,867	26,645
Liberty	0	0	0	8,365	0
Madison	0	0	0	10,553	8,671
Manatee	4,849	14,032	20,278	171,894	111,780
Marion	0	0	0	45,980	285,318
Martin	0	0	17,752	95,554	33,012
Miami-Dade	89,865	137,904	168,936	1,167,648	928,774
Monroe	49,345	14,453	3,548	5,744	0
Nassau	0	12,311	7,980	48,964	4,059
Okaloosa	0	0	0	141,294	39,528
Okeechobee	0	0	0	30,627	9,369
Orange	0	0	0	24,945	1,121,011
Osceola	0	0	0	7,194	261,491
Palm Beach	0	1,683	14,521	956,024	347,234
Pasco	1,487	8,141	16,134	50,114	388,821
Pinellas	0	27,854	95,871	377,269	415,548
Polk	0	0	0	0	602,095
Putnam	0	0	9,421	49,578	15,365
Santa Rosa	0	4,266	4,996	127,972	14,138
Sarasota	0	6,331	8,425	253,376	111,316
Seminole	0	0	7,396	77,961	337,361
St. Johns	0	6,822	17,256	142,915	23,046
St. Lucie	5,841	3,686	4,520	198,634	65,108
Sumter	0	0	0	0	87,023
Suwannee	0	0	0	25,419	16,132
Taylor	0	0	0	13,097	9,473
Union	0	0	0	0	15,535
Volusia	0	15,470	53,573	180,162	245,388
Wakulla	0	0	0	30,776	0
Walton	0	0	0	34,262	20,781
Washington	0	0	0	16,682	8,214

State Summary

Eight Florida counties have residents at some extreme risk (i.e., census tracts with > 75% land below specified elevation to even the lowest prediction of sea level rise being investigated here), with Desoto, Lee, and Monroe counties exhibiting the highest levels of risk to even the lowest predicted sea level rise of 28.5cm (not shown) of SLR. In these counties, at least 50% of the land area in some census tracts is below this elevation. However, the lowest predicted risk is not realistic. Using a middle estimate of 66.9 cm of sea level rise, shows that 16 counties contain tracts with greater than 50% of their land area (affecting more than 168,000 people) in a high risk zone, with immense flooding and erosion risks to coastal and inland areas.

Extreme Events

Here we discuss, albeit briefly due to space limitations, Florida's most important climate-related extreme events — hurricanes. From 1871-1993, there were almost 1,000 hurricanes in the Florida Atlantic, Gulf of Mexico, and Caribbean Sea; 74 made landfall in Florida with hurricane force winds (74 mph) and 105 with tropical storm force winds (up to 73mph) (Doehring et al. 1994). Warming sea surface temperatures from climate change are expected to increase the strength, but not necessarily the frequency, of hurricanes in the Atlantic (Elsner et al. 2008; Webster et al. 2005; Emanuel 2005).

Florida has suffered more than $450 billion in hurricane-related damage since the beginning of the 20th century, and the average loss continues to increase given that "Florida hurricanes are getting more powerful over time" (Malmstadt et al. 2009, 121). The extent of this damage is expected to worsen, not just because there is increasing value of assets on the coast, but also because the effects of hurricanes will likely be more severe, as well. For example, together with SLR, the increases in storm intensity may increase storm surge 25-47% (Balaguru et al. 2016). In places like Miami-Dade, Broward, and Palm Beach where more than 5 million people reside, damage from strong storms will likely be in the tens of billions of dollars unless measures are taken to protect these areas.

Hurricanes present many human dimension problems. At the individual level, they can cause post-traumatic stress disorder or depression (Fullerton et al. 2015), induce pre-term labor and delivery (Grabich et al. 2016), or result in direct or indirect injury. Hurricanes can damage energy infrastructure such as power lines, and infrastructure damage is expected to increase (Grabich et al. 2016). Hurricanes press the government entities charged with managing community risks (see the section on Emergency Management later in this chapter) as well as vulnerable social groups in those communities, necessitating public engagement to adequately and fairly adapt to collective risks (Susskind et al. 2015).

Determinants of Social Vulnerability in Florida's Population

Measures of vulnerability, or the potential for harm, help us understand how social and ecological systems interact and how such interactions might lead to disastrous consequences for people (O'Keefe et al. 1976). Understanding social vulnerability facilitates preparation, response, and recovery in the face of environmental threats and has been widely addressed in literature about global climate change, risks, hazards, and disasters to describe and understand differential impacts on lives and livelihoods (Birkmann 2006; Eakin and Luers 2006; Füssel 2007; Pelling 2003; Polsky et al. 2007; Wisner et al. 2004).

The Social Vulnerability Index (SoVI®) is an empirical measure of pre-event social vulnerability to environmental hazards in the U.S. Initially applied at the county level, SoVI® is a comparative measure for the entire United Statesthat uses a long history of disaster case studies to understand socioeconomic and demographic characteristics that obstruct adequate disaster response (National Research Council 2006; Cutter et al. 2003; Heinz Center 2002). The index combines these population characteristics into a single score for each unit through factor analysis. The measure is closely related to concepts of social well-being, but is focused on characteristics that specifically hinder adequate disaster response. For example, economic status (poverty or wealth) influences a community's ability to absorb losses. On one hand, wealthier communities are better prepared, live in better-constructed homes, and often the residents have "rainy day" funds to cope with unexpected setbacks or catastrophe; whereas, communities in poverty deal with a constant lack of access to resources, safe and secure homes, insurance, and other basic necessities (Table 1.2).

Understanding how social vulnerability manifests across the state provides planners and decision-makers with detailed, geographic information about who and, more importantly, where pockets of the most marginalized people reside. Coupling this information with an empirical understanding of present and future hazards creates opportunities to address deficiencies in social support, identifies targets for governmental and non-governmental disaster preparedness initiatives, and provides a scientific basis for real hazard mitigation aimed at protecting more than just buildings and other infrastructure. Identifying vulnerabilities prior to a disaster can protect lives and livelihoods.

Twenty-seven variables were used in the SoVI-FL 2010-2014 computation (Table 1.2). To facilitate comparisons across counties, all data were from the US Census rolling five-year American Community Survey (ACS) (2010-2014). Each variable was standardized and input into a principal components analysis (PCA) to 1) determine which variables were the most important and 2) condense the number of variables to a smaller set of multi-dimensional attributes or components. Positive values are associated with increasing vulnerability, and negative values associated with decreasing vulnerability.

Six distinct components explain 67.73% of the variances in the data (i.e., the extent of vulnerability) for the SoVI-FL2010-14 (Table 1.3). These components include: 1) Class and race (driven by % in poverty, % black, and % with no access to an automobile); 2) Age (dominated by % social security beneficiaries, % an age-dependent populations); 3) Wealth (characterized by median house value, % earning more than $200K, and per capita income); 4) Ethnicity (represented by % speaking English not well or not at all and % Hispanics); 5) Gender (described by % female an % female labor force participation); and 6) People Per House(hold).

In Fig. 1.3, we see the social vulnerability scores, ranging from + 8.88 (indicating the most vulnerable tract, which is located in Broward County) to – 18.21 (the least vulnerable tract, which is located in Brevard County), mapped by five statistically-created categories (using standard deviation). The darker shades represent the more extreme ends of the scale and tells us where the least and most social vulnerability exists in Florida.

Overall, social vulnerability at the tract level is driven by a combination of socioeconomic and demographic conditions distinct to specific places at the local level. Each county is comprised of a dynamic mix of residents, often with very different demographics. What makes one place vulnerable (e.g. a high number of mobile homes and service sector employment) might not be the same characteristics or variables that influence vulnerability elsewhere (e.g., low educational attainment or unemployment), despite the fact that the resulting SoVI® classification (high) is the same. However, all places with high vulnerability will require additional assistance before, during, and after a shock or stressor such as a hurricane, in order to adequately avoid the long-term effects of the disaster's impact.

The SoVI-FL2010-14 is comprised of the six factor components outlined above. Tract level SoVI® scores by county are detailed in Tables 1.4 and 1.5. Table 1.4 shows the percentage of each county's total population in reference the SoVI® classification of the composite census tracts. For instance, 14.53% of people in Alachua County reside in census tracts with low vulnerability while nearly 20% of Okeechobee County residents reside in tracts with high social vulnerability. Table 1.5 provides an actual count of populations within these same zones for comparative purposes. Using both tables, one can easily see that although nearly 50% of the DeSoto County population resides in areas with elevated vulnerability (medium-high to high), this percentage represents less than 17,000 people; while Miami-Dade's nearly 34% located in the medium-high SoVI® class represents more than 871,000 residents.

Areas with elevated social vulnerability across the state of Florida are concentrated in three main regions. The first is within urban areas located in the southeast part of the state, north from Miami-Dade County, through Broward County, and into Palm Beach County; in these counties, we see that 86%, 28%, and 30% of the respective populations live in areas with elevated vulnerability (Table 1.4 and that social vulnerability is the product of a diverse set of drivers particular to each region. For example, the most vulnerable tracts (medium-high and high SoVI®) within these counties, while primarily driven by ethnicity (component 4) followed closely by class/race (component 1) and gender (component 5), in most cases is not solely an urban vs. rural phenomenon (Table 1.6). Of particular interest is the most vulnerable tract in Miami-Dade, which contains both low and high vulnerability populations, where vulnerability is driven up by age (component 2) and ethnicity (component 4) and down by wealth (component 3) and people per household (component 6).

The second area of elevated SoVI® is comprised of tracts located in south-central Florida, through the I-4 corridor up into the area from Hillsborough County to Orange County and throughout the periphery of Orlando, FL. Here, more than 26% of the population resides in areas with the most extreme vulnerability scores in the state (Table 1.7). In Hillsborough County, nearly 250,000 individuals are situated within 67 census tracts characterized as medium-high to high vulnerability. Nineteen tracts in Osceola County, home to more than 163,000 people, are characterized by at least medium-high vulnerability. Nearly 226,000 people (almost 20%) reside in 69 Orange County census tracts with elevated vulnerability, while Polk County has more than 30% (191,000) of people living in the medium-high to high socially vulnerable tracts. Overall, the I-4 corridor is home to more than 650,000 people within its 152 tracts characterized as high vulnerability. Again, the drivers of social vulnerability are diverse both within each county and between constituent tracts (Table 1.7). Class and Race (component 1), Gender component 5), and People Per Household (component 6) all serve to increase vulnerability in most of the 30 most vulnerable tracts within this zone, while Ethnicity-Hispanic (component 4) and Wealth (component 3) do not serve as major contributors to increased vulnerability.

The third cluster of extreme social vulnerability exists in Central Florida, trending from Pasco County in the south, north to Levy County, and east to Flagler County. Here, 133 census tracts with nearly 613,000 residents are characterized by either medium-high or high vulnerability. One of the main drivers of vulnerability (in 23 of the top 30 census tracts) is the age component. Gender also plays an important role in defining social vulnerability across this area, while class/race and ethnicity are less predictive of high vulnerability. Table 1.8 provides a breakdown of populations for the most vulnerable tracts within each county with respect to an overall social vulnerability score.

Table 1.2. Known correlates of social vulnerability and variables used to compute SoVI-FL2010-14.*

Population Characteristic and Specific Variables	Influence on Social Vulnerability
Race and ethnicity % African American % Native American % Asian or Pacific Islander % Hispanic	Imposes language and cultural barriers for disaster preparedness and response; affects access to pre- and post-disaster resources; there is a tendency for minority groups to occupy high-hazard areas; non-white and non-Anglo populations are often most vulnerable.
Socioeconomic Status Per capita income % households earning more than $200,000 % poverty	Affects a community's ability to absorb losses; wealth enables communities to recover more quickly using insurance, personal resources; poverty makes communities less able to respond and recover quickly.
Gender % females in labor force % female population % female headed household, no spouse present	Women often have a more difficult time coping after disasters than men due to role in the employment sector (care taking), lower wages, and family care responsibilities.
Age Age depended populations (% population under 5 years old and % population over 65) Median age	Age extremes (the elderly and very young) increase vulnerability; parents must care for children when daycare facilities are not available; elderly may have mobility or health problems
Renters % renters Median Gross Rent	Renters are often transient populations with limited ties to the community; they often lack shelter options when lodging becomes uninhabitable after disasters or too costly; they typically lack insurance and often lack savings.
Residential property Median value of owner occupied housing % housing units that are mobile homes % Unoccupied Housing Units	The value, quality, and density of residential construction affects disaster losses and recovery; expensive coastal homes are costly to replace; mobile homes are easily damaged
Occupation % employed in farming, fishing, forestry % employed in service occupations	There is a greater likelihood that some occupations, especially those involving resource extraction (fishing, farming), will be adversely affected by disasters; service sector jobs suffer as disposable income declines; infrastructure employment (transportation, communications, utilities) is subject to temporary disruptions post-disaster.
Family Structure Average number of people per household % children in single parent households	Families with a large number of dependents or single-parent households may be more vulnerable because of the need to rely on paid caregivers.
Employment % civilian labor force unemployed	Communities with high numbers of unemployed workers (pre-disaster) can be more vulnerable, because jobs are already difficult to obtain; this slows the post-disaster recovery.
Education % population over 25 with no high school diploma	Limited educational levels influence ability to understand pre-disaster warning information and likely disaster impacts; knowledge about and therefore access to post-recovery resources

Population Growth % ESL (poorly or not at all)	New immigrant populations often lack language skills and are unfamiliar with state and federal bureaucracies with regards to obtaining disaster relief; may not be permanent or legal residents; unfamiliar with the range of hazards common to an area.
Social Dependency and Special Needs Populations % collecting social security benefits % nursing home residents % no automobile	Residents totally dependent on social services for survival are often economically marginalized; special needs populations (e.g., the infirmed) require more time for evacuation and recovery is often difficult.

*Source: (Cutter et al. 2003; Heinz Center 2002)

Table 1.3. Components of the Social Vulnerability Index-Florida (SoVI-FL2010-2014).

Florida Tract Level 2010-14 Social Vulnerability Component Summary

Component	Cardinality	Name	% Variance Explained	Dominant Variables	Component Loading
1	+	Class and Race (Black)	16.831	% POVERTY	0.786
				% BLACK	0.784
				% WITH NO AUTOMOBILE	0.720
				% FEMALE HEADED HOUSEHOLDS	0.693
				% EDUCATION LESS THAN 12TH GRADE	0.601
				% RENTERS	0.540
				% EMPLOYED IN SERVICE SECTOR	0.520
				% KIDS IN 2 PARENT FAMILIES	-0.674
2	+	Age (Elderly)	14.586	% SOCIAL SECURITY BENFICIARIES	0.916
				% AGE DEPENDENT POPULATIONS	0.886
				MEDIAN AGE	0.847
				% UNOCCUPIED HOUSING UNITS	0.554
3	-	Wealth	13.087	MEDIAN HOUSE VALUE	0.897
				% EARNING MORE THAN $200k	0.881
				PER CAPITA INCOME	0.828
				MEDIAN GROSS RENT	0.529
4	+	Ethnicity (Hispanic)	8.939	% SPEAKING ENGLISH NOT WELL OR NOT AT ALL	0.914
				% HISPANICS	0.910
5	+	Gender (Female)	7.933	% FEMALE	0.791
				% FEMALE LABOR FORCE	0.756
				% EMPLOYED IN EXTRACTIVE INDUSTRY	-0.519
6	+	People Per House	6.353	PEOPLE PER UNIT	0.730
		Cumulative Variance Explained	**67.729**		

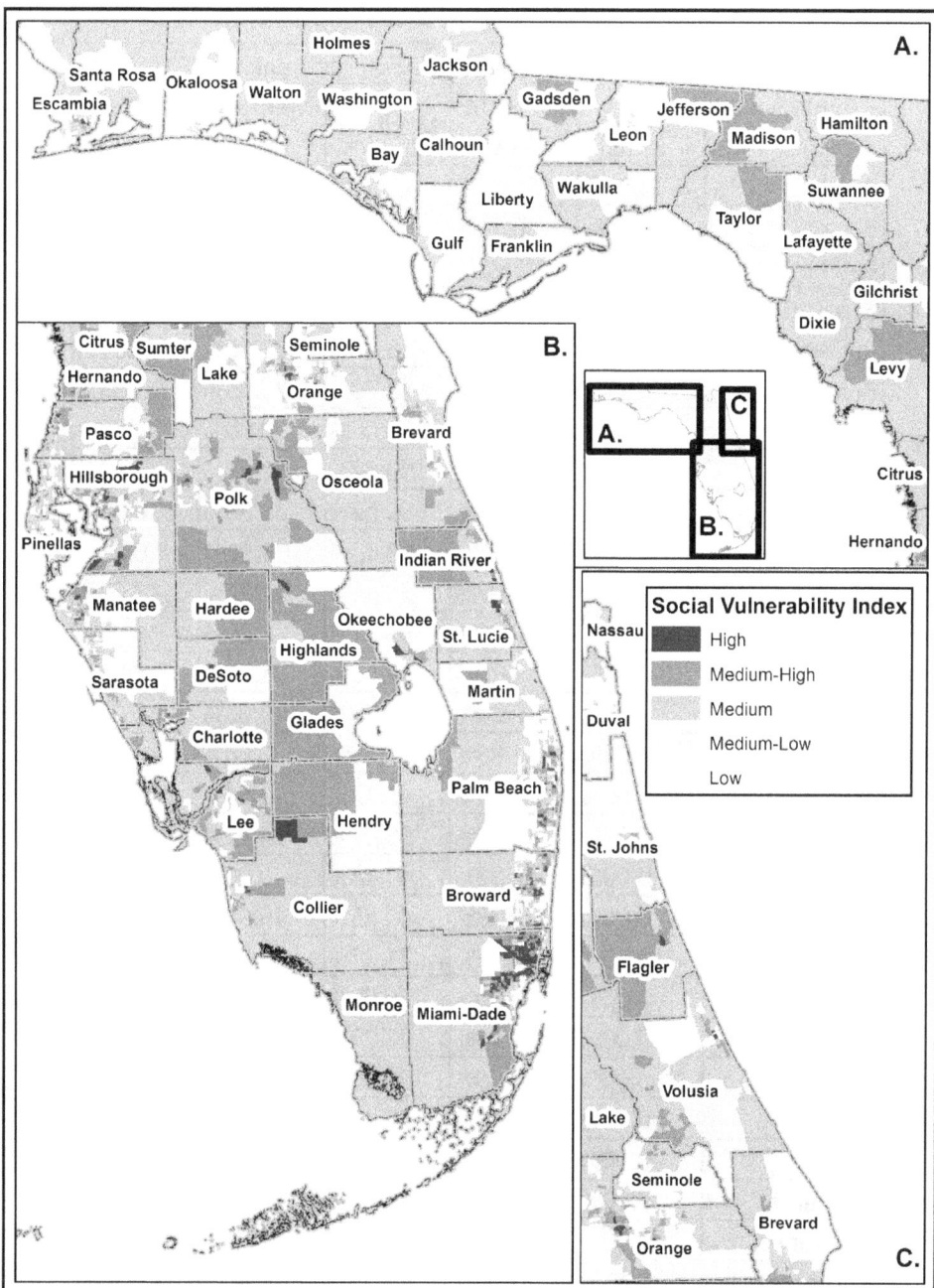

Figure 1.3. SoVI-FL2010-14 Tract-level social vulnerability in Florida.

Table 1.4. Percentage of county population by vulnerability class (SoVI-FL2010-14).

County Name	Social Vulnerability Rank				
	Low	Medium Low	Medium	Medium High	High
Alachua	14.53%	56.04%	20.75%	6.95%	1.72%
Baker	0.00%	46.18%	53.82%	0.00%	0.00%
Bay	1.84%	45.28%	47.39%	5.49%	0.00%
Bradford	0.00%	0.00%	100.00%	0.00%	0.00%
Brevard	0.23%	30.74%	55.07%	13.96%	0.00%
Broward	6.18%	22.99%	42.70%	22.79%	5.33%
Calhoun	0.00%	0.00%	100.00%	0.00%	0.00%
Charlotte	2.02%	2.07%	50.67%	42.58%	2.65%
Citrus	0.00%	3.55%	55.57%	40.88%	0.00%
Clay	0.00%	37.91%	58.13%	3.96%	0.00%
Collier	2.70%	9.07%	55.94%	21.26%	11.03%
Columbia	0.00%	0.00%	95.11%	4.89%	0.00%
DeSoto	0.00%	13.70%	37.50%	41.71%	7.09%
Dixie	0.00%	0.00%	100.00%	0.00%	0.00%
Duval	7.28%	42.88%	38.45%	7.68%	3.72%
Escambia	5.60%	39.61%	43.12%	10.54%	1.13%
Flagler	0.00%	0.00%	59.57%	34.47%	5.96%
Franklin	0.00%	52.82%	47.18%	0.00%	0.00%
Gadsden	0.00%	9.18%	43.85%	46.97%	0.00%
Gilchrist	0.00%	28.73%	71.27%	0.00%	0.00%
Glades	0.00%	28.20%	0.00%	71.80%	0.00%
Gulf	0.00%	77.23%	22.77%	0.00%	0.00%
Hamilton	0.00%	0.00%	100.00%	0.00%	0.00%
Hardee	0.00%	0.00%	55.46%	44.54%	0.00%
Hendry	0.00%	14.31%	28.13%	57.56%	0.00%
Hernando	0.00%	0.99%	60.91%	37.12%	0.98%
Highlands	0.00%	0.00%	16.52%	77.13%	6.35%
Hillsborough	11.44%	32.04%	36.95%	16.47%	3.10%
Holmes	0.00%	0.00%	100.00%	0.00%	0.00%
Indian River	0.00%	15.16%	46.96%	35.21%	2.67%
Jackson	0.00%	28.62%	58.90%	12.48%	0.00%
Jefferson	0.00%	44.40%	28.66%	26.94%	0.00%
Lafayette	0.00%	32.60%	67.40%	0.00%	0.00%
Lake	0.00%	13.58%	60.28%	24.73%	1.41%

County Name	Social Vulnerability Rank				
	Low	Medium Low	Medium	Medium High	High
Lee	0.17%	15.35%	53.63%	27.34%	3.52%
Leon	10.76%	52.88%	30.45%	5.90%	0.00%
Levy	0.00%	0.00%	66.25%	33.75%	0.00%
Liberty	0.00%	100.00%	0.00%	0.00%	0.00%
Madison	0.00%	31.53%	37.84%	30.63%	0.00%
Manatee	0.00%	25.33%	48.32%	21.09%	5.26%
Marion	0.61%	10.72%	43.88%	42.44%	2.34%
Martin	0.00%	19.05%	75.06%	4.63%	1.26%
Miami-Dade	4.75%	6.33%	18.60%	33.50%	36.82%
Monroe	1.72%	60.70%	35.25%	2.34%	0.00%
Nassau	2.02%	47.21%	50.77%	0.00%	0.00%
Okaloosa	7.07%	67.10%	25.83%	0.00%	0.00%
Okeechobee	0.00%	28.54%	39.78%	12.40%	19.27%
Orange	9.26%	35.01%	36.94%	16.43%	2.37%
Osceola	0.00%	5.12%	38.49%	56.38%	0.00%
Palm Beach	3.03%	26.99%	40.09%	24.44%	5.45%
Pasco	0.00%	20.46%	58.90%	19.80%	0.84%
Pinellas	3.60%	36.79%	46.25%	11.20%	2.16%
Polk	0.01%	11.41%	57.50%	27.20%	3.89%
Putnam	0.00%	9.40%	38.65%	51.95%	0.00%
Santa Rosa	5.07%	68.04%	26.89%	0.00%	0.00%
Sarasota	0.00%	18.92%	62.50%	18.59%	0.00%
Seminole	4.63%	40.53%	51.28%	3.56%	0.00%
St. Johns	7.35%	61.53%	28.94%	2.17%	0.00%
St. Lucie	0.00%	0.79%	79.24%	11.84%	8.14%
Sumter	8.48%	0.00%	23.26%	62.97%	5.29%
Suwannee	0.00%	19.07%	53.05%	27.89%	0.00%
Taylor	0.00%	35.73%	47.17%	17.10%	0.00%
Union	23.51%	48.18%	28.31%	0.00%	0.00%
Volusia	1.14%	13.09%	66.81%	17.51%	1.46%
Wakulla	0.00%	58.19%	41.81%	0.00%	0.00%
Walton	0.00%	30.53%	69.47%	0.00%	0.00%
Washington	0.00%	28.37%	71.63%	0.00%	0.00%
State Total	**7.46%**	**4.20%**	**42.78%**	**21.31%**	**24.25%**

Table 1.5. Total county population by vulnerability class (SoVI-FL2010-14).

County Name	Social Vulnerability Rank				
	Low	Medium Low	Medium	Medium High	High
Alachua	36,591	141,082	52,249	17,509	4,328
Baker	0	12,495	14,562	0	0
Bay	3,190	78,325	81,984	9,489	0
Bradford	0	0	27,552	0	0
Brevard	1,240	168,754	302,258	76,639	0
Broward	112,234	417,368	775,177	413,670	96,820
Calhoun	0	0	14,657	0	0
Charlotte	3,294	3,384	82,673	69,471	4,329
Citrus	0	4,959	77,669	57,143	0
Clay	0	73,867	113,281	7,720	0
Collier	9,026	30,335	187,106	71,119	36,888
Columbia	0	0	64,352	3,310	0
DeSoto	0	4,765	13,044	14,510	2,466
Dixie	0	0	16,137	0	0
Duval	64,077	377,687	338,637	67,619	32,730
Escambia	17,023	120,453	131,116	32,062	3,445
Flagler	0	0	58,877	34,076	5,890
Franklin	0	6,146	5,490	0	0
Gadsden	0	4,302	20,550	22,013	0
Gilchrist	0	4,870	12,078	0	0
Glades	0	3,719	0	9,471	0
Gulf	0	12,187	3,594	0	0
Hamilton	0	0	14,466	0	0
Hardee	0	0	15,278	12,271	0
Hendry	0	5,489	10,792	22,079	0
Hernando	0	1,720	105,863	64,510	1,699
Highlands	0	0	16,232	75,790	6,239
Hillsborough	146,372	409,991	472,810	210,785	39,710
Holmes	0	0	19,741	0	0
Indian River	0	21,367	66,178	49,617	3,756
Jackson	0	14,055	28,923	6,127	0
Jefferson	0	6,365	4,108	3,862	0
Lafayette	0	2,876	5,945	0	0
Lake	0	41,416	183,866	75,417	4,311

County Name	Social Vulnerability Rank				
	Low	Medium Low	Medium	Medium High	High
Lee	1,078	99,404	347,263	177,010	22,799
Leon	30,235	148,543	85,519	16,585	0
Levy	0	0	26,539	13,518	0
Liberty	0	8,302	0	0	0
Madison	0	5,959	7,153	5,789	0
Manatee	0	85,064	162,269	70,841	17,666
Marion	2,052	35,898	146,914	142,083	7,824
Martin	0	28,506	112,330	6,933	1,889
Miami-Dade	123,598	164,624	483,683	871,269	957,687
Monroe	1,292	45,648	26,509	1,759	0
Nassau	1,510	35,372	38,036	0	0
Okaloosa	13,353	126,814	48,817	0	0
Okeechobee	0	11,246	15,674	4,886	7,592
Orange	111,119	420,167	443,325	197,192	28,438
Osceola	0	14,823	111,421	163,205	0
Palm Beach	41,117	366,865	544,838	332,149	74,105
Pasco	0	96,746	278,460	93,588	3,951
Pinellas	33,336	340,306	427,819	103,597	19,972
Polk	31	70,454	354,957	167,881	24,000
Putnam	0	6,879	28,291	38,021	0
Santa Rosa	8,021	107,670	42,549	0	0
Sarasota	0	73,193	241,829	71,922	0
Seminole	20,010	175,137	221,592	15,396	0
St. Johns	14,954	125,163	58,872	4,413	0
St. Lucie	0	2,234	225,024	33,622	23,108
Sumter	8,792	0	24,126	65,302	5,488
Suwannee	0	8,275	23,024	12,105	0
Taylor	0	8,107	10,704	3,879	0
Union	3,587	7,352	4,319	0	0
Volusia	5,665	65,302	333,368	87,382	7,264
Wakulla	0	18,048	12,967	0	0
Walton	0	17,653	40,168	0	0
Washington	0	6,997	17,663	0	0
State Total	1,444,394	812,797	8,283,267	4,126,606	4,694,728

Table 1.6. Driving forces of the most vulnerable tracts in Southeast Florida.

County	Population	Housing Units	Components of Social Vulnerability							SoVI Score
			Class and Race (Black)	Age (Elderly)	Wealth	Ethnicity (Hispanic)	Gender (Female)	People Per Household		
Miami-Dade	88	66	1.2179	4.5794	-2.4018	4.5935	-2.5372	-1.4099		8.8454
Broward	7,843	6,255	0.8120	3.1029	-0.9002	2.4288	2.5168	-1.4726		8.2882
Miami-Dade	4,469	1,556	0.8128	0.8368	-0.7876	3.9164	0.8990	0.1206		7.3732
Miami-Dade	3,234	1,512	4.1127	0.8009	0.0694	0.4265	1.5747	0.3201		7.1655
Palm Beach	1,193	1,326	1.0047	3.7280	-0.5712	0.0549	2.3042	-0.5413		7.1218
Miami-Dade	4,940	1,541	1.8853	0.2252	-0.2061	1.3642	1.1489	1.9619		6.7916
Miami-Dade	3,801	1,352	4.2050	0.0982	0.3033	-0.7599	1.9085	1.6076		6.7561
Miami-Dade	5,510	1,502	0.3633	0.9087	-0.5276	3.8080	-0.4286	1.1867		6.3657
Palm Beach	1,963	2,313	0.6703	3.5523	-0.7563	0.5058	2.1626	-1.2931		6.3542
Miami-Dade	5,856	1,901	0.7396	0.6810	-0.4642	4.2680	0.2370	-0.0612		6.3285
Miami-Dade	8,289	2,786	0.3599	0.5737	-0.6671	4.1120	0.5118	-0.0154		6.2092
Miami-Dade	3,630	1,715	3.5189	0.4355	0.2544	-0.7202	2.0929	1.1330		6.2057
Palm Beach	1,802	1,779	-0.2285	2.8070	-0.8385	0.6007	3.0698	-0.9569		6.1306
Miami-Dade	4,014	2,138	0.8865	1.4334	-0.6721	4.0492	0.6981	-1.6755		6.0636
Miami-Dade	4,820	1,235	-0.5343	0.8093	0.0774	3.3765	0.3832	2.0479		6.0053
Miami-Dade	6,276	1,773	-0.2156	0.3694	-1.4943	2.9608	-0.2238	1.5954		5.9805
Miami-Dade	5,166	1,492	2.3073	-0.1479	-0.3674	0.7555	1.3845	1.3128		5.9797
Miami-Dade	3,936	1,287	0.3930	0.9913	-0.6135	3.4766	0.5120	-0.0189		5.9675
Miami-Dade	4,432	1,250	1.3060	0.3074	-0.3404	1.9449	-0.3133	2.3503		5.9357
Miami-Dade	5,117	1,455	0.5912	0.3161	-0.3222	3.6855	0.3673	0.6512		5.9335
Palm Beach	3,266	1,201	3.7131	-0.0261	0.1260	-1.1593	1.9815	1.5496		5.9328
Miami-Dade	4,743	2,008	0.2250	0.7563	-0.8020	4.3913	0.5249	-0.8144		5.8851
Miami-Dade	3,813	1,318	4.3339	-0.1709	0.3042	-0.7122	1.5071	1.2079		5.8616
Miami-Dade	5,565	2,257	1.7737	0.5730	-0.4958	4.3141	0.0336	-1.4471		5.7432
Miami-Dade	5,128	1,493	-0.0003	0.7560	-0.3095	3.5597	0.0549	1.0282		5.7079
Palm Beach	1,645	2,320	0.9547	3.7866	-0.7031	-0.1342	1.7208	-1.3246		5.7064
Broward	4,682	1,611	2.5791	-0.0820	-0.0325	-0.4243	2.0231	1.5523		5.6807
Miami-Dade	5,573	1,908	0.6237	0.2449	-0.6297	3.5361	0.9211	-0.3124		5.6432
Miami-Dade	7,879	3,127	0.7526	0.4651	-0.4743	3.9420	0.5395	-0.5416		5.6320
			Vulnerability Driver			Vulnerability Detractor				

Table 1.7. Driving forces of the most vulnerable tracts in south central and central Florida.

| County | Population | Housing Units | Components of Social Vulnerability ||||||| SoVI Score |
			Class and Race (Black)	Age (Elderly)	Wealth	Ethnicity (Hispanic)	Gender (Female)	People Per Household	
Hillsborough	3,749	1,282	5.5028	-0.0159	0.2392	0.0778	1.5443	0.3552	7.2250
Orange	1,597	549	4.5859	-0.1379	0.3820	-0.7103	2.1612	1.0722	6.5890
Hillsborough	3,547	1,202	3.5645	-0.2142	0.0495	-0.8347	2.1604	1.7238	6.3503
Highlands	5,079	1,842	2.2010	0.7052	-0.3365	0.3124	-0.2242	2.1511	5.4819
Orange	4,268	1,622	3.5390	0.2234	0.0320	-0.9335	1.5974	0.9827	5.3771
Lee	6,903	1,812	2.2318	-0.6297	-0.2523	1.5480	0.1694	1.6941	5.2659
Orange	4,355	1,637	2.3729	0.1808	0.0036	-0.8126	1.8225	1.5804	5.1405
Hillsborough	4,252	1,626	1.0956	0.4405	-0.4032	2.1935	0.7362	0.2000	5.0690
Hillsborough	3,273	2,402	-0.1722	2.9633	-0.2489	0.3772	2.0890	-0.5678	4.9384
Polk	3,284	1,283	2.2267	0.3987	-0.5189	-0.0576	0.2108	1.5705	4.8679
Polk	4,239	2,965	-0.6859	2.3720	-1.8200	-0.2855	0.8241	0.7307	4.7753
Hillsborough	4,055	1,752	3.2071	0.0539	0.0563	-0.8651	1.2306	1.1308	4.7011
Orange	5,167	1,937	-0.0736	0.7876	-1.6029	1.0978	-0.0498	1.2908	4.6558
Hillsborough	2,414	1,844	-0.0114	2.9158	-0.7606	-0.2088	1.7467	-0.5721	4.6308
Hillsborough	2,528	1,957	0.1497	3.4625	-0.3644	-0.1049	1.3907	-0.6374	4.6249
Lee	4,701	1,514	3.2213	-0.1851	0.0489	-0.8772	0.5037	1.8280	4.4418
Hillsborough	2,205	1,681	0.7211	3.3791	0.0487	0.8741	1.2246	-1.7498	4.4004
Lee	3,951	1,259	2.3066	-0.0201	-0.1127	0.5615	-0.2570	1.5958	4.2995
Hillsborough	3,461	1,438	2.8186	-0.3909	-0.2217	-0.4322	1.7687	0.2772	4.2630
Lee	4,414	2,932	-0.9073	3.1801	-1.1557	-0.3881	0.0890	1.0829	4.2123
Hillsborough	54	26	-1.3027	1.3280	-1.7727	-0.9094	3.1393	0.1758	4.2037
Polk	6,382	1,883	1.1911	-0.0714	-0.8297	2.1336	-1.1291	1.1824	4.1362
Hillsborough	1,528	1,137	-0.6433	2.2920	-1.9926	-0.5360	0.3012	0.6820	4.0886
Lee	2,830	1,220	3.3533	0.3663	-0.1391	-0.9024	0.8571	0.2723	4.0856
Hillsborough	2,579	1,139	3.4869	-0.3645	0.4230	-1.1463	1.6308	0.8217	4.0054
Hillsborough	820	550	-0.0260	3.3768	0.3217	0.1805	1.2906	-0.5189	3.9813
Orange	7,046	2,727	2.6517	-0.8453	-0.0532	-0.4778	1.7117	0.8439	3.9374
Orange	6,005	2,416	2.3273	-0.5481	-0.1450	-0.0463	1.0965	0.9344	3.9088
Hillsborough	5,172	1,929	0.6797	-0.1728	-0.3851	1.1672	0.9417	0.8227	3.8236

Vulnerability Driver | Vulnerability Detractor

Table 1.8. Driving forces of the most vulnerable tracts in Central Florida.

County	Population	Housing Units	Components of Social Vulnerability							SoVI Score
			Class and Race (Black)	Age (Elderly)	Wealth	Ethnicity (Hispanic)	Gender (Female)	People Per Household	
Hernando	1,699	1,288	0.2864	3.5500	-0.5571	-0.4180	1.6437	-0.0993	5.5198
Sumter	4,533	3,235	0.0903	3.3919	-0.5514	0.0208	1.8602	-0.4580	5.4565
Marion	1,484	792	4.5145	0.5212	0.2639	-1.4485	0.5967	1.2996	5.2197
Pasco	1,636	1,465	1.0979	2.3097	-1.4662	-0.8829	0.5196	0.2222	4.7327
Marion	6,340	4,526	0.0901	3.2195	-0.6428	-0.1093	1.2391	-0.4918	4.5904
Pasco	2,315	1,704	-0.4419	2.6609	-1.3033	-0.6072	1.1174	0.4387	4.4712
Lake	4,311	3,609	0.0526	3.6739	-1.2826	-0.4053	0.2493	-0.6315	4.2216
Sumter	955	400	3.5249	-0.4418	0.1597	-1.5814	1.4770	1.3868	4.2059
Flagler	5,890	2,951	-0.4619	1.2459	-0.4608	-0.0266	1.5690	0.9188	3.7060
Putnam	4,412	1,659	2.5998	0.2626	-0.2585	-0.9054	0.6123	0.8415	3.6693
Pasco	1,625	1,620	-1.1043	3.5444	-2.2348	-0.9361	-0.6508	0.5298	3.6177
Pasco	2,923	1,790	0.8256	2.0970	-1.4225	0.3053	0.5561	-1.5900	3.6165
Sumter	10,736	6,687	-0.8232	2.7351	-0.6006	-0.2321	1.2832	-0.0741	3.4894
Marion	14,839	8,866	-0.4982	2.2914	-0.9034	-0.2109	0.9680	0.0346	3.4884
Citrus	5,087	2,897	0.5754	1.0065	-0.9511	0.0718	1.1755	-0.3693	3.4111
Pasco	2,755	1,510	-0.0495	2.0205	-1.6526	-0.2754	-0.7949	0.7495	3.3028
Marion	6,776	2,939	0.2380	0.8337	-0.5015	-0.2170	1.0674	0.8281	3.2518
Hernando	4,750	2,885	-1.1681	2.2436	-1.8801	-0.4512	0.1537	0.5891	3.2472
Sumter	5,186	3,158	-0.7210	2.7810	-0.9187	-0.0893	0.7426	-0.4026	3.2294
Hernando	2,457	1,712	-0.2461	3.1665	-0.2692	-0.0787	0.8523	-0.7437	3.2195
Hernando	1,393	1,075	-0.3955	2.6499	-0.8558	-0.2395	1.2934	-0.9610	3.2029
Citrus	2,927	1,874	-0.2764	1.3940	-1.1696	-0.6701	1.0875	0.4605	3.1651
Hernando	3,467	2,335	-0.4464	2.1977	-1.5732	-0.7456	0.0238	0.5510	3.1537
Pasco	3,536	2,249	-1.0511	2.7709	-1.8268	-0.4926	-0.3353	0.4025	3.1213
Citrus	5,796	3,297	0.1345	1.0962	-0.9458	-0.7428	1.3828	0.2575	3.0739
Pasco	3,801	1,327	0.8291	0.7294	-1.1545	0.6400	-1.3576	1.0715	3.0670
Marion	13,701	5,713	0.5960	0.3076	-0.5528	0.1944	0.6660	0.6598	2.9766
Putnam	2,673	1,397	2.8233	0.0381	0.2469	-1.1383	1.2660	0.2037	2.9459
Marion	6,933	4,082	0.2152	1.8160	-0.6425	-0.5971	0.7077	0.1515	2.9357
Marion	10,495	6,107	-0.6963	2.1008	-0.9606	-0.1564	0.5890	0.1176	2.9152

Vulnerability Driver   Vulnerability Detractor

Economic Impacts

Here, space permits only a very brief outline of the potentially massive economic impacts of climate change on Florida that will affect nearly every economic sector -- from transportation to tourism and agriculture. These impacts will not only cost billions in hurricane damage alone, but also alter economic opportunities throughout the state into the future.

Officials in Miami-Dade County are charged with protecting nearly $9 trillion in infrastructure from climate-related threats including SLR. Thus in the city of Miami Beach, they have planned to spend between $400-500 million to upgrade stormwater drains that allow for saltwater intrusion and cause what is referred to as "sunny day" flooding of low-lying neighborhoods (Cocchiarella 2016). Of course, Florida tourism is deeply tied to beach tourism and the comparative attractiveness of Florida may change with threats such as SLR, storms, or the spread of vector-borne diseases (such as Zika) (Agnew and Viner 2001). Unfortunately, beach erosion increases under even minor SLR and can erode away the barrier islands where so many Florida beaches are located (FitzGerald et al. 2008). Since 1998, the Florida Beach Management Funding Assistance Program has paid approximately $626 million of an estimated $2 billion spent to mitigate beach erosion; that represents about a third of the total cost, with the rest of the funds coming from federal and local governments (Florida Department of Environmental Protection).

More flooding, higher heat extremes, and stronger tropical storms and hurricanes are anticipated across the southeast region of Florida, all of which can have significant economic impacts. Increased CO_2 can aid photosynthesis of tree crops (increasing revenues) as well as weeds (adding costs) (Asseng et al. 2013). Roads engineered to handle SLR exceed typical road construction costs by $2-3 million per lane mile (Bloetscher et al. 2013), adding to the state's transportation costs. And, of course, energy costs will be. For example, much literature already exists to demonstrate that household consumption goes up (studies vary on how severely) with the need for increased use of air conditioning, although it is also the case that such an increase can ultimately lead to the adoption of long-term, energy cost-saving measures such as the purchase of solar panels or energy-efficient appliances (Auffhammer and Mansur 2014). Other climate-related effects such as ozone increases due to warming air, more harmful algae blooms due to warmer waters, and vector-borne diseases such as dengue, which can be affected by warming temperatures (Schramm 2013); all of these will increase health costs for individuals, and potentially affect work productivity.

These are but a few examples of how wide-ranging the economic impacts of climate change will be on Florida, and this does not even include those costs associated with mitigating greenhouse gases. Policies to guard against these trends require investing in programs that increase options, such as coastal conservation, distance people from harm, open discussions, and increase education.

"Loss and Damage"

Loss and damage is an Intergovernmental Panel on Climate Change (IPCC) category for economic and non-economic impacts people cannot adapt to but are forced to pay for, such as the costs of internally displaced residents who must flee their homes for safer ground to areas that are not ready for them. Given the SLR analysis above, some of Florida's coastal areas will ultimately have to be abandoned, which will result in both economic (e.g., infrastructure and buildings) and non-economic losses (e.g., ecological and cultural losses). The options for dealing with SLR range from installation of concrete sea walls that often lead to even worse erosion, to restoring dunes and protecting coastal habitats (e.g. salt marshes) through ecological engineering. Ecosystem restoration may be "the most important for reducing exposure to hazards" (Arkema et al. 2013). — But protecting all of Florida's threatened coastlines is probably not practical. Thus, there will be land and property lost, people displaced, and infrastructure investments "sunk."

On the other end of the spectrum, refugees from climate-vulnerable areas such as the Caribbean and Haiti may choose to relocate to Florida, having been pushed out of their homes due to climate-related events. This is one example of inequities that will affect not only people vulnerable to direct effects but poor people who will not have capital or other resources to adapt. And the cost to Florida for welcoming these affected refugees will be high. As a frame of reference, as of 2000, Florida was paying $250 million a year to assist Haitian refugees, many of whom had been pushed out of their homes as a result of a "grand-scale rundown of the environmental resources — soil, water and trees—that underpin their agricultural economy" (Myers 2002, 610).

Human Dimensions of Adaptation to Climate Risks

Florida is the most susceptible state in the US to tropical cyclones and flooding. Florida is also vulnerable to drought, heat events, storms, vector-borne diseases, wildfires, and SLR, all of which have the potential to become more intense and/or more frequent as climate change unfolds (Melillo et al. 2014). As a result, adaptation must be taken seriously. In fact, if the global community of nations were to aggressively cut emissions, 900 fewer municipalities will be submerged due to SLR. One city in Florida with over 100,000 people in this category is Jacksonville (Strauss et al. 2015). Thus, Floridians are dependent on the global community to cut (mitigate) these emissions lest we experience severe damage in the future. That said, some impacts will occur regardless of actions taken by the local, regional, and global communities, so we must also consider how Florida can adapt to inevitable changes. There are multiple definitions of adaptation (Smit and Wandel 2006), for example, the IPCC (2001) defines it as adjustments in ecological, social, or economic systems aimed at alleviating the negative effects and/or taking advantage of emerging opportunities that result from observed or expected changes in climate.

Certainly, to effectively respond to climate change, Floridians must modify their behavior: Adger, Arnell, and Tompkins (2005) wrote, "Adaptation is made up of actions throughout society, by individuals, groups, and government" across social sectors, from businesses to city councils. Fortunately, Florida does have control over some adaptive strategies at our disposal including coastal development arrangements, inland migration policies, institutions, and conservation. Unfortunately, significant obstacles exist for many of these adaptive measures.

Planning Context for Climate Change Adaptation in Florida

Given that climate change impacts will be felt at multiple scales, federal, state, and local initiatives can help the state adapt to coming changes. This section reviews state, regional, and local responses to climate change adaptation, with a particular focus on SLR adaptation planning at each of these levels.

State Climate Change Planning Initiatives

In 2007, Florida's governor established the Action Team on Energy and Climate Change to develop a climate action plan for the state. The plan, which was completed in October 2008, included recommendations for adapting to temperature changes, SLR, extreme storm events, and precipitation. It called for the creation of an eponymous commission to oversee implementation (Georgetown Climate Center 2014). However, in 2010 the Florida Legislature abolished the commission and the newly-elected governor apparently directed state officials not to use the terms "climate change," "global warming," and "sustainability" (Korten 2015).

Nevertheless, several state agencies continue their work on climate-related specific initiatives funded by federal grants. For example, the Department of Economic Opportunity (DEO) established the Community Resiliency Initiative in 2011 to provide technical support to local governments facing SLR. DEO planning staff have worked with the US Department of Environmental Protection, the National Oceanic and Atmospheric Administration (NOAA), and the US Environmental Protection Agency to inform planning/time horizons for SLR determine the scientific needs for SLR projections, develop a legal framework for action, and implement Adaptation Action Areas under the state's amended land use planning and growth management policy (Butler et al. 2013; Deyle et al. 2013; Markell 2016). The Florida Department of Health (DoH), funded by the US Centers for Disease Control, convened a technical advisory team to conduct climate- and health-related vulnerability assessments throughout the state. Florida State University now oversees this program, funding local efforts to address climate change-related health vulnerabilities. Similar efforts to develop robust data, analyses, and model building to increase our understanding of these vulnerabilities have also been undertaken in Florida's Department of Transportation, Department of Environmental Protection, Division of Emergency Management, and the Florida Fish and Wildlife Conservation Commission among others, along the same lines as the original SoVI® analysis above (Butler et al. 2013).

Sea Level Rise (SLR) Planning at the Local Level

While climate change is a global phenomenon, local and regional areas are differentially impacted and much adaptation depends on local governments' land use planning and policies, locally-relevant education and outreach, and capital investments in infrastructure improvements.

While SLR is already evident, we still must and can plan for future SLR because we understand the basic mechanics and are familiar with how to deal with the related impacts such as coastal erosion, coastal flooding, and saltwater intrusion into both surface and groundwater (Church et al. 2013; Butler et al. 2016; Nicholls and Cazenave 2010; Wong et al. 2014). However, many of Florida's coastal communities have been slow to respond to this inevitable threat. As discussed earlier, the topic of SLR remains rife with uncertainty and complexity. This has led to what Butler et al. (2016) characterized as a "low-regrets incrementalism" approach to adaptation planning in Florida's coastal communities, where only around half of Florida coastal counties and 15% of coastal municipalities addressed SLR in their planning documents, often in non-binding planning documents such as sustainability or adaptation plans. Of those communities that did include SLR in their binding policy documents, the majority called for tentative planning in future community infrastructure, development regulations, land use amendments, or beach and inlet management. A few communities, mostly in South Florida, called for establishing Adaptation Action Areas, investing in storm water infrastructure, or raising sea wall height restrictions.

Interviews with planners in communities with more progressive responses revealed that political will to act could be influenced by high quality information and SLR models from reputable sources, along with visible impacts that could obviously be attributed to SLR (such as sunny day flooding) (Butler et al. 2016). Community attitudes to under-adaptation, tolerance of economic opportunity costs, and tolerance of uncertainty seem to determine how aggressive a community will choose to be in developing responses to SLR (Deyle et al. 2013), and most localities have not overcome the political barriers to action.

Regional Collaboration in Adaptation Planning

Where locals seem to be hesitating in many parts of the state, regional agencies and collaborative groups in the Southeast, Northeast, Southwest and Tampa Bay regions are leading the way by calculating regional sea level rise projections, supporting or developing local scale and regional scale vulnerability assessments, and convening local, regional, state and federal agencies and stakeholders to determine appropriate paths forward. An exemplar regional effort is the Southeast Florida Climate Change Compact (SEFCCC), an agreement and commitment to work together among Monroe, Broward, Palm Beach and Miami-Dade counties (Southeast Florida Regional Climate Change Compact 2016), see http://www.southeastfloridaclimatecompact.org/. This collaborative approach is important because many local governments lack the capacity to develop the complex models necessary to guide robust and adequately flexible actions that allow learning and continuous adaptation (Deyle and Butler 2013). Also, communities need to coordinate their actions — for instance, a sea wall in one community can undermine nature-based restoration of dunes or mangroves in another. A recent analysis of the SEFCCC (Vella et al. 2016) found that while it has no regulatory authority or funding to offer, it is well-regarded in the region and has influenced the policies, investments, and initiatives of many of the communities in the area. In particular, the counties involved in the SEFCCC worked with regional, state, and federal agencies, as well as other scientists and experts, to develop a Unified Sea Level Rise Projection (SEFCCC 2015) and a Regional Climate Action Plan (SEFCCC 2012) that contains 110 mitigation and adaptation action items for implementation, many of which have been undertaken by local governments in the region. Moreover, compact members share technical expertise among high level public officials in the counties and municipalities throughout the region. All of the counties and many municipalities have adopted climate adaptation elements into their comprehensive plans, relying on the Unified Sea Level Rise Projection for setting adaptation policies. Moreover, the exchange of information among key professionals from government, nonprofit and private sectors has generated new ideas for policy experimentation and adoption in many communities. This voluntary regional collaborative approach holds promise for a state where state-level planning and action has largely stalled (Vella et al. 2016).

Inland Migration and Managed Retreat

In 2015, over 20 million people were estimated to live in Florida, most of them in counties directly on the coast. Since the 1970s, the proportion of Florida population on the coast has ranged between 75-80% (U.S. Bureau of the Census 2008).

However, the threats to coastal residents from SLR are so serious that the Swiss RE Group, the largest company in the world that insures other insurance companies, official testified before the US Senate that parts of Florida may not be insurable by 2100 (Staletovich 2014). Inland migration from the coast and "managed retreat" are two important options. Managed retreat is the removal of buildings and infrastructure while also restoring the ecosystem to allow these ecosystems to protect areas from threats such as coastal flooding. Policy tools include fixed setbacks (how close a building is allowed to be to the shoreline), land acquisition, zoning options for hazardous areas, conservation easements, and immanent domain; but each of these face serious opposition from property owners (Dyckman et al. 2014), which means these tools require just and thoughtful community engagement (Susskind et al. 2015).

Inland migration faces several related challenges. Because inland areas are limited, they create what biologists call a "coastal squeeze" (i.e., as the tide rises, organisms typically move inland, but cannot if they are blocked by inland development) (Doody 2004). For plants and animals, this can cause local and regional extinctions (Luisa Martínez et al. 2014). For people, the impacts are pressures on inland land use (and prices), infrastructure, and services such as schools and first responders.

Beyond that, there are psycho-social reasons that inland migration will face resistance despite its obvious practical utility. People develop an allegiance to the place they live, as evidenced by the fact that very vulnerable places such as coastal Louisiana are continually re-inhabited after devastating storms (see for example Chamlee-Wright and Storr 2009). Furthermore, individuals and groups are often resistant to abandoning a doomed project when they have "sunk costs" into the project. Sunk costs – costs already incurred that cannot be recovered – are a prime example of how irrational decision-making should not be underestimated. For example, Janssen et al. (2003) showed empirically that pre-Columbian Pueblo societies in the southwest failed to adapt to existential civilization threats even when threats become known, including a changing climate that affected agriculture, because they had invested in the construction of buildings.

Emergency Management Response

Federal Initiative

Since the terrorist attacks of September 11, multiple policy and program efforts have focused on restructuring the Federal Emergency Management Agency (FEMA) mission and organizational structure to build capacity of local jurisdictions to prepare for, respond to, recover from, and mitigate all hazards (see Hu et al. 2014 for summary of policy changes and implementation efforts). One of the most recent efforts involves planning for future hazard risks, including climate change, at the state and local levels.

As of March 6, 2016, FEMA requires states to include these risks in the state hazard mitigation plans in order to qualify for federal disaster funding (e.g., Hazard Mitigation Assistance Mitigation Project and Public Assistance Grants Categories C-G), though states do not need to use the exact words "climate change" In their mitigation plans. Under Title 44 Code of Federal Regulation Part 201, FEMA requires all states to include a risk assessment of future hazard events and changing conditions, which aligns with the original intent of the Stafford Act. The Robert T. Stafford Disaster Relief and Emergency Assistance Act (42 U.S.C. 5121-5207; Public Law 93-288) is the largest source of federal funding to state and local governments for disaster recovery. Signed into law on November 23, 1988, the act provides up to 75% reimbursement to local and state governments recovering from a disaster. FEMA started reviewing the state hazard mitigation plans (SHMP) (as per State Guide Appendix A) for these future events and conditions. Plans need to include: …a summary of the probability of future hazard events that includes projected changes in occurrences for each natural hazard in terms of location, extent, intensity, frequency, and/or duration and considerations of changing future conditions, including the effects of long-term changes in weather patterns and climate on the identified hazards (Hazard Mitigation Planning 2016, ¶ 11).

Most states have not included climate risks in their SHMP because FEMA had not previously required it and the agency only mandates plan updates once every five years to remain eligible for this mitigation funding (Bagley 2015).

FEMA has provided approximately $1 billion annually for state and territory hazard mitigation efforts since 2010, of which Florida has received nearly $52 million annually. Between 2010 and 2014, Florida ranked sixth in the nation for obtaining FEMA's mitigation funding. Governors of states who either do not approve a plan with climate risks or who approve a plan without including these risks are ineligible for this funding (Bagley 2015).

State Plan

Currently, Florida has begun collecting information about SLR and climate change per Appendix K of the 2013 State Enhanced Hazard Mitigation Plan. Climate change is directly referenced in: Objective 2.4 of that and the 2016 plan, "Assist in the integration of climate change and SLR research into state, local and regional planning efforts;" and, Objective 4.5, "Participate in climate change and SLR research that will further the state and local government's ability to plan for and mitigate the impacts of future vulnerability." (State of Florida Enhanced Hazard Mitigation Plan 2016, objective 4.5).

Local Hazard Mitigation Strategies

Federal or state requirements to create, adopt, and implement local land use plans increases the likelihood of compliance and higher quality plans (Peacock et al. 2009; Berke 1996; Berke and French 1994; Berke et al. 2014; Burby 2006; Burby and May 1997).

Many coastal jurisdictions in the US have begun to include climate change in local hazard mitigation plans; however, these plans are not typically included in comprehensive land use plans, even though communities incorporating hazard reduction mechanisms in their land use plans experience less damage from a disaster (e.g., Nelson and French 2002). Still, not all hazards are included in land use plans equally (Srivastava and Laurian 2006). Without this link to land use plans, a hazard mitigation plan lacks regulatory power. As a stand-alone plan, there is no legal requirement for implementation or compliance and local jurisdictions allowing, at times even incentivizing, planning in hazards areas can *increase* a disaster's damaging effects by making development in these areas less expensive through government hazard reduction subsidies (Burby 2006).

Conclusion

Florida faces multiple serious threats from climate change dominated by human-emitted greenhouse gasses (GHGs). Florida is a major economy in the world and is likewise a significant emitter of GHGs. In order to reduce these emissions, it is important to address the structural causes and not simply individual behaviors that can result in GHG emissions. Structural determinants are things that organize a society. These include economic sectors, institutions, infrastructure, social stratification, and political-economic architecture, which all guide the behavior of large numbers of individuals and groups. Often these structures hinder behavioral changes. For example, individuals in Florida who wish to reduce their GHG footprint may not have access to the tools to do so, such as effective and reliable mass transit. Other structural elements, such as the state's pro-development stance that has favored land development, has reduced the primary production that allows for CO_2 sequestration and is also a precursor to highways and other features of the built-environment that provide future pathways for GHG emissions. Meanwhile, at the individual level, communicating the dynamics of climate change is not merely a matter of informing citizens of the problem, partly because there is an organized effort to cast doubt on well-established climate science basics thus confusing members of the general public, and also because many people carry varied cognitive biases that make it very difficult for them to think and act rationally to avoid climate change risks that evolve slowly, are fairly invisible, and may contradict individual beliefs or sense of reality. For example, we are far more prone to believe a risk is real if it is consistent with our preexisting beliefs, but to believe otherwise even when that risk (e.g., slowly rising seas) is a serious threat to our well-being and active behavioral change would protect our welfare.

Meanwhile, the impacts of climate change will not affect everyone equally. This chapter has detailed the geographic organization of areas most vulnerable to SLR as well as the most socially vulnerable areas throughout Florida. Eight Florida counties have census tracts where 75% of the land is under the submergence elevation – the elevation level that would flood and be inundated – even in a case of the smallest, and quite frankly unrealistic, SLR projections. DeSoto, Lee, and Monroe counties have the most extreme exposure to this risk. Worse yet, under more realistic SLR scenarios, 16 counties have census tracts where more than half of the land is under the submergence elevation (with the potential to affect more than 168,000 people living in these high risk zones), with the potential for immense SLR impacts to coastal and inland areas. Florida is arguably the state most likely to be affected by SLR in the US and with the most to lose. SLR threatens our state's revenue-generating beaches, trillions of dollars in coastline infrastructure, the income and investments of residents who may be forced to migrate inland (putting pressure on the inland areas, as well), insurance losses and the designation of more areas as uninsurable, as well as the need for massive government expenditures to solve problems such as erosion, relocation, and "sunny day flooding" in areas such as Miami-Dade County. In short, SLR in Florida has the potential to be a powerful force toward social disorganization and instability.

Unfortunately, some members of Florida's population are especially vulnerable to the risks associated with climate change, such as the effects of extreme storm events. Using the Social Vulnerability Index (SoVI®), this chapter shows that vulnerability is clustered geographically and driven by different demographic components in different areas. For example, in the I-4 corridor, components 1 (Class and Race), five (Gender), and 6 (Persons Per Household) of the

SoVI® influence the vulnerability of most of the 30 most-vulnerable tracts. Among the most vulnerable areas in Florida, 86% of Miami-Dade residents live with elevated risk to the impacts of climate change. Understanding how risk is distributed across Florida provides decision-makers with critical information to inform how they can plan ahead and spend resources most effectively.

Planning efforts that address climate change in Florida has been mixed across the state. Federal programs and resources for emergency management planning already exist, but local level planning for climate change has been slow and incremental. At the same time, Florida boasts the Southeast Florida Climate Change Compact, which has earned national attention for its effective use of science and local-regional coordination. Further, other regional efforts are underway in Florida at various stages of assessment and planning for climate change adaptation. Still, some planning for climate change has been scuttled for political reasons, as have critical conservation efforts. For example, the 2014 Land and Water Conservation Amendment ("Amendment 1") passed with more than 60% of the vote (2.8 million votes), indicating a strong bipartisan majority preference of voters to invest in critical conservation; but, it was derailed by the state's Governor who won re-election by only 1%, (Jones 2015). The prior administration had been more proactive on climate issues, but many of these previous efforts were also aborted. This demonstrates the power of state leadership. Meanwhile, climate has become a polarizing issue in the state and nationwide, sharply dividing Democrats who tend to see climate as real and as an important threat to act on, and Republicans who are more likely to see climate change science as exaggerated or even as a fabrication and scientific deception. Research indicates some of the climate change sceptics' beliefs are fueled not by an understanding of climate science basics, but by fears of governmental abuse of power, loss of traditional energy sources, and increased taxes (Dunlap 2016; Jacques and Knox 106). If and as climate denial is normalized, state leadership will have an increasingly heavy hand in determining how proactive Florida will be on mitigating and adapting to climate change.

There are many resources in Florida to approach the problems of climate change in the state, from a robust scientific infrastructure in the university system and networks of federal agencies to the genuine commitment to conservation of land and coastline by the public. All of these resources will need to be employed to overcome the serious macro and micro obstacles that challenge our ability to address a warming world in the coming century.

References

Adger, W.N., et al., *Cultural dimensions of climate change impacts and adaptation.* Nature Clim. Change, 2013. **3**(2): p. 112-117.

Adger, W.N., N.W. Arnell, and E.L. Tompkins, *Successful adaptation to climate change across scales.* Global Environmental Change, 2005. **15**(2): p. 77-86.

Adger, W.N., Social capital, collective action, and adaptation to climate change, in Der klimawandel. 2010, Springer. p. 327-345.

Agnew, M.D. and D. Viner, *Potential impacts of climate change on international tourism.* Tourism and Hospitality Research, 2001. **3**(1): p. 37-60.

Agrawal, A., *Local institutions and adaptation to climate change.* Social dimensions of climate change: Equity and vulnerability in a warming world, 2010: p. 173-197.

Allison, I., et al., The Copenhagen Diagnosis: Updating the world on the latest climate science. 2011: Elsevier.

Arkema, K.K., et al., Coastal habitats shield people and property from sea-level rise and storms. Nature Clim. Change, 2013. **3**(10): p. 913-918.

Asseng, S., et al., Agriculture and Climate Change in the Southeast USA, in Climate of the Southeast United States: Variability, Change, Impacts, and Vulnerability, K.T. Ingram, et al., Editors. 2013, Island Press/Center for Resource Economics: Washington, DC. p. 128-164.

Atran, S. and J. Henrich, The evolution of religion: How cognitive by-products, adaptive learning heuristics, ritual displays, and group competition generate deep commitments to prosocial religions. Biological Theory, 2010. **5**(1): p. 18-30.

Auffhammer, M. and E.T. Mansur, Measuring climatic impacts on energy consumption: A review of the empirical literature. Energy Economics, 2014. **46**: p. 522-530.

Awosika, L., et al. The impact of sea level rise on the coastline of Nigeria. in Proceedings of IPCC Symposium on the Rising Challenges of the Sea. Magaritta, Venezuela. 1992.

Bagley, K., FEMA to States: No Climate Planning, No Money, in Inside Climate News. 2015.

Balaguru, K., D.R. Judi, and L.R. Leung, Future hurricane storm surge risk for the U.S. gulf and Florida coasts based on projections of thermodynamic potential intensity. Climatic Change, 2016. **138**(1): p. 99-110.

Baumgartner, F.R., et al., *Punctuated Equilibrium in Comparative Perspective.* American Journal of Political Science, 2009. **53**(3): p. 603-620.

Beamish, T., Silent Spill: The Organization of Industrial Crisis. 2002, Cambridge: MA: MIT Press.

Berke, P.R. and S.P. French, *The influence of state planning mandates on local plan quality.* Journal of planning education and research, 1994. **13**(4): p. 237-250.

Berke, P.R., Enhancing plan quality: evaluating the role of state planning mandates for natural hazard mitigation. Journal of environmental planning and management, 1996. **39**(1): p. 79-96.

Berke, P.R., W. Lyles, and G. Smith, *Impacts of federal and state hazard mitigation policies on local land use policy.* Journal of Planning Education and Research, 2014: p. 0739456X13517004.

Biesbroek, G.R., et al., *On the nature of barriers to climate change adaptation.* Regional Environmental Change, 2013. **13**(5): p. 1119-1129.

Birkmann, J., Measuring vulnerability to natural hazards: towards disaster resilient societies. 2006, New York: United Nations Publications.

Bloetscher, F., et al., Climate Change and Transportation in the Southeast USA, in Climate of the Southeast United States: Variability, Change, Impacts, and Vulnerability, K.T. Ingram, et al., Editors. 2013, Island Press/Center for Resource Economics: Washington, DC. p. 109-127.

Boarnet, M.G., R.B. McLaughlin, and J.I. Carruthers, *Does state growth management change the pattern of urban growth? Evidence from Florida.* Regional Science and Urban Economics, 2011. **41**(3): p. 236-252.

Bolstad, E., *The seas may be rising, but Florida keeps building*, in *Climate Wire.* 2016, Environment and Energy Publishing.

Broad, K., et al., *Misinterpretations of the" Cone of Uncertainty" in Florida during the 2004 Hurricane Season.* Bulletin of the American Meteorological Society, 2007. **88**(5): p. 651.

Bullard, R.D., Addressing urban transportation equity in the United States. Fordham Urb. LJ, 2003. **31**: p. 1183.

Burby, R.J. and P.J. May, *Making governments plan: State experiments in managing land use.* 1997, Baltimore, MD: John's Hopkins University Press.

Burby, R.J., Hurricane Katrina and the paradoxes of government disaster policy: Bringing about wise governmental decisions for hazardous areas. The Annals of the American Academy of Political and Social Science, 2006. **604**(1): p. 171-191.

Bureau of Economic and Business Research, *Categories / Transportation and Infrastructure / Airports.* 2016: Gainesville, FL.

Bureau, U.C., Population Estimates March 2016. 2015

Burke, W.W. and G.H. Litwin, *A Causal Model of Organizational Performance and Change.* Journal of Management, 1992. **18**(3): p. 523-545.

Burke, W.W., *Organization change: theory and practice*. Fourth edition. ed. Foundations for organizational science. 2014, Los Angeles: SAGE. xviii, 425 pages.

Butler, W.H., et al., *Sea Level Rise Projection Needs, Capacities and Alternative Approaches*. 2013, Florida Planning and Development Lab: Tallahassee, FL.

Butler, W.H., R.E. Deyle, and C. Mutnansky, *Low-Regrets Incrementalism Land Use Planning Adaptation to Accelerating Sea Level Rise in Florida's Coastal Communities*. Journal of Planning Education and Research, 2016: p. 0739456X16647161.

Camber, G., Global Climate Change and the Rising Challenge of the Sea. Assessment of the Vulnerability of Coastal Areas, in Global Climate Change and the Rising Challenge of the Sea, I.P.o.C.C. Coastal Zone Management Subgroup, Editor. 1992, Environmental Protection Agency and National Ocean Service: Washington, DC.

Carter, L.J., The Florida experience: Land and water policy in a growth state. Second ed. 2013, New York: Earthscan.

Center for Research on Environmental Decisions, The Psychology of Climate Change Communication: A Guide for Scientists, Journalists, Educators, Political Aides, and the Interested Public. 2009, Center for Research on Environmental Decisoins at Columbia University: New York.

Chamlee-Wright, E. and V.H. Storr, "There's No Place Like New Orleans": Sense of Place and Community Recovery in the Ninth Ward after Hurricane Katrina. Journal of Urban Affairs, 2009. **31**(5): p. 615-634.

Chmura, G.L., et al., *Global carbon sequestration in tidal, saline wetland soils*. Global Biogeochemical Cycles, 2003. **17**(4): p. n/a-n/a.

Church, J.A., et al., *Sea level change*. 2013, PM Cambridge University Press.

Cocchiarella, D., Fighting Sunny Day Flooding in Miami Beach, in Forward Florida. 2016.

Costanza, R., et al., *The Value of Coastal Wetlands for Hurricane Protection*. AMBIO: A Journal of the Human Environment, 2008. **37**(4): p. 241-248.

Cutter, S.L., B.J. Boruff, and W.L. Shirley, *Social vulnerability to environmental hazards*. Social science quarterly, 2003. **84**(2): p. 242-261.

Damasio, A.R., *Descartes' Error: Emotion, Reason, and the Human Brain*. 1994, New York: Avon Books.

Dasgupta, S., et al., The impact of sea level rise on developing countries: a comparative analysis. Climatic change, 2009. **93**(3-4): p. 379-388.

DeConto, R.M. and D. Pollard, *Contribution of Antarctica to past and future sea-level rise*. Nature, 2016. **531**(7596): p. 591-597.

DeFries, R., et al., Combining satellite data and biogeochemical models to estimate global effects of human-induced land cover change on carbon emissions and primary productivity. Global Biogeochemical Cycles, 1999. **13**(3): p. 803-815.

Derr, M., Some kind of paradise: A chronicle of man and the land in Florida. 1989: William Morrow & Company.

Dewey, J. and D. Denslow, *Tougher Choices: Shaping Florida's Future*. 2014, Leroy Collins Institute: Tallahassee, FL.

Deyle, R., W. Butler, and L. Stevens, *Planning Time Frames for Coastal Hazards and Sea Level Rise*. 2013, Report prepared for the FL Department of Economic Opportunity.

Deyle, R.E. and W. Butler, Resilience Planning in the Face of Uncertainty: Adapting to Climate Change Effects on Coastal Hazards. Disaster Resilience: Interdisciplinary Perspectives, in Disaster Resiliency: Interdisciplinary Perspectives, N. Kapucu, C.V. Hawkins, and F.I. Rivera, Editors. 2013, Routledge: New York. p. 178-206.

Diaz, H.F. and R.J. Murnane, *The significance of weather and climate extremes to society: an introduction*. Climate extremes and society. Cambridge University Press, Cambridge, UK, 2008: p. 1-7.

Doehring, F., I.W. Duedall, and J.M. Williams, Florida hurricanes and tropical storms: 1871-1993, an historical review. 1994.

Doody, J.P., *'Coastal squeeze'— an historical perspective*. Journal of Coastal Conservation, 2004. **10**(1): p. 129-138.

Dunbar, J.B., L. Britsch, and E.B. Kemp, *Land loss rates. Report 3. Louisiana Coastal Plain*. 1992, US Army Corps of Engineers: New Orleans, LA.

Dunlap, R.E. and R.J. Brulle, eds. *Climate Change and Society: Sociological Perspectives*. 2015, Oxford University Press: Oxford.

Dunlap, R.E., A.M. McCright, and J.H. Yarosh, *The Political Divide on Climate Change: Partisan Polarization Widens in the U.S.* Environment: Science and Policy for Sustainable Development, 2016. **58**(5): p. 4-23.

Dyckman, C.S., C. St. John, and J.B. London, *Realizing managed retreat and innovation in state-level coastal management planning.* Ocean & Coastal Management, 2014. **102, Part A**: p. 212-223.

Eakin, E., I feel, therefore I am, in The New York Times. 2003.

Eakin, H. and A.L. Luers, *Assessing the vulnerability of social-environmental systems.* Annual Review of Environment and Resources, 2006. **31**(1): p. 365.

Elliott, J.R. and M.T. Clement, Urbanization and carbon emissions: a nationwide study of local countervailing effects in the United States. Social Science Quarterly, 2014. **95**(3): p. 795-816.

Elsner, J.B., J.P. Kossin, and T.H. Jagger, *The increasing intensity of the strongest tropical cyclones.* Nature, 2008. **455**(7209): p. 92-95.

Emanuel, K., Increasing destructiveness of tropical cyclones over the past 30 years. Nature, 2005: p. 436, 686-688.

Engelen, G., et al., *Using cellular automata for integrated modelling of socio-environmental systems.* Environmental monitoring and Assessment, 1995. **34**(2): p. 203-214.

Ewing, R.H., et al., Growing cooler: the evidence on urban development and climate change. 2007: Smart Growth America.

Faught, M.K., The underwater archaeology of paleolandscapes, Apalachee Bay, Florida. American Antiquity, 2004: p. 275-289.

Federal Emergency Management Agency, F., *Guidelines & Specifications for Flood Hazard Mapping Partners.* 2013, U.S. Dept. of Homeland Security: Washington, D.C.

FEMA, *Projected Impact of Relative Sea Level Rise on the National Flood Insurance Program.* 1991, Federal Emergency Management Agency's Federal Insurance Administration: Washington, DC.

Ferrario, F., et al., The effectiveness of coral reefs for coastal hazard risk reduction and adaptation. Nature communications, 2014. **5**.

FitzGerald, D.M., et al., *Coastal impacts due to sea-level rise.* Annu. Rev. Earth Planet. Sci., 2008. **36**: p. 601-647.

Florida Department of Agriculture and Consumer Services. *Florida Agriculture Overview and Statistics.* 2016 9/26/2016]; Available from: http://www.freshfromflorida.com/Divisions-Offices/Marketing-and-Development/Education/For-Researchers/Florida-Agriculture-Overview-and-Statistics.

Florida Department of Environmental Protection. *Beach Management Funding Assistance (BMFA) Program* 2016 9/27/2016];
Available from: http://www.dep.state.fl.us/BEACHES/programs/becp/index.htm.

Florida Public Service Commission, *Facts and Figures of the Florida Utility Industry.* 2016: Tallahassee, Florida.

Florida Public Service Commission, *Florida Energy Efficiency and Conservation Act Fact Sheet.* 2014: Tallahassee, Florida.

Florida Solar Energy Center, Effectiveness of Florida's residential energy code: 1979-2009. 2009.

Florida Trend, 2015 Economic yearbook: gaining momentum, in April. 2015. p. 44-45.

Florida, U.o.C., *Florida 2060: We can do better.* 2007, Metropolitan Center for Regional Studies at the University of Central Florida.

Fullerton, C.S., et al., Depressive Symptom Severity and Community Collective Efficacy following the 2004 Florida Hurricanes. PLoS ONE, 2015. **10**(6): p. e0130863.

Füssel, H.-M., Vulnerability: a generally applicable conceptual framework for climate change research. Global environmental change, 2007. **17**(2): p. 155-167.

Georgetown Climate Center, *"Florida Climate and Energy Profile."* 2014: Washington, D.C.

Grabich, S.C., et al., *Hurricane Charley Exposure and Hazard of Preterm Delivery, Florida 2004.* Maternal and Child Health Journal, 2016: p. 1-9.

Gramsci, A., *Prison Notebooks*, J.A. Buttigieg, Editor. 2011, Columbia University Press: New York.

Heinz Center, *Human links to coastal disasters.* The H. John Heinz III Center for Center for Science, Economics, and the Environment, 2002.

Hillier, B., The new science of space and the art of place: toward a space-led paradigm for researching and designing the city, in New urbanism and beyond: designing cities for the future, T. Haas, Editor. 2008, Rizzoli: New York. p. 30-39.

Hoffman, J.S., D. Keyes, and J.G. Titus, *"Projecting Future Sea Level Rise: Methodology, Estimates to the Year 2100.* 1983, US Environmental Protection Agency Office of Policy and Resource Management: Washington, DC.
Hu, Q., C.C. Knox, and N. Kapucu, What have we learned since September 11, 2001? A network study of the Boston marathon bombings response. Public Administration Review, 2014. **74**(6): p. 698-712.
Imhoff, M.L., et al., The consequences of urban land transformation on net primary productivity in the United States. Remote Sensing of Environment, 2004. **89**(4): p. 434-443.
IPCC, I.P.o.C.C., Climate change 2001: Impacts, Adaptation, and Vulnerability. Summary for Policy Makers. 2001, Geneva: World Meteorological Organisation.
Ives, A.R. and S. Carpenter, *Stability and Diversity of Ecosystems.* Science, 2007. **317**: p. 58-62.
Jacques, P.J. and C.C. Knox, Hurricanes and hegemony: A qualitative analysis of micro-level climate change denial discourses. Environmental Politics, 2016. **25**(5): p. 831-852.
Jacques, P.J., *"Emerging Issues: Civil Society in an Environmental Context."*, in *The Guide to US Environmental Policy*, S. Fairfax and E. Russell, Editors. 2014, CQ Press: Washington, D.C. p. 409-419.
Jacques, P.J., *Autonomy and Activism in Civil Society*, in *New Earth Politics*, S. Jinnah and S. Nicholson, Editors. 2016, MIT Press: Cambridge MA. p. 221-246.
Jacques, P.J., R.E. Dunlap, and M. Freeman, *The Organization of Denial: Conservative Think Tanks and Environmental Scepticism.* Environmental Politics, 2008. **17**(3): p. 349 — 385.
Janssen, M.A., T.A. Kohler, and M. Scheffer, *Sunk-Cost Effects and Vulnerability to Collapse in Ancient Societies1.* Current anthropology, 2003. **44**(5): p. 722-728.
Jones, M., Environmentalism and Environmental Constitutional Ballot Initiatives in Florida: The Elements of Support for Amendment One in 2014 in the Context of Current Environmental Attitudes, in Political Science. 2015, University of Central Florida: Orlando, FL.
Kahan, D.M., et al., The polarizing impact of science literacy and numeracy on perceived climate change risks. Nature climate change, 2012. **2**(10): p. 732-735.
Kahneman, D., P. Slovic, and A. Tversky, *Judgment under uncertainty: Heuristics and biases.* 1982, Cambridge University Press: Cambridge.
Kahneman, D., *Thinking, fast and slow.* 2011, New York: Farrar, Straus and Giroux.
Kana, T., B. Baca, and M. Williams, Potential impacts of sea level rise on wetlands around Charleston, South Carolina. US Environmental Protection Agency. 1986, EPA 230-10-85-014.
Kana, T.W., B.J. Baca, and M.L. Williams, *Charleston Case Study*, in *Greenhouse Effect, Sea Level Rise, and Coastal Wetlands. US Environmental Protection Agency*, J. Titus, Editor. 1988, EPA 230-05-86-013. 186 pages. Available at http://epa.gov/climatechange/effects/coastal/SLRLandUse.
Kana, T.W., et al., *The physical impact of sea level rise in the area of Charleston, South Carolina.* Greenhouse Effect and Sea Level Rise: A Challenge for This Generation. New York: Van Nostrand Reinhold Company, 1984: p. 105-150.
Klas, M.E., Who's operating renewable energy in Florida?, in Tampa Bay Times. 2016.
Korten, T., In Florida, Officials Ban Term 'Climate Change', in Florida Center for Investigative Reporting. 2015: St. Petersburg.
Leiserowitz, A. and K. Broad, *Florida: Public opinion on climate change.* A Yale University/University of Miami/Columbia University Poll. New Haven, CT: Yale Project on Climate Change, 2008.
Levin, J., Art Basel: Miami Dade College exhibit provokes thought on sea-level rise, in The Miami Herald. 2015: Miami, FL.
Li, X., et al., *GIS analysis of global impacts from sea level rise.* Photogrammetric Engineering & Remote Sensing, 2009. **75**(7): p. 807-818.
Lind, E.A., Fairness heuristic theory: Justice judgments as pivotal cognitions in organizational relations. Advances in organizational justice, 2001. **56**(8).
LiPuma, E. and T. Koelble, Cultures of circulation and the urban imaginary: Miami as example and exemplar. The Urban Sociology Reader, 2012: p. 370.
Loewenstein, G.F., et al., *Risk as feelings.* Psychological bulletin, 2001. **127**(2): p. 267.
Luisa Martínez, M., et al., Land use changes and sea level rise may induce a "coastal squeeze" on the coasts of Veracruz, Mexico. Global Environmental Change, 2014. **29**: p. 180-188.
Malmstadt, J., K. Scheitlin, and J. Elsner, *Florida Hurricanes and Damage Costs.* Southeastern Geographer, 2009. **49**(2): p. 108-131.

Maniates, M.F., *Individualization: Plant a Tree, Buy a Bike, Save the World?* Global Environmental Politics, 2001. **1**(3): p. 31-52.

Markell, D., *Sea Level Rise and Changing Times for Florida Local Governments.* Columbia Journal of Environmental Law, 2016. **42**.

Marlen, J., et al., Detecting Local Environmental Change: The role of experience in shaping risk judgements about global warming, in Global Environmental Change. under review.

Mazria, E. and K. Kershner, *Nation under siege: sea level rise at our doorstep.* 2007: 2030 Research Center.

McCoy, J., K. Johnston, and E.s.r. institute, *Using ArcGIS spatial analyst: GIS by ESRI.* 2001: Environmental Systems Research Institute.

McCright, A. and R.E. Dunlap, Challenging Global Warming as a Social Problem: An Analysis of the Conservative Movement's Counter Claims. Social Problems, 2000. **47**(4): p. 499–522.

Melillo, J.M., T.T. Richmond, and G. Yohe, Climate change impacts in the United States: The Third National Climate Assessment, in Third National Climate Assessment. 2014, U.S. Global Change Research Program: Washington, D.C.

Meyer, R., et al., Dynamic simulation as an approach to understanding hurricane risk response: insights from the Stormview lab. Risk analysis, 2013. **33**(8): p. 1532-1552.

Mormino, G.R., Land of sunshine, state of dreams: A social history of modern Florida. 2005: University Press of Florida.

Moser, S.C., *Communicating climate change: history, challenges, process and future directions.* Wiley Interdisciplinary Reviews: Climate Change, 2010. **1**(1): p. 31-53.

Mulkey, S., Ed, Towards a Sustainable Florida: A review of Environmental, Social and Economic Concepts for Sustainable Development in Florida. 2006, Report for the Century Commission for a Sustainable Florida.

Myers, N., *Environmental refugees: a growing phenomenon of the 21st century.* Philosophical Transactions of the Royal Society of London B: Biological Sciences, 2002. **357**(1420): p. 609-613.

National Research Council, *Facing hazards and disasters: understanding human dimensions.* 2006: National Academies Press.

Nelson, A.C. and S.P. French, Plan quality and mitigating damage from natural disasters: A case study of the Northridge earthquake with planning policy considerations. Journal of the American Planning Association, 2002. **68**(2): p. 194-207.

Neumann, J.E., et al., Assessing sea-level rise impacts: a GIS-based framework and application to coastal New Jersey. Coastal management, 2010. **38**(4): p. 433-455.

Nicholls, R.J. and A. Cazenave, *Sea-level rise and its impact on coastal zones.* Science, 2010. **328**(5985): p. 1517-1520.

O'Keefe, P., K. Westgate, and B. Wisner, *Taking the naturalness out of natural disasters.* Nature, 1976. **260**: p. 566-567.

Oppenheimer, M., *Climate change impacts: accounting for the human response.* Climatic Change, 2013. **117**(3): p. 439-449.

Oreskes, N. and E. Conway, Merchants of Doubt: how a handful of scientists obscured the truth on issues from tobacco smoke to global warming. 2010, New York: Bloomsbury Press.

Pachauri, R.K., et al., Climate change 2014: synthesis Report. Contribution of working groups I, II and III to the fifth assessment report of the intergovernmental panel on climate change. 2014: IPCC.

Parris, A., et al., *Global sea level rise scenarios for the United States National Climate Assessment.* NOAA Tech Memo OAR CPO-1. 2012, Washington, D.C.: NOAA

Peacock, W.G., et al., *An assessment of coastal zone hazard mitigation plans in Texas.* A Report Prepared for the Texas General Land Office and The National Oceanic and Atmospheric Administration, 2009.

Pelling, M., The vulnerability of cities: natural disasters and social resilience. 2003: Earthscan.

Peterson, T.D. and A.Z. Rose, Reducing conflicts between climate policy and energy policy in the US: The important role of the states. Energy Policy, 2006. **34**(5): p. 619-631.

Pittman, C. and M. Waite, *Paving paradise: Florida's vanishing wetlands and the failure of no net loss.* 2009: University Press of Florida Gainesville, FL.

Polski, C., et al., *Adaptation Pathways 1.0: A Guide for Navigating Sea-Level Rise in the Built Environment*, A. Edwards, et al., Editors. 2016, Florida Atlantic University.

Polsky, C., R. Neff, and B. Yarnal, *Building comparable global change vulnerability assessments: The vulnerability scoping diagram.* Global Environmental Change, 2007. **17**(3): p. 472-485.

Poulter, B. and P.N. Halpin, *Raster modelling of coastal flooding from sea-level rise.* International Journal of Geographical Information Science, 2008. **22**(2): p. 167-182.

Rahmstorf, S., A semi-empirical approach to projecting future sea-level rise. Science, 2007. **315**(5810): p. 368-370.

Rogers, E.M., *Diffusion of innovations.* 5th ed. 2003, New York: Free Press. xxi, 551 p.

Rowley, R., et al., Risk of rising sea level to population and land area. Eos, 2007. **88**(9): p. 105.

Schramm, P.J., Human Health and Climate Change in the Southeast USA, in Climate of the Southeast United States: Variability, Change, Impacts, and Vulnerability, K.T. Ingram, et al., Editors. 2013, Island Press/Center for Resource Economics: Washington, DC. p. 43-61.

Slovic, P., *Perception of Risk.* Science, 1987. **236**(4799): p. 280-285.

Smit, B. and J. Wandel, *Adaptation, adaptive capacity and vulnerability.* Global Environmental Change, 2006. **16**(3): p. 282-292.

Smith, J.B. and D.A. Tirpak, *The potential effects of global climate change on the United States.* Vol. 1. 1989, Washington, D.C.: Office of Policy, Planning and Evaluation, US Environmental Protection Agency.

Smith, S.K. and M. House, *Snowbirds, Sunbirds, and Stayers: Seasonal Migration of Elderly Adults in Florida.* The Journals of Gerontology Series B: Psychological Sciences and Social Sciences, 2006. **61**(5): p. S232-S239.

Smith, S.K., *Florida Population Growth: Past, Present and Future.* Bureau of Economic and Business Research, University of Florida, Gainesville, 2005.

Solar Energy Industries Association, Solar power purchase agreements (PPAs) Fact Sheet. 2012.

Solar Energy Industries Association, *Solar Spotlight: Florida.* 2016.

South Florida Regional Climate Change Compact, *A unified sea level rise projection for Southeast Florida, Southeast Florida Regional Climate Change Compact.* 2015, A document prepared for the Southeast Florida Regional Climate Change Compact Steering Committee.

Southeast Florida Regional Climate Change Compact, A Region Responds to a Changing Climate: Southeast Florida Regional Climate Change Compact Counties Regional Climate Action Plan. 2012, Southeast Florida Regional Climate Change Compact.

Southeast Florida Regional Climate Change Compact. *Home page.* 2016; Available from: http://www.southeastfloridaclimatecompact.org/.

Srivastava, R. and L. Laurian, *Natural hazard mitigation in local comprehensive plans: The case of flood, wildfire and drought planning in Arizona.* Disaster Prevention and Management: An International Journal, 2006. **15**(3): p. 461-483.

Staletovich, J., Task force on rising seas says Miami-Dade County needs step-by-step plan, in The Miami Herald. 2014: Miami.

Staletovich, J., With conservation money, Florida lawmakers aim to foot other bills, in The Miami Herald. 2016: Miami, Fl.

State of Florida Hazard Mitigation Plan, S., 2016.

Steffen, W., P.J. Crutzen, and J.R. McNeill, *The Anthropocene: are humans now overwhelming the great forces of nature.* AMBIO: A Journal of the Human Environment, 2007. **36**(8): p. 614-621.

Steiner, R.L., et al., VMT-Based Traffic Impact Assessment: Development of a Trip Length Model. 2010.

Stocher, T., et al., *Ipcc Workshop on Sea Level Rise and Ice Sheet Instabilities: Workshop Report.* 2010, Intergovernmental Panel on Climate Change: Kuala Lumpur.

Stone Jr, B. and M.O. Rodgers, *Urban form and thermal efficiency: how the design of cities influences the urban heat island effect.* Journal of the American Planning Association, 2001. **67**(2): p. 186-198.

Strauss, B.H., S. Kulp, and A. Levermann, *Carbon choices determine US cities committed to futures below sea level.* Proceedings of the National Academy of Sciences, 2015. **112**(44): p. 13508-13513.

Susskind, L., et al., Managing Climate Change Risks in Coastal Communities: Strategies for Engagement, Readiness, and Adaptation. 2015, London, UK: Anthem Press.

Titus, J.G. and V. Narayanan, *The Probability of Sea Level Rise.* 1995, US Environmental Protection Agency: Washington, DC.

Titus, J.G., *Greenhouse effect, sea level rise, and coastal zone management.* Coastal Management, 1986. **14**(3): p. 147-171.

Trope, Y. and N. Liberman, *Construal-level theory of psychological distance.* Psychological review, 2010. **117**(2): p. 440.

U.S. Bureau of the Census, *2015 Population Estimates.* 2016.

U.S. Bureau of the Census, *Coastline Population Trends in the United States: 1960 to 2008*, U.S.B.o.t. Census, Editor. 2008, Department of Commerce: Washington, D.C.

U.S. Environmental Protection Agency, Technical Support Document 2014: Survey of existing state policies and programs that reduce power sector CO2 emissions. 2014: Washington, D.C.

U.S. National Map. 2013, United States National Map Accuracy Standards. U.S. Bureau of the Budget.

University of Florida Bureau of Economic and Business Research, Florida's population grows again after first decline since mid-1940s, in University of Florida News. 2010.

US Bureau of Labor Statistics, *Economy at a Glance*. 2015.

US Bureau of the Census Urban and Rural Population 1900 to 1990. 1995.

US Energy Information Administration, Energy Consumption Estimates per Capita by End-Use Sector Ranked by State. 2014.

US Energy Information Administration. *Florida State Profile and Energy Estimates*. 2016 10/10/2016]; Available from: http://www.eia.gov/state/?sid=FL.

van Aalst, M.K., T. Cannon, and I. Burton, Community level adaptation to climate change: The potential role of participatory community risk assessment. Global Environmental Change, 2008. **18**(1): p. 165-179.

Vella, K., et al., Voluntary Collaboration for Adaptive Governance The Southeast Florida Regional Climate Change Compact. Journal of Planning Education and Research, 2016: p. 0739456X16659700.

Warner, L.A., S. Galindo-Gonzalez, and M.S. Gutter Building Impactful Extension Programs by Understanding How People Change. 2014.

Weber, E.U., Experience-based and description-based perceptions of long-term risk: Why global warming does not scare us (yet). Climatic change, 2006. **77**(1-2): p. 103-120.

Webster, P.J., et al., Changes in Tropical Cyclone Number, Duration, and Intensity in a Warming Environment. Science, 2005. **309**: p. 1844-1846.

Wise, R.M., et al., Reconceptualising adaptation to climate change as part of pathways of change and response. Global Environmental Change, 2014. **28**: p. 325-336.

Wisner, B., et al., At Risk: Natural hazards, people's vulnerability and disasters. 2004, New York: Routledge.

Wong, P.P., et al., *Coastal systems and low-lying areas.* Climate change, 2014: p. 361-409.

Yohe, G., et al., The economic cost of greenhouse-induced sea-level rise for developed property in the United States. Climatic Change, 1996. **32**(4): p. 387-410.

Yohe, G., The cost of not holding back the sea: Toward a national sample of economic vulnerability. Coastal Management, 1990. **18**(4): p. 403-431.

Zaval, L., et al., *How warm days increase belief in global warming.* Nature Climate Change, 2014. **4**(2): p. 143-147.

Zhao, T., et al., Vegetation productivity consequences of human settlement growth in the eastern United States. Landscape ecology, 2012. **27**(8): p. 1149-1165.

Zhao, T., M.W. Horner, and J. Sulik, A geographic approach to sectoral carbon inventory: examining the balance between consumption-based emissions and land-use carbon sequestration in Florida. Annals of the Association of American Geographers, 2011. **101**(4): p. 752-763.

Zwick, P.D. and M.H. Carr, *Florida 2060: a population distribution scenario for the State of Florida.* A research project prepared for, 2006. **1000**.

CHAPTER 2

Florida Land Use and Land Cover Change in the Past 100 Years

Michael I. Volk[1], Thomas S. Hoctor[1], Belinda B. Nettles[2], Richard Hilsenbeck[3], Francis E. Putz[4], and Jon Oetting[5]

[1]*Center for Landscape Conservation Planning, Department of Landscape Architecture, University of Florida, FL;* [2]*Department of Urban and Regional Planning, University of Florida, Gainesville, FL;* [3]*The Nature Conservancy – Florida Chapter, Saint Augustine, FL;* [4]*Department of Biology, University of Florida, Gainesville, FL;* [5]*Florida Natural Areas Inventory, Tallahassee, FL*

This chapter provides an overview of land use and land cover change in Florida over the past 100 years and a summary of how it may change in the future. We begin by providing a baseline description of Florida's pre-1900 land cover, natural resource distribution, and biodiversity. This is followed by a description of major land use changes and trends related to transportation, agriculture, mining, urbanization, tourism, disruption of natural processes, and conservation from 1900 to the present. We also describe changes in land use and land cover caused by climate change. The chapter concludes with a discussion of current land use and land cover patterns, and the potential impacts of climate change and continued human population growth on the remaining natural and rural landscapes in Florida. Much has changed in Florida over the last century due to a combination of wetland draining, agriculture conversion, urban development, and establishment of several dominant exotic plant species, as well as accelerating sea level rise and shifting climate zones due to climate change.

Key Messages

- Land cover and land use within Florida have changed dramatically since pre-settlement times, primarily due to human activities, with significant impacts on ecosystems and biodiversity.
- Climate-related impacts on land cover, resulting from human-caused climate change, have also been documented in Florida.
- Patterns of historic land use and land cover change are important to quantify and visualize so that we can assess the degree to which natural systems have been impacted and changed by human activities.
- Florida still has highly significant cultural and natural landscapes, which provide important services to people, in addition to possessing intrinsic values separate from their value to humans.
- As future changes continue to occur as a result of climate change and population growth, it will be more important than ever to conduct careful land use planning and management so that we can preserve natural and cultural resources, and maintain the qualities that make Florida the special place that it is today.

Keywords

Land use; Land cover; Climate change; Transportation; Tourism; Agriculture; Mining; Urbanization; Population growth; Natural processes; Conservation

Historical Overview

Florida has a diverse history of land use and human settlement, coupled with a wide range of natural communities, high biodiversity, and abundant natural resources. Land use trends throughout the state's history have been directly influenced by the natural resources, geomorphology, and climate that exist within the state. In turn land use change caused by human populations has altered the natural features that existed prior to human settlement. In this chapter we define land cover as simply the physical characteristics of the earth's surface including natural communities and altered land cover types (e.g., rocks, water, ice, forest, wetlands, rangeland, desert, etc.) whereas land use refers to specific ways that humans are using land (e.g., pastures, crops, residential, commercial, industrial, mining, transportation, utilities, etc.). (NOAA 2015).

Since 1900, Florida has seen substantial changes in land use patterns and land cover. Even though people had lived in Florida for thousands of years prior to 1900, their overall impact had been minimal. The Native Americans altered the land by building settlements, cultivating fields, building mounds, establishing transportation routes, grading causeways, and digging canals and fishponds (Derr 1998). European explorers and settlers arrived in the 1500s, but much of Florida, particularly the central and southern regions, remained relatively undeveloped until the last decades of the 19th century. Significant increases in population and tourism were coincident with new development, facilitated by new railroads and highways, and inspired by an aggressive marketing campaign for new residents and visitors to come to the state (Derr 1998). In creating the ideal Florida community, destination, or attraction, developers directly and indirectly caused significant changes to the natural landscape and resources of the state, fragmenting and degrading natural landscapes, introducing invasive species, and exploiting natural resources.

In addition to development and tourism, Florida's agriculture and extraction industries also led to land cover changes. Agriculture, Florida's second largest industry, led to land clearing, drainage projects, the introduction of invasive species, and pollution. By the early 20th century, the lumber industry had cleared most of North Florida's old growth forests (Florida Natural Areas Inventory 2005). Mining removed natural land cover, altered soil composition, and often left behind large abandoned excavations in the landscape (Shukla et. al. n.d.). Large-scale crop farming operations significantly altered drainage patterns and impacted water resources, particularly in South Florida (Stone and Legg 1992). In response to the environmental degradation that was occurring, Florida started to implement more environmental protection and growth management policies beginning in the 1970s and 1980s (Davis 2009). Efforts were also made to set aside conservation areas and to create wildlife corridors (Florida Department of Environmental Protection 2015; Hoctor et al. 2015). In recent years (as of 2016), state support for these efforts has weakened, but many people and organizations are still actively working to maintain the natural heritage and resources that remain in Florida. The history of land use change and development within the state is particularly important to understand when making future land use decisions and choices about how to adapt to climate change. The land use decisions that we make today will affect the ability for natural systems to adapt to climate change tomorrow.

Pre-1900 Conditions

Much of Florida was sparsely developed until the late 1800s, and settlers built many of the state's early towns along the coasts and rivers in areas with natural ports. In 1900, four cities in Florida had populations greater than 10,000. These were Jacksonville, Pensacola, Key West, and Tampa (U.S. Census 1910). Jacksonville was a well-established port town. Despite having two yellow fever epidemics in the 1880s that drove away nearly half the population, Jacksonville still had a population of more than 28,000 in 1900 (U.S. Census 1910). Pensacola, initially settled by the Spanish in the 1600s, was a thriving town due to its lumber industry and harbor. Key West was briefly settled in the 1500s and resettled in the 1800s. The town was a major source of salt during the first half of the 1800s. Other important industries included salvage, fishing, and turtling. Tampa rapidly grew to more than 15,000 people as the result of a development boom that began in the 1880s with the arrival of Henry Plant's railroad (U.S. Census 1910). The phosphate industry, cigar industry, and the influx of Spanish, Cuban, and Italian immigrants contributed to the growth of Tampa and neighboring Ybor City and West Tampa.

Orlando and Miami, which are now both major metropolitan areas and tourist destinations, were relatively small in 1900. Orlando had a population of less than 2,500, which still ranked it as one of the top 15 largest cities in Florida (U.S. Census 1910). Orlando had been the hub of the citrus industry in the late 1800s, but the Great Freeze of 1895-1896 caused many citrus growers to move further south. Miami was just beginning to boom at the turn of the century. It had been a small frontier town with a population of about 400 when Henry Flagler's railroad reached the area in the 1896 (City of Miami 2016).

Forests, including longleaf pine (*Pinus palustris*) forests, covered much of North and Central Florida prior to development. Longleaf pine forests are characterized by widely spaced trees, a wiregrass understory (*Aristida stricta*), and a very high level of biodiversity (Myers and Ewel 1990). These forests were logged extensively and used for naval stores. Sawmills operated in North Florida as early as the 1830s, and the industry was well-established by the 1850s although old growth forests remained in North Florida until the 1920s.

In 1900, South Florida remained relatively natural. Native Americans and early settlers had been in the area, but they had minimal impact as they lived primarily on subsistence farming and small-scale extraction. The Everglades dominate this region, and it contains seven ecosystems: cypress, freshwater marl prairie, freshwater slough, coastal lowlands, mangrove, pinelands, and hardwood hammock (National Park Service 2016). Attempts to drain the Everglades began in the 19th century, but these efforts had not yet made a significant impact.

Florida was (and still is) a biologically diverse state, though loss and fragmentation of habitat, introduction of non-native species, overexploitation of resources, pollution, and disease have reduced biodiversity. Since development had been limited, few species were lost prior to 1900. The four vertebrate species lost were the Passenger Pigeon (*Ectopistes migratorius*), Carolina Parakeet (*Conuropsis carolinensis*), Red Wolf (*Canis rufus*), and Plains Bison (*Bison bison bison*) (Endries et al. 2009). One major impact to the state's biodiversity prior to 1900 was plume hunting, which devastated bird populations during the last decades of the 1800s. During the Victorian era, hats adorned with feathered plumes were fashionable and the high price of feathers led to millions of bird deaths each year. In Florida, the decimation of bird populations occurred first in the northern areas of the state and later in the Everglades. By the turn of the century, 95% of the state's shore birds had been killed (Burns 2009). This led to legislation in 1891 and 1901 to protect plume birds, and other bird and wildlife protection laws passed in the early 20th century that greatly reduced the impact of market hunting on birds and other species (Palmer 1902).

Major Post-1900 Land Use Changes

Not much of Florida's growth occurred during the last decades of the 19th century and the 20th century (Mohl 1996). New transportation infrastructure, land development, and tourism partnered to bring people to Florida. The major railroads and highways followed the Atlantic Coast, traversed the Panhandle from east to west, and ran north-south through the center of the state before curving to Tampa (Derr 1998). Early development typically followed transportation, so it primarily occurred along the Atlantic Coast and in the Tampa Bay region. Later development occurred in the Orlando region, the southern Gulf Coast, and along new transportation corridors. Land booms during the early and mid-20th century resulted in the development of new communities and the expansion of low-density suburbia across many parts of the state. Automobiles, window screens, the yellow fever vaccine, ice factories and refrigerators contributed to the first major boom in the 1920s (Derr 1998). The next major boom occurred in the 1950s following World War II. Affordable financing options and new construction techniques made homes more affordable for the middle class contributing to a nationwide housing boom (Jackson 1995; Rome 2001). In Florida, the increasing use of air conditioning provided more year-round comfort. Even though they were not extensively used in residential homes until the late 1960s, the use of air conditioning in hotels, apartments, and commercial buildings increased in the 1950s (Derr 1998).

Agricultural and extraction activities played a major role in shaping the state's current remaining natural land cover and resources. Major agricultural crops included tung oil, citrus (*Citrus spp.*), and sugarcane (*Saccharum spp.*) in addition to row crops such as strawberries and tomatoes. Drainage projects were used to dry up wetlands in an effort to create new agricultural land. The cattle industry is also important in Florida and led to land clearing to create pastureland as well as using (with some conversion to improved pasture) the natural prairies of South Central

Florida for cattle production. The lumber industry removed most of the old growth forest, including the longleaf pine forests and cypress (*Taxodium spp.*), though people replanted some pineland areas with weaker species of slash (*Pinus elliottii*) and loblolly pines (*Pinus taeda*). Cypress was logged out later because it was spread through much of the state and was harder to reach (Harris 1999). Mining operations cleared large tracts of land and altered the terrain and soil composition. In response to these changes, subsequent growth management policies have attempted to more carefully guide and manage development and protect the state's natural resources, though persistent urban and suburban development continues to convert rural lands across the state.

Transportation Development

Transportation networks helped drive the development and growth patterns in the state. Until the late 1800s and early 1900s, inland transportation through the state was difficult and many areas were accessible primarily by boat. However, shipping and travel by water could be unreliable due to weather conditions. Just prior to the Civil War, David Yulee completed the first cross-state railroad from Fernandina (now Fernandina Beach) to Cedar Key, two of the state's major ports at the time. Yet, further railroad developments were delayed until after Reconstruction (Turner 2003).

Beginning in the 1880s, two men developed extensive railroad lines throughout the state. Henry Flagler constructed the East Coast Railroad along the Atlantic Coast with his line reaching the Florida Keys in 1912 (Willing 1957). Henry Plant's extensive railroad system primarily connected northeast Florida to the state's West Coast (Johnson 1966). Railroads improved transportation for people, produce, and goods because it was faster and more reliable than transportation by water (Derr 1998).

In the early 1900s, automobile travel was increasing, but roadways were limited. Trail associations formed to select, improve, and promote interstate roadways, creating two major trails in Florida. The Old Spanish Trail connected St. Augustine to San Diego. The Dixie Highway, promoted heavily by Carl Fisher of Miami, ran north-south from Michigan to Miami. These named trails posed challenges to travelers because some trails, like the Dixie Highway, were a series of roads, all with the same name, that allowed travelers to take alternate routes through the state. To alleviate the problem, the government adopted the U.S. Highway numbering system in 1926 (Weingroff 2015).

Since the mid-20th century, turnpikes and interstate highways have been constructed throughout the state. Florida's Turnpike system began in the mid-1950s and is comprised of a number of toll roads. The main line runs from Wildwood through Orlando to Miami. Another line, the Suncoast Parkway runs north-south through Hernando and Pasco counties into Tampa. Shorter routes are located near Ft. Lauderdale, Orlando, Lakeland, and Tampa. Florida is still considering additional toll road projects including a new highway between Orlando and Melbourne, Tampa to Jacksonville, and across parts of the Florida Panhandle. However, these projects are controversial as development patterns begin to change, transportation infrastructure costs increase, and the social and environmental impacts are considered (Warren 2016).

The Federal-Aid Highway Act of 1956 legislated a federal-state partnership to build interstate highways. These highways followed transportation corridors in the state similar to those of earlier highways and railways. I-95 runs along the East Coast. I-75 enters Florida near Jasper and runs north-south through the center of the state before turning westward to Tampa. Tampa was the initial terminus for I-75, but the route was extended down the Gulf Coast and across the Everglades to Miami. I-4 connects Tampa to Daytona Beach, and I-10 traverses east-west along the Panhandle to Jacksonville. The Interstate System connected Florida's major cities and facilitated development along its corridors. These multi-lane, high-speed roads also created barriers for wildlife and fragmented habitat, though some recent wildlife crossing structure projects have attempted to mitigate these impacts (Buford 2015; Land and Lotz n.d.). One of the last sections to be completed was the I-75 extension across the Everglades. This section of highway was designed with extensive bridges to provide for hydrologic flow through the Everglades, and it included wildlife underpasses in an attempt to reduce the number of animals killed, particularly the Florida Panther.

Tourism
Tourism in Florida began growing in the 1870s and remained strong until the Great Depression (Youngs 2005). Early tourists were often invalids that came to Florida for the salubrious climate and springs. The state's natural attractions also drew tourists to hunt, fish, and stroll. Steam boating along the rivers was one of the main modes of travel and a favorite pastime until the 1890s, and tourists often killed native wildlife from the boats (Noll 2004).

At the turn of the century, Henry Flagler and Henry Plant built luxury hotels along their railroad lines that drew wealthy tourists to Florida to spend the winter season. Flagler's eight hotels were located along the East Coast in Atlantic Beach, St. Augustine, Ormond Beach, Palm Beach, and Miami (Braden 2002). Plant built or acquired nine hotels in the center of the state and near the West Coast. His two main luxury hotels were the Tampa Bay Hotel (now part of the University of Tampa) and the Bellview-Biltmore in Clearwater, of which only a small portion remains (Braden 2002).

As the Depression waned in the late 1930s, developers built roadside attractions to draw visitors back to the state. Many early attractions focused on the state's 'natural' elements. These attractions entertained visitors with water shows, animal acts, and lush gardens. Some of the earlier parks included Cypress Gardens in Winter Haven, Jungle Gardens in Sarasota, and Marineland near St. Augustine. While some attractions retained more natural features, others made significant changes to the landscape. For example, the 'natural' beauty of Cypress Gardens was created by digging canals and planting thousands of flowering plants (Branch 2002).

In 1971, one of the world's most visited attractions opened in Orlando – Walt Disney World. This changed Florida forever, including changing Orlando from a small town into a massive metropolitan area. During the 1960s, Walt Disney purchased over 40 square miles of land in Central Florida, just south of Orlando, to build his East Coast theme park. Disney World also drew many new tourists to Florida, which increased the state's exposure to a new wave of immigrants attracted to the climate and the economy. Other large attractions, such as Sea World and Universal Studios, were developed near Walt Disney World making Orlando the largest tourist destination in the U.S. Hotels, restaurants, shopping, and smaller attractions were also built nearby contributing to the area's sprawl and congestion with suburbs and populations expanding to support the tourist industry. In 2014, Orlando became the first U.S. city to have over 60 million total visitors in one year, a number which includes in-state visitors (Dineen 2016).

Agricultural Development
Agriculture is Florida's second largest industry, and it has helped shape land use patterns and influence natural land cover. Some of the major crops include tung oil, citrus, and sugarcane. Livestock and timber are also significant industries.

Timber for logging and naval stores became a major industry in Florida in the 1830s with Florida as the world's leading producer of naval store in the early 1900s. However, the industry's practice of abandoning deforested land without replanting depleted most of the forests' old growth by 1930. As a result, mills closed down, towns were deserted, and the deforested land was abandoned. Around the same time, the Florida Forest Service began to promote reforestation with faster growing trees that could be used for pulp, such as loblolly and slash pines. Most of Florida's timber land is in the northern half of the state, and many of those counties have at least 50% of their land covered in pine forests (Florida Forestry Association 2016).

After the longleaf pine forests had been cut down, locals began looking for new industries. They tried satsuma oranges (*Citrus unshiu*), but freezes and fungus decimated the groves. Tung oil seemed like a good option with trees brought to the United States from China. The oil is used in products such as paint, ink, and linoleum, and it is used for waterproofing. The tree was introduced in Florida in the 1920s, and by the 1930s, 90% of the tung oil produced in the U.S. came from Alachua County (Robb and Travis 2013). However, one problem with tung oil is that the tree is invasive, and its leaves and seeds are poisonous. Due to alternative products and a series of hard freezes in the late 1960s, the tung oil industry is now nearly gone in Florida.

The Spanish brought sugarcane to St. Augustine, however, early attempts to grow it at a large scale in St. Augustine and New Smyrna failed due to freezes and soil conditions. In the 1920s, growers planted sugarcane in South Florida, and the industry grew after the U.S. embargoed Cuban sugar. Sugarcane is grown commercially south of Lake Okeechobee in Palm Beach, Hendry, Martin, and Glades counties. Florida is now the largest producer of sugarcane in the U.S., and it produces over 50% of the nation's cane sugar (Baucum and Rice 2009). However, the crop has significant impacts on surrounding land cover and water regimes because it requires water management to control seasonal flooding of the fields. Also, run-off from fertilizers contributes to algae blooms and the growth of invasive aquatic species, which choke out other native plants and grasses.

Spanish settlers also brought oranges to Florida in the 16th century. These plants eventually became naturalized to Florida and could be found growing amidst other trees. The citrus industry boomed in the 1870s with many groves along the St. Johns River. Yet, the freeze of 1895-96 destroyed groves and farmers moved further south. Additional freezes have occurred during the 20th century, and the industry continues to move further south (Davis 1937).

The Spanish settlers also brought cattle, with early cattle ranches located near Tallahassee, Gainesville, St. Augustine, and the St Johns River. The industry declined during the Civil War and did not recover until the 1920s due to problems with ticks and nutrition. After decades of research on nutrition, ranchers started relying more upon improved pastures versus pastures seeded with native grasses. Ranchers also began to depend more on maintenance of their own pastures after Florida passed a fence law in 1949 that ended open grazing.

Mining
The extraction of Florida's mineral resources contributed to the growth and decline of towns and impacted land cover and natural resources in the 20th century. Florida's main mineral resources include phosphate, limestone, and sand (Florida Department of Environmental Protection [FDEP] 2014). The state also contains deposits of heavy minerals that include zircon, leucoxene, ilmenite, and rutile (FDEP 2014).

Phosphate mining is a major industry in Central Florida, and the state produces about one-quarter of the world's phosphate (FDEP 2014). Phosphate was initially discovered in Alachua County in the late 1880s, but the first phosphate boom was in the Dunnellon area (Florida Industrial and Phosphate Research Institute [FIPR] 2016a). The industry later moved further south to the Polk County area, and Dunnellon's last mine closed in the 1960s. The City of Dunnellon continues to exist today, but other phosphate towns such as Romeo, LeRoy, Brewster, and Parkersburg do not. Early mining was done by hand, but this practice was later replaced by strip mining. Mining removes vegetation, alters drainage patterns and recharge, changes soil profiles, and destroys habitat. Processing phosphate is also water intensive, which has caused springs to dry up (Derr 1998). By 2000, more than 460 square miles of Florida had been mined for phosphate (FIPR 2016b).

The state's limestone, sand, and gravel are primarily used for road and building construction. Limestone has been quarried in Marion County since the early 1900s, and even though mines are located throughout the state, concentrations are still located in Marion and Miami-Dade County. Sand is mined throughout the state, but many mines are located in the Panhandle and Central Florida. Heavy minerals are mined in northeast Florida. Heavy minerals mining began in 1916 near the present day city of Ponte Vedra Beach. Two of these minerals, ilmenite and rutile, are used as pigments in manufacturing items such as paints, plastics, and paper (FDEP 2014).

Urbanization

Land booms have occurred multiple times in Florida resulting in rapid population growth and development. One of the first major land booms occurred after World War I. By this time, middle class people had the time, money, and means to travel to Florida. Automobiles and improving roadways made travel more accessible for middle class families, and Florida became a popular tourist destination. Cities developed to attract tourists, but also to meet the needs of visitors that were interested in buying homes. Developers built new communities to meet demand, and they altered the land to do so. For example, Carl Fisher cleared mangroves to build Miami Beach (George 1981). D.P. Davis dredged nearly 100 million cubic feet of sand to merge two small islands near Tampa into one, now known as Davis Islands, by covering the mudflats (History 2014). George Merrick designed and built the Mediterranean Revival community of Coral Gables, which included the construction of canals that offered gondola rides (Parks 2015). However, a few seasons of bad weather helped end the post-World War I boom by 1925.

Following World War II, development in Florida boomed again. New home financing options and improved construction techniques that lowered costs made homeownership possible for more people. Additionally, retirees were drawn to Florida for its climate and lower housing costs. Developers once again set out to build new communities in Florida. These developers utilized economies of scale to create large suburban communities that sometimes included shopping, schools, parks, and community centers (Nettles 2015). Once again, developers transformed the landscape by completely clearing large tracts of lands during construction. Developers also created new canals to maximize the amount of waterfront property. Some of these large housing developments were designed as new towns, such as Spring Hill or Beverly Hills, and other developments catered solely to retirees, such as Sun City. This boom slowed in the late 1960s due to an economic recession.

Since the 1970s, Florida has been a growth management state and has sought to regulate new development in an attempt to minimize infrastructure costs and environmental impacts. Florida's earliest land use regulation was the Zoning Enabling Act of 1928, which allowed local governments to control development by enacting and enforcing zoning codes (Arrant n.d.). No further regulation occurred until after Florida's mid-century boom, but in 1972 and 1973 Florida passed two planning statutes. The first created Regional Planning Councils (RPCs) to address regional land use issues and the impacts of large-scale developments. The other created Developments of Regional Impact (DRIs) and Areas of Critical State Concern (Arrant n.d.). DRIs are large development projects that impact more than one county, and are required to undergo an approval process that considers and mitigates the impacts. The DRI process has been scaled back since its inception, and the types of development it addresses was reduced in 2011. Areas of Critical State Concern are significant areas and natural resources that the state protects by overseeing local approvals for development. The state currently has five Areas of Critical State Concern: Big Cypress, Green Swamp, Florida Keys, Key West, and Apalachicola (Florida Department of Economic Opportunity 2016).

The next step in growth management was the Local Government Comprehensive Planning Act enacted in 1975, requiring local governments to have comprehensive land use plans. Nearly a decade later, in 1984, Florida adopted a State Comprehensive Plan with planning goals and action steps. The following year, Florida enacted the Growth Management Act. This revised the 1975 act by requiring local government plans and amendments to be adopted by ordinance and approved by the state. This act also required local governments to have Future Land Use Maps (FLUMs) and Land Development Regulations (LDRs) (Stroud 2012). Florida revised its comprehensive plan requirements again in 2011, this time significantly reducing the process for state review of local plans and generally relaxing local planning requirements (Shelley and Brodeen 2011). Although growth management policies have helped facilitate a coordinated land use planning process throughout the state, Florida is still highly impacted by rising populations and policies that incentivize development, making careful land use planning more important than ever.

Disruption of Natural Processes
Throughout Florida's history, people have disrupted natural processes to 'improve' the land. These efforts have included draining wetlands, converting forests to farm fields and citrus, introducing exotic species, and suppressing fires. Before people understood the causes of malaria or yellow fever, which were once prevalent in the state, they linked the diseases to swamp gases or miasma that came from standing water. Swamps were considered undesirable places with deleterious effects on health, and draining these areas was considered beneficial. Additionally, people believed that swamps and marshes, once drained, would make good agricultural land. In other cases, filling in marshes was a way to create more land for development. One of the largest of these projects was the draining of the Everglades. Early efforts began in the 1800s, but the initiative intensified in 1906 under Governor Napoleon Bonaparte Broward (Davis 2009). To drain the Everglades, a series of canals were dug to channelize and drain the water. The new canals often expanded or altered existing rivers, such as the Miami and Kissimmee rivers, but efforts to manage water and control flooding had limited success (Davis 2009). By the late 1970s and early 1980s, the state undertook plans to restore hydrology within the watershed where feasible. Efforts have also included dechannelizing and restoring the natural flow of the Kissimmee River, which serves as the headwaters of the Everglades and flows into Lake Okeechobee. Water treatment reservoirs have been built in several areas south of the Everglades Agricultural Area (EAA), and others are planned in areas throughout the Everglades watershed as part of the Comprehensive Everglades Restoration Plan (CERP) (U.S. Department of the Interior [DOI] 2016). Figure 2.1 provides a comparison of historic and current hydrology in the Everglades, as well as future hydrology as proposed to be restored under CERP, with primary flow patterns indicated by blue arrows.

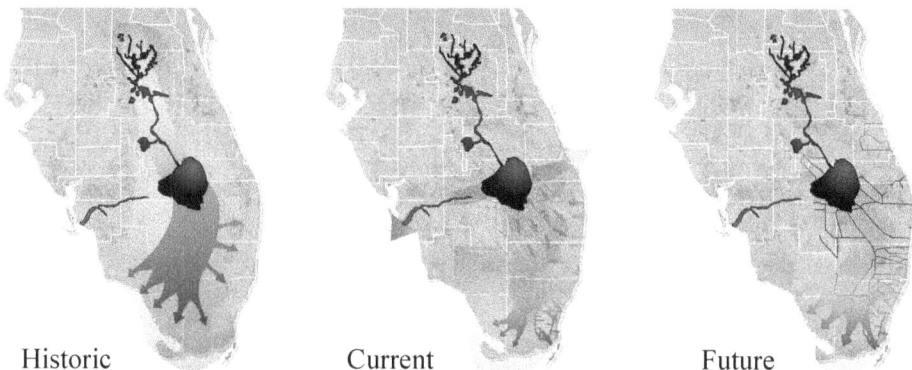

Historic   Current   Future

Figure 2.1. Historic, current, and future hydrologic patterns in the Everglades watershed. Future hydrologic patterns are those anticipated under the Comprehensive Everglades Restoration Plan (CERP). Primary flow patterns are shown by blue arrows, with canals indicated by blue and red lines. Image credit: Jacksonville District, U.S. Army Corps of Engineers.

Since the Spanish settlers arrived in the 1500s, people have been introducing new exotic species to Florida. As mentioned earlier, the Spanish brought citrus, sugarcane, and cattle as well as hogs. Some of the species that have been introduced are invasive and have led to changes in land cover. For example, melaleuca (*Melaleuca quinquenervia*) was introduced in the early 1900s and used as an ornamental, for erosion control, and in efforts to drain the Everglades (Serbesoff-King 2003; Silvers 2004). However, melaleuca often outcompetes native plants and does not provide habitat for most native species (Silvers 2004). Another example is the fast-growing Australian pine (*Casuarina spp.*), which mariners introduced to create windbreaks. Like melaleuca, these trees shaded out native vegetation and created areas that primarily contained one species (Pernas et al. 2013). The Brazilian pepper (*Schinus terebinthifolius*), introduced as an ornamental in the 19th century, has also invaded over 700,000 acres in Central and South Florida creating dense shrublands that shade out many other plants (Florida Fish and Wildlife Conservation Commission, 2016). The impacts of this exotic, invasive, and ecosystem-transforming species will be touched upon later in this chapter.

Many Florida ecosystems are dependent on fire including sandhills, flatwoods, and scrub. Frequent landscape-scale fires clear away undergrowth and help maintain open pine-dominated forests with high biodiversity. Fire suppression in Florida began in the 1930s to facilitate forest regeneration and protect areas of timber production. However, suppressing fires led to landscape-scale alterations in forest structure, and species dependent on fire-maintained forest, shrub, and grassland ecosystems declined precipitously. In recent years, natural and prescribed burns have been used to improve the health of fire-dependent ecosystems, but fire suppression is still a major issue across Florida (Florida Department of Agricultural and Consumer Services [FDACS] 2016).

Conservation

The formal conservation of lands and waters within Florida spans over a century, coinciding well with the land cover and land use changes detailed in this chapter. While the federal government created many of these protected areas, including the earliest, the sheer number, size, and natural resources conserved is impressive. Florida has been a magnet for conservation action by various governmental and private conservation entities because of its subtropical location, its peninsular geography, its many endemic and imperiled species, and its rapid development since the beginning of the 20th century.

President Theodore Roosevelt established Florida's first national wildlife refuge (NWR) and national forest. Roosevelt created the Pelican Island NWR in 1903 to protect wading bird populations from decimation by plume hunters. There are now 29 NWRs in Florida protecting hundreds of thousands of acres. Roosevelt also established the Ocala National Forest in 1908. Florida now supports three national forests covering over 1.2 million acres and oversees the 1,400-mile-long congressionally-authorized Florida National Scenic Trail. These national forests help protect at least 145 species of endangered, threatened and sensitive plant species and 51 such animal species. The National Park Service also manages large and diverse conservation areas including Everglades National Park and Big Cypress National Preserve in South Florida, plus national seashores and monuments across Florida.

The formation of Everglades National Park, recognized as a World Heritage Site and International Biosphere Reserve, began in 1915 when the Florida legislature gave 960 acres of land encompassing Royal Palm Hammock in Dade County to establish Royal Palm State Park. The legislature added 2,080 acres to the park in 1921. Congress authorized Everglades National Park in 1934, which included Key Largo and the Big Cypress Swamp. An additional 1.3 million acres was transferred to the federal government by Florida and donated or sold by several private landowners with Everglades National Park dedicated in December of 1947. The park now encompasses over 1.5 million acres and helps protect numerous federally imperiled species. However, water flows from Lake Okeechobee through the historic River of Grass to the park have been severely compromised with solutions still possibly decades away.

The Water Conservation Areas (WCAs) south and east of Lake Okeechobee were initially designated in the early 1900s by the Everglades Drainage District on state lands deeded to Florida by Roosevelt in 1904. The Central and South Florida Flood Control District expanded and formalized these areas during the development of the Central and South Florida Flood control project in the 1940s and 1950s. Additional lands were acquired by the state in the 1990s as a Save Our Rivers project. Covering nearly 850,000 acres, they are a critical component of South Florida's water management system. They are also extremely important for helping to recharge the Biscayne Aquifer, the major source of urban South Florida's drinking water supply.

Efforts at the state level to conserve Florida's biodiversity and water resources are likewise significant. Florida's first state park was initially acquired with private funds in 1929, and opened to the public two years later. Florida's State Park system now includes 161 state parks covering more than 800,000 acres. With over 100 miles of beach habitat and providing protection for thousands of Florida's plant and wildlife species, the park system attracts more than 25 million visitors annually.

Pine Log State Forest, located just north of Panama City, was established as Florida's first state forest in 1936. Now, 37 state forests, managed by the Florida Forest Service, protect nearly 1.1 million acres of productive habitats. Although it is difficult to pin down when the first state wildlife management area (WMA) was established, there are now over 150 Florida WMAs covering approximately 5.8 million acres. While natural resource and habitat management are important components of these areas, the hunting of game animals is one of the reasons for their popularity. Combined, the WMAs in Florida generate over 220,000 jobs and a $25 billion economic impact.

In more recent times, Florida has sought a more focused and science-based effort toward the conservation of its natural resources. In the 20 years between 1969 and 1989, the state protected approximately one million acres of land, with most acquired through fee simple (outright land purchases) during the latter half of that period. The major programs involved were the Environmentally Endangered Lands (EEL), Conservation and Recreation Lands (CARL), Save Our Coasts, and Save Our Rivers programs. In 1972, the Florida legislature enacted the Land Conservation Act, which created the EEL program, specifically designed to protect environmentally unique and irreplaceable lands (Knight et al. 2011). Governor Bob Graham (Democrat) is credited with creation of the CARL program, which was crafted by the Florida legislature in 1979 to acquire and manage public lands and to conserve and protect environmentally unique lands and areas of critical state concern.

In 1990, the CARL program was largely replaced by the Preservation 2000 (P2000) program, which was a 10-year, $3 billion land acquisition program funded annually through the sale of bonds (Farr and Brock 2006). At the time, it was the largest conservation land acquisition program in the country. The program's state governmental agency recipients, often with the help of private conservation organizations, purchased more than 1.7 million acres of new conservation lands under the program, including many of Florida's most important conservation holdings.

The state's second robust land protection program, Florida Forever (FF), succeeded P2000 with another 10-year, multi-billion-dollar commitment. The FF program was created under the leadership of Governor Jeb Bush (Republican) by the Florida legislature in 1999. It authorized the issuance of up to $3 billion in bonds for land acquisition, water resource development, preservation and restoration of open space, greenways and trails, and outdoor recreation (Farr and Brock 2006). The legislature also mandated public land protection agencies to focus on using alternatives to fee simple acquisition. Since 2000, FF has protected more than 1.3 million acres of water resources, environmentally sensitive lands, and parks (Department of Environmental Protection [DEP] 2016). Many of these areas have been protected through less-than-fee arrangements, involving purchase of conservation easements. Conservation easements allow lands to stay in private ownership but legally restrict what activities can occur on the land. At the same time, the land stays on the local tax rolls and under private management. Figure 2.2 shows conservation lands in Florida classified by date of protection based on the Florida Natural Areas Inventory Florida Conservation Lands dataset. This data includes both public and private lands that are either protected primarily for conservation or where conservation is an important activity (e.g., various military installations across Florida).

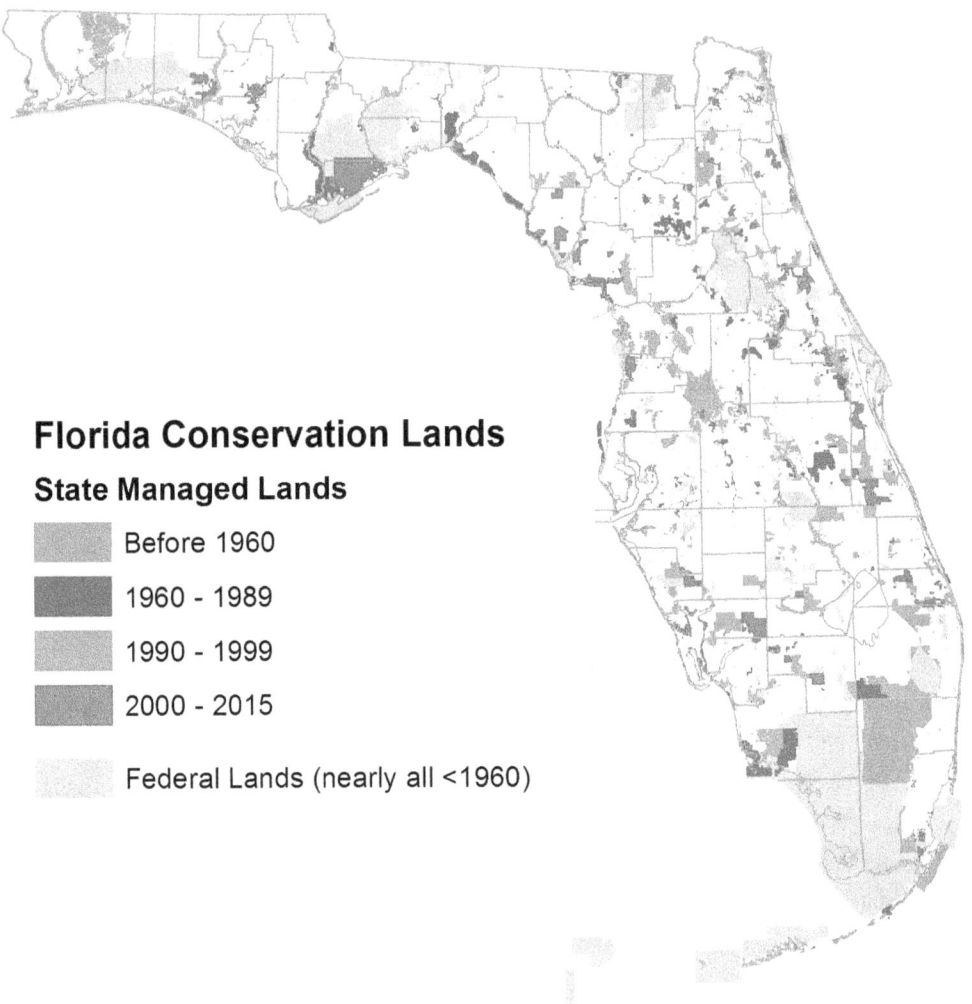

Figure 2.2. Florida conservation lands and date of protection. Data Source: Florida Natural Areas Inventory Florida Conservation Lands (Florida Natural Areas Inventory 2016)

Today Florida has a substantial acreage of its lands and waters in some kind of conservation designation. Combined local, state, and federal conservation holdings equate to 29.4% of the state, with 9,447,419 acres held in fee simple ownership and another 760,400 acres under conservation easements (this figure is slightly higher if private conservation lands and lands in private mitigation banks are included). These figures also include lands managed by the Department of Defense (e.g., Eglin Air Force Base) for conservation benefits (667,200 acres) and land under conservation easements held by the federal Natural Resources Conservation Service (121,122 acres). Figure 2.3 shows conservation and managed lands in Florida classified by managing entity.

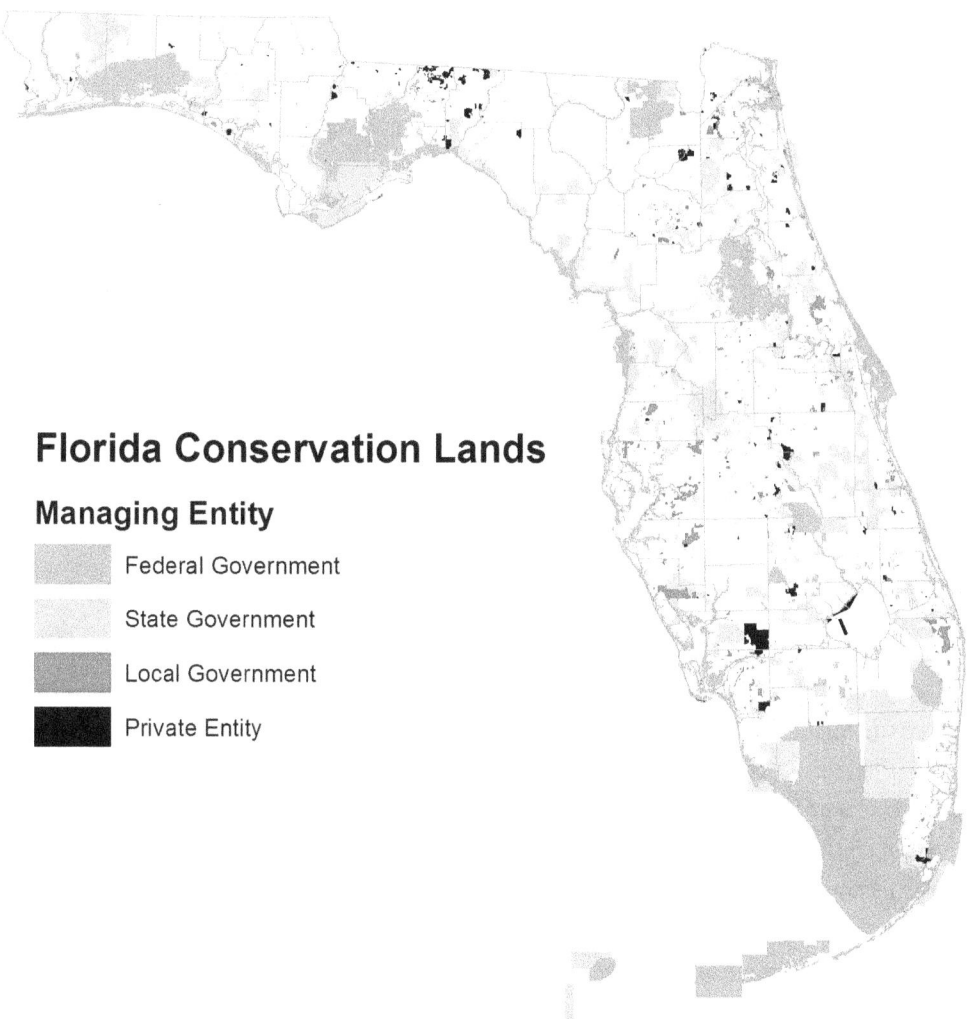

Figure 2.3. Florida conservation lands and managing entities. Data Source: Florida Natural Areas Inventory Florida Conservation Lands (Florida Natural Areas Inventory 2016).

While Florida's history of conservation achievement has been impressive, it has not been enough to prevent wholesale changes to and conversion of the Florida landscape – and its ecology – on a massive scale, and additional conservation land is needed to effectively conserve the still significantly, but quickly vanishing, natural resources that are still unprotected.

Climate Change Impacts on Land Use and Land Cover To-Date

It is important to understand how changes in climate are already impacting land cover within the state, as an indicator of what types of future changes may occur. Significant shifts in land cover are already being seen in both upland and coastal areas as a result of sea level rise, precipitation, and temperature change. Some of these are summarized here, though others are undoubtedly occurring or have been documented.

Coastal Land Cover Changes along the Big Bend Coast
Florida's Big Bend Coast, which stretches along the Gulf of Mexico from just north of Tampa to south of Tallahassee, remains the least developed coastal region in the lower 48 states of the U.S. The region's forests, salt marshes, and near coastal marine environments are legendary for their productivity due to a combination of factors including massive fresh water inputs from the Suwannee River and several first-magnitude springs. That said, it has certainly not been spared from human use, which is intensifying, while it is being rapidly inundated again by rising seas (Williams et al. 1999; de Santis et al. 2007; Raabe and Stumpf 2016).

An understanding of the area's geography is necessary to understand current changes. Like most of Florida, the area is low in elevation and topographically nearly flat; the highest points along the coast are the paleodune on Sea Horse Key and nearby Shell Mound, which started as a dune but was enhanced by Amerindian engineers more than 1000 years ago — neither of those peaks exceeds 20 meters in elevation. The flatness of the coastal region extends far out into the Gulf of Mexico.

Another distinctive characteristic of the Big Bend region relates to the abundant water, both fresh and somewhat salty. Tidal amplitudes in the Gulf of Mexico are small, less than half of those in the Atlantic. Massive influxes of fresh water keep the near-shore salinities in Gulf water at half of "normal" sea water. While all the rivers and spring runs moderate the Gulf's salinity, their waters do not deliver much sediment. Trapped sediments help build land, but the amount deposited on coastal areas of the Big Bend is a small fraction of that carried by rivers.

The natural ecosystems along the Big Bend reflect the impacts of sea salt, as determined by centimeters of elevational change and tempered by distance from the coast. As elevation rises from the open water of the Gulf one crosses sea grass flats, some of which are exposed at especially low tides, mud flats, and oyster bars. Salt marshes are next, often dominated by black needle rush (*Juncus roemerianus*) with saltmarsh cordgrass (*Spartina spp.*) on the depositional banks of meandering tidal streams. At slightly higher elevations, salt marsh shrubs flourish often under the dying crowns of the red cedars (*Juniperus virginiana*) and cabbage palms (*Sabal palmetto*) they replace as sea level rises. Healthy forested islands of cabbage palms and cedars surrounded by salt marsh are next at elevations of 50-60 cm where inundation is only by the highest of high tides as well as by storm surges, which can be several meters high. Inland and slightly uphill from the coastal hammocks of cedar and palms there can be slash pine flatwoods on sandy soils where fires are frequent or, more often, swamp forests referred to as hydric hammocks (Vince et al., 1989). These long hydroperiod wetlands can have as many as 25 species of canopy trees including several species of ash (*Fraxinus spp.*), oak (*Quercus spp.*), and elm (*Ulmus spp.*), with only scattered cabbage palms.

Big Bend ecosystems have been shaped by humans since they arrived some 14,000 years ago. For example, early as well as recent occupants harvested the bountiful shell and bony fish (McCarthy 2006). While modern clam farms benefit from new technologies, farming the sea has been underway for millennia, as indicated by archeological discoveries of massive fish weirs and managed oyster bars. Less than a century ago, pencil slat and brush factories near Cedar Key gobbled up thousands of red cedar and cabbage palm trees every month. Up until a few decades ago when the furniture mills closed, hardwoods were harvested from the hydric hammocks. To this day, stands of slash pine are clear-cut for pulp and saw timber, hunters seek deer, turkeys, bear, and other wildlife throughout the region while crabbers, oyster harvesters, and fisherman ply the coastal waters and up into the tidal creeks.

In addition to the direct effects of sea level rise, ecosystems in the Big Bend Region are being influenced by the decreasing frequency of the hard freezes that set the northern limits to the distributions of many plant species including mangrove trees (Williams et al. 2014). For reasons that are not yet clear, no tree species can withstand both high salinity and cold freezes. This means that where there are hard freezes, salt marshes predominate, whereas the coasts of warmer areas typically support mangrove forests. Black mangrove (*Avicennia germinans*), the most common species of mangrove tree in the region, can withstand super-high salinities but is killed by freezing temperatures. The northern limit of black mangrove is currently about halfway up the Big Bend Coast, but that limit continues to shift northwards as will be described in more detail in the next section. This massive switch from marsh to forest has numerous implications for the biota, biogeochemistry, and the effects of storm surges, which are blocked better by dense forests than by low-growing herbaceous vegetation.

The other big change underway along the Big Bend Coast is also related to global change, but is driven by Brazilian pepper, an invasive species described earlier in this chapter. Brazilian pepper is top-killed by hard freezes but, unlike mangrove trees, re-sprouts afterwards. The fact that it tolerates fairly high soil salinities allows it to proliferate where cabbage palms and cedar trees are succumbing to salt stress (Ewe and Sternberg 2005). Because it can grow taller than the live oak (*Quercus virginiana*) and other trees with which it competes, and can itself grow as a tree, a shrub, or a woody vine (Spector and Putz 2006), it crowds out native tree species in coastal hammocks up to about the same latitude as where mangroves stop.

Rising sea levels have driven the inland and uphill migration of Big Bend ecosystems since the end of the last glacial period some 14,000 years ago as they did during previous interglacials. The current rate of rise is by no means unprecedented; sea levels were rising more than twice as fast when paleoindians occupied the area. What is different now is that the uphill and inland migration is often impeded by the infrastructure of humans who are less willing to move than our coastal predecessors. Given the low human population density and sparsity of development along the Big Bend Coast, the financial consequences of sea level rise are modest in the aggregate while devastating for the people who do live in the region.

Changes in Mangrove Distribution within the Florida Peninsula

Mangrove forests consisting of black mangrove, red mangrove (*Rhizophora mangle*), and white mangrove (*Laguncularia racemose*) are a common coastal community on both the low energy Gulf and Atlantic shorelines in Florida. Along with tidal marshes and other coastal ecosystems, they provide a number of important services including carbon storage, shoreline protection and sediment accretion, water quality improvement, habitat for a number of important fish and wildlife species, as well as recreational opportunities (Osland et al. 2013).

The northern extent of each of the three mangrove species endemic to Florida varies due to differing resilience to freeze events. Precise range boundaries are difficult to determine, but over the past century, black and white mangroves have been found as far north as the Guana Tolomato Matanzas National Estuarine Research Reserve on the East Coast (Wunderlin and Hansen 2008; Zomlefer et al. 2006) and as far north as Cedar Key on the West Coast of Florida. Typically red mangroves are found further south than the other species due to a greater sensitivity to cold temperatures.

However, these ranges are not static, and as already noted change continues to occur. For example, in the Ten Thousand Islands, between 1927 and 2005 mangrove encroachment occurred upstream into salt and brackish marshes resulting in a roughly 35% increase in mangrove coverage (Krauss et al. 2011). Within the Tampa Bay region, Raabe et al. (2012) have documented conversion of marsh to mangrove habitat by comparing digitized nineteenth century topographic and public land surveys with 2005 digital land cover. Though specific conversion rates varied in different locations, the average ratio of non-mangrove to mangrove habitat over a 125-year period reversed from 86:14 to 25:75 across the four sites that they examined.

Depending on location, there are varying and interrelated reasons for these shifts that have been cited, including construction of waterways and interruption/reduction of freshwater flows (Krauss et al. 2011; Raabe et al. 2012), sea level rise (Krauss et al. 2011; Raabe et al. 2012), and changes in temperature resulting from climate change (Raabe et al. 2012; Williams et al. 2014) Storm disturbances have been a historic driver of change in forest structure (Doyle et al. 1995), and future changes may also be driven by precipitation (Ward et al. 2016),

South to north shifts in mangrove ranges seem particularly telling of the influence of climate change because, at least in Florida, the northern distribution of mangroves is limited by temperature. When mangroves begin to migrate further north, it is an indication that freeze events are no longer limiting colonization of mangroves in places where they have not recently existed. Northern migration of mangroves along the Atlantic Coast is now being documented and attributed to climate change. The frontline of this change is the Guana Tolomato Matanzas National Estuarine Research Reserve. In a 2013 study, Williams et al. (2014) surveyed the northernmost locations of black, red, and white mangroves, and compared those locations with historical data identifying the northern extent of these species. In the case of black mangroves, they found an occurrence 27 km north of the prior most northerly occurrence documented in 2007 (Wunderlin and Hansen 2008). They found a red mangrove 26 km north of the previous outlier documented in 2006 (Zomlefer et al. 2006) and a white mangrove occurrence approximately 67 km north of historic observations.

The overall future trend may be a gradual intrusion and northern expansion of mangroves into areas that have historically been dominated by saltmarshes or other types of coastal habitat. Whether this trend will continue, at what rate, and with what effect remains to be seen. At the very least, the changes that have occurred to date underscore the importance of minimizing human influences on these systems, including alteration of hydrology and freshwater flows. The ultimate impacts from climate change on coastal ecosystems is uncertain, but minimizing human impacts will help natural systems remain as resilient as possible to the changes that will occur.

Land Cover Changes in the Florida Keys
The Florida Keys are one of the most sensitive and at-risk regions in Florida with regard to climate change, and especially sea level rise, and changes to land cover to date are already being documented. One example is the loss of South Florida slash pine forest (*Pinus elliottii* var *densa*) as described by Alexander (1976) and Ross et al. (1993) in the Lower Keys. In a study on Key Largo, Alexander (1976) proposed that sea level rise was the cause of this loss, where flooded low-lying freshwater dependent pine communities had been replaced by more salt tolerant mangroves. A second and later study by Ross et al. (1993) reached the same conclusion through an examination of aerial photos and field evidence to compare historic and current distribution of pines on Sugarloaf Key. Ross et al. (1993) estimated the historical extent of pines on Sugarloaf Key to be approximately 217 acres prior to 1935. At the time of their study in 1991, it had been reduced to approximately 74 acres, with the earliest mortalities in areas with the lowest elevations. The areas of early pine mortality had been populated by new salt tolerant species. They also found that groundwater and soil water salinity were higher in areas of rapid pine forest reduction, and that pines in those areas exhibited higher physiological stress. At the time of their study, local sea levels had increased by 15 cm over the past 70 years, with the implication that further sea level rise would only increase the loss of upland pine communities. Ultimately, the entire Florida Keys as an upland ecosystem is endangered by projected sea level rise in the next century, which will necessitate consideration of various conservation strategies including potentially translocation of the many endangered species and subspecies found here (Noss et al. 2014).

The Impacts of Land Cover and Land Use Change on Florida's Climate
It is important to understand that the land cover and land use changes that have occurred in Florida have affected the climate — certainly in their contribution to the greater phenomenon of

global climate change, but also most likely at a regional and local level. These changes in turn affect land cover and land use in the future. As described elsewhere in this chapter, the pre-1900 landscape of Florida has been significantly altered by agriculture and urbanization. One impact of dense urbanization can be the "heat island effect," where urban areas actually cause an increase in local temperatures due to the absorption and re-radiation of solar heat by buildings and paved surfaces. Within an urban or suburban environment, local temperatures can vary based on the amount of tree cover and density of buildings and paved surfaces. For example, a study conducted by Sonne et al. (2000) in Melbourne, Florida showed average summer temperatures to be as much as 1.3 degrees cooler in an undeveloped, forested site when compared to an adjacent residential site with 4.6 houses/hectare and significant tree canopy. Average temperatures were up to 2.9 degrees cooler when compared to a residential site without trees and 10.1 houses/hectare.

At the peninsular scale, Marshall et al. (2004) conducted a series of simulations that found urbanization and agricultural conversion during the 20th century has contributed to a regional increase in summer temperatures and an average decrease in precipitation of 10-12% compared to the pre-1900 total. An earlier study by Pielke et al. (1999) in South Florida found similar results. Changes modeled by Marshall et al. (2004) were particularly apparent in portions of the interior peninsula that had been drained and converted to agricultural land, and land cover changes were also found to have significantly impacted sea-breeze circulation and strength. An important note is that this study was based on 1993 land cover data. It would be useful to know how land cover change since 1993 has affected temperature and precipitation since then, particularly given the continued rate of urbanization. Modeling studies have also shown that drainage and conversion of wetlands to agricultural uses has likely increased the frequency, severity, and duration of freezes in South Florida (Marshall et al. 2003). These simulations were also conducted via a comparison of models that used pre-1900 land cover and 1993 land cover, which showed that wetlands exhibit a moderating effect on sub-freezing temperatures.

Current Land Use and Land Cover in Florida

A comparison of historic land cover data and current land cover/land use is useful to provide a quantitative understanding of changes and potential impacts to date from land use change. Kautz (1998) provides the most recent source of a detailed comparison of long-term land use change within the last 100 years in Florida. Kautz (1998) describes patterns of land cover and land use change between 1936 and 1995, driven in large part by population growth, urbanization, and agricultural conversion. He notes that between 1936 and 1995, Florida's population grew from 1.7 million to 14.1 million residents, resulting in significant declines in natural land cover. This included a 60% increase in agricultural lands and a 632% increase in urban lands. Forest area overall decreased by 22%, with herbaceous wetlands decreasing by 51%. By 1995, longleaf pine forests had decreased by 90% from 1936 levels, slash pines had become the dominant pine species in Florida, and non-commercial forests were only 3% of the remaining forest lands in Florida. Interestingly, between 1980 and 1995 some of the trends described above began to reverse, with herbaceous wetland area actually increasing, and agricultural land area decreasing — likely due to urban conversion.

A careful comparison between the data provided by Kautz (1998) and current land use/land cover needs to be conducted to identify more recent land use trends. Since 1995, the population of Florida has increased to over 19 million, resulting in the conversion of more than 18% of land in Florida to urban land uses as of 2010 (Carr and Zwick 2016). Without a doubt, continued population growth within Florida has only exacerbated the conversion of natural and semi-natural lands to urban land uses, and expanded the loss of biodiversity and ecosystem services. However, in some cases there are projects that could result in restoration of certain natural communities to the extent that statewide statistics could be affected. This includes the Kissimmee River restoration and various other wetland restoration projects in the upper Everglades watershed (as well as some in other watersheds). In addition, there is some momentum for restoration of longleaf pine flatwoods, sandhill, and upland pine forests in North Florida that could result in significant increases in acreage of several upland natural communities in the near future (Regional Working Group 2009).

As a means of visualizing these changes, Figures 2.4 and 2.5 compare the extent of major natural land cover types prior to European settlement based on Davis (1967), with a 2003 version of land use/land cover data. Current land use data is frequently updated and there have certainly been land use changes in the state since the data used for Figure 2.5 was created, including additional expansion of urban land uses, so this comparison should be updated in the future.

The basis for Figure 2.5 is an early version of the Cooperative Land Cover (CLC) dataset. This dataset has become the most comprehensive and up-to-date source of Florida land cover spatial information, and is a starting point for identifying more recent statistics on current land use and land cover. It is currently produced cooperatively by the Florida Fish and Wildlife Conservation Commission and the Florida Natural Areas Inventory, with the latest version completed in October 2015. In the following section, we have created several tables of current land cover and land use categories based on the October 2015 version of the CLC and other relevant data sources to characterize the current Florida landscape. The statistics make clear that although the majority of Florida is still rural, much of that rural land is agriculture or other disturbed categories (including due to land clearing and fire suppression) and natural uplands have become increasingly rare (Table 2.1). Other than freshwater herbaceous and forested wetlands, the top 10 land cover categories combined from the CLC source data are dominated by urban (which lumps all intensive to low intensive developed land uses in this table), agriculture, tree plantations, or land cover classes that are most often indicators of fallow agriculture or disturbed natural communities including shrubs and other rural and mixed hardwood-coniferous. Mixed hardwood-coniferous forests could be considered "natural" and in some limited locations are natural communities, but in the current Florida landscape they are primarily the product of either oldfield succession on former farmlands or fire suppression of various fire-adapted natural communities including flatwoods, sandhill, upland pine, and scrub (Myers and Ewel 1990). The only other exception, and by far the largest of the natural upland natural community classes, is flatwoods, which have been largely replaced by tree plantations but still occur on public and private lands across the state.

Table 2.2 provides more detail on current remaining natural communities in Florida also based on the Florida Cooperative Land Cover Data from 2015. With the clearing of uplands for agriculture and development, it is not surprising that 7 out of the top ten natural communities based on remaining acres are wetland types. Table 2.3 provides statistics regarding acres of protected natural and semi-natural land cover. Protected is defined here as occurring in any area included in the Florida Natural Areas Inventory Florida Conservation Lands database. Conservation lands are disproportionately wetlands, which is not surprising given their lower development potential and the dominance of large wetlands in South Florida conservation lands.

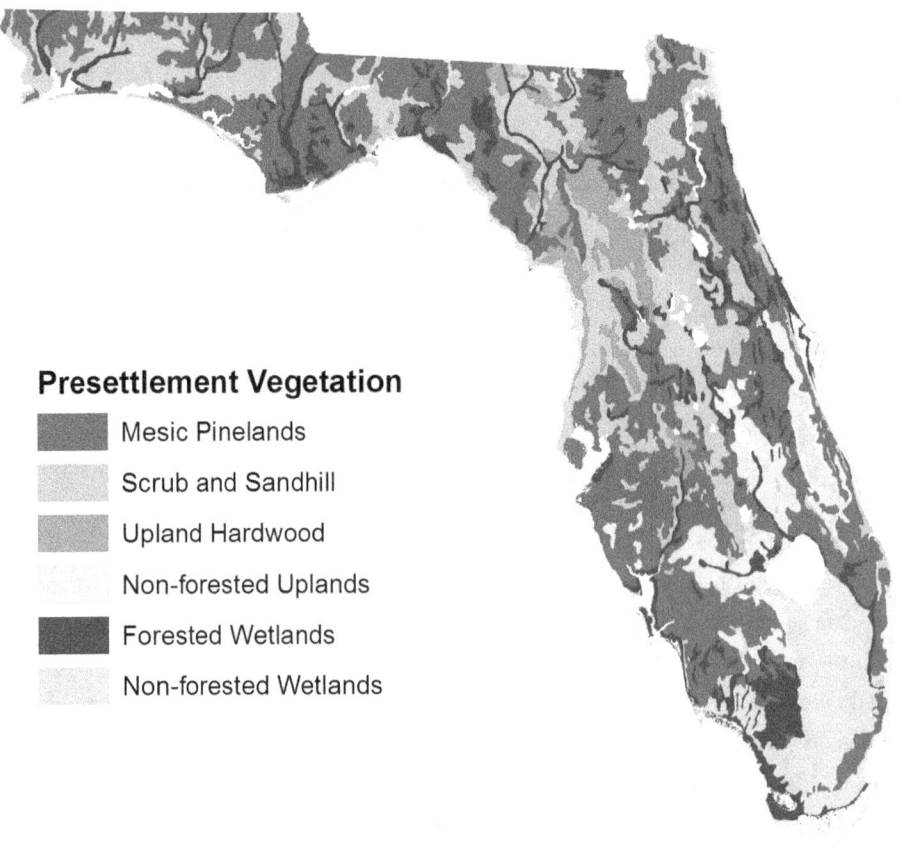

Figure 2.4. Pre-settlement vegetation map. Data Source: Davis (1967).

Figure 2.5. Current (2003) land cover. Data Source: FWC/FNAI Cooperative Land Cover, Version 1.0.

Table 2.1. Current major land cover classes based on Florida Cooperative Land Cover data (2015).

Land Cover Class	Acres	% of Statewide
Urban	5,664,034	15.76%
Freshwater Herbaceous Wetlands	4,637,696	12.91%
Freshwater Forested Wetlands	4,563,153	12.70%
Tree Plantations	4,516,626	12.57%
Pasturelands	4,094,759	11.40%
Crops, Groves, Nurseries	2,839,100	7.90%
Flatwoods	2,219,596	6.17%
Shrubs and Other Rural	1,923,632	5.35%
Mixed Hardwood-Coniferous	1,329,657	3.70%
Freshwater (Lakes, Ponds, Rivers, Streams)	1,310,344	3.64%
Sandhill and Upland Pine	943,053	2.62%
Scrub	784,757	2.18%
Mangroves	614,098	1.71%
Upland Hardwood Forest/Hammock	516,640	1.43%
Salt Marsh	378,678	1.05%
Extractive	256,978	0.71%
Dry Prairie	177,022	0.49%
Coastal Uplands	85,834	0.23%
Exotic Plants	66,089	0.18%
Rockland Forests	36,186	0.10%

Table 2.2. Current natural community types based on Florida Cooperative Land Cover data (2015).

Land Cover Class	Acres	% of Statewide	Land Cover Class	Acres	% of Statewide
Freshwater Forested Wetlands	2,676,694	17.21%	Cypress/Tupelo	92,145	0.59%
Marshes	2,435,732	15.66%	Isolated Freshwater Swamp	74,557	0.48%
Wet Prairies and Bogs	1,736,441	11.16%	Floodplain Marsh	49,974	0.32%
Mixed Hardwood-Coniferous	1,329,657	8.55%	Dome Swamp	48,862	0.31%
Mesic Flatwoods	1,325,011	8.52%	Strand Swamp	44,236	0.28%
Sandhill	775,755	4.99%	Tidal Flat	43,950	0.28%
Wet Flatwoods	761,947	4.90%	Scrub Mangrove	42,388	0.27%
Cypress	637,310	4.10%	Maritime Hammock	29,654	0.19%
Mangrove Swamp	571,710	3.68%	Other Coniferous Wetlands	26,800	0.17%
Floodplain Swamp	421,270	2.71%	Sand Beach (Dry)	24,386	0.16%
Salt Marsh	378,678	2.43%	Xeric Hammock	24,211	0.16%
High Pine and Scrub	290,829	1.87%	Other Hardwood Wetlands	23,022	0.15%
Isolated Freshwater Marsh	276,763	1.78%	Palmetto Prairie	21,131	0.14%
Hydric Hammock	240,562	1.55%	Coastal Scrub	19,554	0.13%
Upland Hardwood Forest	224,388	1.44%	Rockland Hammock	19,320	0.12%
Sand Pine Scrub	220,967	1.42%	Pine Rockland	16,866	0.11%
Basin Swamp	192,634	1.24%	Coastal Uplands	16,570	0.11%
Upland Pine	164,839	1.06%	Non-vegetated Wetland	13,828	0.09%
Scrub	159,788	1.03%	Keys Tidal Rock Barren	8,519	0.05%
Dry Prairie	155,891	1.00%	Coastal Strand	6,703	0.04%
Freshwater Non-Forested Wetland	138,786	0.89%	Slope Forest	5,875	0.04%
Mesic Hammock	126,285	0.81%	Dry Flatwoods	2,459	0.02%
Baygall	111,861	0.72%	Outcrop Communities	507	0.0033%
Pine Flatwoods/Dry Prairie	105,838	0.68%	Upland Glade	34	0.0002%
Scrubby Flatwoods	93,619	0.60%	**Total**	15,553,254	100.00%

Table 2.3. Acres of natural and semi-natural wetlands and uplands in conservation lands.

Description	Acres	Percent of Category
Natural Uplands in Conservation Lands	2,397,767	44.3%
Natural Uplands not in Conservation Lands	3,010,844	55.7%
Wetlands in Conservation Lands	6,155,458	56.2%
Wetlands not in Conservation Lands	4,798,293	43.8%
Semi-natural Uplands in Conservation Lands	1,167,281	14.9%
Semi-natural Uplands not in Conservation Lands	6,675,938	85.1%

Possible Future Changes in Land Use and Land Cover

Although the focus of this chapter is on historical land use and land cover changes to date, it would not be complete without a brief discussion of potential future changes from population growth and development. Climate change may also have significant impacts on land cover and land use throughout the state, but those topics are discussed in other chapters.

Though the rate of growth fluctuates, as of 2015 the population of Florida was increasing by approximately 1,000 people per day (O'Donnell 2015), requiring additional housing and infrastructure be built to accommodate them. A 2016 study of future population allocation and development scenarios for 2070 (Carr and Zwick 2016) showed the extensive impacts that continued development and land use change will have on existing agricultural and natural landscapes if current growth rates and development trends continue. In addition to creating a scenario that mapped future development at current trends and densities, Carr and Zwick (2016) created an alternative scenario that showed how future population growth in the state might be accommodated using higher densities and rates of infill. A baseline scenario that identifies 2010 land use patterns within the state was also created for comparison. Table 2.4, adapted from Carr and Zwick (2016), provides an acreage comparison of the 2010 baseline, a future "trend" scenario, and an alternative scenario for 2070, which shows the significant reduction in acreage of agricultural and other undeveloped lands that will occur if current population growth and development trends are maintained.

Coastal development will, at least in some places, be required to relocate inland in response to sea level rise, compounding the development pressure and impacts on existing undeveloped agricultural and natural landscapes. Where this occurs, the character and ability of inland landscapes to provide important agricultural and ecosystem services will be altered. Vargas et al. (2014), and Noss et al. (2014) have provided scenarios that show the potential impacts from "in-migration" of human population, combined with additional development to accommodate continued population growth. Figure 2.6 shows one such scenario for 2060, where future population growth was allocated throughout the state based on current trends and densities, and coastal populations impacted by a 1 m sea level rise were forced to relocate elsewhere within the state. Where coastal development remains in place, coastal protection and hardening structures may be used to stabilize the shoreline, which has been shown to cause significant damage to coastal ecosystems (Pilkey et al. 2009). Specific studies on the combined impacts of future development and sea level rise on biodiversity, natural communities, and landscape-level ecological priorities have been conducted by Noss et al. (2014) and the Florida Fish and Wildlife Conservation Commission (2008), with results showing that it is more critical than ever to carefully conduct future land use planning in a way that protects the resources critical to our state.

Table 2.4. An acreage comparison of Florida 2070 alternative population allocation scenarios not considering sea level rise (adapted from Carr and Zwick 2016).

	Baseline	% of Land	Trend	% of Land	Alternative	% of Land
Developed	6,412,000	18.56%	11,647,716	33.72%	9,777,000	28.30%
Protected (including protected agriculture)	10,870,000	31.47%	10,870,000	31.47%	18,647,664	53.98%
Agriculture (croplands, livestock, aquaculture)	7,518,267	21.76%	5,520,237	15.98%	4,827,759	13.98%
Other (mining, timber, unprotected natural areas)	9,742,733	28.20%	6,505,047	18.83%	1,290,577	3.74%
Totals (excluding open water)	34,543,000	100.00%	34,543,000	100.00%	34,543,000	100.00%

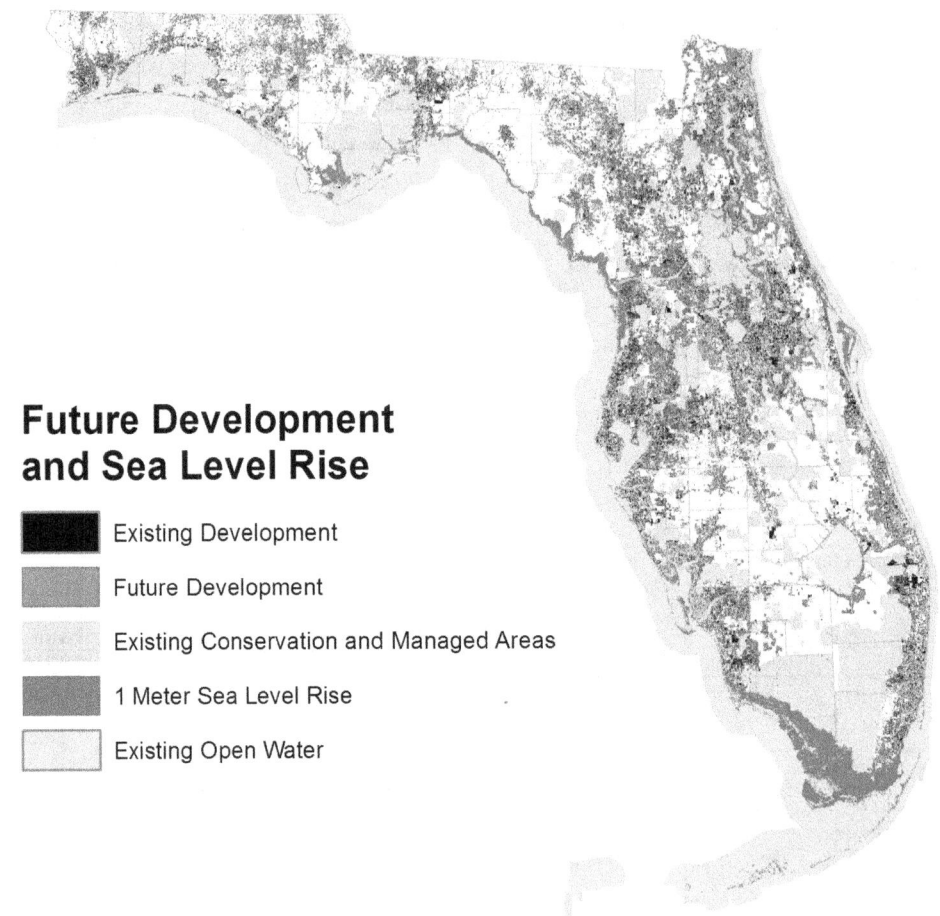

Figure 2.6. Future population allocation for Florida for 2060 with a one-meter sea level rise. Data Source: Noss et al. (2014), Florida Natural Areas Inventory Florida Conservation Lands (Florida Natural Areas Inventory, 2016)

Conclusion

It's clear from an assessment of historic land use and land cover patterns in Florida that significant changes have occurred since the beginning of the 20th century, primarily caused by human population growth, development, and activities such as agricultural production. Other natural phenomena have also caused shifts in the landscape, but the primary drivers have been humans. However, even today Florida still has highly significant cultural and natural landscapes and biodiversity that provide important services to the people that live in the state, in addition to possessing intrinsic values separate from their value to humans. As future changes continue to occur as a result of climate change and population growth, it will be more important than ever to conduct careful land use planning and management so that we can preserve these resources, and maintain the qualities that make Florida the special place that it is today.

References

Alexander, Taylor R. 1974. Evidence of Recent Sea Level Rise Derived from Ecological Studies on Key Largo, Florida. In *Environments of South Florida: Present and Past*, edited by Patrick. J. Gleason, 219-222. Miami, Fla.: Miami Geological Society.

Arrant, Tony A. n.d. Planning and Growth Management. In *Florida County Government Guide*, edited by Florida Association of Counties. Retrieved from http://factor.fl-counties.com/themes/bootstrap_subtheme/sitefinity/documents/growth-management-chapter.pdf

Baucum, Leslie E. and Ronald W. Rice. 2009. An Overview of Florida Sugarcane. IFAS Extension University of Florida. SS-AGR-232. Retrieved from http://ufdcimages.uflib.ufl.edu/IR/00/00/34/14/00001/SC03200.pdf

Braden, Susan R. 2002. *The Architecture of Leisure: The Florida Resort Hotels of Henry Flagler and Henry Plant*. Gainesville, Fla.: University Press of Florida.

Branch, Stephen E. 2002. The Salesman and His Swamp: Dick Pope's Cypress Gardens. *The Florida Historical Quarterly, 80*(4): 483-503. Retrieved from http://www.jstor.org/stable/30146373

Buford, Daniel. 2015. Wildlife and Highways: An Overview. U.S. Department of Transportation. Federal Highway Administration. Retrieved from http://www.fhwa.dot.gov/environment/critter_crossings/overview.cfm

Burns, Ken. 2009. The National Parks: America's Best Idea: Parks – Everglades. Retrieved from http://www.pbs.org/nationalparks/parks/everglades/

Carr, Margaret and Paul D. Zwick. 2016. Florida 2070 Mapping Florida's Future – Alternative Patterns of Development in 2070. Technical Report. 1000 Tallahassee, Fla.: 1000 Friends of Florida.

City of Miami. 2016. City of Miami: History. Retrieved from http://www.miamigov.com/home/history.html

Davis, Frederick. 1937. Early Orange Culture in Florida and the Epocal Cold of 1835. *The Florida Historical Quarterly 15*(4): 232-241.

Davis, John H. 1967. General Map of Natural Vegetation of Florida. Institute of Food and Agricultural Sciences. Gainesville: University of Florida.

Davis, Jack E. 2009. *An Everglades Providence*: Marjory Stoneman Douglas and the American Environmental Century. Athens, GA: University of Georgia Press.

Derr, Mark. 1998. Some Kind of Paradise: A Chronicle of Man and the Land in Florida. Gainesville, Fla.: University Press of Florida.

De Santis Larisa R. G., Smriti Bhotika, Kimberlyn Williams, and Francis E. Putz. 2007 Sea-level rise and drought interactions accelerate forest decline on the Gulf Coast of Florida, USA. *Global Change Biology 13*: 2349-2360.

Dineen, Caitlin. 2016, February 18. "Florida sees record tourism numbers in 2015." *Orlando Sentinel*. Retrieved from http://www.orlandosentinel.com/business/consumer/os-florida-record-tourism-2015-20160218-story.html

Doyle, Thomas W., Thomas J. Smith III, and Michael. B. Robblee. 1995. Wind Damage Effects of Hurricane Andrew on Mangrove Communities along the Southwest Coast of Florida, USA. *Journal of Coastal Research 21*:159–168.

Edenfield, Gray. 2014, June 11. David Levy Yulee. *From the Jailhouse*. Retrieved from http://ameliamuseum.blogspot.com/2014/06/david-yulees-history.html

Endries, Mark, Beth Stys, Gary Mohr, Georgia Kratimenos, Susan Langley, Karen Root, and Randy Kaut. 2009. Wildlife Habitat Conservation Needs in Florida: Updated Recommendations for Strategic Habitat Conservation Areas. FWRI Technical Report TR-15. Retrieved from http://myfwc.com/media/1205682/TR15.pdf

Ewe, Sharon M. L. and Leonel da Silveira Lobo Sternberg. 2005. Growth and gas exchange responses of Brazilian pepper (Schinus terebinthifolius) and native South Florida species to salinity. *Trees 19*: 119-128.

Farr, James A. and O. Greg Brock. 2006. Florida's Landmark Programs for Conservation and Recreation Land Acquisition. *Sustain 14*: 35–44.

Florida Department of Agriculture and Consumer Services. 2016. The Natural Role of Fire. Retrieved from http://www.freshfromflorida.com/Divisions-Offices/Florida-Forest-Service/Wildland-Fire/Prescribed-Fire/The-Natural-Role-of-Fire

Florida Department of Economic Opportunity. 2016. Areas of Critical State Concern Program. Retrieved from http://www.floridajobs.org/community-planning-and-development/programs/community-planning-table-of-contents/areas-of-critical-state-concern

Florida Department of Environmental Protection. 2014. Florida Minerals: Making Modern Life Possible. Retrieved from http://www.dep.state.fl.us/geology/geologictopics/minerals.htm

Florida Department of Environmental Protection. 2015. State Lands History. Retrieved from http://www.dep.state.fl.us/lands/history_more.htm

Florida Department of Environmental Protection (FDEP). 2016. Florida Forever. Retrieved from http://www.dep.state.fl.us/lands/fl_forever.htm

Florida Fish and Wildlife Conservation Commission. 2016. Brazilian Pepper. Retrieved from http://myfwc.com/wildlifehabitats/invasive-plants/weed-alerts/brazilian-pepper/

Florida Forestry Association. 2016. About Florida Forestry. Retrieved from http://floridaforest.org/about-us/fl-forests-facts/

Florida Industrial and Phosphate Research Institute. 2016a. Discovery of Phosphate in Florida. Retrieved from http://www.fipr.state.fl.us/about-us/phosphate-primer/discovery-of-phosphate-in-florida/#

Florida Industrial and Phosphate Research Institute. 2016b. Florida's Phosphate Deposits. Retrieved from http://www.fipr.state.fl.us/about-us/phosphate-primer/floridas-phosphate-deposits/

Florida Natural Areas Inventory. 2005. History: Timbering in North Florida. Retrieved from http://www.fnai.fsu.edu/arrow/almanac/history/history_forestry.cfm

Florida Natural Areas Inventory. 2016. Florida Conservation Lands Dataset. Retrieved from http://www.fnai.org/gisdata.cfm

George, Paul S. 1981. Passage to the New Eden: Tourism in Miami from Flagler through Everest G. Sewell. *The Florida Historical Quarterly 60*(4): 440-463.

Harris, L.D. 1999. Remembering Florida's Ancient Forests: Old Florida in Words and Pictures. Gainesville, Fla.: Florida Defenders of the Environment Special Bulletin.

History of Davis Islands. 2014, August 19. *South Tampa Magazine.* Retrieved from http://southtampamagazine.com/the-history-of-davis-islands/

Hoctor, Thomas, Reed Noss, Richard Hilsenbeck, Joe Guthrie, and Carlton Ward. 2015, November 11. The History of Florida Wildlife Corridor Science and Planning Efforts. Retrieved from http://floridawildlifecorridor.org/wp-content/uploads/2011/12/FWC_History_11_09_2015.pdf

Jackson, Kenneth T. 1985. *Crabgrass Frontier: The Suburbanization of the United States*. New York: Oxford University Press.

Johnson, Dudley S. 1966. Henry Bradley Plant and Florida. *The Florida Historical Quarterly 45*(2): 118-131.

Kautz, Randy S. 1998. Land Use and Land Cover Trends in Florida 1936-1995. *Florida Scientist 61*(3-4): 171-187.

Knight, Gary R., Jon B. Oetting, and Lou Cross, eds. 2011. *Atlas of Florida's Natural Heritage – Biodiversity, Landscapes, Stewardship, and Opportunities.* Tallahassee, FL: Institute of Science and Public Affairs, Florida State University.

Krauss, Ken W., Andrew S. From, Thomas W. Doyle, Terry J. Doyle, and Michael J. Barry. 2011. Sea-level Rise and Landscape Change Influence Mangrove Encroachment onto Marsh in the Ten Thousand Islands Region of Florida, USA. *Journal of Coastal Conservation 15*: 629. doi:10.1007/s11852-011-0153-4

Land, Darrell and Mark Lotz. n.d. Wildlife Crossing Designs and Use by Florida Panthers and Other Wildlife in Southwest Florida. Retrieved from http://fwcg.myfwc.com/docs/wildlife_crossings_large_mammal_crossings_swflorida.pdf

Marshall, Curtis H., Roger A. Pielke, and Louis T. Steyaert. 2003. Wetlands: Crop Freezes and Land-use Change in Florida. *Nature 426*: 29-30. doi:10.1038/426029a

Marshall, Curtis H., Roger A. Pielke, Louis T. Steyaert, and Debra A. Willard. 2004. The Impact of Anthropogenic Land-Cover Change on the Florida Peninsula Sea Breezes and Warm Season Sensible Weather. *Bulletin of the American Meteorological Society. 132*(1): 28-52. doi: http://dx.doi.org/10.1175/1520-0493(2004)132<0028:TIOALC>2.0.CO;2

Mohl, Raymond A. 1996. Asian Immigration to Florida. *The Florida Historical Quarterly 74*(3): 261-286.

Myers, Ronald L. and John J. Ewel. 1990. Ecosystems of Florida. Orlando: University of Central Florida Press.

National Oceanic and Atmospheric Administration (NOAA). 2015. What is the difference between land cover and land use? Retrieved from http://oceanservice.noaa.gov/facts/lclu.html

National Park Service. 2016. Natural Features & Ecosystems. Retrieved from https://www.nps.gov/ever/learn/nature/naturalfeaturesandecosystems.htm

Nettles, Belinda B. 2015. Suburban Landscape Evaluation Method: A Methodology Using GIS for Identifying and Evaluating Post-World War II Suburban Cultural Landscapes. Terminal Project. Gainesville, Fla: Department of Landscape Architecture, College of Design, Construction and Planning, University of Florida.

Noll, Steven. 2004. Steamboats, Cypress, and Tourism: An Ecological History of the Ocklawaha Valley in the Late Nineteenth Century. *The Florida Historical Quarterly 83*(1): 6-23.

Noss, Reed, Joshua Reece, Thomas Hoctor, Michael Volk, and Jonathan Oetting. 2014. *Adaptation to Sea-level Rise in Florida: Biological Conservation Priorities*. Final Report. Kresge Foundation, Troy, MI. Retrieved from http://conservation.dcp.ufl.edu/Project-Downloads.html

O'Donnell, Christopher. 2015, December 25. Squeezing Them In: Florida's Growing Population Will Come at a Price. *TBO.com.* Retrieved from http://www.tbo.com/news/politics/squeezing-them-in-floridas-growing-population-will-come-at-a-price-20151225/

Osland, Michael J., Nicholas Enwright, Richard H. Day, and Thomas W. Doyle. 2013. Winter Climate Change and Coastal Wetland Foundation Species: Salt Marshes vs. Mangrove Forests in the Southeastern United States. *Global Change Biology 19*: 1482–1494. doi: 10.1111/gcb.12126

Palmer, T. S. 1902. *Legislation for the Protection of Birds Other than Game Birds*. U.S. Department of Agriculture Division of Biological Survey Bulletin No 12, Revised Edition. Washington: Government Printing Office.

Parks, Arva Moore. 2015. *George Merrick, Son of the South Wind: Visionary Creator of Coral Gables*. Gainesville, Fla.: University Press of Florida.

Pernas, Tony, Greg Wheeler, Ken Langeland, Elizabeth Golden, Matthew Purcell, Jonathan Taylor, Karen Brown, D. Scott Taylor, and Ellen Allen. 2013. *Australian Pine Management Plan for Florida*. Florida Exotic Pest Plant Council. Retrieved from
http://www.fleppc.org/Manage_Plans/Casuarinamgmntplan_FINAL-05-13-13.pdf

Pielke, Roger A., Robert L. Walko, Louis T. Steyaert, Pier L. Vidale, Glen E. Liston, Walter A. Lyons, and Thomas N. Chase. 1999. The Influence of Anthropogenic Landscape Changes on Weather in South Florida. *Bulletin of the American Meteorological Society 127*(7): 1663-1673. doi: http://dx.doi.org/10.1175/1520-0493(1999)127<1663:TIOALC>2.0.CO;2

Pilkey, Orrin H., William J. Neal, and David M. Bush. 2009. Coastal Erosion. In *Coastal Zones and Estuaries*, edited by Federico Ignacio Isla and Oscar Iribarne, 32-42, Oxford, UK: EOLSS Publishers.

Raabe, Ellen A., Laura C. Roy, and Carole C. McIvor. 2012. Tampa Bay Coastal Wetlands: Nineteenth to Twentieth Century Tidal Marsh-to-Mangrove Conversion. *Estuaries and Coasts 35*:1145-1162. doi:10.1007/s12237-012-9503-1

Raabe Ellen and Richard P. Stumpf. (2016) Expansion of Tidal Marsh in Response to Sea-level Rise: Gulf Coast of Florida, USA. *Estuaries and Coasts 39*: 145-157.

Regional Working Group for America's Longleaf. 2009. Range-wide Conservation Plan for Longleaf Pine. Retrieved from http://www.americaslongleaf.org/media/86/conservation_plan.pdf

Robb, Jeffrey B. and Paul D. Travis. 2013. The Rise and Fall of the Gulf Coast Tung Oil Industry. *Forest History Today*. Retrieved from http://www.foresthistory.org/publications/FHT/FHTSpringFall2013/TungOil.pdf

Rome, Adam. 2001. *The Bulldozer in the Countryside: Suburban Sprawl and the Rise of American Environmentalism*. Cambridge: Cambridge University Press.

Ross, Michael S., Joseph J. O'Brien, Leonel da Silveira, and Lobo Sternberg. 1994. Sea-Level Rise and the Reduction in Pine Forests in the Florida Keys. *Ecological Applications 4*(1): 144-156.

Serbesoff-King, Kristina. 2003. Melaleuca in Florida: A Literature Review on the Taxonmy, Distribution, Biology, Ecology, Economic Importance and Control Measures. *Journal of Aquatic Plant Management 41*: 98-112.

Shelley, Linda Loomis and Karen Brodeen. (2011). Home Rule Redux: The Community Planning Act of 2011. *Florida Bar Journal 85*(7): 49-54.

Shukla, Manoj K., Rattan Lal, and M. H. Ebinger. n.d. Soil Physical Quality of Reclaimed Mine Soil Using FGD Products. Retrieved from https://www.netl.doe.gov/publications/proceedings/03/carbon-seq/PDFs/190.pdf

Silvers, Cressida. 2004. A Century of Melaleuca Invasion in South Florida. Retrieved from http://pesticide.ifas.ufl.edu/courses/pdfs/melaleuca/Melaleuca.pdf.

Sonne, Jeffrey K. and Robin K. Viera. 2000. "Cool Neighborhoods: The Measurement of Small Scale Heat Islands." Proceedings of 2000 Summer Study on Energy Efficiency in Buildings, American Council for an Energy-Efficient Economy, 1001 Connecticut Avenue, Washington, DC.

Spector, Tova. and Francis.E. Putz. 2006. Biomechanical Plasticity Facilitates Invasion by Brazilian Pepper (*Schinus terebinthifolius*). *Biological Invasions* 8: 255-260.

Stone, J. A. and D. E. Legg. 1992. Agriculture and the Everglades. *Journal of Soil and Water Conservation* 47(3): 207-215.

Stroud, Nancy. 2012. A History and New Turns in Florida's Growth Management Reform. *The John Marshall Law Review* 45(2): 397- 415.

Turner, Gregg. 2003. *A Short History of Florida Railroads*. Charleston, S.C.: Arcadia Publishing.

U.S. Census Bureau. 1910. Supplement for Florida. Retrieved from https://www2.census.gov/prod2/decennial/documents/41033935v9-14ch01.pdf

U.S. Department of the Interior. n.d. EvergladesRestoration.Gov. Retrieved from http://evergladesrestoration.gov/index.htm

Vargas, Juan C., Michael Flaxman, and Barry Fradkin. 2014. Landscape Conservation and Climate Change Scenarios for the State of Florida: A Decision Support System for Strategic Conservation. Summary for Decision Makers. GeoAdaptive LLC, Boston, MA and Geodesign Technologies Inc., San Francisco CA.

Vince, Susan W., Stephen R. Humphrey, and Robert W. Simons. 1989. The Ecology of Hydric Hammocks: A Community Profile. Technical Report. Washington D.C.: Fish and Wildlife Service. Retrieved from: http://www.nwrc.usgs.gov/techrpt/85-7-26.pdf

Ward, Raymond D., Daniel A. Friess, Richard H. Day, and Richard A. MacKenzie. 2016. Impacts of Climate Change on Mangrove Ecosystems: A Region by Region Overview. *Ecosystem Health and Sustainability* 2(4):e01211. doi:10.1002/ehs2.1211

Warren, April. 2016, February 13. Efforts to relieve jammed I-75 could face rocky road. *Gainesville.com*. Retrieved from http://www.gainesville.com/news/20160213/efforts-to-relieve-jammed-i-75-could-face-rocky-road/1

Weingroff, Richard F. 2015. U.S. Route 80 the Dixie Overland Highway. Retrieved from https://www.fhwa.dot.gov/infrastructure/us80.cfm

Williams, Asher A., Scott F. Eastman, Wendy E. Eash-Loucks, Matthew E. Kimball, Michael L. Lehmann, and John D. Parker. 2014. Record Northernmost Endemic Mangroves on the United States Atlantic Coast with a Note on Latitudinal Migration. *Southeastern Naturalist.* 13(1):56-63. doi: 10.1656/058.013.0104

Williams, Kimbelyn, Katherine C. Ewel, Richard P. Stumpf, Francis E. Putz, Thomas W. Workman. 1999. Sea-level Rise and Coastal Forest Retreat on the West Coast of Florida, USA. *Ecology* 80(6): 2045-2063.

Williams, Kimberlyn, Michelina MacDonald, Kelly McPherson and Thomas H. Mirti. 2007. Ecology of the Coastal Edge of Hydric Hammocks on the Gulf Coast of Florida. In *Ecology of Tidal Freshwater Forested Wetlands of the Southeastern United States*, edited by William H. Conner, Thomas W. Doyle, and Ken W. Krauss, 255-289. Netherlands: Springer.

Willing, David L. 1957. Florida's Overseas Railroad. *The Florida Historical Quarterly* 35(4): 287-302.

Wunderlin, Richard P., Bruce F. Hansen, Alan R. Franck, and F. B. Essig. 2016. *Atlas of Florida Plants*. http://florida.plantatlas.usf.edu/. USF Water Institute.] Institute for Systematic Botany, University of South Florida, Tampa.

Youngs, Larry R. 2005. The Sporting Set Winters in Florida: Fertile Ground for the Leisure Revolution, 1870-1930. *The Florida Historical Quarterly,* 84(1): 57-78.

Zomlefer, Wendy B., Walter S. Judd, and David E. Giannasi. 2006. Northernmost Limit of *Rhizophora mangle* (Red Mangrove; Rhizophoraceae) in St. Johns County, Florida. *Castanea* 71(3):239–244. doi: http://dx.doi.org/10.2179/05-33.1

CHAPTER 3

Implications of Climate Change on Florida's Water Resources

Jayantha Obeysekera[1], Wendy Graham[2], Michael C. Sukop[3], Tirusew Asefa[4], Dingbao Wang[5], Kebreab Ghebremichael[6], and Benjamin Mwashote[7]

[1]*South Florida Water Management District, West Palm Beach, FL;* [2]*Water Institute and Department of Agricultural and Biological Engineering, University of Florida, Gainesville, FL;* [3]*Department of Earth and Environment, Florida International University, Miami, FL;* [4]*Tampa Bay Water, Clearwater, FL;* [5]*Department of Civil, Environmental, and Construction Engineering, University of Central Florida, Orlando, FL;* [6]*Patel College of Global Sustainability, University of South Florida, Tampa, FL;* [7]*School of the Environment, Florida Agricultural and Mechanical University, Tallahassee, FL*

Water resources systems in Florida are unique and exhibit significant diversity in hydrogeologic characteristics and in rainfall and temperature patterns. In many parts of the state, both surface and groundwater systems are complex, highly interconnected, and any change in hydrologic drivers such as rainfall or temperature has the potential to impact the water resources of the urban, agricultural, and ecological systems. Because of this diversity, it is not possible to present a single overall outlook regarding the implications of climate change on the water resources of the state. This chapter presents brief summaries of individual studies that are available for major water resources systems in the state, which include the Everglades, the Tampa Bay region, the St. Johns River watershed, and the Suwannee River and Apalachicola River basins. Available climate models and their downscaled versions have varying degrees of bias and lack of skill that need to be considered in impact analyses. In all regions, projected changes in rainfall, temperature, and sea level may have significant impacts on water supply, water levels in environmentally sensitive areas, flood protection, and water quality.

Key Messages

- Water resources are an integral contributor to Florida's economy, but there is increasing competition for water supply among the urban, agricultural, and environmental sectors due to population growth in the state.
- Climate change along with rising sea levels will exacerbate the competition for water and it is extremely important to understand the potential impacts on this vital resource through actionable science that is relevant to this region.
- Although different climate models predict a consistent increase in future temperatures, future precipitation is not yet consistently predicted and could be higher or lower. Differences in precipitation propagate into significant differences in future streamflow, groundwater levels, and ET predictions.
- The range of future hydrologic conditions predicted by climate models allows an evaluation of the spectrum of possible future risks, but does not provide actionable information because the uncertainty is so high. Improvement in the ability of the climate models to simulate both retrospective and future rainfall patterns will be required before their projections can reliably be used for water resource planning and management
- Impact assessment to date on large-scale, regional basins in the state demonstrates that future climate change has a significant potential to impact both water quantity and quality, and as a consequence, additional research is necessary to develop standardized climate projections and conduct impact assessment on the water resources systems on a statewide basis.

- Potential increases in temperature, and the variations in precipitation patterns may degrade water quality, exacerbate algae problems, and cause eutrophication of important water bodies.

Keywords

Rainfall; Temperature; Groundwater; Sea level rise; Water quality; GCM, Downscaling; Uncertainties

Introduction

The state of Florida includes more than 1,700 streams and rivers, 7,800 freshwater lakes, 700 springs, 11 million acres of wetlands, and numerous freshwater aquifers, all of which play an important role in meeting water needs of both humans and the environment (Marella 2015). Water use data in the state, which have been collected by the US Geological Survey, show that the combined fresh and saline water withdrawals have increased 465% (over 12,000 million gallons a day) between 1950 and 2010. During the same period, the population has increased by 580% (16 million) (Fig. 3.1 reproduced from Marella 2014). Increased withdrawal of freshwater for human use is triggering significant challenges for maintaining the water supply to environmentally sensitive areas, which are experiencing significant pressure from urbanization. The potential decrease in water supply due to future climate change, along with the contamination of freshwater aquifers from sea level rise, will exacerbate the challenges in meeting the water supply needs of Florida's urban, agricultural, and environmental sectors.

In Florida, 64% of the state's freshwater supply is from groundwater and it is a vital resource essential to public and private water supply, irrigation, aquaculture, and industrial use. Groundwater is recharged by infiltration of precipitation and seepage from canals, lakes, and streams. Groundwater flows down a hydraulic gradient and ultimately into wells, canals, streams, lakes, or to the ocean through seeps and springs, thus closing the continuous water cycle between land, ocean, and atmosphere (Mwashote et al. 2010, 2013). As groundwater use has increased in coastal areas, so has the recognition that groundwater supplies are vulnerable to overuse, contamination, and climate change impact. Any change in recharge and withdrawal from groundwater aquifers due to climate change has the potential to change the water budgets of the various parts of the state.

Man-made and natural water resources systems in Florida are unique, complex, and diverse. The landscape of Florida varies significantly from north to south, with different patterns and extents of urban, agricultural, and natural systems. The state's hydrologic systems are influenced by changes in rainfall patterns, evapotranspiration (ET), and sea level—the primary hydrologic drivers of both surface water and groundwater conditions (both quality and quantity). In addition, the supply and demand of urban, agricultural, and environmental sectors vary from one part of the state to another. Consequently, it is not possible to discuss the implications of climate change on water resources in the state as a single entity. There is currently no comprehensive statewide assessment of the potential impacts of climate change on water resources. Implications of potential changes to climate are being investigated by numerous institutions, and the number and quality of such studies are evolving rapidly. There have been some pilot efforts to assess the impacts of climate change on some regions of the state and those are considered to be the best available investigations to date. This chapter presents a summary of such investigations, focusing on some large water resources systems in the state and concluding with a general assessment of climate change implications on water quality in Florida.

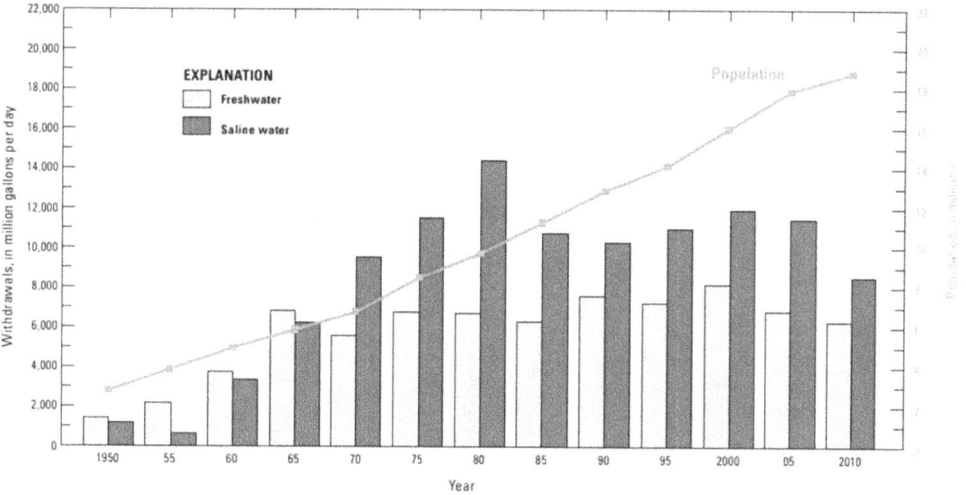

Figure 3.1. Historic total population, freshwater, and saline water withdrawals in Florida 1950–2010 (Marella 2014).

Major Water Resources Systems in Florida

Management of water resources in Florida is delegated to five water management districts (WMDs) that include the (Fig. 3.2): (a) South Florida Water Management District (SFWMD); (b) Southwest Florida Water Management District (SWFWMD); (c) St. Johns River Water Management District (SJRWMD); (d) Suwanee River Water Management District (SRWMD); and (e) Northwest Florida Water Management District (NWFWMD). In general, the WMDs administer flood protection, water supply, water quality, and environmental protection through a variety of functions including planning, operations, and regulation. There are numerous watersheds of varying size in the state; however, this chapter will only cover the current state of knowledge regarding climate change investigations associated with four large and important systems in the state. They include the following regions (maps showing them are provided later in this chapter):

- Greater Everglades Ecosystem

- Tampa Bay Region
- St. Johns River Region
- Suwannee River and Apalachicola River Basins

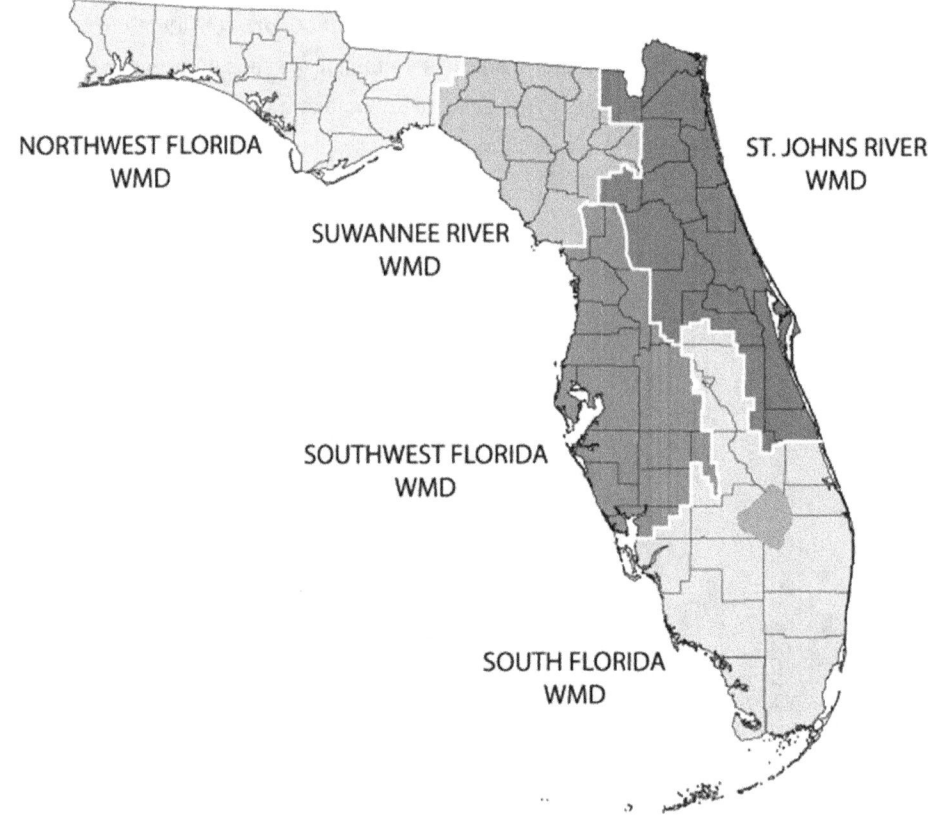

Figure 3.2. Five water management districts in Florida.

Greater Everglades Ecosystem

Background

The Greater Everglades Ecosystem spans from the Kissimmee River Basin, north of Lake Okeechobee all the way south to Florida Bay; and it includes a national treasure, America's Everglades. The climate of this ecosystem is strongly seasonal and exhibits significant interannual variability with prolonged drought and wet periods. The water resources system originates in the upper chain of lakes in the Kissimmee River Basin, and includes Lake Okeechobee, the Everglades Agricultural Area, the Water Conservation Areas, and Everglades National Park. It is bordered by heavily urbanized areas of Florida's lower East Coast (Fig. 3.3 showing areas below Lake Okeechobee). The natural, urban, and agricultural systems in the region are strongly interconnected in terms of supply and demand for water, and any change in rainfall patterns, evapotranspiration (ET), or sea level can have a significant impact on water quantity and quality in any or all of the systems. The urbanized lower east coast area has a unique geology due to the presence of the highly porous and permeable limestone aquifer known as the Biscayne Aquifer. Projected sea level rise will significantly influence saltwater intrusion along the coast affecting numerous wellfields that supply water to the rapidly expanding population.

Implications of Changing Climate

There have been several attempts to assess the implications of climate change and sea level rise on the Greater Everglades Ecosystem (e.g., Obeysekera et al. 2011). The general approach that has been used for such investigations is shown in Fig. 3.4 and it includes the detection of historical trends from observations, understanding the role of teleconnections, skill testing of global and regional climate models, and finally, the assessment of impacts on water resources systems.

Irizarry et al. (2013) evaluated the observed data with the objective of detecting trends in both temperature and precipitation in the entire state. The observations consisted of long-term (1892–2008) precipitation and raw temperature records at 32 stations distributed throughout the state. They used several climate metrics based on both averages and extremes. The trend detection techniques included the non-parametric Mann-Kendall Trend Test (Kendall 1976), Sen-Theil Regression (Sen 1968), and the nonstationary Generalized Extreme Value distribution fitting for the extremes. The results showed a general decrease in wet season precipitation, most evident for the month of May and possibly tied to a delay in the onset of the wet season. The number of wet days during the dry season, especially during November through January, were found to have increased over the period of record. The number of "dog days" (temperature above 26.7 °C during the wet season) per year has increased in many locations. A decrease in the daily temperature range was also observed and it was attributed to an increase in daily minimum temperature. Although there was no attempt to attribute the trends to climate change or anthropogenic causes, "urban heat island" effects were conjectured to have caused observed trends at some locations. In addition, climate teleconnections due to phenomena such as the Atlantic Multi-decadal Oscillation (AMO), the Pacific Decadal Oscillation (PDO), and the El Niño-Southern Oscillation (ENSO) have significant effects that are tied to seasonal and decadal trends (see Chapter 16), and it is difficult to separate observed trends into natural versus anthropogenic causes.

Following the approach in Fig. 3.4, there have been several attempts to evaluate the skills of the general circulation models (GCMs) of both the CMIP3 and CMIP5 suites of models (IPCC 2007, 2013). Comparison of simulation results of the late 20th century GCM output to historical data has shown that the GCMs do not capture the statistical characteristics of the regional rainfall patterns and temperature adequately. Due to the coarse resolution of most present-day GCMs, the region of south central Florida is not well represented. Some models do not adequately represent large areas of the land mass of Florida (Obeysekera et al. 2011). As a consequence, the models are unable to mimic temporal and spatial patterns resulting from mesoscale phenomena such as sea and lake breezes. Furthermore, it is not clear how well such models are able to simulate teleconnections to global phenomena such as the AMO, the ENSO, and the PDO (Trimble et al. 2006). Such teleconnections include, but are not limited to, wetter (drier) than normal precipitation during winter months of El Niño years (La Niña years), and wetter (drier) conditions during warm (cold) phase of both AMO and PDO.

There have been several attempts to downscale GCMs to produce higher-resolution historical rainfall and temperature records for water resources investigations. They have included both statistical downscaling (Maurer et al. 2007) and dynamical downscaling (Mearns et al. 2012; Stefanova et al. 2012) of temperature, precipitation and other climatic variables. Obeysekera et al. (2011) document the evaluation of 112 finer-resolution (1/8 degree), statistically-downscaled ensemble datasets based on 15 climate models of the CMIP3 scenarios B1, A1B, and A2 for the period 1950–2009 (Maurer et al. 2007). The analysis of bias-corrected, statistically-downscaled data showed that the simulation of climatology and the variability of temperature are adequate. However, the precipitation values still showed some biases. The dynamically-downscaled data (NARCCAP) were not much better, as they exhibited significant spatial biases although they mimicked the seasonal patterns of both precipitation and temperature well. The conclusion was that, even for dynamically-downscaled data, further bias correction may be necessary. A careful review of the downscaled products for Florida indicate that a reasonable range for percent change in annual rainfall is ± 5% for 2040 and ± 10% for 2070. For temperature, the corresponding range for 2040 is +0.5° to +1.5° C (+0.8° to +2.4° F) with a median value of +1° C (+1.6° F). For 2070, a reasonable planning range is +1° to +3° C (+1.6° to +4.8° F) with a median value of +2° C (+3.2° F) (Obeysekera et al. 2011; Dessalegne et al. 2016)

Efforts are underway to assess the CMIP5 suite of GCM models and the corresponding downscaled datasets. The bias-corrected constructed analogs (BCCA, Maurer et al. 2010) of precipitation and daily minimum and maximum temperature projections at 12 km resolution have been analyzed (Dessalegne et al. 2016). The analysis included identification of future trends in precipitation and temperature based on a total of 119 models covering three RCP scenarios for the period 1950–2099. In an attempt to identify trends in precipitation and temperature, percent change in precipitation for near future (2025–2055) and far future (2055-2085) as compared to change in mean annual temperature were computed. Spatial trends in temperature and precipitation as a function of latitude are shown in Fig. 3.5. The results show that the there is a robust increase in temperature as expected. However, trends in precipitation are scenario-dependent, with RCP85 showing the largest average increase up to about 10%. Some models do show a reduction in precipitation (Fig. 3.5b) and, as with CMIP3 models, precipitation change is more uncertain than change in temperature. However, in all cases, temperature increases are expected in the future.

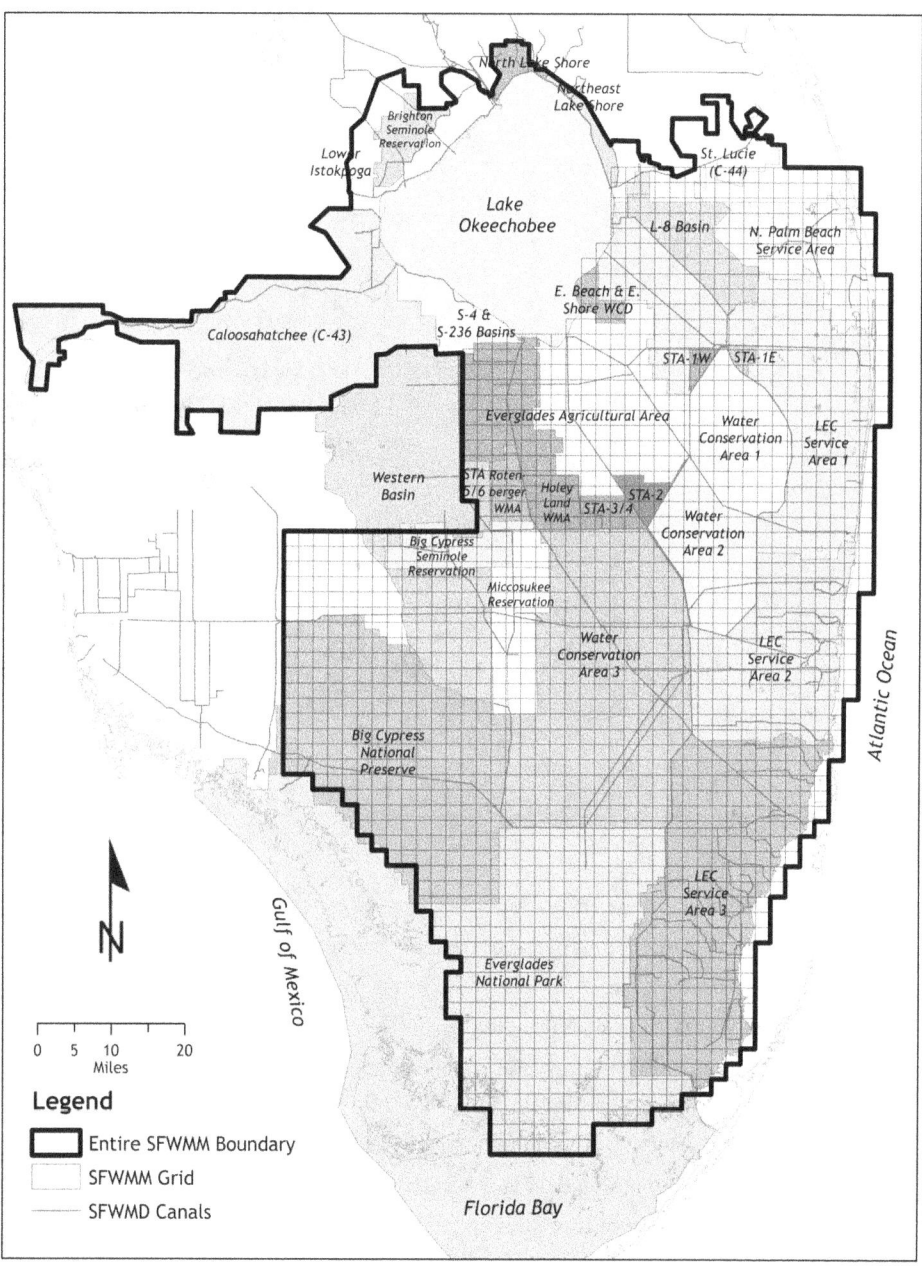

Figure 3.3. Map of South Florida with primary hydrologic regions and domain simulated by the South Florida Water Management Model (thick black outline); the southeast sub-region of the model domain below Lake Okeechobee is modeled using a distributed hydrologic model with a mesh of 3.2 km x 3.2 km (2 mile x 2 mile) cells. This figure shows the region of the Greater Everglades Ecosystem, south of Lake Okeechobee.

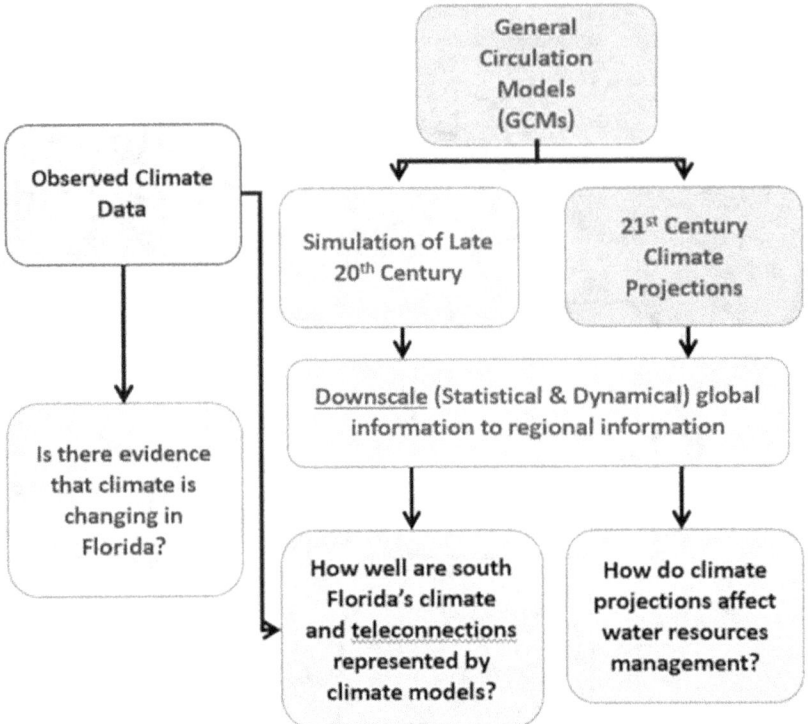

Figure 3.4. General approach for using climate data and projections for water resources investigations in the Greater Everglades Ecosystem Region of South Florida (Obeysekera et al. 2011).

Sea Level Rise

Tide gage records show that relative sea level is rising along the entire Florida coastline. This can have significant implications for coastal areas with low relief and highly permeable geology. The implications may include direct flooding of coastal landscape during storms and high-tide (including what is known as "nuisance flooding"), inefficiencies in coastal water control systems affecting flood protection, saltwater intrusion into water supply wells, and inundation of natural systems such as the Southern Everglades (SFWMD 2009). Sea level estimates based on the Unified Sea Level Rise Projections of the Southeast Florida Regional Climate Change Compact agencies are used for planning purposes (Fig. 3.6). They include at least three scenarios covering a planning range and a high curve that is intended for evaluating high-risk projects (SFRCCC 2015).

Evaluation of Climate Scenarios

Obeysekera et al (2014) focused on general implications of potential changes in future temperature, and associated changes in ET, precipitation, and sea levels within the regional boundary of Southeast Florida. Using a Bayesian approach known as the reliability ensemble average (REA) (Tebaldi et al. 2005), Obeysekera et al. (2011) provided probabilistic projections of both precipitation and temperature that are used to define scenarios for the water resources impact assessment. Based on this analysis, and the analysis of CMIP5 data sets (Dessalegne et al. 2016), +/-10% for precipitation scenarios and the single scenario of +1.5 °C for temperature were selected for an assumed planning horizon of 2060. The sea level projection, assumed to be coincident with temperature increase, was assumed to be 1.5 ft. Based on the above information, seven modeling scenarios were developed (Table 3.1).

Table 3.1. Water resources modeling scenarios based on temperature, precipitation, and sea level rise projections used for evaluation of climate change impacts on water resources in the Greater Everglades Ecosystem Region of South Florida.

Scenario Name	Temperature Change	Precipitation Change	Sea Level Rise	Coastal Canal Levels Increased?
BASE	No change	No change	No change	No
-RF	No change	-10%	No change	No
+RF	No change	+10%	No change	No
+ET	+1.5 °C (2.7 °F)	No change	0.46 m (0.81 ft)	Yes
-RF+ET	+1.5 °C (2.7 °F)	-10%	0.46 m (0.81 ft)	Yes
-RF+ETnoC	+1.5 °C (2.7 °F)	-10%	0.46 m (0.81 ft)	No
+RF+ET	+1.5 °C (2.7 °F)	+10%	0.46 m (0.81 ft)	Yes

ET = Evapotranspiration; RF = Rainfall; noC = No change in canal maintenance levels.

The hydrologic implications of the above scenarios were investigated using the South Florida Water Management Model (SFWMD 2005). This model simulates groundwater and surface water movements, including the complex operations and water management, over the entire Greater Everglades Ecosystem including the heavily urbanized areas of the Lower East Coast (Fig. 3.3), on a gridded mesh with a cell size of 2 miles × 2 miles.

The extreme rainfall scenarios together with warming show that the water budget of South Florida could be altered significantly, affecting the performance of all sectors (agricultural, ecosystems, and urban). In particular, the -RF+ET scenario would dry out the Everglades significantly, which would greatly alter its ecosystems and water supply function. One of the major implications of the reduction in rainfall and the increased ET is that tributary inflows (e.g. from the Kissimmee River basin) would be reduced by a large percentage, causing a significant lowering of Lake Okeechobee levels. The only positive aspect of this scenario would be the significant reduction in damaging high flows to estuaries. The infrastructure could handle increased rainfall, but this may cause considerable harm to the estuarine and wetland ecosystems in terms of too much water. Depending on the rainfall and ET scenario, the agricultural and urban demands would be increased or decreased by a significant percentage. In the worst case scenario, the demands not met by the agricultural service areas would increase significantly (by as much as 50–60 percent). A thorough analysis of the scenarios are available in a series of published papers (Aumen et al. 2015).

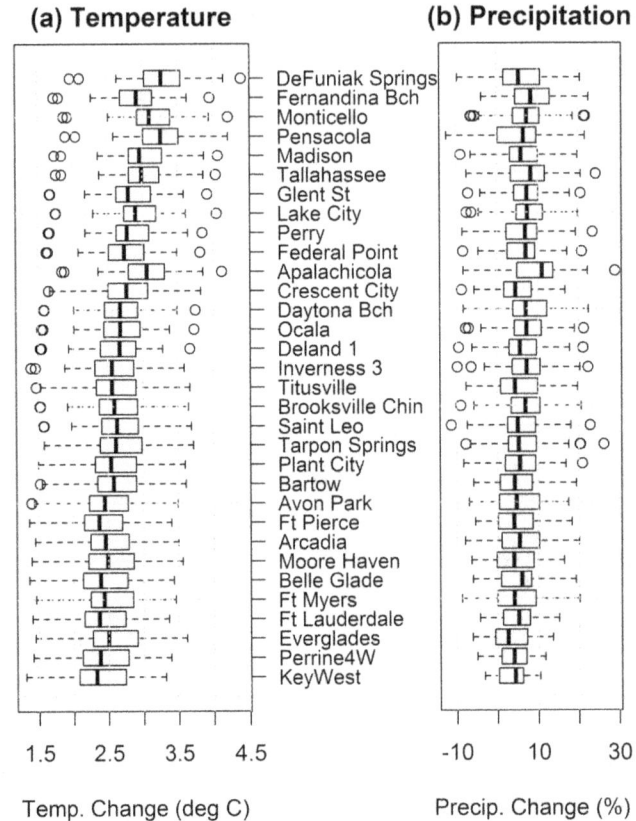

Figure 3.5. Box and whisker plots of temperature and precipitation change from 1970–2000 to 2055–2085 sorted by increasing latitude for the RCP 8.5 Scenario (Obeysekera et al. 2011; Dessalegne et al. 2016).

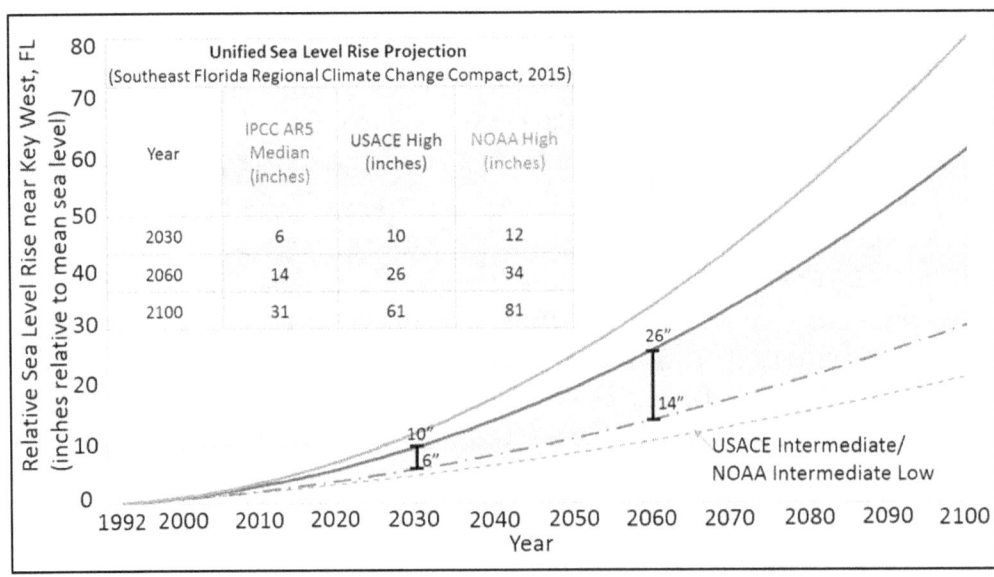

Figure 3.6. Unified sea level rise projections for regional planning purposes (SFRCCC 2015).

In summary, there are still considerable uncertainties in climate projections of hydrologic drivers such as precipitation and ET affecting the Greater Everglades region. Consequently, most studies have employed a scenario approach for impact assessment of climate change. Over the next 50 years, such scenarios to date cover 1.6 to 4° F (0.5 to 2.2° C) of temperature increase, ±10% change in rainfall, and 13 to 26 inches (0.33 m to 0.66 m) in sea level rise. Hydrologic modeling using the scenarios demonstrate that if such changes materialize, significant impact to the water resources of South Florida could occur.

Tampa Bay Region

Background

Tampa Bay is the largest estuary in Florida and extends about 50 km (30 miles) inland from the Gulf of Mexico. Tampa Bay Water, a wholesale regional drinking water supplier, operates a diverse water supply system in this region that includes regional well fields, river withdrawals from the Hillsborough and Alafia rivers, and a seawater desalination plant.

Within the Tampa Bay region, there are eight major watersheds: the Pithlachascotee, Anclote, Hillsborough, Alafia, Little Manatee, Withlacoochee, Peace, and Manatee river watersheds. In addition, an anthropogenically-altered surface conveyance system, the Tampa Bypass Canal system, operates as both flood protection and water supply for the City of Tampa and Tampa Bay Water. Besides the major river watersheds and the Tampa Bypass Canal, there are various smaller creeks in the region including Bullfrog, Delaney, Pinellas County Coastal creeks, Northwest Hillsborough Creek, and various minor coastal creeks. All of these creeks discharge either to the Gulf of Mexico or to Tampa Bay (Geurink and Basso 2013). A variety of land cover types are present in the study area, including urban, grassland, forest, agricultural, mined land, water, and wetlands.

Average annual rainfall for region from 1989 to 2001 was approximately 1230 mm (48.5 inches). The region typically has eight drier months followed by four months of wet summer season when 50-70% total annual rainfall occurs. Currently, depending on the rainfall and prevailing conditions, ET accounts for 30-90% of the water balance (Geurink and Basso 2013). Accurate prediction of seasonal, interannual, and decadal climate variability, as well as potential long-term climate change, are crucial to water resources planning and management in the region. Changes in rainfall frequency, seasonal shifts such as the onset and end of rainfall, as well as increases in temperatures have important implications for surface and groundwater availability. Higher than average temperatures can significantly change the hydrologic water balance, increasing water losses to the atmosphere through higher ET.

The Tampa Bay region is one of the most urbanized regions in Central Florida, and public water supply is one of the largest water users in the region. During the 1980s and 1990s, the

Tampa Bay region experienced severe drought. This problem was further exacerbated by a 108% increase in population between 1970 and 2000 to over 2 million people. Lack of adequate rainfall and continuous reliance on only the Upper Floridan aquifer for public water supply led to dewatering of the region's wetlands that resulted in significant environmental impact. The ensuing regional conflict and need to balance municipal and agricultural water use with natural system needs resulted in the creation of Tampa Bay Water as a regional water supply utility (see Asefa et al. 2015 for history). Since then, Tampa Bay Water has diversified its public water supply sources to include surface water (50-60%), groundwater (30-40%), and desalinated sea water (0-10%) in its portfolio.

The shift from an all-groundwater source supply to significant surface water reliance changed the risk profile for the agency and led to the need to understand the potential impacts of climate change on water supply and demand. Changes in climate could have important implications for utilities like Tampa Bay Water; e.g., changes in public water supply operations due to changes in the magnitude and seasonality of surface and groundwater availability, changes in wetland and lake ecosystems and associated regulatory programs due to change in ET and precipitation, and impacts on asset management programs for infrastructure than might be affected by rising temperatures. In response to these needs, dynamically downscaled CMIP3 GCMs were bias-corrected for the Tampa Bay region and used as climatic input for the integrated hydrologic model developed for the region by Tampa Bay Water and the Southwest Florida Water Management District.

Hydrologic Model

There are strong interactions among surface, subsurface, and ET processes in the Tampa Bay area due to the complex geology and relatively flat topography in the region (Geurink and Basso 2012). In order to understand and predict the dynamic surface–groundwater interactions in this complex system, two regional water management agencies, Tampa Bay Water and the Southwest Florida Water Management District, jointly developed an integrated surface/subsurface hydrologic model for the area. The Integrated Hydrologic Model (IHM) couples the EPA Hydrologic Simulation Program-Fortran (HSPF; Bicknell et al. 2001) and the USGS MODFLOW96 (Harbaugh and McDonald 1996) for surface and groundwater modeling, respectively (Geurink et al. 2006). The model is characterized as deterministic, semi-distributed, and semi-implicit with variable time steps and spatial discretization (Ross et al. 2004). Subsequently, Tampa Bay Water developed the Integrated Northern Tampa Bay (INTB) model application using the IHM to improve hydrologic assessment capabilities of West Central Florida. The hydrologic model domain for INTB is bordered by the Gulf of Mexico on the west (Fig. 3.7), by the Floridan aquifer flow lines on the north and east, and by a general head boundary condition at the southern boundary, which located far enough from the area of interest for this study to minimize the influence of the boundary condition (Geurink et al. 2006). The INTB model was calibrated and verified for the Northern Tampa Bay Region using hydrologic observations from 1989 to 2006 (Geurink and Basso 2012).

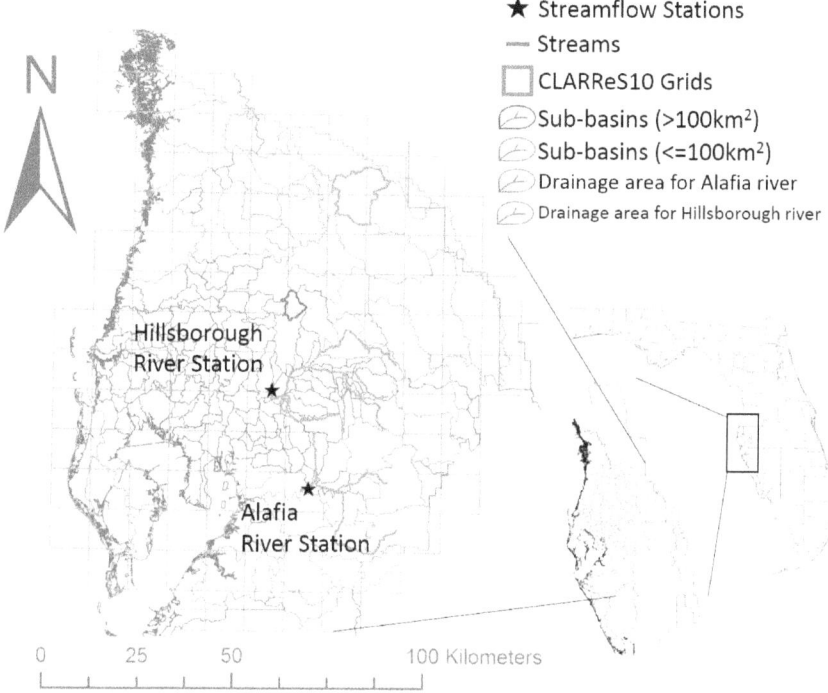

Figure 3.7. Map of the Tampa Bay region including the extent of the integrated hydrologic model, the locations of the streamflow predictions shown, and the CLARReS10 grid. Colored areas indicate the contributing areas for the streamflow prediction locations. (1 km=0.62 miles)

CLARREnCE10 Data Products

The CLARREnCE10 dataset was developed by the Florida State University (FSU) Center for Ocean Atmospheric Prediction Studies (http://floridaclimateinstitute.org/resources/datasets/regional-downscaling). The data includes retrospective predictions (historical climate conditions, 1969–2000) and future climate scenario projections (A2 scenario for years 2039–2070) from three CMIP3 GCMs that were dynamically downscaled to 10 km resolution using the FSU Regional Spectral Model (RSM) (see Fig. 3.7). The three GCMs selected by FSU for downscaling were the Community Climate System Model (CCSM), version 3 of the Hadley Centre Coupled Model (HadCM3), and the Geophysical Fluid Dynamics Laboratory (GFDL). Emission scenarios were generated by the Intergovernmental Panel on Climate Change (IPCC) and are described in the IPCC Special Reports on Emission Scenarios (IPCC 2000). Scenarios were developed that describe different storylines about possible future social, economic, technological, and demographic developments. The emission scenarios have internally consistent relationships that were used to describe future pathways of greenhouse gas emissions. The A2 scenario describes a very heterogeneous world and represents a "high future CO_2 emissions" scenario. Projected CO_2 concentrations affect the Earth's radiative energy budget, and thus are the key forcing input used in global climate model simulations of future conditions.

Methodology

GCMs are run at coarse resolution (100- to 200-km (60 to 120-mile) grid cells) to make them computationally tractable. As a result, GCMs typically show bias in the model outputs. Regional climate models (RCMs) are run at much finer scale (10 to 50-km (6 to 30-mile) grid cells) in order to capture local scale processes that may not be well-represented by large-scale GCM models. However, previous studies have shown there is still a need to bias-correct even high-resolution RCM output for the Tampa Bay region in order to accurately reproduce historic rainfall totals and predict historic streamflows using hydrologic models (Hwang et al. 2013). Therefore in this analysis, the daily precipitation and temperature data for the retrospective predictions from each GCM were bias-corrected using a monthly cumulative distribution function (CDF) mapping approach (Panofsky and Brier 1968; Wood et al. 2002; Hwang and Graham 2013).

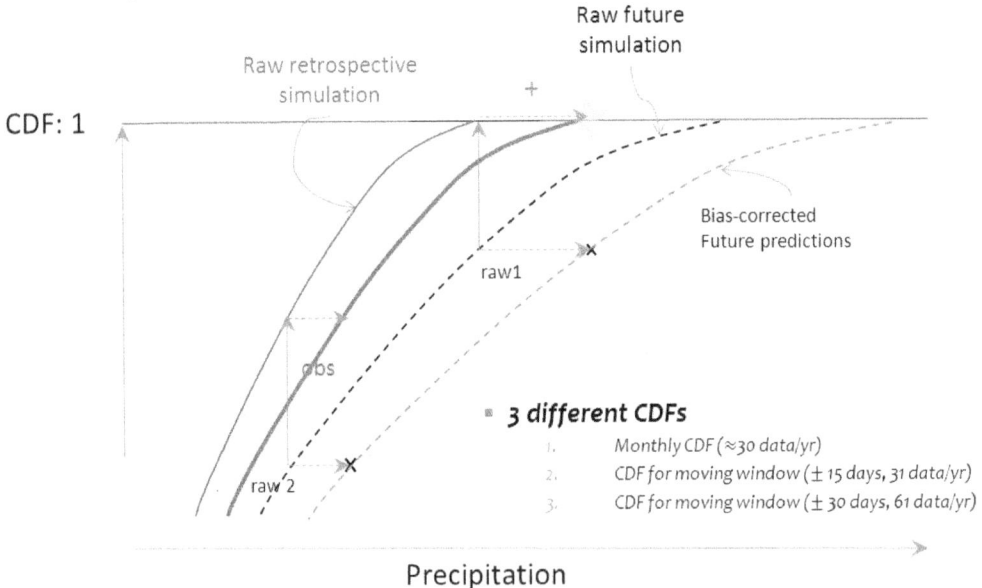

Figure 3.8. Schematic representation of bias-correction procedure (CDF mapping) used in this study. The process is conducted for each monthly cumulative distribution function (CDF).

For future scenarios, the bias for a particular daily value of precipitation or temperature was assumed to be the same in the retrospective and future periods. Thus, for each daily future projection, the bias-correction for the retrospective prediction with that same value was applied (Fig. 3.8). This method assumes GCM biases at a given temperature or rainfall amount stay the same in future simulations. The bias-corrected, dynamically-downscaled retrospective and future daily precipitation and temperature data were used as inputs for the Tampa Bay Water INTB model. All other parameters, forcing terms, and initial boundary conditions for hydrologic simulation were identical to those used in the calibrated model.

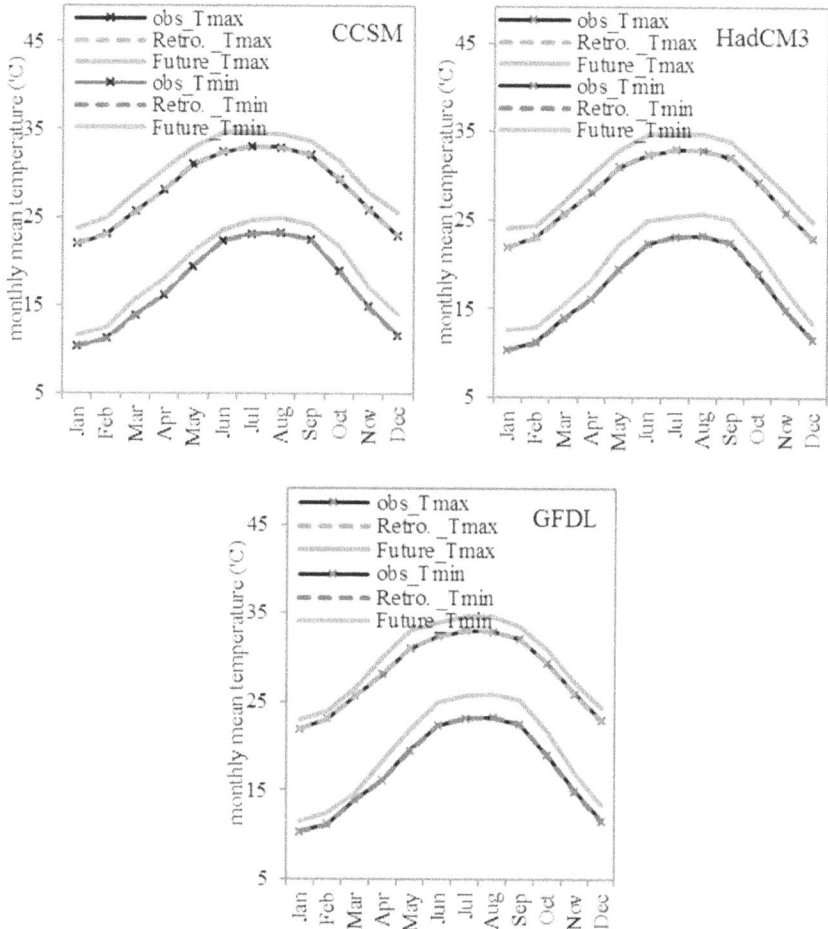

Figure 3.9. Monthly mean of T_{max} and T_{min} of bias-corrected CLAREnCE10 data for CCSM (top left), HadCM3 (top right), and GFDL results (bottom) using monthly CDFs for retrospective (1969–2000) and future (2039–2070) periods. (°C= (°F-32)*5/9)

Results

Temperature

Fig. 3.9 compares monthly mean T_{max} and T_{min} for observed and bias-corrected CLAREnCE10 data for the retrospective and future periods. This figure indicates that the annual cycle of observed mean T_{max} and T_{min} were accurately reproduced by all three GCMs after bias-correction, and that all bias-corrected GCMs predict a systematic increase in T_{max} and T_{min} over the entire annual cycle. Fig. 3.10 compares the predicted change in future monthly mean T_{max} and T_{min} (future–retrospective) for each GCM in the CLAREnCE10 experiment. Bias-corrected

CLAREnCE10 results predict that the average monthly increase of temperature will range from 1–3 °C (1.8–5.4° F), with some variation among different GCMs.

Precipitation

Fig. 3.11 compares monthly mean precipitation for observed and bias-corrected CLAREnCE10 data for the retrospective and future periods. Fig. 3.12 compares the predicted change in future precipitation (future–retrospective) for the three bias-corrected GCMs. The bias-corrected CCSM predicts a decrease in precipitation for all months in the future. The bias-corrected HadCM3 shows a slight increase in precipitation in the winter months and a decrease in the summer months. GFDL shows a significant decrease in July precipitation but increases in precipitation for most months of the year.

Streamflow

Fig. 3.13 compares the annual cycle of mean monthly streamflow predicted by the IHM-INTB using bias-corrected retrospective predictions and future projections to both historic streamflow observations and the calibrated IHM-INTB results for the Alafia and Hillsborough rivers. Differences between retrospective and future predicted mean monthly streamflow for each GCM are plotted in Fig. 3.14. These results show that predicted changes in the annual cycles of future streamflow for each GCM generally follow the predicted mean monthly precipitation change pattern (Fig. 3.12). The differences among the streamflows for different GCMs are significant, with the CCSM predicting much lower mean monthly streamflow throughout the entire year, HadCM3 predicting a slight decrease in mean monthly streamflow in July and August, and GFDL predicting an increase in streamflow throughout most of the wet season (June through October).

Fig. 3.15 compares retrospective and future mean annual ET and the ET-to-precipitation ratio to the calibrated IHM-INTB estimate (all averaged over the study area). The retrospective ET predicted by the hydrologic model using the bias-corrected GCM precipitation and temperature is similar to the ET predicted by the calibrated model for all GCMs. The future HadCM3 and GFDL results predict an increase of ET compared to the retrospective results due to projected increases in temperature, but a decrease in the ET to precipitation ratio, indicating that more excess precipitation is available for groundwater recharge or streamflow generation. In contrast, the CCSM results predict a significant decrease of mean annual ET and a significant increase in the ET to precipitation ratio. For the CCSM future, projected ET decreases because available water in the soil zone decreases due to the decrease in precipitation in all months (Figs. 3.10, 3.11, and 3.12). Thus, the ET becomes more moisture-limited for the CCSM future scenarios, whereas the ET remains largely energy-limited for the HADCM3 and GFDL future scenarios.

Tampa Bay Region Summary

The results of this investigation show that although each of the GCMs predicts a consistent increase in future temperature, differences among future precipitation estimates propagate into significant differences in future streamflow and ET predictions. In other words, the precipitation signal overwhelms the temperature signal in predicting hydrologic implications of projected future changes. Due to the large variation in precipitation and thus streamflow and ET estimates across the three GCMs considered here, this analysis does not provide actionable information for water resource planning. Additional GCM model projections (using multiple greenhouse gas emission scenarios and the next generation of GCM models) must be examined to more rigorously evaluate the expected magnitude of, and variation among, future hydrologic projections from the existing generation of GCMs. Improvement in the ability of the GCMs to simulate both retrospective and future rainfall patterns will be required before their projections can reliably be used for water resource planning and management in the Tampa Bay region. This is the same conclusion arrived at for the Greater Everglades region.

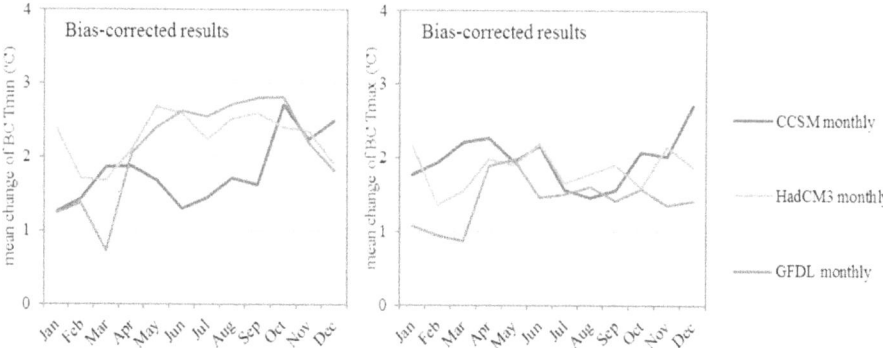

Figure 3.10. Predicted change in future bias-corrected monthly mean T_{min} (left) and T_{max} (right) for each bias-corrected GCM. (°C= (°F-32)*5/9)

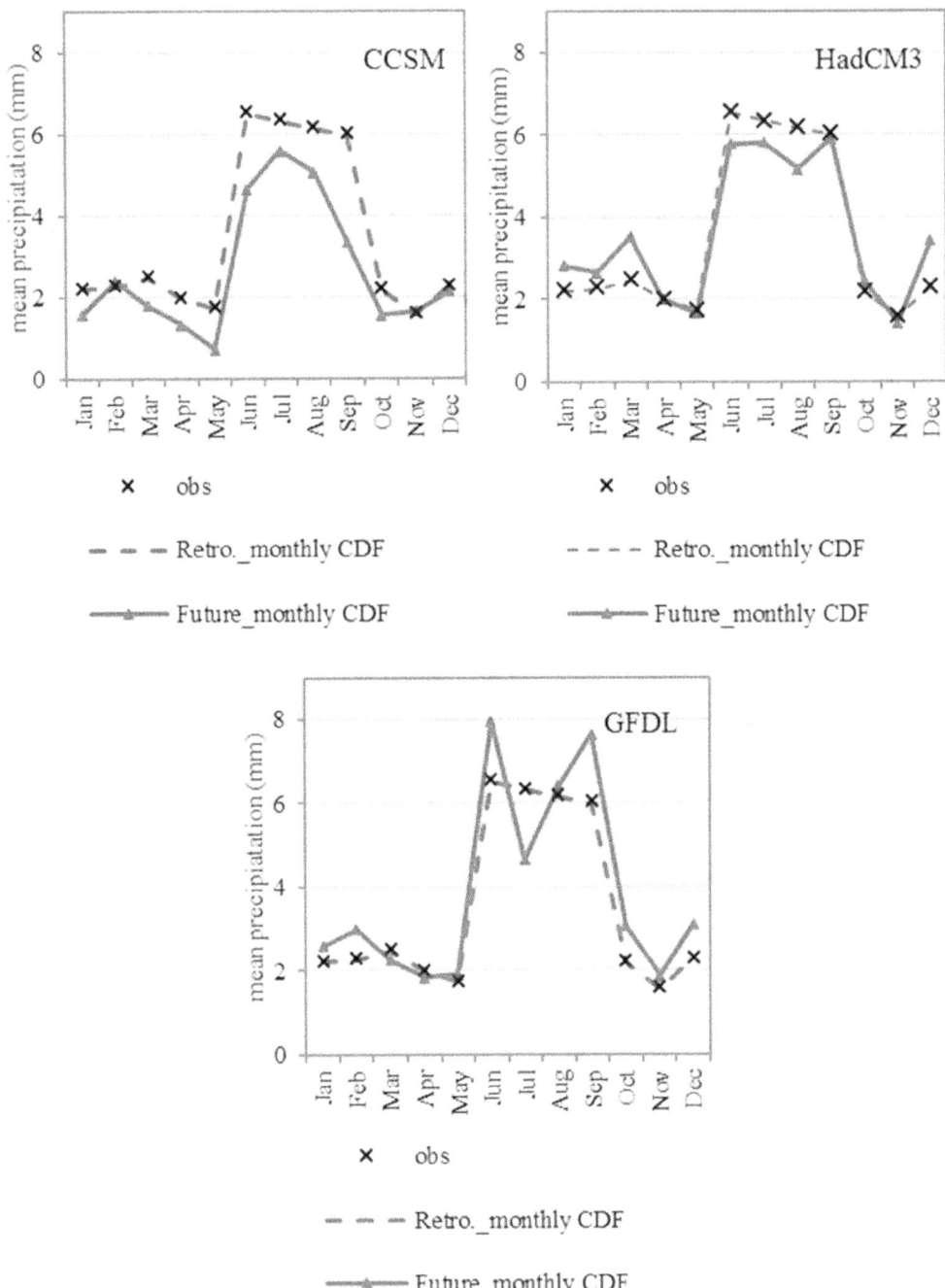

Figure 3.11. Daily mean precipitation of bias-corrected CLAREnCE10 precipitation data for CCSM (first column), HadCM3 (second column), and GFDL results (third column) for retrospective (1969-2000) and future (2039-2070) periods.

IMPLICATIONS OF CLIMATE CHANGE ON FLORIDA'S WATER RESOURCES • 101

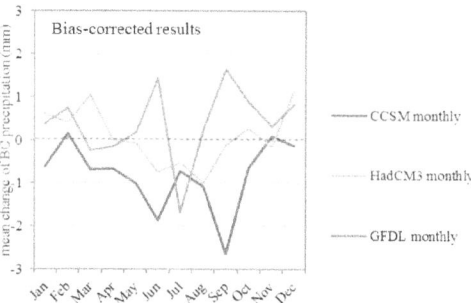

Figure 3.12. Predicted change in future bias corrected mean precipitation for each GCM. (1 mm = 0.04 in).

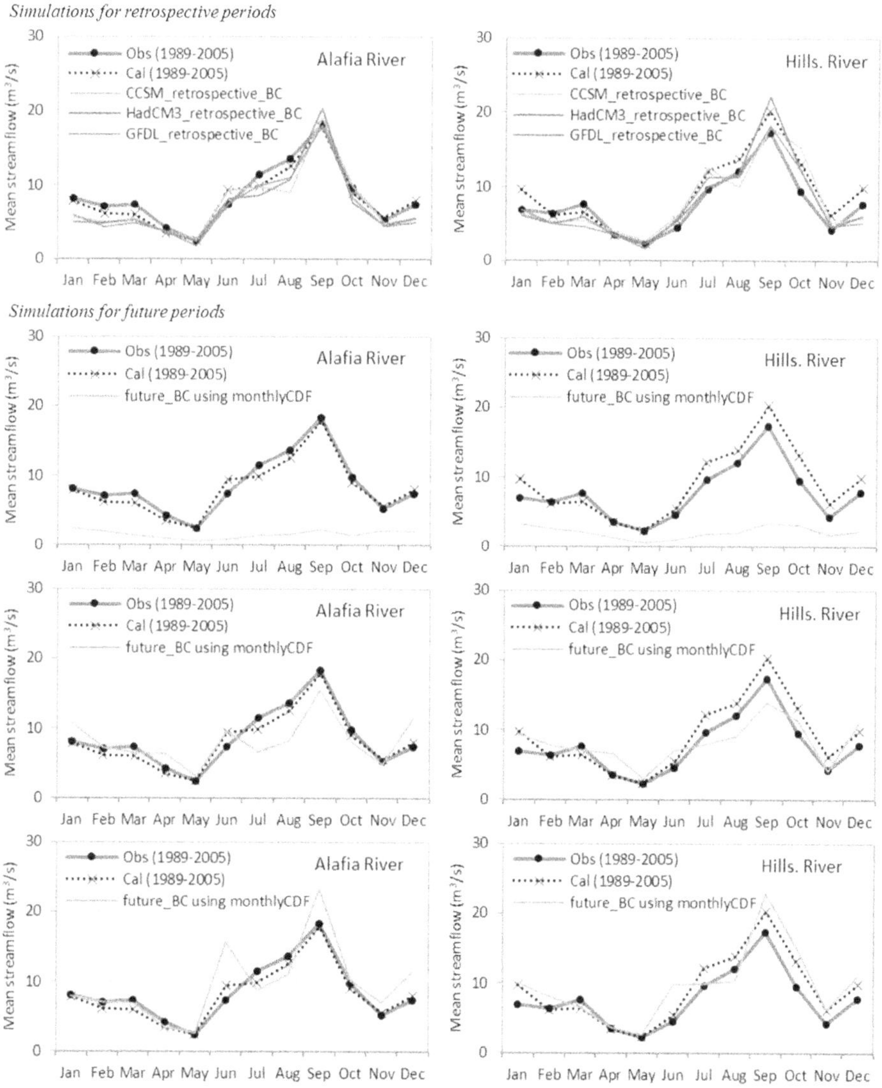

Figure 3.13. Simulated daily mean streamflow using bias-corrected retrospective CLAREnCE10 data (1969–2000, first row) and future data (2039–2070: CCSM (second row), HadCM3 (third row), and GFDL (fourth row) for Alafia River (left column) and Hillsborough River (right column). (1 m³/s = 35.3 cfs)

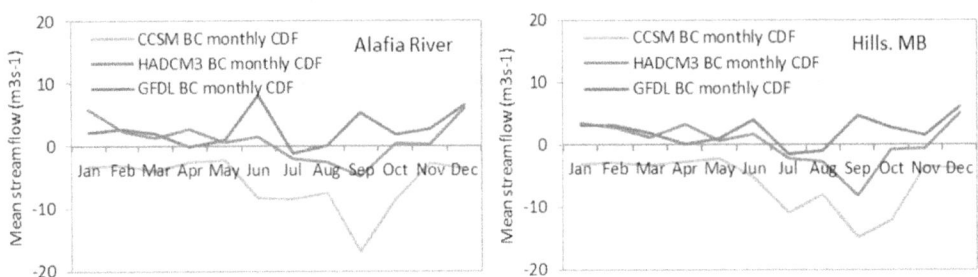

Figure 3.14. Predicted change in future streamflow simulations for each GCM for Alafia River (left column) and Hillsborough River (right column). (1 m^3/s = 35.3 cfs)

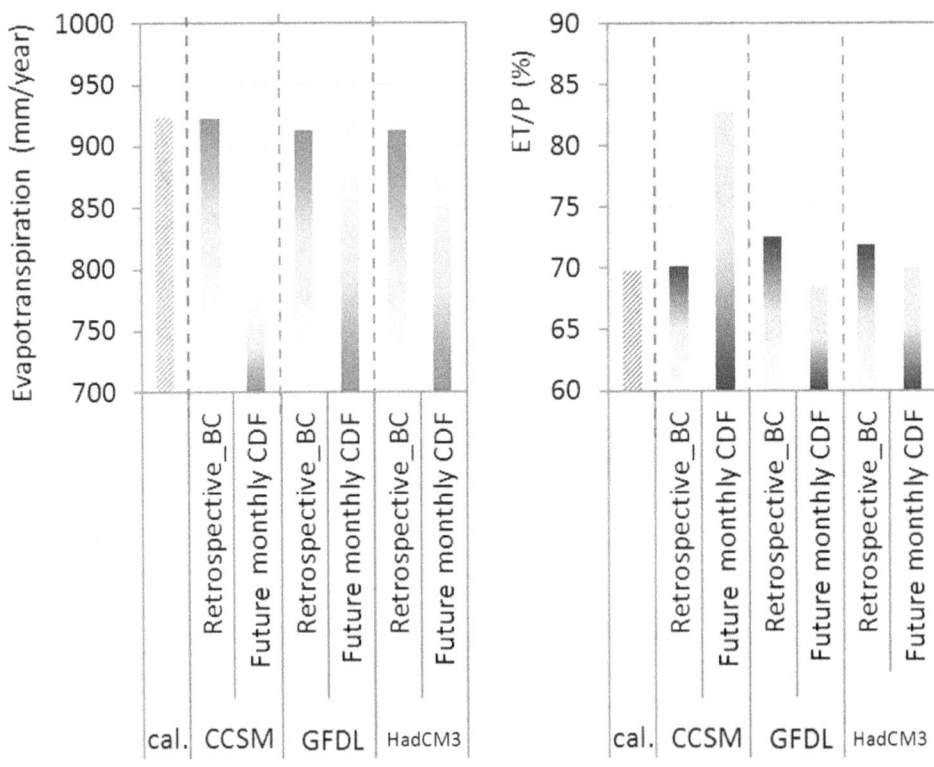

Figure 3.15. Comparison of calibrated, retrospective and future mean annual evapotranspiration (left column) and evapotranspiration ratio to precipitation (right column), averaged over the study area. (1 mm = 0.04 inches).

St. Johns River Region

Background

The watershed of the lower St. Johns River Basin encompasses 7000 km² and extends from Lake George to the mouth of the river at the Atlantic Ocean near Jacksonville. The landscape of the lower St. Johns River Basin is relatively low and flat, with surface elevations ranging from a maximum of 77 m near the western boundary to a minimum of sea level (0 m) at the river mouth. The lower St. Johns River is a low-gradient, lake-like river (Bacopoulos et al. 2009). The average bed slope of the river is only 0.000022 (Toth 1993), which allows tidal effects to extend up to Lake George, although the tidal range there is only a few centimeters (Giardino et al. 2011).

Hydrologic Modeling and Climate Change Impacts

The water level and flooding extent of the lower 200 km of the St. Johns River in northeast Florida under a 100-year flood event are evaluated by coupling hydrologic and hydrodynamic models. The 1% annual exceedance probability flood (i.e., 100-year flood event) is the basis for the National Flood Insurance Program. Therefore, Tropical Storm Fay in 2008 is selected in this analysis since it is an approximate 100-year return period rainfall event (Bacopoulos et al. 2017). Tropical Storm Fay passed over Cuba and the Florida Keys into the Gulf of Mexico (August 18, 2008), steered into Naples, Florida (August 19, 2008), crossed over the state, and emerged into the Atlantic Ocean off the east-central Florida coast (August 20, 2008). Then, changes in atmospheric conditions set up a broad flow pattern causing Fay to turn north and track at slow speeds of 1.5 to 2 m/s (3.4 to 4.5 mph), which allowed heavy rain bands to continually pass over northeast Florida for several hours. As the broad flow pattern weakened, a high pressure ridge set in north of Fay causing it to turn westward making a third Florida landfall near Flagler Beach (August 21, 2008). The westward motion was maintained across the northern Florida peninsula and Fay emerged into the extreme northeastern Gulf of Mexico (August 22, 2008), later making a fourth and final Florida landfall near Carrabelle in the Florida Panhandle (August 23, 2008).

The coupled model integrates a hydrologic model (SWAT) and a hydrodynamic model (ADCIRC). SWAT (Arnold et al. 1998) is a **S**oil and **W**ater **A**ssessment **T**ool used for prediction of water, sediment, nutrient, and pesticide yields with reasonable accuracy on large, ungauged river basins. SWAT has been successfully applied to many areas around the world at annual, monthly, and daily scales (Wang et al., 2011). ADCIRC is an **AD**vanced **CIRC**ulation numerical code for simulating shallow water flow (tides and surge) in shelves, coasts, and estuaries. ADCIRC solutions consist of time-dependent variables of water surface deviation from a datum and depth-integrated velocities in the longitudinal and latitudinal directions for all nodes of the computational mesh (Luettich et al. 1992).

The flows of tributaries and the upper main river stem affect the hydrodynamics of the lower main river stem. To accurately model this effect, runoff from the upper St. Johns River and tributaries of the lower St. Johns River is integrated with ADCIRC as inflow boundary conditions. Due to the limited number of streamflow gauges at the outlets of tributaries, a spatially distributed hydrologic model is applied to the lower St. Johns River Basin to provide simulated inflows. This integration of hydrologic and hydrodynamic models is set up such that the hydrologically computed inflows from any tributary are incorporated into the hydrodynamic simulation directly within the domain of flooding. Therefore, the constraints of limited observed flows for boundary conditions of ADCIRC, especially under extreme rainfall events, are overcome by the model integration.

In summary, it was shown that most of the flooding due to Tropical Storm Fay occurred in the upstream parts of the lower St. Johns River Basin (river km 130–160) where the incoming storm surge combined with high influx of watershed runoff, causing water levels to rise above the river banks and inundate the local floodplain. River flooding due to Tropical Storm Fay and the associated large amount of watershed runoff was shown to increase well beyond that of tidal conditions, and the filling and draining of water within watershed basins adjacent to the river during and after the peak of the local surge was shown to vary both spatially and temporally. The results indicate that the ADCIRC model can accurately capture storm surge if boundary conditions are complete and reliable.

The impacts of climate change and sea level rise on flood inundation were assessed based on the developed coupled model. Fig. 3.16 shows the inundation map for the simulated 100-year rainfall event represented by Tropical Storm Fay. Fig. 3.17 shows the inundation map under sea level rise and climate change impacts. The sea level rise is set to 0.51 m and the rainfall intensity is increased by 10% from Tropical Storm Fay.

Climate Change and Sea Level Rise Impact on Groundwater

In coastal aquifers, saline and fresh groundwater are in a dynamic equilibrium, and a landward shift of the equilibrium can cause landward encroachment of saline groundwater, resulting in the occurrence of saltwater intrusion (SWI). The low-lying alluvial plains and barrier islands located in coastal portions of the St. Johns River Basin are also vulnerable to flooding from rising water tables driven by sea level rise (SLR), because the water table depth is usually shallow and can even breach the land surface during and after a heavy rainfall. Water quality of the shallow coastal aquifer is also vulnerable to SLR-induced SWI, especially during prolonged drought. Hence, the low-lying coastal alluvial plains and barrier islands are dynamically influenced by climate change, and the negative effects include, but are not limited to, shoreline erosion, wetland inundation and migration, SWI, and alterations of the distribution and productivity of vegetation communities.

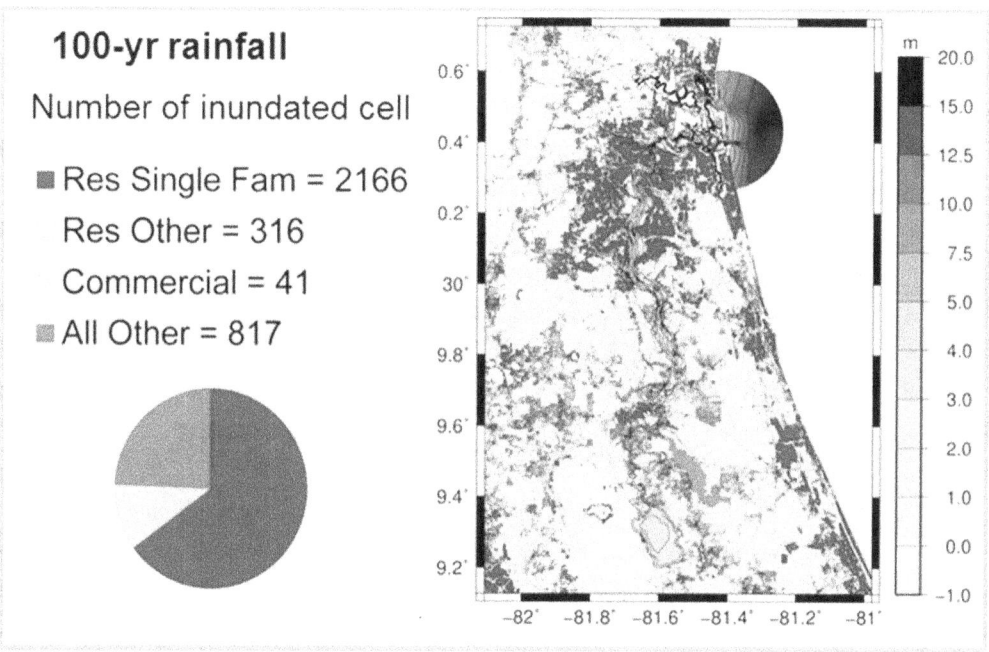

Figure 3.16. Inundation map of the lower St. Johns River under an 100-year extreme rainfall event (Tropical Storm Fay). The pink color represents simulated flooding.

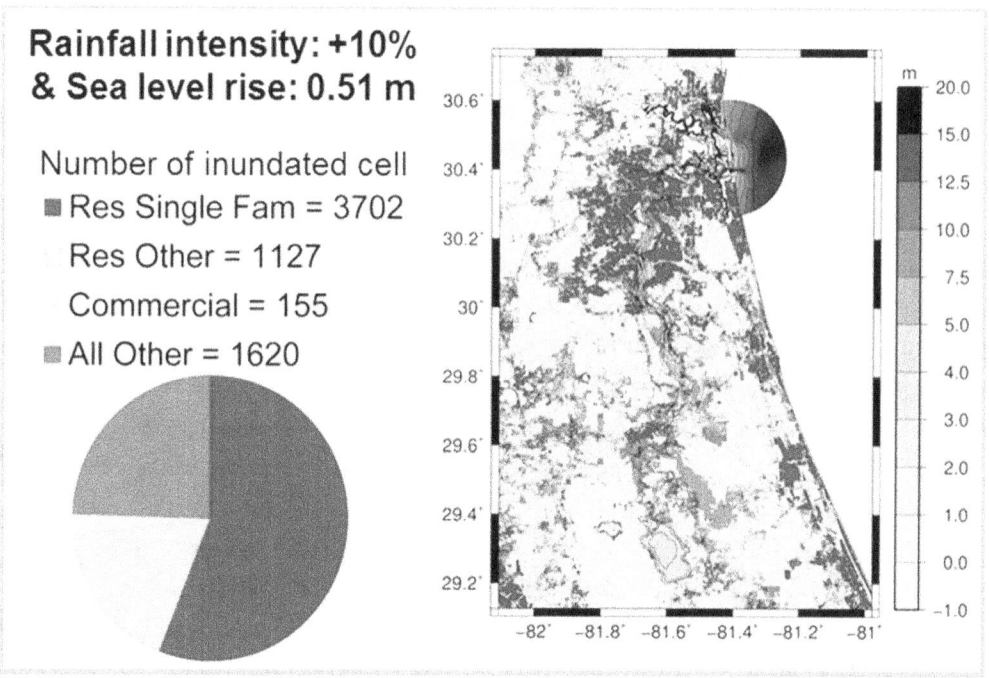

Figure 3.17. Inundation map of the lower St. Johns River under 100-year extreme rainfall event (Tropical Storm Fay) +10% and a 0.51 m increase in sea level. The pink color represents simulated flooding.

The SEAWAT model (Langevin et al. 2008) was applied to simulate the spatial variation of water table depth and salinity in the surficial aquifer system at Cape Canaveral Island and Merritt Island under steady-state 2010 hydrologic and hydrogeologic conditions. The model was calibrated against the field-measured groundwater levels monitored from 2006 to 2014. The calibrated model was used to evaluate climate change and SLR impact on the surficial aquifer system (Xiao et al., 2016).

Compared to 2010, precipitation is estimated to vary from a 7% decline to a 17% increase, while SLR is estimated to be 13.2 cm, 31.0 cm, and 58.5 cm for the low, intermediate, and high ice melt projections, respectively. These downscaled projections for 2050 are developed using data provided by Radley Horton and Daniel Bader, Center for Climate Systems Research, Earth Institute, Columbia University as part of the NASA Climate Adaptation Science Investigators Program (Rosenzweig et al. 2014).

Fig. 3.18 shows the simulation results, and the 'sensitive' areas are highlighted in yellow-brown. The predictions indicate that the effects of SLR and precipitation change are significant in west Merritt Island. This area is particularly vulnerable because of its low-lying coastal areas with flat topography and shallow water table depth having a high risk of being inundated during and after an extreme rainfall event. Also, low land surface elevation corresponding with low potential for freshwater recharge due to shallow water table result in low fresh groundwater pressure head and low hydraulic head gradient between inland fresh groundwater and coastal saline groundwater, which further results in a low rate of submarine groundwater discharge. This reduces the protection from SLR-induced SWI offered by freshwater groundwater recharge. In west Merritt Island, the land cover is mainly composed of fresh marsh, intermediate marsh (less saline than brackish), brackish marsh, and saline marsh.

Landward migration of saline water into the traditionally freshwater areas can cause degradation of ecologic systems and alter the distribution and productivity of vegetation communities. Increased rainfall can contribute to flushing, while prolonged drought can intensify salinity problems. Salt tolerance of plant communities is dependent on vegetation type, duration of exposure to saline water, rate of salinity increase, mineral content of soil, and degree of submergence. Some species can tolerate a wide range of salinity and can recover quickly once the salinity declines. However, some species die off quickly and cannot recover. Potential consequences of exposure to salinity include, but are not limited to, shift of wetland from fresh or less saline marsh to brackish or saline marsh, vegetation species dieback and limited recovery, shift in vegetation species composition from less salinity-tolerant species to more salinity-tolerant species, and reduction in biomass production. SWI not only affects marshes/wetlands, but also affects agriculture. Citrus is the main agricultural product in this area, and a reduction in citrus production due to increased groundwater salinity might be a big problem. Currently, no consumptive use wells operate in this area, and SWI does not have a negative effect on public drinking water supply.

The rates of growth of the aquifer areas affected by SWI were computed. The intruded area growth rate is faster at first and then slightly declines. At the beginning, the growth rate is high because west Merritt Island is vulnerable to even a small amount of SLR due to its low elevation and flat topography. Afterwards, the growth rate slightly declines because the amount of SLR assumed for the 2050 time horizon is not large enough to affect Cape Canaveral Island and east Merritt Island significantly.

In order to prevent saltwater intrusion, it is very important to minimize the effects of SLR, since its effects are clearly more influential than the effects of precipitation change. In order to 'balance' SLR, it is necessary to increase the inland fresh groundwater pressure head by artificial recharge. Recharge wells could be constructed close to the coastline, along with detention ponds designed for flood control to avoid inland flooding, since the region is humid subtropical with plenty of precipitation especially in the rainy season. The designed detention ponds could be used to temporarily hold the excess rainwater while slowly draining to the coastal recharge wells. Artificial recharge is even more important in the dry season because of less precipitation. It is not necessary to 'shut off' the two pumping wells that are used occasionally for lawn irrigation, since the pumping rates are very low and the effect would be tiny.

In summary, the increased inundation area due to intensified rainfall and rising sea levels is significant in the lower St. Johns River Basin, especially residential and commercial areas. In terms of groundwater, the predictions indicate that the effects of SLR and precipitation change will not be significant in Cape Canaveral Island and east Merritt Island by 2050. Both areas serve as the primary recharge area due to its high elevation, deep water table depth, land cover (forest and pasture), and soil type (sand). However, it is estimated that the negative effects could be noticeable if SLR and precipitation change turn out to be greater than projected.

Suwannee and Apalachicola River Basins

Suwannee River Basin (SRB)

The SRB (Fig. 3.19) covers approximately 11,042 square miles and is located entirely within the coastal plain physiographic region of the southeastern U.S.— extending from Cordele, Georgia to Cedar Key, Florida at the Gulf of Mexico (Katz and Raabe 2005). The SRB extends from its eastern headwaters in the Okefenokee Swamp to the Gulf of Mexico, and it is considered one of the most pristine and undeveloped river systems in the United States (Fig. 3.18). The SRB typically entails a unique mix of subtropical and temperate forests, swamps, fresh and tidal wetlands, and a rich habitat for aquatic and terrestrial wildlife. This significantly expansive, grassy estuary provides one of the most scenic nearshore habitats within the northeastern Gulf of Mexico.

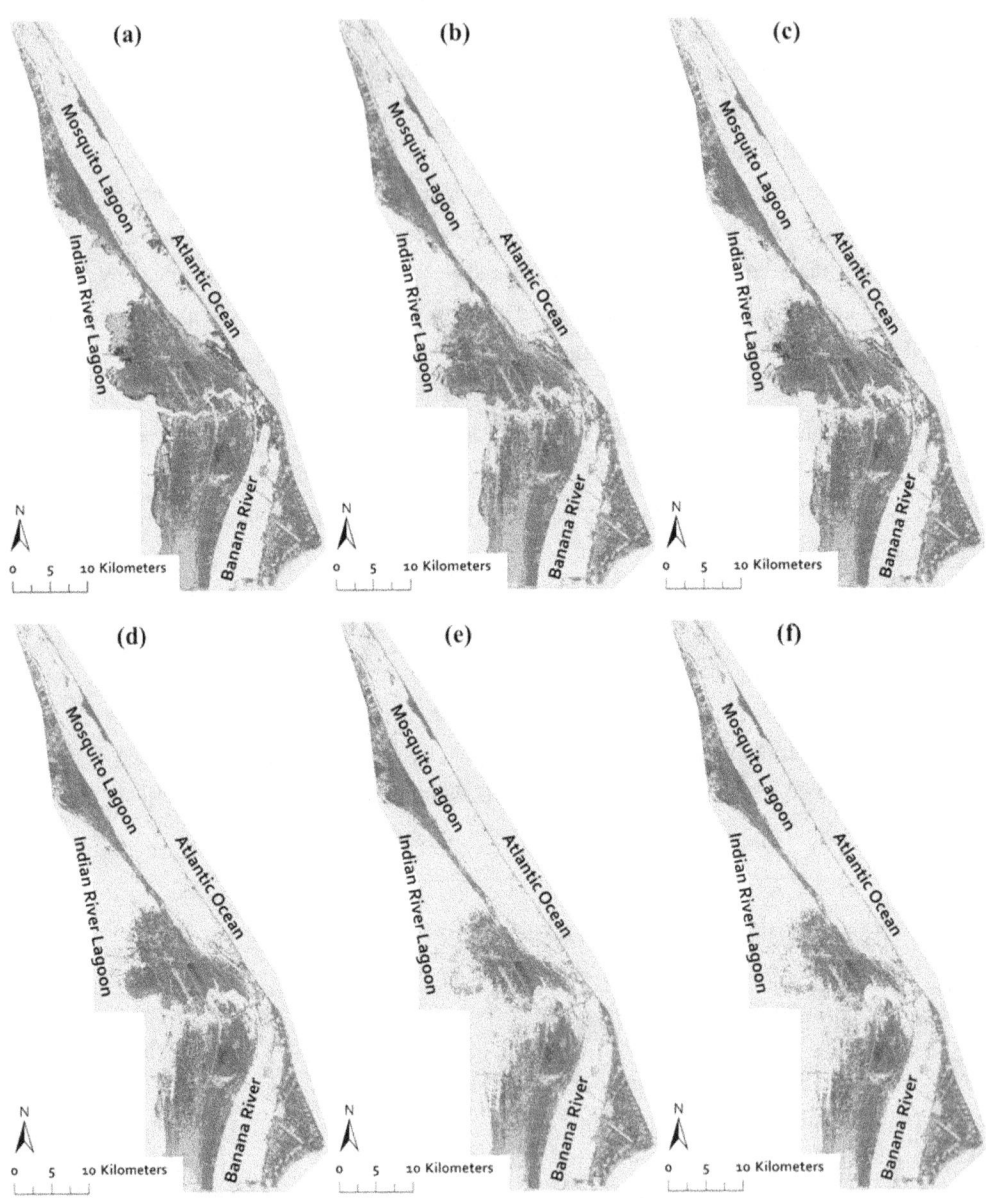

Figure 3.18. The area, which is sensitive to sea level rise and/or rain-induced flooding, is represented by yellow-brown color: (a) Case 0; (b) Case 1 (13.2 cm SLR, +17% precipitation); (c) Case 2 (13.2 cm SLR); (d) Case 3 (31.0 cm SLR); € Case 4 (58.5 cm SLR); and (f) Case 5 (58.5 cm SLR, -7% precipitation).

Groundwater resources in the SRB are supplied by the Floridan aquifer system, which is one of the most productive groundwater resource reservoirs in the United States. The system is a major water source throughout much of Florida and most of South Georgia (Lindsey et al. 2009). The Floridan aquifer underlies the rest of the southern portion of the basin. It is overlain by approximately 25–125 feet of sandy clay residuum derived from chemical weathering of the underlying rock. The total thickness of the Floridan aquifer in the basin ranges from a few tens of feet in the north to more than 400 feet to the southern portion of the basin.

Apalachicola River Basin (ARB)

The Apalachicola River Basin (ARB), which is part of the larger Apalachicola-Chattahoochee-Flint (ACF) river basins, is a basin in the Florida Panhandle that drains a watershed of some 20,000 square miles (Fig. 3.20). The northern reaches of the basin include a dramatic picturesque landscape of steep bluffs and deep ravines that are some of the most significant natural features of the southeastern coastal plain. The river and its surrounding forests, prairies, and coastal habitats are recognized as one of six biodiversity hotspots in the United States. The river basin has the highest species diversity of reptiles and amphibians in the U.S. and Canada, with more than 40 species of amphibians and 80 species of reptiles (Couch et al. 1996). The Apalachicola National Forest, which borders the river, is one of the largest contiguous public lands east of the Mississippi River.

The geology of the Chattahoochee River Basin largely determines the groundwater characteristics of the ARB area. The Chattahoochee River makes its way to the coastal plain. Aquifers in the coastal plain consist of porous sands and carbonates, and include alternating units of sand, clay, sandstone, dolomite, and limestone that dip gently and thicken to the southeast. Most of these aquifers remain reliable and consistent prolific producers of groundwater. The aquifers in the coastal plain typically comprise of two types: unconfined and confined. The unconfined aquifers are hydraulically interconnected to surface water bodies. The confined or artesian aquifers are buried and hydraulically isolated from surface water bodies. Confining units between these aquifers are mostly silt and clay. The five major aquifers that underlie the Chattahoochee River Basin include: 1) the Floridan aquifer system (one of the most productive aquifers worldwide). The complex hydrogeology of the Floridan aquifer system is reflected by highly variable transmissivities ranging from 2,000 to 1,300,000 ft^2 per day (Miller, 1986); 2) the Claiborne aquifer – an important source of water in part of southwestern Georgia (McFadden and Perriello, 1983); 3) the Clayton aquifer, another important source of water in southwestern Georgia; 4) the Providence aquifer system, the deepest of the principal aquifers in South Georgia that serves as a major source of water in the northern one-third of the coastal plain (McFadden and Perriello 1983); and 5) the Crystalline rock aquifer, a bedrock aquifer that underlies the Chattahoochee River Basin.

The Apalachicola River receives streamflow and sediment from the Chattahoochee and Flint rivers, and flows through the Florida Panhandle eventually draining into the Gulf of Mexico. As the source of 90% of Florida's oyster production, Apalachicola Bay is an important marine nursery area. Streamflow and sediment load from the Apalachicola River have a direct impact on the ecosystem, particularly commercial oyster production in Apalachicola Bay. It is important to assess the impact of climate change on the Apalachicola River's streamflow and sediment load to identify potential ecological effects.

Hydrologic Modeling

A SWAT model developed for the Apalachicola River region was used to simulate runoff and sediment loading under present and future conditions (Chen et al. 2014). The model was calibrated and validated for historical conditions (1984–1994) and then used to simulate 30 years of daily discharge and sediment load under present (circa 2000) and future (2100) conditions. Future scenarios incorporated changes in climate, land use and land cover (LULC), and coupled climate and LULC. Wang et al. (2013) assessed projected climate change impact on extreme rainfall events in the ARB based on RCMs. Downscaled climate data for three GCMs and LULC projections detailing changes for 16 distinct land classes are characterized by the IPCC Special Report on Emissions Scenarios (SRES) scenarios A2, A1B, and B1 for 2100 (IPCC 2000). The Long Ashton Research Station Weather Generator (LARS WG) was selected for downscaling GCM data due to its ability to generate site specific, daily, stochastic temperature and precipitation (Semenov and Stratonovitch 2010). In keeping with IPCC carbon emission scenarios, projected 2100 A2, A1B, and B1 LULC provided by the United States Geological Survey Earth Resources Observation and Science (USGS EROS) Center were selected to assess LULC change impacts (USGS EROS Center 2014). Detailed procedures for model calibration and validation, as well as climate change and LULC projections, are described in Hovenga et al. (2016).

The coupling of both future climate and LULC change was simulated for each GCM. Climate based on different GCMs plus LULC change produced large variation in flow outputs. That is, the HADCM3 model predicted higher high flows and lower low flows, IPCM4 indicated overall lowered runoff, and MPEH5 produced generally increased flow. The general seasonal behaviors for sediment loading were very similar to the climate-only simulation results.

Climate change predicted by individual GCMs showed noticeable differences in future rainfall seasonal patterns, with no single carbon emission scenario resulting in higher values, emphasizing structural differences among GCMs. The consensus on temperature is that it will increase, with the A2 scenario predicting the highest values and B1 predicting the lowest. When incorporating climate change into the SWAT model, output in terms of runoff and sediment loading showed large distinctions between GCMs, implying that these parameters might have been driven more by rainfall than temperature. Both the runoff and sediment loading responded to future climate change, yet the ways in which they responded may be conflicting between

GCMs. Under present conditions, high flows occur around October to December. Results from all GCMs are in agreement that flow increased for the months of September and October, indicating the current wet season may occur earlier in the year and with greater magnitude. Sediment loadings were also predicted to increase for these months. Further, sediment loading for the baseline period was at its minimum from July to September, yet a seasonal shift may occur with minimal loading occurring earlier in the year, around April to June. This response may be driven by lowered future precipitation predicted during these months.

LULC change had little effect on the runoff response. Surface runoff was computed using the Soil Conservation Service (SCS) curve number method. While individual curve number values assigned to the distinct land classes varied within the model, the weighted cumulative curve numbers for the circa 2000, 2100 A2, A1B, and B1 land cover datasets within the study domain were 46.9, 48.7, 45.7, and 44.4, respectively. The slight variability between curve number values might explain why runoff was so minimally affected by LULC change. An alternative future LULC class assignment than the one used in this study may result in a more significant response. Further, response to alterations in LULC may be more appropriately assessed using a finer temporal resolution capable of addressing peak runoff. The slight increase that does occur from August to October for the A2 scenario land cover might be explained by the plant growth model and consequent ET that is incorporated into SWAT.

Sediment loading was more evidently impacted by changes made to land cover than runoff. The loading increase observed from the A2 LULC may be a result of the increase in agricultural lands and loss of forested area. Sediment loading decreased for all months as a result of the B1 coverage. Compared to the circa 2000 coverage, B1 had more forested regions and less agriculture. It is inferred that the amount of agricultural and forested lands was directly related to sediment loading and that an increase in agriculture and/or loss of forest may have caused loading quantities to increase.

Runoff response for simulations that coupled climate and LULC change produced runoff values that were very similar to those produced by incorporating climate change only, suggesting future climate change may affect flow more than LULC change. Regardless of the behavior of increased or decreased runoff predicted by the individual models, the patterns of amplification and de-amplification were alike, demonstrating a homogenous interaction occurring within the simulations that was not affected by the GCM data. Sediment loading response was more reactive. When climate and LULC both independently modeled sediment as increasing or as decreasing, the coupled response resulted in sediment values that were overall larger than would be estimated from the individual, added deviations from the baseline. This suggests climate and LULC change effects amplify one another, resulting in larger loadings than if estimated by the separately modeled responses.

Many of the biological species in the Apalachicola region are sensitive to salinity and total suspended solids levels, which can affect both their productivity and distributions (Scavia et al. 2002). The increase in high flow magnitude and seasonal shifts for runoff and sediment caused by climate change have implications for threatening the phenology of the system by affecting migration, breeding, and distributions. Additionally, as land types change, adverse effects may become amplified. Given the time horizon of this study, results may provide guidance in establishing both short- and long-term monitoring activities and mitigation strategies. Short-term efforts may focus on responses with less uncertainty, i.e., those in agreement among all GCMs. Long-term strategies, with more flexibility to adapt, can help coastal managers adjust sustainability efforts and regulatory procedures as more knowledge is acquired, e.g., how the region is changing and region-specific performance of GCMs. The suite of SRES scenarios provides yet another dimension to adaptation planning.

Climate Change Factors Affecting Groundwater within ARB and SRB

The threats to the water and associated natural resources of the SRB and ARB require a better understanding of the sources and effects of contamination, water withdrawals, and climate change, as well as interactions among these stressors. Climate variation alone can result in significant impacts on these resources. For instance, changes in rainfall patterns alone can cause far-reaching impacts on surface water and groundwater supply. Natural fluctuation in water supply, coupled with water consumption, can place added stress on biological communities. Intermittent droughts in Florida over the last two decades have heightened concerns about management of water resources within the watershed. It therefore seems desirable that future consideration be focused on improving interagency communication and coordination for effective water resource monitoring covering both groundwater and surface water, including development of better predictive measures.

SWI (which is the movement of saline water into freshwater aquifers) is an especially significant potential threat to the potable water supply in areas along the SRB. Based on historical data and hydrogeological principles, the water-level fluctuations can affect the position of the freshwater and saltwater interface.

Increased demands for groundwater from intensifying urban development and extensive agricultural activities in this part of the basin have resulted in increased withdrawals of water from the Upper Floridan aquifer. As an example of water–quantity-related problems in the recent decade, some springs within the ARB and SRB are increasingly being depleted and essentially stop flowing at times of the year due to lowering of the water table. Increased water withdrawals have also caused a secondary deleterious impact on the groundwater resource by salt water intrusion. This phenomenon will be exacerbated by potential decreases in precipitation in the future due to global climate change.

Although studies in the basins indicate generally good overall water quality, there are other ever-increasing threats to the water resources within the ARB and SRB areas. These include nitrogen and phosphorus contamination of groundwater from fertilizers, animal waste, sewage effluent (septic tanks and land application of treated sewage effluent), and atmospheric deposition. These threats are raising concerns regarding both human and ecosystem health. Elevated nitrate concentrations in rivers can cause eutrophication, which can result in algal blooms and depletion of oxygen that can lead to fish kills (Bledsoe and Phlips 2000). Increases in nitrate concentrations from human activities may cause adverse ecological effects, indicated by an increase in periphyton biomass along the middle and lower reaches of the Suwannee (Hornsby and Mattson 1998).

Several human health concerns are also associated with elevated nitrate concentrations in groundwater used for drinking. A typical example is for infants under six months of age who are susceptible to methemoglobinemia when they ingest nitrate in drinking water, which can lead to reduced blood oxygen levels that can result in death. For these health concerns, the U.S. Environmental Protection Agency (EPA) established a maximum contaminant level for nitrate of 10 mg/L as nitrogen for drinking water. Recent studies have also shown that pharmaceuticals, endocrine-disrupting chemicals (hormones), and other organic wastewater contaminants are present in streams throughout the U.S. (Kolpin et al. 2002). Although present in generally very low concentrations, little is known about the potential effects on human health and the health of aquatic organisms that may occur from complex mixtures of organic wastewater contaminants and their metabolites in surface waters.

During prolonged wet periods, when river floodwaters flow into the karstic aquifer system along the Lower Suwannee River corridor, there is an opportunity for waterborne pathogens (such as Cryptosporidium and Giardia oocysts) to enter the aquifer system. These waters also contain very high concentrations of naturally occurring organic matter that could react with disinfectants such as chlorine and produce harmful trihalomethanes and haloacetic acids.

In addition to influencing temperature and precipitation around the world (Chiew et al. 1998; Roy 2006; Keener et al. 2007), the ENSO phenomenon also has impacts on groundwater resources (McCabe and Dettinger 1999; Gurdak et al. 2007). Studies have found strong correlation between ENSO, precipitation, and streamflow (Berri and Flamenco 1999; Simpson and Colodner 1999). These studies have shown that ENSO can have strong correlations with temperature and precipitation in the North America and reported that the northern United States experiences less precipitation and warmer winters during El Niño events.

Possible Adaptation Measures and Research Opportunities

The most practical mitigation measures for water resource issues within the ARB–SRB area should broadly focus on basin-wide optimization of water resource information and management. Suggested measures include, but are not limited to, the development and proper application of: consistent and comparable data collection and analysis methods; improved techniques and their coordination among agencies and across jurisdictions; integrated land use and land cover databases that comprise past, present, and future; and improved groundwater/surface water interaction workable models that will enhance predictive capabilities. Past advances in

understanding the ARB–SRB call for continued and sustained research priorities within the ARB–SRB area that should address the following:
- Extent and significance of SWI on freshwater systems
- Fate of nitrate in drinking water and aquatic systems
- Radionuclide occurrence in drinking water and associated health issues
- Pathogens and bacteria influx to karstic groundwater during flood periods
- Elevated natural organic material and the formation of disinfection byproducts
- Endocrine disruptor chemicals and organic wastewater compounds
- Mercury methylation and other toxic elements accumulation in edible fishery

Water Quality Impacts

Climate change impacts on water resources have been studied extensively from the perspective of changes in quantity, but far fewer studies consider potential changes in water quality and their implications for water and wastewater utilities. Understanding impacts on water quality is important to assess the full implications for water resources because climate change is expected to exacerbate existing water quality problems and create new problems. The principal impacts on water quality are often related to temperature increases, variations in precipitation, SLR, and deposition of gases and particulates from the atmosphere.

At the global scale, increases in temperature and SLR are expected, while precipitation patterns would vary depending on geographical location; some may experience increased precipitation while others may experience reduced precipitation or drought (IPCC 2014). In general, Florida is expected to experience increasing air temperatures (Florida Oceans and Coastal Council 2009). Carter et al. (2014) reported that temperatures in the southeastern U.S. have risen by an average of 2° F since 1970.

Temperature Increase

Increased temperature impacts water quality in a number of ways. It increases ET levels that may lead to higher concentration of pollutants, and it can change water chemistry and biochemical reaction kinetics that affect dissolution, complexation, and biological degradation processes. Increased temperature will also result in lower dissolved oxygen (DO) concentration due to lowering of saturation levels. Reduced DO and increased temperature can cause changes in water chemistry, such as increased mobility and bioavailability of heavy metals (John and Leventhal 1995). In addition to effects on chemical characteristics, increased temperature also leads to higher pathogen levels as microorganisms will remain viable longer in the environment. To help maintain stable water quality and minimize water loss due to increased evapotranspiration and temperature, utilities may consider aquifer storage and recovery, which has regulatory and cost implications and possibly its own water quality issues.

IMPLICATIONS OF CLIMATE CHANGE ON FLORIDA'S WATER RESOURCES • 115

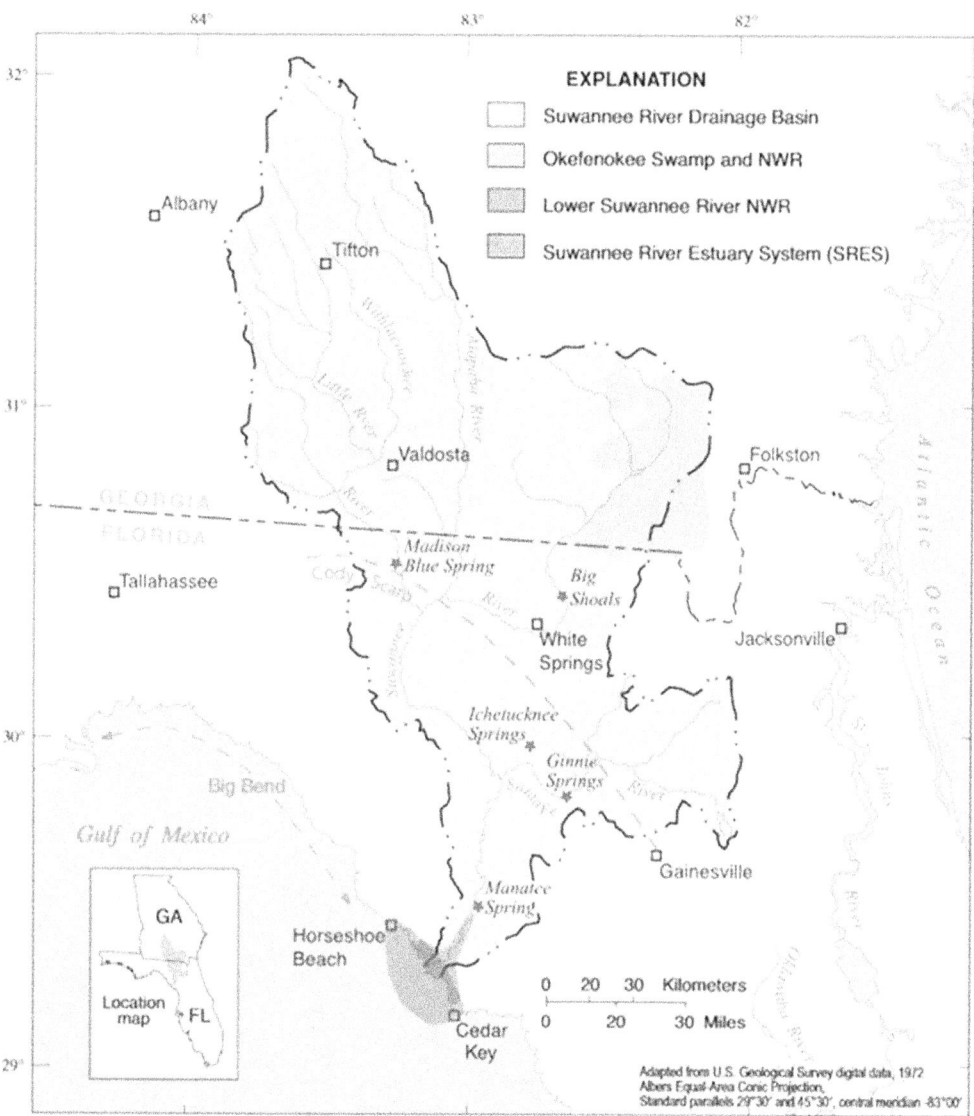

Figure 3.19. Location map showing the Suwanee River Basin and estuary system (Katz et al. 2005).

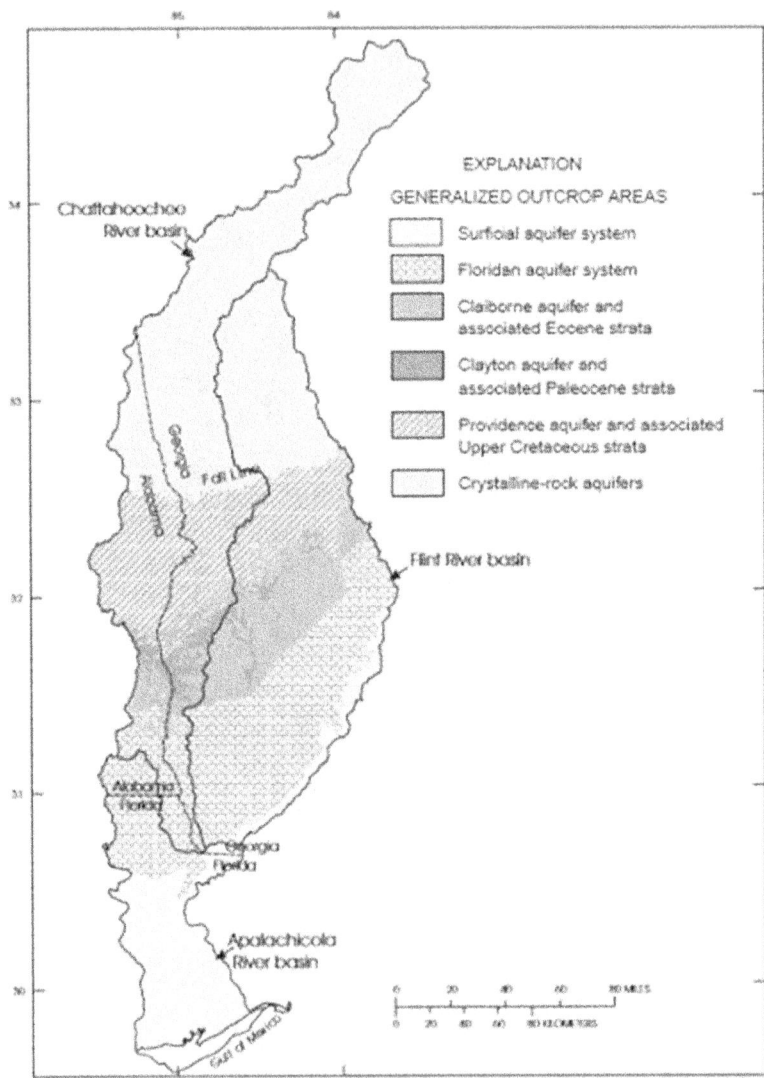

Figure 3.20. Hydrological units underlying the Apalachicola-Chattahoochee-Flint River Basin (Couch et al. 1996).

In the case of higher temperatures coupled with high nutrient loads, lakes, reservoirs, and low flow streams would experience more intense eutrophication and algal blooms. This results in increased levels of cyanobacteria biomass and development of anoxic conditions that affect water quality negatively, and hence can increase water treatment requirements or cost. Based on model simulation of some lakes under higher temperatures, some studies predict extended lake stratification periods with low DO in the hypolimnion layer in summer and consequent phosphorus and heavy metals dissolution, and increased lake turbidity. (Sahoo et al. 2011; Taner et al. 2011; Dupuis and Hann 2009).

Elevated algal blooms also result in higher levels of organic loads that lead to increased disinfection byproducts that are thought to be carcinogenic. Based on experimental study, Kovacs et al (2013) reported that a 1 to 5°C temperature increase above normal values resulted in higher organic load that increased disinfectant needs by up to 15% and resulted in a corresponding increase in disinfection byproducts formation. In wastewater systems, low DO levels can create septic situations in sewers that may corrode pipes and result in offensive odors and other toxic volatile gases.

Precipitation Variations

Changes in the timing, intensity, and duration of precipitation can negatively affect water quality. In places where lower precipitation is expected, lower velocities and hence higher water residence times in rivers and lakes will enhance the potential for toxic algal blooms and reduced DO levels. Lower water levels in lakes or rivers also lead to the release of sediments, organic carbon, and other contaminants into intake structures for water treatment. On the other hand, increased rainfall duration and intensity can result in higher runoff and subsequently higher level of salts, pathogens, heavy metals, and nutrients that will complicate water quality and treatment. Increased precipitation and consequent flooding can also overload combined sewer and wastewater treatment plants, resulting in the direct discharge of untreated or partially treated wastewater to water bodies. The timing of these changes may also have some undesirable consequences. Whitehead et al (2009) reported that storms that terminate drought periods can generate acid pulses in acidified catchments. Such events also flush nutrients from urban and rural areas, which would increase risk of eutrophication in lakes. However, increased runoff also has the potential to reduce such risks because nutrients could be flushed from lakes by more frequent storms and hurricanes. Overall, increased runoff results in negative impacts on water quality due to increased pollutant loads.

Sea Level Rise (SLR)

As sea level continues to rise, the extent of SWI will increase, especially during periods of drought in areas where aquifers are mainly recharged by rainfall and surface water flows. Coastal aquifers in Florida such as the Biscayne aquifer have long experienced SWI due to pumping, seawater movement up canals during high tides and storms, the lowering of water tables by drainage, and reduced flows from the Everglades (Miller et al. 1989; Prinos, et al. 2014). Preventing such impacts has become an expensive endeavor for the South Florida Water Management District involving the construction of salinity control structures, some of which must now be actively pumped to move freshwater to the sea and avoid inland flooding. It is expected that the Comprehensive Everglades Restoration Plan will increase freshwater flow to the southern Everglades, which will help offset the effect of SLR. SLR would also increase saltwater inflows to sewer collection systems and cause changes to the salinity of wastewater, which may impact biological treatment processes.

The IPCC report (2014) stated that expected climate change phenomena will, in general, impact raw water quality and pose risks to drinking water quality. In addition, climate change will also affect discharge permits of wastewater utilities due to more complex environmental conditions and reduced dilution effect of receiving water bodies in reduced precipitation areas. The EPA has identified the drinking water, surface water, discharge permits, and Total Maximum Daily Loads (TMDL) programs that could face potential impacts from air and water temperature increases.

It is important that the linkages between observed effects on water quality and climate be interpreted cautiously for different localities and water bodies due to the nonuniformity of climate change and water quality.

In summary, climate change can have a range of impacts on water quality for utility operation and management that could challenge meeting the regulations of the Safe Drinking Water and Clean Water acts. Additional processes or adaptation measures in water and wastewater treatment systems and stormwater management would likely be required in some cases. Conventional water treatment processes would need to be upgraded to advanced processes that might involve costly operations to handle increased levels of contamination and to comply with stringent standards that may be associated with impacts of climate change.

Some of the important adaptation measures for water and wastewater utilities include:

- Consideration of options for modular systems to provide additional capacity and improved performance or to add flexibility to treatment processes
- Careful consideration of climate change uncertainties in developing asset management strategies
- Enhanced long-term monitoring of temporal and spatial water quality information and development of more precise methods of projection of water quality changes
- Development of robust watershed management systems and holistic approaches to water quality and quantity management that take potential climate change impacts into consideration
- Identification of threshold water quality parameters for upgrading and planning new facilities

Conclusion

Florida has a diverse portfolio of water resources. Groundwater and surface water are intimately linked in many parts of of the state. Urban, agricultural, and ecological systems in Florida depend on water resources and the state relies heavily on groundwater for many of its water supply needs. Large uncertainties in climate projections for Florida make assessment of climate change impacts on water resources challenging. Only a few early assessments of likely future climate conditions in portions of the state have been completed so far. There is good agreement among the available climate models that temperatures will be higher in the future, but there is less consensus about future precipitation. A limited number of global climate models have been downscaled for Florida using both statistical and dynamical modeling approaches and the assessments to date have been used to determine the sensitivity of the water resources system to potential changes in precipitation, temperature, and SLR.

Since there is no statewide, comprehensive assessment of water resources impacts due to climate change, this chapter has provided summaries of several studies associated with four major regions including the Greater Everglades ecosystem, the Tampa Bay region, the St. Johns River basin, and the Suwanee and Apalachicola river basins. Although such studies have been diverse, they can be used to make several general and specific conclusions until more comprehensive, statewide assessment using standardized data and methods become available.

Climate change is likely to impact the drivers of hydrologic cycle in Florida, which primarily include precipitation, temperature, ET, and SLR. Because climate in Florida is also influenced by natural phenomena such as the ENSO, AMO, and others, it is important to understand how the interactions between climate and the teleconnections due to such natural phenomena may change in the future. Unfortunately, the available climate projections do not appear to have sufficient capacity to provide accurate predictions of hydrologic variables that will ultimately influence water resources systems. For this reason, much of the work in Florida has taken a scenario approach based on potential magnitude of the changes or pathways that have been identified for global change.

Impact assessments in various regions have demonstrated that the corresponding impact on water resources could be significant, particularly if the precipitation amounts decrease in the future. Water supplies are vulnerable to decreased recharge from higher ET associated with almost certain higher temperatures in the future. More precipitation could partially or wholly compensate for more ET, but less precipitation coupled with expected higher ET could place severe stresses on water supply systems. There is a great need to enhance the ability to predict future precipitation.

SLR threatens coastal areas with flooding and SWI into water supply aquifers. Near term, this is especially problematic in Southeast Florida, but other areas are beginning to be affected or will be in the more distant future.

There are numerous water quality issues associated with climate change and SLR. Exacerbation of algae and eutrophication problems due to higher temperatures may be one of the most immediate challenges. Water and wastewater treatment are likely to be affected by water quality changes due to climate change and SLR.

Acknowledgements

As coordinating lead authors, Jayantha Obeysekera, Wendy Graham, and Mike Sukop developed the chapter outline, prepared the introductory section, and provided continuous reviews of other sections. Jayantha Obeysekera wrote the Greater Everglades Ecosystem section and would like to acknowledge Jenifer Barnes, Tibebe Dessalegne, and Sashi Nair at SFWMD for their contributions to this section. Wendy Graham and Tirusew Asefa were the contributing authors of the Tampa Bay section, which is based on Dr. Syewoon Hwang's post-doctoral research. St. John River Region section was contributed by Dingbao Wang. Benjamin Mwashote was the contributing author for the Suwannee and Apalachicola River Basins section. Water Quality Impacts section was contributed by Kebreab Ghebremichael.

References

Arnold, J.G., Srinivasan, R., Muttiah, R.S., Williams, J.R., 1998. Large-area hydrologic modeling and assessment, I: Model development. Journal of the American Water Resources Association 34, 73–89.

Asefa, T., A. Adams, and I. Kajtezovic-Blankenship, 2014, A tale of integrated regional water supply planning: Meshing socio-economic, policy, governance, and sustainability desires together, Journal of Hydrology, doi:10.1016/j.jhydrol.2014.05.047, vol. 519, Part C, pp. 2632- 2641

Asefa, T., Adams, A., 2013. Reducing bias corrected precipitation projection uncertainties: a Bayesian based indicator weighting approach. J. Regional Environ. Change (special issue, March 9, 2013).

Aumen N.G., Havens K.E., Best, G.R., Berry, L., 2015. Predicting Ecological Responses of the Florida Everglades to Possible Future Climate Scenarios: Introduction. Environmental Management 55, 741- 748.

Bacopoulos, P., Funakoshi, F., Hagen, S.C., Cox, A.T., Cardone, V.J., 2009. The role of meteorological forcing on the St. Johns River (northeastern Florida). Journal of Hydrology 369, 55–70.

Bacopoulos, P., Tang, Y., Wang, D., Hagen S. C., 2017. Integrated hydrologic-hydrodynamic modeling of estuarine-riverine flooding (Tropical Storm Fay, 2008), Journal of Hydrologic Engineering, 22(8): 04017022.

Berri, J. G., and Flamenco, E. A., 1999. Seasonal volume forecast of the Diamante River, Argentina, based on El Niño observations and predictions. Water Resources Resesearch, 35(12), 3803-3810

Bicknell, B. R., Imhoff, J. C., Kittle, Jr., J. L., Jobes, T. H. and Donigian, Jr., A. S.: Hydrological Simulation Program-Fortran: HSPF Version 12.2 User's Manual., 2005.

Bledsoe, E.L., and Phlips, E.J., 2000, Relationships between phytoplankton standing crop and physical, chemical, and biological gradients in the Suwannee River and Plume region: Estuaries, v. 23, p. 458-473.

Carter, L. M., J. W. Jones, L. Berry, V. Burkett, J. F. Murley, J. Obeysekera, P. J. Schramm, and D. Wear, 2014: Ch. 17: Southeast and the Caribbean. Climate Change Impacts in the United States: The Third National Climate Assessment, J. M. Melillo, Terese (T.C.) Richmond, and G. W. Yohe, Eds., U.S. Global Change Research Program, 396-417. doi: 10.7930/J0NP22CB.

Chen, X., K. Alizad, D. Wang, and S. C. Hagen (2014), Climate Change Impact on Runoff and Sediment Loads to the Apalachicola River at Seasonal and Event Scales, Journal of Coastal Research, Special Issue (68), 35-42.

Chiew, F.H.S., Piechota, T.C., Dracup, J.A., and McMahon, T.A., 1998. El Niño/Southern Oscillation and Australian rainfall, streamflow, and drought: links and potential for forecasting. Journal of Hydrology, 204, 138–149.

Couch, C.A., E.H. Hopkins, and P.S. Hardy. 1996. Influences of Environmental Settings on Aquatic Ecosystems in the Apalachicola-Chattahoochee-Flint River Basin. Water-Resources Investigations Report 95-4278. U.S. Geological Survey, Atlanta, GA.

Dausman, Alyssa, and Langevin, C.D., 2005, Movement of the Saltwater Interface in the Surficial Aquifer System in Response to Hydrologic Stresses and Water-Management Practices, Broward County, Florida: U.S. Geological Survey Scientific Investigations Report 2004-5256, 73 p.

Dessalegne, T., J. Obeysekera, S. Nair, J. Barnes. (2016). "Assessment of CMIP5 Multi-Model Datase to Evaluate Impacts on the Future Regional Water Resources of South Florida. World Environmental and Water Resources Congress.

Dupuis, A. P. and Hann, B. J. (2009) Warm spring and summer water temperatures in small eutrophic lakes of the Canadian prairies: potential implications for phytoplankton and zooplankton. *J. Plankton Res.*, 31, 489–502.

Field, M.S., 2002, A lexicon of cave and karst terminology with special reference to karst hydrology: Washington, D.C., National Center for Environmental Assessment, Office of Research and Development, U.S. Environmental Protection Agency, EPA/600/R-02/003, 214 p.

Florida Oceans and Coastal Council (2009). The effects of climate change on Florida's ocean and coastal resources. A special report to the Florida Energy and Climate Commission and the people of Florida. Tallahassee, FL. 34 pp.

Geurink, J., and R. Basso, 2012: Development, calibration, and evaluation of the Integrated Northern Tampa Bay Hydrologic Model. Tampa Bay Water and the Southwest Florida Water Management District. Clearwater, FL.

Geurink, J., R. Basso, P. Tara, K. Trout, and M. Ross, 2006: Improvements to integrated hydrologic modeling in the Tampa Bay, Florida region: Hydrologic similarity and calibration metrics. Proceedings of the Joint Federal Interagency Conference 2006, Reno, NV.

Giardino, D., Bacopoulos, P., Hagen, S.C., 2011. Tidal spectroscopy of the lower St. Johns River from a high-resolution shallow water hydrodynamic model. International Journal of Ocean and Climate Systems 2, 1–18.

Gurdak, J. J., Hanson, R. T., McMahon, P. B., Bruce, B. W., McCray, J. E., Thyne, G. D., and Reedy, R. C., 2007. Climate Variability Controls on Unsaturated Water and Chemical Movement, High Plains Aquifer, USA. Vadose Zone Journal, 6, 533–547.

Harbaugh, a. and McDonald, M.: User's Documentation for MODFLOW-96, an update to the U.S. Geological Survey Modular Finite-Difference Ground-Water Flow Model, Open-File Report, US Geol. Surv., 96–485, 1996.

Hornsby, H.D., and Mattson, R., 1998, Surface water quality and biological monitoring network: Suwannee River Water Management District Annual Report WR-98-02.

Hovenga, P. A., D. Wang, S. C. Medeiros, S. C. Hagen, and K. Alizad (2016), The response of runoff and sediment loading in the Apalachicola River, Florida to climate and land use land cover change, Earth's Future, 4, 124–142, doi:10.1002/2015EF000348.

Hwang, S. and Graham, W. D.: Development and comparative evaluation of a stochastic analog method to downscale daily GCM precipitation, Hydrol. Earth Syst. Sci., 17(11), 4481–4502, doi:10.5194/hess-17-4481-2013, 2013.

Hwang, S., Graham, W. D., Adams, A. and Geurink, J.: Assessment of the utility of dynamically-downscaled regional reanalysis data to predict streamflow in west central Florida using an integrated hydrologic model, Reg. Environ. Chang., 13(S1), 69–80, doi:10.1007/s10113-013-0406-x, 2013.

Intergovernmental Panel on Climate Change (IPCC), 2007. Climate Change 2007: The Physical Science Basis. Contribution of Working Group I to the Fourth Assessment Report of the Intergovernmental Panel on Climate Change [Solomon, S., D. Qin, M. Manning, Z. Chen, M. Marquis, K.B. Averyt, M. Tignor and H.L. Miller (eds.)]. Cambridge University Press, Cambridge, United Kingdom and New York, NY, USA, 996 pp.

Intergovernmental Panel on Climate Change (IPCC), 2013. "Climate Change 2013: The Physical Science Basis. Contribution of Working Group I to the Fifth Assessment Report of the Intergovernmental Panel on Climate Change, Stocker, T.F. and Qin, D. and Plattner, G.-K. and Tignor, M. and Allen, S.K. and Boschung, J. and Nauels, A. and Xia, Y. and Bex, V. and Midgley, P.M., Cambridge University Press

IPCC (2000), Special Reports: Emissions ScenariosRep., IPCC, UK.

IPCC . 2007. Climate Change 2007 – The Physical Science Basis, Contribution of Working Group I to the Fourth Assessment Report of the IPCC. Cambridge University Press

IPCC (2014) Climate Change 2014: Synthesis Report. Contribution of Working Groups I, II and III to the Fifth Assessment Report of the Intergovernmental Panel on Climate Change [Core Writing Team, R.K. Pachauri and L.A. Meyer (eds.)]. IPCC, Geneva, Switzerland, 151 pp.

Irizarry Michelle, Jayantha Obeysekera, Joseph Park, Paul Trimble, Jenifer Barnes, Winnie Said, Erik Gadzinski (2013). Historical Trends in Florida Temperature and Precipitation, Hydrological Processes Journal, 27(16), 2225-2382

John, D.A and Leventhal, J.S (1995). Bioavailability of Metals. Ch. 2 in: Preliminary compilation of descriptive geoenvironmental mineral deposit models Edward A. du Bray, Editor. U.S. Department of the Interior U.S. Geological Survey Open-File Report 95-831, Denver, Colorado.

Karl T.R., Melillo, J.M. and Peterson, T.C. (2009) Global Climate change impacts in the United States. Cambridge University Press, Cambridge.

Katz, B. G. and Raabe, E. A., 2005, Suwannee River Basin and Estuary: An Integrated Watershed Science Program. U.S. Geological Survey Open File Report 2005-1210, 19p.

Keener, V.W., Ingram, K.T., Jacobson, B., and Jones, J.W., 2007. Effects of El-Niño/ Southern Oscillation on simulated phosphorus loading. Transactions of the ASABE, 50(6), 2081–2089.

Kendall MG. 1976. *Rank Correlation Methods*. 4th Ed. Charles Griffin, 210 pp.

Kolpin, D.W., Furlong, E.T., Meyer, M.T., Thurman, E.M., Zaugg, S.D., Barber, L.B., and Buxton, H.T., 2002, Pharmaceuticals, hormones, and other organic wastewater contaminants in U.S. streams, 1999-2000: A national reconnaissance: Environmental Science and Technology, v. 36, p. 1202-1211.

Kovacs, M.H., Ristoiu, D., Voica, C. and Moldovan, Z. (2013) Climate Change Influence on Drinking Water Quality, Processes in Isotopes and Molecules, AIP conference proceedings 1565, 298-303.

Langevin CD, Thorne DT, Dausman AM, Sukop MC, Guo W (2008) SEAWAT Version 4: A Computer Program for Simulation of Multi-Species Solute and Heat Transport. U.S. Geological Survey Techniques and Methods Book 6.

Li H, Sheffield J, Wood EF (2010) Bias correction of monthly precipitation and temperature fields from Intergovernmental panel on climate change AR4 models using equidistant quantile matching. J Geophys Res 115:D10101. doi:10.1029/2009JD012882

Lindsey, B.D., Berndt, M.P., Katz, B.G., Ardis, A.F., and Skach, K.A., 2009, Factors affecting water quality in selected carbonate aquifers in the United States, 1993–2005: U.S. Geological Survey Scientific Investigations Report 20085240, 117 p. https://pubs.er.usgs.gov/publication/sir20085240

Luettich, R.A., Westerink, J.J., Scheffner, N.W., 1992. ADCIRC: An advanced three-dimensional circulation model for shelves, coasts and estuaries, I: Theory and methodology of ADCIRC-2DDI and ADCIRC-3DL. Technical Report DRP-92-6, United States Army Corps of Engineers, 144p.

Marella, R. L., 2014. Water withdrawals, use and trends in Florida, 2010. U.S. Geological Survey Scientific Investigations Report 2014-5088.

Marella, R.L., 2015. Water withdrawals in Florida, 2012. U.S. Geological Survey Open-File Report 2015-1156, 10p.

Maupin, M.A., and Barber, N.L., 2005, Estimated withdrawals from principal aquifers in the United States, 2000: U.S. Geological Survey Circular 1279, 46 p.

Maupin, M.A., Kenny, J.F., Hutson, S.S., Lovelace, J.K., Barber, N.L., and Linsey, K.S., 2014, Estimated use of water in the United States in 2010: U.S. Geological Survey Circular 1405, 56 p., http://dx.doi.org/10.3133/cir1405.

Maurer EP, Brekke L, Pruitt T, Duffy PB (2007) Fine-resolution climate projections enhance regional climate change impact studies. Eos Trans AGU 88(47):504. doi:10.1029/2007EO470006

Maurer, E.P., Hidalgo, H.G., Das, T., Dettinger, M.D., Cayan, D.R. 2010. "The utility of daily large-scale climate data in the assessment of climate change impacts on daily streamflow in California", Hydrology and Earth System Sciences, 14, 1125-1138, doi:10.5194/hess-14-1125-2010. McCabe, G. J., and Dettinger, M. D., 1999. Decadal variations in the strength of ENSO teleconnections with precipitation in the western United States. International Journal of Climatology, 19(13), 1399-1410.

McFadden, S.S. and P.D. Perriello. 1983. Hydrogeology of the Clayton and Claiborne aquifers in southwestern Georgia. Georgia Geologic Survey Information Circular 55.

Mearns, L.O., et al. 2012. The North American Regional Climate Change Assessment Program dataset, National Center for Atmospheric Research Earth System Grid data portal, Boulder, CO. Data downloaded 2017-10-23. doi:10.5065/D6RN35ST

Melinda Haydee Kovacs, Dumitru Ristoiu, Cezara Voica, Zaharie Moldovan, Processes in Isotopes and molecules, AIP Conf. Proceedings 1565, 298-303)

Miller, J.A. 1986. Hydrogeologic Framework of the Floridan Aquifer System in Florida and Parts of Georgia, South Carolina, and Alabama. U.S. Geological Survey Professional Paper 1403-B.

Miller, T., Walker, JC., Kingsley, G.T. and Hyman, W. A. 1989. Impacts of global Climate change on Urban infrastructure. In J.B Smith and D.A. Tirpak, eds., Potential effects of Global Climate change on the Unites States: Appendix H, Infrastructure, Washington, D.C.: US Environmental Protection Agency

Mwashote, B. M., Burnett, W. C ., Chanton, J, Santos, I. R., Dimova, N. and Swarzenski, P. W., 2010, Calibration and use of continuous heat-type automated seepage meters for submarine groundwater discharge measurements, Estuarine Coastal Shelf Sci., 87, 1–10, doi:10.1016/j.ecss.2009.12.001.

Mwashote, B. M., Murray, M., Burnett, W. C ., Chanton, J., Kruse, S. and Forde, A., 2013, Submarine groundwater discharge in the Sarasota Bay system: Its assessment and implications for the nearshore coastal environment, Continental Shelf Research, 53, 63-76.

NARCCAP. 2011 *The North American Regional Climate Change Assessment Program (NARCCAP)*. Research Data Archive at the National Center for Atmospheric Research, Computational and Information Systems Laboratory. http://rda.ucar.edu/datasets/ds317.0/. Accessed† dd mmm yyyy.

Obeysekera J, J. Barnes, M. Nungesser. (2014). "Climate sensitivity runs and regional hydrologic modeling for predicting the response of the greater Florida Everglades ecosystem to climate change," Environ Manage. 2015 Apr; 55(4):749-62.

Obeysekera, Jayantha, J. Park, M. Irizarry-Ortiz, P. Trimble, J. Barnes, J. VanArman, W. Said, and E. Gadzinski (2011) Past and Projected Trends in Climate and Sea Level for South Florida. South Florida Water Management District, Hydrologic and Environmental Systems Modeling Technical Report (peer reviewed), 148 pp.

Panofsky HA, Brier GW (1968) Some application of statistics to meteorology. Pennsylvania State University, University Park, Pa

Prinos, S.T., Wacker, M.A., Cunningham, K.J., and Fitterman, D.V., 2014, Origins and delineation of saltwater intrusion in the Biscayne aquifer and changes in the distribution of saltwater in Miami-Dade County, Florida: U.S. Geological Survey Scientific Investigations Report 2014–5025, 101 p., http://dx.doi.org/10.3133/sir20145025.

Quinlan, J. F. and Ewers, R. O.: 1989, 'Subsurface drainage in the Mammoth Cave Area', in: W. B. White and E. L. White (Eds.), Karst Hydrology: Concepts from the Mammoth Cave Area: Van Nostrand Reinhold, New York, 65–103.

Rosenzweig C, Horton RM, Bader DA, Brown ME, DeYoung R, Dominguez O, … Toufectis K (2014) Enhancing Climate Resilience at NASA Centers: A Collaboration between Science and Stewardship, Bull. Amer. Meteorol. Soc. 95(9):1351-1363.

Ross, M., Geurink, J., Said, A., Aly, A. and Tara, P.: Evapotranspiration Conceptualization in the HSPF-MODFLOW Integrated Models, J. Am. Water Resour. Assoc., 41(5), 1013–1025, doi:10.1111/j.1752-1688.2005.tb03782.x, 2005.

Roy, S.S., 2006. The impacts of ENSO, PDO, and local SSTs on winter precipitation in India. Physical Geography, 27(5), 464-474.

Sahoo, G., Schaldow, S.G., Reuter, J.E. and Coats, R (2011). Effects of climate change on thermal properties of lakes and reservoir, and possible implications. *Stochastic Environmental Research and Risk Assessment*, 25(4): 445-456.

Scavia, D., et al. (2002), Climate change impacts on U.S. Coastal and Marine Ecosystems, Estuaries, 25(2), 149-164, doi: 10.1007/BF02691304.

Semenov, M. A., and P. Stratonovitch (2010), Use of multi-model ensembles from global climate models for assessment of climate change impacts, Climate Research, 41, 1-14, doi: 10.3354/cr00836.

Simpson, H. J., and Colodner, D. C., 1999. Arizona precipitation response to the Southern Oscillation: A potential water management tool. Wat. Resources Res., 35(12): 3761-3769.

SFRCCC, 2015. A Unified Sea Level Rise Projection for Southeast Florida, Southeast Florida Regional Climate Change Compact, http://www.southeastfloridaclimatecompact.org/wp-content/uploads/2015/10/2015-Compact-Unified-Sea-Level-Rise-Projection.pdf.

South Florida Water Management District. 2009. Climate change and water management in south Florida. Interdepartmental Climate Change Group. South Florida Water Management District, West Palm Beach, FL .

South Florida Water Management District. 2005. Documentation of the South Florida Water Management Model, Version, 5.5. South Florida Water Management District, FL.

Stefanova, L, V. Misra, S. C. Chan, M. Griffin, J. J. O'Brien and T. J. Smith III. 2012. A Proxy for High-Resolution Regional Reanalysis for the Southeast United States: Assessment of Precipitation Variability in Dynamically Downscaled Reanalyses. Clim Dyn, DOI: 10.1007/s00382-011-1230-y.

Stewart, S.R., Beven, J.L.II, 2009. Tropical Storm Fay, 15–26 August 2008. Tropical Cyclone Report, National Hu Wang, D., Hejazi, M., Cai, X., Valocchi, A.J., 2011. Climate change impact on meteorological, hydrological and agricultural drought in central Illinois. Water Resources Research 47, W09527, doi: 10.1029/2010WR009845.

Taner, M.T, Carleton J. N. and Wellman, M (2011) Integrated model projections of climate change impacts on a North American lake. *Ecological Modeling*, 222(18), 3380-3393.

Tebaldi C, Smith RL, Nychka D, Mearns LO (2005) Quantifying uncertainty in projections of regional climate change: A Bayesian approach to the analysis of multimodel ensembles. J Clim 18(10):1524-1540. doi:10.1175/JCLI3363.1

Toth, D.J., 1993. Volume 1 of the Lower St. Johns River Basin reconnaissance: Hydrogeology. Technical Report SJ93-7, St. Johns River Water Management District, 28p.

Trimble, P.J., Obeysekera, T.B., Cadavid, L.G., Santee, E.R. 2006. Applications of climate outlooks for water management in south Florida. In: Garbrecht JD, Piechota TC (ed) Climate variations, climate change, and water resources engineering. ASCE/EWRI, Reston, VA

USGS EROS Center (2014), CONUS Modeled annual land-cover maps of the A1B, A2, and B1 scenario from 2006-2100, edited, U.S. Department of the Interior, U.S. Geological Survey.

Wang, D., S. C. Hagen, and K. Alizad (2013), Climate change impact and uncertainty analysis of extreme rainfall events in the Apalachicola River basin, Florida, Journal of Hydrology, 480, 125-135, 10.1016/j.jhydrol.2012.12.015.

Wang D., Hejazi, M., Cai,X., Valocchi, A.J. 2011. Climate change impact on meteorological, hydrological, and agricultural drought in central Illinois, Water Resources Research, 47, W09527, doi:10.1029/2010WR009845.

Watershed Academy Web (n.d.) The Effect of Climate Change on Water Resources and Programs, Distance Learning Modules on Watershed Management, https://cfpub.epa.gov/watertrain/pdf/modules/Climate_Change_Module.pdf, retrieved on November 29, 2016

Weary, D.J., 2006, Compilation of national-scale karst data by the U.S. Geological Survey with emphasis on the southeastern United States: Geological Society of America Southeastern Section meeting, Abstracts with Programs, v. 38, no. 3, p. 84.

Whitehead, P.G., Wilby, R. L., Battarbee, R. W., Kernan, M. and Wade, A. J (2009). A review of potential impact of climate change on surface water quality. *Hydrological Sciences Journal*, 54:101-123.

Wood AW, Maurer EP, Kumar A, Lettenmaier D (2002) Long-range experimental hydrologic forecasting for the eastern United States. J Geopys Res 107(D20):4429. doi: 10.1029/2001JD000659

Xiao, H., Wang, D., Hagen, S. C., Medeiros, S. C., Hall, C. R., 2016. Assessing the impacts of sea-level rise and precipitation change on the surficial aquifer in the low-lying coastal alluvial plains and barrier islands, east-central Florida (USA), Journal of Hydrogeology, doi: 10.1007/s10040-016-1437-4.

CHAPTER 4

Climate Change Impacts on Human Health

Song Liang[1,5], Kristina Kintziger[2], Phyllis Reaves[3], and Sadie J. Ryan[4,5]

[1]*Department of Environmental and Global Health, College of Public Health and Health Professions, University of Florida, Gainesville, FL;* [2]*Department of Public Health, College of Education, Health & Human Sciences, The University of Tennessee, Knoxville, TN;* [3]*Division of Physical Therapy, School of Allied Health Sciences, Florida Agricultural & Mechanical University, Tallahassee, FL;* [4]*Department of Geography, University of Florida, Gainesville, FL;* [5]*Emerging Pathogens Institute, University of Florida, Gainesville, FL*

Climate change poses major challenges to human society and to Earth systems, influencing the functioning of many ecosystems and thereby affecting human health. Many climate change/variability- and extreme weather-associated events, such as sea level rise, hurricanes, and storm surge, as well as other weather extremes, including excessive precipitation and heatwaves, have direct and/or indirect impacts on human health. These impacts include death/injury, cardiovascular and respiratory diseases, environmentally-mediated infectious diseases, and mental health, among others. Due to its unique geography, Florida is particularly vulnerable to these environmental impacts, which have important health implications for the state's more than 20 million residents. In this chapter, we review the health impacts of climate change and associated weather events, with an emphasis on those relevant to Florida, and environmental hazards, including hurricanes and storms, lightning, sea level rise, excessive precipitation, extreme heat, and drought. There is clear evidence for significant climate-sensitive hazards and human health impacts in the state, despite uncertainties associated with the assessment of some effects. To address health impacts and challenges, policies focused on mitigation and adaptation strategies, health surveillance, and research that could close knowledge gaps on human exposures to the climate-sensitive hazards and health impacts are needed.

Key Messages

- Florida is highly vulnerable to climate-sensitive hazards (e.g. sea level rise, heat waves, storm surge, and hurricanes), which have a wide range of human health effects.
- The health effects can be direct, such as storm/temperature related illnesses, injuries, and deaths; or indirect, such as waterborne, food-borne, and vector-borne diseases; or take social and economic pathways, such as stress and mental illness.
- The health effects exhibit substantial regional disparities across the state.
- Policies focused on health surveillance and research on knowledge gaps between human exposure to the hazards and health effects are much needed.

Keywords

Climate change; Environmental hazards; Human health; Florida

Introduction

Climate change is well-characterized at the global level using metrics such as the global mean temperature and sea level. But the manifestations of global climate change and variability occur at varying geographical scales, causing a variety of weather events, such as excessive rainfall, drought, severe storm/flooding, sea level rise, and heat waves (Field et al. 2014), all of which are likely to have direct and/or indirect human health impacts (McMichael and Haines 1997; Colwell et al. 1998; Frumkin et al. 2008). Indeed, there is clear and increasing evidence that many health outcomes (most adverse) exhibit high sensitivities to these varying weather events. The interface between the climate change-associated weather events and human health exhibits a complex web of relationships involving both natural and social environments through direct or indirect impact pathways, social institutional disruption, or a combinations of these things (Fig. 4.1). Globally, substantial disease burdens are attributable to hazards that are associated with climate change and related weather events. According to the World Health Organization's 2008 estimate, climate change-associated hazards were responsible for more than 150,000 deaths and 5,517,000 disability-adjusted life years, which were highly likely to be underestimated as the estimates were only based on selected risk factors and illnesses associated with climate change and related weather events (WHO 2008). The actual disease burden might be greater and the adverse health impacts are likely to increase in the years to come, as the Intergovernmental Panel on Climate Change (IPCC) and US Global Change Research Program have clearly indicated that the accelerated changing climate poses a substantial threat to global human health and the risk will continue to become severe if no remediation action is taken (Field et al. 2014; Balbus et al. 2016).

Due to its unique geography, Florida is highly vulnerable to a variety of hazards associated with climate change and variability, and related weather events. Sea level rise, hurricane and storm surge, excessive precipitation, and heatwaves all pose threats to Florida's agriculture, ecosystems, tourism, and public health. For example, Florida has experienced a greater number of hurricane landfalls than other states in the country (Knight and Davis 2009), and more intense hurricanes are expected in the future (Knutson et al. 2013). In addition, rising sea levels are influencing and expected to continue to have significant impacts on communities and residents along Florida's coastal areas, particularly in the southern part of the state, making Floridians exceptionally vulnerable to these environmental hazards (Elsner et al. 2008).

This chapter is organized into sections addressing impact pathways and specific environmental hazards of particular concern in Florida (Fig. 4.1). For direct impacts, we focus on extreme heat, flooding, storms and lightning, and sea level rise. For indirect impacts on public health, we focus on water, food, and vector-borne diseases. And lastly, we conclude with our discussion of impact pathways through social disruption on mental and community health impact associated with climate change and related weather events. Throughout this chapter, health impacts and implications, particularly those pertinent to Florida, are reviewed primarily based on the *Building Resilience Against Climate Effects* (BRACE) project reports from the Florida Department of Health (FDH) (FDH 2015b, a) and other relevant publications.

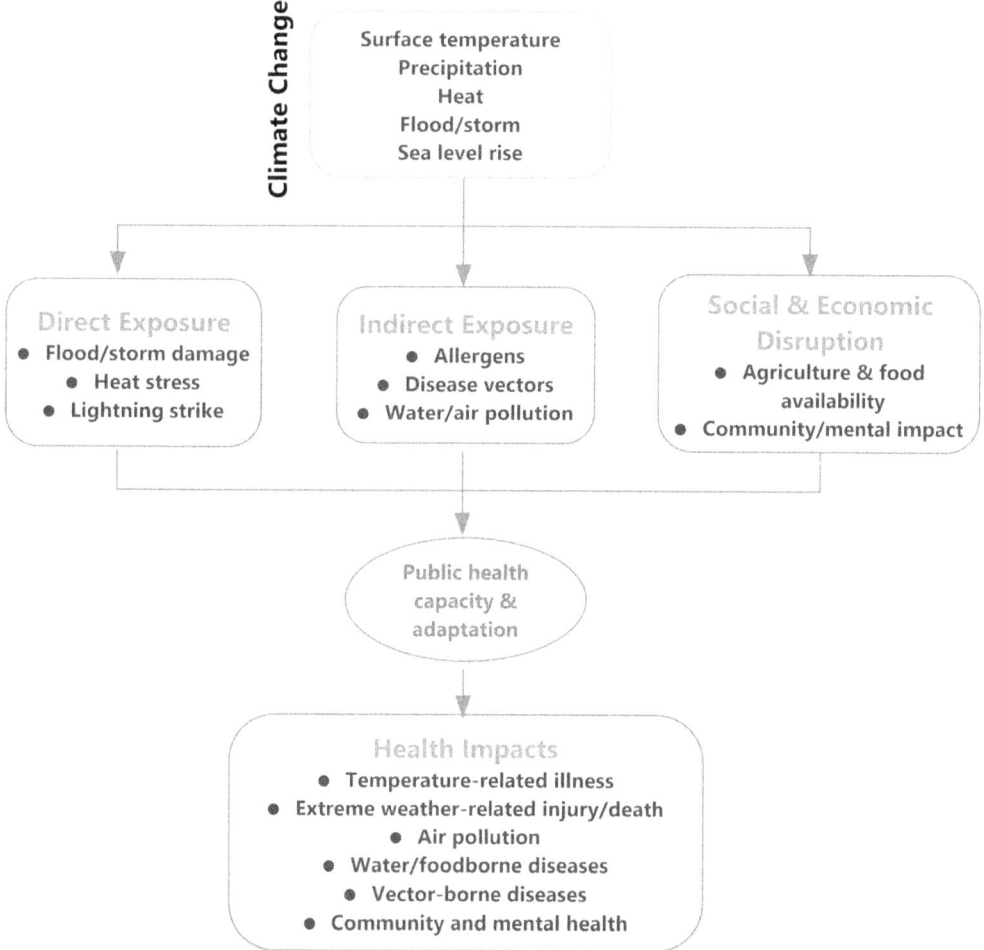

Figure 4.1. Impact pathways by which climate change and variability and related weather events affect human health.

Direct Impact Pathways – Environmental Hazards and Human Health

A number of human health impacts come from direct exposure to environmental hazards associated with climate change and related weather events. Among those of particular concern in Florida are extreme heat, flooding, storms and lightning, and sea level rise.

Extreme Heat and Health Impacts

Extreme climate and weather events (e.g., heat, cold, storms, and floods) are expected to occur more frequently worldwide due to climate change. These extreme events, such as heatwaves and cold spells (e.g. frequent very hot days and possibly fewer very cold days), have a direct impact on human health by compromising the body's ability to regulate its internal temperature. Increased human morbidity and mortality have been associated with extreme temperatures, both hot and cold, and documented in both developing and developed countries worldwide (Center for Disease Control and Prevention 1994; Keatinge et al. 2000a; Keatinge et al. 2000b; Basu and Samet 2002; Medina-Ramon et al. 2006; Kovats and Hajat 2008; Bandyopadhyay et al. 2012). The mechanistic effects of extreme temperature on death and illness are briefly summarized below (Sarofim et al. 2016):

- Extreme heat may induce heat cramps, heat exhaustion, heatstroke, and hyperthermia.
- Extreme cold may induce hypothermia and frostbite.
- Extreme heat/cold may exacerbate chronic conditions related to cardiovascular disease, respiratory disease, cerebrovascular disease, and diabetes-related illnesses.
- Extended exposure to high temperatures may induce increased hospital admissions for cardiovascular, kidney, and respiratory disorders.
- Extreme heat may induce or exacerbate mental health and behavioral disorders.

A number of studies have reported associations between high temperatures and human deaths or illnesses around the world. For example, a large number of heat-related deaths have been reported in the United States (Center for Disease Control and Prevention 1994; Curriero et al. 2002; Hoshiko et al. 2010), Europe (Keatinge et al. 2000a; Keatinge et al. 2000b; Huynen et al. 2001), and Asia (Qian et al. 2008; Chung et al. 2009). A recent study by Berko et al. examined deaths attributed to extreme weather events from 2006 to 2010 in the United States and found that each year about 2,000 U.S. residents died from weather-related causes, and among those 31% died due to exposure to excessive heat, heat stroke, and/or sun stroke (Berko et al. 2014). Mounting evidence has suggested that substantial increases in mortality from respiratory and cardiovascular diseases, in particular among the elderly and youth, are associated with high temperatures in the United States (Basu and Samet 2002; Medina-Ramon et al. 2006; Kovats and Hajat 2008). During the 2006 California heat wave, 16,166 excess emergency department visits and 1,182 excess hospitalizations were reported statewide. This included a significant increase in emergency department visits for acute renal failure, cardiovascular diseases, and diabetes (Knowlton et al. 2009). The association between heat and respiratory diseases has also been reported in many studies. In New York state, a significant number of respiratory hospitalizations were attributable to excessive heat, and a projection suggested that excess respiratory admissions in the state due to extreme heat will be two to six times higher in 2080–2099 than what was seen in 1991–2004 (Lin et al. 2012). In another study conducted in Greater London, UK, an analysis of historical data suggested that heat-related increases in emergency admissions for respiratory and renal diseases had been observed in children under five, and for respiratory disease in the 75+ age group (Kovats et al. 2004).

Extreme Heat and Public Health in Florida

Florida has a warm subtropical and tropical climate. Due to continuous summer heat and extreme weather events, temperatures around the state can reach levels potentially harmful to human health. Through analyses of historical data (for the periods 1895–2009 and 1970–2009), a significant trend in Florida's mean temperature has been found (Martinez et al. 2012). The average summer temperature (June–August) has been showing an overall increasing trend, particularly after the 1980s, which shows a greater upward trend (FDH 2015a), likely due to a variety of factors ranging from rapid urban development (Martinez et al. 2012; FDH 2015a) to climate change (Ji et al. 2014). Within the state, substantial spatial heterogeneities in mean, minimum, and maximum temperature have been reported (Martinez et al. 2012). For example, the distribution of annual mean number of days with a maximum temperature greater than or equal to 95 °F and greater than or equal to 75 °F showed that more days with hottest daytime temperatures occurred in the northern and interior portions of the state (FDH 2015a).

Using statewide information on specific health outcome-related emergency department visits that occurred May to October from 2005–2012, the BRACE project examined the effects of daily maximum temperature or daily maximum heat index on emergency department visits through a two-stage analysis—the first stage involved the analysis on regional (National Weather Service regions) scales using the Poisson regression model; the second stage used a meta-analysis technique to integrate regional estimates into a statewide estimate (FDH 2015a). Rate ratios, based on the comparison between the exposure (e.g., hotter days) and the reference (e.g. 88 °F for temperature and 94 °F for heat index), were used to compare emergency department visits for specific health conditions (FDH 2015a).

a) *Heat and heat-related illness.* A strong association between heat-related illness and heat exposure was observed; the significant increase in emergency department visit rates for heat-related illness was associated with increasing temperatures and a strong dose-response relationship across all regions in Florida was found (FDH 2015a).

b) *Heat and cardiovascular disease.* Two types of cardiovascular disease were considered—myocardial infarction and ischemic stroke. No statistically significant association was found between temperature and ischemic stroke at the state level; however, significant regional variations were identified. For example, the Tampa region showed a significant positive association between heat index (above the reference level at 94 °F) and stroke (FDH 2015a). Furthermore, a positive, statistically significant relationship was observed between myocardial infarction and heat index despite remarkable geographical variations (FDH 2015a).

c) *Heat and respiratory disease.* Overall, increases in the number of emergency department visits for asthma were associated with higher temperatures. Although no significant association was found at the state level between temperature and emergency department visits for asthma, the regional analysis suggested that all regions except Tallahassee showed a statistically significant positive association between temperature and asthma (FDH 2015a).

Flooding, Storms, Lightning, and Sea Level Rise

Excessive precipitation, hurricanes, coastal storms, sea level rise, and thunderstorm-related lightning, which are all typically accompanied by coastal and inland flooding, have the potential for substantial direct human health impacts (e.g. injury and death, maternal and child health, and mental health issues). Natural disasters related to flooding and storms are a significant cause of mortality and morbidity. For example, according to the International Federation of Red Cross and Red Crescent Societies (IFRCRCS), more than 3,448 climatic, hydrological and meteorological disasters (including 1,751 major floods and 988 storms) resulting in 339,710 deaths (59,092 and 177,685 deaths specifically tied to flooding and storms) were reported worldwide between 2005 and 2014 (IFRCRCS 2015). Lightning strike is the second leading cause of weather-related mortality at the global scale, with an estimated 0.2-1.7 deaths per million people, and those who survive lightning strikes often suffer from significant injuries (Aslar et al. 2001; Ritenour et al. 2008). The magnitude and impacts of these extreme weather events, their severity, and the extent of the effects are influenced by many different factors. For example, short- and long-term averages and variability of weather conditions and physical impacts of associated extreme events, as well as social environmental factors (e.g. infrastructure, social and individual vulnerability), are considered important (Bouma et al. 1997, Kovats 2000, Kovats et al. 2003, Miranda 2004). For projections of future flooding under different climate change scenarios, climate models have consistently suggested that episodes of severe flooding may become more frequent in inland river systems including flood plains (Christensen and Christensen 2003, Booij 2005), urban and coastal environments in various parts of the world (Schreider et al. 2000; Douglas et al. 2008; Kirshen et al. 2008; Thompson et al. 2009, Diez et al. 2011; Lyle and Mills 2016). Meanwhile, climate models also predict positive correlations between lightning and global temperatures, suggesting the likelihood of a greater number of and more severe lightning episodes (Price and Rind 1994; Reeve and Toumi 1999; Kochtubajda et al. 2006). These episodes are expected to increase the risk of extreme environmental hazards and adverse public health impacts.

a) *Natural disasters in Florida.* Due to its geography (extensive coastline and a peninsular shape) and its tropical/subtropical climate, Florida is particularly vulnerable to tropical storms, hurricanes, and lightning strikes. Historically the occurrence of hurricanes in the US has been clustered in Florida and along the Atlantic coastline (Ellis et al. 2015). A total of 67 known Florida hurricanes have occurred over the 108-year period (1900-2007) with four different hurricanes occurring in one year (2004) in Florida. The climate change model suggests a 46% chance that Florida will be hit by at least one hurricane each year in the future (Malmstadt et al. 2009), and recent experience seems to support this. Fig. 4.2 illustrates the distribution of average return time of hurricane landfalls and populations along the coastal areas of Florida, showing spatial heterogeneities in vulnerability to hurricanes.

b) *Public health impacts.* Key direct public health impacts associated with extreme weather events include injury and death due to trauma, drowning, destructive forces of wind, collapsed building and trees, and lightning strike and carbon monoxide poisoning related to power outages. Historically, Florida experienced significantly higher mortality in the early 20th century; approximately 3,000 deaths were attributed to extreme weather-related events (Winsberg 2003) and the largest number of deaths due to lightning strike in the US were reported during this time (Duclos et al. 1990; Ritenour et al. 2008). During 2004 and 2005, eight hurricanes hit Florida resulting in 213 deaths, over half of which were caused by trauma, followed by drowning, other injury, electrocution, and carbon monoxide poisoning (Ragan et al. 2008). Injury is common cause of hurricane-related morbidity and mortality. For example, in the aftermath of Hurricane Katrina between September 8 and October 14, 2005, 7,543 non-fatal injuries among residents and relief workers were recorded in the surveillance system (Sullivent et al. 2006). During a 2006 flood event in El Paso County in Texas, 43% of individuals (out of 475 surveyed) were reported having physical health issues related to the flooding episode (Collins et al. 2013).

c) *Injury.* The BRACE study compared injury-related emergency department visits and hospitalizations during impact periods (periods covering tropical cyclone landfall) vs. control periods (periods with no such impact) between 2004 and 2012. The study found that emergency department visit and hospitalization rates were significantly higher during the impact periods, with a rate ratio of 1.03 (95% confidence interval (CI): 1.02, 1.05) for the emergency department visit rate and 1.4 (95% CI: 1.02, 1.05) for the hospitalization rate. The most common types of injuries during the impact periods included falls, being struck by an object, being cut or pierced, and motor vehicle transport accidents (FDH 2015b).

d) *Carbon monoxide poisoning.* Carbon monoxide (CO) is a colorless, odorless, poisonous gas that can be harmful when inhaled in a large amount. Inhaled CO enters the blood stream and reduces the delivery of oxygen to the body's critical organs, such as the heart and brain. The gas is primarily generated through incomplete combustion. The greatest sources of CO outdoors are usually related to motor vehicles, machinery that burn fossil fuels. Indoors, some major sources include unvented kerosene and gas space heaters, leaking chimneys and furnaces, and gas stoves. Exposure to CO may cause weakness, headache, dizziness, nausea, shortness of breath, confusion, and even death at very high concentration levels, which is more likely in indoor environments. A recent systematic review of the health impacts of power outages due to extreme events indicated that CO poisoning is an important health concern (Klinger et al. 2014). In Florida, the BRACE study reported that during the study period 2004–2012, the rate of CO exposure calls during impact periods was 6.59 times (95% CI: 4.48, 9.7) the rate of CO exposure calls to poison control centers during the control periods. The study also found the rate of CO-poisoning related emergency department visits was significantly higher for the impact periods than during the control periods with the rate ratio 3.44 (95% CI: 2.07, 5.72). Finally, the BRACE study revealed that rates of CO poisoning-related hospitalizations were also significantly higher during the impact periods with the rate ratio 4.0 (95% CI: 2.9, 5.51) (FDH 2015b).

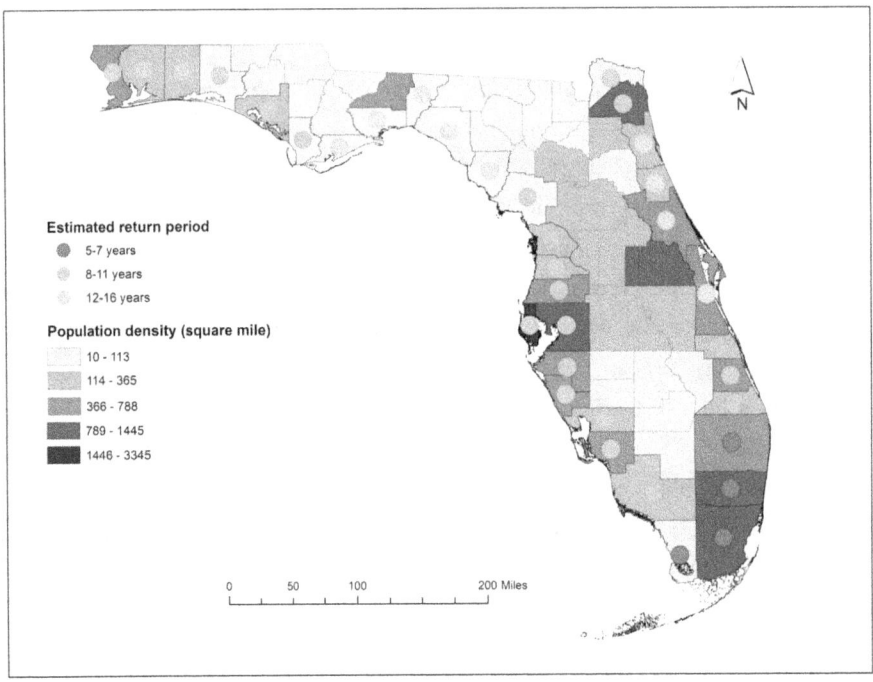

Figure 4.2. Historical distribution of hurricane landfalls in Florida.

Indirect Impact Pathways–Infectious Diseases

Climate change and variability and associated variable weather conditions can affect human health through indirect pathways (Fig. 4.1). These indirect impacts are modulated through biological and ecological processes that can influence infectious diseases (in particular, some water/food- and vector-borne diseases) and air quality. In this section, we focus on water/food-, and vector-borne diseases.

Disease Transmission Groupings

The impact of climate change and variability and associated weather events on infectious diseases has been receiving much attention in the past several decades. Across the globe, high human mortality rates and disability have been attributed to infectious diseases; climate change is intensifying and will continue to exacerbate transmission of infectious diseases, poses increasing concerns (Patz et al. 2005). Indeed, the transmission of many infectious diseases is closely linked to physical environmental factors, such as water and temperature. Many water-associated and zoonotic pathogens, for instance, which account for more than 75% of known pathogens causing human diseases, are associated with hydrological process (WHO 2003). From a transmission perspective, infectious diseases can be broadly classified into two groups: 1) directly transmitted (i.e., those spread directly from person to person via direct contact or droplet exposure, such as influenza and tuberculosis); and 2) indirectly transmitted, including those mediated through an intermediate vector organisms (vector-borne) or through environmental media such as water and soil (environmentally-mediated) (Eisenberg et al. 2007). Because the diseases in the indirectly transmitted group are either directly or indirectly related to water, this group can be further broken down into the following sub-groups based on water's role in disease transmission: waterborne (e.g., cholera typically via fecal-oral transmission), water-based (e.g. part of the pathogen life cycle requires aquatic environment, such as schistosomiasis), and water-related (e.g., malaria and dengue, which need water for breeding of insect vectors to fulfill the transmission cycle) (Yang et al. 2012; Mordecai et al. 2013; Ryan et al. 2015; Johnson et al. 2015; Mordecai et al., 2017).

Climate-Sensitivity of Microbial Agents and Their Associated Diseases

Table 4.1 lists some of the major pathogens and their associated diseases that are climate sensitive and associated with water. These disease-causing pathogens, either through infection or toxin-generating (e.g. caused by protozoa, bacteria, virus, helminth, or algae), and their associated vectors (e.g., snails, mosquitoes, sand flies) are very small in size and lack thermostatic mechanisms (i.e. lack of function to maintain body temperatures). Their biological (e.g., survival, reproduction, development/growth, and infectivity) and environmental (e.g., contamination and movement) processes are determined by environmental conditions, such as rainfall, storm runoff, temperature, and sunlight. Climate change and variability and associated weather events are expected to affect both fresh and marine water environments thereby changing humans' exposure to these water-related contaminants or pathogens that cause ill health.

a) Waterborne Diseases

Waterborne diseases remain an important contributor to the global burden of disease (Pruss et al. 2002; Yang et al. 2012; Pruss-Ustun et al. 2014). Many different viral, bacterial, and parasitic diseases have been associated with waterborne transmission (Table 4.1). Human exposure to waterborne pathogens is primarily through drinking water, recreational water (e.g. via accidental drinking and, for some pathogens, via dermal contact), or foods. The distribution of these pathogens exhibits significant geographical variations, depending on both physical environmental and socio-economic conditions (Yang et al. 2012). For example, waterborne cholera, viral hepatitis, and some other water-based and water-related diseases (e.g., schistosomiasis and onchocerciasis) are largely confined to certain tropical areas where the prevalence of such diseases is largely due to lack of access to improved drinking water, sanitation and hygiene (Pruss et al. 2002), while some pathogens, such as *Cryptosporidium parvum*, *Giardia duodenalis*, and *Campylobacter* spp. have a much wider geographical distribution (Yang et al. 2012). Regardless of the geographical region, heavy rainfall, flooding, and temperature are among the most important factors associated with transmission and outbreaks of these waterborne diseases.

There is mounting evidence that weather events are often an important factor triggering waterborne outbreaks. In developed countries, extreme weather events such as excessive rainfall and flooding can overwhelm water treatment plants and/or increase runoff into recreational water, leading to water contamination that causes outbreaks (Kistemann et al. 2002). Some recent studies have clearly indicated that excessive rainfall has been a significant contributor to waterborne outbreaks. An analysis of historical records of 548 waterborne outbreaks in the US reported between 1948 and 1994 suggested a significant association between the outbreaks and rainfall—51% of waterborne disease outbreaks were preceded by precipitation events above the 90th percentile and 68% by events above the 80th percentile; the strongest association was between the outbreaks and surface water contamination due to extreme precipitation (Curriero et al. 2001). In Canada, Thomas et al. examined extreme rainfall and spring snowmelt in relation to 92 Canadian waterborne disease outbreaks that occurred between 1975 and 2001. They found that warmer temperatures and extreme rainfall were significant contributing factors to waterborne disease outbreaks in Canada (Thomas et al. 2006). Similar studies have been conducted in England (Nichols et al. 2009), the Netherlands (Schijven and de Roda Husman 2005), Finland (Miettinen et al. 2001), Denmark (Laursen et al. 1994), and Taiwan (Chen et al. 2012). In many developing countries, lack of access to improved drinking water and sanitation exacerbates the impact of these extreme precipitation weather events on waterborne diseases. Diarrhea, caused by many waterborne pathogens, remains a top killer of children under five in the developing world, particularly in Africa (Pruss-Ustun et al. 2014). A large number of studies have been carried out in the developing world. For example, cholera outbreaks caused by toxigenic Vibrio cholerae have been consistently shown to be correlated with excessive rainfall, flooding, and high temperatures in the epidemic areas of West Africa and Bangladesh (Mhalu et al. 1987; Hashizume et al. 2008; Luque Fernandez et al. 2009; Ngwa et al. 2016), although the exact mechanistic relationships that result in this correlation remain elusive (Colwell et al. 1998). Significant (positive) relationships have been reported between excessive rainfall and/or high temperatures and diarrhea in the Pacific Islands (Singh et al. 2001), Ecuador (Carlton et al. 2014), and Sub-Saharan Africa (Bandyopadhyay et al. 2012), while evidence also suggests that low rainfall and even drought can also be associated with diarrhea, as reported in Bangladesh (Hashizume et al. 2008) and Denmark (Senhorst and Zwolsman 2005).

b) Waterborne Diseases and Climate Impact in Florida

Across the US, climate change and variability and associated weather events are affecting and are expected to continue to affect both marine and freshwater resources. These effects extend to some water-associated pathogens and related diseases. In Florida, waterborne and foodborne diseases of major public health concerns are highlighted in Table 1. Enteric bacteria, protozoan parasites, enteric viruses including *Salmonella enterica*, *Campylobacter* spp., toxigenic *Escherichia coli*, *Vibrio* bacteria species, *Cryptosporidium* and *Giardia* enteroviruses, rotaviruses, noroviruses, and hepatitis A and E, are among those closely related to drinking and recreational waters, and shellfish. These pathogens, while in environmental stages, are sensitive to temperature, precipitation, and water flow. Several important toxin producers of water sources are of particular public health importance in Florida (Table 4.1) including toxins from harmful algal blooms, toxigenic marine species of *Alexandrium* (causing paralytic shellfish poisoning), *Karenia brevis* (causing neurotoxic shellfish poisoning), and *Gambierdiscus* spp. (causing ciguatera fish poisoning). Human exposure pathways are primarily through the consumption of contaminated shellfish and fish, and in recreational waters. Cyanobacteria, consisting of multiple species, can produce toxins (including microcystin), and the primary human exposure is from drinking water and recreational water. For all of these harmful algal blooms species, temperature, and precipitation are among the important factors affecting their reproduction, growth, and distribution.

Fig. 4.3 shows yearly distribution of total reported outbreaks and cases of food-borne and waterborne diseases in Florida from 1989 to 2011 and Fig. 4.4 shows the monthly distribution of reported outbreaks and cases for 2011, exhibiting marked annual variations and distinct seasonal distribution. According to the BRACE report for the period 2004–2012, approximately 1,150 cases of campylobacteriosis were reported annually, with the majority of cases reported between May and September. From that same report, ~426 cases of cryptosporidiosis were reported annually, with the majority of cases reported between June and October. Approximately 940 cases of giardiasis were reported annually, with an annual incidence rate of 5.1 cases per 100,000 population and the majority of cases reported between May and October. Salmonellosis accounted for the greatest number of foodborne illnesses reported, with an average of 5,438 cases reported each year and ~70% reported between June and November. Finally, ~101 cases of vibriosis were reported annually, with 81% of the cases reported between April and October (FDH 2015b). Focusing on the five diseases—campylobacteriosis. cryptosporidiosis, giardiasis, salmonellosis, and vibriosis—the BRACE study identified a total of 1,231 follow-up days of interest including 775 control days and 456 impact days—42 associated with hurricanes and 414 with tropical storms (FDH 2015b); the result suggested that the occurrences of cryptosporidiosis and salmonellosis were significantly associated with the tropical cyclones, with the risk ratio being 1.26 (95% CI, 1.04, 1.52) and 1.35 (95% CI, 1.29, 1.42), respectively (FDH 2015b).

Table 4.1. A list of climate-sensitive, water-associated agents and related illnesses.

Transmission Grouping	Organism	Disease	Main Exposure Pathways	Main Illness & Symptoms
	Protozoa			
Waterborne	*Cryptosporidium parvum*	Cryptosporidiosis	Fecal-oral via water (D,R)	Diarrhea, stomach pain, weight loss, and fatal for immunocompromised people
	Giardia duodenalis	Giardiasis	Fecal-oral via water (D,R)	Diarrhea, stomach pain, weight loss
	Cyclospora cayetanensis	Cyclosporiasis	Fecal-oral via water (D)	Diarrhea, stomach pain, weight loss
	Entamoeba histolytica	Amebiasis	Fecal-oral via water (D)	Diarrhea, stomach pain, dysentery
	Toxoplasma gondii	Toxoplasmosis	Water (D, F), congenital	Fever, damage on brain, eye, fetal damage in pregnant woman
Water-related (Vector-borne)	*Plasmodium* SPP	Malaria	Bite by mosquitos	Fever, chills, headache, sweats, fatigue, nausea
	Trypanosoma SPP	African trypanosomiasis	Bite by tsetse fly	Chancre, sleeping sickness, fever, headache, fatigue, can be fatal
	T. cruzi	Chagas disease	Bite by triatomine bug, contaminated (F), and others	Fever, fatigue, headache, rash, diarrhea, vomiting
	Leishmania SPP	Leishmaniasis	Bite by sand fly	Fever, weight loss, spleen/liver enlargement (visceral), skin sores (cutaneous)
	Viruses			
Waterborne	Hepatitis A, E virus	Viral hepatitis	Contaminated water (D, R, F)	Hepatitis
	Norovirus	Viral gastroenteritis	Contaminated water (D, R,F)	Vomiting and diarrhea
	Enteroviruses	Various	Contaminated water (D,R,F), contact	Various
Water-related (Vector-borne)	DENV 1, 2, 3, 4	Dengue hemorrhagic fever (DHF)	Bite by mosquito	Fever, severe headache, severe joint, muscle and bone pain, rash, bleeding
	Japanese encephalitis virus	Japanese encephalitis	Bite by mosquito	Fever, headache, vomiting, movement disorder, seizures
	West Nile virus	West Nile virus disease	Bite by mosquito	Most cases asymptomatic, febrile illness, encephalitis or meningitis
	St. Louis encephalitis virus	St. Louis encephalitis	Bite by mosquito	Most cases asymptomatic, fever, headache, dizziness, nausea, and malaise
	Chikungunya virus	Chikungunya	Bite by mosquito	Fever, joint pain and swelling, muscle pain, rash
	Eastern equine encephalitis virus	Eastern equine encephalitis	Bite by mosquito	Rare in humans, headache, high fever, chills, and vomiting in severe cases
	La Crosse encephalitis virus	La Crosse encephalitis	Bite by mosquito	Most cases asymptomatic, fever, headache, nausea, vomiting, and tiredness

Transmission Grouping	Organism	Disease	Main Exposure Pathways	Main Illness & Symptoms
	Bacteria			
Waterborne	Vibrio cholerae	Cholera	Contaminated (D, F)	Severe diarrhea, vomiting, muscle cramps
	Vibrio SPP	Vibriosis	Contaminated Shellfish, open wound	Diarrhea, septicemia
	Salmonella spp.	Salmonellosis	Contaminated (F, less D)	Diarrhea, abdominal pain, fever
	Salmonella typhi	Typhoid	Contaminated (D, F)	Fever, malaise, abdominal pain, can be fatal
	Shigella spp.	Shigellosis	Contaminated (D, F)	Diarrhea, fever, and stomach cramps
	Campylobacter spp.	Campylobacteriosis	Contaminated (D, F, R)	Diarrhea, fever, and stomach cramps
	Enterotoxigenic E. coli	Diarrheal illness	Contaminated (D,F)	Diarrhea, abdominal cramping, fever, nausea, muscle aches
	Enterohaemorrhagic E. coli	Diarrheal illness	Contaminated (D,F)	Bloody diarrhea, and haemolytic-uraemic syndrome
	Yersinia spp.	Yersiniosis	Contaminated (D,F)	Diarrhea, fever, abdominal pain
Water-related (Vector-borne)	*Francisella tularensis*	Tularaemia	Bites of tick or deer fly, contact with infected animals, can be through air, water	Mild skin lesion to life-threatening pneumonic
	Helminths			
Water-based	*Schistosoma* spp.	Schistosomiasis	Dermal contact with water	Diarrhea, urinary and intestinal damage
	Dracunculus medinensis	Dracunculiasis	Contaminated (D)	Painful ulcer and inflammation
	Ancylostoma duodenale and Necator americanus	Hookworm infection	Dermal contact with soil	Diarrhea, loss of appetite, fatigue, and anemia
	Ascaris lumbricoides	Ascaris	Fecal-oral (D, F)	Abdominal discomfort, growth retardation in children
	Trichuris trichiura	Whipworm infection	Fecal-oral (D,F)	Bloody and watery stool, growth retardation
Water-related (Vector-borne)	*Wuchereria bancrofti*	Lymphatic filariasis	Bite by mosquito	Most asymptomatic, lymphedema, pulmonary eosinophilia
	Onchocerca volvulus	Onchocerciasis	Bite by blackfly	Skin rash, eye disease, and nodules under the skin
	Loa loa	Loiasis	Bite by deerfly	Itchy, muscle and joint pain, fatigues, eye worm
	Algae			
Water-related	Cyanobacteria	Various	Exposure to toxins via D and direct contact	Dermatitis, intestinal illness, liver damage, neurotic reactions
	Alexandrium spp	Paralytic shellfish poisoning	Contaminated (F) by toxin	Circumoral and extremity paresthesia, respiratory paralysis
	Karenia brevis	Neurotoxic shellfish poisoning	Contaminated (F) by toxin	Gastrointestinal symptoms, paresthesia, respiratory and eye irritation
	Gambierdiscus spp.	Ciguatera fish poisoning	Contaminated (F) by toxin	Diarrhea, vomiting, malaise, extremity paresthesia

Note: 1. Abbreviations for main exposure pathways: drinking water (D), recreational water (R), and foods (F); 2. Highlighted in red indicates pathogens diseases of major public health significance in Florida.

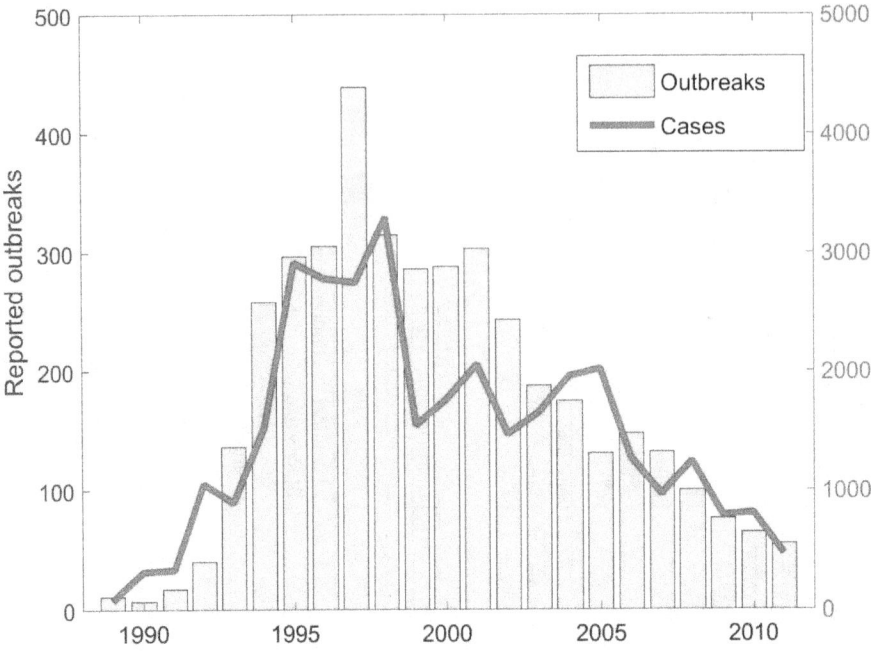

Figure 4.3. Reported outbreaks (bar) and human cases (line) of food-borne and waterborne diseases in Florida each year from 1989 to 2011 (Source: Florida Department of Health). Toxigenic V. cholerae infections are highly unusual in Florida and the first V. cholerae O75 outbreak was detected and reported in Florida between March 23 and April 13, 2011.

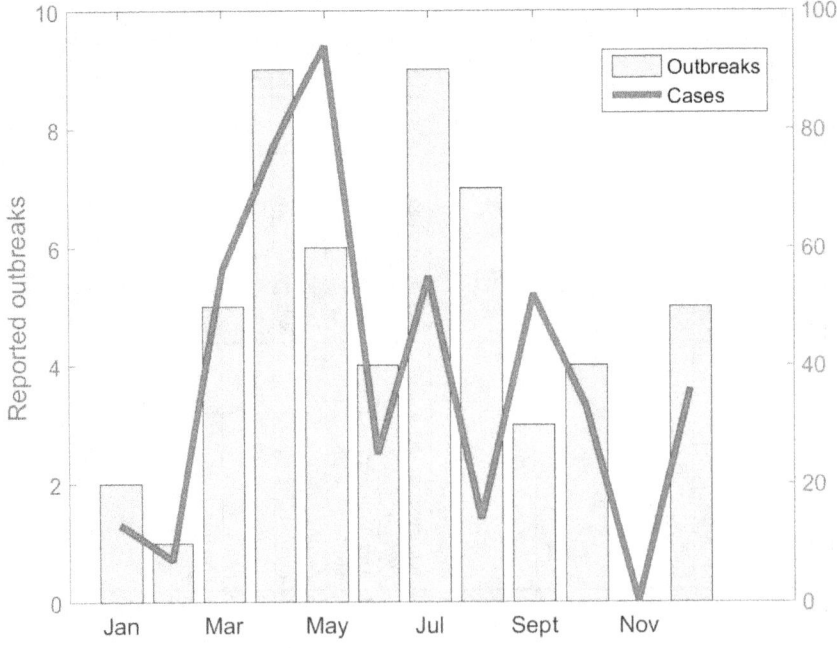

Figure 4.4. Monthly distribution of reported outbreaks and cases of food-borne and waterborne diseases in Florida, 2011 (Source: Florida Department of Health)

c) Vector-Borne Diseases

Vector-borne diseases pose serious public health threats throughout the world. According to the World Health Organization, vector-borne diseases account for more than 17% of infectious diseases, causing more than one million deaths annually from diseases such as malaria, dengue fever, yellow fever, Japanese encephalitis, and schistosomiasis (WHO 2016). Vector-borne diseases are transmitted through vector organisms including mosquitoes, ticks, flies, fleas, and snails. The transmission of vector-borne diseases takes two primary transmission pathways. The first is the human-vector pathway, with transmission through a vector biting a host; pathogens survive outside hosts in arthropod vectors and humans are typically the only host (Eisenberg et al. 2007). Diseases in this group include malaria, dengue fever, onchocerciasis, trypanosomiasis, and filariasis (Table 4.1). The second pathway involves zoonotic transmission, with a vector biting a nonhuman host as the transmission route and humans as the dead-end host (Eisenberg et al. 2007). Diseases in this group include Lyme disease, Yellow fever, West Nile virus, and Japanese encephalitis (Table 4.1). The transmission and spread of vector-borne diseases are determined by complex interactions between the host (either human or nonhuman), the vector (e.g., mosquitos, flies, snails, and ticks), and various pathogens ranging from protozoans, viruses, bacteria, and helminthes (Table 4.1). Important biological properties underlying the transmission of vector-borne diseases include survival, development, and reproduction of the vector; also the pathogens in the vector, the vector's biting rate, and the behavior of hosts (human and nonhuman), all of which are associated with climatic conditions. The following sections discuss some specific aspects of the effects of rainfall, temperature, and flooding with regard to vector-borne diseases.

Temperature Sensitivity

Through the modulating effects on physiological activities and tolerance limits of vectors and infectivity of pathogens, temperature can affect distribution of the vector and transmission of the pathogens between vectors and hosts. Some key mechanistic effects of temperature on vectors, vector-borne pathogens and their transmission are summarized by Gulber et al. (2001).

- Change in survival of vectors (e.g. increase or decrease depending on species)
- Change in susceptibility of vectors to pathogens
- Change in vector population growth
- Change in feeding behavior and host contact
- Change in incubation period of pathogens in vectors (e.g. decreased extrinsic incubation period at higher temperatures)
- Change in pathogen replication and infectivity
- Change in transmission season

A large number of studies on vector-borne diseases at different scales have provided empirical evidence and predictions on the climatic factors and vector-borne disease links. For example, studies have shown significant relationships between ambient temperature and malaria transmission for both mean temperature (Loevinsohn 1994; Lindblade et al. 2000; Thomson et al. 2005; Tian et al. 2008; Wardrop et al. 2013) and its daily fluctuations (Paaijmans et al. 2010). Laboratory studies have shown clear temperature-dependent effects on dengue virus infection in mosquitoes (Alto and Bettinardi 2013). Fluctuations at a low mean temperature accelerate dengue virus transmission in *Aedes aegypti* mosquitoes (Carrington et al. 2013). A recent systematic review and meta-analysis based on 137 studies on the relationship between dengue risk and temperature suggests that the dengue transmission and risk are sensitive to temperature, with a positive relationship between them for a minimum range (18.1–24.2 °C) to a maximum temperature range (28.0–34.5 °C) (Fan et al. 2014). A regional predictive study reported that increased temperature increases the likelihood of many vector-borne diseases, in particular malaria, dengue fever, and Lyme disease in new areas (Githeko et al. 2000). Similar findings are also reported for West Nile virus (Soverow et al. 2009). These empirical findings have been used to parameterize mechanistic models of temperature driven transmission for vector-borne diseases, which can be coupled to projected climate scenarios to anticipate shifting geographies of transmission risk, e.g. for malaria (Mordecai et al. 2013; Johnson et al. 2014; Johnson et al. 2015; Ryan et al. 2015; Ryan et al. 2015) and for *Aedes* spp transmitted diseases of concern for Florida: dengue, chikugunya, and zika viruses (Moedecai et al. 2017; Ryan et al. 2017).

Rainfall
Rainfall can supply water to both transient and persistent environments that can serve as habitats or breeding sites for vectors and/or for part of pathogens' lifecycles, critical for transmission of vector-borne diseases. Meanwhile, rainfall can also have negative impacts on the habitats (for example, washing away), thereby reducing the transmission risk. Variations in rainfall may have direct and indirect impacts on the distribution and timing of transmission of vector-borne diseases. Some key mechanistic effects of rainfall on vectors, vector-borne pathogens, and their transmission are summarized by Gubler et al. (2001).

- Increased rain may increase larval habitat and vector population.
- Excess rain (e.g. causing flooding) may eliminate habitat for both vector and hosts.
- Increased humidity may increase vector survival and biting rate.
- Increased rainfall may facilitate transport of pathogens for certain distance (Liang et al. 2007).
- Increased rain may increase vertebrate host size through increased food availability (e.g. vegetation).
- Flooding may decrease vertebrate host size (e.g. reduced food availability) but increase interactions (e.g. contacts) with humans.

Such mechanistic links have been reported by many studies including for diseases such as malaria (van der Hoek et al. 1997; Thomson et al. 2005; Galardo et al. 2009; Gao et al. 2012; Yang et al. 2012), dengue fever (Li et al. 1985; Su 2008; Johansson et al. 2009; Banu et al. 2011;

Hii et al. 2012), West Nile virus (Shaman et al. 2005; Soverow et al. 2009), and Lyme disease (Ogden et al. 2006; Ostfeld et al. 2006).

Vector-Borne Diseases in Florida

In the United States, 14 vector-borne diseases are of primary public health concern. Table 4.2 lists key notifiable vector-borne diseases and reported cases between 2013 and 2015 in Florida. Major mosquito-borne diseases include the St. Louis encephalitis virus, West Nile virus, Eastern equine encephalitis virus, Western equine encephalitis virus, Venezuelan equine encephalitis virus, Everglades virus, and the California serogroup viruses including La Crosse encephalitis virus, all of which are transmitted by mosquitoes. More recently, concern about *Aedes spp* transmitted flavivirus and alphaviral disesaes have become more prevalent: dengue (DENV), chikungunya (CHIKV), and zika (ZIKV). The 2015-2016 global public health emergency for Zika declared by the WHO focused attention on Florida as an introduction and establishment gateway for the U.S.

Table 4.2. A summary of reported case counts of notifiable vector-borne diseases in Florida. (Source: Florida Department of Health).

Diseases	2013-2015 Reported Cases	Number of Counties Reported
Tick-borne		
Lyme disease	459	46
Spotted Fever Rickettsia	74	18
Anaplasmosis/Ehrlichiosis	82	30
Tularemia	2	2
Mosquito-Borne		
West Nile virus	37	14
Malaria	146	24
Dengue	331	31
California serogroup viruses	2	2
Eastern equine encephalitis	3	3
St. Louis encephalitis	2	1

St. Louis Encephalitis. The St. Louis encephalitis virus, a flavivirus, was the most common mosquito-transmitted human pathogen in the U.S. prior to the introduction of the West Nile virus in 1999 (FDH 2016). In Florida, the principal vector is *Culex nigripalpus*, a ubiquitous species throughout the state. No human case has been reported in Florida since 2003. However, sentinel chickens testing positive for antibodies to the St. Louis encephalitis virus have been reported in several counties in Central and South Florida, suggesting the potential for a possible resurgence (FDH 2014, 2016).

West Nile virus. The peak period of transmission in Florida for the West Nile virus is July through September. The natural cycle of West Nile virus involves *Culex* mosquitoes and wild birds. Between 2001 and 2013, 318 human West Nile virus cases were reported in Florida and

West Nile virus activities have been consistently reported in many counties around the state (FDH 2014).

Eastern Equine Encephalitis. Eastern equine encephalitis occurs in natural cycles involving birds and *Culiseta melanura* in freshwater swampy areas, with a peak in transmission occurring between May and August. Historical and current evidence indicates limited human Eastern equine encephalitis epidemic potential in Florida (FDH 2014, 2016).

Dengue Fever. In Florida, a historical dengue fever epidemic occurred in 1934-1935 and then it ceased. It re-emerged in 2009, and since then a small number of cases have been reported each year among individuals who had previous travel to dengue-endemic countries. In the summer of 2009, local dengue transmission was identified in Key West, Florida. Since then, sporadic local transmission has been identified in other Central and South Florida counties during 2010, 2011, 2012, and 2013 (FDH 2014, 2016). The presence of the *Aedes spp* mosquito, both *Aedes aegypti*, also known as the Yellow Fever mosquito, which has been a persistent invasive mosquito in Florida for a couple of centuries, and the more recently introduced Asian Tiger Mosquito, *Aedes albopictus*, maintain transmission potential in Florida. *Aedes* mosquitoes are capable of breeding in as little as a teaspoon of water, and are able to adapt well to urban environments, and exhibit transmission of dengue, chikungunya, and zika, at temperature ranges found almost year-round in most of Florida (Mordecai et al. 2017). The optimal transmission temperature range for *Aedes aegypti*, in particular, is higher than that for *Aedes albopictus*, suggesting that future climates will promote *Ae aegypti* transmission in Florida, before it becomes too hot for transmission (Ryan et al. 2017).

Malaria. Malaria is caused by the mosquito-borne parasite *Plasmodium falciparum*. Endemic malaria was eliminated in Florida in the late 1940s. However, imported cases—either from travelers returning to the state from malaria-endemic regions or tourists carrying the pathogen—have been reported. The *Anopheles* mosquitoes responsible for transmitting the malaria parasite to humans are common in the state so establishment of local transmission is still possible (FDH 2014, 2016).

Tick-Borne Diseases. The most common tick-borne diseases are ehrlichiosis, anaplasmosis, Lyme disease, Rocky Mountain spotted fever, and other spotted fever illnesses. Lyme disease is the most commonly reported vector-borne disease in the United States. It is caused by a bacteria known as *Borrelia burgdorferi*. Lyme disease is reported in Florida year-round. An estimate suggests that about 23% of cases were acquired in Florida and about 77% were acquired while travelling to other states or countries (FDH 2016). *Ehrlichiosis and Anaplasmosis*—several pathogenic species in the genus *Ehrlichia* and *Anaplasma*—can cause human illness. The human illness caused by *Ehrlichia chaffeensis* is called Human Monocytic Ehrlichiosis, and the illness caused by *Anaplasma phagocytophilum* is called Anaplasmosis or Human Granulocytotropic Anaplasmosis. In Florida, the majority of Human Monocytic Ehrlichiosis cases (73%) are acquired in Florida, primarily in the northern and central parts of the state, and Human Granulocytotropic Anaplasmosis cases are even more likely to be acquired in Florida. Cases are reported year-round, with a peak occurring during the spring and summer months (FDH 2016). *Rocky Mountain Spotted Fever* is a disease is caused by the bacterium *Rickettsia richerrsii* and, in Florida, it is transmitted primarily by the American dog tick. It can also be caused by *Dermacentor variabilis* and cases of this transmission are reported year-round with more than 70% acquired in Florida, and the majority of cases are reported in the northern and central regions of the state (FDH 2016).

Drought

Drought is considered a meteorological anomaly characterized by a prolonged and abnormal moisture deficiency. Climate change and variability are progressively increasing the severity and frequency of drought events. Drought has many possible public health implications (Kalis et al. 2009; Kalis and Curtiss 2016). Drought may:

- Compromise quantity and quality of water for drinking, sanitation, and hygiene
- Compromise air quality (e.g. increased particulate matters in the air)
- Reduce crop yield, food availability and nutrition
- Increase or decrease risks to infectious pathogens (e.g. waterborne, air-borne, and vector-borne)
- Increase risk of non-communicable illness (e.g. respiratory infections and/or illness)

Drought has been linked to deterioration of water quality (Golladay and Battle 2002; Al-Kharabsheh and Ta'any 2003) and air quality (Taylor and Davies 1990; Field et al. 2009) in various settings in both the developed and developing world. Drought has been implicated for infectious diseases—waterborne (Lipp et al. 2002; Schuster et al. 2005; Senhorst and Zwolsman 2005), vector-borne (Chretien et al. 2007; Erlanger et al. 2009; Medlock and Leach 2015), and airborne (Polymenakou et al. 2008). Drought conditions have also been linked to increased incidence of respiratory diseases (Smith et al. 2014).

In Florida, the BRACE study examined associations between monthly drought conditions (using the Standardized Precipitation Index), disease rates and emergency department visits for specific health conditions (e.g. allergic rhinitis, asthma, and all respiratory diseases excluding asthma) among Florida residents from 2005 to 2012. The results suggested that drought was significantly associated with emergency department visits for respiratory diseases. It is worth noting that the relationship between the most extreme drought conditions and emergency department visits tended to be protective (e.g. less emergency department visits for respiratory illness), while more moderate drought appeared to be associated with an increase in emergency department visits for the health conditions. A similar pattern was also observed for asthma (FDH 2014b).

Indirect Impact Pathway – Mental Health, Well-Being, and Community Health

Mental health disorders and well-being are an important aspect of the human health impacts associated with climate change and variability, and associated weather events. Many mental health consequences, ranging from simple stress and distress symptoms to some severe conditions including anxiety, distress, posttraumatic stress disorder (PTSD), and even suicidal tendency have been the focus of much of the research and public health responses in the past few decades. Here are some key mental disorders and community health issues that may be related to climate change and associated weather events, based on the U.S. Global Change Research Program report (Crimmins et al. 2016).

Experience with extreme weather events, such as flooding, drought, and hurricanes, may cause the following mental disorders:

- Increased stress, anxiety, depression, grief, and even suicidality;
- Increased tensions on social stress and relationships;
- Increased substance abuse; and
- Increased risk of PTSD.

At the community level (Crimmins et al. 2016), potential impacts may include:

- Increased interpersonal aggression,
- Increased violence and crime,
- Increased social instability, and
- Decreased community cohesion.

Assessing the impacts of natural disasters on mental or community health and well-being of affected populations, as well as mediation techniques, has been a focus of much research, community and public health responses. In Florida, studies have suggested that anxiety disorder, major depressive episodes, and PTSD among affected populations could be significantly associated with storm exposure and displacement during the occurrence of Hurricane Andrew in 1992 (David et al. 1996), as well as a number of hurricanes in 2004 (Acierno et al. 2007) and 2012 (Neria and Shultz 2012). Studies have also indicated that social support would help to alleviate such negative impacts (Acierno et al. 2007). Similar impacts have been reported for community and public health workers who responded to these weather events. For example, during the 2004 Florida hurricane season, four hurricanes (Charley, Frances, Ivan, and Jean) as well as one tropical storm (Bonnie) hit Florida during a period of seven weeks. Increased PTSD, other mental health issues, and substance abuse were reported among public health workers during and after these events (Fullerton et al. 2013; Fullerton et al. 2015). The studies also suggested that these mental health outcomes were influenced by multiple community characteristics including the collective efficacy of neighborhood populations in the community.

Significantly lower depressive symptoms were associated with communities that had sufficient resources and received social support (Acierno et al. 2007, Fullerton et al. 2013, Fullerton et al. 2015).

Impacts on children following natural disasters are of particular concern. For example, Hurricane Andrew was a devastating category 5 Atlantic hurricane that struck South Florida in 1992. Studies showed significant impacts of hurricane exposure, stressors occurring during the hurricane, and recovery periods on children's persistent posttraumatic stress. In the Miami-Dade County area, 35% to 60% of children surveyed reported moderate to very severe levels of posttraumatic stress symptoms (Vernberg et al. 1996; La Greca et al. 2010), with hurricane-related stressors influencing children's persistent posttraumatic stress symptoms and other life events in later stages of children's life (La Greca et al. 2010; Weems et al. 2010; La Greca et al. 2013). These findings have important community health implications for identifying and potentially helping youth in the aftermath of natural disasters.

Conclusion

The current available evidence has clearly indicated increases in average temperature, total annual precipitation, frequency of extreme temperature conditions and heavy precipitation in the United States in the past five decades, as well as increases in the tropical cyclone activity in regions along the Atlantic Ocean, the Caribbean, and the Gulf of Mexico. Similar patterns are expected in the future if no remediation is conducted (USEPA 2016). Florida is likely to face even greater effects than the rest of the US, particularly effects associated with sea level rise, heat waves, storm surge, hurricanes, and others (FOCC 2010) that affect human health in a variety of different ways. Some effects are through direct exposure pathways, such as hurricane/storm and temperature-related illnesses, injuries, and deaths. Some are through indirect exposure pathways such as waterborne and vector-borne diseases. And still others are through social and economic pathways, such as stress and mental illness. In addition to these direct and indirect effects, we see coupled and compounded effects of climate change, which are anticipated to be exacerbated in the future. For example, Florida is home to a large elderly population; the warm climate and housing development capacity made this ideal. However, the elderly are more vulnerable to the effects of heatwaves, less mobile in the event of hurricanes, and suffer greater impacts of the symptomology of febrile vectorborne diseases. This puts the state in a position higher public health urgency by virtue of an interaction of demographic profile and increased impact of climate change due to geography. Given the increase in hurricances and storm impacts expected under climate change, we are likely to also see the added impact of vectorborne disease risk anticipated in the aftermath of such disasters. For example, the additional potential vector breeding sites provided by debris and damaged infrastructure after hurricanes have been noted as concerns for Zika transmission in the 2016-2017 hurricane season. An additional component of this is the unanticipated exposures post-disaster, in which both displaced residents and rescue workers have radically increased outdoor exposure, reduced access to air conditioned spaces, and to basic protection measures such as insect repellent and appropriate protective clothing. These are just two of the compounded scenarios of public health concern that the state of Florida faces in a changing climate. Despite the comprehensive information reviewed in the chapter, many things are still unknown or are not well understood about the health impacts of Florida's changing climate. For instance, what are the impacts of sea level rise, flooding, and other extreme weather events on water contamination that will ultimately influence human health? How will future warming trends and other climate-sensitive hazards influence the transmission and spread of existing and/or new waterborne and vector-borne diseases? What are the disease burdens and their distributions attributable to known climate-sensitive hazards in the state? To address these challenges, policies focused on mitigation, adaptation strategies, health surveillance, and research that could close knowledge gaps on human exposures to these climate-sensitive hazards and associated health impacts are much needed.

Acknowledgments

The authors thank Melissa Jordan for her instrumental discussion on this chapter. Song Liang was supported in part by grants from NSF/EEID (EF-1015908) and NSF/WSC-Category 3 (1360330). Sadie J. Ryan was supported in part by grants from NSF/EEID (DEB-1518681), NSF RAPID (DEB-1641145), and an Early Career Fellowship from the Florida Climate Institute.

References

Acierno, R., K. J. Ruggiero, S. Galea, H. S. Resnick, K. Koenen, J. Roitzsch, M. de Arellano, J. Boyle, and D. G. Kilpatrick. 2007. Psychological sequelae resulting from the 2004 Florida hurricanes: implications for postdisaster intervention. Am J Public Health **97 Suppl 1**:S103-108.

Al-Kharabsheh, A., and R. Ta'any. 2003. Influence of urbanization on water quality deterioration during drought periods at South Jordan. Journal of Arid Environments **53**:619-630.

Alto, B. W., and D. Bettinardi. 2013. Temperature and dengue virus infection in mosquitoes: independent effects on the immature and adult stages. Am J Trop Med Hyg **88**:497-505.

Aslar, A. K., A. Soran, Y. Yildiz, and Y. Isik. 2001. Epidemiology, morbidity, mortality and treatment of lightning injuries in a Turkish burns units. International Journal of Clinical Practice **55**:502-504.

Balbus, J., A. Crimmins, J. L. Gamble, D. R. Easterling, K. E. Kunkel, S. Saha, and M. C. Sarofim. 2016. Ch. 1: Introduction: Climate Change and Human Health. Pages 25–42 The Impacts of Climate Change on Human Health in the United States: A Scientific Assessment. U.S. Global Change Research Program, Washington, DC.

Bandyopadhyay, S., S. Kanji, and L. M. Wang. 2012. The impact of rainfall and temperature variation on diarrheal prevalence in Sub-Saharan Africa. Applied Geography **33**:63-72.

Banu, S., W. Hu, C. Hurst, and S. Tong. 2011. Dengue transmission in the Asia-Pacific region: impact of climate change and socio-environmental factors. Trop Med Int Health **16**:598-607.

Basu, R., and J. M. Samet. 2002. Relation between elevated ambient temperature and mortality: A review of the epidemiologic evidence. Epidemiologic Reviews **24**:190-202.

Berko, J., D. D. Ingram, S. Saha, and J. D. Parker. 2014. Deaths attributed to heat, cold, and other weather events in the United States, 2006-2010. Natl Health Stat Report:1-15.

Booij, M. J. 2005. Impact of climate change on river flooding assessed with different spatial model resolutions. Journal of Hydrology **303**:176-198.

Bouma, M. J., R. S. Kovats, S. A. Goubet, J. S. Cox, and A. Haines. 1997. Global assessment of El Nino's disaster burden. Lancet **350**:1435-1438.

Carlton, E. J., J. N. S. Eisenberg, J. Goldstick, W. Cevallos, J. Trostle, and K. Levy. 2014. Heavy Rainfall Events and Diarrhea Incidence: The Role of Social and Environmental Factors. American Journal of Epidemiology **179**:344-352.

Carrington, L. B., M. V. Armijos, L. Lambrechts, and T. W. Scott. 2013. Fluctuations at a Low Mean Temperature Accelerate Dengue Virus Transmission by Aedes aegypti. Plos Neglected Tropical Diseases **7**.

Center for Disease Control, and Prevention. 1994. Heat-related deaths--Philadelphia and United States, 1993-1994. MMWR Morb Mortal Wkly Rep **43**:453-455.

Chen, M. J., C. Y. Lin, Y. T. Wu, P. C. Wu, S. C. Lung, and H. J. Su. 2012. Effects of extreme precipitation to the distribution of infectious diseases in Taiwan, 1994-2008. PLoS One **7**:e34651.

Chretien, J. P., A. Anyamba, S. A. Bedno, R. F. Breiman, R. Sang, K. Sergon, A. M. Powers, C. O. Onyango, J. Small, C. J. Tucker, and K. J. Linthicum. 2007. Drought-associated chikungunya emergence along coastal East Africa. Am J Trop Med Hyg **76**:405-407.

Christensen, J. H., and O. B. Christensen. 2003. Climate modelling: Severe summertime flooding in Europe. Nature **421**:805-806.

Chung, J. Y., Y. Honda, Y. C. Hong, X. C. Pan, Y. L. Guo, and H. Kim. 2009. Ambient temperature and mortality: an international study in four capital cities of East Asia. Sci Total Environ **408**:390-396.

Collins, T. W., A. M. Jimenez, and S. E. Grineski. 2013. Hispanic health disparities after a flood disaster: results of a population-based survey of individuals experiencing home site damage in El Paso (Texas, USA). J Immigr Minor Health **15**:415-426.

Colwell, R. R., P. R. Epstein, D. Gubler, N. Maynard, A. J. McMichael, J. A. Patz, R. B. Sack, and R. Shope. 1998. Climate change and human health. Science **279**:968-969.

Crimmins, A., J. Balbus, J. L. Gamble, C. B. Beard, J. E. Bell, D. Dodgen, R. J. Eisen, N. Fann, M. D. Hawkins, S. C. Herring, L. Jantarasami, D. M. Mills, S. Saha, M. C. Sarofim, J. Trtanj, and L. Ziska. 2016. Executive Summary. Pages 1–24 The Impacts of Climate Change on Human Health in the United States: A Scientific Assessment. U.S. Global Change Research Program, Washington, DC.

Curriero, F. C., K. S. Heiner, J. M. Samet, S. L. Zeger, L. Strug, and J. A. Patz. 2002. Temperature and mortality in 11 cities of the eastern United States. Am J Epidemiol **155**:80-87.

Curriero, F. C., J. A. Patz, J. B. Rose, and S. Lele. 2001. The association between extreme precipitation and waterborne disease outbreaks in the United States, 1948-1994. American Journal of Public Health **91**:1194-1199.

David, D., T. A. Mellman, L. M. Mendoza, R. Kulick-Bell, G. Ironson, and N. Schneiderman. 1996. Psychiatric morbidity following Hurricane Andrew. J Trauma Stress **9**:607-612.

Diez, J. J., M. D. Esteban, R. Paz, J. S. Lopez-Gutierrez, V. Negro, and J. V. Monnot. 2011. Urban Coastal Flooding and Climate Change. Journal of Coastal Research:205-209.

Douglas, I., K. Alam, M. Maghenda, Y. McDonnell, L. McLean, and J. Campbell. 2008. Unjust waters: climate change, flooding and the urban poor in Africa. Environment and Urbanization **20**:187-205.

Duclos, P. J., L. M. Sanderson, and K. C. Klontz. 1990. Lightning-related mortality and morbidity in Florida. Public Health Rep **105**:276-282.

Eisenberg, J. N., M. A. Desai, K. Levy, S. J. Bates, S. Liang, K. Naumoff, and J. C. Scott. 2007. Environmental determinants of infectious disease: a framework for tracking causal links and guiding public health research. Environ Health Perspect **115**:1216-1223.

Ellis, K. N., L. M. Sylvester, and J. C. Trepanier. 2015. Spatiotemporal patterns of extreme hurricanes impacting US coastal cities. Natural Hazards **75**:2733-2749.

Elsner, James B., James P. Kossin, and Thomas H. Jagger. "The increasing intensity of the strongest tropical cyclones." Nature 455.7209 (2008): 92.

Erlanger, T. E., S. Weiss, J. Keiser, J. Utzinger, and K. Wiedenmayer. 2009. Past, Present, and Future of Japanese Encephalitis. Emerging Infectious Diseases **15**:1-7.

Fan, J., W. Wei, Z. Bai, C. Fan, S. Li, Q. Liu, and K. Yang. 2014. A systematic review and meta-analysis of dengue risk with temperature change. Int J Environ Res Public Health **12**:1-15.

FDH. 2014a. Surveillance and control of selected mosquito-borne diseases in Florida. Florida Department of Health, Tallahassee, Florida.

FDH. 2014b. Health effects of precipitation abundance and deficits in Florida. Tallahassee, Florida.

FDH. 2015a. Health effects of summer heat in Florida. Florida Department of Health, Tallahassee, Florida.

FDH. 2015b. Health effects of tropical storms and hurricanes in Florida. Florida Department of Health, Tallahassee, Florida.

FDH. 2016. Diseases and Conditions, Florida Department of Health, Tallahassee, Florida.

Field, C. B., V. R. Barros, and Intergovernmental Panel on Climate Change. Working Group II. 2014. Climate change 2014: impacts, adaptation, and vulnerability: Working Group II contribution to the fifth assessment report of the Intergovernmental Panel on Climate Change. Cambridge University Press, New York, NY.

Field, R. D., G. R. van der Werf, and S. S. P. Shen. 2009. Human amplification of drought-induced biomass burning in Indonesia since 1960. Nature Geoscience **2**:185-188.

FOCC. 2010. Climate change and sea-level rise in Florida. An update of the effects of climate change on Florida's ocean & coastal resources. Page 36 *in* T. F. O. a. C. Council, editor, Tallahassee, Florida

Frumkin, H., A. J. McMichael, and J. J. Hess. 2008. Climate change and the health of the public. Am J Prev Med **35**:401-402.

Fullerton, C. S., J. B. A. McKibben, D. B. Reissman, T. Scharf, K. M. Kowalski-Trakofler, J. M. Shultz, and R. J. Ursano. 2013. Posttraumatic Stress Disorder, Depression, and Alcohol and Tobacco Use in Public Health Workers After the 2004 Florida Hurricanes. Disaster Medicine and Public Health Preparedness **7**:89-95.

Fullerton, C. S., R. J. Ursano, X. Liu, J. B. McKibben, L. Wang, and D. B. Reissman. 2015. Depressive Symptom Severity and Community Collective Efficacy following the 2004 Florida Hurricanes. PLoS One **10**:e0130863.

Galardo, A. K. R., R. H. Zimmerman, L. P. Lounibos, L. J. Young, C. D. Galardo, M. Arruda, and A. A. R. D. Couto. 2009. Seasonal abundance of anopheline mosquitoes and their association with rainfall and malaria along the Matapi River, Amapi, Brazil. Medical and Veterinary Entomology **23**:335-349.

Gao, H. W., L. P. Wang, S. Liang, Y. X. Liu, S. L. Tong, J. J. Wang, Y. P. Li, X. F. Wang, H. Yang, J. Q. Ma, L. Q. Fang, and W. C. Cao. 2012. Change in rainfall drives malaria re-emergence in Anhui Province, China. PLoS One **7**:e43686.

Githeko, A. K., S. W. Lindsay, U. E. Confalonieri, and J. A. Patz. 2000. Climate change and vector-borne diseases: a regional analysis. Bull World Health Organ **78**:1136-1147.

Golladay, S. W., and J. Battle. 2002. Effects of flooding and drought on water quality in gulf coastal plain streams in Georgia. Journal of Environmental Quality **31**:1266-1272.

Gubler, D. J., P. Reiter, K. L. Ebi, W. Yap, R. Nasci, and J. A. Patz. 2001. Climate variability and change in the United States: potential impacts on vector- and rodent-borne diseases. Environ Health Perspect **109 Suppl 2**:223-233.

Hashizume, M., B. Armstrong, S. Hajat, Y. Wagatsuma, A. S. Faruque, T. Hayashi, and D. A. Sack. 2008. The effect of rainfall on the incidence of cholera in Bangladesh. Epidemiology **19**:103-110.

Hii, Y. L., H. P. Zhu, N. Ng, L. C. Ng, and J. Rocklov. 2012. Forecast of Dengue Incidence Using Temperature and Rainfall. Plos Neglected Tropical Diseases **6**.

Hoshiko, S., P. English, D. Smith, and R. Trent. 2010. A simple method for estimating excess mortality due to heat waves, as applied to the 2006 California heat wave. International Journal of Public Health **55**:133-137.

Huynen, M. M., P. Martens, D. Schram, M. P. Weijenberg, and A. E. Kunst. 2001. The impact of heat waves and cold spells on mortality rates in the Dutch population. Environ Health Perspect **109**:463-470.

IFRCRCS. 2015. World disasters report: focus on local actors, the key to humanitarian effectiveness International Federation of Red Cross and Red Crescent Societies.

Ji, F., Z. H. Wu, J. P. Huang, and E. P. Chassignet. 2014. Evolution of land surface air temperature trend. Nature Climate Change **4**:462-466.

Johansson, M. A., D. A. T. Cummings, and G. E. Glass. 2009. Multiyear Climate Variability and Dengue-El Nino Southern Oscillation, Weather, and Dengue Incidence in Puerto Rico, Mexico, and Thailand: A Longitudinal Data Analysis. Plos Medicine **6**.

Johnson, L. R., K.D. Lafferty, A. McNally, E. Mordecai, K. Paaijmans, S. Pawar, S.J. Ryan. 2014. "Mapping the distribution of Malaria: Current Approaches and Future Directions". in Analyzing and Modeling Spatial and Temporal Dynamics of Infectious Diseases. D. Chen, B. Moulin, J. Wu (Eds). John Wiley & Sons. pp 189-209.

Johnson, L.R. Ben-Horin, T., Lafferty, K.D., McNally, A., Mordecai, E., Paaijmans, K.P., Pawar, S., Ryan, S.J. 2015. Understanding uncertainty in temperature effects on vector-borne disease: A Bayesian approach. Ecology 96(1): 203-213.

Kalis, M., and E. Curtiss. 2016. CDC's Drought Guidance: Your Public Health Resource for Understanding and Preparing for Drought in Your Community. J Environ Health **78**:34-35.

Kalis, M. A., M. D. Miller, and R. J. Wilson. 2009. Public health and drought. J Environ Health **72**:10-11.

Keatinge, W. R., G. C. Donaldson, K. Bucher, G. Jendritzky, E. Cordioli, M. Martinelli, K. Katsouyanni, A. E. Kunst, C. McDonald, S. Nayha, I. Vuori, and G. Eurowinter. 2000a. Winter mortality in relation to climate. Int J Circumpolar Health **59**:154-159.

Keatinge, W. R., G. C. Donaldson, E. Cordioli, M. Martinelli, A. E. Kunst, J. P. Mackenbach, S. Nayha, and I. Vuori. 2000b. Heat related mortality in warm and cold regions of Europe: observational study. BMJ **321**:670-673.

Kirshen, P., C. Watson, E. Douglas, A. Gontz, J. Lee, and Y. Tian. 2008. Coastal flooding in the Northeastern United States due to climate change. Mitigation and Adaptation Strategies for Global Change **13**:437-451.

Kistemann, T., T. Classen, C. Koch, F. Dangendorf, R. Fischeder, J. Gebel, V. Vacata, and M. Exner. 2002. Microbial load of drinking water reservoir tributaries during extreme rainfall and runoff. Appl Environ Microbiol **68**:2188-2197.

Klinger, C., O. Landeg, and V. Murray. 2014. Power outages, extreme events and health: a systematic review of the literature from 2011-2012. PLoS Curr **6**.

Knight, D. B., and R. E. Davis. 2009. Contribution of tropical cyclones to extreme rainfall events in the southeastern United States. Journal of Geophysical Research-Atmospheres **114**.

Knowlton, K., M. Rotkin-Ellman, G. King, H. G. Margolis, D. Smith, G. Solomon, R. Trent, and P. English. 2009. The 2006 California heat wave: impacts on hospitalizations and emergency department visits. Environ Health Perspect **117**:61-67.

Knutson, T. R., J. J. Sirutis, G. A. Vecchi, S. Garner, M. Zhao, H. S. Kim, M. Bender, R. E. Tuleya, I. M. Held, and G. Villarini. 2013. Dynamical Downscaling Projections of Twenty-First-Century Atlantic Hurricane Activity: CMIP3 and CMIP5 Model-Based Scenarios. Journal of Climate **26**:6591-6617.

Kochtubajda, B., M. D. Flannigan, J. R. Gyakum, R. E. Stewart, K. A. Logan, and T. V. Nguyen. 2006. Lightning and fires in the Northwest Territories and responses to future climate change. Arctic **59**:211-221.

Kovats, R. S. 2000. El Nino and human health. Bull World Health Organ **78**:1127-1135.

Kovats, R. S., M. J. Bouma, S. Hajat, E. Worrall, and A. Haines. 2003. El Nino and health. Lancet **362**:1481-1489.

Kovats, R. S., and S. Hajat. 2008. Heat stress and public health: a critical review. Annu Rev Public Health **29**:41-55.

Kovats, R. S., S. Hajat, and P. Wilkinson. 2004. Contrasting patterns of mortality and hospital admissions during hot weather and heat waves in Greater London, UK. Occup Environ Med **61**:893-898.

La Greca, A. M., B. S. Lai, M. M. Llabre, W. K. Silverman, E. M. Vernberg, and M. J. Prinstein. 2013. Children's Postdisaster Trajectories of PTS Symptoms: Predicting Chronic Distress. Child Youth Care Forum **42**:351-369.

La Greca, A. M., W. K. Silverman, B. Lai, and J. Jaccard. 2010. Hurricane-Related Exposure Experiences and Stressors, Other Life Events, and Social Support: Concurrent and Prospective Impact on Children's Persistent Posttraumatic Stress Symptoms. Journal of Consulting and Clinical Psychology **78**:794-805.

Laursen, E., O. Mygind, B. Rasmussen, and T. Ronne. 1994. Gastroenteritis: a waterborne outbreak affecting 1600 people in a small Danish town. J Epidemiol Community Health **48**:453-458.

Li, C. F., T. W. Lim, L. L. Han, and R. Fang. 1985. Rainfall, abundance of *Aedes aegypti* and dengue infection in Selangor, Malaysia. Southeast Asian J Trop Med Public Health **16**:560-568.

Liang, S., E. Y. Seto, J. V. Remais, B. Zhong, C. Yang, A. Hubbard, G. M. Davis, X. Gu, D. Qiu, and R. C. Spear. 2007. Environmental effects on parasitic disease transmission exemplified by schistosomiasis in western China. Proc Natl Acad Sci U S A **104**:7110-7115.

Lin, S., W. H. Hsu, A. R. Van Zutphen, S. Saha, G. Luber, and S. A. Hwang. 2012. Excessive heat and respiratory hospitalizations in New York State: estimating current and future public health burden related to climate change. Environ Health Perspect **120**:1571-1577.

Lindblade, K. A., E. D. Walker, A. W. Onapa, J. Katungu, and M. L. Wilson. 2000. Land use change alters malaria transmission parameters by modifying temperature in a highland area of Uganda. Tropical Medicine & International Health **5**:263-274.

Lipp, E. K., A. Huq, and R. R. Colwell. 2002. Effects of global climate on infectious disease: the cholera model. Clin Microbiol Rev **15**:757-770.

Loevinsohn, M. E. 1994. Climatic warming and increased malaria incidence in Rwanda. Lancet **343**:714-718.

Luque Fernandez, M. A., A. Bauernfeind, J. D. Jimenez, C. L. Gil, N. El Omeiri, and D. H. Guibert. 2009. Influence of temperature and rainfall on the evolution of cholera epidemics in Lusaka, Zambia, 2003-2006: analysis of a time series. Trans R Soc Trop Med Hyg **103**:137-143.

Lyle, T. S., and T. Mills. 2016. Assessing coastal flood risk in a changing climate for the City of Vancouver. Canadian Water Resources Journal **41**:343-352.

Malmstadt, J., K. Scheltin, and J. Elsner. 2009. Florida hurricanes and damage costs. Southeastern Geographer **49**:108-131.

Martinez, C. J., J. J. Maleski, and M. F. Miller. 2012. Trends in precipitation and temperature in Florida, USA. Journal of Hydrology **452**:259-281.

McMichael, A. J., and A. Haines. 1997. Global climate change: the potential effects on health. BMJ **315**:805-809.

Medina-Ramon, M., A. Zanobetti, D. P. Cavanagh, and J. Schwartz. 2006. Extreme temperatures and mortality: assessing effect modification by personal characteristics and specific cause of death in a multi-city case-only analysis. Environ Health Perspect **114**:1331-1336.

Medlock, J. M., and S. A. Leach. 2015. Effect of climate change on vector-borne disease risk in the UK. Lancet Infect Dis **15**:721-730.

Mhalu, F. S., W. M. Mntenga, and F. D. Mtango. 1987. Seasonality of cholera in Tanzania: possible role of rainfall in disease transmission. East Afr Med J **64**:378-387.

Miettinen, I. T., O. Zacheus, C. H. von Bonsdorff, and T. Vartiainen. 2001. Waterborne epidemics in Finland in 1998-1999. Water Sci Technol **43**:67-71.

Miranda, J. J. 2004. El Nino and health. Lancet **363**:247-248.

Mordecai, E.A., Cohen, J., Evans, M.V., Gudapati, P., Johnson, L.R., Lippi, C.A., Miazgowicz, K., Murdock, C.C., Rohr, J.R., Ryan, S.J., Savage, V., Shocket, M., Stewart Ibarra, A.M., Thomas, M.B., Weikel, D.P. 2017. Detecting the impact of temperature on transmission of Zika, dengue, and chikungunya using mechanistic models. PLOS NTDs. 11(4): e0005568

Mordecai, E.A., Paaijmans, K.P., Johnson, L.R., Balzer, C.H., Ben-Horin, T., deMoor, E., McNally, A., Pawar, S., Ryan, S.J., Smith, T.C., Lafferty, K.D., 2013. Physiological constraints dramatically lower the expected temperature for peak malaria transmission. Ecology Letters 16(1):22-30

Neria, Y., and J. M. Shultz. 2012. Mental health effects of Hurricane Sandy: characteristics, potential aftermath, and response. JAMA **308**:2571-2572.

Ngwa, M. C., S. Liang, I. T. Kracalik, L. Morris, J. K. Blackburn, L. M. Mbam, S. F. Ba Pouth, A. Teboh, Y. Yang, M. Arabi, J. D. Sugimoto, and J. G. Morris, Jr. 2016. Cholera in Cameroon, 2000-2012: Spatial and Temporal Analysis at the Operational (Health District) and Sub Climate Levels. PLoS Negl Trop Dis **10**:e0005105.

Nichols, G., C. Lane, N. Asgari, N. Q. Verlander, and A. Charlett. 2009. Rainfall and outbreaks of drinking water related disease and in England and Wales. J Water Health **7**:1-8.

Ogden, N. H., A. Maarouf, I. K. Barker, M. Bigras-Poulin, L. R. Lindsay, M. G. Morshed, C. J. O'Callaghan, F. Ramay, D. Waltner-Toews, and D. F. Charron. 2006. Climate change and the potential for range expansion of the Lyme disease vector Ixodes scapularis in Canada. International Journal for Parasitology **36**:63-70.

Ostfeld, R. S., C. D. Canham, K. Oggenfuss, R. J. Winchcombe, and F. Keesing. 2006. Climate, deer, rodents, and acorns as determinants of variation in Lyme-disease risk. Plos Biology **4**:1058-1068.

Paaijmans, K. P., S. Blanford, A. S. Bell, J. I. Blanford, A. F. Read, and M. B. Thomas. 2010. Influence of climate on malaria transmission depends on daily temperature variation. Proc Natl Acad Sci U S A **107**:15135-15139.

Polymenakou, P. N., M. Mandalakis, E. G. Stephanou, and A. Tselepides. 2008. Particle size distribution of airborne microorganisms and pathogens during an Intense African dust event in the eastern Mediterranean. Environ Health Perspect **116**:292-296.

Price, C., and D. Rind. 1994. Possible Implications of Global Climate-Change on Global Lightning Distributions and Frequencies. Journal of Geophysical Research-Atmospheres **99**:10823-10831.

Pruss-Ustun, A., J. Bartram, T. Clasen, J. M. Colford, O. Cumming, V. Curtis, S. Bonjour, A. D. Dangour, J. De France, L. Fewtrell, M. C. Freeman, B. Gordon, P. R. Hunter, R. B. Johnston, C. Mathers, D. Mausezahl, K. Medlicott, M. Neira, M. Stocks, J. Wolf, and S. Cairncross. 2014. Burden of disease from inadequate water, sanitation and hygiene in low- and middle-income settings: a retrospective analysis of data from 145 countries. Tropical Medicine & International Health **19**:894-905.

Pruss, A., D. Kay, L. Fewtrell, and J. Bartram. 2002. Estimating the burden of disease from water, sanitation, and hygiene at a global level. Environmental Health Perspectives **110**:537-542.

Qian, Z. M., Q. C. He, H. M. Lin, L. L. Kong, C. M. Bentley, W. S. Liu, and D. J. Zhou. 2008. High temperatures enhanced acute mortality effects of ambient particle pollution in the "Oven" city of Wuhan, China. Environmental Health Perspectives **116**:1172-1178.

Ragan, P., J. Schulte, S. J. Nelson, and K. T. Jones. 2008. Mortality surveillance: 2004 to 2005 Florida hurricane-related deaths. Am J Forensic Med Pathol **29**:148-153.

Reeve, N., and R. Toumi. 1999. Lightning activity as an indicator of climate change. Quarterly Journal of the Royal Meteorological Society **125**:893-903.

Ritenour, A. E., M. J. Morton, J. G. McManus, D. J. Barillo, and L. C. Cancio. 2008. Lightning injury: a review. Burns **34**:585-594.

Ryan, S.J., McNally, A., Johnson, L.R., Mordecai, E.A., Ben-Horin, T., Paaijmans, K.P., Lafferty, K.D. 2015. Mapping physiological suitability limits of malaria in Africa under climate change. Journal of vector borne and zoonotic diseases 15(12): 817-725.

Ryan, S.J., Ben-Horin, T., Johnson, L.R. 2015. Malaria control and senescence: the importance of accounting for the pace and shape of ageing in wild mosquitoes. Ecosphere 6(9):170.

Ryan, S.J. et al. 2017. Climate change drives uncertain global shifts in potential distribution and seasonal risk of Aedes-transmitted viruses. BioRXiv doi: https://doi.org/10.1101/172221.

Sarofim, M. C., S. Saha, M. D. Hawkins, D. M. Mills, J. Hess, R. Horton, P. Kinney, J. Schwartz, and A. St. Juliana. 2016. Ch. 2: Temperature-Related Death and Illness. Pages 43-68 The Impacts of Climate Change on Human Health in the United States: A
Scientific Assessment. U.S. Global Change Research Program, Washington, DC.

Schijven, J. F., and A. M. de Roda Husman. 2005. Effect of climate changes on waterborne disease in The Netherlands. Water Sci Technol **51**:79-87.

Schreider, S. Y., D. I. Smith, and A. J. Jakeman. 2000. Climate change impacts on urban flooding. Climatic Change **47**:91-115.

Schuster, C. J., A. G. Ellis, W. J. Robertson, D. F. Charron, J. J. Aramini, B. J. Marshall, and D. T. Medeiros. 2005. Infectious disease outbreaks related to drinking water in Canada, 1974-2001. Can J Public Health **96**:254-258.

Senhorst, H. A., and J. J. Zwolsman. 2005. Climate change and effects on water quality: a first impression. Water Sci Technol **51**:53-59.

Shaman, J., J. F. Day, and M. Stieglitz. 2005. Drought-induced amplification and epidemic transmission of West Nile virus in southern Florida. J Med Entomol **42**:134-141.

Singh, R. B., S. Hales, N. de Wet, R. Raj, M. Hearnden, and P. Weinstein. 2001. The influence of climate variation and change on diarrheal disease in the Pacific Islands. Environ Health Perspect **109**:155-159.

Soverow, J. E., G. A. Wellenius, D. N. Fisman, and M. A. Mittleman. 2009. Infectious Disease in a Warming World: How Weather Influenced West Nile Virus in the United States (2001-2005). Environmental Health Perspectives **117**:1049-1052.

Su, G. L. S. 2008. Correlation of climatic factors and dengue incidence in Metro Manila, Philippines. Ambio **37**:292-294.

Sullivent, E. E., 3rd, C. A. West, R. S. Noe, K. E. Thomas, L. J. Wallace, and R. T. Leeb. 2006. Nonfatal injuries following Hurricane Katrina--New Orleans, Louisiana, 2005. J Safety Res **37**:213-217.

Taylor, G., and W. J. Davies. 1990. Root-Growth of Fagus-Sylvatica - Impact of Air-Quality and Drought at a Site in Southern Britain. New Phytologist **116**:457-464.

Thomas, K. M., D. F. Charron, D. Waltner-Toews, C. Schuster, A. R. Maarouf, and J. D. Holt. 2006. A role of high impact weather events in waterborne disease outbreaks in Canada, 1975 - 2001. Int J Environ Health Res **16**:167-180.

Thompson, K. R., N. B. Bernier, and P. Chan. 2009. Extreme sea levels, coastal flooding and climate change with a focus on Atlantic Canada. Natural Hazards **51**:139-150.

Thomson, M. C., S. J. Mason, T. Phindela, and S. J. Connor. 2005. Use of rainfall and sea surface temperature monitoring for malaria early warning in Botswana. American Journal of Tropical Medicine and Hygiene **73**:214-221.

Tian, L., Y. Bi, S. C. Ho, W. Liu, S. Liang, W. B. Goggins, E. Y. Chan, S. Zhou, and J. J. Sung. 2008. One-year delayed effect of fog on malaria transmission: a time-series analysis in the rain forest area of Mengla County, south-west China. Malar J **7**:110.

USEPA. 2016. Climate change indicators in the United States. Page 96 *in* U. S. E. P. Agency, editor.

van der Hoek, W., F. Konradsen, D. Perera, P. H. Amerasinghe, and F. P. Amerasinghe. 1997. Correlation between rainfall and malaria in the dry zone of Sri Lanka. Annals of Tropical Medicine and Parasitology **91**:945-949.

Vernberg, E. M., A. M. LaGreca, W. K. Silverman, and M. J. Prinstein. 1996. Prediction of posttraumatic stress symptoms in children after Hurricane Andrew. Journal of Abnormal Psychology **105**:237-248.

Wardrop, N. A., A. G. Barnett, J. A. Atkinson, and A. C. A. Clements. 2013. Plasmodium vivax malaria incidence over time and its association with temperature and rainfall in four counties of Yunnan Province, China. Malaria Journal **12**.

Weems, C. F., L. K. Taylor, M. F. Cannon, R. C. Marino, D. M. Romano, B. G. Scott, A. M. Perry, and V. Triplett. 2010. Post Traumatic Stress, Context, and the Lingering Effects of the Hurricane Katrina Disaster among Ethnic Minority Youth. Journal of Abnormal Child Psychology **38**:49-56.

WHO. 2008. Global Burden of Disease 2004 update. World Health Organization.

Winsberg, M. D. 2003. Florida weather. 2nd edition. University Press of Florida, Gainesville, Fla.

Yang, K., J. LeJeune, D. Alsdorf, B. Lu, C. K. Shum, and S. Liang. 2012. Global distribution of outbreaks of water-associated infectious diseases. PLoS Negl Trop Dis **6**:e1483.

CHAPTER 5

Climate Change Impacts on Florida's Energy Supply and Demand

Wendell A. Porter[1] and Hal Knowles III[2]

[1]*Department of Agricultural & Biological Engineering, University of Florida, Gainesville, FL;* [2]*Program for Resource Efficient Communities, University of Florida, Gainesville, FL*

Florida's unique location in the contiguous United States ensures that the effects of climate change will be significant and persistent across the state. Florida's current economy and its population have developed energy use patterns based on fully developed fossil fuel industries. These industries and Florida's consumption patterns are presented and analyzed. Location of Florida's electricity generating facilities are shown and a significant proportion of these facilities are literally at the water's edge. Future actions to protect the state's energy supply may need to include costly moving of significant fossil fueled facilities and/or outright replacement by newer, cheaper renewable energy power plants. The current status of energy consumption in Florida is presented in this chapter, along with disruptive technologies in energy efficiency, renewable energy, and the electrical grid. World photovoltaic (PV) and wind power adoption rates are used to explore the possible time frames for renewable energy transformation.

Key Messages

- Currently, Florida has very few sources of energy within its borders. There are no appreciable coal mines, natural gas fields, or oil fields.
- Florida has very little diversity in its sources of energy. More than two-thirds of Florida's electricity is produced from natural gas, and almost all of the oil used in Florida is used for transportation.
- Significant portions of Florida's energy infrastructure are located at the water's edge and may be exposed to the effects of sea level rise much sooner than we expect.
- Florida's dependence on electricity to run its economy is inefficient and not competitive in comparison with leading U.S. states and other fully developed countries.
- A significant push to incorporate energy efficiency and renewable energy into Florida's energy mix can rapidly make the state more competitive and bring down the cost of energy for all of Florida residents.

Keywords

Power generation; Energy efficiency; Conservation behaviors; Renewable resources; Building performance; Consumer financing; Solar photovoltaics; Fuel economy

Introduction

Since the burning of fossil fuels is the single largest contributor to the greenhouse gas emissions (EPA 2016) that are forcing anthropogenic climate change, the issue of Florida's energy supply and demand is vital to the state's future. In addition to the stresses associated with rising temperatures, Floridians of the near future will also have to deal with sea level rise. Florida has the second longest coastline of all states in the US, and its relatively low coastal elevation means that these climate effects will be felt in Florida first. A study by The World Bank (2013) ranks cities in relation to potential for damage associated with increased flooding brought on by climate change. Of the top ten cities worldwide, five are in the US and two (Miami and Tampa) are in Florida; Miami is ranked second overall in the world. Florida also happens to be the third largest state in the US in terms of population, with just over 20 million residents as of 2016. Taking all of these facts together, one could argue that Florida is the state with the most dramatic and direct link between its location, its people, and the impending effects of climate change.

While climate change and our future scenarios have become familiar topics to many over the years, the energy systems that power modern society are often less familiar. There are many databases around the world that tabulate national energy use in primary terms to correlate the amount of energy pulled from the earth in the form of coal, oil, and natural gas. The United States uses a term called quads, which is short for quadrillion British Thermal Units (BTUs), while the Australians tabulate their energy use in peta-joules. While informative to energy analysts, these terms are probably unfamiliar to the general public and may fail to report energy consumption patterns in ways that can positively influence climate-friendly behavior changes. For the purposes of this chapter, oil is described in terms of barrels, each of which contains 42 US gallons. Natural gas is sold in USD per million BTUs, which is approximately 1,000 cubic feet, and production quantities are in cubic feet. Coal is sold in the United States by the short ton, which is 2,000 pounds; its energy quantity varies by the mine and general geographic location, with eastern coal having a higher energy content than western coal. Consumption patterns include transportation use and electricity consumption; we find that these two areas alone make up the majority of fossil fuel use in Florida. Thus, by focusing on consumption patterns, actionable decisions by Floridians are easier to for the general public to understand and to make.

Additionally, once Florida's consumption patterns are described, the source of these fuels can be analyzed. Broadly stated, many of the fuels we need in Florida are brought in by water, and many of the power plants that produce the electricity Floridians consume are located at the water's edge for cooling purposes. This is of critical concern when faced with even the most benign sea level rise predictions. The predictions described in Chapter 19 related to the most current thinking on this topic require us to start planning for the relocation of these energy-producing assets or their outright replacement with new, renewable technology located away from the changing coastline.

In fact, new and potentially disruptive technologies are already entering the state's energy market. Long-planned power plants will not be needed due to new energy efficient technologies that are actually leading to a decrease in total electricity demand within the state. Along with these new technologies will be the latest in renewable energy sources, especially solar photovoltaic (PV) electricity, which is rapidly becoming the cheapest new power source and is particularly well-suited to the "Sunshine State." This combination of new energy-efficient technologies and solar PV electricity will remake the energy landscape as we know it, not just here in Florida but around the world.

Implementation of these new approaches to energy supply and demand will require careful planning if we are to seamlessly transition from a modern economy fueled by fossil fuels to one operated by more renewable forms of energy. While this might seem a daunting task, it is worth noting that as of 2015, approximately 55 countries worldwide produced more than 50% of their electricity from renewable sources (EIA 2016).

Fossil Fuel Consumption and Status

The largest single source of greenhouse gases comes from the burning of fossil fuels to operate our modern societies. From the gasoline in our automobiles, the natural gas furnaces in our homes, and the coal-fired electric power plants that light our homes and power our offices, the use of fossil fuels permeates every aspect of our lives. This tends to make any discussion too big of a hill for most people to climb. It is hard, in the staggering amounts of energy consumed by today's society, to see where a single person could make a difference. To do so, we need to break down the larger problem into smaller concepts with more direct explanations and solutions. What we discover by investigating energy use patterns in Florida can help us make better decisions for the future.

Florida's agricultural- and tourism-based economy helps to simplify our analysis. The following sections will show that nearly all of the oil used in the state is consumed in transportation systems. In a similar manner, nearly all coal and natural gas burned in Florida is combusted in the state's electric power plants. These very good approximations will help us design effective mitigation strategies specifically for Florida.

Florida Oil Consumption

During 2015, Florida consumed approximately 832,000 bbl/day of oil products (EIA, 2016). A barrel (bbl) of oil is equal to 42 US gallons. The direct relation between personal consumption of gallons of fuel and the marketplace in barrels makes it easier for the reader to comprehend these relationships. Unlike some of the more industrial states, Florida uses nearly all of its oil in the transportation sector: 489,000 bbl/day of gasoline consumption, 134,000 bbl/day of diesel fuel consumption, and 89,000 bbl/day of jet fuel use (EIA 2016). The remaining oil use consists of small amounts of propane, mainly for space heating and fuel oil for heating, and oil-fired power plant operations. Individually, these are only a couple of percent each, so for general purposes we will focus on transportation when we are discussing oil use in Florida.

Florida Natural Gas Consumption

Nationally, natural gas has three main uses: natural gas-fired electric power plants, space and water heating in the built environment, and industrial process heating. Natural gas consumption for electricity production is about 40% of the total use nationally, with the other two categories splitting the remainder. However, Florida's warm climate and lack of heavy industry make it much easier to track total natural gas use in the state. More than 90% of the natural gas use in Florida is consumed in power plants to produce electricity. Natural gas consumption in Florida during 2014 totaled just over 1.22 trillion cubic feet (EIA 2016), and the use of natural gas for electricity production is steadily rising in Florida. Natural gas was used to produce 62% of total electricity in 2014, 66% of total electricity in 2015, and 68% in the first four months of 2016 (EIA 2016).

Florida Coal Consumption

Coal is also consumed in Florida to produce electricity. During 2014, approximately 26 million short tons were burned in Florida power plants to produce approximately 52 million MWh (EIA 2016). Nationally, approximately 90% of US coal production was used to produce electricity while the remainder was either exported or used for industrial purposes, mainly steelmaking. However, Florida is different in that there is an absence of heavy industry that would use coal.

The use of coal to produce electricity in Florida has been steadily decreasing. In 2000, coal was used to produce nearly 73 million MWh, representing 38% of Florida's total electricity consumption. In 2015, this had dropped to 43 million MWh, representing only 18% of Florida's electricity use.

Florida Nuclear Power Production

As of the end of 2015, there were four nuclear reactors operating located at three different power plants in Florida. Two reactors operate at the St. Lucie Nuclear Power Plant in Jensen Beach and two additional reactors operate south of Miami at the Turkey Point Nuclear Generating Station. An additional reactor has been permanently shut down at Crystal River. Florida Power & Light has permission to start building two new reactors at Turkey Point, but construction has not yet begun. Plans called for two more reactors to be built in Levy County, but construction has been halted. The four existing, operating reactors combine to produce 28 million MWh or 12% of Florida's annual electricity needs (EIA 2016).

Florida Energy Systems Locations

Florida's long coastline provides ample opportunity to locate electric power plants close to cooling water sources. Our state's many port facilities also provide strategic locations to receive, stage, and deliver fuel sources such as coal and oil close to population centers. However, locating power plants near the shoreline greatly increase their risks as the sea level rises. It might seem that the risks posed to these facilities are far in the future, but a few very real examples involving actual power plant sites show that this is not true.

The independent organization, Climate Central, has developed an interactive online tool that allows a user to explore the effect of rising sea levels on the detailed shoreline. This tool, Surging Seas Risk Zone Map (Surging Seas 2016), combines sea level rise predictions based on the "business as usual" world greenhouse gas emissions scenario with the specific coastal elevation map of Florida. Using this tool in conjunction with an online mapping program enables anyone to correlate specific locations with risk. Florida's three nuclear power plants provide compelling examples of the relatively immediate risk posed by sea level rise.

Anyone with access to a computer and the internet can see for themselves that some of Florida's power plant locations represent a serious problem with regards to sea level rise. For example, using Google Maps, a satellite view of the Turkey Point power plant site can easily be seen (Google Maps 2016). The power plant is the very visible structure right on the coast, approximately five miles southeast of the Homestead Air Reserve Base. The long canal structure to the south of the plant is a series of cooling water canals designed to limit thermal pollution to Biscayne Bay. This power plant site includes the two, 693 MW nuclear reactors, two 400 MW oil/natural gas-fired units, and one 1,150 MW combined cycle natural gas-fired unit. Turkey Point is one of the largest power plant sites in the US. This same section of coastline can be found and highlighted using Climate Central's Surging Seas risk mapping program. One of the most notable features of this program is that the sea level rise can be set in one foot increments from zero to ten feet.

Recent projections (Rahmstorf, Perrette, and Vermeer 2012) predict that the seas will rise a little more than three feet by 2100. Other researchers incorporating new data concerning the stability of the West Antarctic Ice Sheet warn that the sea level rise might be twice that by 2100 (DeConto and Pollard 2016). The Surging Seas risk mapping tool allows users to see what roughly one meter and two meters of sea level rise will look like. So, considering the location of the Turkey Point power plant site and using the Surging Seas Risk Zone Map tool (Surging Seas 2016), one can see that approximately one meter of sea level rise will result in fully flooded wetlands that extend all the way from the power plant to the runway for Homestead Air Force Base located nearly five miles to the northwest. In this scenario, the Turkey Point power plant site will be surrounded by water and essentially miles from dry land. And while the built up base of the power plant will still be well above the water level, the property itself will be well off shore. Side-by-side images of the current coastline and a future coastline with a three foot increase in sea level is shown in Fig. 5.1. And an extra one meter rise associated with the West Antarctic Ice Sheet melting added would be nothing short of catastrophic for southeast Florida (Surging Seas 2016).

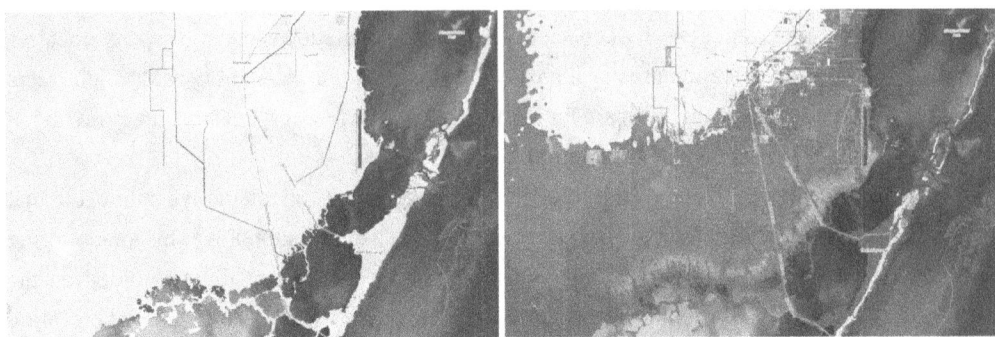

Figure 5.1. Existing Florida coastline compared to a future coastline scenario with a three foot rise in sea level (Harrison, adapted from Surging Seas 2016).

These sea level rise predictions are based on greenhouse gas emissions governed by a business-as-usual approach in relation to our state's collective carbon footprint. The ice sheet predictions are based on a rapidly evolving understanding of a very complex interrelationship between many climate variables, the complexity of which might lead some to delay in decision-making until a greater understanding is obtained. However, the Turkey Point site has a replacement value in excess of $20 billion and is fixed in place at the water's edge. Even a one foot rise puts this site in peril, and not too far into the future.

The Surging Seas Risk Zone Map tool can be used in a similar manner to look at the St. Lucie Nuclear Power Plant, which is located on Hutchinson Island, a barrier island with direct exposure to the Atlantic Ocean. Like Turkey Point, the power plant property in St. Lucie has been extensively elevated. However, even choosing a minimum one foot rise sea level on the Surging Seas mapping tool demonstrates how rising waters will isolate the island's main access road, A1A, in both directions, with water on both sides of the road bed (Surging Seas 2016). At this point, it should be noted that it can take ten years or more to safely decommission a nuclear power plant, and losing road access to a plant severely limits the ability to decommission the nuclear aspects of the plant in a timely manner. Choosing three feet of sea level rise on the mapping tool will show that access to the mainland through causeways north and south is compromised, and looking at a six foot sea level rise scenario results in a completely isolated nuclear power plant site located on the remains of a barrier island surrounded by the ocean.

Crystal River is a third nuclear power plant site in Florida, but at the time of publication the reactor has been permanently shut down, although it has not been fully decommissioned. The Crystal River site is also home to approximately 2200 MW of coal-fired electricity production, with a planned addition of 1,640 MW of combined cycle gas turbine-based power production. With development halted for the Levy County nuclear power complex and a natural gas power plant addition, some portion of the coal-fired complex will remain in the near future. Just a one-foot rise in sea level would dramatically change the shoreline around the Crystal River power plant property. This one-foot rise would bring the shoreline to the landward side of the power plant property, and the property itself is a peninsula jutting out into the Gulf of Mexico. A three-foot rise would bring the shoreline to the east of the power plant property, almost completely isolating the plant site (Surging Seas 2016). And a six-foot sea level rise, a possibility introduced by the recent publications, would have the site completely surrounded by the Gulf of Mexico.

A final example would be an oil import facility located in Tampa Bay. The vast majority of liquid fuels used for transportation purposes is delivered to Florida through its port facilities. The Port of Tampa oil terminal is located on the lower western shore of the peninsula that roughly bisects Tampa Bay. This peninsula is also the location of the Central Command MacDill Air Force Base, which contains runways that are easily recognized. Locating the Port of Tampa fuel terminal using the Surging Seas Risk Map tool (Surging Seas 2016) shows that with just one foot of sea level rise the southern portion of the facility's tank farm will be below sea level. Many additional tanks and some of the docks go below sea level with three feet of sea level rise (Surging Seas), and with a sea level rise of six feet the entire tank farm will be flooded, as will the docks. Under this scenario, what remains of the terminal will be entirely separated from the mainland.

These are a few examples in just our state. And unfortunately, none of these sites could be protected individually by a system of pumps and levees from the encroaching seas. And the truth is that we will not have the luxury of facing these problems individually; they will occur all along Florida's shoreline at roughly the same time. Sometime in the very near future, one such site will be slated for an upgrade, significant maintenance, or a service life extension. Somebody in the long line of permitting offices will ask the simple question: "Should we invest this much money for a facility that will soon be under water?" At this point, the choices will be:

- Do nothing and watch as an expensive energy facility is stranded by the rising seas.
- Upgrade the facility with levees and pumps to protect it for the near future.
- Move the legacy fossil fuel or nuclear fueled facility to a new location at considerable expense.
- Decommission the facility and transfer energy needs to new renewable energy facilities located inland.

Current conservative predictions show that the scenarios presented here could start happening as early as 2060, and this does not take into account new research describing the instability of the West Antarctica ice sheet referenced earlier. Newer predictions include estimates of six feet of sea level rise by 2100. As of this publication, the owner of the Turkey Point facility has decided to delay the construction start of the two new reactors for at least four years. With four more years of climate science and analysis behind us, it is doubtful that these new reactors will ever be built.

Any analysis of the choices that we, as a society, will be faced with brings us to the conclusion that our only viable long-term option is to transform our energy systems from fossil-fueled to a system powered by renewable energy.

Energy Efficiency and Disruptive Technologies

Nationally, about 74% of all electricity consumed in the US is used in the built environment, with the remainder consumed by industrial processes (EIA 2016). At present, only small amounts of electricity are used in domestic transportation systems. In 2014, 93% of all electricity consumed in Florida was used in residential (52%) and non-residential (41%) buildings (EIA 2016). These residential structures include single-family homes, townhomes, apartments, and condominiums; non-residential structures include retail businesses, office complexes, municipal buildings, schools, and universities.

It is important to quantify electrical consumption by totals, end use, and by sector. Knowledge of these quantities allows us to begin to benchmark our consumption and construct state and international level comparisons. It also helps to think of all energy consumption as a parasitic economic cost. By using less energy to run our society, we become more efficient and more of our output can be used to provide for our direct needs. With this in mind, let's look at the comparative data from a few key states and countries (Table 5.1).

Table 5.1. Energy consumption benchmarks for selected states and countries.

Country/State	Population, millions	GDP, Trillions ($ ppp)	Electricity consumption, Billion MWh/yr	MWh/yr, ca	$GDP/MWh
California	39.1	2.45	0.261	6.68	9,387
United Kingdom	64	2.68	0.335	5.23	8,000
Germany	81	3.84	0.585	7.23	6,560
Japan	126	4.83	0.960	7.62	5,031
USA	321	17.95	4.05	12.62	4,432
Florida	20.3	0.89	0.237	11.69	3,768

At first glance, the ordering of the countries and states seems random; unfortunately for Florida, it is not. Each one of these governmental entities has a population that is organized and works to produce the easily identified benchmark of gross domestic product (GDP). To make the comparison effective, the purchase power parity (ppp) category for GDP is also used. This equalizes the effects of currency value differences. Annual electricity consumption is tabulated for each entry. With this last entry, the benchmarks in the last two columns can be produced. The first is a per capita electricity consumption figure that simply divides all of the year's electricity consumption into the total population. This column generally increases down the list, with the US and Florida being at the bottom. The last column relates the creation of wealth (GDP) to the consumption of electricity (MWh). Again, the US and Florida are at the bottom of the list. If Germany, Japan, and the United Kingdom are assumed to be fully developed nations that the US competes with on a daily basis, our inefficient use of electricity puts us at a significant disadvantage. The inclusion of California shows that even within the United States, it is possible to be a world leader in the efficient use of electricity.

What if both Florida and the USA used electricity as efficiently as California to produce wealth? If Florida was as efficient as California, Florida would use 95 million MWh/yr instead of 237 million MWh/yr and the USA would only use 1.9 billion MWh/yr instead of 4.05 billion MWh/yr. Simply put, if the USA were as efficient as California, about 85% of our electricity would currently be produced from non-fossil fuel sources!

Energy Efficiency in the Built Environment

Stating that the United States and Florida should be more efficient is one thing, making it so is quite another. Many organizations promote ideas that show homeowners, renters, and commercial businesses how to become more energy efficient in economically viable ways. One of the more useful sites specifically built for Florida is *My Florida Energy Home* (2016). The online site introduction breaks down the energy use in a typical Florida home as follows:
- Air conditioning and heating is 36%
- Domestic hot water is 14%
- Appliances, electronics, and lighting consumes 50%

Nationally, lighting is about 15% of home energy consumption, which would make appliances and electronics approximately 35%. The actual energy consumed in each of these areas is related to the characteristics of each end use device and, just as importantly, how each device is operated. A simple example would be a light bulb. Operating a 60-watt incandescent light bulb four hours each day instead of eight hours each day will reduce overall energy consumption by 50%. This is easily accomplished by always turning a light off when not in use. Examples of this are seen everywhere in our lives—from our homes, to our schools, and in our businesses. Take this example one step further and replace the 60-watt bulb with an 8-watt LED light bulb that operates four hours each day. This would reduce the total energy used by 93%.

A near-term goal for both business competitiveness and climate change mitigation would be to make every effort to increase the energy efficiency of all economic sectors in our state and the nation. Fortunately for Floridians, there are ample opportunities to make quick progress towards this goal. The following are just a few ideas that cost literally nothing to implement, and yet will significant reduce a consumer's bottom line:
- Phantom loads: Many devices used in homes have instant-on, or standby features that consume power as long as they are plugged into an electrical receptacle. Entertainment centers house televisions, cable boxes, DVRs, receivers, powered sub-woofers, game boxes, etc. All of these devices consume power when not in use. When totaled over an entire year, many devices consume more energy when they are off than when they are actually being

used. An easy fix for this problem is to plug all of these devices into a surge-protected power strip. When at work or asleep, the power strip can simply be switched off.
- Lights: A good amount can be saved on an electric bill just by turning things off when they are not being used. When you leave a room, turn out the lights.
- Ceiling fans: Another device that wastes energy by being on when it isn't needed is ceiling fans. The truth is that fans cool people, not rooms. When a room is not in use, turn the fan off. It's just that simple.
- Thermostats: Thermostats are directly linked to the size of a utility bill. The balance between comfort and expense is different for everybody, but the recommended set points for summer and winter are 78 °F and 68 °F, respectively. For every degree away from these recommended settings (i.e., cooler set points in summer, warmer set points in winter), air conditioning and heating costs are increased by about 4%. If current use settings are different than those recommended, consumers can adjust set points slowly, one degree at a time over a period of months and even seasons. The human body will adjust very well to warmer and colder temperatures, just not very quickly.

Combined, these simple, no-cost actions can typically reduce an electric bill by as much as 20% with a capital cost of exactly zero. These examples are meant to show that the manner in which many of us operate our residences is far from efficient. Taking these steps can be done without any degradation of lifestyle while also returning some hard-earned money back to consumers to be used for other needs. Some people might construe a different temperature setting for heating and air conditioning as a lifestyle change. However, there are other benefits to doing so. Indoor temperatures below the prevailing summer dew point temperature of 77-79 °F can cause condensation on interior surfaces, which in turn produces an excellent environment for mold. Over air conditioning can produce a cold and clammy sensation and a situation that negatively affects indoor air quality.

At present, all of these suggestions for reducing energy consumption can be accomplished by everyone in their own residences. By doing these, the electricity consumption in the residential sector could be decreased by 20% overnight, which would amount to 24 million MWh/yr in Florida and approximately 300 million MWh/yr in the entire country. Technology trends support this transition. New standards for electronic devices have significantly dropped standby power consumption and instant-on power drain. Thus, as equipment gets replaced, more and more energy will be saved. New LED lights come with Wi-Fi connectivity that allows users to turn them off from anywhere. The same is true for new ceiling fans. One of the fastest selling home accessories relatively new to the market is the smart thermostat. Sales of smart thermostats already number in the millions, with growth rates in double digits. at this rate, the market will be saturated with smart thermosets within ten years.

It should also be noted that the primary new lighting technology, LED lights, will grow even more efficient in the near future. Currently, a quality, economically viable LED will produce between 100 and 120 lumens per watt. Laboratory results have already demonstrated a lamp efficiency of about 300 lumens per watt. The disruptive transition of LED lights over the next few years will cause the portion of our residential electrical bills related to lighting to drop from about 15% to about 1%.

The previous recommendations consist entirely of things that can be done without any capital expense, assuming that most houses currently have a power strip or two that can be used to manage phantom loads. There are also a number of things that households can do to reduce utility bills at very minimal expense. Some of these are:

- Replace one incandescent light with an LED version every month or two. A 60-watt incandescent can be replaced with an 8.5-watt LED. Electricity for lighting typically comprises 10–15% of a home's electrical consumption. The combination of upgrading to LEDs and always turning lights off when not in use can almost eliminate this category of consumption.
- Add faucet aerators and low-flow shower heads to all sinks and showers in a house or apartment, and remember that the length of shower is directly related to the energy consumed. In Florida the majority of water heaters use electricity. By reducing the use of hot water, electricity consumption is also reduced.
- Add pipe insulation to the accessible portions of the hot and cold water lines directly adjacent to the water heater. Also, add insulation to the larger of the two lines that go from the outside air conditioning unit into a house.
- Install a smart thermostat that will help keep the home at its most efficient temperature when residents are away at work, without the need to remember to do anything. Learning how to program an old style thermostat is no longer necessary.

Nearly every family in Florida and the US can afford to implement both sets of these no-cost, low-cost recommendations. The return on investment by saving 20-40% on an electricity bill can be used to pay for more capital intensive upgrades to a home, such as adding attic insulation, upgrading appliances, installing double-pane windows, and new energy efficient heating and air conditioning systems.

Financing Energy Efficiency and Renewable Energy in the Built Environment

If society is to both mitigate and adapt to climate change, it must be affordable. Fortunately, consumers hold a massive amount of power through their purchasing choices. Additionally, new financing options empower Floridians to support the energy efficiency and renewable energy (EERE) sectors while meeting their daily needs. However, no financial product is a panacea for people. Every household and business will have unique EERE goals, dependent on individualized credit worthiness. Thus, the financing options described in this section should be evaluated with critical minds and a balanced approach. The *My Florida Energy Home* website (2016) includes more complete coverage of this topic, including the SMART approach to achieving consumer financial goals.

While subjected to the political winds and industry whims, a variety of local, state, and federal incentives, manufacturer rebates, and EERE programmatic resources are often available to Floridians. A great place to start an evaluation of consumer options is the North Carolina Clean Energy Technology Center's Database of State Incentives for Renewables and Efficiency (DSIRE; 2016).

The DSIRE online repository documents dozens of policy and incentive categories. However, some of these incentives may be out of date and others may not be reported on DSIRE. Thus, savvy consumers will also investigate the current options directly advertised through their local government agencies, their utility providers, and the manufacturers of any EERE products and services they may be considering. The most common EERE incentive programs target the following major building systems: (1) structural, including thermal and air barriers; (2) mechanical, including ductwork and space conditioning; (3) appliances; (4) lighting; (5) distributed renewable energy generation, including solar PV and solar water heating; and (6) electrical grid load management. Additional details and resources can be found at the My Florida Energy Home website (www.myfloridahomeenergy.com).

Florida Statue 163.08 establishes guidelines and requirements enabling counties, municipalities, and special districts to implement property assessed financing (PAF) programs, also called property assessed clean energy (PACE) programs. These PAF/PACE programs may be used to finance qualifying building energy efficiency retrofits, renewable energy generation installations, and wind resistance improvements. Typically, local governments enter into agreements with one or more private sector financial partners who originate and service the loans. The governmental role comes in during the repayment process and occurs through non-ad valorem property assessments for participating property owners.

While other existing financial products may be used to fund EERE, the perceived benefits of PAF/PACE programs come primarily in their stimulus to local economies and their property assessment and repayment structures, which stay with the properties, regardless of the owner. Some evidence suggests measurable economic benefits. However, it is unclear whether PAF/PACE merely replicates the equivalent local impacts that might exist in conventional consumer financing. In theory, PAF/PACE may enable leveraged property owners with little to no equity available in their existing mortgage to receive the financing necessary to implement EERE and storm hardening property improvements. In reality, most PAF/PACE programs have underwriting standards similar to conventional consumer financing and they may offer less competitive rates due to programmatic overhead.

Unfortunately, property renters often lack direct access to PAF/PACE programs. Thus, in situations where property owning landlords have neither the means (e.g., lack credit worthiness), nor the motivation (e.g., lack awareness, lack interest), PAF/PACE programs may do little to address the "split incentive (principal agent) problem" commonly constraining EERE innovations within the rental marketplace.

While their benefits, constraints, and legal challenges are sometimes debated, the addition of PAF/PACE programs to the suite of climate change mitigation adaptation financing strategies is likely a positive trend, as they increase the diversity of options available to residential and non-residential property owners. A more detailed explanation and analysis of PAF/PACE programs is available at http://www.myfloridahomeenergy.com (My Florida Energy Home-Financing 2016)

Disruptive Technologies in the Built Environment

Disruptive technologies that are more energy efficient are starting to affect the entire US electricity market. Total electricity generation in 2004 was approximately 3.97 billion MWh. The 2016 data available as of this chapter's publication show the US on track to generate approximately 4.0 billion MWh (EIA 2016), in spite of an additional 27 million more people since 2004 and an increase in the GDP from $12.3 trillion to $18.3 trillion (projected for 2016).

An even more extreme example would be to compare Florida to California. Florida has approximately half the population of California and just over one third of the GDP, yet Florida consumes significantly more electricity. California made a concerted effort after the 1974 Oil Embargo to change the way Californians consume energy. The biggest change was to slowly evolve building codes over the years, eventually allowing California to develop some of the most energy efficient buildings in the United Sates, if not the entire world. The other significant change made in California was to de-couple the state's electric power producers' profits from generation. As a result, electric utilities were able to direct their customers to many new and existing technologies that could provide the same service with much lower energy consumption. As grid-wide electricity consumption fell, the utility was allowed to raise rates to cover the cost of transmission. At first glance, it might look as though the customer would lose because of the higher rates; however, the opposite is true. California residential customers pay higher electrical rates than their Florida counterparts but the average residential bill is lower (EIA 2016). Simply put, California saves more than the higher rates cost. It has taken California nearly 40 years to make the necessary changes and begin to reap the benefits. How does a state like Florida catch up, and catch up quickly?

The most disruptive technologies might be financial rather than physical. For example, many homeowners would like to upgrade their homes to consume less energy but are stymied by a lack of working capital. This lack of capital is one of the main reasons why the majority of homes in Florida and the nation have outdated home systems in the following key areas: air conditioning and heating systems, attic insulation, and ductwork; windows; lighting systems, and water heating

Conventionally designed and built, residential duct systems in Florida can wastefully leak between 20 to 40% (My Florida Energy Home 2016). Sealing duct systems can almost entirely eliminate this loss and increase the overall efficiency of a system. Another answer is to eliminate a duct system entirely by using what is called a mini-split design. Including new direct current motors and inverter designs allows mini-split design systems to be nearly twice as efficient as existing technology. Coupled with new smart thermostats, these new designs and controls offer significant savings as well as greater comfort.

While attic insulation might not seem to be an example of a disruptive technology, an energy efficient building envelope is certainly a critical component. Nationally, approximately 40% of homes have levels of attic insulation that do not meet current regional and national standards for design. Home insulation is quantified in terms of R-value, with a higher number reflecting better insulation levels that minimize the transfer of heat. The attic insulation level recommended by the Department of Energy for Florida is between R-30 and R-60 (EnergyStar 2016). A major Florida electric utility still recommends R-19 as the proper level for a home built in Florida. It is easy to see why better attic insulation could be seen as a disruptive technology by that utilty.

When it comes to the issue of windows, only 25% of the residential windows in Florida are double pane (EIA 2016). Single frame, aluminum frame windows still make up the majority of window systems in Florida homes. While nearly all of new construction incorporates double pane windows, many with even more advanced selective coatings, replacing existing windows with newer, more efficient designs is costly. However, as already mentioned, PACE programs can provide the necessary funds to speed this transition.

Lights are another area where updates are needed. Most homes built in Florida incorporate linear fluorescent lights in common areas (e.g., the garage, kitchen, laundry and bathrooms). Manufacturers of lighting products now produce linear LEDs that are drop-in replacements for typical 4 ft fluorescent fixtures. LED lamps use between 35 and 50% less electricity and the price is dropping dramatically. While existing fluorescent and incandescent technologies are as efficient as they can be, current LED technology is predicted to become almost three times as efficient as they are currently.

And finally, another area faced with a new, disruptive technology is electric water heating. Just as a heat pump for a home transfers heat from outside to inside when in heating mode, a heat pump water heater can do the same. Depending on the location and other home characteristics, a heat pump water heater can also remove heat from a garage or utility room and transfer it to a water heater tank. As a result, a home actually gets a small amount of additional air conditioning. While this "waste" cooling can be detrimental to households during the winter in heating-dominant climates, Floridians are fortunate to benefit from this heat pump water heater side effect for the vast majority of our year. The current generation of heat pump water heater units are typically between two to three times more efficient than typical electric water heaters.

Table 5.1 shows that more efficient states and countries can have robust economies and while still dramatically reducing their energy use. By making the state's electricity consumption as efficient as other top ranked states and countries, Florida can address significant decreases in the consumption of both natural gas and coal.

Before the subject of renewable energy transformation can be explored, a few benchmarks related to an energy-efficient economy can be established. The reasons for which are two-fold. An energy efficient economy is a more cost effective economy. Think of energy costs as parasitic, or bottom line costs. Retail establishments must turn the lights, computers and air conditioning systems must go on before any sales can take place. Lower utility bills can translate into greater profits. The same can be said for residences. As they become more efficient, residents have more money to spend on other items. The second reason is that an energy-efficient economy can be transformed easier and faster into a renewable energy economy.

As was mentioned earlier in this chapter, 52% of Florida's electricity is consumed in the residential sector and 41% in the commercial sector. Using the list of trends and upgrades already described, our built environment, encompassing both the residential and commercial sectors can be made twice as efficient as they are currently. This would result in Florida's electricity consumption dropping from 237 million MWhr/yr to approximately 127 million MWhr/yr.

Another way to arrive at a similar conclusion would be to establish energy efficiency goals similar to what California has done. On a per capita basis, California consumes just at 5,000 kWhr/year for each resident. Using the same rate, Florida would use just about 103 million MWhr/yr. This is significantly less than if the goal was to make the built environment twice as efficient.

Disruptive Technologies in Transportation

What about oil, the third fossil fuel? As stated earlier, almost all of the oil consumed in Florida is used for transportation. So where are the ideas for energy efficiency and disruptive technologies in relation to our transportation systems? Following are several major trends in transportation that will positively affect Florida's transportation energy consumption in both the near and long term:

- The US automobile fleet average miles per gallon is required to be 54.5 mpg by 2025, a rate that is more than twice as efficient as today.
- The move toward electric cars and plug-in hybrids is growing rapidly.
- Mass transit usage continues to climb, and more cities are adding mass transit systems around the country.
- Driverless cars and companies such as Uber and Lyft have demonstrated the potential benefits of these technologies and new business models penetrating the marketplace. Cars sit idle in parking spaces 95% of the time. Decreasing this to just 90% translates into a need for just half as many cars. The use of rideshare services will eliminate the need for parking spaces. Clients can choose to ride alone or decrease costs by traveling with more than one passenger. Traffic jams will become just a memory.

- New heavy duty 18 wheeler truck models have been developed that provide twice the miles per gallon as previous models.
- Railroads will compete for more road cargo with fewer rail cars being loaded with coal.

Taken in total, all of these trends will transform the use of oil in the US. Some of these changes will happen faster than others, but the end effect will be that the age of oil as we know it will start to end. The US consumed nearly 22.2 million barrels per day at its peak in late 2005, but despite adding approximately 25 million people since then, this level has not been topped. In part this is because by the end of 2016, the US automobile fleet will include more than 550,000 electric vehicles, including plug-in hybrids and total electric models (Inside EVs 2016). Lately, annual growth rates have been almost 30% and they are expected to increase as prices drop and capabilities increase. In less than ten years, the US market will have close to ten million electric cars on the road, remaining gasoline car models will have much better mileage figures, and driverless car services will allow many to forgo automobile ownership altogether.

How will these trends of electrification and efficiency affect our state? Florida drivers travel a total of just over 200 billion miles per year (EIA 2016). Converting half of these miles to electric cars that use 0.25 kWh/mile requires the production of 25 million MWh/yr, which represents about 11% of Florida's current electricity production. At the same, doubling the efficiency of the remaining automobiles reduces the remaining gasoline consumption by half. All in all, this yields a 75% reduction in oil use for automobiles in Florida.

Similarly, upgrades to truck fleets with new (existing) designs that can double the fuel efficiency and moving a portion of truck cargo back to the rails will also yield a 75% reduction in diesel fuel use. However, fuel use by the railroads will not increase because they will no longer be transporting coal, thereby saving about half of their current fuel consumption. And finally, many city truck systems including delivery trucks, buses and garbage trucks will convert to electric powered vehicles.

Renewable Energy Systems

Many areas in the United States have ready access to multiple sources of renewable energy. California is the site of the largest geothermal power plant on earth. The West Texas plains have the largest collection of wind turbines in the US. The hydroelectric power plants on the Columbia and Colorado rivers are among the largest in the world. And, California is home to several of the largest photovoltaic power plants on the globe.

The state's lack of steady, powerful wind patterns leaves wind power currently off the table in Florida, and a lack of elevation means that our state is a poor site for hydroelectric power plants. Florida's geothermal potential is also quite limited. Since there are some sources of waste biomass available within the state associated with forestry operations, these could be important in several areas of the state but will not be a major player based on cost alone. But there is one viable renewable energy source that is more than well-suited to Florida—solar energy. It would appear that photovoltaic (PV) power (commonly referred to as solar power) could be the main source of renewable energy for Florida's future. It involves the direct conversion of sunlight into electricity and offers many advantages:

- The direct conversion from sunlight to electricity requires no thermodynamic cycle as is required in all combustion and nuclear power plants. Therefore, no water source is needed for cooling.
- PV panels have no moving parts and are much simpler to maintain. They typically enjoy tremendous longevity; in fact, some of the original retail market panels made by ARCO Solar more than 40 years ago are still producing power.
- The price curve for solar panels has generally followed the semiconductor industry pattern that results in steady price drops as production grows. Prices were $76.67/watt in 1977 and are currently in the $0.35 to $0.37/watt range (Energy Trend PV 2016), and prices are still dropping.
- Solar panels are made from readily available materials with no foreseeable bottlenecks.
- Electric power produced by the very latest systems can do so at rates that are cheaper than all other sources, renewable or fossil-fueled (Lazard 2015).

The only disadvantage that PV power systems present is that they can only operate when the sun is out. They are not unpredictable or intermittent, at least not when aggregated over a region. Regional weather forecasts can provide utilities with meaningful predictions hours or even days in advance. That said, PV systems do have the lowest capacity factor of any grid-level system currently in use today. The capacity factor is defined as that portion of a year that a given power plant operates at its full rated capacity. For example, a 100 MW biomass power plant with a capacity factor of 0.7 would be able to operate at rated capacity for 6,132 hours per year and produce 613,200 MWhr during a year. In the same manner, a 100MW PV power plant located in Florida would be expected to operate with a capacity factor of 0.21, or 1,840 hours per year at rated capacity and produce about 184,000 MWhr/yr.

This capacity factor analysis does not mean that a given power plant only operates at 100% capacity for the defined number of hours each year. Especially with regards to PV plants, the production of power begins as soon as the early morning sun hits the solar panels and continues through late afternoon or early evening. Power produced both early and late is a fraction of full output, but the long-term averages result in a capacity factor for each region of the globe. This basic analysis can be used to estimate how much PV capacity needs to be added to transform Florida's energy systems.

Renewable Energy Systems Costs

One of the biggest hurdles to adopting renewable energy systems is the issue of cost. A popular way to look at costs is to quote the price on a per watt basis for a panel. However, it is more important to quote complete system prices on a per watt basis. There are three distinct markets for PV systems in the U.S.: utility, commercial, and residential.

Table 5.2. PV System Costs by Market (Solar Energy Industries Association 2016).

System scale	System Cost, $/Watt, Q1 2016
Utility, fixed axis	1.24
Utility, one axis tracking	1.41
Commercial	1.90
Residential	3.21

These capital costs can be used to estimate the complete system costs on an average, national basis. But what about the wholesale cost of delivered power in comparison to current fossil-fueled power plants? A series of annual reports have been produced by the Lazard Company, an international business firm, analyzing the unsubsidized costs of virtually all power-producing systems currently on the market. The reports compare each system based on the levelized cost of electricity (LCOE) in terms of US dollars per megawatt-hour produced ($/MWhr). For example, a cost of $43/MWhr would be equivalent to $0.043/kWhr, remembering that this represents the wholesale cost of power and not what individual consumers pay on their utility bill. The major points in the report (Lazard 2015) are as follows:

1. Energy efficiency upgrades are the best investment across the entire spectrum of electricity sources, but in some cases are beat out by the very cheapest PV and wind systems.
2. Utility-scale crystalline and thin-film PV systems using 2017 pricing models produce electric power cheaper than all fossil-fueled and nuclear electric systems.
3. Onshore wind power systems are cheaper than all fossil-fueled and nuclear powered electric systems.
4. Biomass and geothermal electric power systems are competitive, but they deliver power more expensively than the cheapest coal-fired and natural gas combined cycle systems.

The Lazard report presents a snapshot in time that puts a very positive spin on the future of renewable energy. However, it can't be emphasized enough that the cost structure for PV and wind power is still falling. The same cannot be said for coal, natural gas, and nuclear power systems. The cheapest of these, combined cycle natural gas systems, is very dependent on continued low prices for natural gas. The combination of US natural gas producers beginning to export liquefied natural gas and all-time lows for active natural gas drilling rigs make any prediction of continued low natural gas prices risky.

Renewable Energy Storage Systems

Exciting press releases from companies such as Tesla and Sonnen means that people often think of energy storage only in terms of batteries, and typically the next step in this thought process is to think of price, as in dollars per kWh stored. However, there is a bigger picture to be explored first. What consumers really want is increased capability to use energy that is created by various natural, renewable systems, but often the renewable energy is not collected when it is needed for use. Two examples would be 1) wind energy that is produced in the middle of the night in the Great Plains region of the US and 2) solar energy collected in the middle of the day.

The second largest use of electricity in US residences is for the production of hot water. Modern electric water heaters store 40 gallons or more of heated water and are surrounded by an excellent foamed-in-place insulation product that will keep water in a tank hot for hours. More than 90% of the water heaters in Florida are electric, as opposed to natural gas or propane. So based on an estimate of 2.4 people per Florida household and 90% electric water heaters, Florida has approximately 7.6 million water heaters. Programming each of these to be heated 10 °F each day by solar energy is equivalent to deploying 1,860 MW of storage capability that can be operated for four hours, which is equal to a total storage capacity of about 7,440 MWhr. This is significant in that it represents more deployed storage in Florida alone than what is expected for the entire US battery storage market by 2020. The only additional technology required for this type of move toward solar power is a Wi-Fi-connected or programmable water heater, and these are available in the marketplace right now. With Florida's naturally occurring "hard water," the lifetime of most electric water heaters is about ten years. So with suitable incentives, Florida could convert its entire residential water heating system within a decade.

There are several other trends that will also contribute to a vastly different grid load profile in the future. Just a few are listed here:
- Replacement of incandescent lights with LEDs will dramatically flatten the evening grid load peak.
- Emphasis on adequate attic insulation will also contribute to a lower evening peak load.
- Programmable dishwashers that will turn on in the middle of the day when solar energy is available.
- Charging electric cars during the day when renewable solar energy is available.

Renewable Energy Transformation

Unfortunately, Florida is not even ranked in the top ten states when looking at states' renewable electricity production. In fact, Florida is ranked 48th (EIA 2016), with only 2.2% of its electricity produced from renewable energy sources. Six states produce more than 50% of their electricity from renewable energy sources, and ten additional states produce from approximately 20% to 40% of their electricity from renewable energy sources. These examples within our own country show that this transformation can be accomplished. While there are a number of promising renewable energy technologies, the focus in Florida is mainly on solar or PV electric power. As mentioned already, Florida winds speeds, although high at times, do not have long term averages great enough to support wind power with current technology. When the offshore wind power market expands in the US and wind energy companies have access to turbines designed for lower wind mean conditions, this will change and wind power will become a viable alternative in Florida. Florida is also not a hot spot for geothermal electricity production. So what will it take to transform Florida's current energy systems into a more renewable future?

Let's first take a look at the status of PV power in the US and the world as a whole. Fig. 5.2 shows the history of the cumulative capacity of solar power in the US from 200-2016 and Fig. 5.3 shows a similar history for the entire world. The first thing that stands out is that the growth rates have been tremendous. Growth curves such as these are similar to or even greater than the market penetration rates seen in the case of automobiles, radios, color television, and smart phones. As prices continue to drop and more markets are opened, there appears to be no national or international barriers for continued expansion of this promising energy source.

Florida's lack of progress in the renewable energy arena stems from a number of factors. First, Florida is one of a few states that presently does not have a legislative or voluntary renewable portfolio standards. These state-level standards require utilities to implement renewable energy.

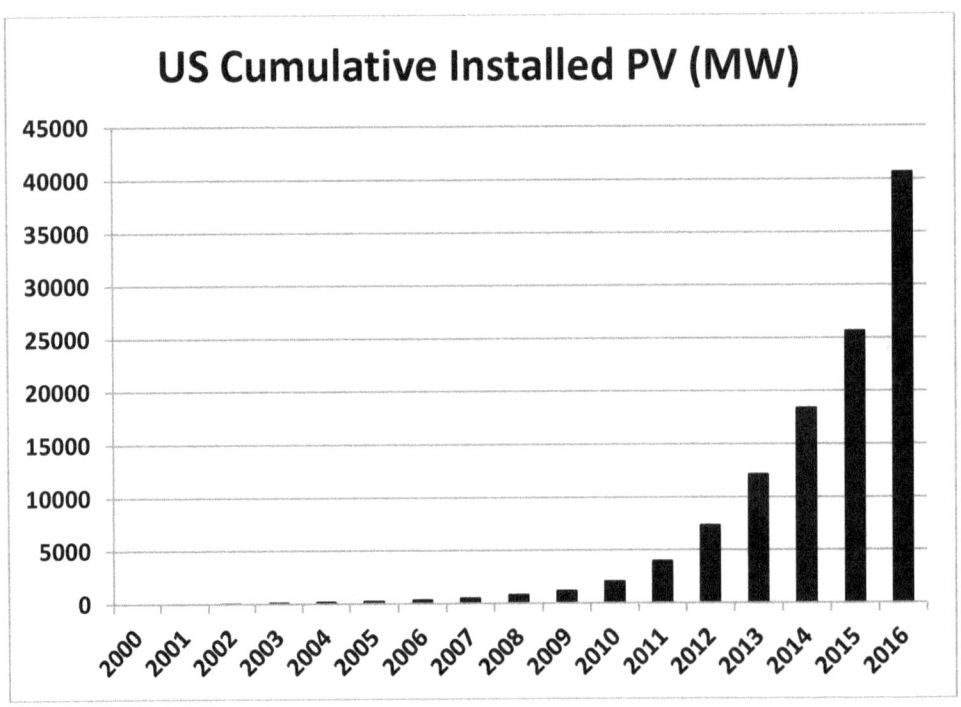

Figure 5.2. US cumulative installed PV capacity, adapted from Solar Energy Industries Association (2016).

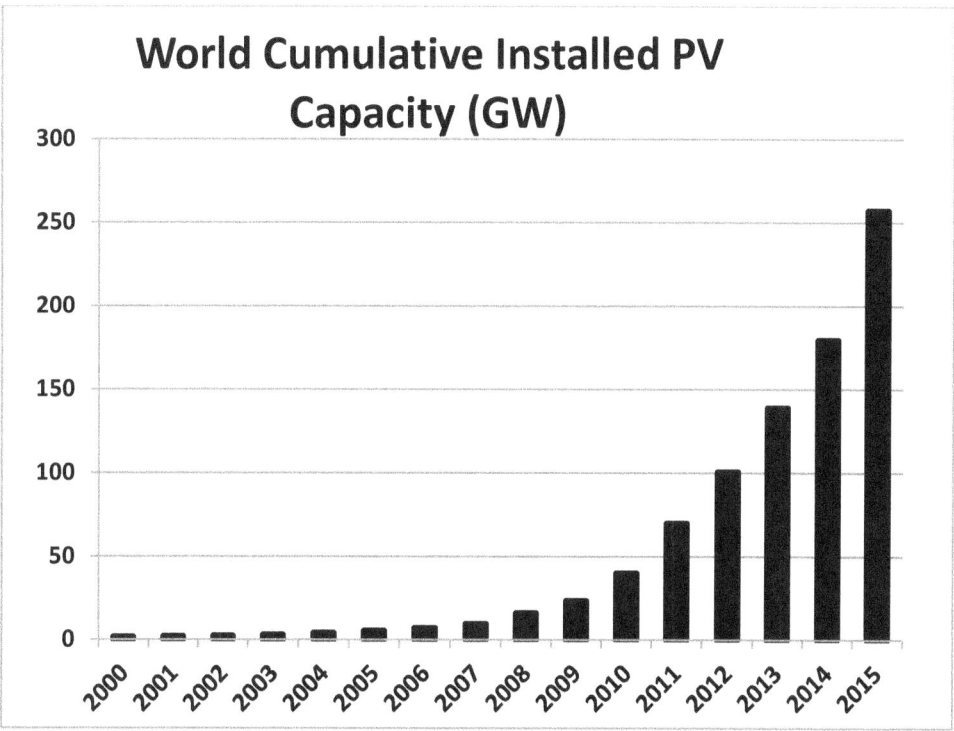

Figure 5.3. World cumulative installed PV capacity, adapted from Solar Energy Industries Association (2016).

For most renewable portfolio standards, the amounts and types are quite flexible, as are the time frames for attainment. For example, California's renewable portfolio standards requires the state to produce 50% of its electricity from renewable sources by the year 2030. Texas requires 10,000 MW of renewable electricity capacity be installed by 2025. Adoption of renewable portfolio standards is quite effective in that they produce a demand for a market, yet remain fairly flexible so that different regions can follow the paths that are most suitable for that geographic area. So, back to the question of what it will take to transform Florida.

A summary of the current status and the trends described in this chapter will provide an answer.

- Florida currently consumes 237 million MWh/yr of electricity, 52% in the residential sector and 41% in the commercial sector
- Doubling the efficiency of our built environment allows the state to better compete in the national and international market places. This also brings down the need for electricity in Florida to 122 million MWhr/yr.
- Florida currently produces nearly 30 million MWhr/yr from four nuclear reactors located at the Turkey Point Nuclear Generating Station and the St. Lucie Nuclear Power Plant. It can be assumed that, for near future of ten years, this will not change.
- The combination of biomass power plants, landfill gas power plants, and a small amount of remaining natural gas power plants can produce 20 million MWhr/yr.

Totally eliminating the use of coal to produce electricity in Florida and reducing the use of natural gas by about 90% would result in the need to produce 72 million MWhr/yr from renewable energy sources. With an overall capacity factor in Florida of 0.21, the state would have to add 39,000 MW of PV systems in order to reach this goal. Furthermore, producing the additional amount of electricity needed to charge electric vehicles for half of the existing fleet will require an additional 13,600 MW for a total capacity needed of 52,600 MW. While this may seem daunting, spread over a ten-year time span, just over 5,000 MW would have to be installed each year. This number is only slightly more than what California will have installed in 2016 in a market that is still growing at double digit rates each year.

Conclusion

This chapter has presented three broad areas of thought concerning Florida's energy systems and their relation to climate change. Every attempt has been made to use the most up-to-date information in order to provide an accurate picture of the positive and negative aspects of the transformation that we face as a state, a nation, and world.

- The current status of fossil-fueled power and transportation systems and consumption patterns was described. Comparisons were made between Florida, other states and nations to illustrate the relative inefficiency of Florida's energy landscape.
- Florida's current energy systems, such as power plants and coastal fuel terminals, were discussed in relation to sea level rise. While estimates of three feet of sea level rise by the year 2100 are well established in the scientific literature, the most recent research concerning the West Antarctic Ice Shelf suggests that planning for twice that, or six feet of sea level rise, might be necessary. Decommissioning a nuclear power plant can take ten or more years, which means that decisions about what to do with the reactors at Turkey Point, Crystal River, and St. Lucie must be put at the top of our state's to-do list. This issue of time needed to transition is not just related to nuclear power plants. Most of the oil products used in Florida are delivered via ship and barge, and coastal oil terminals will have to be moved if we continue to insist on an oil-powered transportation system.
- In comparison to other nations and states, Florida is considerably underachieving and lagging in both energy efficiency and renewable energy implementation. Proceeding down a more efficient path would make the transformation from a fossil-fueled economy to a renewable-powered one much faster and less expensive. The battery storage market does not have to be fully developed before the state can incorporate large amounts of renewable electricity with lower capacity factors into the grid. Solar power and wind energy are now the cost leaders for electric power, and their annual market growth rates reflect that.

How then does Florida rise from its 48th position for production of renewable energy electricity to a top five national ranking more representative of its position as the third most populous state? At the policy level, there are several successful strategies being used in other states that could be transferable to Florida. The renewable portfolio standards policy adopted by many other US states has led to innovations and cost reductions in the renewable energy marketplace. Recent open electric power auctions in the US and worldwide have resulted in the provision of solar PV and wind at the least expense, even considering no subsidies. Florida needs to adopt a significant renewable portfolio standard immediately, one that would challenge the state to produce at least 20% more of its electricity from renewable sources by 2025.

Florida could also create standards and policies to ensure that the state's built environment keeps up with a changing world. California created a Long-Term Energy Efficiency Strategic Plan (http://www.cpuc.ca.gov/General.aspx?id=4125) in September 2008 that established zero-net energy use goals for residential buildings by 2020 and for commercial buildings by 20430. These goals were motivated by the global Architecture 2030 Challenge (http://architecture2030.org/2030_challenges/2030-challenge/), which could provide an internationally-inspired platform for Florida's energy transition and climate change mitigation strategies, as well. The dual objectives of creating a more efficient built environment and large-scale deployment of solar PV energy capabilities would allow Florida to stop using coal and nearly end the use of natural gas to meet its energy needs.

Even more significantly, Florida could open up access to the electrical grid to additional providers, much like what Texas has done. Purchase power agreements can be signed between renewable energy providers and businesses or municipalities, and these arrangements often result in lower utility bills or lower rates of increases.

Moving away from oil for Florida's transportation needs would be fairly straightforward: Whether it is personal automobiles or the trucking industry, Florida could double the efficiency of existing systems by moving half of passenger vehicle miles to electric vehicles and half of the state's truck freight back to railroads. At the same time, fuel efficiencies for both modes will double. By simple ratios, this represents a 75% reduction in oil use in the state of Florida.

Rising seas will force Florida to confront our legacy fossil fuel energy systems much sooner than anticipated. Energy efficiency and renewable energy trends will rapidly shrink the need for coal and natural gas and begin a massive transformation of transportation systems. Florida's unique geographic location and history as an ever-shifting peninsula separating the hurricane-prone Atlantic Ocean and Gulf of Mexico make it a flashpoint in terms of facing the challenges of an already changing climate. So, shouldn't Florida also serve as a beacon for the 21st century cultural changes necessary to foster more resilient and adaptive communities from coast to coast? Promoting the policies and practices critical to energy transition and carbon-neutral urban renewal represent a promising first step. There is no more time left to delay. The future will not be forced, but rather chosen. May Floridians choose wisely.

Acknowledgements

Both authors would like to thank their spouses, their children, the students they've had the pleasure to teach, and all of those yet to be taught. You are our hope for a better future.

References

Database of State Incentives for Renewables and Efficiency. 2016. "DSIRE." Accessed October 4. http://www.dsireusa.org/

DeConto, Robert M., and David Pollard. 2016. "Contribution of Antarctica to past and future sea-level rise." *Nature International weekly journal of science* 531:591-597. Accessed September 30, 2016. doi:10.1038/nature17145.

Energy Trend PV. 2016. "PV Spot Prices." Accessed September 30. http://pv.energytrend.com/pricequotes.html

EnergyStar. 2016. "Recommended Home Insulation R-Values." Accessed September 28. https://www.energystar.gov/index.cfm?c=home_sealing.hm_improvement_insulation_table

Environmental Protection Agency. 2016 "Global Greenhouse Gas Emissions Data." Accessed September 30. https://www.epa.gov/ghgemissions/global-greenhouse-gas-emissions-data

Google Maps. 2016. Accessed Sept 30. https://www.google.com/maps/@25.4468079,-80.347613,15941m/data=!3m1!1e3

Inside EV's. 2016. "Monthly Plug-In Sales Scorecard." Accessed September 30. http://insideevs.com/monthly-plug-in-sales-scorecard/

Lazard, 2015. "Lazard's Levelized Cost of Electricity, Version 9.0." Accessed September 18. https://www.lazard.com/perspective/levelized-cost-of-energy-analysis-90/

My Florida Energy Home. 2016. "My Florida Energy Home – Financing." Accessed October 4. http://www.myfloridahomeenergy.com/help/library/financing-incentives/financing/

My Florida Energy Home. 2016. "My Florida Energy Home – Financing." Accessed October 4. http://www.myfloridahomeenergy.com/help/library/financing-incentives/property-assessed-financing/

My Florida Energy Home. 2016. "My Florida Energy Home." Accessed September 29. http://www.myfloridahomeenergy.com/

My Florida Energy Home. 2016. "My Florida Energy Home." Accessed September 29. http://www.myfloridahomeenergy.com/help/library/hvac/duct/#sthash.OCg233qS.dpbs

My Florida Energy Home. 2016. "My Florida Energy Home-Incentives." Accessed October 4. http://www.myfloridahomeenergy.com/help/library/financing-incentives/incentives/

Rahmstorf, S., Perrette, M. & Vermeer, M. 2012. "Testing the robustness of semi-empirical sea level projections." *Climate Dynamics*, 39:861. Accessed September 30, 2016. doi: 10:1007/s00382-011-1226-7.

Solar Energy Industries Association. 2016. "Solar Market Insight Report 2016 Q2." Accessed September 19. http://www.seia.org/research-resources/solar-market-insight-report-2016-q2

Surging Seas. 2016. "Risk Zone Map." Accessed September 30. http://sealevel.climatecentral.org/

Surging Seas. 2016. "Risk Zone Map." Accessed September 30. http://ss2.climatecentral.org/#12/25.4447/-80.3245?show=satellite&projections=0-RCP85-SLR&level=3&unit=feet&pois=hide

Surging Seas. 2016. "Risk Zone Map." Accessed September 30. http://ss2.climatecentral.org/#12/25.4447/-80.3245?show=satellite&projections=0-RCP85-SLR&level=6&unit=feet&pois=hide

Surging Seas. 2016. "Risk Zone Map." Accessed September 30. http://ss2.climatecentral.org/#15/27.3436/-80.2403?show=satellite&projections=0-RCP85-SLR&level=1&unit=feet&pois=hide

Surging Seas. 2016. "Risk Zone Map." Accessed September 30. http://ss2.climatecentral.org/#13/28.9542/-82.6692?show=satellite&projections=0-RCP85-SLR&level=3&unit=feet&pois=hide

Surging Seas. 2016. "Risk Zone Map." Accessed September 30. http://ss2.climatecentral.org/#16/27.8570/-82.5397?show=satellite&projections=0-RCP85-SLR&level=3&unit=feet&pois=hide

The World Bank. 2013. "Which Coastal Cities Are at Highest Risk of Damaging Floods? New Study Crunches the Numbers." August 19th, 2013. http://www.worldbank.org/en/news/feature/2013/08/19/coastal-cities-at-highest-risk-floods

U.S. Energy Information Administration. 2016. "Electric Power Monthly, February 2016." Accessed and adapted September 18. http://www.eia.gov/electricity/monthly/current_year/february2016.pdf

U.S. Energy Information Administration. 2016. "Independent Statistics & Analysis, Petroleum & Other Liquids." Accessed and adapted September 30.
http://www.eia.gov/dnav/pet/pet_cons_prim_dcu_SFL_a.htm

U.S. Energy Information Administration. 2016. "Independent Statistics & Analysis, Florida Electricity Profile." Accessed and adapted Sept 30. https://www.eia.gov/electricity/state/florida/

U.S. Energy Information Administration. 2016. "Independent Statistics & Analysis, Florida Electricity Profile." Accessed and adapted Sept 30. https://www.eia.gov/electricity/state/florida/

U.S. Energy Information Administration. 2016. "Independent Statistics & Analysis, Profile Data." Accessed and adapted September 30. http://www.eia.gov/state/data.cfm?sid=FL#EnergyIndicators

U.S. Energy Information Administration. 2016. "Independent Statistics & Analysis, Electric Power Monthly." Accessed and adapted, September 30.
http://www.eia.gov/electricity/monthly/epm_table_grapher.cfm?t=epmt_5_1

U.S. Energy Information Administration. 2016. "Independent Statistics & Analysis, Electricity, State Electricity Profiles." Accessed and adapted, September 30.
https://www.eia.gov/electricity/state/florida/

U.S. Energy Information Administration. 2016. "Independent Statistics & Analysis, Electric Power Monthly." Accessed and adapted, September 30.
https://www.eia.gov/electricity/monthly/epm_table_grapher.cfm?t=epmt_1_1

U.S. Energy Information Administration. 2016. "Independent Statistics & Analysis, Electric Power Monthly." Accessed and adapted, September 30. U.S. Energy Information"
http://www.eia.gov/electricity/sales_revenue_price/pdf/table5_a.pdf

U.S. Energy Information Administration. 2016. "International Energy Statistics." Accessed and adapted September 21. http://www.eia.gov/cfapps/ipdbproject/IEDIndex3.cfm

U.S. Energy Information Administration. 2016. "U.S. States." Accessed and adapted September 19. http://www.eia.gov/state/data.cfm?sid=FL

U.S. Energy Information Administration. 2016. "U.S. States." Accessed Sept 30.
http://www.eia.gov/state/data.cfm?sid=FL#ConsumptionExpenditures

U.S. Energy Information Administration. 2016. "U.S. States." Accessed and adapted September 19. https://www.eia.gov/state/data.cfm?sid=FL

U.S. Energy Information Administration. 2016. Accessed September 30, 2016.
https://www.eia.gov/consumption/residential/reports/2009/state_briefs/pdf/fl.pdf

CHAPTER 6

Climate Change Impacts on Insurance in Florida

Lorilee Medders[1]

[1]*Department of Finance, Banking and Insurance, Appalachian State University, Boone, NC*

Climate change presents added risks as well as related opportunities for the insurance industry and financial sector. Implications must be evaluated for property, casualty and life insurance industry segments as well as for the financial sector more broadly. While climate change exacerbates the existing volatility of these markets, it also inherently creates opportunities for product development. Florida is a unique contributor to both the risk and opportunity since the state is the world's largest insured catastrophe region. The state of Florida itself is heavily leveraged as insurer for much of the cost of extreme weather in the form of hurricanes and other tropical storms. Unlike other insurance risk bearers, however, this state cost of risk cannot be offset by commensurate market opportunity. Increased volatility in insurance, reinsurance, and capital markets are all challenges for Florida, with potentially adverse collateral effects on residual insurance market pressures, policyholder assessments, state debt, and tax strategies. Insurance industry initiatives, to the extent they are successful, can have a balancing effect on these challenges.

Key Messages

- The potential for substantial changes in the climate make the risk assessment, underwriting, and pricing of insurance and insurance-linked securities more complex.
- Catastrophe loss models incorporate scientific assumptions about climate change into the risk assessment capabilities for multiple disaster perils, including hurricane, flood, and wildfire, among others.
- Insurers and reinsurers are at the forefront of research regarding the effects of climate and climate change on future loss costs, loss uncertainty, and opportunity.
- Florida is and will continue to be one of the world's largest insured, catastrophe markets, and as such is highly vulnerable to changes in the climate as well as changes in the markets for protecting against financial risk.
- It is especially important for Florida and Floridians to implement smart ways of adapting to climate change and its effects in order to protect our economic sustainability.

Keywords

Disaster risk; Catastrophes; Insurance; Reinsurance; Economic loss; Risk reduction; Insurability

Introduction

During recent decades, natural catastrophes as well as man-made disasters have posed a rising threat to societies and the world economy. In 2015 alone, there were 353 recorded catastrophe events worldwide, resulting in estimated total economic losses of $92 billion (Swiss Re 2016) Of these events, 198 were deemed natural catastrophes, the most ever recorded in a single year, with estimated overall losses of $80 billion (Swiss Re 2016). A continuation of changes in environmental factors is expected to have an upward effect on the magnitude as well as on the geography of future loss events. The private (re)insurance sector recognizes climate change as a factor in strategic planning for the future of the business as major reinsurers, insurers, and industry consortiums are placing substantial resources on research and development to respond appropriately to the emerging changes in risk.

This chapter provides an overview of the possible insurance impacts of such changes. The implications for insurance and more specifically insurance in Florida are potentially far-reaching and are subject to several important, but often little understood, aspects of the marketplace for risk and Florida's place in it.

Beginning with Hurricane Andrew in 1992, Florida has been at the forefront of discussions related to natural catastrophes and insurance markets. Florida's geographic, or physical, exposure to catastrophic weather, especially in the form of tropical windstorms, is clear and is discussed thoroughly in other chapters. Socioeconomic factors directly influence Florida's level of economic losses resulting from catastrophic events. Florida's increasing overall population, the increasing migration of its population to coastal areas, and the rise in total insured property values at risk in these areas combine to substantially increase Florida's concentration of insurance exposure to catastrophes.

The population of Florida in 2015 is estimated at approximately 20 million, according to the U.S. Census Bureau, making it the third most populated state in the nation behind California and Texas. This represents a drastic increase since 1950 when there were a mere 2.8 million inhabitants. Even just since 1980, Florida's overall population has doubled. Furthermore, while the nation's 673 coastal counties make up only 17% of the U.S. land area they account for 55% of the nation's population (AIR Worldwide 2013; U.S. Census Bureau 2010). Florida serves to highlight this trend, with 61 of its 67 counties listed as coastal by the National Oceanic and Atmospheric Administration (NOAA), comprising over 75% of the state's total population (NOAA 2013).

Exacerbating the risk, Florida construction values have risen sharply during this period of time as well (Florida Catastrophic Storm Risk Management Center 2011). A portion of this rise is easily explained by the need for housing stock as the population has risen. But two additional factors have contributed to the rise. Individual property values have risen, even after adjusting for inflation (Florida Catastrophic Storm Risk Management Center 2011). Land values have risen as populations (and thus demand for land) have risen. And newer homeowners who either ignore the disaster risk, or can affordably insure against it, have built homes more expensive to construct than the typical Florida home of the 1950s and 1960s (Florida Catastrophic Storm Risk Management Center).

Businesses and commercial real estate have flocked to Florida since the 1960s as the state grew as a tourism economy with low taxes. Florida also has a high density of property insurance coverage, with most houses protected against windstorm losses and about one-third insured against floods (Florida Catastrophic Storm Risk Management Center 2011). Indeed, it was estimated that in 2012 nearly 80% of insured real estate assets in Florida were located in coastal counties (AIR Worldwide 2013). This represents $2.86 trillion of insured residential and commercial exposure located in Florida coastal areas. The exposure of Florida to natural hazards, particularly tropical storms, along with the state's high level of insurance penetration combine to make Florida arguably the world's greatest insured natural catastrophe region.

Not surprisingly, and for the reasons given, the cost to insurers of Florida weather catastrophes has risen far faster than inflation during the past 30 years (Kunreuther et. al. 2012). These factors will continue to have a major impact on the level of insured losses from natural catastrophes. Given the growing concentration of exposure on both the Atlantic and Gulf coasts of the state, future disaster events would be likely to inflict significant property damage and business interruption losses on Florida even without changing environmental conditions. And Florida's insurers can be expected to foot much of the bill.

Assessing the potential impacts of climate change on insurance is complex, and depends on the temporal and spatial scales over which one is concerned, the entities of interest, judgment criteria, and the desired level of certainty. Moreover, climate change—given its potential for systemic impact—can dramatically alter the risk management landscape more broadly than just insurance. We are challenged to understand climate change in a world where all climates contain a spectrum of extremes and related catastrophes. This spectrum can be quite broad with no easily discernable trends (Muir-Wood 2016). It is in such a world the insurance sector attempts to find long-term profitability (private sector) and financial stability (public sector).

This chapter addresses potential impacts of climate change on insurance within the framework of how healthy insurance markets function. The first section is comprised of a broad conversation on the insurance and financial markets, their functions and roles, as well as their financial volatility and problems. Second is a discussion of why and how governments serve as public insurers or as sponsors of insuring entities. In the third section, the chapter addresses the assessment of the risk, both from an insurance standpoint and a climate change perspective. Fourth, insurance underwriting and pricing are examined, followed by a fifth section on the potential financial impacts of climate changes. The sixth and last section examines recent and current efforts—both public and private—to abate the financial impacts of potential climate changes.

The Insurance and Financial Markets

The American insurance market is generally robust, competitive, and innovative. Citizens expect insurance to be available and affordable for a wide range of risks. Floridians are no exception. One could argue Floridians have higher than average expectations of their insurance. It has been asserted the growth of Florida since 1960 is attributable primarily to two factors: air conditioning and affordable property insurance (Florida Catastrophic Storm Risk Management Center 2011). Whether the statement is true, it is undeniable that Floridians have come to depend on both for quality of life purposes. The dependence on affordable insurance is a factor that takes on increasing pressure as insured losses increase. It is important then to discuss how this market for risk financing works in order to understand how it may be impacted by climate change.

Functioning Insurance Markets

Private insurance markets function well only if necessary conditions are met. Private insurers generally only insure pure risks, meaning risks that do not include any speculative "upside outcome" for the person or entity facing the risk (i.e., risks wherein only loss or status quo are possible). Beyond this first condition, however, six additional characteristics of a risk are critical to making it ideally insurable in the private market (Rejda 2011).

First, there should be a large number of roughly similar (although not necessarily identical) individual risk exposures subject to the same peril or group of perils. For instance, a large number of masonry dwellings subject to hurricane exposure can be grouped together (pooled) for purposes of providing property insurance on them. This enables an insurer to predict losses, with reasonable accuracy, based on the law of large numbers and to spread loss costs over all insureds in the group.

Second, losses should be accidental, or unintentional, in nature, so that the insured cannot affect the probability that a loss will occur. The law of large numbers is based on the random occurrence of events. Prediction of future loss experience may be highly inaccurate if intentional, or otherwise nonrandom, losses occur.

Third, losses should be determinable and measurable, meaning the loss should be definite as to cause, time, place and amount. The basic purpose of this requirement is to enable an insurer to determine if a loss is covered, and if it is, how much should be paid.

Fourth, losses should not be catastrophic. This means that a large portion of exposed units should not incur losses at the same time. Although insurers ideally wish to avoid all catastrophic losses, practically most natural catastrophes are covered and reinsurance is purchased to limit the insurer's catastrophe potential.

Fifth, the chance of loss must be calculable, both in terms of frequency and severity. Some losses are difficult to insure because the chance of loss cannot be accurately estimated, and the potential for a catastrophic loss is present.

Sixth, the premium must be economically feasible, meaning the insured must be able to afford the premium. To have an economically feasible premium, the chance of loss must be relatively low.

When individual exposures are independent and numerous, and losses are accidental in nature, they become more predictable than if these conditions are not met. An insurer can compute actuarial projections of the probability of such losses sufficient to assess the premium it must charge to be able to insure the risk. So long as the premium (based on the fair price of the risk) is not infeasible for insureds to pay, a market for insuring such risks can function well.

Insurance in theory is supposed to be an *ex ante* financial instrument, where insurers over time build up capital to use for paying damages that occur later. But today, the vast majority of property insurance policies carry 12-month terms. Thus, insurers strongly prefer risks that can be pooled over a short period of time (i.e., the law of large numbers works over multiple exposure units in one short term period rather than requiring a long term average for statistical accuracy) so that one year's premiums cover one year's losses. Furthermore, insurance pricing is optimized by competitive markets, so that each insured pays the cost of adding that insured to the risk pool and no insurer can survive in the short term by underpricing the expected true cost of the risk.

Insurance in practice has a tradition of serving as a risk averse financial instrument and particularly averse to catastrophe losses and potential. Because of this, insurers are prone to restrict contract terms as they raise prices of their products and services whenever and wherever catastrophic loss potential is estimated to exist. Regulation—self-regulation, industry regulation and public regulation—with an interest in long-term stability and availability serves to temper the propensity to behave in wholly reactionary fashion to climate extremes, catastrophic losses and climate change.

The Role of Insurance within the Financial Markets

Insurance is a specialty sector of the financial markets where, as previously mentioned, pure risks are the focus. Because these risks inherently should provide no opportunity for financial gain by the insured, profitability for the insurer is contingent on accurate information for every risk insured and the ability to convert this information into pricing that adequately and fairly represents the risk the insurer takes.

The reinsurance market serves as back stop to losses from the insurance market. Most reinsurance aggregates risks (by books of exposure units subject to similar perils) for which insurers have already obtained credible information on each individual risk (by exposure unit). Buying reinsurance, insurers can take on larger and more volatile risks than would otherwise be financial feasible as reinsurance not only spreads losses across its insurer clients (through its pricing scheme) but also infuses additional capital that indirectly increases insurers' financial capacity to bear large losses (Rejda 2011; Kunreuther et. al. 2012).

The broader capital markets play a role as well in two ways: first, these markets may participate directly in risks and losses through insurance-linked securities (ILS), which are available to large investors; and second, they may participate indirectly as investors infuse capital into publicly traded insurers and reinsurers as well as investment funds.

Volatility and Market Problems

Based on the preceding requirements of an ideally insurable risk, even with participation by reinsurers and ILS investors, private insurance markets struggle to provide sufficient coverage to meet society's needs for some risks (Klein and Kleindorfer 2003; Cummins 2006; Medders, Nyce and Karl 2013). These risks—commonly called extreme or catastrophic risks—are uninsurable through conventional insurance markets because they defy the conditions private markets require for operation. Insurance markets can face problems in providing coverage for truly large events; the size and rarity of insured events can make them difficult to predict. Losses may be intentional (as in terror attacks) or affected by trends that render losses nonrandom. The infrequency of large loss insured events may also require risk pooling across several time periods for insurers to expect to break even. In the case of catastrophic risk, private insurers may not be able to adequately address the information problems they encounter in attempts to price and underwrite. Where they are able to arrive at rates in which they have confidence, buyers may have insufficient income to afford to pay (Skipper and Kwon 2007). And financial markets can be disrupted when an extremely large insured loss occurs, complicating the rebuilding of capital after a large payout.

Markets that include high potential for catastrophic insurance industry losses are prone to a variety of market problems. Private insurers may choose to decrease market exposure, and thus decrease capacity in the highest-risk zones. For example, Cummins (2006) states, "Insurance markets tend to respond adversely to mega-catastrophes. They respond to large events ... by restricting the supply of insurance and raising the price of the limited coverage available." The Florida property insurance market experienced such market problems in 1993 after Hurricane Andrew and again in 2006, on the heels of the brutal 2004-2005 hurricane seasons. In both market years, reinsurers restricted reinsurance supply to insurers thus creating a cascading effect on property insurance availability, especially for homeowners and commercial residential property owners.

In Florida, given the size of the catastrophe risk as well as its volatility, the availability of private capital to support catastrophic windstorm exposure is contingent upon regulatory and legislative directives intended to ease market pressures and stabilize pricing. Negative externalities, and ultimately market failures, can result from insurance rate suppression (regulatory decreases in the price ceiling) within this catastrophe-prone market.[1] Public policy considerations in the state, however, have in recent years dictated suppressed pricing and subsidized coverage, rather than having each insured necessarily pay the fair cost (based on the law of large numbers) of being added to the risk pool (Medders et.al. 2013).

In a state such as Florida, where development of the built environment has revolved around population and business (particularly tourism) growth, it makes sense the political economy

---

[1] Insurance rate regulation has three purposes: ensure adequate rates, ensure non-excessive rates and ensure rates are not unfairly discriminatory. Protection against insurance rate inadequacy is critical since it reduces the likelihood that an insurer will not have sufficient funds from premiums collected to pay claims.

would will people to continue to live and work in certain geographic areas where there are high risks, however infrequent the losses may be. Thus, in addition to restricting the competitive environment within the private insurance market, the state chose to develop a residual market for property insurance and reinsurance as well, with implications for the private insurance industry.

It is important to note that climate extremes and losses due to climate change may be felt most quickly and strongly by the reinsurance sector and the public insurance markets since these are the "back stops" for exposures, risks and losses determined extreme, catastrophic or otherwise difficult to insure. It therefore makes sense to begin a Florida-centered discussion of insurance participation with its residual (public) market for insurance.

State of Florida and Federal Government as Florida "Insurers"

Based on the principles for well-functioning insurance markets, government plans would ideally use risk-based premiums in setting the price charged for each individual risk and not offer subsidies in setting the premiums on individual risks (at the least not subsidize the riskiest locations). To do otherwise is to actively induce people to put themselves in harm's way. Such subsidies also risk crowding out any potential private market initiatives. Of course, to follow this advice, the government must have the will to reject requests to provide special help to affected industries and/or regions.

A review of government programs providing coverage for extreme or catastrophic events shows that these programs do not function like insurance; where premiums are charged, they may be explicitly subsidized or set based on incomplete measures of the risks involved, resulting in an implicit subsidy (Cummins 2006; Grace and Klein 2009). There may not even be a clear statutory intent to subsidize coverage. The government's insurance commitment may extend over multiple time periods, allowing the government to recoup past losses through future premiums or other revenues. The federal government may provide back-up coverage financed not through premiums paid by insureds but through general government revenues. In short, these programs bear less resemblance to insurance than to targeted public spending or risk management programs aimed at meeting the government's responsibilities of providing economic security and economic stabilization (Kunreuther et.al. 2012).

The Florida homeowners insurance market consists of a unique combination of private and residual insurers grappling with increasing demand- and supply-side economic pressures in the face of high-density development near high-risk coastlines. Insurance for Florida's residential property insurance market includes both private insurers and several quasi-governmental property insurance mechanisms. In 1970 the Florida Windstorm Underwriting Association (FWUA) was enacted by the Florida Legislature to offer "wind only" coverage in Monroe County and the Florida Keys. The FWUA was gradually expanded to provide wind coverage in 29 of Florida's coastal counties. Since this initial attempt to provide a public policy response to catastrophic windstorm risk, three entities have evolved with expressly different purposes: Citizens Property Insurance Corporation, the Florida Hurricane Catastrophe Fund and the Florida Insurance Guaranty Association. These statewide programs are backed implicitly by the State of Florida itself, although none of the programs has explicit backing. Flood insurance has not historically been provided by the private insurance market but rather has been covered under a Federal program, the National Flood Insurance Program. (Florida Catastrophic Storm Risk Management Center 2011).

Citizens Property Insurance Corporation[2]

After Hurricane Andrew in 1992, the Florida Legislature met in a special session to address problems in the residential insurance market. Several insurers had become insolvent, and others were concerned about increased insolvency risks. The Legislature addressed the need for residential property insurance policies that provided "full" (multi-peril) coverage rather than wind-only policies offered by the FWUA. The Florida Residential Property and Casualty Joint Underwriting Association (FRPCJUA) or (JUA) was created in 1992, and later combined with the residual market mechanism that insured commercial residential or condominium and apartment buildings (the Florida Property Casualty Joint Underwriting Association).

The Florida Legislature merged the FWUA with the FRPCJUA, creating Citizens Property Insurance Corporation (Citizens) effective August 1, 2002. Citizens has three distinct accounts; the Personal Lines Account, the Commercial Lines Account, and the Coastal (formerly High-Risk) Account. Citizens has policies available to cover multiple perils (excluding flood and other perils considered uninsurable), the wind peril only, and multiple perils excluding wind.

When Citizens experiences a financial deficit due to losses, it may levy assessments. These assessments are not only against its policyholders but also against the policyholders of private insurers in almost all lines of property-casualty insurance in Florida.

The Florida Hurricane Catastrophe Fund[3]

The Florida Hurricane Catastrophe Fund (FHCF) was created by the Florida Legislature in 1993 to provide additional insurance capacity and help stabilize the property insurance market in Florida (Fla. Stat. s. 215.555(1)). The FHCF provides reimbursement for a portion of a property insurer's hurricane losses above the amount retained by the insurers. Insurers enter into contracts with the FHCF and pay a premium. The FHCF is able to accumulate premium payments on a tax-free basis as it is exempt from federal income taxation. Except for certain exemptions, all admitted insurers writing residential property insurance in Florida, including Citizens Property

---

[2] All information here was adapted from Florida Catastrophic Storm Risk Management Center, 2011.
[3] Ibid.

Insurance Corporation, are required by Section 215.555, Florida Statutes, to obtain FHCF reimbursement coverage.

In the event that the FHCF's losses exceed its surplus, the FHCF is authorized to collect assessments on policyholders in almost all lines of property-casualty insurance. The amount of coverage available from the FHCF, the cost of the coverage, and the potential assessments are significant factors in the state of the insurance market. The maximum obligation of the FHCF for a given contract year is specified by statute. The current maximum is $17 billion. Each insurer's reimbursement coverage is limited to its share of the $17 billion maximum obligation.

Because for now the FHCF coverage is for one peril only—hurricane. And because reimbursements are made for direct losses only, not for indirect losses associated with claims payments (such as attorney fees), the FHCF loss experience is a fairly simple barometer for the level of pure losses generated by one type of extreme climate event. The possibility of increases in event intensity due to climate changes means the capacity of the FHCF will continue to be under pressure to stabilize the Florida residential property insurance market.

Florida Insurance Guaranty Association[4]

The Florida Insurance Guaranty Association (FIGA) was created by the Florida Legislature in 1970 to address concerns about the adverse effects of insolvent insurers. Its specific purpose is to "provide a mechanism for the payment of covered claims under certain insurance policies to avoid excessive delay in payment and to avoid financial loss to claimants or policyholders because of the insolvency of an insurer." (Section 631.51(1), F.S.) Thus, FIGA is the state entity that pays the claims of insolvent insurers and has the ability to assess in the event of insolvencies related to catastrophic storms. FIGA does not accumulate funds in advance of an insurer's insolvency, but similar to Citizens and the FHCF, it obtains funds through pro-rata assessments levied by the Office of Insurance Regulation on insurers subject to assessment. Its use is limited primarily to protecting the state's policyholders against potential insolvencies of private insurers since the public insurers—Citizens and FHCF—as discussed above have their own respective assessment capabilities in the event of large losses.

National Flood Insurance Program

The private insurance market, through its standard homeowners policy, generally does not pay for flood losses due to the market failures concerns previously discussed. The lack of private insurance availability led to government intervention. In 1968, Congress created the National Flood Insurance Program (NFIP) to help provide a means for property owners to financially protect themselves from water loss (Kunreuther and Michel-Kerjan 2012). The NFIP offers flood insurance to homeowners, renters, and business owners if their community participates in the

---

[4] Ibid.

NFIP. Participating communities agree to adopt and enforce ordinances that meet or exceed the Federal Emergency Management Agency (FEMA) requirements to reduce the risk of flooding.

Floods are the most common and most destructive natural disaster in the United States. Ninety percent of all natural disasters involve flooding, and all 50 states have experienced floods or flash floods in the past five years (NAIC 2016). In 2014, global sea level was 2.6 inches above the 1993 average and continues to rise at a rate of about one-eighth of an inch per year (NOAA 2017). Higher sea levels mean that deadly and destructive storm surges push farther inland than they once did, which also means more frequent nuisance flooding. Nuisance flooding is estimated to be from 300 percent to 900 percent more frequent within U.S. coastal communities than it was just 50 years ago (NOAA 2017). Furthermore, rising sea levels threaten the integrity of infrastructure—roads, bridges, ports, utility plants, sewage treatment plants, etc.

Since Hurricane Katrina, the NFIP has been in a financial deficit state. Super Storm Sandy increased the deficit substantially. According to the Government Accountability Office, the NFIP "likely will not generate sufficient revenues to repay the billions of dollars borrowed from the Department of the Treasury to cover claims from the 2005 and 2012 hurricanes or potential claims related to future catastrophic losses. This lack of sufficient revenue highlights what have been structural weaknesses in how the program is funded."

The losses generated by the NFIP to date, as well as the potential for future losses, leave the federal government and taxpayers in a highly exposed financial state. The Biggert-Waters Flood Insurance Reform Act of 2012 (Biggert-Waters Act) extended the NFIP for five years and contained provisions to help strengthen the financial solvency of the program, including phasing out almost all discounted insurance premiums (for example, subsidized premiums). The extent to which its changes would have reduced NFIP's financial exposure is unclear since in 2014 the Homeowner Flood Insurance Affordability Act of 2014 was enacted, which reinstated some premium subsidies and slowed some premium rate increases that were included in the Biggert-Waters Act.

The maximum elevation above sea level in Florida is less than 400 feet (NOAA 2017). There are over 2 million flood insurance policies in effect in Florida, representing more than one-third of the approximate 5.6 million policies nationally (OIR 2016). Because of the relative importance of flood insurance in Florida, the state's response to the potential effects of the Biggert-Waters Act was negative and strongly so. Florida Insurance Commissioner, Kevin McCarty, issued a press release in spring of 2013 warning homeowners of the expected price increases that would be taking effect in fall of that year (OIR 2016). Florida Governor, Rick Scott, wrote an open letter to Florida's U.S. Senators Nelson and Rubio in the fall of 2013, urging them to resist the Biggert-Waters Act changes. Also in the fall of 2013, the Florida Office of Insurance Regulation (OIR) issued an informational memorandum to insurance companies interested in writing private flood insurance to serve as an alternative to the NFIP (OIR 2016). As of year-end 2016, there were five private insurers writing both primary and excess-of-NFIP insurance coverage, six offering primary policies only, four offering excess policies only, and one writing surplus flood insurance to homeowners in the state of Florida (OIR 2016).

Risk Assessment

Given insurance market problems and public policy pressures that can arise in catastrophe-prone regions, along with the critical importance of information to a well-functioning insurance market, members of the insurance industry have been paying attention to changes in environmental factors and the science of climate change for years. Yet it remains a challenge to take concrete measures to address climate change. One reason is because the there is pressure to price for profitability each policy period (usually a year) insurers are under pressure to focus on what may happen in the short term than about the impacts of climate change that will likely occur over the long term. Nevertheless, insurers are increasingly asked by regulators and rating agencies to explain what they are doing to manage the risk of climate change (Climate Working Group 2012).

Traditionally, insurers relied on historical loss data to assess their property risks. As losses from atmospheric perils such as windstorms and flood grew in the 1980s and 1990s, however, insurers began to develop and use catastrophe loss models that are based on simulation analysis. These models take into account expert assumptions about present and future events as well as historic data. Catastrophe models can either be stochastic, randomly generating loss events based on data inputs, or deterministic, running loss event scenarios. Both types of catastrophe models can be utilized to inform insurers about the possible effects of the climate and weather patterns on insured losses. Concerns about climate change effects on loss expectations has increased the demand for quantification of risk and uncertainty and so has increased the demand for use of these models (Climate Working Group 2012).

Physical Risk Effects and Catastrophe Loss Models

Catastrophe loss models inherently reflect the climate since they are by definition models meant to represent reality. For example, flood models estimate storm surge losses based on today's sea levels, not those of 50 or 100 years ago. As the models are updated, they capture the most recent seasons of activity. So whatever impact a climate change has had to date is in essence already present in the models.

Estimation of losses is key to modeling for insurance purposes. Catastrophe models produce not only estimates of average annual losses, but also probable maximum loss (PML) and tail value at risk (TVaR). The modeled 100-year PML represents a 1% (1 in 100) chance losses will reach a particular value or greater, based on an exceedance probability curve (in turn based on thousands of simulated loss years). It is easy to forget, however, that the "or greater" part of the sentence is important; statistically, the probability that losses will be much greater is significant. Estimation of the TVaR becomes critical to understanding the risk profile of insurable events for low-probability loss levels. The catastrophe TVaR, similar to value-at-risk measurements for other applications, estimates the average amount of loss expected above a critical loss level—in the catastrophe modeling, above the PML. These calculations are particularly critical to pricing reinsurance and ILS products.

All of Florida is exposed to hurricane events, so the markets require PML estimates that include the entire state as well specific books of business individual insurers may underwrite. Not surprisingly, Florida has the highest estimates of any state. Table 6.1 below indicates estimates of Florida PMLs for 2016, at 0.4%, 1%, and 2% probabilities.

Table 6.1. Probable Maximum Loss Estimates for the State of Florida Due to 2016 Hurricanes.

Return Period (in Years)	Critical Probability	Aggregate Gross PML (in Billions)
250	0.004	$80.6
100	0.01	$53.9
50	0.02	$36.0

Interpretation: A one-in-hundred loss year (associated with a one% probability) would produce estimated $53.9 billion or greater in gross loss to all Florida residential policyholders, including loss adjustment expenses. Source: State of Florida Financial Services Commission 2016b.

Table 6.1 indicates a 0.4% likelihood that insured residential policyholders in Florida would experience at least $80.6 billion in hurricane wind losses, given what the catastrophe model(s) utilized are assuming about climate effects to date. These modeled results are based on wind losses only since flood losses are primarily covered by the NFIP, not the private insurance market or Citizens. Even without further environmental changes that exacerbate loss events and losses, hurricanes clearly pose an enormous risk to the state. If considering hurricane-related flood and other categories of loss not related to hurricanes, the Florida PMLs are substantially higher. Future environmental changes would impact the estimates as well, although it is not yet clear precisely how, or by how much.

Climate-Sensitive Modeling

Several commercial loss modelers offer "climate conditioned" stochastic models and deterministic models that presume extreme disaster scenarios (Climate Working Group 2012). In order to develop scientific scenarios showing the long-term impacts from climate change, the potential paths for development of environmental changes must be fed into the models. A global community of scientists investigate the potential impacts of climate change on the frequency and intensity of natural disasters. For instance, the Fifth Assessment Report of the United Nations Intergovernmental Panel on Climate Change (IPCC) is the expectation that hydroclimatic intensity will be exacerbated by continuing climate change (Niehörster et. al. 2013). This means that regions that are already humid today will become even more humid, and areas that are dry today will become even drier (and in some cases warmer too). Such a trend can have a significant impact on agriculture and forestry and also increase the frequency of extreme hydroclimatic events, such as long heatwaves and periods of drought.[5] More specific to catastrophe events, the frequency of hurricane activity in the Atlantic basin may stay constant, or even decrease, while storm intensity may increase. Climate model results for extratropical cyclones, or winter storms, show a similar pattern. Floods, on the other hand, may increase both in number and intensity (Niehörster et. al. 2013).

But there is significant uncertainty in understanding relationships between various climate signals and the frequency of occurrence of natural disasters, especially where regional-scale expectations are concerned. There is a wide scientific consensus that climate variability may increase, but the current climate is already highly variable—witness the 10-year absence of Florida hurricane landfalls. Because the natural climate variability is so large, detecting a clear signal due to climate change remains a challenge. Given all the natural fluctuations in the climate, it could take decades before establishing whether the frequency of extreme conditions is increasing. The main problem is the rarity of extreme events and a lack of data about them. That alone makes it difficult to assess the situation.

Two recent studies show the danger of jumping to conclusions in the short term about long-term climate effects. In these studies, researchers looked at hurricanes and tornadoes, in turn, and found that if the same number of housing units had been built at the time, then the then the amount of damage caused by storms actually would have decreased to today rather than increased. The first study discovered the most damaging single hurricane would have been the Great Miami hurricane of 1926, followed by Katrina in 2005, and two hurricanes in Galveston, Texas in 1900 and 1915 (Pielke et.al. 2008). The report did not find any trend in increasing intensity in hurricanes. The second study conducted an analysis of 56,457 tornadoes since 1950, and showed that if the same number of housing units had existed over the past six decades, there would be a decreasing trend in tornado loss damages between 1950 and today (Simmons et.al. 2013). These efforts to replicate past weather patterns and events yield results that counter the notion of increasing extreme events. But more such studies are needed. Once past weather patterns can be modeled consistently, they can then be used to more confidently forecast future patterns. For (re)insurers, a prudent approach is to use scenario testing (via deterministic event models) in addition to stochastic models to evaluate climate impacts.

Model Volatility and Technological Changes

Volatility in loss estimation, especially PML and TVaR estimation, has been a driver in the changing and updating of model assumptions as new data and scientifically validated approaches are discovered.[6] Catastrophe models increasingly emphasize their consideration of the inherent

---

[5] For instance, the IPCC indicates Florida and the Southeast U.S. are expected to experience lengthier heatwaves in the foreseeable future than in the past.

[6] The variability in modeled loss outcomes is a well-known challenge in the insurance markets. In fact, the State of Florida has a public commission, the Florida Commission on Hurricane Loss Projection

uncertainty in loss outcomes. This consideration is critical to capturing the true loss potential from an event and must address both primary and secondary uncertainty concerns. Primary uncertainty lies in the random event generation (i.e., the factors impact probability of an event), while secondary uncertainty lies in estimation of event intensity, damage and financial loss. Representing the full distribution of potential event outcomes reduces bias in loss estimates. For reinsurance and ILS instruments, model developments in this area are critical for events resulting in mean modeled losses that lie near the trigger point of a transaction. Credible consideration of the full range of potential loss outcomes from the event reduces the worst form secondary uncertainty can take—that the loss potential to the transaction would be underestimated (Guin 2016).

Today's catastrophe models can express model uncertainty both probabilistically (through expert assumptions and sophisticated simulation testing) and deterministically (through extensive "what if" scenario testing). Different models may achieve these tail values using different methodologies, and each methodology has drawbacks that contribute to model risk. Nevertheless, they provide important tools for gaining information about the tail and the tail risk profile.

Vast improvements in computing power and underlying risk information have resulted in considerable refinement of catastrophe loss modeling as well. Increased resolution, specification and online platforms have made the use of models faster and friendlier and the modeled outcomes increasingly informed. Also affording additional improvements in the modeling environment are the technologies that allow for real-time event and loss information as well as predictive analytics. High-performance computing has made many catastrophe modeling advances possible, particularly on the predictive analytics front. Predictive analytics are concerned with determining whether an event will occur as opposed to estimating losses over a specified time period. As a potential market disruptor, predictive analytics may also alter what is offered in the catastrophe finance market. For now, the technologies at a minimum improve the intelligence with which insurance operates in a volatile environment.

(Re)insurers can use updated information from the models for core business – risk selection and pricing – improvements as well as for better investment decisions. Risk selection, commonly referred to as underwriting, is the process of evaluating whatever is potentially insured (properties, businesses, people, etc.) for the perils to which they are exposed and the hazards to which they are susceptible (Rejda 2011).

The Special Case of Flood Risk Assessment

While private catastrophe risk modelers have been working diligently on the assessment of U.S. wind risk and other perils, they have been slower to develop models that assess flood risk. This

---

Methodology, which reviews the reasonableness of model assumptions and the modeling process of commercial models before they can be used in Florida to set residential property insurance rates.

relative slowness by commercial modelers to tackle flood risk assessment is attributable to at least two important factors, both of which are market driven. First, the fact that the NFIP has been the virtually exclusive insurer of U.S. flood risk for 50 years and so there has previously been little or no demand by private insurers for commercial risk modelers to spend resources to develop flood models. Second, the NFIP is already aided by the FEMA Risk Mapping, Assessment and Planning. FEMA identifies flood hazards, assesses flood risks, and partners with states and communities to provide accurate flood hazard and risk data to guide them to mitigation actions. FEMA flood hazard mapping is the basis of NFIP regulations and flood insurance requirements (FEMA 2017).

As FEMA and the NFIP struggle to update flood maps, the flood mapping process and flood insurance rates to more accurately reflect the risk, and as private insurers indicate interest in writing U.S. flood insurance coverage, commercial catastrophe risk modelers have increasingly begun to pursue the modeling of U.S. flood risk. An indication of the uptake in research and development by modelers in this area is the fact that three commercial modelers have invited the Florida Commission on Hurricane Loss Projection Methodology (Modeling Commission) professional team to review their flood modeling efforts on site in 2017 (SBA 2017). The Modeling Commission began to develop standards for acceptability of flood models in 2014.

Meanwhile, FEMA is simultaneously making strides to improve its flood risk assessment structure and processes. The Technical Mapping Advisory Council is a federal advisory committee established to review and make recommendations to FEMA on matters related to the national flood mapping program authorized under the Biggert-Waters Act. The Technical Mapping Advisory Council is charged with reviewing the national flood mapping process and making recommendations to FEMA regarding flood mapping as well as the "impacts of climate sciences and future conditions and how they may be incorporated into the mapping program" (FEMA 2016). In a June 2016 report issued by Technical Mapping Advisory Council, it is clear the group is prioritizing structural and process improvements in existing FEMA mapping, and then will focus more heavily on the potential for climate change effects on the maps.

Scientific researchers outside of the modeling industry and FEMA have been specifically looking at sea level rise as a contributor to past and future flooding. While such research does not directly attempt to tie the sea level rise directly to insured loss amounts, the (re)insurance industry can incorporate the estimates generated from this research into their internal financial models to estimate the potential impact of future flooding due to sea level rise.

Underwriting and Pricing Impacts

Underwriting is the first line of defense an insurer has to protect its profitability. Proper underwriting results in books of business that are not only insurable, but are pooled according to the risk each brings to the company. Thus, proper pricing of each risk pool of similarly-exposed units, is contingent on proper underwriting as well as on the quality of information used directly for rate and premium calculations.

If insurers fail to respond to changes in the risks they have selected, negative externalities almost certainly occur. Two negative externalities are especially dangerous to profitability—moral hazard and adverse selection (Rejda 2011). Moral hazard occurs when those with insurance can influence the chance of loss by taking on more risk than they would in the absence of insurance. Adverse selection occurs when those at higher risk of loss are more likely to seek coverage, or seek more coverage, than those at a lower risk.

The insurance industry has a long history of modifying underwriting guidelines, tightening contract terms and parsing out classes of risk more granularly so as to accurately reflect expected losses whenever risk and/or market conditions change unfavorably. If climate change is expected or observed to change the loss landscape, it is in insurers' best interests to respond with changes in the risk transfer landscape.

Multiple Lines of Insurance Impacted

Natural disasters can destroy homes, cars, businesses and crops, leading to an increase in the number and severity of property-casualty insurance claims in multiple lines of coverage—residential, commercial, auto, business interruption, crime and more. Environmental changes that lead to larger losses therefore impact multiple property-casualty insurance lines as well.

The potential issue of liability under law for the consequences of climate change has not been tested adequately but the risk is real. Attempts by individuals and groups of people to sue industrial groups have so far generally failed (Niehörster 2013). Thus far, there has been no case that has really tested whether and how liability claims based on the consequences of climate change would be settled. The insurance industry, however, follows trends in court decisions, and responds to case precedent with changes in their business decisions.

Furthermore, climate change is not just a property and liability issue. It also has potential ramifications for consumers' health. Poor air quality can lead to an increase in the number of people with asthma, which can also lead to an increase in health insurance claims.

Have insurers in some parts of the country stopped offering certain types of insurance coverage or limited the types of coverage they offer, based on sustained rises in their loss costs? Yes. And still more have charged higher insurance premiums that are often unaffordable for consumers (Kunreuther and Michel-Kerjan 2012). As a result, some consumers buy policies that do not provide as much coverage as they need, while others go without insurance. But to what extent are these market decisions based on climate change?

Effects on Underwriting, Terms, and Prices

Insurance is the only product in the world that is sold before the cost of producing the "goods" is known with certainty. Thus, insurers are inherently in the business of information. The better the information, the better the business decision. In a marketplace of changing dynamics, insurers are confronted daily with new information and must cull it for credibility and reliability before incorporating it into business operations. And if good business were not enough reason to proceed cautiously, regulatory scrutiny over underwriting and pricing requires judicious treatment of information.

If insurers were to review the current scientific information for decision purposes, it would be difficult to respond confidentially from an underwriting and pricing standpoint. For instance, the IPCC relays information that could justify either increasing or decreasing increasing premiums. The IPCC has for years cautioned about an increase in heat waves, torrential rains and floods. According to a recent IPCC report, however, there may be fewer cold weather disasters and storms in the future in some places.[7]

It is currently difficult, given the current state of climate science and catastrophe models, to take the output of climate-sensitive models and apply them directly to insurance pricing. Direct application, however, is not necessary for capture of climate change information. Indeed, insurers do not need to take "climate change," per se, into account at all. If probable loss, frequency, and/ or severity are changing, they do not have to be labeled as climate changes for insurers to respond appropriately. Even times of heightened extreme weather activity do not change industry fundamentals: to assess the risk based on the current modeling assumptions and price it accordingly.

From an underwriting perspective, expect to see increasing granularity and fragmentation. Granularity is the extent to which insurers drill down within a risk pool to the individual exposure unit for purposes of deciding insurability and contract terms (Florida Catastrophic Storm Risk Management Center 2011). The more they drill down the more risk pools they create, and the smaller the pools get, all else the same. In the extreme, residential property insurers could drill down to a point where each house is its own risk pool. The problem with this is insurers would not be able to use the law of large numbers to accurately determine expected losses. Realistically, they can only get as granular as makes statistical and financial sense. Fragmentation is another way in which insurers can respond to changes in its risks, and/ or the marketplace. This refers to comparing risk pools holistically, according to risk as well as customer preferences, and providing different products, services and contract terms to different market segments, or fragments, based on multiple profitability factors (Swiss Re 2016).

The property insurance business in Florida is already underwritten, and priced, more granularly than in most other regions. The catastrophe loss models used by insurers all, to one degree or another, utilize global positioning satellite (GPS) data to precisely locate homes and businesses, and build vulnerability curves into the modeling process to incorporate information

---

[7] On the whole, it seems likely that insurance premiums for houses and buildings in flood-prone coastal regions will rise, based on current flood models. On the other hand, because of the complex impacts of climate change, risks (and premiums) may fall in other areas. There is evidence, for example, that snowmelt-driven springtime floods will become less frequent in the future.

about property improvements that are known to affect susceptibility to perils and hazards. Market fragmentation is occurring, too, albeit more at the reinsurance than primary insurance level. Primary insurers are limited by the Florida OIR as to how much they can incorporate differences in customer preferences into their underwriting and pricing algorithms. Reinsurers, not limited directly by the same rules, can utilize profitability analytics to fragment the insurance base to optimize expected underwriting profits and capital allocation decisions.

The insurance industry appears to be taking a watchful approach to environmental change. Reinsurers, who write policies that help insurance companies pay catastrophic claims, may not be anxious about short-term loss impacts of environmental changes since insurers and reinsurers can deal with that year to year by raising prices following catastrophic events. What they are likely more concerned about is that if risks become too big, they may become incalculable, and thus uninsurable. For instance, if sea levels continue to rise in some regions then properties, or even entire cities, on those affected coasts could become uninsurable at some point.

Underwriting and pricing are complicated for events that are low-frequency and catastrophic. A significant portion of the complexity is driven by political pressures. Regulators, legislators and other public policymakers are under tremendous pressure, particularly in the months that follow extreme loss events, to ensure insurance is both available and affordable for their constituents. Insurers do not make underwriting and pricing decisions in a vacuum; they must conform to the regulations and legislation to which their lines of business are subject. Regulators and legislators, in turn, desire a healthy insurance market. To the extent private insurers cannot provide sufficient insurance capacity profitably, policymakers are under pressure to utilize the state insurance entities at a minimum for insurance availability, and may also use them for insurance affordability (Medders et al. 2013).

In the state of Florida, if environmental changes produce more extreme loss events, Citizens' policyholder count will likely rise as private insurer profits get squeezed (at a time when Citizens would likely be facing large claims payouts itself). The FHCF will be under pressure to increase its capacity (and thus its financial liability). The adverse financial impacts on the state of Florida could be substantial.

State of Florida Financial Impacts

The ability of Florida's state-sponsored insurance entities (residual insurers)—Citizens, FHCF and FIGA—to pay losses is vital to the state's ability to respond adequately to weather disasters.

Due to the magnitude and volatility of catastrophic losses, it is virtually impossible to finance all of the potential losses in any single time period. This leaves two options: prefund all potential losses or utilize some form of post-loss funding. The state of Florida has chosen to finance a significant portion of its catastrophic risk exposure through post-loss assessments. In Florida, these assessments are levied on most property-casualty insurance policyholders by the state's residual insurers (Florida Catastrophic Storm Risk Management Center 2011).

Citizens Financial Impacts and Policyholder Assessments

Citizens is smaller than it was five years ago due to multiple factors. A decade of no Florida-landfalling hurricanes and an influx of ILS capital combined to soften the Florida property reinsurance and primary insurance markets. Plus, a vigorous Citizens depopulation program intentionally reduced Citizens' market share.

Annually, Citizens reports its aggregate PMLs, potential assessments, and financing options to the state of Florida (Florida Financial Services Commission 2016a). Largely due to Citizens' reduced policyholder base, a 100-year (one% likelihood) or 50-year return period (two% likelihood) in 2016 would result in no financial shortfall. A 250-year return period (0.4% likelihood) would, however, result in an estimated shortfall, an amount in excess of $2.8 billion (Florida Financial Services Commission, 2016a).

Based on the 2016 estimated Citizens policyholder base and combining the effects on the Personal Lines Account, the Commercial Lines Account, and the Coastal Account, under a one-in-250-year loss scenario, there would be an estimated one-time Citizens policyholder surcharge of $136 million (15% surcharge) during the first year, an estimated one-time assessment of non-policyholders of $799 million (2%), and an estimated emergency assessment of nearly $1.9 billion, with estimated annual assessment of $122 million, representing a 0.3% assessment (Florida Financial Services Commission 2016a).[8]

The insurance impact of environmental changes in the short-term are inherently embedded as much as is practicable in these estimates. Long-term changes are not necessarily embedded in the modeling used to arrive at these numbers. Suffice to say if in the long-term catastrophic losses are substantially higher than estimated today, the effect on Citizens could be substantial. Higher losses, resulting in higher assessment potential over an almost-certainly-higher policyholder base are the likely results. Since assessments are subject to a maximum percentage, however, there is a limit to the direct financial impact on Citizens. What then the impact on the FHCF and FIGA?

FHCF Financial Impacts and Policyholder Assessments[9]

Once an FHCF-participating insurer's hurricane losses exceed its share of the aggregate industry retention (deductible), it triggers FHCF coverage.[10] The claims-paying resources of the FHCF

---

[8] These numbers assume annual assessment for 30 years using an interest rate of 5%.

[9] Discussion of the FHCF estimates is based on Financial Services Commission 2016b and the Florida Catastrophic Storm Risk Management Center.

[10] An insurer's FHCF reimbursement coverage is triggered after it meets its retention (the functional equivalent of a deductible). For the contract year that began on June 1, 2015 and ends on May 31, 2016, the aggregate retention for all participating insurers is $6.9 billion. Aggregate retention for the contract year beginning on June 1, 2016 is projected to be $7.0 billion.

include cash available from current and past accumulation of reimbursement premiums and investment income[11]; proceeds from pre-event financing and post-event debt; and risk transfer (reinsurance and ILS) recoverables.[12] Cash is used before any of the other claims-paying resources are used. Pre-event debt is repaid from FHCF investment income. In situations involving large losses that must be paid quickly the FHCF would likely attempt to use post-event bonds to finance its reimbursement payments to participating insurers based on the losses generated by the hurricane or hurricanes. As stated earlier in the chapter, these bonds would be repaid using emergency assessments.

Post-event debt is repaid from emergency assessments on most Florida property and casualty premiums of both admitted and non-admitted lines of business (the exceptions are workers' compensation, medical malpractice, accident and health, and federal flood insurance). Post-event resources could also include funds from assessments levied without the issuance of post-event debt. The maximum assessment percentage is 6% with respect to any one contract year's losses and 10% with respect to all contract years' losses combined. No such post-event debt is outstanding as of the date of this writing; there are currently no assessments.

Similar to Citizens, the FHCF reports annually its aggregate PMLs, potential assessments, and financing options. Table 6.2 shows the estimated annual assessment impact from various hurricane loss scenarios for 2016.

Table 6.2. Potential FHCF assessment impact of PML scenarios.

Return Time (in Years)	Potential Post-Event Bonding (in Millions)	Annual Assessment (in Millions)	Annual Assessment %
250	$3,146	$205	0.50%
100	$2,593	$169	0.41%
50	$429	$28	0.07%

Interpretation: Based on a one-in-100 loss year (associated with a 1% probability) would produce estimated $53.9 billion or greater in gross loss, the bond financing need is estimated at $3,146 in order to make insurer reimbursement payments. (Assumes annual assessment for 30 years using an interest rate of 5% and an assessment base of $40.9 million.) Source: State of Florida Financial Services Commission 2016b.

As with Citizens, if in the long-term catastrophic losses are substantially higher than estimated today, the FHCF would face additional financial pressures with limited claims-payment sources, thus likely placing a larger share of the financing burden ultimately on Florida policyholders. Although the FHCF has no statutory obligation to pay after its financial resources are exhausted, and thus insurers could be left without full reimbursement for their losses, it is not likely Florida policymakers would allow such an outcome. First, without reimbursement, many Florida property insurers would be unable to meet their claims liabilities and would face

---

[11] The FHCF is projected to collect $1.2 billion in reimbursement premium, net of expenses, mitigation and debt service, for the 2016-2017 contract year and the total projected cash balance as of December 31, 2016 is $13.8 billion.

[12] The FHCF purchased $1 billion of reinsurance for the first time ever for the 2015 season. No such risk transfer products are in place for the 2016 season.

bankruptcy. Second, although the primary purpose of Florida's third state-sponsored insurance entity, FIGA, is to pay claims in the event private insurers cannot, it is not intended or financially prepared to backstop the private insurance industry in the event of a disaster (or set of disasters) so large it impairs a large portion of the industry.

FIGA Financial Impacts and Policyholder Assessments[13]

FIGA does not accumulate funds in advance of an insurance company's insolvency. Therefore, when a company insolvency occurs, FIGA must obtain the funds it needs through pro-rata assessments levied by the Office of Insurance Regulation on insurance companies subject to assessment.[14] These insurers must then recoup the cost through their policyholders. Depending on the number and size of property insurance companies that become insolvent following future hurricane strikes (or other disasters) in Florida, FIGA may need to levy its own FIGA Regular Assessments and FIGA Emergency Assessments to meet its hurricane claims payment obligations under Florida law.

FIGA has three separate accounts (Section 631.55(2), Florida Statutes): (1) the automobile liability account; (2) the automobile physical damage account; and (3) the account for all other insurance required to be part of FIGA. Only insurers writing business in the lines of insurance included in the account in which the insolvent company was writing business can be assessed. The "all other" account is relevant since it includes the property insurance lines of business.

FIGA has three sources of income to pay claims other than through assessments: distributions from estates in receivership, recoveries from the FHCF and investment income. It can be assumed that in a worst-case weather loss year, FHCF recoveries may not be forthcoming and investment income may be negative. Clearly, FIGA operates in a cascading-effect environment with the other two state insurance entities. If they are under extreme financial pressure, FIGA, by its very definition is under financial pressure as well in attempting to pay claims.

This third leg of the state's residual insurance stool receives the least attention, at least publicly, for the financial risk it represents to insurance policyholder, state government and Florida taxpayers. Assessments are a quasi-tax on insurance policyholders. But there also exists the risk of direct taxation implications of extreme weather.

Florida Taxpayer Financial Impacts

The state of Florida total tax revenues were nearly $33 billion for 2012 (Unites States Census Bureau 2015). It is notable that at the state's current taxation level, state tax income would hardly be adequate to pay the gross losses of a one-in-50-year storm (or set of storms) in a single year,

---

[13] Discussion of FIGA impacts is based largely on review of FIGA 2016 and on the work of the Florida Catastrophic Storm Risk Management Center.

[14] Insurance companies may be required to pay these assessments in as little as 30 days.

which is estimated at $36 billion, according to FHCF figures (Financial Services Commission 2016b).

In a worst-case year, without any special consideration for climate-related losses, weather-related losses could leave the state in a financially awkward position. Without a "nuclear briefcase" (aka a financial plan for a worst-case situation), the state would be left with potentially few options. These involve a combination of maximum policyholder assessments along with issuing debt, necessitating the likely imposition of additional other taxes at a time when the taxpayer income base would be necessarily impaired already, due to the non-tax financial burdens resulting from the event(s).

Maximum assessments could increase policyholder premiums by up to 75%. Putting this into perspective, Florida homeowners already pay the highest homeowners insurance premiums, on average, in the nation, at over $1,100 per year (based on annually-updated data from the National Association of Insurance Commissioners). State of Florida workers today pay no state income tax. Arguably, Florida's growth and economic vitality have come to depend on the appeal of the zero individual income tax. The state would virtually be forced to tax the business and tourism base more heavily to make up the shortfall—a risky option. In some worst-worst case scenarios, Florida could face wide-scale population emigration, sustaining a heavy loss of its tax base.

Better for the state is to find ways to avert the possibility of such a crisis and/ or promote insurance industry efforts to do so. The balance of this chapter discusses insurance innovations and initiatives underway, designed in part to reduce the insurance risks of environmental changes.

NFIP Financial Impacts

Storm surge will be exacerbated by sea level rise, and the impacts of increases in storm surge will be felt long before the effects of static sea level rise in Florida (Florida Oceans and Coastal Council 2010). The estimates of climate change on Florida insurance markets and taxpayers provided above have not included flood at all, much less the potential increases in flood losses due to sea level rise. Flood loss estimates due to sea level rise have not been released by catastrophe modelers although we know modeler efforts are underway to build models that generate such estimates.

We must not forget Florida is the NFIP's largest coverage region, representing one-third of the program's insurance policies. Even if we set aside increases in tropical storm-related loss potential, the threat of sea level rise to Florida is sufficient to exacerbate the financial challenges the NFIP already faces. Based on NOAA's four global mean sea level rise scenarios—Low, Intermediate Low, Intermediate High and High—state, regional, county and municipal area planners can estimate the impact of sea level rise on a particular locality and its population (NOAA 2016). The Tampa Bay Climate Science Advisory Panel recently completed a study of these scenarios and concluded that the Tampa Bay region might experience sea level rise between 0.5 to 2.5 feet by 2050 (Hillsborough City-County Planning Commission 2016). This is especially important information given pockets of the Hillsborough-Tampa Bay area would have been especially hard hit by the effects of Biggert-Waters on NFIP flood insurance rates (OIR 2016).

When considering the potential combined impact of the NFIP's existing financial distress, the changes in technical mapping taking place within FEMA and the potential for significant sea level rise within the next 20-50 years, it is difficult to imagine a financially healthy NFIP unless the program is entirely revamped, and frankly, is not available or affordable to all Florida property owners.

Mitigation and Risk Transfer Solutions

The insurance-reinsurance sector does not exactly have a reputation for innovation, but this does not mean insurers and reinsurers are not innovative. As discussed earlier in the chapter, insurers must proceed with caution to protect their financial solvency and their policyholders' exposures. Innovation must be well researched and vetted before (re)insurers move forward on significantly new or altered products or pricing techniques. But there are several risk transfer solutions underway that can help to mitigate catastrophe risk.

Closing the Protection Gap

Insurance penetration in Florida is high, both in raw terms and relative to most other regions in the world. The simplest insurance solution to reduce the relative risk faced by Florida is to close the insurance gap. Much of the world is underinsured, according to the world's leading reinsurers (Swiss Re 2015). Meanwhile, the U.S. is heavily insured, and Florida is the arguably the most heavily insured of all relative to its catastrophe potential (Swiss Re 2015; Kunreuther et.al. 2012). Fig. 6.1 indicates, for instance, 45% of North America catastrophe losses in 2013 were insured while only 5% of Asia's were insured. The percentages were even lower for South America and Africa, at 1% insured losses each.

Fortification, Building Standards, and Zoning

Insurability requires accidental, unforeseen losses and a feasible insurance price. If real estate development is unlimited and/ or building standards are lax, many property owners are not likely able to afford the premiums to insure their properties. Florida boasts among the best building codes in the U.S. and the world.[15] Although the state's record on fortification of existing building stock is spotty, it is a world leader in property fortification. The insurance sector supports responsible building and zoning through risk-based pricing.

---

[15] The Insurance Institute for Business and Home Safety (IBHS) reported in 2013 and 2015 their rankings of U.S. Gulf and Atlantic Coastal states by strength of building code and effectiveness of building code enforcement. According to IBHS, Florida ranked 1st in 2013 and 2nd (behind Virginia) in 2015.

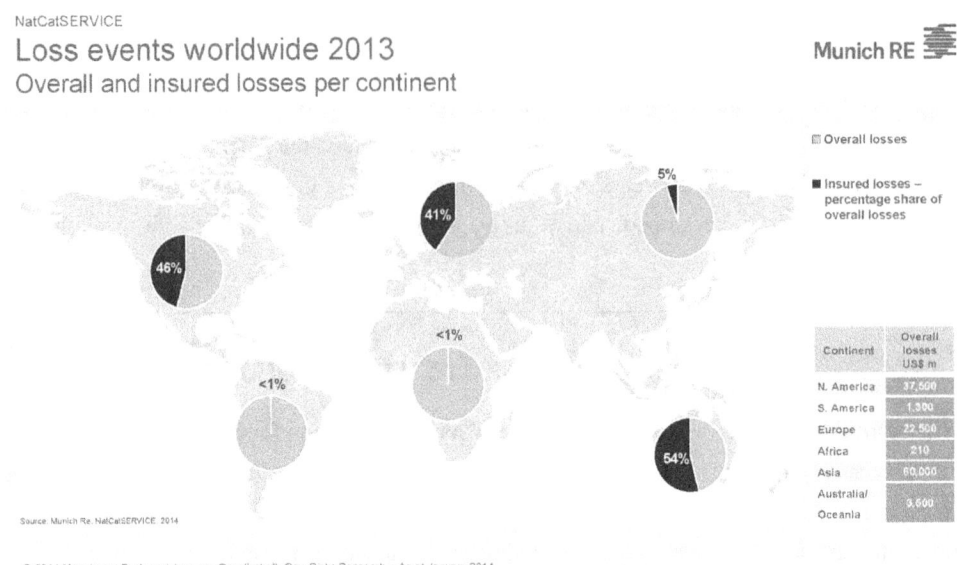

Figure 6.1. The best measures to close the gap address the root causes of underinsurance.

Mandatory Insurance Programs

In the U.S., including Florida, insurance is often mandated, either directly by statute or indirectly by lenders who require insurance in lieu of collateral for obtaining needed financing (Kunreuther et al. 2012). U.S. residential and commercial properties, if financed, must be insured, with few exceptions. Given Florida's high property and construction values, the state's asset base is a volatile insurance and reinsurance risk. Mandating insurance for properties in other regions would help to spread the risk.

Microinsurance

The property protection gap is exacerbated by poverty and lack of insurance availability in other regions of the world. Microinsurance can come in many forms, but is basically community-based insurance provided to individuals and businesses with such small financials they do not ordinarily meet the financial minimums for profitability (Swiss Re 2015).

Parametric Insurance

Insurance payout triggers and benefits that are based on parameter indexes, such as wind speeds, temperatures, or industry-level loss amounts. Because it does not require historical loss data or current loss measurement to determine claims payouts or premiums, index-based insurance has the advantage of simplicity over traditional indemnity-based insurance.

Other Innovative Reinsurance Solutions

(Re)insurance for Renewables

Insurance makes possible the growth in use of new economic developments, including new technologies, by providing protection against loss (Skipper and Kwon 2007). A challenge for insurers posed by new technologies is that there are no long-term statistics on loss experience, which is the most important factor for risk assessment. Thus, insurers closely monitor the development of new technologies so as to understand and assume new risks. In this way, insurers who are first movers make an important contribution to removing barriers to the use of new types of technology for investors and operators. Reinsurance companies are at the forefront of encouraging (or discouraging, in some cases) new renewables technologies, such as wind energy, hydroelectric, biofuel, and geothermal power. Besides this, insurers can provide further insurance solutions that help protect the environment, such as low-cost policies for climate-friendly vehicles.

Holistic Earnings and Capital Protection Covers

Insurers and corporations can efficiently combine multiple risks and/ or interdependent triggers so long as the joint costs of these is less than the sum of their separate costs. In so doing, overall costs of catastrophes to both the insured and the insurer are reduced.

Parametric Solutions

Discussed previously, these index-based insurance solutions can be used to enhance the insurability of any difficult-to-insure risk (London School of Economics and Political Science 2015; Swiss Re 2016).

Integrated Reinsurance and ILS Programs

When the sheer size of a risk is challenging, increasingly, panels of reinsurers combine their capital with alternative sources of capital (ILS) to fund upper layers of potential losses on a collaborative, integrated basis.

Retrospective Covers

These programs allow insurers to transfer the risk of liabilities from past underwriting years, and are especially helpful to insurers who face potentially significant but unknown long-tail losses and need to release capacity to underwrite new business.[16]

---

[16] Long-tail losses are any losses that tend to be slow to develop over time.

Industry Initiatives

In 2016, the World Economic Forum declared "failure of climate change mitigation and adaptation" as a top five global-level risk facing the world over the next 10 years, based on the collective judgment of 742 surveyed experts and decision makers drawn from business, academia, civil society, and the public sector (World Economic Forum 2016). Extreme weather events, water crises, food crises, and profound social instability round out the "top 5" list and can be linked directly to climate risks. The insurance industry and its regulators, as participants in the Forum's Global Risk Survey, clearly recognizes the gravity of climate change as a potential global economic game changer. Insurers have led initiatives like reducing greenhouse-gas emissions, increasing their use of renewable energy, and being more energy efficient.

(Re)insurance Industry Research Efforts

The insurance industry is aware of possible catastrophic ramifications of environmental changes. It responds to the risk potential by undertaking research initiatives, apart from its direct risk transfer innovations. Reinsurers, such as Munich Re, Renaissance Re and Swiss Re, have invested in human capital to aid in finding answers to the scientific questions and discover solutions. These companies are among the several that have hired in-house mathematicians, scientists and legal counsel to help problem solve new risks before the risks emerge as trends. Industry groups are active in the research as well. For instance, the Casualty Actuarial Society, Canadian Institute of Actuaries, Society of Actuaries, and the American Academy of Actuaries' Property/Casualty Extreme Events Committee have responded to climate risks by collaboratively commissioning committees to recommend, support, and perform research on climate change and assess the potential risk management implications for the insurance industry.

Regulatory Research Efforts

The FEMA, Technical Mapping Advisory Council, and NFIP research efforts were mentioned earlier in this chapter. It is incumbent upon regulators and policy makers to understand and respond to the threat of sea level rise and increased storm surge that may arise from climate change and result in greater flood losses. Recognizing the need to account for any potential effect climate-driven risks might have on the marketplace and the availability and affordability of insurance, state insurance regulators and other stakeholders are undertaking their own separate research efforts. The National Association of Insurance Commissioners (NAIC) has a working group on climate risk. The objectives of the group are to review risk management efforts by insurers regarding the effects of climate change and investigate the use of modeling by insurers and reinsurers regarding climate change impacts.

The NAIC adopted the Insurer Climate Risk Disclosure Survey (survey) in 2010. The survey is comprised of eight questions that assess insurer strategy and preparedness in the areas of investment, mitigation, financial solvency (risk management), emissions/carbon footprint and engaging consumers. The survey results provide trends, vulnerabilities and best practices related to insurers' response to climate change. Disclosure allows policymakers to gain an insight into needed public policy changes.

The NAIC also adopted revisions to the 2013 Financial Condition Examiners Handbook at the end of 2012. These revisions incorporated risk-focused examination questions that provide examiners with needed guidance on what questions to ask insurers regarding any potential impact of climate change on solvency. They were specifically designed to help examiners identify unmitigated risks and to provide a framework for them when examining such risks and their impact on how an insurer invests its assets and prices its products. The participating states for the latest reporting year include California, Connecticut, Minnesota, New Mexico, New York, and Washington.

Florida Mitigation and Adaptation

Given Florida's susceptibility to natural disasters from wind and water, it is wise for Florida to take the potential catastrophe threats of climate change seriously. Climate change mitigation and adaptation opportunities for Florida are covered thoroughly in another chapter. Nevertheless, it is worthwhile to address the issue here, even if only briefly.

Mitigation of potential losses via capacity restrictions on public insurance entities (e.g., Citizens and FHCF) is critical as is adaptation to loss patterns. There is also increasing pressure to harden shorelines and implement engineering strategies to protect structures and infrastructure from loss.

The Hillsborough City-County Planning Commission is an example of a local government effort to be informed about the potential risks and consider adaptation efforts that may reduce the financial (and other) impacts of sea level rise. The group's 2016 report points out the most at-risk areas as well as the at-risk populations with discussion about the need for infrastructure improvements and modifications to adapt to the looming threat. For instance, the report indicate more than 80 percent of the affected area is publicly owned and thus is both 1) critically important for the local government to address directly; and 2) within the local government's control to practice adaptation methods, such as coastal retreat, to reduce the future impact of sea level rise on the public budget.

The Florida Oceans and Coastal Council (2010) similarly has evaluated potential impacts of sea level rise on areas of Florida and has reported on opportunities for mitigation and adaptation. One example in this report is that almost 60% of the land along the U.S. Atlantic Coast that is one meter or less above sea level is expected to be developed, and less than 10% has been set aside for conservation. In addition to hardening shorelines and fortifying structures, a strategy of coastal retreat from key high risk locales may be a wise course of action with catastrophe risk management as well as conservation benefits.

Conclusion

As this chapter has discussed, the potential for climate change as risk is real. Some of these risks are insurable today but may become uninsurable over the long term unless sufficiently mitigated. Low-frequency, high-severity events, within feasible financial limits, generally are considered the most ideally insurable, but insurability also presumes such events occur randomly. The non-randomness issue must be addressed before insurers and reinsurers can fully capitalize on related opportunities for the insurance industry and financial sector. Systemic environmental effects that could have implications must be evaluated for property, casualty and life insurance industry segments as well as for the capital markets.

While potentially risky consequences of changes in the climate exacerbate the baseline volatility of the financial sector, they also inherently create opportunities for insurers, reinsurers and other "investors in risk" to develop new products and services. Florida is a unique contributor to both the risk and opportunity since the state is the world's largest insured catastrophe region. The state of Florida itself is heavily leveraged as pseudo-insurer for much of the cost of its extreme weather exposure. Unlike private insurers, however, Florida's cost of bearing risk has no market opportunity offsets. Increased volatility in insurance, reinsurance and capital markets are all challenges for Florida, with potentially adverse collateral effects on residual insurance market pressures, policyholder assessments, and taxation. Insurance industry risk transfer solutions and research initiatives, are especially important to the state in balancing the need for reliant economic development with insurance availability and life safety.

References

AIR Worldwide. 2013. "The Coastline at Risk: 2013 Update to the Estimated Insured Value of U.S. Coastal Properties," Boston: AIR Worldwide.

Casualty Actuarial Society (CAS) Climate Index Working Group. 2012. "Determining the Impact of Climate Change on Insurance Risk and the Global Community," Solterra Solutions.

Cummins, J. David. 2006. "Should the Government Provide Insurance for Catastrophes?" Federal Reserve Bank of St. Louis Review, July/August 2006, 88(4), pp. 337-79.

Federal Emergency Management Agency (FEMA). 2016. *TMAC: National Flood Mapping Program Review*.

Federal Emergency Management Agency (FEMA). 2017. National Flood Insurance Program: Flood Hazard Mapping. https://www.fema.gov/national-flood-insurance-program-flood-hazard-mapping

Financial Services Commission. February, 2016a. Citizens Property Insurance Corporation: Annual Report of Aggregate Net Probable Maximum Losses, Financing Options and Potential Assessments, Tallahassee, FL: Citizens Property Insurance Corporation and State of Florida Financial Services Commission.

Financial Services Commission. February, 2016b. Florida Hurricane Catastrophe Fund Annual Report of Aggregate Net Probable Maximum Losses, Financing Options and Potential Assessments, Tallahassee, FL: State Board of Administration and State of Florida Financial Services Commission.

Florida Catastrophic Storm Risk Management Center. 2011. *The State of Florida's Property Insurance Market*, The Florida State University, Tallahassee, Florida, Submitted to the Florida State Legislature.

Florida Insurance Guaranty Association, Inc. 2016. *Florida Insurance Guaranty Association, Inc. Annual Report 2015,* Tallahassee, FL: Florida Insurance Guaranty Association, Inc.

Florida Oceans and Coastal Council. 2010. *Climate Change and Sea-level Rise in Florida*, Tallahassee, FL.

Florida Office of the Governor. 2016. http://www.flgov.com/governor-rick-scott-urges-senators-to-avert-increase-in-flood-insurance-premiums-2/

Florida Office of Insurance Regulation (OIR). 2016.
http://www.floir.com/sections/pandc/floodinsurance/floodinsurance.aspx
http://www.floir.com/Sections/PandC/FloodInsurance/FloodInsuranceWritersFL.aspx

Florida State Board of Administration (SBA). 2017. Florida Commission on Hurricane Loss Projection Methodology: Commission Documents.
https://www.sbafla.com/methodology/CommissionDocuments.aspx

Grace, M. F., and R. W. Klein. 2009. "The Perfect Storm: Hurricanes, Insurance and Regulation," *Risk Management and Insurance Review*, 12(1), 81-124.

Guin, Jay. February, 2016. "Does Climate Change Matter to the Insurance Industry?" White paper report, Boston: AIR Worldwide.

Hillsborough County City-County Planning Commission, 2016. Sea Level Rise Vulnerability: Assessment for the City of Tampa.

Klein, Robert W., and Paul R. Kleindorfer. 2003. "Regulation and Markets for Catastrophe Insurance," in: M. R. Sertel and S. Koray, Eds., *Advances in Economic Design*, Berlin: Springer-Verlag.

Kunreuther, Howard C., & Erwann O. Michel-Kerjan, eds., with Neil A. Doherty, Martin F. Grace, Robert W. Klein, & Mark V. Pauly. 2009. Marquardt, William H., & Karen J. Walker. 2012. *At War with the Weather: Managing Large-Scale Risks in a New Era of Catastrophes*, 29–61. Cambridge: MIT Press.

London School of Economics and Political Science. 2015. "Evaluating the Economics of Climate Risk and Opportunities in the Insurance Sector."

Medders, Lorilee A., Charles M. Nyce and J. Bradley Karl. 2013. "Market Implications of Public Policy Interventions: The Case of Florida's Property Insurance Market," *Risk Management and Insurance Review*, 17(2), 117-132.

Muir-Wood, Robert. 2016. The Cure for Catastrophe: How We Can Stop Manufacturing Natural Disasters, 233-280. New York: Basic Books.

National Association of Insurance Commissioners (NAIC). 2016. Center for Insurance Policy and Research Brief on Flood Insurance.

Niehörster, Falk, Markus Aichinger, Richard Murnane, Nicola Ranger and Svenja Surminski. 2013. "Warming of the Oceans and Implications for the (Re)insurance Industry: A Geneva Association Report," Geneva: Geneva Association.

NOAA. 2013. *National Coastal Population Report*. http://stateofthecoast.noaa.gov/features/coastal-population-report.pdf

NOAA. 2017. National Ocean Service Ocean Facts. http://oceanservice.noaa.gov/facts/sealevel.html

Pielke, Roger A., Joel Gratz, Christopher W. Landsea, Douglas Collins, Mark A. Saunders and Rade Musulin. 2008. "Normalized Hurricane Damage in the United States: 1900-2005," *Natural Hazards Review*, 9(1), 29-42.

Rejda, George E. 2011. *Principles of Risk Management and Insurance*, 11th Edition, 13-37. Boston: Prentice Hall.

Simmons, Kevin M., Daniel Sutter and Roger Pielke. 2013. "Normalized Tornado Damage in the United States: 1950-2011, *Environmental Hazards*, 12(2).

Skipper, Harold D. and Wook J. Kwon. 2007. *Risk Management and Insurance: Perspectives in a Global Economy*, 110-127. London: Blackwell Publishing.

Swiss Re Economic Research and Consulting. 2015. "Underinsurance of Property Risk: Closing the Gap," *sigma* report 5/2015, 25-32.

Swiss Re Economic Research and Consulting. 2016. "Strategic Reinsurance and Insurance: The Increasing Trend of Customized Solutions," *sigma* report 5/2016, 1-4.

U.S. Census Bureau. 2010. *Coastline Population Trends in the United States: 1960 to 2008*. https://www.census.gov/prod/2010pubs/p25-1139.pdf

U.S. Government Accountability Office (GAO). 2015. *High Risk Report*.

World Economic Forum, 2016. *The Global Risks Report 11th Edition*, Geneva, Switzerland.

CHAPTER 7

Climate Change Impacts on Law and Policy in Florida

Thomas Ruppert[1] and Erin L. Deady[2]

[1]*Florida Sea Grant, University of Florida, Gainesville, FL;* [2]*Erin L. Deady, P.A., West Palm Beach, FL*

Climate change and sea level rise have made obsolete the notion that law and policy develop in the context of a relatively stable natural environment. The need of communities to adapt to climate change and sea level rise reflects the need for laws and policies governing those communities to facilitate rather than undermine such adaptation. This chapter provides an overview of law and policy issues at three levels of government—state, local, and federal. It highlights changes in state law and policy in Florida that relate to climate change and sea level rise. The chapter also focuses on local governments, and includes sections about regional collaborations of local governments, financial issues and climate change/sea level rise at the local level, examinations of impacts on infrastructure, and impacts on the public's use of beaches in Florida. The chapter concludes with discussion of a policy change related to climate change and sea level rise at the federal level that impacts local governments.

Key Messages

- The state of Florida engagement with climate change began early, with energy law in 2006. Since then the focus of engagement in climate and sea level rise has shifted from energy to disaster planning and flooding. This shift to focus on flooding and resilience resulted from local government experience of roads and drainage being the first types of infrastructure to suffer from rising seas.
- Local governments, with much of their focus on infrastructure, have been some of the greatest centers of action on climate change, with many adopting extensive comprehensive plan policies that are increasingly being implemented through ordinances. Collaboration among local governments has resulted in increased focus on climate change and sea level rise as well as harmonized approaches to the challenges.
- Among the challenges that Florida faces is protecting the state's beaches, which are the lifeblood of Florida's tourism industry.
- Federal actions (from federally-supported research and data to federal policy changes to the National Environmental Protection Act and the new Federal Flood Risk Management Standard) have been both drivers and supporters of state and local activities on climate change and sea level rise, though recent changes at the federal level have eliminated some of these drivers.

Keywords

Climate change; Sea level rise; Infrastructure; Flooding; Local government; Policy; Law; Planning; Resiliency; Adaptation

Introduction

Law has typically developed on the assumption that as much as society and law may change, the natural world around us is, for the most part, a world of basic stationarity.[1] (Craig 2010) Climate change and sea level rise, however, mean that this fundamental assumption about the natural world does not apply any longer, and law and policy need to enable necessary adaptations to changes in physical and related conditions. This chapter reviews areas of law and policy in Florida impacted by climate change and sea level rise, including related changes in state law, provides a focus on local government action and regional collaborations in Florida, and includes a brief overview of federal actions and policy with greater emphasis on federal policy changes that affect local governments. Local governments form the real focus of this chapter since most of the impacts and most of the adaptations to climate change and sea level rise directly implicate local governments as the seat of comprehensive land use planning, infrastructure, public finance, and the level of representational government closest to people when they feel the impacts of climate change or sea level rise.

Climate change and sea level rise can and will impact law and policy in almost any substantive area and discipline in the long run. However, this chapter maintains a relatively narrow focus for two reasons. First, the authors wish to keep this summary short and readable. Second, a number of issues that could fall under the rubric of climate change or sea level rise policy already appear in other chapters of this book. For example, land use and land cover discussions appear in Chapter 2; water policy related to climate change and sea level rise is included in Chapter 3; some discussion of urban infrastructure policy occurs in Chapter 11; energy policy and impacts appear in Chapter 5; impacts on insurance, both the National Flood Insurance Program and the Florida's Citizens Insurance, are analyzed in Chapter 6.

State Law, Climate Change, and Sea Level Rise

At the level of state government, Florida has had a relationship with climate change and sea level rise for longer than most might think. Florida was an early adopter when then-governor Jeb Bush signed into law the Renewable Energy Technologies and Energy Efficiency Act in 2006. A major component of the Act was the creation of the new Florida Energy Commission in an advisory role related to state energy policies. The first report of the Commission (Florida Energy Commission 2007)[2] was required to include recommended steps and a schedule for the development of a state climate action plan.

---

[1] See, e.g. Robin Kundis Craig, "Stationarity is Dead"—Long Live Transformation: Five Principles for Climate Change Adaptation Law, 34 Harv. Envt'l L.R. 9 (2010).

[2] 2007 Florida Energy Commission, Report to the Legislature. Available at http://www.dms.myflorida.com/content/download/54452/228726/file/2007.

The Commission's report noted the scientific community's consensus that human-caused increases in greenhouse gases need to be addressed and recommended setting targets to reduce them. This would require an inventory of greenhouse gases but it would put the state in a position to lead by example through education and unification of Florida's energy governance. In 2007, under the administration of then-Governor Charlie Crist, the focus on climate continued at the state level, with executive orders setting targets and actions to reduce greenhouse gas emissions statewide.[3]

Climate change activity continued in 2008 and throughout the Crist administration. One bill created a cap and trade program for utilities, set up a renewable portfolio standard for energy, and addressed automobile efficiency and emissions.[4] Another addressed issues such as green building, efficient land use patterns, energy conservation, greenhouse gas emissions in planning, and provided for the Florida Building Commission to make recommendations on energy efficiency.[5] In 2009, the statutorily-created "Florida Energy & Climate Commission" began meeting.

Under the subsequent administration of Governor Rick Scott, state agencies continue working on climate change-related issues. Rather than concentrating resources on policy development through more publicly-focused commissions or task forces, the administration has turned its attention more to the disaster planning and recovery aspects of climate change. The Florida Fish and Wildlife Conservation Commission is doing a significant amount of data collection and monitoring related to habitat and species impacts.[6] The Department of Economic Opportunity, both through its own statutory mission and with funding from the federal government, has been doing extensive work on sea level rise, including pilot planning efforts in several communities.[7] Its approach has been to also provide technical assistance for local governments and to offer review and comment on compliance with legislation passed in 2015 related to addressing "Peril of Flood" issues in comprehensive plans. The Department of Economic Opportunity has also created numerous guides and compilations of resources for local governments that want to start addressing sea level rise in their policy framework.[8] The Florida Department of Environmental

---

[3] This discussion of climate change history and policy in Florida owes much to Erin Deady, Esq., and the article "The Link Between Future Flood Risk and Comprehensive Planning," in The Environmental and Land Use Law Section Reporter of the Florida Bar, Vol. 37, No.2, 7-14, (Sept. 2015). Thomas Ruppert was co-author with Erin Deady on this article.

[4] Laws of Florida, 2008-227.

[5] Laws of Florida, 2008-191.

[6] *See, e.g.* Florida Fish and Wildlife Conservation Commission, Climate Change, available at http://myfwc.com/conservation/special-initiatives/climate-change/.

[7] *See, e.g.* Florida Dept. of Econ. Oppt'y, Adaptation Planning, available at http://www.floridajobs.org/community-planning-and-development/programs/community-planning-table-of-contents/adaptation-planning.

[8] Various products of the work of the Department of Economic Opportunity can be found on their "Adaptation Planning" website at http://www.floridajobs.org/community-planning-and-development/programs/community-planning-table-of-contents/adaptation-planning.

Protection has primarily been limited to work related to climate change impacts on coral reefs.[9]

In 2011, HB 7207, a nearly 200-page bill, became law and essentially overhauled the state's growth management policy, which is laid out in Chapter 163, Florida Statutes (F.S.). Included in the changes were many viewed as contentious by planners and conservationists, with a fundamental reorganization of the Florida Department of Community Affairs into a new Florida Department of Economic Opportunity; a shift that retained some planning and growth management functions but clearly focusing more on the state's economic development policy. In addition, the law reduced state oversight of local planning decisions and actions, concentrating more on resources and issues with statewide significance. Relative to climate change, the law eliminated many, but not all, of the energy efficiency and greenhouse gas reduction provisions from Chapter 163, F.S.; these had been added only three years before in 2008 with the passage of HB 697. Although many of these provisions were eliminated, there are several that remain even today to consider in the planning context:

- Comprehensive planning standards on data in Section 163.3177(1)(f), F.S. "All mandatory and optional elements of the comprehensive plan and plan amendments shall be based upon relevant and appropriate data and an analysis by the local government that may include, but not be limited to, surveys, studies, community goals and vision, and other data available at the time of adoption of the comprehensive plan or plan amendment."
- A project could be considered "sprawl" if: (VIII) plan or plan amendment allows for land use patterns or timing which disproportionately increase cost in time, money or energy of providing and maintaining facilities/services, including roads, potable water. Section 163.3177(6)(a)9.a.(VIII), F.S.
- "Discourage the proliferation" of sprawl if: project incorporates a development pattern or urban form that achieves four (4) or more of the following ... promotes conservation of water and energy. Section 163.3177(6)(a)9.b.(IV).
- Conservation element: must contain principles, guidelines, and standards for conservation that provide long-term goals to protect air quality. Section 163.3177(6)(d)2., F.S.

The 2011 law eliminated Chapter 9J-5 from the Florida Administrative Code (F.A.C.), which was an implementing rule providing detailed guidance on comprehensive plan implementation pursuant to Chapter 163, F.S. While the implementing rule was eliminated, some of that guidance was incorporated into Chapter 163, F.S. The stated goal was to eliminate duplication between the Chapter itself and the implementing rule by consolidating the policies into one location. The sections of the F.A.C. that were eliminated included details and requirements related to post-disaster and resilience planning, evaluation of erosion and accretion trends and their impacts, public access, infrastructure, and other considerations. The new requirements are more flexible

---

[9] *See, e.g.* Florida Dept. of Environmental Protection, Climate Change and Coral Reefs, available at http://www.dep.state.fl.us/coastal/programs/coral/climate_change.htm.

than what was required pursuant to Chapter 9J-5, F.A.C.[10] How this flexibility ultimately impacts local government post-disaster redevelopment planning is still an evolving matter.

The 2011 law also opened some doors. It added the option for local governments to address sea level rise adaption as part of the Coastal Management Element of local government comprehensive plans through the establishment of optional adaptation action areas. Potential criteria for such an area are broad and include, but are not limited to: areas for which the land elevations are below, at, or near mean higher high water; areas with a hydrologic connection to coastal waters; or areas that are designated as evacuation zones for storm surge.[11] This addition is reinforced by a definition for "adaptation action area" or "adaptation area," which is "a designation in the coastal management element of a local government's comprehensive plan which identifies one or more areas that experience coastal flooding due to extreme high tides and storm surge, and that are vulnerable to the related impacts of rising sea levels for the purposes of prioritizing funding for infrastructure needs and adaptation planning."[12]

Finally, the 2011 law did not prohibit longer timeframes for planning,[13] but the minimum required planning horizons in Florida's comprehensive planning law (5 and 10 years) remain too short to effectively include consideration of climate change and sea level rise impacts, even when taking into account infrastructure or development with usable life spans of many decades.[14]

In summary, the changes to comprehensive planning law in 2011 do not prevent Florida's local governments from engaging in detailed, proactive efforts to increase their resilience to coastal hazards, such as erosion, storms, and sea level rise; but they allowed for a more discretionary function than a prescriptive one. However, the flexible and discretionary mechanisms for resilience and coastal planning may be politically challenging at the local level. One view is that without the "stick" of mandatory requirements, local governments will choose not to address them at all. Another view is that incrementally providing local governments tools to address these issues within the state's comprehensive planning law provides a basis for doing so for those that want to, and allows greater local control and self-determination. Several local governments have proactively utilized provisions in Chapter 163, F.S. to create optional elements of comprehensive plans far exceeding the requirements in Chapter 163 and addressing a range

---

[10] Section 163.3178(2)(f), F.S. contains language somewhat similar to the former requirement of Chapter 9J-5 of the Florida Administrative Code requirement for post-disaster redevelopment plans. The statute provides that the coastal management element shall contain a "redevelopment component which outlines the principles which shall be used to eliminate inappropriate and unsafe development in the coastal areas when opportunities arise."

[11] Laws of Florida, 2011-139, codified at Section 163.3177(6)(g)10, F.S. (2016).

[12] Laws of Florida, 2011-139, codified at Section 163.3164(1), F.S. (2016).

[13] The new law allowed "additional planning periods for specific components, elements, land use amendments, or projects." Section 163.3177(5)a, F.S. (2016).

[14] Section 163.3177((5)(a), F.S. stating, "Each local government comprehensive plan must include at least two planning periods, one covering at least the first 5-year period occurring after the plan's adoption and one covering at least a 10-year period."

of topics including climate change, adaptation, energy, or combinations of all of these on some level.[15]

As discussed in chapter 6 of this book, changes to the National Flood Insurance Program in 2012 and 2014 hit Florida extremely hard. As part of its efforts to deal with the impacts of such changes, the 2015 Florida Legislature passed a law entitled "An Act Relating to the Peril of Flood."[16] While important parts of the law directly address flood insurance issues, other portions focus on flooding issues, disaster planning and recovery, and pre-disaster mitigation. Under the 2015 law, coastal management elements must include "A redevelopment component that outlines the principles that must be used to eliminate inappropriate and unsafe development in the coastal areas when opportunities arise."[17] While the redevelopment component itself is not new, what is required to be addressed in the component has been enhanced. The requirements include:

1. Employing development and redevelopment principles, strategies, and engineering solutions that reduce the flood risk in coastal areas that result from high-tide events, storm surge, flash floods, stormwater runoff, and the related impacts of sea level rise.
2. Encouraging the use of best practices development and redevelopment principles, strategies, and engineering solutions that will result in the removal of coastal real property from flood zone designations established by FEMA.
3. Identifying site development techniques and best practices that may reduce losses due to flooding and claims made under flood insurance policies issued in the state.
4. Being consistent with, or more stringent than, the flood-resistant construction requirements in the Florida Building Code and applicable flood plain management regulations set forth in 44 C.F.R. part 60.
5. Requiring that any construction activities seaward of the coastal construction control lines established pursuant to Section 161.053, F.S. be consistent with Chapter 161, F.S.
6. Encouraging local governments to participate in the National Flood Insurance Program Community Rating System administered by FEMA to achieve flood insurance premium discounts for their residents.

Because of the law passed in 2011, Section 163.3178(2)(f)1., F.S. now includes sea level rise as one of the root causes that must be addressed in the "redevelopment principles, strategies, and engineering solutions" to reduce flood risk. Local governments required to have coastal management elements in their comprehensive plans appear to have broad discretion as to how they comply with this new mandate. The law does not specify a date by which local governments must comply. Section 163.3191(1), F.S. still requires local governments to evaluate their plans at least once every seven years to determine if amendments are necessary to reflect relevant

---

[15] Section 163.3177(1)(a), F.S. stating, "The comprehensive plan shall consist of elements as described in this section, and may include optional elements."

[16] Laws of Florida, 2015-69.

[17] Laws of Florida, 2015-69, section 1, codified at Section 163.3178(2)(f), F.S. (2016).

changes in state law. That said, a local government also has the authority pursuant to Section 163.3191(2), F.S. to make a determination that amendments are necessary sooner than that seven-year requirement. With that, local governments do have discretion in how they want to comply with these new future flood risk requirements and could do so sooner than their next required evaluation and appraisal report, if they chose to.

While explicitly now requiring flood risk from sea level rise to be addressed in comprehensive plans (coastal management elements), this shouldn't be considered the single driving force for local governments to address climate change and sea level rise. As outlined previously, policy development based on solid data, infrastructure risk planning, limiting expenditures in coastal areas susceptible to storm damage, protection of air quality, elimination of sprawl, and other notions of balanced planning should be considered holistically when updating comprehensive plans. The importance of considering the state's policy and regulatory structure as a floor, not a ceiling, when dealing with climate and sea level rise issues is a shift that is already starting to occur at the local level.

Another area of state law that should be integrating the reality of sea level rise but has largely failed to do so is Florida's "Coastal Construction Control Line" (CCCL) permitting program.[18] Part of the CCCL program's stated goal is "to preserve and protect [Florida's beaches] from imprudent construction which can jeopardize the stability of the beach-dune system, accelerate erosion, provide inadequate protection to upland structures, endanger adjacent properties, or interfere with public beach access."[19] Building requirements associated with the CCCL program have increased the resistance of coastal construction to storm events generally. However, the CCCL program has many shortcomings that, as demonstrated by the risks to and loss of structures permitted under it, do seem to allow imprudent construction.[20] The CCCL program statutes never mention sea level rise, and the regulations that implement it do not account for sea level rise in permitting of new "major habitable structures." In fact, regulations implementing the CCCL program only mention sea level rise once: regulations allow consideration of sea level rise when determining whether to issue a permit to allow armoring or hardening of the shoreline.[21] Thus, state laws and regulations do not allow the CCCL program to account for sea level rise when

---

[18] The CCCL program appears in statute at Chapter 161, Part I (161.011 – 161.242) (2016).

[19] FLA. STAT. § 161.053 (2016).

[20] Florida's experience with loss of and damage to many coastal homes during Hurricane Matthew's glancing blow to Florida in 2015 demonstrated shortcomings of the CCCL program. Many homes or structures damaged or lost due to coastal erosion were permitted under the CCCL program. *See also, e.g.* Thomas Ruppert, Eroding Long-Term Prospects for Dynamic Beach Habitat in Florida, 1 Sea Grant Law & Policy Journal 65 (2008), available online at http://www.olemiss.edu/orgs/SGLC/National/SGLPJ/SGLPJ.htm. This cited article was based on a more comprehensive document that reviewed several structures permitted under the CCCL program and includes photos of the structures. This is available as the link "White Paper on Dynamic Turtle Nesting Habitat Accommodation in Florida" and "Appendices to the White Paper on Dynamic Turtle Nesting Habitat Accommodation in Florida" at https://www.law.ufl.edu/academics/dynamic-habitat-accommodation-the-policy-framework-for-migrating-shorelines.

[21] Fla. Admin. Code 62B-41.005(7)(c) (2016).

deciding what types of development are permissible, where development may be sited, or how it is designed but do allow for consideration of sea level rise when deciding whether to allow coastal armoring.

Florida has extensive state requirements for stormwater and wetlands permitting through the state's Environmental Resource Permit (ERP) program.[22] The ERP program regulates activities that alter the flow of surface waters, such as construction creating stormwater; the ERP also regulates dredging and filling in wetlands or surface waters. However, the rules for implementing the ERP program do not currently consider future conditions specifically related to changing rainfall patterns or increasing sea levels. Thus, the ERP program, when permitting construction in wetlands in coastal areas, may be permitting construction that will soon experience significant impacts from sea level rise. Likewise, without taking changing rainfall patterns into account, such as the increased incidence of heavy rainfall events, design requirements for stormwater systems will, in the future, likely not achieve the level of service intended.

Finally, in the 2017 legislative session, SB 464/HB 181 (now Section 252.3655, F.S.) passed creating an interagency workgroup to share information on the current and potential impacts of natural hazards throughout the state. The goal is to coordinate the ongoing efforts of state agencies in addressing the impacts of natural hazards, and collaborate on statewide initiatives to address the impacts of natural hazards. "Natural hazards" was specifically defined to include, extreme heat, drought, wildfire, sea-level change, high tides, storm surge, saltwater intrusion, stormwater runoff, flash floods, inland flooding, and coastal flooding. The workgroup is to meet quarterly to share information, leverage agency resources, coordinate ongoing efforts, and provide information for inclusion in the annual progress report to the Governor, the President of the Senate, and the Speaker of the House of Representatives starting January 2019.

Local Government: The Real Seat of Climate Change and Sea-Level Rise Action in Florida

Evolving approaches to climate and sea level rise policy at the state level, depending on what administration is in place, have left some local governments wondering how to address this important challenge. Whether driven by local politics or efforts to comply with state or federal planning requirements, there is recognition that, on some level, policy discussion must occur. Some of the most impacted communities have also begun to work together regionally to address the issue regardless of what legislative or administrative requirements may or may not exist.

Local governments have a visceral connection to climate change and sea level rise; when exceptionally heavy rainfall events occur—when heat waves hurt people, when droughts limit

---

[22] State permitting is also conducted in conjunction with U.S. Army Corps of Engineers permitting. *See* Florida Department of Environmental Protection, "What is the Environmental Resource Permit (ERP) Program?", available at http://www.dep.state.fl.us/WATER/wetlands/erp/index.htm.

water supplies, and when sea level rise causes flooding—more often than not the first ones to hear from the residents impacted are local governments. As one person put it: "If the road in front of your house is flooded, you will probably call the local government before you call the state or federal government." Some local governments have responded by taking the lead in developing planning language and local resolutions and ordinances that account for climate change and sea level rise.

In terms of implementation of the 2015 Peril of Flood legislation, 195 local governments are required to have a coastal management element in their comprehensive plans (161 municipalities and 34 counties). As of May 2017[23]:

- 43 (22%) explicitly address sea level rise in their comprehensive plans
- Eleven mention Adaptation Action Areas (AAAs) in their comprehensive plans (six of these are located in southeast Florida)
- Six have a physical designation:
    - Satellite Beach designates coastal high hazard areas as AAAs
    - Village of Pinecrest designates AAAs
    - Broward County sand bypass project at Port Everglades
    - Ft. Lauderdale 16 areas 38 stormwater projects
    - Yankeetown
    - Fernandina Beach

The following local governments have addressed the new Peril of Flood requirements in Section 163.3178, F.S. within their comprehensive plans or updates to them. Several others are in process (Levy and Santa Rosa). But as of May 2017, the following now have some language, either proposed or final and adopted, related to redevelopment principles, plans, or strategies that address future flood risk including sea level rise.

- North Miami
- Miami Beach
- Lake Park
- Ponce Inlet
- Sunny Isles Beach
- St. Petersburg
- Boynton Beach
- Jupiter Inlet Colony
- West Palm Beach
- Jupiter
- Yankeetown

---

[23] Erin L. Deady, Why the Law of Climate Change Matters: From Paris to a Local Government Near You, Florida Bar Journal (November 2017), available at: https://www.floridabar.org/news/tfb-journal/.

- Palm Beach
- Clearwater
- Broward County
- Pinecrest

Some local governments also have optional elements of their comprehensive plans addressing adaptation, sea level rise, energy, or a combination. For instance, Broward County has a "Climate Change Element" and Monroe County has an "Energy and Climate Element." Both have similar characteristics in that they address things such as greenhouse gas reductions on the mitigation side through energy policies and otherwise; but they also address linkages between a changing climate and the built environment, such as infrastructure and land use. Using the optional element approach to address these issues within a comprehensive plan probably affords a local government the widest latitude to address their issues at the most individual level through direct policies and commitments to further develop data to support future policy or both.

Additionally, several local governments in Florida are using their own resources or grant funding to develop datasets upon which to develop new climate-related policies. Some are doing this in furtherance of the Peril of Flood legislation, and some just because it makes good planning sense. Partnerships between not-for-profit organizations and local governments, or just local governments themselves are a driving factor in opportunities local governments are seizing to address these issues. A dual dynamic is also evolving whereby the policy structure is being put into place, but concurrently, local governments are starting to implement actual strategies to build resiliency into their communities. This implementation aspect is taking on many forms including stormwater master planning; actual project development for roads or other drainage infrastructure; identification, analysis and retrofits of actual vulnerable facilities; beach and shoreline protection; and the raising or flood-proofing of structures.

Regional Collaboration

In 2010, southeast Florida created what has become one of the best-known regional collaborations on climate change in the United States. Palm Beach, Broward, Miami Dade, and Monroe counties joined together to officially form the Southeast Florida Regional Climate Change Compact (Compact). The Compact had its genesis in part from a realization that parallel lobbying efforts by the counties could be strengthened by working from similar baselines.[24] While the official voting members of the Compact's Steering Committee include two members from each county and one member from a municipality from each county, several other organizations have actively worked in coordination with the Compact, including the U.S. Army

---

[24] David L. Markell, Sea-Level Rise and Changing Times for Florida Local Governments, 42 Colum. J. Envtl. L. (forthcoming 2016).

Corps of Engineers, the National Oceanic and Atmospheric Administration, the U.S. Environmental Protection Agency, the South Florida Regional Planning Council, the South Florida Water Management District, and The Nature Conservancy, among others.

One of the first activities of the Compact was development by the "Sea Level Rise Technical Ad Hoc Working Group" of a white paper on sea level rise projections for Compact members in their planning efforts. This document was updated in the Compact's 2015 "Unified Sea Level Rise Projection" paper.[25] While the Compact has generated dozens of valuable documents, this one has played an important role in helping bring together Compact members and others on the committee that developed and updated it.

Another crucial document, both in development and implementation, is the Compact's Regional Climate Action Plan, or RCAP. Developed in 2012, the RCAP was again a broadly collaborative project involving nearly 100 subject-matter experts.[26] The RCAP contains recommendations in seven broad areas: sustainable communities and transportation planning; water supply, management and infrastructure; natural systems; agriculture; energy and fuel; risk reduction and emergency management; and outreach and public policy. The RCAP also contains 110 recommendations intended to be implemented "through existing local and regional agencies, processes and organizations."[27] The RCAP has resulted in several workshops as well as an implementation guide, and the Compact has completed surveying of municipalities to determine levels of implementation of RCAP recommendations. The RCAP is, as of late 2017, going through an update process scheduled for completion in December 2017.

The extensive collaboration at the highest level of county government represented by the Southeast Florida Regional Climate Change Compact is not the only model for regional work to address climate change. In northeast Florida, an early effort of the Northeast Florida Regional Council to promote discussion by the business community about the potential risks of sea level rise led to members of the business community beginning to address not only the risks of sea level rise but also of climate change. The Northeast Regional Council and its related Regional Council Institute have been facilitating the work of Public/Private Regional Resiliency Committee.[28]

Florida's Tampa Bay region has also been working towards greater regional collaboration with the Tampa Bay Regional Planning Council and its "One Bay Resilient Communities"

---

[25] Available at http://www.southeastfloridaclimatecompact.org/wp-content/uploads/2015/10/2015-Compact-Unified-Sea-Level-Rise-Projection.pdf

[26] Southeast Florida Regional Compact Counties, A Region Responds to a Changing Climate: Regional Climate Action Plan (Oct. 2012), http://www.southeastfloridaclimatecompact.org/wp-content/uploads/2014/09/regional-climate-action-plan-final-ada-compliant.pdf. *See also,* David L. Markell, Sea Level Rise and Changing Times for Florida Local Governments, 42 Colum. J. Envtl. L. (forthcoming 2016).

[27] Southeast Florida Regional Compact Counties, A Region Responds to a Changing Climate: Regional Climate Action Plan vi (Oct. 2012).

28 http://www.rcinef.org/P2R2.html.

working group; together with Florida Sea Grant, they facilitated creation of the Tampa Bay Climate Science Advisory Panel. In August 2015, this panel released a report entitled "Recommended Projection of Sea Level Rise in the Tampa Bay Region." This effort, focused on developing a regional sea level rise projection scenario similar to the one produced by the Southeast Florida Regional Climate Change Compact, served as a way to bring together a plethora of local actors, including local governments, state and federal agencies, academic institutions, and non-profit entities. The resulting sea level rise projections are now being integrated into policy and codes by some local governments in the region.

Sea level rise and climate change do not respect political borders at the local level any more than at the global level. Developing regional approaches to climate change can help build momentum for efforts to reduce the causes of climate change, while also making it possible to more rationally address the impacts that sea level rise will have on our communities.

Drainage and Road Infrastructure—the Canaries in the Coal Mine

Climate change and sea level rise were once thought to be issues for the future. But communities in Florida and elsewhere have seen that climate change and sea level rise are already impacting them.[29] Local governments are struggling with water supply issues brought on by more severe dry seasons and increased saltwater intrusion, heavier rainfall and flooding, warmer temperatures, and flooding exacerbated by rising seas. This subsection looks at how the challenges of roads and drainage are an example of the challenges that local governments—and all of us that live in them—face. This subsection focuses primarily on sea level rise rather than the full spectrum of climate change impacts.

Roads and drainage systems are among the most basic infrastructure that makes areas inhabitable. Development of an area requires road access. And for decades, most development in Florida has required dealing with issues related to stormwater.[30] Roads and drainage, typical of infrastructure, are most conspicuous in how little people think or talk about them when everything is working well.

But roads and drainage have long presented challenges in certain parts of Florida. Drainage has been a challenge due to the very flat nature of much of Florida, resulting in efforts to drain land that is miles from the ocean but just a few feet above sea level. Roads may present problems either due to lack of drainage or, along the coast, hazards such as erosion.

While not specifically about sea level rise per se, an instructive case about the challenges that local governments face regarding roads and other infrastructure presented itself in 2011 northeast

---

29 See, e.g. 2014 National Climate Assessment, Overview Introduction, available at http://nca2014.globalchange.gov/highlights/overview/overview#intro-section-2.
30 Programs impacting stormwater regulation include: Florida's implementation the National Pollution Elimination Discharge System, the Florida Department of Environmental Protection's Nonpoint Source Management Program, and the Environmental Resource Permit (ERP) Program managed by Florida's five water management districts.

Florida.[31] Property owners sued St. Johns County for a "taking" of private property because the Atlantic Ocean had been undermining the only road access to their homes since long before most of the homes were built; the County had been unable to keep the road in an equivalent condition to other county roads. After the property owners lost at the trial court level, an appellate court ruled that "[t]he County must provide a reasonable level of maintenance that affords meaningful access, unless or until the County formally abandons the road."[32] However, in many areas where the road used to be there was nothing left but wet sand beach, and part had been entirely washed away by a new inlet. Estimates of the cost at that time for a beach nourishment project, which would have been a precursor to reconstructing the road, were over $13 million up front and $5.7-8.5 million every 3-5 years for maintenance. This would have been a major financial burden for a county that then had a population of about 170,000.

Ultimately the case settled with the property owners and the county in very similar positions to the ones they were in prior to several years of litigation. The county had, however, expended nearly $1 million just in litigation costs to defend against an action by owners of a handful of homes that were built or purchased on land that already had almost 50 years' history of obvious erosion problems. St. Johns County, during the litigation, passed a new ordinance that would have would have required building permit applicants to sign a "hold harmless" agreement in order to receive a building permit.[33] The trial court judge called this policy "repugnant," but an almost identical policy has been consistently used for over a decade in California when issuing permits to build along the coast. During Hurricane Matthew in 2016, the road and properties at issue in the St. Johns case were heavily damaged. Virtually all the remaining road that had been repaired and repaved under the settlement agreement three years earlier was wiped out. A breach in the spit of sand created a new inlet that cut off three homes from the mainland and left them almost in the waters of the Atlantic Ocean.

The settlement agreement bound the county, in case of a "catastrophic weather event," to "make timely and good faith efforts to obtain state, federal, and/or other available funds to restore, to the greatest extent reasonably possible, [the road]." Additionally, the county and property owners came to an agreement on levels of service for the road in the future, recognizing the environmental challenges impacting the quality of the road. As part of the settlement, the following were agreed to:

- County agreed to use "good faith" efforts to maintain Old A1A in "As Is" condition;
- County agreed to use "timely and good faith efforts" to keep access open;
- County agreed to include the existing paved portion of Old A1A in the pavement management schedule and repave it as needed;

---

[31] Jordan v. St. Johns County, 2011 Fla. App. LEXIS 11337 (Fla. Dist. Ct. App. 5th Dist., June 22, 2011); St. Johns County v. Jordan, 2011 Fla. LEXIS 2819 (Fla., Dec. 5, 2011). The case settled in 2013.

[32] *Id.* at 838.

[33] St. Johns County Ordinance 2008-45.

- County agreed to resurface a 0.3-mile portion of Old A1A to create a connection with New A1A;
- County agreed to remove diminished road access as an impediment to obtaining building permits;
- Property owners agreed to give the county notice and an opportunity to buy properties along this roadway before selling to others;
- County agreed to repeal the requirement that prospective home builders sign "hold harmless" agreements to get building permits;
- Property owners agreed to grant easements to restore access to parcels outside of the existing paved area;
- Agreed to allow transit over county-owned parcels to facilitate access to parcels outside of the existing paved area;
- Agreed to consider recommendations of the Summer Haven Municipal Services Taxing Unit regarding the use of funds; and
- County agreed to pay $75,000 to partially reimburse plaintiff-owners' costs.

Ultimately, the case in St. Johns County points out a new reality that local governments need to consider: the need to provide and maintain infrastructure services to existing or future development needs to be considered as a potentially massive legal and financial liability looming over local governments. This case in St. Johns County indicates a potential for local government liability when a road cannot, in the local government's assessment, realistically be repaired due to some combination of environmental challenges and cost. Currently, this court decision is binding on all trial courts in Florida, but local governments may seek to take a proactive approach to try to minimize potential liability in similar situations.[34] It is also important to note that St. John's County has had an ordinance in place since 2012 that indicates when forces of nature and environmental conditions create difficulties in maintaining a road according to usual design and maintenance standards, the county will apply different criteria.[35] The 2012 ordinance also specifies that in some cases "roads in environmentally-challenging locations" may have unpaved surfaces; substandard lane widths or single lanes; vehicle type, size, and weight limitations; periods of time when the roads may be submerged, be buried by soil, covered by sand or blocked by vegetative debris; and no assurance that emergency vehicles can use or routinely use the road for access. Finally, the 2012 ordinance notes several other things, including that property owners with existing improvements that are accessed by roads that are in environmentally-challenging locations may encounter access issues; that access may be limited by naturally occurring

---

[34] *See, e.g.* Thomas Ruppert, John Fergus & Alex Stewart, Environmentally Compromised Road Segments—A Model Ordinance (October 2015), available at https://www.flseagrant.org/wp-content/uploads/Envirntly_Comp_Rds-FINAL_10.20.15_1.pdf.

[35] St. Johns County Ordinance No. 2012-35 (December 2012), available at http://www.sjccoc.us/minrec/OrdinanceBooks/2012/ORD2012-35.pdf.

conditions beyond the control of the county; and that the county has no obligation to build or improve any roads areas designated as "environmentally challenging."

In Monroe County, a case study is currently underway (now in its implementation phase) to "pilot" a methodology in two neighborhoods to determine a basis for a future sea level rise level of service for road improvements and construction. Completed in January 2017, the effort has analyzed the current levels of inundation routinely seen in these neighborhoods, which were severely impacted king tides in 2015 and 2016. The effort has developed numerous options based on elevation targets (with stormwater features) to ensure that alternatives can be permitted and implemented.[36] From a policy perspective, Monroe County is selecting a future scenario of sea level rise and basing design criteria to reduce flooding to a specified return frequency. The County adopted a Resolution that includes design standards capturing these concepts and incorporating the consideration of future sea level rise for the useful life of the project.[37] The concept is to apply typical "level of service" approaches to account for future flood risk. The county has also planned a more comprehensive analysis of the same methodology countywide to develop a phased approach to retool roads based on level of vulnerability in the future.

In terms of flooding, drainage presents similar challenges to that of roads, but the impacts to drainage typically occur first. Problems in the lowest areas begin with sea water rising to the level of outfall pipes which leads to a slowing of drainage. Next, pipes fill with sea water, and as sea levels continue to rise, eventually water backs up out of storm drains and into streets. Many local governments experiencing such drainage problems begin by placing one-way valves that prevent sea water from moving backward into the system. Once that no longer works, pumps are the next logical option. Finally, water quality requirements at the state level and local permitting will like require new or enhanced stormwater features accompany the rehabilitation or retrofit of a road. These stormwater features will be necessary not only to address water quality but also to mitigate any potential impacts to adjacent property owners. The need for such stormwater infrastructure could have a large influence on the ultimate design and cost of a road improvement project.

So far, Florida law does not yet impose the level of liability for drainage whereby sea level rise and legal liability could potentially impact local governments financially the way the St. John's County case discussed above could. Under current law and jurisprudence, it appears that Florida local governments might only be liable for flooding if they fail to maintain their stormwater systems.[38] With regard to services, Florida courts distinguish between upgrading and

---

[36] Monroe County, FL, "Monroe County Pilot Roads Project: The Sands and Twin Lakes Communities" (January 2017), available at: http://monroecountyfl.iqm2.com/Citizens/FileOpen.aspx?Type=1&ID=1038&Inline=True.

[37] Monroe County, FL, Resolution 028-2017.

[38] Thomas Ruppert and Carly Grimm, Drowning in Place: Local Government Costs and Liabilities for Flooding Due to Sea-Level Rise, Florida Bar Journal, Vol 87, No. 9 (2013). However, as this article points out, a lack of local government liability hinges on whether courts find that flooding due to SLR results from

maintenance of infrastructure. Courts have held that the decision to upgrade infrastructure is considered a planning level function to which absolute immunity applies.[39] This should be contrasted with courts holding that failing to maintain infrastructure is an operational activity that exposes the government to potential liability.[40] Therefore, when a local government upgrades, there is now a duty to maintain and operate the system so that it will properly function.[41] That said, liability is a fact-specific inquiry considering the project design, function, history and infrastructure operations.[42] In the face of changing future conditions, such as changing rainfall volumes and tidal inundation, these principles are likely to morph, especially when previously constructed projects can no longer function as designed.[43]

A crucial question for the future is how the law will evolve as local (as well as state and federal) governments find it increasingly expensive and difficult to provide infrastructure services to existing properties in areas more and more subject to coastal hazards. This presents a serious policy conundrum that requires balancing extremely important and fundamental issues, such as rights to access and use of private property, the responsibility of private property owners, fairness to taxpayers, honoring the long-standing admonition in state law to avoid subsidizing development in hazardous areas, and the financial solvency of local governments.

Beaches and Tourism

Florida is the world's number one tourist destination; tourism generates $67 billion of activity in the state. Florida's beaches are the single biggest draw for tourists. Thus, access to beaches for those who do not own coastal property is essential to maintaining the tourism lifeblood in Florida's economic veins. In Florida, many beaches that are considered "public" because the public uses them are actually comprised of private properties with boundaries that reach down to the mean high water line. Under state common law, in some areas the public has established a customary right to use the dry sand beach on private property for recreational purposes if that use has been longstanding, uninterrupted, and without dispute.[44]

Florida's Department of Environmental Protection (DEP) has statutory authority to protect established public use of dry sand beaches when the department authorizes coastal construction.[45]

---

a lack of maintenance, which results in local government liability, or from a need to upgrade the system, which very likely means that there is no local government liability for flooding damage.

[39] Dep't of Transportation v. Konney, 587 So.2d 1292,1296 (Fla. 1991).

[40] Dept't of Transportation v. Neilson, 419 So.2d 1071,1073 (Fla. 1982).

[41] Erin L. Deady, Why the Law of Climate Change Matters: From Paris to a Local Government Near You, Florida Bar Journal (November 2017), available at: https://www.floridabar.org/news/tfb-journal/.

[42] Id.

[43] Id.

[44] City of Daytona Beach v. Tona-Rama, Inc., 294 So.2d 73 (Fla. 1974).

[45] Fla. Stat. §§ 161.021(1), 161.041(1)(a), 161.052(12), 161.053(1)(a), 161.053(4)(e) (2016). Note that the definition in section 161.021(1) protects established customary use rights of the public and states that it

Assessing the extent to which the DEP considers lateral public access in the CCCL permitting process presents challenges since few permits mention lateral public access or include permit conditions to protect it.[46] However, DEP officials have indicated that when lateral public access issues arise, they usually are addressed through design modifications during the permitting process. The most serious challenge remains in that DEP asserts that its current statutory authority to address lateral public access only allows the department to look at the current situation; DEP indicates it lacks authority to look into the future and consider potential or likely erosion or the impacts of sea level rise. This failure to look towards the future of the beach and lateral public access to the beach during the permitting of coastal construction may threaten the future of publicly accessible beaches.

The potential right of the public to customary use of dry sand beach areas has been a point of contention in some areas of Florida. While disputes have arisen around many parts of Florida, they have been particularly acrimonious and continuous for the past few years in Walton County, which is located in Florida's Panhandle.[47]

In July 2016, Lionel and Tammy Alford filed a suit challenging Walton County Ordinance 2016-23 ("Customary Use Ordinance").[48] The suit involved a facial challenge to the Customary Use Ordinance alleging that the County was without authority to enact it. They claimed that certain County regulations pertaining to the dry sand beach on their property violate their free speech rights (Count I) and substantive due process rights (Count II), and they sought declaratory relief (Count III). The Alfords contended that the Obstruction Amendments, which prohibit obstructions on the beach, including "ropes, chains, signs, or fences," effectively prevented them from conveying messages to public beachgoers regarding the boundary of their property, as well as religious and political messages, and precluded them from excluding the general public from their private property.

In October 2016, during the pendency of the case, the County enacted a new ordinance recognizing the public's customary right to use the beaches. Walton Cty. Code Ch. 23, "Customary Use Ordinance" (Ord. No. 2017-10) (amended Mar. 28, 2017, effective April 1, 2017). In December 2016, the Alfords amended their Complaint to add Count IV, challenging the Customary

---

applies to the access referred to in section 161.041(1)(a), but the latter section's definition only specifies access seaward of the mean high water line.

[46] Thomas Ruppert, Eroding Long-Term Prospects for Florida's Beaches: Florida's Coastal Management Policy, 111-13 (2008), available at https://www.law.ufl.edu/_pdf/academics/centers-clinics/clinics/conservation/resources/coastal_management_finalreport.pdf. For two examples of permits that do mention public access, see DEP permit #DA-708 and permit #BO-721.

[47] Customary use battle rages on 30A, nwfdailynews.com (Sept. 11, 2016), available at http://www.nwfdailynews.com/news/20160911/customary-use-battle-rages-on-30a. *See also, e.g.* Erika Kranz, *Sand for the People: The Continuing Controversy Over Public Access to Florida's Beaches*, 83 Fla. Bar J. 10 (June, 2009).

[48] Alford v. Walton County, U.S. District Court for the Northern District of Florida, Case No: 3:16-cv-00362-MCR-CJK

Use Ordinance and seeking a declaration that the ordinance is void ab initio on grounds that customary use is a common law doctrine reserved to the courts for determination on a case-by-case basis, and therefore, the County exceeded its authority and acted ultra vires by legislating customary use on a county-wide basis.

On September 26, 2017 the court entered an order finding for the Plaintiffs stating: "It is declared that the beach obstruction amendments to the Walton County Code, specifically, § 22-54(g)(2)(a)(3), to the extent it defines "obstructions [as] including but not limited to ropes, chains, signs, or fences," and § 22-55 to the extent is states, "Obstructions include, but are not limited to ropes, chains, signs, or fences," are facially unconstitutional in violation of the First Amendment and are STRICKEN. This does not impact any other provision of the Walton County Waterways and Beach Activities Ordinance."

On August 15, 2017 numerous condominium associations and individuals also challenged the Walton County Ordinance in Federal Court as violative of Due Process, Equal Protection and Takings claims.[49] These claims are different than those litigated in the previous litigation with discovery not concluding until February 2018. Given the broader application of these claims, at this time, it is clear Walton County's Customary Use Ordinance remains in contention.

Regardless of how the issues are settled in Walton County or elsewhere, a new combination of court precedent and sea level rise could threaten the public's right to use beaches to which the public currently has a customary right of recreational use, thus potentially undermining Florida's tourism industry. Hurricanes and tropical storms took an extremely heavy toll on beaches in Volusia County during 1999 and 2004. After significant loss of beach sand, Volusia County responded to the eroded, narrower, and more landward position of the dry sand beach by landward adjustment of the driving access lanes on the beach in parts of the county. This readjustment positioned the driving lanes within the confines of some private parcels. Three property owners sued the county for trespass and for a "taking" of their private property that resulted from maintaining parking and driving on the dry sand beach within their property boundaries and allowing public use.[50] The trial court ruled for Volusia County, and the landowners appealed to Florida's Fifth District Court of Appeals.

The appeals court noted that "customary use" might apply to the beach in question. However, the Fifth District Court of Appeals also noted that whether the public's customary use right to the dry sand beach moved landward along with the dry sand beach was unclear. After all, said the court, in instances of "avulsion" or sudden and dramatic loss of beach due to a hurricane or strong storm, property boundaries do not move. But with erosion, property boundaries usually move with the shifting mean high water line. In either case, the court intimated the possibility

---

[49] Seaside Town Council et al v. Walton County, U.S. District Court for the Northern District of Florida, Case No.: 3:17cv682-MCR/CJK

[50] Alfred J. Trepanier, etc. et al. v. County of Volusia, 965 So.2d 276, 278-79 (5th DCA 2007).

that an established customary use easement providing the right of the public to recreational use of the privately owned dry sand beach might not migrate landward with the dry sand beach area.[51]

Due to the case-specific and fact-intensive nature of determining a right to customary use of the dry sand beach by the public, the Fifth District Court of Appeal sent the case back to the trial court for additional fact finding. The trial court then made findings of fact that supported the position that the public had established a customary right to use of the dry sand beach portion of the property owners' parcels.[52] The court noted that over the past century, evidence amply indicated that the location of the dry sand beach had varied dramatically, and that at times in the past the dry sand beach had indeed been located within the property boundaries of the plaintiff property owners' parcels. This finding of fact essentially skirted the key issue raised by the Fifth District Court of Appeals: whether an established public easement for use of the dry sand beach migrates with the dry sand beach onto and could be applied to private property to which it had not previously been applied.

Currently, the ambiguous holding of the Fifth District Court of Appeals in the Volusia County case (*Trepanier et al. v. County of Volusia*), which potentially puts at risk public easements by custom as sea level rise impacts beaches, is the law for all trial courts in Florida; however, if a trial court ruling depending on the *Trepanier* case is appealed in a district outside of Florida's Fifth District Court of Appeals, the case will not bind that appeals court. Ultimately, as the court noted, this issue carries so much significance for Florida that it will eventually have to be decided by the Florida Supreme Court.

Federal Policy and Action Impacting Local Governments

While the heart of adaptation for most people and communities will be their local government, federal law and policy still play important roles that affect communities. As mentioned earlier, major federal changes contributing as drivers for policy movement in Florida are the 2012 and 2014 changes to the National Flood Insurance Program.

The federal government has seen its role, in part, as being a generator of information and data that supports climate change and sea level rise analysis. For example, the National Aeronautics and Space Administration has an extensive network of satellites and other resources that generate

---

[51] The court noted that "if it can be shown that, by custom, use of the beach by the public as a thoroughfare has moved seaward and landward" onto private property, the right of the public remains. However, the court continued by saying that "it is not evident, if customary use of a beach is made impossible by the landward shift of the mean high water line, that the areas subject to the public right by custom would move landward with it to preserve public use on private property that previously was not subject to the public's customary right of use." Alfred J. Trepanier, etc. et al. v. County of Volusia, 965 So.2d 276, 294 (5th DCA 2007). The court may have meant that an established right to customary use *does* move with the dry sand beach but does not if the movement of the beach was due to avulsion. However, due to the court's language, this is not entirely clear.

[52] Trepanier v. County of Volusia, No. 2000-10528-CIDL, ¶¶4-5, 7th Judicial Circuit of Florida (March 30, 2010), 2010 WL 2849823 (Fla.Cir.Ct.).

climate data. The National Oceanic and Atmospheric Administration and the National Weather Service also generate data relevant to climate and sea level rise science. The United States Global Change Research Program is a program in which numerous federal agencies participate and that was mandated by Congress in the 1990 Global Change Research Act to "assist the nation and the world to understand, assess, predict, and respond to human-induced and natural processes of global change." In addition to conducting its own research, federal agencies often support independent researchers in conducting their own work via grant funding.

Additionally, over the last several years, the federal government has pushed federal agencies to consider climate change in their missions.[53] At the federal level, international agreements and policy have also had an impact on federal initiatives. The Paris Agreement of 2015 came into force during November 2016 (on 5 October 2016, the threshold for entry into force of the Paris Agreement was achieved), meaning that countries that are part of the agreement, including the United States, are obligated to meet their "nationally determined contributions" of greenhouse gas reductions.[54]

The Paris Agreement's central aim is to strengthen the global response to the threat of climate change by keeping a global temperature rise this century well below 2 °C above pre-industrial levels and to pursue efforts to limit the temperature increase even further to 1.5 °C. Additionally, the agreement looks to strengthen the ability of countries to deal with the impacts of climate change. To reach these goals, appropriate financial flows, a new technology framework, and an enhanced capacity building framework will be put in place, thus supporting action by developing countries and the most vulnerable countries, in line with their own national objectives. The Paris Agreement also provides for enhanced transparency of action and support through a more robust transparency framework. In 2018, participating parties will assess efforts in relation to the goals set and prepare the nationally determined contributions. There will also be a report card of sorts every five years to assess the collective progress.

Other federal activities that relate to or have contributed to climate change or sea level rise activities at the state and local level in Florida are detailed next.

U.S. Army Corps of Engineers ("Corps")

The Corps has considered sea level change in its planning activities since 1986. This is separate from the regulatory aspects of its mission, but in 2000, sea level change considerations were included within its Planning Guidance Notebook. In 2009 the Corps released its first Engineer Circular (EC), 1165-2-211, "Incorporating Sea-Level Change Considerations in Civil Works Programs," and EC 1165-2-212 "Sea-Level Change Considerations for Civil Works Programs".

---

[53] Executive Order 13653 (Nov. 1, 2013), available at https://www.whitehouse.gov/the-press-office/2013/11/01/executive-order-preparing-united-states-impacts-climate-change.

[54] United Nations / Framework Convention on Climate Change (2015) *Adoption of the Paris Agreement*, 21st Conference of the Parties, Paris: United Nations.

Most recently in December 2013, EC 1100-2-8162 extended this guidance. In July 2014, the Corps created guidance (Engineer Technical Letter 1100-2-1) covering "Procedures to Evaluate Sea Level Change: Impacts, Responses and Adaptation." Guidance is still in effect today.[55] The Corps also has available a tool (called "Beach-fx") to create vulnerability assessments of non-developed natural coastlines or beach protection projects, which was updated for use with the new sea level guidance.[56]

Considered "regulations," these policies establish a framework "for incorporating the direct and indirect physical effects of projected future sea level change across a project lifecycle in managing, planning, engineering, designing, constructing, operating, and maintaining Corps projects and systems of projects."[57] Again, this does not apply to the Corps' regulatory review duties of permits; but rather, the need to take into account changing sea levels only currently applies to projects the Corps is bound to undertake under congressional funding and direction, more commonly known as civil works projects.

National Environmental Policy Act ("NEPA")

On December 24, 2014, the White House's Council on Environmental Quality released revised draft guidance on how federal agencies should evaluate greenhouse gas emissions and the impacts of climate change when conducting reviews pursuant to National Environmental Policy Act (NEPA) evaluation. This guidance updated and expanded previous guidance from 2010 and applied to all proposed federal actions, including land and resource management activities.

Focusing on the climate change and sea level aspects, the new guidance directed agencies to consider the implications of climate change impacts on the proposed action, including potential adverse environmental effects that could result from drought or sea level rise. While agencies have wide discretion in how to consider climate change and sea levels, two key considerations are: 1) reliance on agency experience and expertise to determine whether an analysis of greenhouse gas emissions and climate change impacts would be useful, and 2) application of the "rule of reason" to ensure that the type and level of analysis is appropriate for the anticipated environmental effects of the project. The focus is on the long-term viability of the project, tying design alternatives to climate change effects on a proposed federal action of the useful life of that project. This is especially true in cases when it will be located in a vulnerable area or impact vulnerable populations or resources. With the NEPA guidance, the main message is that while the level of analysis is somewhat flexible, addressing the issue is not.

---

[55] U.S. Army Corps of Engineers, Engineering and Construction Bulletin (September 16, 2016), available at: http://www.iwr.usace.army.mil/Portals/70/docs/frmp/eo11988/EDB_2016_25.pdf

[56] U.S. Army Corps of Engineers, Analyzing Evolution and Cost-Benefits of Shore Protection Projects, available at: http://www.erdc.usace.army.mil/Media/Fact-Sheets/Fact-Sheet-Article-View/Article/476718/beach-fx/

[57] U.S. Army Corps of Engineers, Dept. of the Army, Sea-Level Change Considerations for Civil Works Programs, EC 1165-2-212 (October, 1, 2011).

In August 2016, the Council on Environmental Quality released final guidance[58] for federal agencies on how to consider the impacts of their actions on climate change in their NEPA reviews and how all types of federal actions will impact climate change. The guidance builds from the 2010 draft guidance and 2014 revised draft guidance, and incorporates comments and feedback received. Additionally, finalization of the 2014 revised draft guidance was specifically called for by the State, Local and Tribal Leaders Task Force on Climate Preparedness and Resilience's recommendations to the president. The new final guidance also:

- Advises agencies on how to quantify projected greenhouse gas emissions of proposed federal actions whenever the necessary tools, methodologies, and data inputs are available;
- Encourages agencies to determine the appropriate level (broad, programmatic, or project- or site-specific) and extent of quantitative or qualitative analysis required to comply with NEPA;
- Has agencies consider alternatives that would make the action and affected communities more resilient to the effects of a changing climate; and
- Promotes use of existing information and science when assessing proposed actions.

The guidance is applicable when a federal agency initiates any new NEPA review, and agencies should exercise judgment when considering whether to apply this guidance to the extent practicable to an ongoing NEPA process. Finally, agencies are encouraged to consider applying this guidance to projects in the environmental impact statement or environmental assessment preparation stage. The standard is if it would inform the alternatives analysis or address comments raised through the public comment if (based on science) the review would be incomplete without application of the guidance, and the additional time and resources needed would be proportionate to the value of the information included.

It should be noted that the President's March 28, 2017 Executive Order, "Promoting Energy Independence and Economic Growth," directed the Council on Environmental Quality to rescind this guidance.[59] Given that courts have demanded climate consideration in agency NEPA analyses already, this could cause the administration future legal challenges, despite the directive to rescind it.[60]

The Federal Flood Risk Management Standard

Two significant issues at the federal level that already impact local communities and will continue to even more so in the future include the National Flood Insurance Program and the Federal Flood Risk Management Standard. The National Flood Insurance Program is discussed

---

[58] Final Guidance for Federal Departments and Agencies on Consideration of Greenhouse Gas Emissions and the Effects of Climate Change in National Environmental Policy Act Reviews.

[59] The White House Office of the Press Secretary, Executive Order on Promoting Energy Independence and Economic Growth (March 28, 2017), available at: https://www.whitehouse.gov

[60] *See e.g.*, Ctr for Biological Diversity v. Nat'l Highway Traffic Safety Admin., 538 F.3d 1172, 1217 (9th Cir. 2008).

in detail in Chapter 6, so this section of the book will focus on the Federal Flood Risk Management Standard (FFRMS).

The FFRMS is contained in the 2015 Executive Order 13690 – Establishing a Federal Flood Risk Management Standard and a Process for Further Soliciting and Considering Stakeholder Input. as an amendment to Executive Order 11988 – Floodplain Management, which was issued in 1977. The goal of modifying the 1977 executive order was to ensure that, in light of increased flooding around the country and especially in coastal areas suffering from sea level rise, federal investments do not contribute to flooding and are wisely sited to maximize the utility of the federal investment.

The original 1977 executive order's purpose was "avoiding [federal] actions in or impacting the base floodplain and minimizing potential harm if [federal] action must be located in the base floodplain."[61] To accomplish this, severa steps were delineated for a federal action, including: 1) determining if the project will be in a floodplain; 2) identification of practical alternatives to locating in a floodplain; 3) if location in the floodplain is necessary, identify potential impacts; 4) minimize harm to the floodplain; 5) reevaluation of the proposal in light of previous steps; and 6) release of findings prior to implementation. The 2015 executive order altered these steps as part of a movement beyond "emphasis on flood control and protection to a broader focus on flood risk management.".[62]

The 2015 executive order applies to "federally-funded projects," which means "actions where federal funds are used for new construction, substantial improvement, or to address substantial damage."[63] This is a more limited application than the 1977 executive order, which applied to "federal actions," including when agencies of the federal government engage in (1) acquiring, managing, and disposing of Federal lands, and facilities; (2) providing federally undertaken, financed, or assisted construction and improvements; and (3) conducting federal activities and programs affecting land use, including but not limited to water and related land resources planning, regulating, and licensing activities."[64]

The 2015 Executive Order 13690's FFRMS made three primary changes to Executive Order 11988. First, it modified the way that floodplains are determined. Whereas Executive Order 11988 used the 100-year floodplain as its base, Executive Order 13690 expanded the floodplain used both vertically and horizontally. This occurs in three possible ways. The preferred method,

---

[61] In the context of Executive Order 11988, "base floodplain" refers to the area subject to flooding by the base flood, which, as with the National Flood Insurance Program, (also known as the "100-year" floodplain.

[62] Guidelines for Implementing Order 11899, Floodplain Management, and Executive Order 13690, Establishing a Federal Flood Risk Management Standard and a Process for Further Soliciting and Considering Stakeholder Input, 14 (Oct. 8, 2015).

[63] Guidelines for Implementing Order 11899, Floodplain Management, and Executive Order 13690, Establishing a Federal Flood Risk Management Standard and a Process for Further Soliciting and Considering Stakeholder Input, 16 (Oct. 8, 2015).

[64] E.O. 11988, Section 1 (1977).

if the science is available, is the climate-informed science approach, which consists of customized analysis of the area in question under potential future climate scenarios. Another method is the freeboard value approach, which requires projects be built two feet above the base flood (100-year flood) elevation or three feet above the base flood elevation for critical projects.[65] Finally, agencies may use the "500-year" elevation approach, which is an elevation equivalent to the 0.2%-annual-chance-flood elevation.[66] Whichever method is used, the agency must include not only the increase in *elevation* of the floodplain but also the corresponding *horizontal expansion* of the floodplain.

Second, the 2015 executive order 13690 incorporated the idea of "critical action determinations" by agencies. A critical action "shall mean any activity for which even a slight chance of flooding would be too great."[67]

And third, the 2015 executive order added significant focus on the use of natural features and nature-based approaches "to reduce flood risks, as well as minimize the impacts of Federal actions to natural and beneficial floodplain values and to lives and property."[68]

The 1977 executive order was implemented by regulation for very few federal agencies; most agencies implemented it through policy changes. This is also true for the modifications established by Executive Order 13690; only five agencies—FEMA, Housing and Urban Development, the Army Corps of Engineers, the Tennessee Valley Authority, and the Federal Energy Regulatory Commission—are currently anticipated to adopt new rules to implement the FFRMS. Other agencies will incorporate the FFRMS changes into policy and procedures guiding their work.

The FFRMS is very important to local governments because many projects are built with a contribution of federal funds, and this will trigger use of the FFRMS. One potentially devastating scenario for local governments is that a disaster strikes and the local government begins rebuilding destroyed or substantially damaged infrastructure or buildings. Then, upon seeking eligible reimbursement from the federal government, the local government realizes that it only built facilities to the 100-year floodplain rather than the standards of the FFRMS, and thus is not eligible for federal funds. While, as of late 2016, this scenario has not happened given that no federal agency has yet finalized its implementation of the FFRMS, local governments should be aware that implementation will occur in the near future as (FEMA has completed a draft rule for FFRMS implementation, and other agencies are approaching that mark. In August of 2017,

---

[65] "Critical Action" is defined in the Implementing Guidelines to E.O. 11988.

[66] Exec. Order No. 13,690, 80 FR 6425, § 6(1)(c) (2015).

[67] *Id.* § 2(j) (2015); Guidelines for Implementing Order 11899, Floodplain Management, and Executive Order 13690, Establishing a Federal Flood Risk Management Standard and a Process for Further Soliciting and Considering Stakeholder Input, 35, 38-39 (Oct. 8, 2015).

[68] Guidelines for Implementing Order 11899, Floodplain Management, and Executive Order 13690, Establishing a Federal Flood Risk Management Standard and a Process for Further Soliciting and Considering Stakeholder Input, 41 (Oct. 8, 2015).

President Trump signed an executive order that eliminated the Federal Flood Risk Management Standard.[69]

Great uncertainty surrounds the future of federal involvement and support of climate change and sea level science and adaptation. The evidence is mounting that the Trump administration has taken a very different stance on climate change and sea level rise than the previous administration. Regardless of how the federal response may change, local governments in Florida will continue to increase their activity on adapting to rising seas since they are seeing more flooding each year.

Conclusion

Florida has taken many steps at the state, regional, and local levels to begin addressing the challenges of climate change and sea level rise. However, the current state requirements alone remain far from sufficient to build a more resilient future for Florida. Over the past few years, major rainfall events have caused massive flooding in Florida's Panhandle and in southeast Florida; and after a decade without a major hurricane, we have seen the paths of destruction left by Hurricane Matthew off Florida's East Coast and Hurricane Irma through much of the peninsula.

While events of the past have slowly pushed Florida and its local governments towards more resilience, such as stronger building codes, the state must now confront two realities that merge to create a tremendous policy challenge. First, even after past events, Florida has, as a state, been largely unwilling to address where we build; the focus after disasters has been almost entirely on how we build. While strengthening building standards is good, we need further discussion about what types of land uses may or not be appropriate in certain hazardous areas. Second, the past is not a good indicator of the future. Currently, permitting programs and planning take place almost exclusively based on data of past trends. Whether it be rainfall, calculation of five-year, 25-year, or 100-year storm events, or erosion rates, we look to a past that is no longer indicative of the future we need to face.

To address these weaknesses, permitting programs should look to the future. This raises many questions. For example, should water supply and storage be expanded to account for the possibility of increasingly severe droughts? And even if there are more severe droughts, should permitting of drainage systems be modified to account for heavier rainfall events when the rains do come? Amid the many questions there are also some clear needs. Construction of drainage systems needs to account for future sea levels rather than being engineered solely for today's sea level. Sea level projections should be incorporated into both local planning decisions as well as

---

[69] Presidential Executive Order on Establishing Discipline and Accountability in the Environmental Review and Permitting Process for Infrastructure, Executive Order ----, (August 15, 2017), available at https://www.whitehouse.gov/the-press-office/2017/08/15/presidential-executive-order-establishing-discipline-and-accountability.

into the state of Florida's Coastal Construction Control Line permitting program to inform what is built, perhaps allowing some uses in at-risk areas but not allowing more sensitive uses. Some policy options most appropriately fit in state programs and others at the local level, but we need to work together at the local and state level to effectively incorporate climate change and sea level rise into all relevant areas of policy and planning in Florida.

Finally, in the aftermath of hurricanes Harvey, Irma, and Maria in 2017, it remains to be seen if Federal and state responses will shift to better link the future of flood risk with climate and sea level rise considerations. After Hurricane Sandy, there were some shifts in policy and a noticeable rise in legal cases that shaped some recovery and resiliency strategies. The Florida House has established a Select Committee on Hurricane Response and Preparedness. The extent to which that committee delves into the issues of future flood risk related to climate change, for both coastal and inland communities, remains to be seen.

CHAPTER 8

Climate Change Impacts and Adaptation in Florida's Agriculture

Young Gu Her[1], Kenneth J. Boote[2], Kati W. Migliaccio[3], Clyde Fraisse[3], David Letson[4], Odemari Mbuya[5], Aavudai Anandhi[6], Hongmei Chi[7], Lucy Ngatia[5], and Senthold Asseng[3]

[1]*Agricultural and Biological Engineering Department / Tropical Research and Education Center, Institute of Food and Agricultural Sciences, University of Florida, Homestead, FL;* [2]*Agronomy Department / Agricultural and Biological Engineering Department, Institute of Food and Agricultural Sciences, University of Florida, Gainesville, FL;* [3]*Agricultural and Biological Engineering Department, Institute of Food and Agricultural Sciences, University of Florida, Gainesville, FL;* [4]*Marine Ecosystems and Society Department, Rosenstiel School of Marine and Atmospheric Science, University of Miami, Miami, FL;* [5]*Center for Water and Air Quality, College of Agriculture and Food Sciences, Florida Agricultural and Mechanical University, Tallahassee, FL;* [6]*Biological Systems Engineering Department / Center for Water and Air Quality, College of Agricultural and Food Sciences, Florida Agricultural and Mechanical University, Tallahassee, FL;* [7]*Computer and Information Sciences Department, College of Science and Technology, Florida Agricultural and Mechanical University, Tallahassee, FL*

In this chapter, we describe Florida's agriculture, the vulnerability of its crops and livestock to climate change and possible adaptation strategies. Much of Florida's agricultural success is linked to its moderate climate, which allows vegetable and fruit crop production during the winter/spring season as well as the production of perennial crops such as citrus and sugarcane. In addition, there is a substantial livestock industry that uses the extensive perennial grasslands. While rising CO2 is generally beneficial to crop production but detrimental to nutritional quality, increase in temperature will cause mostly negative effects on yield. Florida's agriculture faces additional challenges from climate change characterized by sea level rise and intensified extreme climate events, affecting land and irrigation water availability, livestock productivity and pest and disease pressure. New technologies and adaptation strategies are needed for sustainable agricultural production in Florida, including increased water and nutrient use efficiency in crops, crop and livestock breeding for heat stress, pest and disease resistance and reduced exposure of livestock to high temperature. Irrigation is a favored adaptation, but places an even greater burden or potential conflict between agriculture and community use of water resources.

Key Messages

- Florida's agricultural industries provide over $120 billion in economic revenue to the state, second only to tourism, and support more than two million jobs.
- Florida's diverse climate conditions make it suitable for many crops, fruits, livestock, and seafood, although these are vulnerable to climate variations that occur from year to year.
- Florida's agriculture has a long history of successful adaptations to the vagaries of weather and climate, but climate change poses a challenge that is unprecedented in magnitude and rates of change.
- Although current temperatures are near optimal for growing many of our crops, yields are lower during the hotter seasons that occur now, and additional increases in future temperatures will lead to lower crop yields, creating challenges to the competitiveness of current production systems.
- Florida's agriculture faces additional challenges from climate change characterized by sea level rise and intensified extreme climate events, affecting land and irrigation water

availability, crop yield and quality, livestock productivity, as well as pest and disease pressures.
- The known increases in atmospheric CO_2 concentration can stimulate growth in some crops but will reduce the nutritional value of many food crops. Higher atmospheric CO_2 concentration will also increase canopy temperatures and could add to the adverse effects of temperature.
- New technologies and adaptation strategies needed for sustainable agricultural production in Florida include increased water and nutrient use efficiency in crops, crop and livestock breeding for heat stress, pest and disease resistance, and reduced exposure of livestock to high temperatures.
- Knowledge gaps include an understanding of climate change impacts on growth and nutritional value of vegetable and fruit crops, the dynamics of pests and diseases, and direct and indirect effects (the latter via pasture growth) on livestock and livestock-crop systems.
- New experiments and development of modeling and analysis tools are needed for many of the economically-important agricultural systems in order to better estimate climate change impacts on Florida's diverse agricultural production systems.

Keywords

Florida's agriculture; Climate change; Crops; Fruits; Livestock; Sea level rise; Irrigation; Water resources; Elevated carbon dioxide; Increased air temperature; Rainfall change; Salt water intrusion; Salinity; Climate change adaptation; Cover crop; Conservation tillage; Sod-based rotation; Plastic mulch; Drought tolerant crops; Heat tolerant corps, Drought tolerant forage; Heat tolerant livestock; Livestock facility renovation; Livestock genomic selection; Mixed crop-livestock systems; Decision support systems; Crop modeling

Introduction

Florida is one of the largest crop producers in the United States. The state's agricultural industries contribute over $120 billion to the economic revenue of the state (Putnam 2015a) and the annual market value of Florida's crop products is $7.8 billion (Putnam 2015c). In fact, agricultural products from Florida have increased in export value by about $2 billion US dollars between 2004 and 2014; the state of Florida ranked eighth in the US, with $4 billion in agricultural exports in 2014 (Sleep and Gitzen 2015). The largest importers of Florida products are Canada, Bahamas, The Netherlands, Dominican Republic, Mexico and Colombia (Sleep and Gitzen 2015). Export products from Florida include meat products, orange juice, grapefruit juice, fruits, and nuts. Hence, Florida significantly contributes to the local, regional and global food supply via fresh market vegetables and fruits (Putnam 2015c); and the effects of climate change on agricultural yields, nutritional value and prices in Florida as well as follow-on impacts on food processing, storage, transportation and retailing have implications for food security beyond the state's borders.

Agricultural trends in terms of land area and agricultural products sold have changed in Florida since the 1960s, with the market value of agricultural products increasing and the area of farmland decreasing. The number of farms has increased over this same period, with the average farm size being 81 ha (200 acres) in 2015 (Putnam 2015b). The number of farms in Florida has increased from 44,000 in 2002 to 47,300 in 2015, supporting two million full- and part-time jobs (Putnam 2015b, Putnam 2015c, Sleep and Gitzen 2015).

Florida's agriculture is a major consumer of water. Surface and groundwater fed by rainfall is the main source of irrigation for growing crops, and about 12.1 billion liters (3.2 billion gallons) per day of Florida's water resources are used to grow crops (Marella 1999).

The Florida peninsula covers six degrees of latitude between approximately 25° N and 31° N, with a range of climatic regions from Tallahassee to Key West characterized by differences in frost occurrence, chill accumulation, growing degree accumulation, and solar radiation that affects crops. Average annual temperature ranges from 19.8 °C (67.6 °F) in the north (Tallahassee Airport) to 23.8 °C (74.8 °F) in the south (Homestead General Aviation Airport) (NOAA 2017b). Average annual rainfall ranges from 1,475 mm (58.1 in) in the north to 1,458 mm (59.5 in) in the south, with more rainfall occurring in summer than winter. Average relative humidity is 95% during the summer (Gainesville, mornings in September) and 47% in the winter (Orlando and Tallahassee, afternoons in April). Hurricanes can affect all of Florida but have been infrequent in recent years, with the only hurricanes making landfall in Florida being Wilma in 2005 and two in 2016 (NHC 2017). Parts of North Florida can also be affected by tornadoes. North and Central Florida are classified as humid subtropical, while South Florida includes savanna, monsoon, and rainforest (Peel et al. 2007).

Such a diverse and mild climate makes Florida suitable for growing many different crops including oranges, grapefruit, snap beans (fresh market), cucumbers (fresh market), bell peppers, squash, sweet corn, tomatoes (fresh market), watermelons, sugar cane, tangerines, and strawberries (Putnam 2015c) (Fig. 8.1). South Florida is warm enough for growing vegetables such as sweet corn, tomato, strawberry, green beans, and lettuce, even during winter. Florida's warm winters make it possible to grow tropical fruits and vegetables such as avocado, mango, cassava, boniato, and lychee in South Florida (Campbell 1994; Klassen et al. 2002). North Florida climate conditions are less favorable for this type of tropical fruit production, but are favorable for agronomic grain and fiber crops during the summer growing season (April to September). While some crops are regionally-specific in Florida, others are grown throughout the state but with varying production seasons and different market windows. Florida's agricultural land has been declining in recent decades, mostly due to economic drivers but sometimes due to disease pressure (e.g. citrus greening) and partly due to changes in climate (e.g., citrus). However, at the same time, Florida's agricultural production has grown steadily over the past four decades, due to increased efficiency and expanding irrigation.

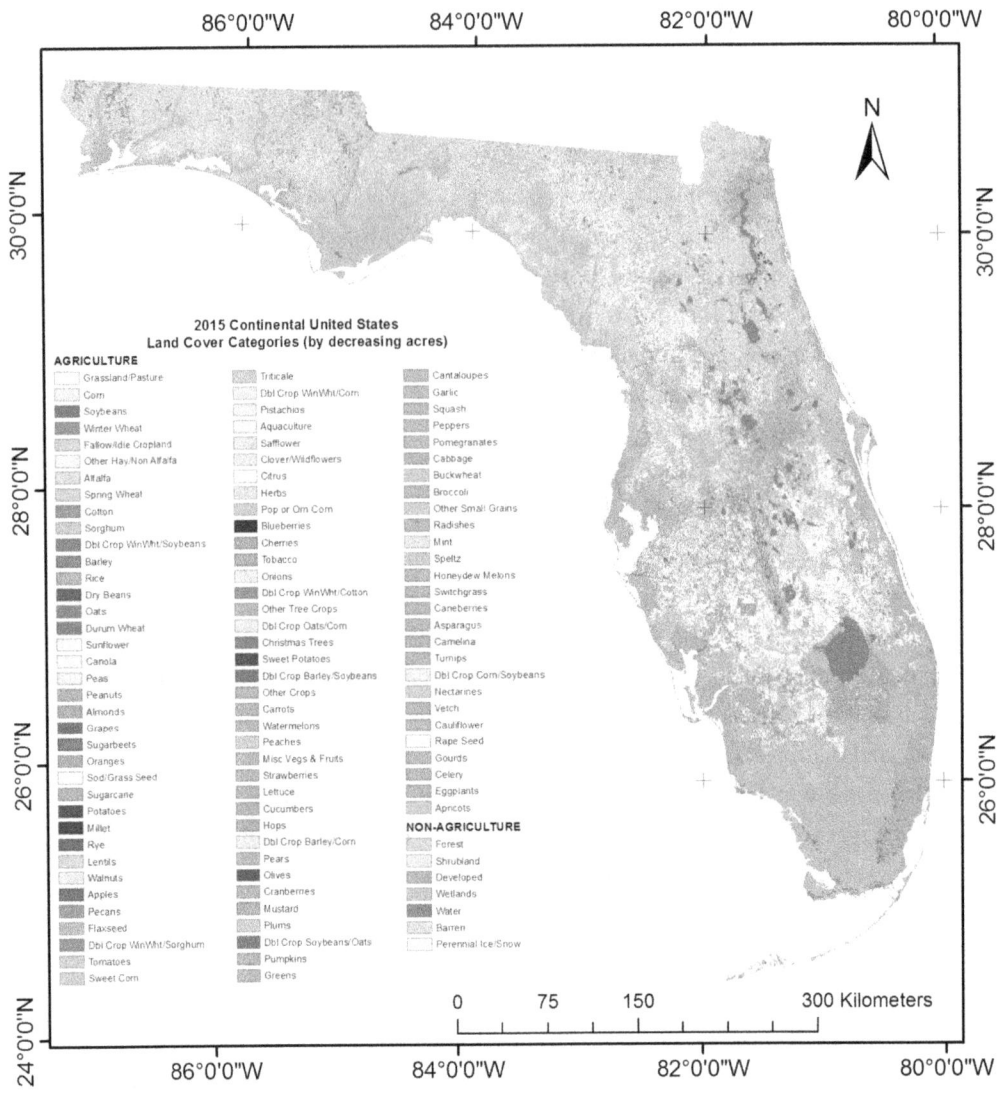

Figure 8.1. Distribution of crop areas in Florida (USDA-NASS 2016).

Crop growth and productivity, as well as the occurrence of pests and diseases, are all influenced by climate factors, including atmospheric CO_2 concentration, temperature, and rainfall. Crops can directly suffer damage from low temperatures (e.g., freezing), high temperatures (e.g., heat stress), strong winds (e.g., tornadoes and hurricanes), periods of low rainfall (e.g., droughts) or intensive rainfall events that can cause runoff, flooding, and/or erosion. Many pests and diseases flourish in high relative humidity conditions. Any change in climate factors will, therefore, affect crops directly or indirectly via pests and diseases (Walthall et al. 2013).

The frequency and intensity of extreme events such as heavy storms, flooding, hurricanes, and drought are expected to increase under projected climate change scenarios due to elevated temperatures and a resulting increase in the moisture-holding capacity of the air (National Academies of Sciences 2016; Anyamba et al. 2014). Agricultural productivity and water resources can be degraded with unfavorable sequences from these weather events, resulting in: substantial losses of soil, nutrients, and fertilizers in agricultural fields; pollutant loadings to waterbodies; and subsequent water quality issues, particularly in agriculturally dominated regions (Whitehead et al. 2009; Johnson et al. 2015). The National Climate Assessment (2014) concluded that climate change "is already affecting the American people in far-reaching ways" and that the future will be unlike the past. In Florida, some impacts of climate change on the state's agricultural economy have already been observed (Maul and Martin 1993; Scavia et al. 2002; Sallenger et al. 2012). For example, studies have found that the crop yields of vegetables such as snap beans, bell peppers, and tomatoes are related to El Niño Southern Oscillation (ENSO) phases. ENSO is an irregularly periodical variation in winds and sea surface temperatures over the tropical eastern Pacific Ocean, affecting much of the tropics and subtropics including Florida. Climate variability characterized by freeze probabilities have also directly affected Florida's citrus production (Hansen et al. 1999; Miller and Glantz 1988). And the fact is that climate change and variability will continue to affect Florida's agricultural productivity in the coming decades through increased or more intense occurrences of extreme events, such as droughts, floods, and storms (Adams et al. 1990; Hansen et al. 1998; Reilly et al. 2003; Gao et al. 2012).

The United States Department of Agriculture (USDA) defines food insecurity as "limited or uncertain availability of nutritionally adequate and safe foods or limited or uncertain ability to acquire acceptable foods in socially acceptable ways" (Anderson 1990). And the USDA's National Food Security Surveys are the main tools to measure food security. The most recent survey in 2015 found that 12.7% (one million) of the eight million Florida households were food insecure, and 5.4% (447,000) of the households had very low food security. Climate change will likely make it more difficult to improve those statistics, thus the poor will suffer more from climate change (Lobell et al. 2008; Mendelsohn et al. 2006). Understanding agricultural implications of climate change is critical in developing the climate change adaptation strategies and methods needed to achieve improved food security and agricultural sustainability in Florida.

In the coming decades, Florida farmers will need to take steps to adapt to climate change; that is, they must choose to make investments today to offset climate changes' negative impacts and take advantage of possible positive impacts in the future. The economic challenge will be for farmers to be able to increase productivity and incomes while they cope with temperature and precipitation patterns that are increasingly likely to be unfavorable. This chapter gives an overview of Florida's agriculture and describes expected agricultural impacts of climate change. In addition, agricultural adaptation strategies to climate change in Florida are discussed, and recommendations are made for future research needs.

Florida's Agriculture

Florida's agriculture is among the most diverse in the U.S., contributing over 300 commodities to national and international markets (Putnam 2011; Putnam 2015c). Land dedicated to agriculture in Florida has slowly decreased from 4.2 million ha (10.3 million acres) in 2002 to 3.8 million ha (9.4 million acres) in 2015 (USDA-NASS 2017; Putnam 2015b; Fig. 8.2). The market value of Florida's agricultural products from the state's 47,600 commercial farms was estimated to be $8.5 billion US dollars in 2013 (Putnam 2015c; Figs. 8.3 and 8.4), and that market value has grown steadily over the past four decades (Fig. 8.3).

Florida agriculture encompasses a variety of commodity groups including grains, fiber, vegetables, fruits, nursery and floriculture, livestock, and aquaculture. Of these, the crops producing the greatest sales are oranges, sugarcane, foliage plants, strawberries, tomatoes, and peppers (Fig. 8.5). In the 2013-2014 season, Florida accounted for 59% of the total U.S. citrus production with 105 million boxes of citrus (Putnam 2015c). And Florida ranked second in the nation in the production of greenhouse and nursery products as well as vegetables including melons and potatoes, with cash receipts totaling over $7.7 billion (USDA-NASS 2017). Livestock also contributes to Florida's commodity receipts, with cattle and calves, milk, poultry, and eggs being the most prominent.

Florida crops are often irrigated due to frequent periods of low rainfall in the state, particularly in winter. In 2012, approximately 11,744 farms (25%) had irrigated crops (USDA-NASS 2012). The total irrigated area in Florida is estimated to be 1.65 million ha (4.08 million acres), which is 21% of the total agricultural areas in Florida, using an average 3.5 million liters per hectare (350 mm or 0.37 million gallons per acre) annually (FDACS 2016; USDA-NASS 2012). A variety of methods are used for irrigation including center pivot, drip, gravity systems, sprinkler, spray, overhead, and traveling irrigation guns. In Florida, water is also applied to prevent crops from freezing (frost damage). Citrus and sugarcane production is estimated to apply 5 billion liters (1.3 billion gallons) of water per day on average, with some of this for freeze-protection; this is 61% of the total irrigation water (8 billion liters per day or 2.1 billion gallon per day) applied in Florida (FDACS 2016). And agricultural water use is expected to increase 17% by 2035 (FDACS 2016).

Each year, Florida dairies produce more than 1.1 billion kilograms (2.5 billion pounds) of milk valued at $560 million, from 122,000 cows (Putnam 2015c; USDA-NASS 2017) (Fig. 8.6). There are 1.7 million head of beef cattle (0.9 million beef cows and 0.8 million calves) on Florida farms and ranches, primarily located in southwest Florida (Okeechobee, Osceola, and Polk counties). The cash receipts from cattle and calf marketing are about $838 million. Nationally, Florida ranked 10th in beef cows and 19th in milk cows. Hens and pullets of laying age on farms amount to about nine million birds that produce 2.4 billion eggs corresponding at a market value of $219 million every year. Florida also produces about 63 million broilers each year, valued at

$170 million. Florida has about 18,000 hogs valued at $3.1 million, and 54,000 goats for milk and meat in Florida.

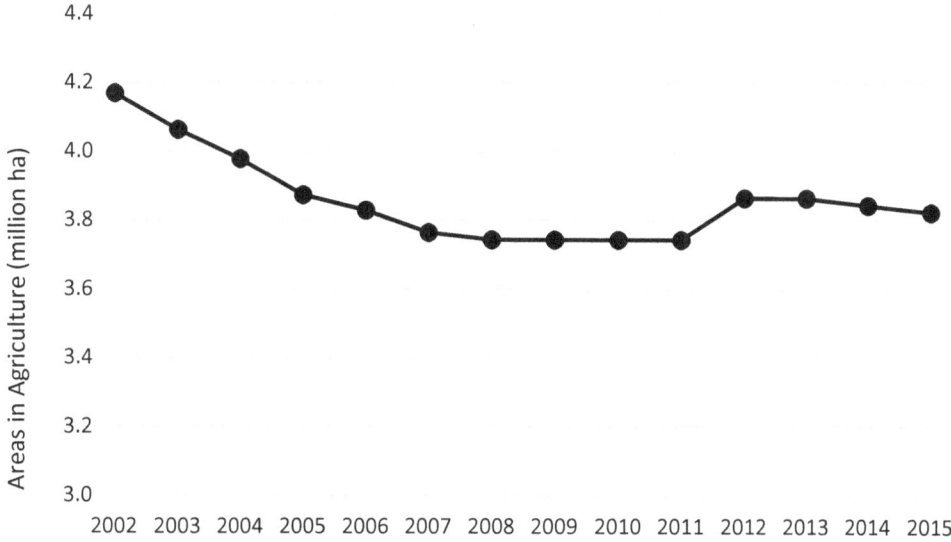

Figure 8.2. Agricultural areas in Florida (Putnam 2015b).

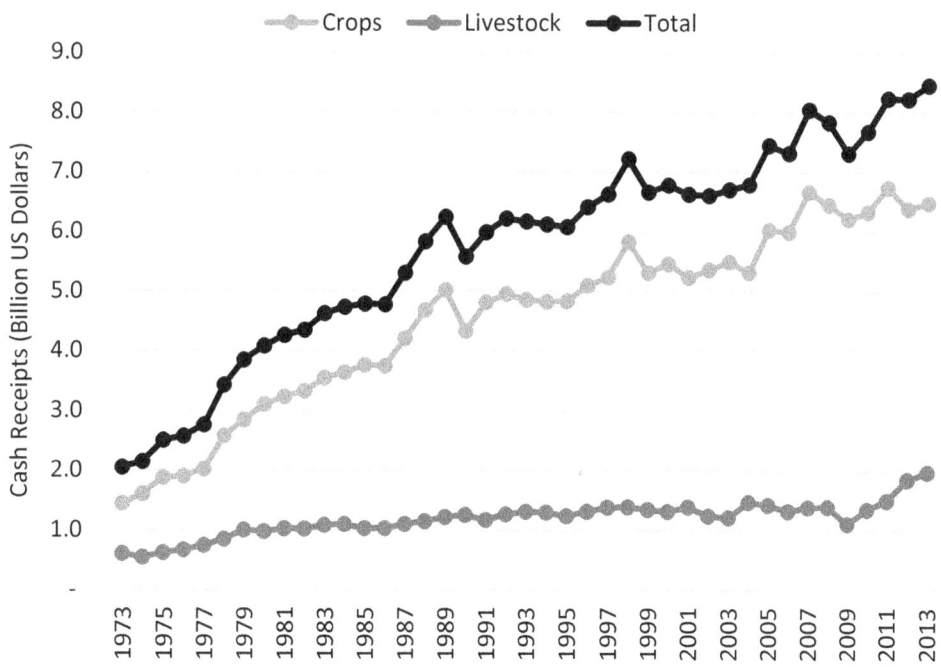

Figure 8.3. Florida's agricultural production in cash receipts (Putnam 2015c).

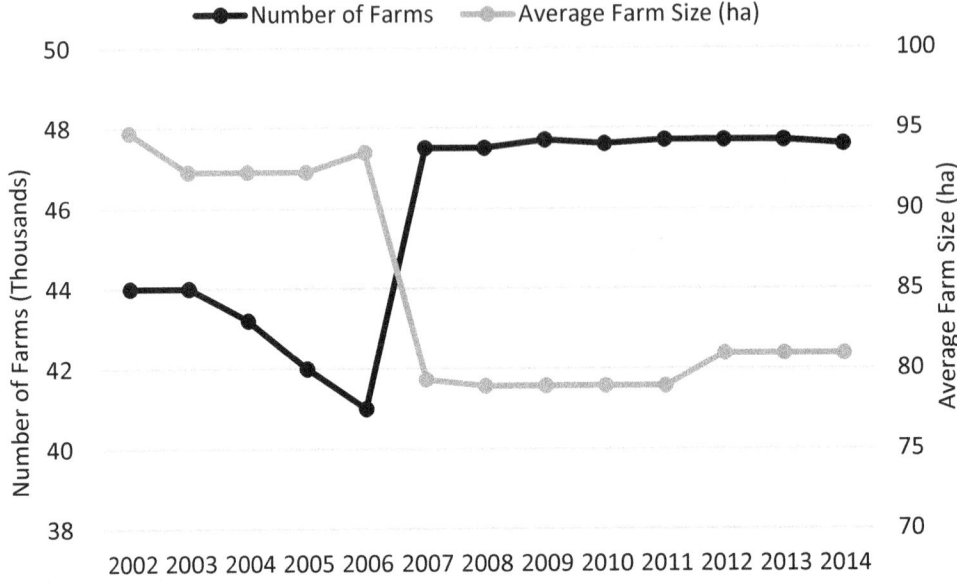

Figure 8.4. Number and average size of farms in Florida (Putnam 2015c).

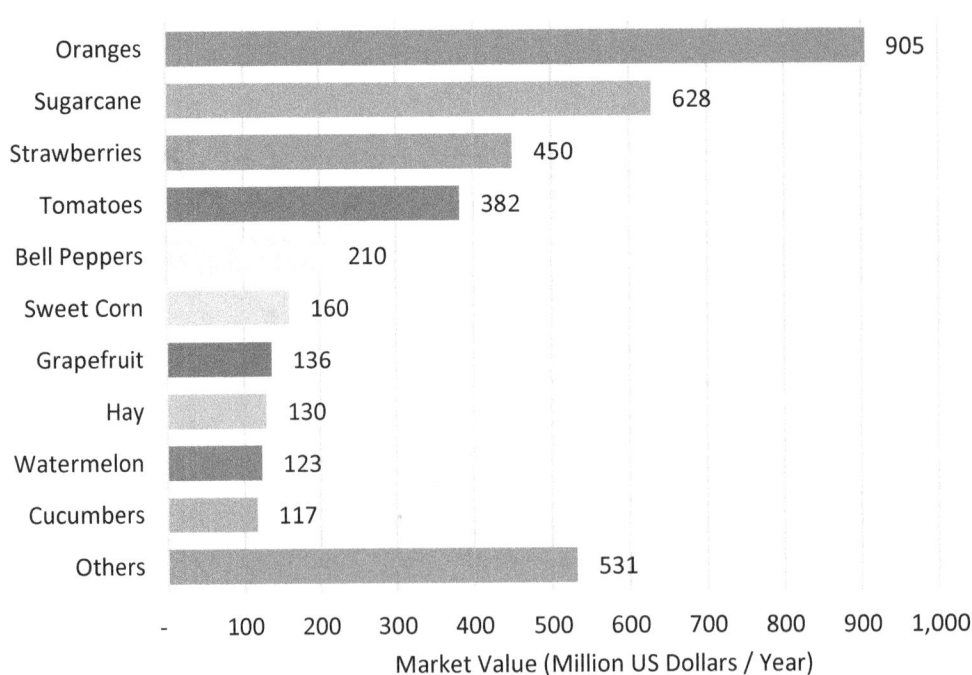

Figure 8.5. Market values of Florida crops (USDA-NASS 2017).

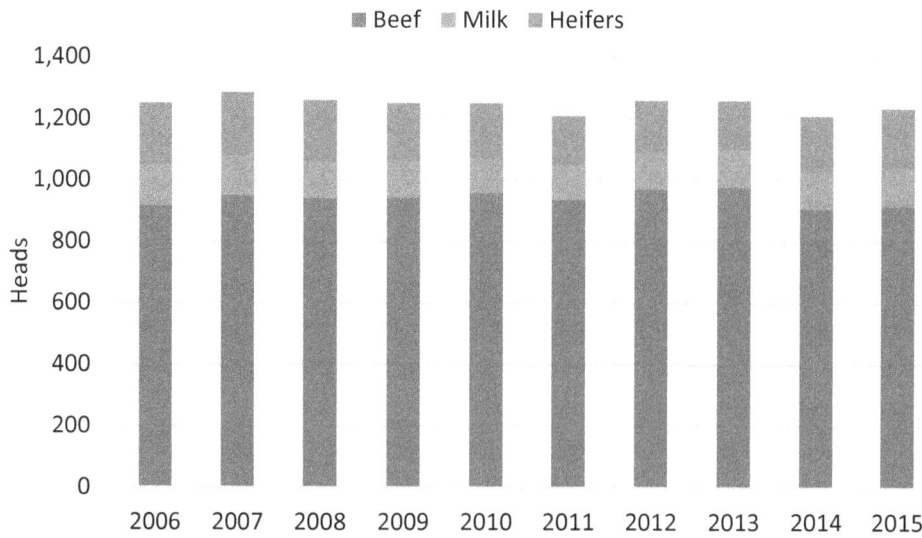

Figure 8.6. Head of beef and dairy cows in Florida (Putnam 2015c).

Climate Change Impacts on Florida's Crop Production

Elevated Atmospheric CO_2

Since accurate CO_2 recording was initiated in 1958 at Mauna Loa, Hawaii, the atmospheric CO_2 concentration has increased from 316 to 404 ppm in 2016, a 28% increase (NOAA 2017a). Projected atmospheric CO_2 concentration will reach more than 500 ppm in this century (Pachauri et al. 2014). An analysis with CO_2-dependent photosynthesis equations in the Decision Support System for Agrotechnology Transfer (DSSAT) crop models (Jones et al. 2003) indicates that yields of crops such as peanuts would have increased by 17% due to the CO_2 increase over that same period. The degree of crop yield increase from elevated CO_2 depends on the photosynthetic features of a crop, which are commonly characterized by C_3 and C_4 pathways. C_3 crops have CO_2 responsive photosynthesis (often called temperate or cool season plants) and include beans, rice, and potatoes. On the other hand, the C_4 crops are less responsive to elevated atmospheric CO_2 concentrations but have higher temperature optimums for photosynthesis (corn and sugarcane) and are called tropical or warm season plants.

In the 1980s when research on crop CO_2 response had just begun, the so-called ambient reference point was about 330 ppm. In later years 350 ppm CO_2 was used as the reference point. Assuming a starting point of 350 ppm CO_2, early research showed that C_3 photosynthesis crop yields (temperate dicots and cereals such as soybean, peanut, dry bean, rice, and wheat) increased about 30% with a doubling of CO_2, from 350 to 700 ppm for rice (Baker et al. 1990; Baker et al. 1992; Baker et al. 1995), peanuts (Prasad et al. 2003), common beans (Prasad et al. 2002), cotton

(Reddy et al. 2000), and tomatoes (Tripp et al. 1991). Except for sweet corn (which is a C_4 plant), all of Florida's vegetable crops, fruits, and citrus are C_3 species and are presumed to have CO_2 responses similar to the other C_3 crops (for details see summaries of measured crop response to CO_2 reviewed by Backlund et al. (2009) and Kimball et al. (2002), and crop model responses to CO_2 reviewed by Boote et al. (2010)). The response of C_3 crops to rising CO_2 is asymptotic, tending toward saturation at about 700 ppm. Thus, the yield improvement above present 400 ppm CO_2 will be occurring at a lesser rate, and the beneficial effects of this aspect of climate change are declining (i.e. saturating). On the other hand, the response to CO_2 is much less for those crops with a C_4 photosynthetic pathway, such as corn, sugarcane, sorghum, millet, and nearly all of Florida's tropical grasses (Ghannoum et al. 2000; Leakey et al. 2006; Manderscheid et al. 2014; Rosenzweig and Hillel 1998; Tubiello et al. 2002).

At the present-day levels of atmospheric CO_2 of 400 ppm, there are recent reports of no response to CO_2 for maize (Ghannoum et al. 2000; Leakey et al. 2006; Manderscheid et al. 2014). In CO_2-Free Air Carbon Dioxide Enhancement (FACE) experiments in Illinois (in which an elevated CO_2 concentration is maintained under field conditions by constantly blowing air with elevated CO_2 concentration into a field experiment to create future CO_2 conditions), maize yields were not increased by elevated atmospheric CO_2 under well-watered conditions (Leakey et al. 2006; Twine et al. 2013). The CO_2-FACE experiments in Germany (Manderscheid et al. 2014) likewise showed a non-existent yield response to elevated atmospheric CO_2 increase from 387 to 550 ppm for irrigated maize, although yields were responsive to CO_2 under water-deficit conditions. Prior to those FACE studies, there was only one experiment on maize grown to maturity (King and Greer 1986) that reported a 6.2% and 2.6% increase with atmospheric CO_2 concentration elevated from 355 to 625 ppm and 875 ppm, respectively (this translates to less than a 2% response for the smaller increment from 376 to 542 ppm CO_2 in the Illinois FACE experiments). There are few studies of yield response to CO_2 for other tropical C_4 species, and sorghum among them was responsive only under water-deficit conditions (Ottman et al. 2001). Bahiagrass (*Paspalum notatum* Flüggé), the most common pasture grass in Florida, presented a 9% response to CO_2 increasing from 360 to 700 ppm (Fritschi et al. 1999; Newman et al. 2001; Newman et al. 2006). Simulated biomass response for corn, sugarcane, and tropical grass (the C_4 species) in the Decision Support System for Agrotechnology Transfer (DSSAT) crop models (Jones et al. 2003) is 4.2% with a doubling of CO_2, going from 350 to 700 ppm, with simulated increases between 1–2% (statistically not detectable in field studies) for a CO_2 increase from 387 to 550 ppm. The response of crops to elevated atmospheric CO_2 concentrations is greater under drought stress and water limitation (Kimball et al. 1995; Sionit et al. 1981), especially for C_4 crops such as sorghum (Ottman et al. 2001) and maize (Manderscheid et al. 2014). However, the response to CO_2 is much less under nutrient, particularly nitrogen, limitations (Sionit et al. 1981; Kimball et al. 1995).

Rising CO_2 is expected to have modest effects to reduce crop transpiration (and hence evapotranspiration), but the exact effect is confounded by the extent to which elevated atmospheric CO_2 concentration increases crop leaf area index (which increases transpiration) as well as compensations in crop energy balance processes by which any decrease in transpiration results in an increased canopy temperature. Leaf conductance is reduced on average by about 38–40% with a doubling of CO_2 from 350 to 700 ppm (Morison 1987); however, whole crop transpiration is only reduced by about 9–10% under the same doubling of atmospheric CO_2 (Backlund et al. 2009). Concurrently, midday foliage temperature increases about 1–1.5 °C (1.8–2.7 °F) with doubling of CO_2 (Prasad et al. 2006) due to stomatal closure and less transpiration causing a warming of the canopy as a way to dissipate energy; but warming, in turn, increases transpiration and possibly enhances the impact of high temperature on crops. Sometimes there is no reduction in transpiration when the elevated atmospheric CO_2 concentration stimulates additional leaf area growth (Allen et al. 2003). The C_3 and C_4 crops differ in their degree of transpiration reduction, with C_3 crops near 8–10% reduction with doubled CO_2 (350 to 700 ppm) (Allen et al. 2003; Bernacchi et al. 2007; Backlund et al. 2009), while C_4 crops such as maize and sorghum show about 18% reduction (Allen et al. 2011; Chun et al. 2011).

Elevated CO_2 concentration has been shown to inhibit protein assimilation in main food crops (Bloom et al. 2010), but less is known about this effect in vegetables and fruits. Similarly, elevated CO_2 concentrations have been shown to reduce zinc (Zn) and iron (Fe) concentrations in major grains (Myers et al. 2014), but less is known how this might affect other crops grown in Florida.

Table 8.1. Climate change impacts on crops.

Changes	Positive Impacts	Negative Impacts
Elevated CO_2 concentration	Increased growth rate Increased water use efficiency	Increased weeds Decreased nutritive products Warmer canopies
Increased temperature	Less frost damage Improved winter growth Earlier planting	Faster phenology Reduced chill hours Increased heat stress Increased water use Increased pest/disease Increased risk of freeze if early flowering Crop water-logging/flooding Decreased arable lands (due to salt water intrusion induced by sea level rise)
Intensified rainfall and prolonged dry period		Increased runoff Increased erosion Increased irrigation requirement Increased chemical leaching

Rising Air Temperature

Many crops in Florida are grown at air temperatures that typically already exceed the optimum for a crop. This applies to maize and soybean, for which optimum temperature conditions are 23–24 °C (73–75 °F), matching temperatures presently experienced in the Midwestern United States, but are often higher in Florida. Some wheat and barley are grown in northern Florida over winter, but temperatures then are already too warm and chill requirements are often not adequately met in some years (compare this to the cool temperatures in northern Europe where wheat yields are often the highest, or to the Midwestern United States, which is generally cooler than Florida). Crops such as snap beans and tomatoes are grown only in winter–spring in Florida, but not in the summer because temperatures are too high and prevent bean and fruit formation. The optimal temperatures for the growth of citrus, tomatoes, and sugarcane are 20–30 °C (68–86 °F), 19–25 °C (66–77 °F), and 26–27 °C (79–81 °F), respectively (Morton 1987; Sato et al. 2000; Ebrahim et al. 1998). Temperatures outside the optimal ranges will decrease crop growth. Thus, increased temperatures may lead to the northward shift of some crops (EPA 1997).

Rising temperatures shorten crop life cycles, thus reducing resource capture and often resulting in a reduced yields (some of this might be overcome by crop breeding to recover the crop cycle). Another response could be slightly reduced daily photosynthesis depending on the crop. Crop respiration increases with rising temperatures but is often not the main reason for large yield reductions. As temperature increases further, fruit set and grain set are reduced, first slowly but eventually reaching a cut-off where zero pollen fertility occurs (this happens to tomatoes and snap beans during the summer in Florida). Crops differ in their failure point temperatures as described in the next paragraph.

The response of a number of Florida crops to increased temperatures has been evaluated in sunlit, controlled-environment chambers by University of Florida researchers. Findings for rice, beans, peanuts, soybeans, and sorghum are summarized here.

The optimum temperature for rice yield is 25 °C (77 °F) mean daily temperature (compare to a T_{max}/T_{min} of 30/20 °C (86/68 °F)) (Baker et al. 1992; Baker et al. 1995). Summer temperatures in Florida are 2–3 °C (3.6–5.4 °F) warmer than that optimum, and yield declines slowly, at first above 25 °C (77 °F) but reaching complete failure at 35 °C (95 °F) (compare to a T_{max}/T_{min} of 40/30 °C (104/86 °F)). The optimum temperature for bean yield is below 23 °C (73 °F), the coolest optimum temperature in that study, and seed-set and seed yield of beans failed completely at 32 °C (90 °F) mean temperature (Prasad et al. 2002).

The peanut, an important Florida crop, was grown from sowing to maturity at a range of temperatures in the same chamber system. The optimum temperature for peanut pod yield was less than 26 °C (79 °F), the coolest treatment of that study (Prasad et al. 2003). By contrast, present average temperatures in Florida's peanut growing season are about 27 °C (81 °F). These studies showed that peanut yield is expected to decline progressively as temperature increases until a failure of yield is projected at about 40 °C (104 °F) mean temperature. Soybeans have a similar yield response to temperatures, and the total failure of seed yield also occurs at about 39 °C to 40 °C (86 to 104 °F) mean temperature (Allen and Boote 2000; Boote et al. 2005; Pan 1996).

Sorghum, while not an important crop in Florida, has a sensitivity to temperature (Prasad et al. 2006) that is very similar to rice (also not an important crop in Florida), with an optimum at or below 25 °C (77 °F) (note: the 32/22 °C (90/72 °F) diurnal temperature cycle was the lowest temperature tested). Sorghum, like rice, had complete failure at 35 °C (95 °F) (40/30 °C (104/86 °F) diurnal cycle). By analogy to these two warm-season cereals, rice and sorghum, we assume that maize has a similar temperature sensitivity.

A common key feature in all of these experiments (rice, beans, peanuts, soybeans, and sorghum) is that the upper failure temperature threshold is associated with progressive failure of pollen viability and seed-set, and subsequently reduced seed harvest index and yield (Boote et al. 2005; Prasad et al. 2002; Prasad et al. 2006; Prasad et al. 2003; Baker et al. 1995). Soybeans and peanuts are the most tolerant, maize-sorghum-rice have moderate tolerance, while bean and tomato crops are very susceptible to rising temperature (see percent seed-set of these crops in Fig. 8.7). Indeed, present high summer temperatures are the cause for tomato and green bean crop failure to produce any tomatoes or beans because of non-viable pollen when planted late (May onward), thus exposing plants to high temperatures during fruiting. Hot conditions of maximum day temperatures above 32 °C (90 °F) and/or minimum night temperatures above 21 °C (70 °F) cause increased pollen sterility and reduced fruit set in tomatoes (Benedictos and Yavari 1997; Sato et al. 2000). There will certainly be increased pressure on tomato and bean breeders to obtain genetic heat tolerance (e.g. via breeding programs) or for growers to sow crops earlier (provided there is no freeze risk).

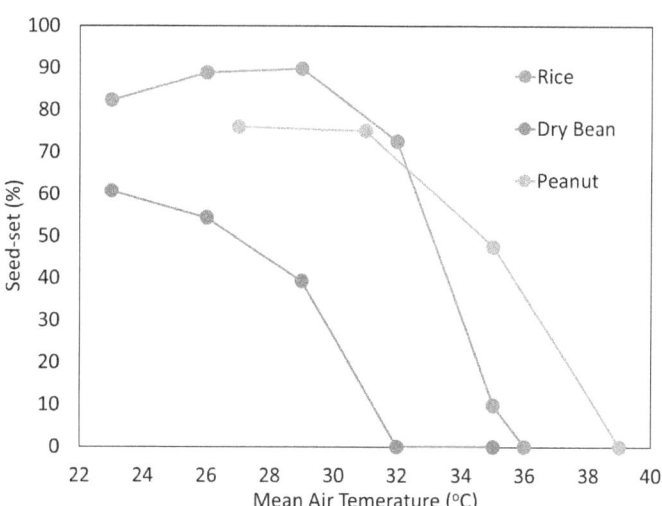

Figure 8.7. Percent seed set of rice, dry beans, and peanuts in response to temperature (Baker et al. 1995; Prasad et al. 2002; Prasad et al. 2003). Seed set is very closely related to yield, and a lower seed set means a lower yield.

Forage crop response to temperature varies and temperature effects are different from the effect on grain crops. The C_3 grasses, such as ryegrass or wheat used for grazing, are responsive to rising temperatures; small increases in temperature during winter season may actually stimulate growth, although also triggering earlier onset of flowering and thus less total forage production. The vegetative growth of C_4 tropical grasses (for which seed-set is not an issue) responds positively to rising temperatures at all times of the year, and experiments on bahiagrass showed increased production with up to 4 °C (7.2 °F) higher temperature, provided water is not limited (Fritschi et al. 1999; Newman et al. 2001; Newman et al. 2006). By analogy, but with no experimental evidence, the same response could potentially be expected for sugarcane, although sugar concentration in stalks is likely to be lower.

Changing Rainfall Pattern and Irrigation Demand

Rainfall is an important climate factor that influences plant growth and may also be supplemented by producers (via irrigation). Agricultural production has long been dependent on rainfall and irrigation for optimum yields. Because of alternating wet and dry seasons, the range of commodities grown in Florida will benefit from supplementary irrigation. Irrigation provides an input for Florida agriculture to grow a variety of crops and target market windows that allow the various commodities to be economically viable. Climate change and climate variability introduce uncertainty into the availability and quality of water for irrigation and the potential for Florida producers to meet the desired market windows.

Current user demands on freshwater supplies are already causing disputes, as is illustrated by the Florida/Georgia conflict in the Apalachicola-Chattahoochee-Flint River Basin (Stephenson 2000). In 2013, Florida sued Georgia in the US Supreme Court over water use in this basin and identified reduced water flow into Florida as a result of increased agricultural irrigation water withdrawals in Georgia. The Tampa Bay conflict is another example of the competing demand of water resources (Regan 2003). Thus, the potential change in freshwater quantities that may result from climate change will contribute additional stress to the water supply system. Climate changes that might influence available freshwater, and thus irrigation, are directly connected to the timing and amount of rainfall received and the rate of evapotranspiration or water losses from the system.

Climate change impacts on the frequency and amount of rainfall have a direct influence on available water for irrigation as well as the amount of irrigation needed to produce an agricultural crop. Climate change predictions for rainfall have shown an increase in extreme rainfall intensity, with greater increases near the coast and lower increases inland for Florida locations (Wang et al. 2013). Rainfall that occurs with greater intensity will result in less total effective rainfall for use by crops because of the low water holding capacity of Florida soils. This, in turn, will lead to greater irrigation needs to maintain crops under current farming practices. Furthermore, agricultural water demand increases due to a warmer and drier climate will compete with other water resource uses (EPA 1997). Model projections have suggested that total rainfall amounts will be reduced with climate change in Florida (Biasutti et al. 2012; Todd et al. 2012). Lower rainfall totals will translate into less recharge to water supplies and therefore less available water for irrigation. In addition, lower rainfall will result in greater irrigation needs depending on the distribution of rainfall events. Some caution is warranted with these potential future scenarios of rainfall change as rainfall projections are uncertain. Further research is needed to better understand how Florida rainfall will change in the future (Misra et al. 2011).

While changes in rainfall have a direct impact on irrigation, changes in evapotranspiration also influence irrigation needs. Climate change models have projected increases in annual atmospheric evapotranspiration demand of 70 to 130 mm (2.76 to 5.12 inches) by 2050 in Florida (Obeysekera 2011). The projected higher atmospheric evapotranspiration demand for future climate scenarios would translate into greater crop water needs for irrigation under current farming conditions. However, the effects of elevated atmospheric CO_2 concentration on stomatal conductance and transpiration may compensate/offset temperature-induced increases in evapotranspiration (Backlund et al. 2008). The potential changes related to evapotranspiration, such as temperature and solar radiation, could have other effects on agricultural production and thus irrigation. For example, temperature changes will influence the timing and rate of crop development. The shifting of a crop season and harvest would create additional irrigation modifications, which could be either an increase or a decrease depending on the timing.

Another climate change concern for agricultural irrigators is the quality of water used for irrigation. Sea level rise will increase the risk of salt water intrusion into aquifers that are used for irrigation purposes (Karl 2009). Thus, some groundwater wells used for irrigation may no longer be viable as freshwater supplies. Alternative water sources or water treatment will be needed for agriculture in areas where salinity concentration exceeds the level that is safe for crops.

Irrigation strategies that could be explored as an adaptation to climate change in Florida agricultural production include primed acclimation, deficit irrigation, drought-resistant crops, and variety (cultivar) improvements. Primed acclimation refers to the practice of providing deficit irrigation amounts during the initial phases of crop development and full irrigation amounts in the later portion of crop development to create a more resilient plant (Rowland et al. 2012). Deficit irrigation refers to providing a less than optimum irrigation for the entire production period of a crop. Thus, primed acclimation is a variation of a deficient irrigation strategy where deficit amounts are applied in the early season of crop development. Reducing overall irrigation inputs is one strategy for addressing the projected change in agricultural systems due to climate change. Another strategy is to grow crops that are more efficient water users either through improved breeding or alternative crops. Irrigated agriculture will likely need a combination of strategies to remain viable under predicted climate change conditions. The future viability of irrigated agriculture in Florida will depend on the ability of producers to successfully implement new strategies that adapt to climate change.

Salt Water Intrusion from Sea Level Rise Affects Crops and Irrigation Needs

Increased air temperature will accelerate melting of ice sheets and glaciers on land, which will increase the volume of water in the oceans and then the sea level (Nicholls and Cazenave 2010). As warmer water takes up more volume (thermal expansion), the sea level will also be raised as seawater is warmed by increased air temperature (Nicholls and Cazenave 2010). The sea level rise will not be uniform spatially, but will be more prominent along the equator due to the centrifugal force of the Earth's rotation; thus, the amount of sea level rise will be higher in Florida than other coastal states. Raised sea level will further the intrusion of salt water into coastal aquifers, which will increase the salinity of groundwater and then soils (Prinos 2016; Ketabchi et al. 2016). In South Florida, for instance, sea level has been rising by 2.32 mm/year (0.0913 in/year, Daytona Beach, 1925–1983) to 2.78 mm/year (0.1094 in/year, Vaca Key, 1971–2006) in the past (Zervas 2009). If sea level rises by 1.0 m (3.28 ft) by 2060, which is close to the National Oceanic and Atmospheric Administration's (NOAA) "High" scenario (1.03-m increase) (Sweet et al. 2017), then about 2,000 km2 (770 mi2) of three counties—Miami-Dade, Broward, and Palm Beach— in South Florida will be below the projected sea level (Zhang 2011) resulting in loss of Florida's agricultural land in that region.

The Floridan aquifer system is an important source of freshwater in Florida and it is shallow and highly permeable, which makes the aquifers more vulnerable to sea level rise and saltwater intrusion. The aquifer systems have been experiencing saltwater intrusion caused by sea level rise, leading to the contamination of wells for agricultural and domestic water supplies and changes in water management practices, especially in South Florida (Blanco et al. 2013; Trimble et al. 1998; Heimlich and Bloetscher 2011). Increased salinity in groundwater will result in increased irrigation costs and then cropping system will decrease in productivity and profitability. Salty irrigation water will lead to an accumulation of salts in soils and increase soil salinity, which will require more freshwater irrigation to wash it out or, in some cases, it will result in soil degradation and loss of arable land.

Climate Change Impacts on Florida's Livestock Production

The increase in air temperatures associated with climate change will affect livestock production directly and indirectly (Thornton 2010; Reynolds et al. 2010). Rising temperature can increase heat stresses, illness, diseases, and mortality, which subsequently reduce the productivity of livestock (Nardone et al. 2010; Rojas-Downing et al. 2017; Das et al. 2016). Increased temperatures will promote the growth of some forage crops but decrease nutrient availability (Rojas-Downing et al. 2017; IPCC 2007). In addition, the feed intake and digestive efficiency of livestock decreases at high temperatures (Mader and Davis 2004; Tankson et al. 2001). Dairy cows produce less milk, and meat production can decrease due to reduced growth rate at increased temperatures (Nardone et al. 2010; Mitlöhner et al. 2001). In addition, reproduction of cows, pigs, and poultry decreases with increases in temperature (Hansen 2007; Nardone et al. 2010; De Rensis and Scaramuzzi 2003; Kunavongkrit et al. 2005). Prolonged exposure to high temperatures can affect livestock health, metabolism, and liver function (Bernabucci et al. 2006). When livestock performance is high, livestock production will be more vulnerable to high temperatures (Hahn 1999). Such adverse impacts of climate change could lead to the northward movement of livestock production in Florida and into other parts of the USA (EPA 1997; Von Lehe 2007).

Table 8.2. Climate change impacts on livestock and their feed sources.

Changes	Positive Impacts	Negative Impacts
Increased temperature	Increased herbage growth	Decreased nutrient quality of feed Decreased feed intake Decreased efficiency of feed conversion Decreased milk production Decreased meat production Decreased reproduction Increased mortality Increased diseases
Intensified rainfall and prolonged dry period		Decreased forage growth and quality Decreased pasture biodiversity Increased flood damage Decreased forage quality Changed optimal growth rate
Elevated CO_2 concentration	Increased herbage growth Reduced transpiration Improved water use efficiency	Decreased nutrient quality of forage/feed

Increased frequency of extreme events such as drought and flood can degrade the quality of forage. Dry periods prolonged by increased air temperatures will affect the growth of forages and feed crops and will reduce their nutrient availability for livestock (Polley et al. 2013). In addition, rising temperatures will increase evapotranspiration and decrease soil water content. Warmer air can hold more moisture, which can result in more intense storm events. Increased heavy storm and flood frequency is another impact we can expect with climate change (Schmidt 2000). Such hydrologic consequences of climate change will lead to a reduction in forage and feed crop production, and subsequently increase feed costs and decrease cattle and poultry products such as milk, meat, and eggs. Combined changes in temperature and rainfall can promote the spread of vector-borne pests such as flies, ticks, and mosquitoes, as well as increase livestock diseases (Thornton et al. 2009; Rojas-Downing et al. 2017; Kurukulasuriya and Rosenthal 2003). For instance, the hydrologic condition (surface wetness and groundwater table depth) of the land surface was found to be associated with the transmission of the West Nile Virus to chickens in South Florida (Shaman et al. 2005; Day and Shaman 2008).

Adaptation Strategies

Overall

Adaptation to expected changes in climate is of considerable importance to the long-term sustainability of agricultural production in Florida. Adapting to climate change can be achieved through a broad range of management alternatives and technological advances. While decision-making in agriculture involves many aspects beyond climate, including economics, social factors, and policy considerations, climate-related risks are a primary source of yield and income variability. Researchers and cooperative extension services must play a proactive role to cogenerate necessary responses and technologies that farmers will need to handle such future challenges. In addition to improving and/or developing management practices and technologies, climate literacy of extension faculty and producers is required, as well as climate information and decision support systems to help the industry mitigate risks associated with climate variability and change.

A wide range of management practices can help producers adapt and increase the resilience of agricultural production systems to climate variability and change. Many aspects related to vulnerability (defined as the degree of sensitivity) and ability to cope with climate variability, and adaptation (defined as adjustments to environmental stresses caused by climate variability) can also be applied to climate change (Fraisse et al. 2009). Existing strategies, mostly developed for row crops—such as the use of high-biomass winter cover crops, conservation tillage, sod-based rotation systems, efficient irrigation technologies, and precision agriculture—can help producers minimize the risks associated with climate variability and change as well as improve their resource-use efficiency.

High Residue Cover Crops

High-residue cover cropping is an adaptation of conservation tillage in which a high-biomass cover crop is grown during the winter and is rolled or cut down prior to no-till or strip-till planting in the spring. Examples of winter cereals used as high residue cover crops include rye (*Secale cereale*), black oats (*Avena strigosa*), wheat (*Triticum*), or triticale (*Triticosecale*). High-residue cover crops and reduced tillage can lessen some negative impacts from climate and weather, such as high-intensity rainfall events, spring and summer dry spells, droughts, and extreme soil temperatures during critical crop reproduction periods. Keeping soil covered year-round with crop residue can reduce soil erosion, improve water infiltration, reduce evaporative moisture loss, and moderate soil temperature. Some benefits depend on the climate and soil types of the system, and these positive impacts can increase with repeated use of high-residue cover crops. The main differences between high-residue cover crops and traditional winter cover crops are the types of crops selected and the amount of fertilizer applied. A high-residue system uses winter cereals with fertilizer applications, resulting in greater production of biomass than a traditional cover crop system. Many producers find the cost of high-residue cover crops are justified in dryland systems because of the improved water management and soil quality that result from greater crop residues (Joel Love 2015).

Conservation Tillage

The U.S. Department of Agriculture, Natural Resources Conservation Service (USDA-NRCS) defines conservation tillage as a system that leaves enough crop residues from cover crops and/or cash crops on the soil surface after planting to provide at least 30% soil cover. Research has identified 30% soil cover as the minimal amount of residue needed to avoid significant soil loss, but greater residue amounts are preferred. The use of cover crops is critical to producing this additional plant residue. In addition to maximizing surface residues, conservation tillage can increase below-ground disruption to eliminate compacted soil layers by maintaining plant roots and soil macropores. While conservation tillage can resolve the occurrence of a shallow plow-compacted layer in some systems, subsoil tillage may be required in some soils to manage compaction from vehicle traffic or from naturally occurring compacted layers. Together with cover crops, conservation tillage has the potential to reduce erosion, increase rainfall infiltration, reduce subsurface compaction, and maximize soil organic carbon accumulation, which positively affects many soil physical and chemical properties. The main way that conservation tillage can reduce risks related to climate variability (particularly droughts and dry spells) is by increasing the water available to plants. Areas where conservation tillage is used have revealed a number of benefits including reduced erosion and runoff, increased water infiltration, more plant-available water, reduced soil water evaporation, and reduced diurnal soil temperature fluctuations (Balkcom et al. 2012).

Sod-Based Rotation

A sod-based rotation incorporates two or more consecutive seasons of a perennial grass into a conventional row-crop rotation. One example of a sod-based rotation is an adaptation of the conventional peanut/cotton rotation that farmers follow in North Florida. In a four-year, sod-based rotation, bahiagrass is grown for two years, followed by a year of peanuts, and then a year of cotton (Fig. 8.8). Such rotation is beneficial to the many soils in Florida that have a high sand content, low organic matter, and compaction layers making them more vulnerable to stresses from variability in climate, namely dry spells and droughts (Wright et al. 2015).

Sod-based rotation can reduce climate-related risks by increasing soil water holding capacity, potentially reducing the negative effects of droughts and dry spells, increasing the water infiltration rate, and reducing soil bulk density. The soil water holding capacity is increased as a result of improved soil organic matter promoted by the sod-based rotation. Increased infiltration rate and reduced soil bulk density results from an increase in soil macropores due to greater root mass and biological activity for soils. Field data from 2002 to 2007 in Quincy, Florida showed that water-use efficiency of peanut crops under sod-based rotation was 15% greater in irrigated fields and 19% greater in dryland fields compared to the water-use efficiency of peanut crops in a conventional rotation (Zhao et al. 2008). Here, water-use efficiency is defined as the ratio of crop yield to the sum of irrigation and rainfall. These data suggest that yield increases have resulted from improvements in soil water-holding capacity. In the very dry years of 2006 and 2007, peanut yields in a sod-based rotation were 13% greater than those under conventional rotation (Zhao et al. 2008).

Figure 8.8. Illustration of conventional and sod-based peanut/cotton rotations. Credits: David Wright (Wright et al. 2015).

Irrigation

The International Panel on Climate Change (IPCC) indicated in their most recent report (Field et al. 2014) that there is medium confidence that drought will intensify in the 21st century in some seasons and areas due to reduced precipitation and/or increased evapotranspiration. With that in mind, irrigation, if available, can be considered as a potential alternative for reducing the risk of yield losses especially in soils with low water-holding capacity, such as those common in Florida. While irrigation requires considerable investment over dryland production, it can also result in considerable increases in yields and profits. However, irrigation management must be as efficient as possible to avoid losses and groundwater contamination.

Center pivot, micro, and subsurface drip irrigation systems require different investment costs and management practices. Micro-irrigation is the slow, frequent application of water directly to relatively small areas adjacent to individual plants through emitters placed along a water delivery line. Water is generally conveyed in low-pressure, flexible plastic tubing. Water must be of high quality to avoid clogging the small emitters; this is often managed with filtration and occasional chemical treatments. A principal advantage of micro-irrigation is that non-beneficial evaporation—meaning evaporation of water from soil surfaces that do not contribute to crop growth—is greatly reduced when compared to sprinkler irrigation (Zotarelli 2015). Subsurface drip irrigation water is applied below the soil surface through driplines that are installed at a depth of 30–46 cm (12–18 in). Tillage, planting, and other field operations are not impeded by driplines because they are established at a sufficient depth to allow for those operations and long-term use. Emitter flow rates for subsurface irrigation are generally less than 11.4 liters/hour (3 gallons/hour). Subsurface drip irrigation can have a useable lifetime of up to 20 years, making it economically competitive with center pivot irrigation used for low-value commodity row crops (Lamm et al. 2010).

Center pivot irrigation consists of a galvanized steel lateral that rotates in a circle around a fixed point (pivot) in the center of a field. The lateral is supported above the crop on A-shaped steel frames using cables and trusses. Sprinklers are used to distribute the water across the field, as the area to be irrigated increases towards the outer end of the lateral, thus varied size or spacing of sprinklers is used to gradually increase the water application rate. Variable-rate irrigation is an innovative technology that enables a center pivot irrigation system to optimize irrigation application. Most fields are not uniform because of natural variations in soil type or topography. When water is applied uniformly to a field, some areas of the field may be overwatered while other areas may remain too dry. Some farmers manage these individual zones by excluding these problematic areas from the acres cropped. However, variable-rate irrigation technology gives farmers an automated method to vary rates of irrigation water based on the individual management zones within a field (Perry et al. 2015). Using a variable-rate irrigation system can reduce the total irrigation water volume required to grow field crops in two ways: first, producers can exclude non-cropped or marginal areas from water application; and second, producers can lower application rates in low-lying areas or in soils with high water-holding capacity.

Plastic Mulch

Plastic mulch applications for vegetable production provide several adaptations relative to climate change and variability. They reduce water loss to evaporation and minimize leaching of soluble nutrients (e.g., nitrogen) that are applied under the mulch and thus are protected from rainfall-induced leaching. In addition to weed control and minimizing aboveground soil-vegetable contact, plastic mulch also can provide warmer conditions for winter-grown vegetables, although various colors can be used to minimize the degree of heating at a later growth period and attacks from pest insects (e.g. aphids). High tunnels can be used to create a more protected environment for certain crops such as strawberries, to protect from wind and rain damage, to reduce diseases associated with dew formation, and to provide some degree of temperature regulation from day to night, and especially to minimize freeze damage during the winter season. In addition, plastic mulch can help control and prevent soil-borne pathogens and diseases from spreading by providing a shield protecting crops from pests and virus (Katan et al. 1976; Espi et al. 2006).

Drought-Tolerant Crops and Forage

Grasses have varying ability to adjust their growth in response to extreme hydrologic events such as flood and drought, so a careful selection of forage type should be made (Baruch 1994). In particular, prolonged dry periods can reduce the cover and productivity of grasses significantly (Evans et al. 2011; Craine et al. 2013). Thus, the selection of drought-tolerant forages in breeding programs and improvements in grazing management to buffer against drought could aid producers in adapting to more frequent drought conditions under projected climate change. A study found that native grasslands have a wide spectrum of drought tolerance, suggesting that a focused breeding effort including native grasses could lead to an effective adaptation option (Craine et al. 2013). Drought tolerance of high-value vegetable and fruit crops is typically not a breeding objective for Florida because production is irrigated.

Livestock Facility Renovation

Warm air can help reduce winter housing requirements for livestock and additional forage for feed in winter. However, higher temperatures will increase needs for cooling in intensive livestock production systems to reduce potential heat stress resulting in lower productivity in summer (Howden et al. 2007; St-Pierre et al. 2003). Livestock, such as poultry and pigs, kept indoors can be directly affected by increased air temperature. Dairy cattle are susceptible to warm summer temperatures at sunny times of the day during much of the year. Shade structures can reduce livestock heating and minimize a reduction in milk production, fertility, and growth during the warm season. Thus, the renovation—of feedlots and barns, including additional shades, spraying, fanning, air conditioning, circulation, and ventilation—can be a way to adapt to projected warmer temperature, but the efficiency and cost of the cooling system are important considerations (Howden et al. 2007; Armstrong 1994).

Livestock Genomic Selection and Breeding Strategy

Increasing the genetic diversity of livestock can be a strategic approach to reducing climate change impacts on food security. Genetic selection has made significant contributions to the improvement of the feed-to-food conversion efficiency (Herrero et al. 2010; Havenstein et al. 2003). Genetic evaluation and new data collection systems have led to improved genetic selection (Zwald et al. 2004a; Zwald et al. 2004b; VanRaden et al. 2004), and its use will increase in the future (Weigel 2006; Schaeffer 2006; Jonas and de Koning 2015). Genetic selection and crossing with tropical breeds of beef animals could be useful to improve their heat tolerance.

Mixed Crop-Livestock Systems

Mixed crop–livestock systems can be more productive than monoculture systems, as the outcome of one system can be beneficial to another (Rojas-Downing et al. 2017; Thornton and Herrero 2015). Two-thirds of the global population is involved in the mixed agricultural systems, which produce more than half of the livestock and crop products in the world, including meat, milk, cereals, millet, rice, and sorghum(Herrero et al. 2012). Mixed crop-livestock systems have supported increasing food demands in sub-Saharan Africa and South Asia (Thornton and Herrero 2014; Thornton and Herrero 2015). Transition to more efficient livestock systems including mixed crop-livestock could be an effective measure to reduce greenhouse gasses such as thermogenic methane and nitrous oxide, while increasing livestock productivity (Havlík et al. 2014).

Climate Information and Decision Support Systems

Climate information and decision support systems can help agricultural operators develop the capacity to detect expected changes early, respond to the changes quickly, and manage cases appropriately. Thus, they can help reduce production risk, and increase resource use efficiency and the profitability of agricultural operations. Information and decision support systems are not just a compilation of data, but learning tools to develop knowledge-based management plans. Simply providing better climate information and forecasts to potential users will not be enough. Climate information has value only when there is a clearly defined adaptive response and a benefit once the content of the information is considered in the decision-making process (Fraisse et al. 2016). AgroClimate (http://agroclimate.org/) and the conceptual model CISTA-A (Conceptual model using Indicators selected by SysTems thinking for Adaptation strategies for Agro-ecosystems) (Anandhi 2017) are examples of how climate information can be prepared and provided to assist producers and stakeholders in making informed climate-related decisions.

AgroClimate is a web-based climate information and decision support system. The website includes seasonal forecasts, expected impacts of management options for different crops and climate scenarios, and a wide variety of interactive tools that help producers monitor current conditions and plan for the season ahead. AgroClimate has been developed to serve agricultural stakeholders in the southeastern states of Florida, Georgia, Alabama, South Carolina, and North Carolina (Breuer et al. 2009). Users can monitor variables of interest such as growing degree-days, chill hours, disease risks for selected crops, and current and projected drought conditions. Users can also learn about the forecast of climate cycles affecting the Southeastern United States, such as the ENSO phenomenon. Water and carbon footprint calculators can provide estimates of how efficiently water and energy are being used. AgroClimate can assist producers to develop a strategy for a coming season and track current climate conditions affecting crop development and yield. Based on the expected seasonal climate outlook or other climate information, producers are better able to adapt to expected conditions by changing crop selection, planting dates, plant population, cover crop management, livestock management, input purchasing, and nutrient management.

CISTA-A is a decision support tool developed to explore how to adapt ecosystems to climate change using a systems-thinking approach to adaptation (Anandhi 2017). CISTA-A allows users to consider abiotic/biotic information (e.g. temperature, rainfall, and crop yield) and employs ecological, agro-hydrological, and climatological indicators (e.g. length of the growing season, growing degree days, and plant failure temperature) that affect the ecosystem in climate change adaptation planning. The translation of information from indicators to adaptation strategies (incremental systems and transformational adaptation) depends on the degree of change and the level of adaptation. For instance, CISTA-A uses temperature change information to predict spring freezes, and then translates the changes to propose early sowing dates as an adaptation strategy.

Knowledge Gaps and Recommendations for Future Research

Researchers have tried to identify the potential impacts of climate change on the agricultural production system and to develop adaptation and mitigation plans to safely accommodate the changes in the system. However, there are still many knowledge gaps to be filled.

Healthy ecosystems are resistant to external changes and able to quickly recover from damages induced by changes. Thus, maintaining and building healthy ecosystems can help agricultural systems be resilient and sustainable (Tompkins and Adger 2004). Plant and animal biodiversity regulates ecosystem health by controlling hydrological processes, nutrient cycles, and microclimate (Altieri 1999). Some effective ways to improve the ecological resistance and resilience of an agricultural system is to restore its functional biodiversity by rotating crops, using cover crops, intercropping, implementing agroforestry, and mixing crops and livestock (Altieri 1999; Verchot et al. 2007).

Mechanistic crop simulation models (e.g. DSSAT; Jones et al. 2003) have been key tools in extrapolating the impacts of climate variability and change from limited field and controlled-environment experiments to other climatic zones, rainfall regions, soil types, management regimes, crops, and climate change scenarios (Chenu et al. 2017). The impact of individual climate change components and the combined effect of climate change scenarios on crop production and externalities have been explored with such models. However, these models mostly exist for main food grain crops but not for many of the vegetables and fruits grown in Florida. Hence, developing crop models for vegetables and fruit crops based on field experimentation will be critical to assessing the impact of climate change on Florida's agriculture and for preparing adaptation and mitigation strategies. Detailed field experiments investigating the impact of CO_2, temperature, and water supply changes will also be important and are recommended for such needed model development.

Interactive effects of elevated CO_2 with temperature, such as reduction of transpiration canopy cooling as a result of elevated CO_2 and stomata closure, are often not considered in crop models but can, for example, increase pollen sterility in rice (Ziska and Bunce 2007) and sorghum (Prasad et al. 2006). As the frequency of high temperatures (> 32 °C (90 °F)) during the growing season will increase with climate change, the interactions with elevated CO_2 need to be better understood and considered (Attri and Rathore 2003), particularly for crops grown in Florida.

Other interacting effects of climate change on flooding and salinity (Ziska and Bunce 2007) need to be considered but are not yet known, not even for main food crops.

Climate factors often affect crop quality, including protein composition and oil content (Kimball et al. 2001) and various minerals in main food crops (Myers et al. 2014); but less is known about how the nutritional value of vegetables and fruits will be affected by climate change. Field experiments on changes in nutritional contents of vegetables and fruits are needed and should be included in crop models.

Plant breeding technology will help to improve heat tolerance of crops under the projected rising temperatures of climate change (Tester and Langridge 2010). Pollen viability and reproductive fertility of heat-sensitive crops (such as tomatoes and snap beans) are limited at high temperatures, but there may be a potential to improve this genetically (Bita and Gerats 2013). It will be important to improve drought tolerance of crops under projected increases in air temperature and with changes in rainfall, particularly for those crops that are typically rainfed, such as peanuts, cotton, maize, and tropical grasses. Factors influencing the genetics of crop diseases have been recently discovered, and specific genetics can be selected for (Bishop and Woolliams 2014; Stear et al. 2001). Genetic selection for disease resistance traits in animals can allow farmers to raise livestock in less preferable climates (Berry et al. 2011). A better understanding of disease pressure under climate change and the ability to breed for improved disease resistance in livestock will be important for adaptation of livestock to climate change.

Florida's growers are part of a global market, and production in competing regions will also be affected by climate change. Not only will Florida's growers be adapting to climate changes, they will also be competing against growers elsewhere making their own adaptations. When considering the economic prospects for example for Florida's citrus and tomato growers in the coming decades, climate change in Florida needs to be considered, but one must also consider climate change and potential adaptations in main competing regions within the US and other countries like Brazil and Mexico.

As population increases in Florida are likely to continue, urban sprawl and demands for water from municipal, energy, and other sectors will increasingly conflict with agricultural irrigation requirements. Policy research will be essential to balance these competing demands for land and water resources. This research should evaluate agricultural competitiveness in Florida, both for its ability to meet food, fiber, and fuel needs of the region as well as to contribute to national and global food, fiber, and fuel production (Marcus and Kiebzak 2008).

Initial research has shown that there is a strong interaction between changes in agricultural land use/land cover and regional climate, a feedback often overlooked and less understood (Shin and Baigorria 2012). Traditional agricultural research has followed a linear approach, from research scientists to extension agents to farmers. To address the complex issues of sustainability in the face of changing and variable climate, research must follow a new paradigm—one that emphasizes the integration of research, teaching, and extension, invites the participation of decision-makers throughout the research process, and assembles the diverse elements of agriculture through a systems approach (Breuer et al. 2009, 2010; Bartels et al. 2012; Roncoli 2006). Crop and livestock breeding and management research are important elements to the overall agricultural research portfolio, but they should be incorporated into integrated approaches to ensure that they contribute to agricultural sustainability.

While researchers are already working with farmers in Florida to develop and assess technologies to mitigate and adapt to climate variability and change, additional research is needed to identify and incorporate adaptive technologies into agricultural systems. Analysis of these technologies should include carbon, energy, water, and nutrient balances as well as life cycle, risk and economic analysis (a systems analysis that is only rarely applied to agricultural research and development).

Agriculture and food production systems are complex and associated with many fields of science including agronomy, biology, crop physiology, soil science, economics, sociology, mathematics, physics, and environmental sciences. Thus, solutions to an agricultural challenge will be multidisciplinary, requiring holistic views and approaches from emerging scientific platforms such as biotechnology, nanotechnology, information science, and cognitive science (Scott et al. 2016; Conway 2012). Recent advances in data and network sciences, sensing and robotics technology, and computing resources are expected to enhance informed decision-making in agriculture by promoting the accurate and quick exchange of information among farmers, researchers, and tool/equipment manufacturers, and by improving precision agriculture and smart farming technology.

Florida's agriculture has a long history of successful adaptations to the vagaries of weather and climate. However, climate change poses a challenge that is unprecedented in its magnitude and pace of onset. As with any major change in global agricultural markets, the winners will be those who are able, with the help of their government and industrial leaders, to cope with these challenges and to recognize and take advantage of opportunities.

Acknowledgments

Carolyn Cox, Florida Climate Institute Coordinator, is recognized for her role in coordinating this chapter.

References

Adams, R. M., C. Rosenzweig, R. Peart, J. T. Ritchie, and B. A. McCarl (1990), Global climate change and US agriculture, Nature, 345(6272), 219.

Allen Jr, L., and K. Boote (2000), Crop ecosystem responses to climatic change: soybean, In Climate Change and Global Crop Productivity, pp. 133-160, University of Florida, Gainesville, FL.

Allen, L. H., V. G. Kakani, J. C. Vu, and K. J. Boote (2011), Elevated CO_2 increases water use efficiency by sustaining photosynthesis of water-limited maize and sorghum, Journal of Plant Physiology, 168(16), 1909-1918.

Allen, L., D. Pan, K. Boote, N. Pickering, and J. Jones (2003), Carbon dioxide and temperature effects on evapotranspiration and water use efficiency of soybean, Agronomy Journal, 95(4), 1071-1081.

Altieri, M. A. (1999), The ecological role of biodiversity in agroecosystems, Agriculture, Ecosystems and Environment, 74(1), 19-31.

Anandhi, A. (2017), CISTA-A: Conceptual model using indicators selected by systems thinking for adaptation strategies in a changing climate: Case study in agro-ecosystems, Ecological Modelling, 345, 41-55.

Anderson, S. A. (1990), Core indicators of nutritional state for difficult-to-sample populations, The Journal of Nutrition.

Anyamba, A., J. L. Small, S. C. Britch, C. J. Tucker, E. W. Pak, C. A. Reynolds, J. Crutchfield, and K. J. Linthicum (2014), Recent weather extremes and impacts on agricultural production and vector-borne disease outbreak patterns, PLoS One, 9(3), e92538.

Armstrong, D. (1994), Heat stress interaction with shade and cooling, Journal of Dairy Science, 77(7), 2044-2050.

Attri, S., and L. Rathore (2003), Simulation of impact of projected climate change on wheat in India, International Journal of Climatology, 23(6), 693-705.

Backlund, P., A. Janetos, and D. Schimel (2008), The effects of climate change on agriculture, land resources, water resources, and biodiversity in the United States. U.S. Climate Change Science Program: Synthesis and Assessment Product 4.3, Report by the U.S. Climate Change Science Program and the Subcommittee on Global Change Research.

Backlund, P., A. Janetos, and D. Schimel (2009), The effects of climate change on agriculture, land resources, water resources, and biodiversity in the United States, Nova Science Publishers, Inc., New York.

Baker, J. T., K. J. Boote, and L. H. Allen (1995), Potential climate change effects on rice: Carbon dioxide and temperature, in Climate Change and Agriculture: Analysis of Potential International Impacts, pp. 31-47, Amerian Society of Agronomy.

Baker, J., L. Allen, and K. Boote (1990), Growth and yield responses of rice to carbon dioxide concentration, The Journal of Agricultural Science, 115(03), 313-320.

Baker, J., L. Allen, and K. Boote (1992), Response of rice to carbon dioxide and temperature, Agricultural and Forest Meteorology, 60(3-4), 153-166.

Balkcom, K., L. Duzy, D. Dourte, C. Fraisse (2012), Conservation Tillage: Agricultural Management Options for Climate Variability and Change. Available: http://agroclimate.org/wp-content/uploads/2016/03/Conservation-tillage.pdf.

Bartels, W., C. Furman, F. Royce, B. Ortiz, D. Zierden, and C. Fraisse (2012), Developing a learning community: Lessons from a climate working group for agriculture in the Southeast USA, Southeast Climate Consortium Technical Report Series, 12-002.

Baruch, Z. (1994), Responses to drought and flooding in tropical forage grasses, Plant and Soil, 164(1), 87-96.

Benedictos Jr, P., and N. Yavari (1997), Optimum sowing date in relation to flower drop reduction in tomato, paper presented at VIII International Symposium on Timing Field Production in Vegetable Crops, ISHS Acta Horticulturae 533.

Bernabucci, U., N. Lacetera, L. Basiricò, B. Ronchi, P. Morera, E. Serene, and A. Nardone (2006), Hot season and BCS affect leptin secretion of periparturient dairy cows, paper presented at Journal of Animal Science, Amer Soc Animal Science 1111 North Dunlap Ave, Savoy, IL 61874 USA.

Bernacchi, C. J., B. A. Kimball, D. R. Quarles, S. P. Long, and D. R. Ort (2007), Decreases in stomatal conductance of soybean under open-air elevation of CO_2 are closely coupled with decreases in ecosystem evapotranspiration, Plant Physiology, 143(1), 134-144.

Berry, D. P., M. L. Bermingham, M. Good, and S. J. More (2011), Genetics of animal health and disease in cattle, Irish Veterinary Journal, 64(1), 5.

Biasutti, M., A. H. Sobel, S. J. Camargo, and T. T. Creyts (2012), Projected changes in the physical climate of the Gulf Coast and Caribbean, Climatic Change, 112(3-4), 819-845.

Bishop, S. C., and J. A. Woolliams (2014), Genomics and disease resistance studies in livestock, Livestock Science, 166, 190-198.

Bita, C. E., and T. Gerats (2013), Plant tolerance to high temperature in a changing environment: scientific fundamentals and production of heat stress-tolerant crops, Frontiers in Plant Science, 4.

Blanco, R. I., G. M. Naja, R. G. Rivero, and R. M. Price (2013), Spatial and temporal changes in groundwater salinity in South Florida, Applied Geochemistry, 38, 48-58.

Bloom, A. J., M. Burger, J. S. R. Asensio, and A. B. Cousins (2010), Carbon dioxide enrichment inhibits nitrate assimilation in wheat and Arabidopsis, Science, 328(5980), 899-903.

Boote, K. J., L. H. Allen Jr, P. V. Prasad, and J. W. Jones (2010), Testing effects of climate change in crop models, in Handbook of Climate Change and Agroecosystems, pp. 109-129, Imperial College Press, London.

Boote, K. J., L. H. Allen, P. V. Prasad, J. T. Baker, R. W. Gesch, A. M. Snyder, P. Deyun, and J. M. Thomas (2005), Elevated temperature and CO2 impacts on pollination, reproductive growth, and yield of several globally important crops, Journal of Agricultural Meteorology, 60(5), 469-474.

Breuer, N. E., C. W. Fraisse, and P. E. Hildebrand (2009), Molding the pipeline into a loop: the participatory process of developing AgroClimate, a decision support system for climate risk reduction in agriculture, Journal of Service Climatology, 3(1), 1-12.

Campbell, C. A. (1994), Handling of Florida-grown and imported tropical fruits and vegetables, HortScience, 29(9), 975-978.

Chenu, K., J. R. Porter, P. Martre, B. Basso, S. C. Chapman, F. Ewert, M. Bindi, and S. Asseng (2017), Contribution of crop models to adaptation in wheat, Trends in Plant Science, 22(6), 472-490.

Chun, J. A., Q. Wang, D. Timlin, D. Fleisher, and V. R. Reddy (2011), Effect of elevated carbon dioxide and water stress on gas exchange and water use efficiency in corn, Agricultural and Forest Meteorology, 151(3), 378-384.

Conway, G. (2012), One billion hungry: can we feed the world?, Cornell University Press, Ithaca, NY.

Craine, J. M., T. W. Ocheltree, J. B. Nippert, E. G. Towne, A. M. Skibbe, S. W. Kembel, and J. E. Fargione (2013), Global diversity of drought tolerance and grassland climate-change resilience, Nature Climate Change, 3(1), 63-67.

Das, R., L. Sailo, N. Verma, P. Bharti, and J. Saikia (2016), Impact of heat stress on health and performance of dairy animals: A review, Veterinary World, 9(3), 260.

Day, J. F., and J. Shaman (2008), Using hydrologic conditions to forecast the risk of focal and epidemic arboviral transmission in peninsular Florida, Journal of Medical Entomology, 45(3), 458-465.

De Rensis, F., and R. J. Scaramuzzi (2003), Heat stress and seasonal effects on reproduction in the dairy cow—a review, Theriogenology, 60(6), 1139-1151.

Ebrahim, M. K., O. Zingsheim, M. N. El-Shourbagy, P. H. Moore, and E. Komor (1998), Growth and sugar storage in sugarcane grown at temperatures below and above optimum, Journal of Plant Physiology, 153(5-6), 593-602.

EPA, (1997), Climate Change and Florida. Available: https://nepis.epa.gov/Exe/ZyPDF.cgi/40000IY6.PDF?Dockey=40000IY6.PDF.

Espi, E., A. Salmeron, A. Fontecha, Y. García, and A. Real (2006), Plastic films for agricultural applications, Journal of Plastic Film and Sheeting, 22(2), 85-102.

Evans, S. E., K. M. Byrne, W. K. Lauenroth, and I. C. Burke (2011), Defining the limit to resistance in a drought-tolerant grassland: long-term severe drought significantly reduces the dominant species and increases ruderals, Journal of Ecology, 99(6), 1500-1507.

FDACS (2016), Florida Statewide Agricultural Irrigation Demand (FSAID). Available: http://freshfromflorida.s3.amazonaws.com/Media%2FFiles%2FAgricultural-Water-Policy-Files%2FAgricultural-Water-Supply-Planning%2FFSAID+II+Final+Report_The+Balmoral+Group_08.31.2015_Corr.1.pdf.

Field, C. B., V. R. Barros, D. J. Dokken, K. J. Mach, and M. D. Mastrandrea (2014), Climate Change 2014: Impacts, Adaptation, and Vulnerability, in Working Group II Contribution to the Fifth Assessment Report of the Intergovernmental Panel on Climate Change.

Fraisse, C. W., N. E. Breuer, D. Zierden, and K. T. Ingram (2009), From climate variability to climate change: Challenges and opportunities to extension, Journal of Extension, 47(2), 2FEA9.

Fraisse, C., J. H. Andreis, T. Borba, , V. Cerbar, E. Gelcer, W. Pavan, D. Pequeno, D. Perondi, X. Shen, C. Staub, and O. Uryasev (2016), AgroClimate - Tools for managing climate risk in agriculture, Agrometeoros Journal, 24(1), 121-129.

Fritschi, F. B., K. J. Boote, L. Sollenberger, and L. Hartwell (1999), Carbon dioxide and temperature effects on forage establishment: tissue composition and nutritive value, Global Change Biology, 5(7), 743-753.

Gao, Y., J. Fu, J. Drake, Y. Liu, and J. Lamarque (2012), Projected changes of extreme weather events in the eastern United States based on a high resolution climate modeling system, Environmental Research Letters, 7(4), 044025.

Ghannoum, O., S. v. Caemmerer, L. Ziska, and J. P. Conroy (2000), The growth response of C4 plants to rising atmospheric CO2 partial pressure: a reassessment, Plant, Cell & Environment, 23(9), 931-942.

Hahn, G. (1999), Dynamic responses of cattle to thermal heat loads, Journal of Animal Science, 77, 10-20.

Hansen, J. W., A. W. Hodges, and J. W. Jones (1998), ENSO influences on agriculture in the southeastern United States, Journal of Climate, 11(3), 404-411.

Hansen, J. W., J. W. Jones, C. F. Kiker, and A. W. Hodges (1999), El Niño–Southern Oscillation impacts on winter vegetable production in Florida, Journal of Climate, 12(1), 92-102.

Hansen, P. (2007), Exploitation of genetic and physiological determinants of embryonic resistance to elevated temperature to improve embryonic survival in dairy cattle during heat stress, Theriogenology, 68, S242-S249.

Havenstein, G., P. Ferket, and M. Qureshi (2003), Growth, livability, and feed conversion of 1957 versus 2001 broilers when fed representative 1957 and 2001 broiler diets, Poultry Science, 82(10), 1500-1508.

Havlík, P., H. Valin, M. Herrero, M. Obersteiner, E. Schmid, M. C. Rufino, A. Mosnier, P. K. Thornton, H. Böttcher, and R. T. Conant (2014), Climate change mitigation through livestock system transitions, Proceedings of the National Academy of Sciences, 111(10), 3709-3714.

Heimlich, B., and F. Bloetscher (2011), Effects of sea level rise and other climate change impacts on southeast Florida's water resources, Florida Water Resources Journal, 63(9), 37-48.

Herrero, M., P. K. Thornton, A. M. Notenbaert, S. Wood, S. Msangi, H. Freeman, D. Bossio, J. Dixon, M. Peters, and J. van de Steeg (2010), Smart investments in sustainable food production: revisiting mixed crop-livestock systems, Science, 327(5967), 822-825.

Herrero, M., P. K. Thornton, A. M. O. Notenbaert, S. Msangi, S. Wood, R. Kruska, J. A. Dixon, D. A. Bossio, J. Van de Steeg, and H. A. Freeman (2012), Drivers of change in crop–livestock systems and their potential impacts on agro-ecosystems services and human wellbeing to 2030: A study commissioned by the CGIAR Systemwide Livestock Programme.

Howden, S. M., J.-F. Soussana, F. N. Tubiello, N. Chhetri, M. Dunlop, and H. Meinke (2007), Adapting agriculture to climate change, Proceedings of the National Academy of Sciences, 104(50), 19691-19696.

IPCC (2007), Synthesis Report. Contribution of Working Groups I, II and III to the Fourth Assessment Report of the Intergovernmental Panel on Climate Change, IPCC, Geneva, Switzerland.

Joel Love, J. D., Kirk Brock, Daniel Dourte, Clyde Fraisse (2015), High-Residue Cover Crops A Management Option for Climate Variability and Change. Available: http://agroclimate.org/wp-content/uploads/2016/03/High-residue-cover-crops.pdf.

Johnson, T., J. Butcher, D. Deb, M. Faizullabhoy, P. Hummel, J. Kittle, S. McGinnis, L. Mearns, D. Nover, and A. Parker (2015), Modeling streamflow and water quality sensitivity to climate change and urban development in 20 US watersheds, Journal of the American Water Resources Association, 51(5), 1321-1341.

Jonas, E., and D. J. de Koning (2015), Genomic selection needs to be carefully assessed to meet specific requirements in livestock breeding programs, Frontiers in Genetics, 6, 49.

Jones, J. W., G. Hoogenboom, C. H. Porter, K. J. Boote, W. D. Batchelor, L. Hunt, P. W. Wilkens, U. Singh, A. J. Gijsman, and J. T. Ritchie (2003), The DSSAT cropping system model, European Journal of Agronomy, 18(3), 235-265.

Karl, T. R. (2009), Global Climate Change Impacts in the United States, Cambridge University Press, Cambridge, UK.

Katan, J., A. Greenberger, H. Alon, and A. Grinstein (1976), Solar heating by polyethylene mulching for the control of diseases caused by soil-borne pathogens, Phytopathology, 66(5), 683-688.

Ketabchi, H., D. Mahmoodzadeh, B. Ataie-Ashtiani, and C. T. Simmons (2016), Sea-level rise impacts on seawater intrusion in coastal aquifers: Review and integration, Journal of Hydrology, 535, 235-255.

Kimball, B. A., P. J. PINTER, R. L. Garcia, R. L. LaMORTE, G. W. Wall, D. J. Hunsaker, G. Wechsung, and F. Wechsung (1995), Productivity and water use of wheat under free-air CO_2 enrichment, Global Change Biology, 1(6), 429-442.

Kimball, B., C. Morris, P. Pinter, G. Wall, D. Hunsaker, F. Adamsen, R. LaMorte, S. Leavitt, T. Thompson, and A. Matthias (2001), Elevated CO_2, drought and soil nitrogen effects on wheat grain quality, New Phytologist, 150(2), 295-303.

Kimball, B., K. Kobayashi, and M. Bindi (2002), Responses of agricultural crops to free-air CO_2 enrichment, Advances in agronomy, 77, 293-368.

King, K., and D. Greer (1986), Effects of carbon dioxide enrichment and soil water on maize, Agronomy Journal, 78(3), 515-521.

Klassen, W., C. F. Brodel, and D. A. Fieselmann (2002), Exotic pests of plants: current and future threats to horticultural production and trade in Florida and the Caribbean Basin, Micronesica Supplement, 6, 5-27.

Kunavongkrit, A., A. Suriyasomboon, N. Lundeheim, T. W. Heard, and S. Einarsson (2005), Management and sperm production of boars under differing environmental conditions, Theriogenology, 63(2), 657-667.

Kurukulasuriya, P., and S. Rosenthal (2003), Climate change and agriculture: A review of impacts and adaptations world bank climate change series paper no. 91, 106 pp., The International Bank for Reconstruction and Development / The World Bank, Washington, D.C.

Lamm, F. R., P. D. Colaizzi, J. P. Bordovsky, T. P. Trooien, J. Enciso-Medina, D. O. Porter, D. H. Rogers, and D. M. O'Brien (2010), Can Subsurface Drip Irrigation (SDI) be a Competitive Irrigation System in the Great Plains Region for Commodity Crops?, paper presented at 5th National Decennial Irrigation Conference Proceedings, 5-8 December 2010, Phoenix Convention Center, Phoenix, Arizona USA, American Society of Agricultural and Biological Engineers.

Leakey, A. D., M. Uribelarrea, E. A. Ainsworth, S. L. Naidu, A. Rogers, D. R. Ort, and S. P. Long (2006), Photosynthesis, productivity, and yield of maize are not affected by open-air elevation of CO_2 concentration in the absence of drought, Plant Physiology, 140(2), 779-790.

Lobell, D. B., M. B. Burke, C. Tebaldi, M. D. Mastrandrea, W. P. Falcon, and R. L. Naylor (2008), Prioritizing climate change adaptation needs for food security in 2030, Science, 319(5863), 607-610.

Mader, T., and M. Davis (2004), Effect of management strategies on reducing heat stress of feedlot cattle: feed and water intake, Journal of Animal Science, 82(10), 3077-3087.

Manderscheid, R., M. Erbs, and H.-J. Weigel (2014), Interactive effects of free-air CO_2 enrichment and drought stress on maize growth, European Journal of Agronomy, 52, 11-21.

Marcus, R. R., and S. Kiebzak (2008), The role of water doctrines in enhancing opportunities for sustainable agriculture in Alabama, Journal of the American Water Resources Association, 44(6), 1578-1590.

Marella, R. L. (1999), Water withdrawals, use, discharge, and trends in Florida, 1995, Geological Survey, Water Resources Div., Tallahassee, FL (United States); Florida State Dept. of Environmental Protection, Tallahassee, FL.

Maul, G. A., and D. M. Martin (1993), Sea level rise at Key West, Florida, 1846-1992: America's longest instrument record?, Geophysical Research Letters, 20(18), 1955-1958.

Mendelsohn, R., A. Dinar, and L. Williams (2006), The distributional impact of climate change on rich and poor countries, Environment and Development Economics, 11(02), 159-178.

Miller, K. A., and M. H. Glantz (1988), Climate and economic competitiveness: Florida freezes and the global citrus processing industry, Climatic Change, 12(2), 135-164.

Misra, V., E. Carlson, R. Craig, D. Enfield, B. Kirtman, W. Landing, S. Lee, D. Letson, F. Marks, and J. Obeysekera (2011), Climate scenarios: a Florida-centric view. Florida Climate Change Task Force, Center for Ocean-Atmospheric Prediction Studies, 14, 1-61.

Mitlöhner, F., J. Morrow, J. Dailey, S. Wilson, M. Galyean, M. Miller, and J. McGlone (2001), Shade and water misting effects on behavior, physiology, performance, and carcass traits of heat-stressed feedlot cattle, Journal of Animal Science, 79(9), 2327-2335.

Morison, J. I. (1987), Intercellular CO_2 Concentration and Stomatal Response to CO_2 James I. L Morison, in Stomatal Function, edited by E. Zeiger, G. D. Farquhar and I. R. Cowan, p. 229, Stanford University Press, Stanford, CA.

Morton, J. F. (1987), Fruits of warm climates, Julia F. Morton, Miami, FL.

Myers, S. S., A. Zanobetti, I. Kloog, P. Huybers, A. D. Leakey, A. Bloom, E. Carlisle, L. H. Dietterich, G. Fitzgerald, and T. Hasegawa (2014), Rising CO_2 threatens human nutrition, Nature, 510(7503), 139.

Nardone, A., B. Ronchi, N. Lacetera, M. S. Ranieri, and U. Bernabucci (2010), Effects of climate changes on animal production and sustainability of livestock systems, Livestock Science, 130(1), 57-69.

National Academies of Sciences, E., Medicine (2016), Attribution of extreme weather events in the context of climate change, National Academies Press.

National Climate Assessment (2014), Available: http://nca2014.globalchange.gov/.

Newman, Y. C., L. E. Sollenberger, K. J. Boote, L. H. Allen, J. M. Thomas, and R. C. Littell (2006), Nitrogen fertilization affects bahiagrass responses to elevated atmospheric carbon dioxide, Agronomy Journal, 98(2), 382-387.

Newman, Y., L. Sollenberger, K. Boote, L. Allen, and R. Littell (2001), Carbon dioxide and temperature effects on forage dry matter production, Crop Science, 41(2), 399-406.

NHC (2017), Hurricanes in History. Available: http://www.nhc.noaa.gov/outreach/history/.

Nicholls, R. J., and A. Cazenave (2010), Sea-level rise and its impact on coastal zones, Science, 328(5985), 1517-1520.

NOAA (2017a), National Centers for Environmental Information - Data Tools: 1981 - 2010 Normals.

NOAA (2017b), Global Greenhouse Gas Reference Network, Earth System Research Laboratory - Global Monitoring Division.

Obeysekera, J., J. Park, M. Irizarry-Ortiz, P. Trimble, J. Barnes, J. VanArman, W. Said, and E. Gadzinski (2011), Past and Projected Trends in Climate and Sea Level for South Florida, South Florida Water Management District, West Palm Beach, FL.

Ottman, M. J., B. Kimball, P. Pinter, G. Wall, R. Vanderlip, S. Leavitt, R. LaMorte, A. Matthias, and T. Brooks (2001), Elevated CO2 increases sorghum biomass under drought conditions, New Phytologist, 150(2), 261-273.

Pachauri, R. K., M. R. Allen, V. R. Barros, J. Broome, W. Cramer, R. Christ, J. A. Church, L. Clarke, Q. Dahe, and P. Dasgupta (2014), Climate change 2014: synthesis report. Contribution of Working Groups I, II and III to the fifth assessment report of the Intergovernmental Panel on Climate Change, IPCC.

Pan, D. (1996), Soybean responses to elevated temperature and doubled CO_2, University of Florida, Gainesville, FL.

Peel, M. C., B. L. Finlayson, and T. A. McMahon (2007), Updated world map of the Köppen-Geiger climate classification, Hydrology and Earth System Sciences, 4(2), 439-473.

Perry, C., C. W. Fraisse, and D. Dourte (2015), Variable-Rate Irrigation: A Management Option for Climate Variability and Change. Available: http://agroclimate.org/wpcontent/uploads/2016/03/Variable-rate-irrigation.pdf.

Polley, H. W., D. D. Briske, J. A. Morgan, K. Wolter, D. W. Bailey, and J. R. Brown (2013), Climate change and North American rangelands: trends, projections, and implications, Rangeland Ecology & Management, 66(5), 493-511.

Prasad, P. V., K. J. Boote, and L. H. Allen (2006), Adverse high temperature effects on pollen viability, seed-set, seed yield and harvest index of grain-sorghum [Sorghum bicolor (L.) Moench] are more severe at elevated carbon dioxide due to higher tissue temperatures, Agricultural and Forest Meteorology, 139(3), 237-251.

Prasad, P., K. J. Boote, L. H. Allen, and J. M. Thomas (2002), Effects of elevated temperature and carbon dioxide on seed-set and yield of kidney bean (Phaseolus vulgaris L.), Global Change Biology, 8(8), 710-721.

Prasad, V. P., K. J. Boote, L. Hartwell Allen, and J. M. Thomas (2003), Super-optimal temperatures are detrimental to peanut (Arachis hypogaea L.) reproductive processes and yield at both ambient and elevated carbon dioxide, Global Change Biology, 9(12), 1775-1787.

Prinos, S. T. (2016), Saltwater intrusion monitoring in Florida, Florida Scientist, 79, 4.

Putnam, A. (2011), Florida Agriculture: By the Numbers., Florida Department of Agriculture and Consumer Services, Tallahassee, FL.

Putnam, A. (2015a), Florida Agriculture: By the Numbers., Florida Department of Agriculture and Consumer Services, Tallahassee, FL.

Putnam, A. (2015b), Florida Agriculture Overview and Statistics.

Putnam, A. (2015c), 2015 Annual Report., Florida Department of Agriculture and Consumer Services, Tallahassee, FL.

Reddy, K. R., H. F. Hodges, and B. A. Kimball (2000), Crop ecosystem responses to climatic change: Cotton, in Climate Change and Global Crop Productivity, edited by K. R. Reddy and H. F. Hodges, pp. 161-187, CABI Publishing, Oxon, UK.

Regan, K. E. (2003), Balancing public water supply and adverse environmental impacts under Florida water law: from water wars towards adaptive management, Journal of Land Use & Environmental Law, 19(1), 123-184.

Reilly, J., F. Tubiello, B. McCarl, D. Abler, R. Darwin, K. Fuglie, S. Hollinger, C. Izaurralde, S. Jagtap, and J. Jones (2003), US agriculture and climate change: new results, Climatic Change, 57(1), 43-67.

Reynolds, C., L. Crompton, and J. Mills (2010), Livestock and climate change impacts in the developing world, Outlook on Agriculture, 39(4), 245-248.

Rojas-Downing, M. M., A. P. Nejadhashemi, T. Harrigan, and S. A. Woznicki (2017), Climate change and livestock: impacts, adaptation, and mitigation, Climate Risk Management, 16, 145-163.

Roncoli, C. (2006), Ethnographic and participatory approaches to research on farmers' responses to climate predictions, Climate Research, 33(1), 81-99.

Rosenzweig, C., and D. Hillel (1998), Climate Change and the Global Harvest, Oxford University Press, New York.

Rowland, D. L., W. H. Faircloth, P. Payton, D. T. Tissue, J. A. Ferrell, R. B. Sorensen, and C. L. Butts (2012), Primed acclimation of cultivated peanut (Arachis hypogaea L.) through the use of deficit irrigation timed to crop developmental periods, Agricultural Water Management, 113, 85-95.

Sallenger Jr, A. H., K. S. Doran, and P. A. Howd (2012), Hotspot of accelerated sea-level rise on the Atlantic coast of North America, Nature Climate Change, 2(12), 884-888.

Sato, S., M. Peet, and J. Thomas (2000), Physiological factors limit fruit set of tomato (Lycopersicon esculentum Mill.) under chronic, mild heat stress, Plant, Cell & Environment, 23(7), 719-726.

Scavia, D., J. C. Field, D. F. Boesch, R. W. Buddemeier, V. Burkett, D. R. Cayan, M. Fogarty, M. A. Harwell, R. W. Howarth, and C. Mason (2002), Climate change impacts on US coastal and marine ecosystems, Estuaries, 25(2), 149-164.

Schaeffer, L. (2006), Strategy for applying genome-wide selection in dairy cattle, Journal of animal Breeding and genetics, 123(4), 218-223.

Schmidt, C. W. (2000), Lessons from the flood: will Floyd change livestock farming?, Environmental Health Perspectives, 108(2), A74.

Scott, N. R., H. Chen, and R. Schoen (2016), Sustainable global food supply, Handbook of science and technology convergence. Springer International, Switzerland.

Shaman, J., J. F. Day, and M. Stieglitz (2005), Drought-induced amplification and epidemic transmission of West Nile virus in southern Florida, Journal of Medical Entomology, 42(2), 134-141.

Shin, D., and G. A. Baigorria (2012), Potential influence of land development patterns on regional climate: a summer case study in the Central Florida, Natural hazards, 62(3), 877-885.

Sionit, N., D. Mortensen, B. Strain, and H. Hellmers (1981), Growth response of wheat to CO2 enrichment and different levels of mineral nutrition, Agronomy Journal, 73(6), 1023-1027.

Sleep, D., and R. Gitzen (2015), Fresh from Florida, 2015 International Report, Available: http://www.freshfromflorida.com/content/download/59882/1184512/2015_International_Report.pdf.

Stear, M., S. Bishop, B. Mallard, and H. Raadsma (2001), The sustainability, feasibility and desirability of breeding livestock for disease resistance, Research in Veterinary Science, 71(1), 1-7.

Stephenson, D. S. (2000), The tri-state compact: falling waters and fading opportunities, Journal of Land Use and Environmental Law, 16, 83.

St-Pierre, N., B. Cobanov, and G. Schnitkey (2003), Economic losses from heat stress by US livestock industries, Journal of Dairy Science, 86, 52-77.

Sweet, W. V., R. E. Kopp, C. P. Weaver, J. Obeysekera, R. M. Horton, E. R. Thieler, and C. Zervas (2017), Global and Regional Sea Level Rise Scenarios for the United StatesRep., National Oceanic and Atmospheric Administration, Silver Spring, Maryland.

Tankson, J., Y. Vizzier-Thaxton, J. Thaxton, J. May, and J. Cameron (2001), Stress and nutritional quality of broilers, Poultry Science, 80(9), 1384-1389.

Tester, M., and P. Langridge (2010), Breeding technologies to increase crop production in a changing world, Science, 327(5967), 818-822.

Thornton, P. K. (2010), Livestock production: recent trends, future prospects, Philosophical Transactions of the Royal Society B: Biological Sciences, 365(1554), 2853-2867.

Thornton, P. K., and M. Herrero (2014), Climate change adaptation in mixed crop–livestock systems in developing countries, Global Food Security, 3(2), 99-107.

Thornton, P. K., and M. Herrero (2015), Adapting to climate change in the mixed crop and livestock farming systems in sub-Saharan Africa, Nature Climate Change, 5(9), 830-836.

Thornton, P. K., J. Van de Steeg, A. Notenbaert, and M. Herrero (2009), The impacts of climate change on livestock and livestock systems in developing countries: A review of what we know and what we need to know, Agricultural Systems, 101(3), 113-127.

Todd, M. J., R. Muneepeerakul, F. Miralles-Wilhelm, A. Rinaldo, and I. Rodriguez-Iturbe (2012), Possible climate change impacts on the hydrological and vegetative character of Everglades National Park, Florida, Ecohydrology, 5(3), 326-336.

Tompkins, E., and W. N. Adger (2004), Does adaptive management of natural resources enhance resilience to climate change?, Ecology and Society, 9(2), 10.

Trimble, P., E. Santee, and C. Neidrauer (1998), Preliminary estimate of impacts of sea-level rise on the regional water resources of southeastern Florida, Journal of Coastal Research, 252-255.

Tripp, K. E., M. M. Peet, D. M. Pharr, D. H. Willits, and P. V. Nelson (1991), CO2-enhanced yield and foliar deformation among tomato genotypes in elevated CO2 environments, Plant Physiology, 96(3), 713-719.

Tubiello, F., C. Rosenzweig, R. Goldberg, S. Jagtap, and J. Jones (2002), Effects of climate change on US crop production: simulation results using two different GCM scenarios. Part I: Wheat, potato, maize, and citrus, Climate Research, 20(3), 259-270.

Twine, T. E., J. J. Bryant, K. T Richter, C. J. Bernacchi, K. D. McConnaughay, S. J. Morris, and A. D. Leakey (2013), Impacts of elevated CO2 concentration on the productivity and surface energy budget of the soybean and maize agroecosystem in the Midwest USA, Global Change Biology, 19(9), 2838-2852.

USDA-NASS (2012), 2012 Census of Agriculture Vol. 1, Ch. 1: State Level Data, Florida.

USDA-NASS (2016), Cropland Data Layer, Available: https://www.nass.usda.gov/Research_and_Science/Cropland/SARS1a.php.

USDA-NASS (2017), 2016 State Agriculture Overview, Available: https://www.nass.usda.gov/Quick_Stats/Ag_Overview/stateOverview.php?state=FLORIDA.

VanRaden, P., A. Sanders, M. Tooker, R. Miller, H. Norman, M. Kuhn, and G. Wiggans (2004), Development of a national genetic evaluation for cow fertility, Journal of Dairy Science, 87(7), 2285-2292.

Verchot, L. V., M. Van Noordwijk, S. Kandji, T. Tomich, C. Ong, A. Albrecht, J. Mackensen, C. Bantilan, K. Anupama, and C. Palm (2007), Climate change: linking adaptation and mitigation through agroforestry, Mitigation and Adaptation Strategies for Global Change, 12(5), 901-918.

Von Lehe, A. (2007), Climate Change and South Carolina's Economy, Southeastern Environmental Law Journal, 16, 359.

Walthall, C., J. Hatfield, E. Marshall, L. Lengnick, P. Backlund, S. Adkins, E. Ainsworth, F. Booker, D. Blumenthal, and J. Bunce (2013), Climate Change and Agriculture in the United States: Effects and Adaptation, USDA, Washington, D.C., Available: https://www.usda.gov/oce/climate_change/effects_2012/CC%20and%20Agriculture%20Report%20(02-04-2013)b.pdf.

Wang, D., S. C. Hagen, and K. Alizad (2013), Climate change impact and uncertainty analysis of extreme rainfall events in the Apalachicola River basin, Florida, Journal of Hydrology, 480, 125-135.

Weigel, K. A. (2006), Prospects for improving reproductive performance through genetic selection, Animal reproduction science, 96(3), 323-330.

Whitehead, P., R. Wilby, R. Battarbee, M. Kernan, and A. J. Wade (2009), A review of the potential impacts of climate change on surface water quality, Hydrological Sciences Journal, 54(1), 101-123.

Wright, D., J. Marois, C. Fraisse, and D. Dourte (2015), Agricultural Management Options for Climate Variability and Change: Sod-Based Rotation, EDIS, UF/IFAS(AE49200).

Zervas, C. (2009), Sea level variations of the United States., National Oceanic and Atmospheric Administration, Silver Spring, MD.

Zhang, K. (2011), Analysis of non-linear inundation from sea-level rise using LIDAR data: a case study for South Florida, Climatic Change, 106(4), 537-565.

Zhao, D., D. Wright, J. Marois, C. Mackowiak, and T. Katsvairo (2008), Yield and water use efficiency of cotton and peanut in conventional and sod-based cropping systems, paper presented at Proc. 30th Southern Conserv. Agric. Syst. Conf. and 8th Ann. Georgia Conserv. Prod. Syst. Trng. Conf., Tifton, Georgia.

Ziska, L. H., and J. A. Bunce (2007), Predicting the impact of changing CO2 on crop yields: some thoughts on food, New Phytologist, 175(4), 607-618.

Zotarelli, L., C. Fraisse, and D. Dourte (2015), Agricultural Management Options for Climate Variability and Change: Microirrigation, EDIS, UF/IFAS(HS1203).

Zwald, N., K. Weigel, Y. Chang, R. Welper, and J. Clay (2004a), Genetic selection for health traits using producer-recorded data. I. Incidence rates, heritability estimates, and sire breeding values, Journal of dairy science, 87(12), 4287-4294.

Zwald, N., K. Weigel, Y. Chang, R. Welper, and J. Clay (2004b), Genetic selection for health traits using producer-recorded data. II. Genetic correlations, disease probabilities, and relationships with existing traits, Journal of dairy science, 87(12), 4295-4302.

CHAPTER 9

Managing Florida's Plantation Forests in a Changing Climate

Timothy A. Martin[1], Damian C. Adams[1], Matthew J. Cohen[1], Raelene M. Crandall[1], Carlos A. Gonzalez-Benecke[2], Jason A. Smith[1], and Jason G. Vogel[1]

[1]*School of Forest Resources and Conservation, University of Florida, Gainesville, FL;* [2]*Department of Forest Engineering, Resources and Management, Oregon State University, Corvallis, OR*

Production forestry provides substantial benefits to the state of Florida, including the provision of ecosystem services, such as regulation of water quantity and quality, provision of wildlife habitat and carbon sequestration, and supporting 80,000 jobs and $16.34 billion/year in economic activity. Climate through the end of the century in the production forestry regions of northern Florida and southern Georgia is predicted to result in substantial increases in potential loblolly pine and slash pine plantation productivity, ranging from 5–35% depending on emissions scenario, species, and location. Climate change is likely to affect the timing and frequency of abiotic disturbances, such as wildfire and windstorms, and will also change the dynamics of forest pests, pathosystems, and forest water resources. But predictions about the nature of these impacts remains uncertain. Regardless, the fact is that plantation forests have been a vital part of protecting regional water quantity and quality, and they will continue to be essential features of healthy productive landscapes, as climate changes and the potential for adverse climate impacts on water resources increases. The key to adapting forest management to changing climate will be the considered application of silvicultural tools, such as competition control, density and fertility management, and proper choice of species for each site. Keeping abreast of research advances related to these tools will be increasingly important for forest managers as climate conditions change. In addition, the development of viable policy options focused primarily on privately owned forests can help protect Florida's existing forests and the benefits they provide, and encourage investment in reforestation of existing forestland and planting new forests on previously unforested land.

Key Messages

- Production forestry provides substantial benefits to the state of Florida, including the provision of ecosystem services, such as regulation of water quantity and quality, provision of wildlife habitat and carbon sequestration, and supporting 80,000 jobs and $16.34 billion/year in economic activity.
- Climate through the end of the century in the production forestry regions of northern Florida and southern Georgia is predicted to warm from 1.5 °C to almost 3.5 °C, with small increases in annual precipitation, and elevated atmospheric CO_2 concentration. Models predict that these changes will result in substantial increases in potential loblolly pine and slash pine plantation productivity, ranging from 5–35% depending on emissions scenario, species, and location.
- Forestry is unique in that it is one of the few industries that sequesters more carbon than it emits. There are opportunities to increase carbon sequestration for mitigation of atmospheric CO_2 through retention or expansion of forested areas, altered forest management, and the use of woody biomass for power generation in place of fossil fuels.
- The frequency and intensity of abiotic disturbances, such as wildfire and windstorms, are likely to be affected by climate change; but predictions remain uncertain about the magnitude of change and their effects on the forest resource.

- Research is underway to better understand how native forest pests and pathosystems may respond to changing climate. The movement of pests or pathogens into previously non-impacted areas is of particular concern.
- Plantation forests have been a vital part of protecting regional water quantity and quality, and they will continue to be essential features of healthy, productive landscapes as climate changes and the potential for adverse climate impacts on water resources increases.'
- The key to adapting forest management to changing climate will be the considered application of silvicultural tools, such as competition control, density and fertility management, and proper choice of species for each site. Keeping abreast of research advances related to these tools will be increasingly important for forest managers as climate conditions change.
- There are several viable policy options for harnessing forests to mitigate climate change and increasing forest resilience and adaptation to climate change. However, since 71% of Florida's forests are privately owned, policy options must align well with landowner needs to have adequate impact. Broadly speaking, policies that improve market conditions, reduce burdens (regulatory and economic), and increase economic sustainability for forest landowners would help protect Florida's existing forests and the benefits they provide, and would encourage investment in reforestation of existing forestland and planting new forests on previously unforested land.

Keywords

Forestry; Natural resources; Water resources; Disturbance; Ecosystem services; Wildfire; Insects; Disease; Invasives

Introduction

Production forestry is a critically important economic resource to the state of Florida. Forests cover nearly half the state (17.3 million acres in 2013), with 15.4 million acres composed of "working forests" that are managed primarily for timber, but also for their economic benefits from other ecosystem goods and services (e.g., hunting). Nearly three-quarters of Florida's forestland is privately owned; the balance is publicly held by state and local or federal entities (FDEP 2016).

Florida is in one of the most productive tree-growing regions in the world–the Southern United States, which produces nearly one-eighth of the world's industrial roundwood and nearly one-fifth of the world's paper and pulp products. Within the U.S., this area is known as the "wood basket" of the country, generating half of the saw log and veneer products, and nearly three-quarters of U.S. pulpwood (Smith et al. 2009). Importantly, this area is expected to become even more critical to U.S. and global wood production, with significant projected increases (+25%–70%) in timber production in the region (Hugget et al. 2013) and losses of timberland elsewhere in the U.S. (e.g., due to mountain pine beetle outbreaks that have devastated western forests).

Florida is an important contributor to forest products markets, with an annual harvest of 472.5 million cubic feet of wood between 2009 and 2013, 90% from private lands (FFS 2015). The associated economic impact on the state is tremendous: forestry contributed $16.34 billion to the state's economy and provided more than 80,000 jobs in 2013 (FDEP 2016; Hodges et al. 2013).

We mostly think of forests as providing much-needed raw materials, such as timber and fiber used for wood products, heat and power generation; but forests are considerably more valuable to society for the ecosystem services that they provide. These include water availability, wildlife habitat, air quality, soil formation, recreation, carbon sequestration and biodiversity. For example, more than one-third of the water supply in the southern United States comes from forested watersheds (Lockaby et al. 2013). In Florida, water quality protection alone provides $154–$230 million in annual average benefits from forests (Kreye et al. 2016). A recent Florida study estimates that the typical acre of non-industrial private forestland annually provides $5,030 of ecosystem services (e.g., timber, carbon storage, water quality, and wildlife habitat), with just 7% of that value from timber (Escobedo et al. 2012).

Florida's residents derive benefits from many types of forests, ranging from conserved forests managed primarily for ecosystem services to very intensively managed planted forests ("plantations") overseen primarily for economic benefits from tree harvesting. This chapter focuses on planted pine forests in northern Florida and southern Georgia, and provides a brief overview of predicted future climate and its likely effects on forest productivity, water quality and quantity, ecosystem services, disturbance by biotic and abiotic agents, and carbon sequestration for mitigation of atmospheric CO_2. Silviculture, the set of techniques used for managing forest structure and composition, will be an important tool for adapting forests to future climate conditions. Accordingly, we outline potential silvicultural approaches for forest management under future climate, and discuss policy options for minimizing future risks to this valuable resource.

Climate Projections for Florida

Climate projections worldwide predict an increase in air temperatures and variability in precipitation. Global circulation models predict increasing air temperatures with a high degree of certainty, with the magnitude and rate of warming varying across the globe (IPCC 2013). There is less certainty around predictions of precipitation, and much more variability in the direction and magnitude of predicted change, with future projected precipitation ranging from drier to wetter depending on region (IPCC 2013). The southeastern U.S. is predicted to have less severe warming and smaller changes in precipitation compared to other regions in North America (Carter et al. 2013). We examined climate model outputs for a range of locations in northern Florida and southern Georgia (Fig. 9.1).

Fig. 9.2 shows projected climate for 2050–2075, under two CO_2 emissions scenarios: Representative Concentration Pathway 8.5 (RCP 8.5), which assumes CO_2 emissions and associated radiative forcing continue to increase through the end of the century, and Representative Concentration Pathway 4.5 (RCP 4.5), which assumes increased CO_2 emissions through mid-century followed by reductions in emissions to approximately 1975 levels by the end of the century (see van Vuuren et al. 2011 for more details on the emissions scenarios). Changes in daily maximum air temperatures vary by emissions scenario and by latitude of location, with increases relative to the 1950–2005 baseline of 1.6 to 1.8 °C (2.9 to 3.2 °F) under the RCP 4.5 scenario, and 2.3 to 2.7 °C (4.1 to 4.9 °F) for the RCP 8.5 scenario (Fig. 9.2). Large decreases in the number of days with frost are projected as well, with reductions of 36 days to 54 days across Florida depending on location and emissions scenario. Projected changes in precipitation across the same locations and emissions scenarios are relatively small, ranging from no change in precipitation to a 4% increase (Fig. 9.2).

Simulated Loblolly and Slash Pine Productivity under Future Climate

The productive potential of planted southern pine underlies most of the economic and many of the ecological benefits derived from managed forests in Florida. To understand how productivity might change under future climate conditions, we used the forest growth model 3-PG (the Physiological Processes Predicting Growth model; Landsberg and Waring 1997), which has been parameterized for the two most important commercial tree species in the region: loblolly pine and slash pine (Gonzalez-Benecke et al. 2014, 2016). We used gridded, interpolated historical climate data (http://metdata.northwestknowledge.net/) as well as the previously described climate projections and CO_2 concentration scenarios as input for 3-PG, and simulated stem wood volume production for 25-year rotations of the two species during a baseline period (1990–2005) and for a future period (2050–2075) at the same locations used for the climate projections (Fig. 9.1). In all simulations, an initial planting density of 1,500 trees per hectare and a site index of 22 m was assumed for unthinned stands.

The 3-PG model predicted increased loblolly and slash pine productivity across all six simulated locations and emissions scenarios during the 2050–2075 time period (Fig. 9.3). This consistent increase in productivity is attributable to the combination of relatively moderate increases in temperature, continued sufficient water availability through precipitation, and increased atmospheric CO_2, which acts as a fertilizer for plants (McCarthy et al. 2010). Relative increases in productivity were largest for slash pine, ranging from about 15% at the southern sites, to greater than 35% at the more northern locations. Loblolly pine also showed greater relative increases in productivity at the more northern sites but the magnitude of increase was smaller, ranging from less than 5% at the Alachua County location to almost 20% at the Jones County site (Fig. 9.3). At each site, productivity increases tended to be larger for the RCP 8.5 scenario than for the RCP 4.5 scenario. While relative increases in productivity were generally larger for slash pine compared to loblolly pine, the absolute productivity of loblolly pine was predicted to be larger than that of slash pine, consistent with current patterns of productivity of the two species (Jokela et al. 2010). It is important to note that the modeling approach used here did not incorporate the effects of disturbances, such as insects, disease, fire, or hurricanes on forest productivity, and as such should be considered an estimate of maximum potential productivity under future climate.

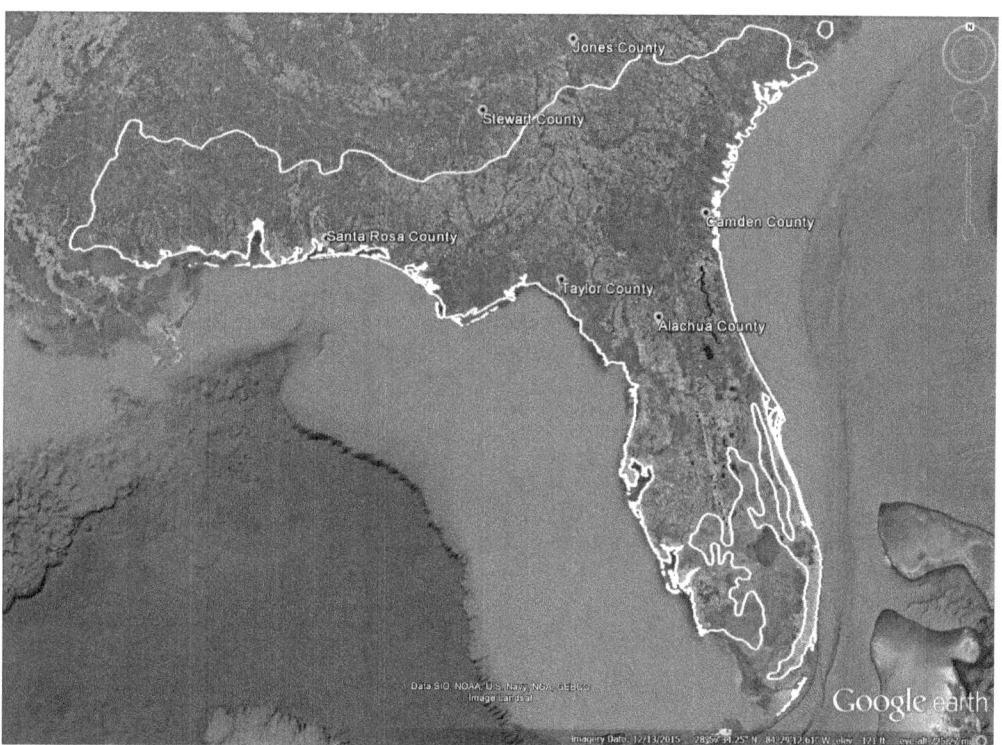

Figure 9.1. Map of the six locations (labeled by counties) in Florida and Georgia used for climate projections and productivity simulations. Current slash pine range is outlined in white.

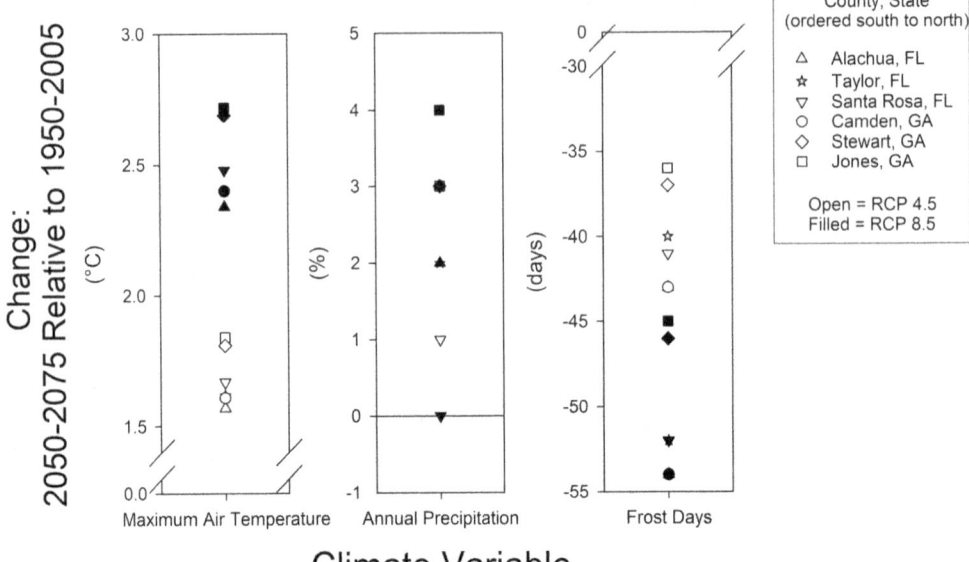

Figure 9.2. Projected change in mean maximum air temperature, annual precipitation, and annual days with frost for six locations in Florida and Georgia under two CO_2 emissions scenarios. Comparisons are for the period 2050–2075 relative to 1950–2005. Projections are the mean of output from 20 downscaled global circulation models (Abatzoglou and Brown 2012; Taylor et al. 2012).

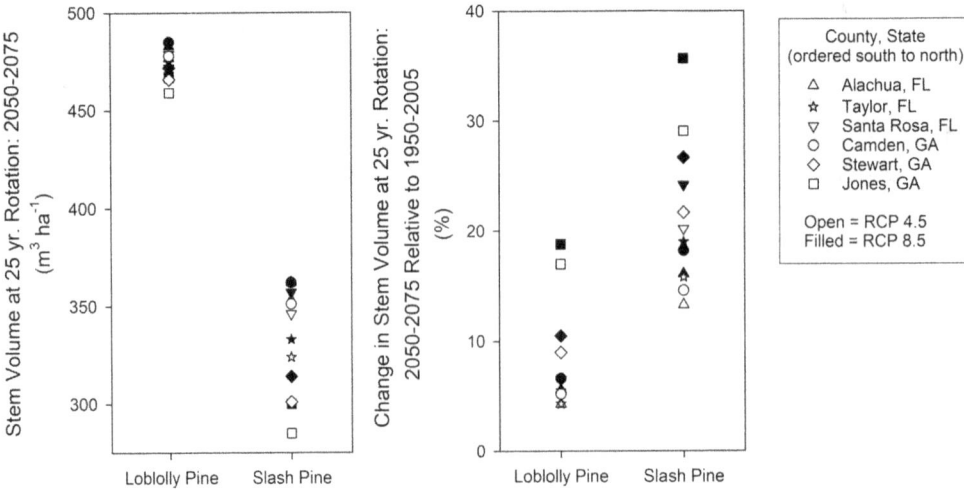

Figure 9.3. Projected stand stem volume at a rotation age of 25 years (left) and relative changes in stem volume (right) for loblolly and slash pine for six locations in Florida and Georgia under two CO_2 emissions scenarios. Stem volume was simulated with the 3-PG model (Gonzalez-Benecke et al. 2014, 2016) using as input the mean of 20 downscaled global circulation models (Fig. 9.2).

Abiotic Disturbance Effects under Future Climate

All pine species used in Florida plantations (longleaf, slash, and loblolly pines) evolved with periodic disturbances including fires, hurricanes, and droughts. These abiotic disturbances are natural components of southeastern U.S. ecosystems, and pines are adapted to survive these perturbations. Although expected to occur periodically, disturbance frequency and intensity have been projected to change in the coming decades as a result of climate change (Dale 2001; Westerling et al. 2011; Becknell et al. 2015; Johnstone et al. 2016); temperatures are rising, the growing season is becoming longer, fires and droughts are becoming more frequent and intense, and hurricane activity is expected to substantially increase. It is currently uncertain exactly how these changes will alter survival, regeneration, and other processes of pine species in plantations. We are, however, confident that current best management practices, such as maintaining proper tree spacing, attending to soil fertility, and controlling excessive competition, is likely to maintain or increase the resilience of forest plantations (Guldin 2014).

Effective understory fuel management can reduce the likelihood of intense and/or frequent fires predicted under climate change scenarios. Modern silvicultural approaches to site preparation and understory competition control using herbicides and mechanical treatments can be quite effective at controlling understory fuel loads, but these approaches can be cost-prohibitive for non-corporate landowners. Frequent, low-intensity fires every two to four years after pine establishment will reduce competing vegetation, be easier to control, and reduce the probability of catastrophic fires (Davis and Cooper 1963; Crow and Shilling 1980; Brose and Wade 2002). Pines have virtually no mortality following low-intensity fires as compared to coexisting vegetation. They survive and benefit from reduced competition as well as nutrient release after fire (Brockway et al. 1997; Mitchell et al. 2006). A large build-up of flammable fuels, which is more likely as growing seasons become longer or if fires are suppressed, increases the probability of a high-intensity fire causing crown scorch and consequentially increasing mortality of planted pines (Mitchell et al. 2009). From an economic standpoint, frequent low-intensity fires that might cause a small loss are preferable to high-intensity fires that are likely to cause a substantial loss.

Pine species are well-adapted to survive fires. Pine foliage is susceptible to fire, but unless all of the needles on a tree are completely scorched, mortality is unlikely. Furthermore, pine bark has good insulating qualities, which protects the aboveground stem from injury. Bark thickness varies considerably between and within pine species; but as a general rule, it increases with age and tree girth. Researchers have found that bark thicker than 12 mm will protect the stem cambium of most pines during prescribed fires (Fahnestock and Hare 1964). Even though pine bark is a good insulator, cambial damage can occur if the fire duration is long, which typically occurs if fuels (e.g., sloughed bark and needles) have accumulated at the base of a tree (Menges and Deyrup 2001; Varner et al. 2005). Damage to either the crown or cambium often results in death of trees months later.

Prescribed fire can be used to enhance the production of some non-timber values in planted forests. For instance, forage for wildlife and recreational opportunities, such as hunting, hiking, picnicking, and horseback riding, all benefit from periodic fires that reduce woody understory vegetation (De Ronde et al. 1990). One incentive to managing for wildlife, in particular, is the opportunity for forest landowners to participate in federal cost-sharing programs designed to offset the cost of improving habitat (Mixon et al. 2009). In addition, managing forests to improve ecosystem services, such as water yield, can be profitable if stands are well-managed and not densely planted (Susaeta et al. 2016a).

The effects of abiotic disturbances on pine plantations differ among species. By many measures, longleaf pine is the most resilient of pines planted in Florida (Wade and Johansen 1986). It can survive fires when young, has high hurricane tolerance (Johnsen et al. 2009), and can grow on dry, low-nutrition sites (Jose et al. 2007). Despite these potential advantages, longleaf pine is rarely planted for timber production because when it is a young tree its productivity is lower than that of slash pine or loblolly pine (Haywood et al. 2015).

In general, exposure time and intensity of drought and hurricane disturbances determine survival of pines, but generally small trees of a given species are easier to kill than large ones. If a disturbance, such as drought, is severe and occurs repeatedly over multiple years, pine growth will slow and some mortality is inevitable. As our climate changes, the frequency and intensity of disturbances will change as will interactions between them. It is predicted that increased drought frequency will increase fire frequency and intensity, as well as the ability of resource managers to use prescribed burns to lessen the intensity of wildfire (Mitchell et al. 2014).

Uncertainty about the future of pine plantations in Florida's disturbance-prone habitats highlights the need for research leading to predictive models that incorporate the effects of disturbance. This will help us forecast how climate change and associated changes in disturbance frequency and intensity, as well as interaction between disturbances, will affect the survival and growth of planted pines. These simultaneous changes in climate and disturbance regimes may require us to rethink how we manage pine plantations in the future (Becknell et al. 2015; Johnstone et al. 2016). Therefore, it is essential that these models help predict future trajectories of forest production and economic profit in ways that can guide policy decisions and management strategies.

Forest Health Impacts under Future Climate

Although the effects of pests and pathogens on conifers under predicted future climate are difficult to forecast, the magnitude and frequency of their impacts are predicted to increase (Garrett et al. 2009; 2013). Even if future climate scenarios are predictable in a given region, the potential effects on tree physiology and interactions with pests and pathogens are less certain (Desprez-Loustau et al. 2009). A useful framework for thinking about changes in forest health risk is the disease triangle (Fig. 9.4), which shows that host susceptibility, pathogen virulence, and conducive environments must all align for pest or disease outbreaks to occur. Climate changes may favor one factor, but not another side of the triangle–or may have counterbalancing effects. Additionally, it is difficult to predict how other exogenous variables, such as predators of pests, changes in silviculture and pesticide use, and other land use changes may influence future outbreaks in forests. Overall, perturbations to climate that challenge pine species' adaptations to current and past conditions will likely result in plant stress. Knowing the severity and duration of the stress can be useful for predicting the types of pests and pathogens that may take advantage of stressed hosts. In general, management for healthier forests in the face of climate change will need to focus on design of resilient, genetically-defined and adapted plantations (Showalter et al. 2016; Coakley et al. 1999).

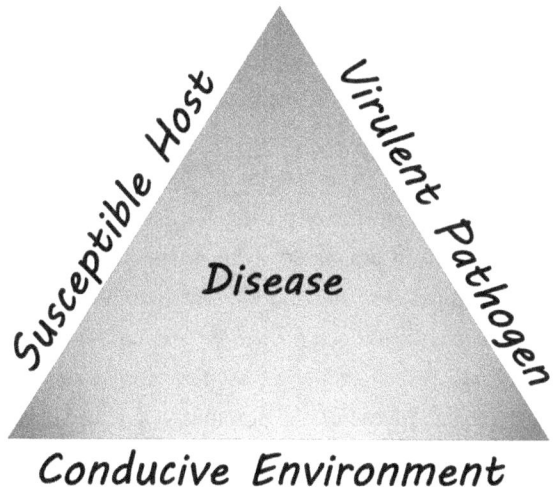

Figure 9.4. The disease triangle showing the three factors necessary for the development of disease. Note that timespans for a conducive environment are skewed for long-lived plants, such as trees.

Despite the uncertainties about how forest health will be affected by future climate, a body of empirical evidence is building that is giving us insight into what to expect. For example, Dothistroma needle blight and Swiss needle cast exemplify foliar diseases that have caused significant damage to plantation conifers worldwide as favorable conditions have aligned with disease biology to enhance the scale and severity of outbreaks. Dothistroma needle blight, caused by the fungal pathogens *Dothistroma septosporum* and *D. pini,* is a major disease of *Pinus* spp. worldwide (Bulman et al. 2013). In many areas, there have been significant increases in outbreak frequency and the disease has recently expanded into new areas (Barnes et al. 2008). Research has implicated strong El Niño Southern Oscillation (ENSO) events coincident with intercontinental disease outbreaks (Woods et al. 2016) and suggests that future climate predictions for the Northern Hemisphere will favor Dothistroma needle blight outbreaks in many areas. Although Dothistroma needle blight is not considered a major threat to southern pines, comparable pathogens occur in the region and could pose a similar threat.

Swiss needle cast, caused by *Phaecryptopus gaeumannii,* is a major foliar disease affecting plantation grown Douglas fir (*Pseudotsuga menziesii*) (Stone et al. 2008). It is particularly damaging to plantations in western Oregon and Washington in the Pacific Northwest that were previously forested with other species, such as western red cedar (*Thuja plicata*), western hemlock (*Tsuga heterophylla*) and Sitka spruce (*Picea sitchensis*) and in New Zealand in wet coastal regions (Stone et al. 2007). The establishment of large numbers of plantations in regions favorable for disease has been a key to why these outbreaks first occurred. However, it is becoming increasingly clear that warmer winters are strongly favoring pathogen reproductive and survival processes and leading to significant increases in severity with growth losses of >50% reported (Stone et al. 2008). Although Swiss needle cast does not pose a direct threat to southern pines, similar foliar diseases, such as brown spot on longleaf pine (caused by the fungus *Mycosphaerella dearnessii*) may behave in a similar way, and with increased efforts to more broadly re-establish longleaf pine, perhaps in uncharacteristic sites at times, this disease needs to be monitored.

In general, little is known about the role climatic factors (e.g., temperature, rainfall, and humidity) play in modulating traits in trees or pathogens that might impact host susceptibility or pathogen virulence. Currently, efforts are underway to assess how pathogen and pest biology and life histories are affected by temperature and humidity. One example includes the important pine disease pitch canker. Caused by the fungus *Fusarium circinatum,* pitch canker disease affects most pine species globally and is responsible for high economic losses in the timber industry. Favored by high temperatures and humidity, the possibility of future outbreaks is high under the environmental conditions predicted for the next 50–100 years. Also, because breeding for disease-resistant trees takes many years to accomplish, it is important to understand the biology of this fungus and be able to predict which disease scenarios would be more likely to thrive under future climate conditions. Assessments of in vitro culture growth as well as sporulation and virulence of isolates collected along north–south gradients in the southeastern U.S. have illustrated isolate-specific preferences for higher or lower temperatures and appear to suggest geographic patterns (Quesada et al. 2016). This information, along with field disease phenology (spore trapping) and host spatial distribution, could be used to develop epidemiological models to predict future outbreaks (Quesada et al. 2016).

One particularly concerning mechanism by which impacts from forest pests may intensify due to climate change is the invasion by native species into new geographical areas where encounters with naïve hosts could lead to devastating effects. A very dramatic example, the mountain pine beetle (*Dendroctonus ponderosae*), has demonstrated this with devastating impacts in the western portion of North America (Carroll et al. 2003). A native species of bark beetle, the mountain pine beetle, typically affected stressed lodgepole (*P. contorta*) and ponderosa pine (*P. ponderosa*) forests in the Rocky Mountains. A combination of fire suppression, limited forest management, and decades of drought led to a massive outbreak of the mountain pine beetle that spanned from northern British Columbia to Guatemala. The effects of drought and climate change on the initial outbreak appear to be compelling, but what was unexpected was the expansion of the mountain pine beetle into previously unaffected high elevation whitebark pine (*P. albicaulis*) and high latitude jack pine (*P. banksiana*) forests (Bentz et al. 2010), presumably due to a lack of low winter minimum temperatures that previously would have kept this pest from these areas. Both host species are now considered new hosts for the mountain pine beetle, allowing the pest to move unchecked through these susceptible, non-co-evolved hosts (Bentz et al. 2009). The effect is the same as a new introduction of an alien pest. There remains uncertainty about how far the mountain pine beetle destruction will go, but there is the possibility of a trans-continental range expansion to eastern North America on jack pine, potentially threatening eastern pine forests in the future. This type of phenomenon is likely to be experienced under future climate scenarios with other native pests and should be emphasized in future efforts to establish resilient silvicultural methods and in assessments of how to manage fire on the landscape (Bentz et al. 2010).

Water Resources under Future Climate

Climate change impacts to water resources are predicted to be significant (Arnell 1999), influencing precipitation and evaporation everywhere, albeit unevenly, and sea levels in coastal areas. Coupled to growing human water demands, these changes are already altering the volume, timing, and quality of fresh water in rivers, wetlands, lakes, aquifers and estuaries, and the availability of water for human needs (Vorosmarty et al. 2000). These changes are likely to impact Florida's commercially harvested forests in both upland and wetland settings, affecting their productive capacity (Sun et al. 2000a), their composition and resilience (Hansen et al. 2001), and their ability to sustain landscape hydrologic services (Sun et al. 2005). Plantation forestry is an extensive enterprise in Florida, however, forests can also be managed to mitigate many of the water resource challenges presented by a changing climate (Ford et al. 2011).

Forests use water (Bosch and Hewlett 1982), with rates of use impacted by composition, density, and understory management (including fire) (Powell et al. 2005). The proportion of precipitation that returns to the atmosphere via evapotranspiration can exceed 90% in Florida's highly productive commercial forests (Gholz and Clark 2002), suggesting that Florida's forests, and indeed forests worldwide, are important regulators of stream flow (Jackson et al. 2005). The links between forest management and landscape hydrology (e.g., streamflow and aquifer recharge) also illustrate opportunities to mitigate climate change impacts, and possibly regional water supply conflict, by connecting landowners willing to manage their plantation forests at lower density with groups willing to pay for enhanced water yield (McLaughlin et al. 2013).

Ecosystem productivity is vulnerable to changes in water availability, particularly for systems like those in Florida where rainfall and evapotranspiration are approximately in balance (Porporato et al. 2006). While forecast mean annual rainfall changes across the southeast are modest, this does not necessarily imply that the hydrologic impacts of these changes are negligible. Forests respond to patterns of rainfall, not just the annual amount (Porporato et al. 2004), potentially creating water stress in sandy soils like those common in Florida if rainfall intensity and frequency changes, even where total rainfall remains unchanged. Predictions strongly support increased incidence and altered timing of extreme rainfall (Wang et al. 2013), with drier summers, wetter winters, and more intense hurricanes (Enfield et al. 2011). Along with increased atmospheric demand for water arising from the 2 to 5 °C forecasted rise in temperature, these changing rainfall patterns can impact growth, fire and disease risk, species invasions (particularly in wetland settings), and nutrient cycling.

Compared to other land uses, plantation forests retain many of the water storage compartments present in natural landscapes, including shallow aquifers, soils with thick surface organic layers, and wetlands (Sun et al. 2000b). Low intensity management to protect soil recharge and storage functions, and best management practices that protect embedded wetlands (FDACS 2008), result in landscapes that persist in their capacity to retain rainfall. This has particularly important implications with increased incidence of extreme events (floods and droughts) since those storages serve the multiple roles of retaining floodwaters under high rainfall conditions (Lane and D'Amico 2010), attenuating downstream risks, and also sustaining flow to streams during drier periods (McLaughlin et al. 2014).

One emerging effect of ongoing carbon dioxide (CO_2) enrichment of Earth's atmosphere is improvement in plant water use efficiency, an effect that is weaker in trees than herbaceous plants (Saxe et al. 1998). With higher CO_2 concentrations, plants satisfy their carbon needs more easily, and thus lose less water for the same production (i.e., they use water more efficiently). In a retrospective modeling study of coastal plain ecosystems, water use efficiency in forests was high compared to other vegetation types but exhibited limited plasticity with rising CO_2 levels, suggesting that long-term regional increases in water use efficiency are not a result of forest CO_2 fertilization (Tian et al. 2010). One reason may be that CO_2 enrichment effects are mitigated by low nitrogen availability, a condition also impacted by increased temperatures (enhancing mineralization and denitrification rates), and reduced soil moisture (decreasing soil nitrogen mineralization rates) (Pastor and Post 1986). In short, while water use efficiency gains are possible, Florida's forests are already high efficiency systems. The implications for forest nutrition and water yield are important, but largely still uncertain.

Impacts to water quality are not frequently part of the global change narrative, except for ocean acidification effects. However, several key water quality attributes are likely to be impacted. Plantation forests are widely observed to protect downstream water quality in Florida and elsewhere (Omernik 1976, U.S. EPA 1995). High rates of primary production, generally low fertilization rates, limited use of agrochemicals, and effective best management practices to limit sediment loading or thermal impacts to streams mean that plantation forests offer a viable option that balances economic production and water quality protection. Climate change is likely to exacerbate existing water quality challenges, elevating the importance of land planning that integrates and incentivizes plantation forests. Increased incidence of extreme rainfall will likely lead to enhanced soil sediment mobilization, particularly from urban and agricultural areas, but also from plantation forests during clearcut and bedding phases (Aust and Blinn 2004). Altered flow and increased temperature are likely to alter landscape delivery of water and concentrations of key constituents, such as dissolved carbon, nitrogen and phosphorus, with some global trends already evident (Evans et al. 2004). Similarly, increased incidence of prolonged dry periods will likely impact stream dissolved oxygen and organic matter dynamics (Mulholland et al. 1997), as well as salinity in coastal forest habitats (Williams et al. 1999). The role of plantation forests in mitigating these water quality challenges follows from the general notion that plantation forests approximate a natural flow regime, that forestry operations, especially in Florida, adhere to long-standing and demonstrably effective best management practices borne of the need to protect water quality, and that flatwoods landscapes, with forests and embedded wetlands, contribute to carbon, nitrogen, and phosphorus cycling that effectively retains these elements.

Predicting the links between water resources and commercial forests in a changing climate is challenged by myriad uncertainties. The magnitude of temperature and rainfall changes, the capacity of planted trees to adjust to these changes and the attendant physiological subsidies and stresses, and the role of management (e.g., fertilization, stand density) together create a complex and contingent problem. However, it is clear that plantation forests have been a vital part of protecting regional water quantity and quality, and that they will continue to be essential features of healthy productive landscapes.

Silvicultural Approaches for Maintaining Pine Plantation Productivity in a Changing Climate

Climate change represents opportunities, threats, and a number of unknowns for land managers practicing silviculture in Florida pine plantations. As mentioned in previous sections, over the last several decades Florida has seen nominal increases in average precipitation and temperature, with most of the changes occurring during the cold season. With projected changes in average temperature and precipitation, silvicultural practices that currently improve plantation productivity will likely interact with climate or CO_2 trends to further increase productivity. Less certain is how silviculture might interact with an increase in the frequency of extreme climatic events (drought, storm events), as these are expected to also increase with climate change (Bell et al. 2016). Here we discuss critical decision points in plantation silviculture and how the importance of these decisions may be affected by climate change in Florida.

Achieving survival targets and rapid early growth in planted seedlings is the first step landowners can take to reduce climate change effects on their plantations. Plantation establishment generally includes site preparation for planting, the planting phase, post-planting release treatments from competing vegetation, and fertilization on low productivity sites. Landowners who use modern establishment and planting practices now average ~90% survival for their plantations in the Southeast (Lang et al. 2016). Land managers who consistently have seedlings survive at less than this average should consider their stand establishment approaches and identify reasons for lower effectiveness, as future conditions might alleviate or make worse the reasons for poor performance.

A stand establishment technique that may be critical for Florida in an era of rapid climate change is the continued use of raised mounds or beds on which to plant seedlings. Beds are widely used on poorly drained soils in the region, as they keep seedling roots out of saturated soil conditions that can slow tree growth or even facilitate mortality (Outcalt 1984). Bedding is likely to remain critical because one climate prediction is for the increased frequency of intense precipitation events (Bell et al. 2016). Land managers can use the geo-located 'Web Soil Survey' internet application, developed by the National Resources Conservation Service, to determine if the soil on their property has poor soil drainage characteristics. Moreover, vegetation in a pre-harvest stand may include indicator plant species (e.g. pitcher plants, wiregrass, and palmetto) that suggest impeded drainage (Jokela and Long 2012). Land managers should avoid relying solely on past experiences as to which sites they do bed or whether double-bedding is required, because a site's bedding requirements may change in response to future, extreme precipitation events or shifts in local hydrology.

Choosing which seedlings to plant is another way that landowners can mitigate potential climate change effects. For example, if bedding is too expensive but an area's soils are prone to flooding or saturation, landowners could consider planting slash pine as it is more resistant than loblolly pine to poorly drained soil conditions (Oucalt 1984). Another decision point is whether to plant bare-root (less expensive) or containerized (more expensive) seedlings. Containerized seedlings have more developed root systems and are generally more resistant to poor soil conditions (drought or saturated soils) than are bare-root seedlings (Grossnickle and El-Kassaby 2016). Containerized seedlings may also be the better choice if temperatures are warmer than optimal during planting, because their greater root density generally reflects greater nutrient reserves. In a period of rapid change, using containerized seedlings on marginal soils may justify their greater expense by reducing uncertainty in seedling survival.

Controlling competing vegetation is an important part of plantation management that could become even more critical with climate change. More intense droughts, in particular, would accentuate the inter-species competition for water that often results in reduced pine growth or even mortality (Zutter et al. 1986). Competing vegetation can also prevent fertilizer from increasing tree growth (Jokela et al. 2010), prolonging the time a plantation remains in a sensitive juvenile state. Mechanical site preparation that turns the soil or a prescribed burn after a harvest offers slight to moderate control of shrubs (e.g. gallberry and palmetto), but herbicide applications are more certain in their control and are needed for herbaceous weeds (Miller et al. 2003). An important unknown is whether competing plants will become more difficult to control in a period of rising atmospheric CO_2 and temperatures, as species may be differentially adapted to these new conditions (Manea and Leishman 2011).

Land managers who practice sound plantation management control the numbers of living trees, or stand density, throughout a rotation. Termed "stand density management," this practice begins at planting with a decision on the spacing of tree seedlings, and then occurs later when the landowner decides on whether and when to thin a stand. Pine seedlings in the southeastern United States are currently planted at an average density of 584 trees per acre (Lang et al. 2016); however, few studies are available to suggest alternative densities that might mitigate climate change effects. In general, increasing planting densities may provide a 'hedge' against increased seedling mortality. This would be particularly important for either droughty or poor drainage soils that have had little site preparation, competition control, or fertilization. However, once tree size increases and crowding causes intense inter-tree competition, pine plantations have an increased risk of suffering mass mortality from southern pine beetles (Nowak et al. 2015)—pests that might increase in virulence as tree stress from drought and heat also increases (Gan 2004). Thinning is recommended as stands approach a level of crowding associated with intense inter-tree competition; but if, for economic reasons, a land manager does not think a stand will be thinned, then planting at a lower tree density (e.g. ~200-450 trees per acre) could help protect it against pathogens later in stand rotation. However, planting at low tree densities may require manual branch pruning as the retention of branches could decrease wood quality.

On soils that have inherently low fertility, fertilization can dramatically increase pine productivity, in particular when it is coupled with inter-species competition control and appropriate site preparation techniques (Jokela et al. 2010). It is possible that future wind and drought effects on plantations could be ameliorated with forest fertilization. Recent research suggests that pine plantations fertilized at higher levels are less sensitive to wind damage than those fertilized at lower rates (Zhai et al. 2015). This may reflect faster growth, increased coarse root development, and faster canopy closure, creating a greater overall resistance to wind effects (Stanturf et al. 2007). In reference to drought, fertilized mature pine plantations apparently grow faster than unfertilized forests under reduced moisture conditions (Maggard et al. 2016). Although more research is needed, the studies currently available suggest that fertilization at recommended rates will increase plantation resistance to some of the negative aspects of climate change, and could potentially increase the positive effect that elevated CO_2 levels have on pine growth (McCarthy et al. 2010).

Perhaps more important than plantation response to climate change will be how land managers respond to the as yet unknown threats and opportunities that will affect managed forests in a rapidly changing world. Land managers who remain engaged with the research community, extension professionals, and their fellow practitioners are the ones most likely to maintain or increase the profitability of their plantations. These individuals will also be critical to reporting any climate-driven changes in plantation function to the forestry community, helping the scientific community mobilize to address threats. Maintaining communication among those invested in pine plantation management will be the key to ensuring the southern pine plantation resource continues to be productive through the upcoming period of rapid change.

Mitigating Atmospheric CO_2 by Storing Carbon in Forests and Wood Products

Plantation forests have the potential to mitigate rising atmospheric CO_2. These forests are unique among agricultural crops in that they are a substantial net sink for CO_2, meaning that they take up more CO_2 than they release during a management cycle. Because of their large production area and high productivity, southern forests are a significant portion of the U.S. carbon budget, containing 36% of the sequestered forest carbon in the conterminous United States (Turner et al. 1995). Forests in the region annually sequester 76 million metric tons of carbon, equivalent to 13% of regional greenhouse gas emissions, and have the potential to sequester more through retention and expansion of forested land area, and improved forest management, which increases productivity and resilience to disturbance (Johnsen et al. 2001, Han et al. 2007).

Forest management can be used to increase sequestration both in forest ecosystems themselves and in harvested wood products. A study by Gonzalez-Benecke et al. (2010) illustrated how forest management can influence carbon pools in slash pine, an important timber species in Florida. This study used models to examine the impacts of different management scenarios on carbon storage in different ecosystem components, as well as in "off-site" pools associated with solid wood products, such as lumber, and pulp used to produce paper and similar products.

Gonzalez-Benecke et al. (2010) showed scenarios that increased rotation length and incorporated periodic thinning accumulated greater amounts of carbon in off-site forest product pools than did scenarios involving shorter rotations and no thinning (Fig. 9.5). These sawtimber-focused scenarios produced larger trees at final harvest that, when used for long-lived products, such as structural lumber and furniture, stored carbon for longer periods and in greater quantities than scenarios that produced smaller trees and more pulpwood, which has a shorter carbon half-life off site. Importantly, this study also demonstrated that the carbon emissions associated with silvicultural activities, such as energy used for fertilizer production, fuel for planting and harvesting equipment, and fuel for transport of logs to the mill, were only a small fraction (about 2%) of the total carbon sequestered by the management system. Additional carbon benefits can be derived if harvested wood or harvest debris is used to generate electricity, since these uses offset emissions that would have been associated with the use of fossil fuel energy sources (Dwivedi et al. 2016). Wear and Greis (2012) pointed to the potential of biomass energy markets as a "game changer" for forestry in the region, which could result in increased demand for productive forests, in turn helping to prevent the conversion of forestland to other land uses while maintaining substantial carbon benefits.

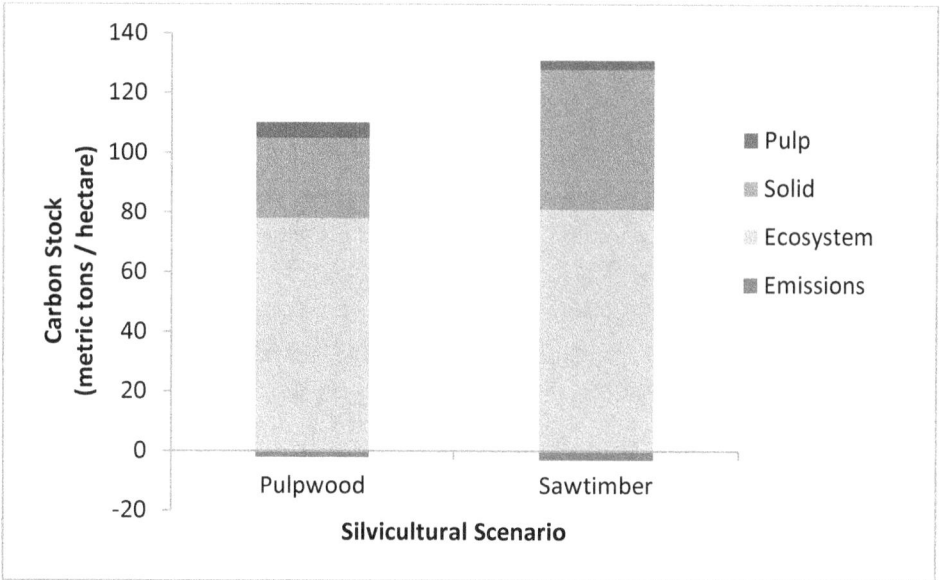

Figure 9.5. Average over five rotations for carbon stock in forest ecosystem pools, off-site pulp and solid wood pools, and average carbon emissions per rotation for two slash pine silvicultural scenarios. The pulpwood scenario involved typical fertilization, no thinning, and clearcut harvest at age 22 years. The sawtimber scenario involved typical fertilization, thinning at age 14 and 22 years, and final clearcut harvest at age 35 years. Adapted from Gonzalez-Benecke et al. (2010).

Forests are dynamic systems that generate ecosystem services that can be defined as tradeoffs (i.e., managing for one service leads to the decrease in another) or bundles (i.e., both services increase). We know that different forest structures and management approaches can affect timber production, carbon sequestration, water quality, water yield, wildlife habitat, and other services in meaningful ways (e.g., Susaeta et al. 2016a-b, 2016a-c). However, the tradeoffs among multiple ecosystem services (e.g., timber, water, carbon simultaneously), which is critical from a policy and forest management perspective, have been largely unexplored, particularly in a climate change context. For instance, increased tree stocking to maximize carbon stores in areas prone to fire or attack of insects and other pests may increase ecosystem vulnerability to natural disturbances, increase water use, and decrease stream flow. Further, these tradeoffs can differ across temporal and spatial scales and depend on interactions between land use and socioeconomic conditions. It is imperative that we gain much better insight into the different management alternatives, tradeoffs, and synergies between the multiple ecosystem services provided by forests, and do so in a climate change context.

Policy Options for Florida's Future Forests

Florida's future forests face increasing pressures from land use change (e.g., due to urbanization and agricultural in-migration from the drought-stricken western U.S.), pests and disease, invasive species, policy (e.g., estate taxes, and protections for endangered species and water resources), and changes in forest ownership (e.g., fragmentation due to estate tax; Butler and Wear 2013) that affect the forest estate. Land use change is by far the largest pressure. By 2060, we expect to lose between 30 and 43 million acres of southern U.S. forest to urbanization (Wear and Greis 2012). Absent strong policy intervention, this will lead to a net reduction in forest carbon stocks by 2060 (Hugget et al. 2013).

There are several viable policy options for harnessing forests to mitigate climate change and increasing forest resilience and adaptation to climate change. However, since 71% of Florida's forests are privately owned, policy options must align well with landowner needs to have adequate impact. Broadly speaking, policies that improve market conditions, reduce burdens (regulatory and economic), and increase economic sustainability for forest landowners would help protect Florida's existing forests and the benefits they provide, and encourage investment in afforestation and reforestation.

Robust, policy-driven markets for forest-based ecosystem services are important considerations. We know that forests are highly effective at sequestering atmospheric CO_2 as biomass and long-lived forest products, and they can provide long-term solutions to offsetting greenhouse gas emissions. In the southern U.S., conservative estimates suggest that forests could offset one-fourth of the region's greenhouse gas emissions (Han et al. 2007), and in a way that is cheap relative to climate change mitigation alternatives (e.g., shuttering coal-fired power plants; Couture and Reynaud 2011; Gren and Carlsson 2013). Their valuable role in mitigating climate change is recognized by programs and policies aiming to reduce greenhouse gases, including the United Nation's Kyoto Protocol, the U.S. Environmental Protection Agency's Clean Power Plan, California's AB32, the Climate Action Reserve, the Voluntary Carbon Standard, and the American Carbon Registry (Soto et al. 2016a). However, carbon sequestration and other ecosystem services from forests are classic examples of market failure–while there exists tremendous economic value associated with the myriad ecosystem services that forests provide (e.g., Costanza et al. 2014), there are few mechanisms available to landowners to capture this value. As a result, forests are largely valued for their timber alone, which may be a relatively small portion of the overall economic value (e.g., Escobedo et al. 2012). Market mechanisms (e.g., incentives through state or federal programs, mitigation markets, and boutique contracts with private conservation groups) have been lauded as an effective class of policy interventions to fix this market failure and protect forestland.

These approaches show considerable promise. For example, payments for ecosystem services is a market-based alternative and has been suggested as a more effective strategy for offsetting landowner costs associated with ecosystem service production (Ingram et al. 2014). Studies have identified non-industrial private forest landowners as receptive to payments for carbon sequestration (Soto et al. 2016a; Dwivedi et al. 2009); but, we still know relatively little about forest landowner preferences for these policy approaches and their use has not been widely subscribed. Also, we lack an understanding of the impact of landowner participation in carbon sequestration payment schemes on jointly produced ecosystem services (e.g., Beach et al. 2005).

Cost-share programs, technical assistance programs, and changes to tax policy (e.g., lower property tax rates for conservation lands) are known to play a key role in supporting working forests, and enjoy broad public support (Kreye et al. 2016). Several existing federal and state programs (e.g., the US Department of Agriculture's Conservation Reserve Program and Environmental Quality Incentive Program, the Southern Pine Beetle Assistance and Prevention Program, and the Longleaf Landowner Incentive Program; FDACS 2016) offer effective conservation cost-share models that could be expanded with a climate change-specific focus. However, landowner enrollment in these programs lags due to perceived program inflexibility and failure to effectively address the needs of individual landowners (Hyde et al. 1996).

Likewise, both conservation easements that allow forest landowners to retain their "working forest" status while either temporarily or permanently limiting their development rights, and outright purchase of forestland by conservation organizations, are approaches to maintaining the forest estate that enjoy broad popular support. These approaches can be particularly relevant for non-industrial private forest landowners who are oriented to activities other than timber production, such as recreation, aesthetics, and wildlife, and may be highly motivated to retain forestland ownership and bequest it to future generations (Smith et al. 2009). With climate change, it may be important to grow the use of these programs to increase forest resilience and/or more fully engage forestland to mitigate climate change effects. To date, these approaches do not seem effective at slowing the pace of forestland conversion (e.g., Kramer and Shabman 1993; Stainback and Alavalapati 2002).

Policies that decrease disturbance risk or increase forest yields also contribute to the economic sustainability of Florida forestland and carbon storage. For example, policy changes to reduce the arrival of destructive forest pests through trade, and funding to provide early detection and rapid reaction to new invasive forest pests have been shown to provide a very high return on investment (e.g., Susaeta et al. 2016d). Similarly, investments in research on improved silvicultural practices and tree breeding (including through the use of genomic information) have led to substantial increases in productivity for Florida forestry (Teskey 2014), although research support has not been consistent.

Finally, more support for and emphasis on generating economic valuation data for forests and forest-based ecosystem services is needed to inform future decisions, both by landowners and policymakers. Recognizing the value of forests, including market (e.g., timber) and nonmarket (e.g., carbon sequestration) benefits, is essential for generating public and policymaker support for forest conservation, and designing durable, effective mechanisms for landowners to capture a portion of the value that their forests provide to society. Lack of adequate information and models regarding economic impacts and ecological tradeoffs associated with managing forests for climate change mitigation or increased resilience (e.g., planting longleaf pine) remains a major barrier for forest landowners in terms of participation, and for policymakers in terms of assessing programs that ensure the sustainable provision of forest-based ecosystem services. It is also essential that we work to understand the social (e.g., distributional impacts and environmental justice) and administrative (e.g., ease of implementation and administration) aspects of forest-related climate change policy, since these factors help define what is "appropriate" climate change policy action for Florida, and are significant drivers of policy choices and their ultimate success or failure (Kreye et al. 2016, Soto et al. 2016b).

Conclusion

Changing climate through the end of this century poses both risks and opportunities for plantation forest management in Florida. Risks include potential alteration in frequency and intensity of abiotic disturbances, such as wildfire and windstorms, modification of pest and pathogen dynamics, and changes in water quantity and quality. Importantly, predictions about the nature and impacts of most of these risks remain uncertain. Despite these uncertainties, changing climate also is likely to open up opportunities for Florida's plantation forests. The productivity of plantation forests is likely to increase substantially through the end of the century, due to relatively moderate changes in temperature and precipitation coupled with increases in atmospheric CO_2 fertilization. There are also important opportunities to increase the sequestration of carbon by plantation forests through retention and expansion of forestland, and appropriate alteration of forest management. These approaches can help to mitigate rising atmospheric CO_2 while simultaneously increasing the many other benefits provided by forests to the state of Florida. Application of appropriate silvicultural technology and the development of supportive public policy will be key factors in adapting future plantation forest management to the risks and opportunities of climate change.

Acknowledgments

Some of the research reported here was a part of the Pine Integrated Network: Education, Mitigation, and Adaptation project (PINEMAP), a Coordinated Agricultural Project funded by the USDA National Institute of Food and Agriculture, Award #2011-68002-30185.

References

Abatzoglou, J.T., and T.J. Brown. 2012. A comparison of statistical downscaling methods suited for wildfire applications. International Journal of Climatology 32:772-780.

Arnell, N.W. 1999. Climate change and global water resources. Global Environmental Change 9:S31-S49.

Aust, W.M, and C.R. Blinn. 2004. Forestry best management practices for timber harvesting and site preparation in the eastern United States: An overview of water quality and productivity research during the past 20 years (1982–2002). Water Air and Soil Pollution 4:5-36.

Barnes, I., T. Kirisits, A. Akulov, D.B. Chhetri, B.D. Wingfield, T.S. Bulgakov, and M.J. Wingfield. 2008. New host and country records of the Dothistroma needle blight pathogens from Europe and Asia. Forest Pathology 38:178-195.

Beach, R.H., S.K. Pattanayak, J. Yang, B. Murray, and R.C. Abt. 2005. Econometric studies of non-industrial private forest management: a review and synthesis. Forest Policy & Economics 7:261-281.

Becknell, J.M., A.R. Desai, M.C. Dietze, C.A. Schultz, G. Starr, P.A. Duffy, J.F. Franklin, A. Pourmokhtarian, J. Hall, P.C. Stoy, M.W. Binford, L.R. Boring, and C.L. Staudhammer. 2015. Assessing interactions among changing climate, management, and disturbance in forests: A macrosystems approach. BioScience 65:263-274.

Bell J.E., S.C. Herring, L. Jantarasami, C. Adrianopoli, K. Benedict, K. Conlon, V. Escobar, J. Hess, J. Luvall, C.P. Garcia-Pando, D. Quattrochi, J. Runkle, and C.J. Schreck, III. 2016. Impacts of extreme events on human health. Chapter 4 *In:* The impacts of climate change on human health in the United States: A scientific assessment. U.S. Global Change Research Program.

Bentz, B. J., J. Regniere, C.J. Fettig, E.M. Hansen, J.L. Hayes, J.A. Hicke, R. Kelsey, J. Negron, and S.J. Seybold. 2010. Climate change and bark beetles of the Western United States and Canada: Direct and indirect effects. BioScience 60:602-613.

Bentz, B., J. Logan, J. MacMahon, C.D. Allen, M. Ayres, E. Berg, A. Carroll, M. Hansen, J. Hicke, L. Joyce, W. Macfarlane, S. Munson, J. Negron, T. Paine, J. Powell, K. Raffa, J. Regniere, M. Reid, B. Romme, S.J. Seybold, D. Six, D. Tomback, J. Vandygriff, T. Veblen, M. White, J. Witcosky, and D. Wood. 2009. Bark beetle outbreaks in western North America: Causes and consequences. Bark Beetle Symposium; Snowbird, Utah; November, 2005. Salt Lake City, UT: University of Utah Press. 42 p.

Bosch, J.M. and J.D. Hewlett. 1982. A review of catchment experiments to determine the effect of vegetation changes on water yield and evapotranspiration. Journal of Hydrology 55:3-23.

Brockway, D.G., and C.E. Lewis. 1997. Long-term effects of dormant-season prescribed fire on plant community diversity, structure and productivity in a longleaf pine wiregrass ecosystem. Forest Ecology and Management 96:167-183.

Brose, P., and D. Wade. 2002. Potential fire behavior in pine flatwood forests following three different fuel reduction techniques. Forest Ecology and Management 163:71-84.

Brown, M.J., and J. Nowak. 2016. Forests of Florida, 2014. Resource Update FS-91. 4 p.

Brown, S. 1981. A comparison of the structure, primary productivity, and transpiration of cypress ecosystems in Florida. Ecological Monographs 51:403-427.

Bulman, L.S., M.A. Dick, R.J. Ganley, R.L. McDougal, A. Schwelm, and R.E. Bradshaw. 2013: Dothistroma needle blight. *In:* Infectious Forest Diseases. Ed. by Gonthier, P.; Nicolotti, G. Boston, MA: CABI, pp. 436-457.

Butler, B.J., and D.N. Wear. 2013. Forest ownership dynamics of southern forests. The Southern Forest Futures Project: Technical Report.

Carter, L.M., J.W. Jones, L. Berry, V. Burkett, J.F. Murley, J. Obeysekera, P.J. Schramm, and D.N. Wear. 2014. Southeast and the Caribbean. *In:* Climate Change Impacts in the United States: The Third National Climate Assessment, edited by J.M. Melillo, T.C. Richmond, and G. W. Yohe. U.S. Global Change Research Program. pp. 396-417.

Clark, K.L., H.L. Gholz, and M.S. Castro. 2004. Carbon dynamics along a chronosequence of slash pine plantations in north Florida. Ecological Applications 14:1154-1171.

Coakley, S. M., H. Scherm, and S. Chakraborty. 1999. Climate change and plant disease management. Annual Review of Phytopathology 37:399-426.

Costanza, R., R. de Groot, P. Sutton, S. van der Ploeg, S.J. Anderson, I. Kubiszewski, S. Farber, and R.K. Turner. 2014. Changes in the global value of ecosystem services. Global Environmental Change 26:152-158.

Couture, S., and A. Reynaud. 2011. Forest management under fire risk when forest carbon sequestration has value. Ecological Economics 70:2002-2011.

Crow, A.B., and C.L. Shilling. 1980. Use of prescribed burning to enhance southern pine timber production. Southern Journal of Applied Forestry 4:15-18.

Currie, D.J. 2001. Projected effects of climate change on patterns of vertebrate and tree conterminous United States. Ecosystems 3:216-225.

D'Amato, A.W., J.B. Bradford, and S. Fraver. 2013. Effects of thinning on drought vulnerability and climate response in north temperate forest ecosystems. Ecological Applications 23:1735-1742.

D'Amato, A.W., J.B. Bradford, S. Fraver, and B.J. Palik. 2011. Forest management for mitigation and adaptation to climate change: insights from long- term silviculture experiments. Forest Ecology and Management 262:803-816.

Dale, V. H., L.A. Joyce, S. McNulty, R.P. Neilson, M.P. Ayres, M.D. Flannigan, P.J. Hansen, L.C. Irland, A.E. Lugo, C.J. Peterson, D. Simberloff, F.J. Swanson, B.J. Stocks, and B.M. Wotton. 2001. Climate change and forest disturbances: climate change can affect forests by altering the frequency, intensity, duration, and timing of fire, drought, introduced species, insect and pathogen outbreaks, hurricanes, windstorms, ice storms, or landslides. BioScience 51:723-734.

Davis, L.S., and R.W. Cooper. 1963. How prescribed burning affects wildfire occurrence. Journal of Forestry 61:915-917.

Davis, S.C., A.E. Hessl, C.J. Scott, M.B. Adams, and R.B. Thomas. 2009. Forest carbon sequestration changes in response to timber harvest. Forest Ecology and Management 258:2101-2109.

De Ronde, C., J.G. Goldammer, D.D. Wade, and R.V. Soares. 1990. Prescribed fire in industrial pine plantations. *In*: Goldammer, J.G., ed. Fire in the tropical biota, Springer Berlin Heidelberg. pp. 216-272.

Desprez-Loustau, M. L., C. Robin, G. Reynaud, M. Deque, V. Badeau, D. Piou, C. Husson, and B. Marcais. 2007: Simulating the effects of a climate-change scenario on the geographical range and activity of forest-pathogenic fungi. Canadian Journal of Plant Pathology 29:101-120.

Dwivedi, P., J.R. Alavalapati, A. Susaeta, and A. Stainback. 2009. Impact of carbon value on the profitability of slash pine plantations in the southern United States: an integrated life cycle and Faustmann analysis. Canadian Journal of Forest Research 39:990-1000.

Dwivedi, P., M. Khanna, A. Sharma, and A. Susaeta. 2016. Efficacy of carbon and bioenergy markets in mitigating carbon emissions on reforested lands: A case study from Southern United States. Forest Policy and Economics 67:1-9.

Enfield, D., S.K. Lee, F. Marks and M. Powell. Mid-Century Expectations for Tropical Cyclone Activity and Florida Rainfall. In Misra, V., E. Carlson, R. K. Craig, D. Enfield, B. Kirtman, W. Landing, S.-K. Lee, D. Letson, F. Marks, J. Obeysekera, M. Powell, S.-l. Shin, 2011: Climate Scenarios: A Florida-Centric View, Florida Climate Change Task Force. [Available online at http://floridaclimate.org/whitepapers/]

Escobedo, F.J., A. Abd-Elrahman, D.C. Adams, A. Frank, N. Kil, M. Kreye, T. Kroeger, T. Stein, and N. Timilsina. 2012. Stewardship Ecosystem Services Project Final Report. School of Forest Resources and Conservation, University of Florida, Gainesville, Florida. http://www.sfrc.ufl.edu/cfeor/SESS.html.

Evans, C.D., D.T. Monteith, and D.M. Cooper. 2004. Long-term increases in surface water dissolved organic carbon: Observations, possible causes, and environmental impacts. Environmental Pollution 137:55-71.

Fahnestock, G.R., and R.C. Hare. 1964. Heating of tree trunks in surface fires. Journal of Forestry 62:799-805.

Florida Department of Agriculture and Consumer Services [FDACS]. 2016. Forestry and wildlife cost share programs. http://www.freshfromflorida.com/Divisions-Offices/Florida-Forest-Service/For-landowners/Programs/Forestry-and-Wildlife-Cost-Share-Programs

Florida Department of Agriculture and Consumer Services [FDEP]. 2016. *Florida Agriculture Overview and Statistics*. http://www.freshfromflorida.com/Divisions-Offices/Marketing-and-Development/Education/For-Researchers/Florida-Agriculture-Overview-and-Statistics

Florida Department of Agriculture and Consumer Services. 2008. Silviculture Best Management Practices. http://www.freshfromflorida.com/Divisions-Offices/Florida-Forest-Service/Our-Forests/Best-Management-Practices-BMP

Florida Forest Service. 2015. Florida Forestry Economic Highlights. FL Forestry News D4: 3.

Ford, C.R., S.H. Laseter, W.T. Swank, and J.M. Vose. 2011. Can forest management be use to sustain water-based ecosystems services in the face of climate change? Ecological Applications 21:2049-2067.

Franklin, J.F., R.J. Mitchell, and B.J. Palik. 2007. Natural Disturbance and Stand Development Principles for Ecological Forestry. General Technical Report NRS-19, USDA-Forest Service.

Gan, J. 2004. Risk and damage of southern pine beetle outbreaks under global climate change. Forest Ecology and Management 191:61-71.

Garren, K.H. 1943. Effects of fire on vegetation of the southeastern United States. The Botanical Review 9:617-654.

Garrett, K. A., A.D.M. Dobson, J. Kroschel, B. Natarajan, S. Orlandini, H.E.Z. Tonnang, and C. Valdivia. 2013. The effects of climate variability and the color of weather time series on agricultural diseases and pests, and on decisions for their management. Agricultural and Forest Meteorology 170:216-227.

Garrett, K. A., M. Nita, E.E. De Wolf, L. Gomez, and A.H. Sparks.2009. Plant pathogens as indicators of climate change. *In*: Climate and Global Change: Observed Impacts on Planet Earth. Ed. by Letcher, T. Oxford: Elsevier Science, pp. 425-437.

Gholz, H.L., and K.L. Clark. 2002. Energy Exchange Across a Chronosequence of Slash Pine Forests in Florida. Agricultural and Forest Meteorology 112:87-102.

Gonzalez-Benecke, C.A., T.A. Martin, W.P. Cropper Jr., and R. Bracho. 2010. Forest management effects on in situ and ex situ slash pine forest carbon balance. Forest Ecology and Management 260:795-805.

Gonzalez-Benecke, C.A., E.J. Jokela, W.P. Cropper, Jr., R.G. Bracho, and D.J. Leduc. 2014. Parameterization of the 3-PG model for *Pinus elliottii* stands using alternative methods to estimate fertility rating, biomass partitioning and canopy closure. Forest Ecology and Management 327:55-75.

Gonzalez-Benecke, C.A., R.O. Teskey, T.A. Martin, E.J. Jokela, T.R. Fox, M.B. Kane, and A. Noormets. 2016. Regional validation and improved parameterization of the 3-PG model for *Pinus taeda* stands. Forest Ecology and Management 361:237-256.

Grace, S.L., and W.J. Platt. 1995. Effects of adult tree density and fire on the demography of pregrass stage juvenile longleaf pine (*Pinus palustris* Mill.). Journal of Ecology 83:75-86.

Gren, M., and M. Carlsson. 2013. Economic value of carbon sequestration in forests under multiple sources of uncertainty. Journal of Forest Economics 19:174-189.

Grossnickle S.C., and Y.A. El-Kassaby. 2016. Bareroot versus container stocktypes: a performance comparison. New Forests 47:1-51.

Guldin, J.M. 2011. Experience with the selection method in pine stands in the southern United States, with implications for future application. Forestry 84:539-546.

Guldin, J.M. 2014. Adapting silviculture to a changing climate in the southern United States. *In:* J.M. Vose and K.D. Klepzig, Eds., Climate Change Adaptation and Mitigation Management Options, A Guide for Natural Resource Managers in Southern Forest Ecosystems. CRC Press, Boca Raton, Florida. pp. 173-192.

Gustafson, E.J., P.A. Zollne, B.R. Sturtevant, H.S. He, and D.J. Mladenoff. 2004. Influence of forest management alternatives and land type on susceptibility to fire in northern Wisconsin, USA. Landscape Ecology 19:327-341.

Guthrie, K., R. Barlow, and F.S. Kush. 2016. Restoring an ecosystem with silvopasture: a short (leaf) story. Ecological Restoration 34:16-19.

Han, F.X., M.J. Plodinec, Y. Su, D.L. Monts, and Z.P. Li. 2007. Terrestrial carbon pools in southeast and south-central United States. Climatic Change 84:191-202.

Hansen, A.J., R.P. Neilson, V.H. Dale, C.H. Flather, L.R. Iverson, D.J. Currie, S. Shafer, R. Cook, and P.J. Bartlein. Global Change in Forests: Responses of Species, Communities and Biomes. BioScience 51:765-779.

Harrington, T.B., and T.A. Harrington. 2016. Early density management of longleaf pine reduces susceptibility to ice storm damage. *In*: Proceedings of the 18th biennial southern silvicultural research conference. General Technical Report SRS-212. Asheville, NC: U.S. Department of Agriculture, Forest Service, Southern Research Station. pp. 313-316.

Haywood, J.D., M.A.S. Sayer, and S.J.S. Sung. 2015. Comparison of planted loblolly, longleaf, and slash pine development through 10 growing seasons in central Louisiana--an argument for longleaf pine. *In*: Holley, A. Gordon; Connor, Kristina F.; Haywood, James D., eds. Proceedings of the 17th biennial southern silvicultural research conference. e–Gen. Tech. Rep. SRS–203, Asheville, NC: U.S. Department of Agriculture, Forest Service, Southern Research Station. pp. 383-390.

Hodges, A.W., M. Rahmani, and T.J. Stevens. 2013. Economic Contributions of Agriculture, Natural Resources, and Food Industries in Florida in 2013. University of Florida, IFAS Extension.

Hodges, A.W., M. Rahmani, and T.J. Stevens. 2015. Economic contributions of agriculture, natural resources, and food industries in Florida in 2013. EDIS report FE969. University of Florida Institute of Food and Agricultural Sciences. 134 p.

Huggett, R., D.N. Wear, R. Li, J. Coulston, and S. Liu. 2013. Forecasts of forest conditions. The Southern Forest Futures Project: Technical Report.

Hyde, W., G.S. Amacher, and W. Magrath. 1996. Deforestation and forest land use: theory, evidence, and policy implications. World Bank Research Observations 11: 223-248.

Ingram, J.C., D. Wilkie, T. Clements, R.B. McNab, F. Nelson, E.H. Baur, H.T. Sachedina, D.D. Peterson, and C.A.H. Foley. 2014. Evidence of payments for ecosystem services as a mechanism for supporting biodiversity conservation and rural livelihoods. Ecosystem Services 7:10-21.

IPCC. 2013. Climate Change 2013: The Physical Science Basis. Contribution of Working Group I to the Fifth Assessment Report of the Intergovernmental Panel on Climate Change. T.F. Stocker, D. Qin, G.-K. Plattner, M. Tignor, S.K. Allen, J. Boschung, A. Nauels, Y. Xia, V. Bex, and P.M. Midgley, Eds. Cambridge University Press, Cambridge, UK. 1535 p.

Jackson, R.B., E.G. Jobbagy, R. Avissar, S.B. Roy, D.J. Barrett, C.W. Cook, K.A. Farley, D.C. le Maitre, B.A. McCarl, and B.C. Murray. 2005. Trading Water for Carbon with Biological Carbon Sequestration. Science 310:1944-1947

Johnsen, K.H., D.N. Wear, R. Oren, R.O. Teskey, F.G. Sanchez, R.E. Will, J.R. Butnor, D. Markewitz, D. Richter, T. Rials, H.L. Allen, J.R. Seiler, D.S. Ellsworth, C.A. Maier, G.G. Katul, and P.M. Dougherty. 2001. Meeting global policy commitments: Carbon sequestration and southern pine forests. Journal of Forestry 99:14-21.

Johnsen, K.H., J.R.Butnor, J.S.Kush, R.C.Schmidtling, and C.D.Nelson. 2009. Hurricane Katrina winds damaged longleaf pine less than loblolly pine. Southern Journal of Applied Forestry 33:178-181.

Johnstone, J.F., C.D. Allen, J.F. Franklin, L.E. Frelich, B.J. Harvey, P.E. Higuera, M.C. Mack, R.K. Meentemeyer, M.R. Metz, G.L.W. Perr, T. Schoennagel, and M.G. Turner. 2016. Changing disturbance regimes, ecological memory, and forest resilience. Frontiers in Ecology and the Environment 14:369-378.

Johnstone, J.F., F.S. Chapin, T.N. Hollingsworth, M.C. Mack, V. Romanovsky, and M. Turetsky, M. 2010. Fire, climate change, and forest resilience in interior Alaska. Canadian Journal of Forest Research 40:1302-1312.

Jokela E.J., and Long A.J. 2012. Using Soils to Guide Fertilizer Recommendations for Southern Pines. Circular 1230, School of Forest Resource and Conservation. University of Florida. 12 pp.

Jokela, E.J., Martin T.A., and Vogel J.G. 2010. Twenty-five years of intensive forest management with southern pines: Important lessons learned. Journal of Forestry 108:338-347

Jose, S., E.J. Jokela, and D.L. Miller. 2007. The longleaf pine ecosystem. In: The Longleaf Pine Ecosystem, Springer, New York. pp. 3-8.

Komarek, E.V. 1974. Effects of fire on temperate forests and related ecosystems: southeastern United States. Fire and Ecosystems 24:251-277.

Kramer, R., and L. Shabman. 1993. The effects of agricultural and tax policy reform on the economic return to wetland drainage in the Mississippi Delta region. Land Economics 69: 85-126.

Kreye, M., D.C. Adams, J. Soto, and F.J. Escobedo. 2016. Does policy process influence public values for forest-water resource protection in Florida? Ecological Economics 129:122-131.

Landsberg, J.J., and R.H. Waring. 1997. A generalised model of forest productivity using simplified concepts of radiation-use efficiency, carbon balance and partitioning. Forest Ecology and Management 260:795-805.

Lane, C.R., and E. D'Amico. 2010. Calculating the ecosystem service of water storage in isolated wetlands using LIDAR in North Central Florida, USA. Wetlands 30:967-977.

Lang, A., S. Baker, and B. Mendell. 2016. Forest management practices of private timberland owners and managers in the U.S. South (2016 Update). Forest Resources Association. Technical Release 16-17.

Law B.E., and M.E. Harmon. 2011. Forest sector carbon management, measurement and verification, and discussion of policy related to climate change. Carbon Management 2:73-84.

Lockaby, G., C. Nagy, J. Vose, C. Ford, G. Sun, S.G. Mcnulty, P.V. Caldwell, E. Cohen, and J. Moore Myers. 2013. Forests and water. In: Wear, D.N., Greis, J. (Eds.), The Southern Forest Futures Project:

Technical Report. U.S. Department of Agriculture Forest Service, General Technical Report SRS-178. USFS, Asheville, NC, 30 p.

Maggard, A.O., R.E. Will, D.S. Wilson, C.R. Meek, and J.G. Vogel. 2016. Fertilization reduced stomatal conductance but not photosynthesis of *Pinus taeda* which compensated for lower water availability in regards to growth. Forest Ecology and Management 381:37-47.

Manea, A., and M.R. Leishman. 2011. Competitive interactions between native and invasive exotic plant species are altered under elevated carbon dioxide. Oecologia 165:735-744.

McCarthy, H.R., R. Oren R., K.H. Johnsen, A. Gallet-Budynek, S.G. Pritchard, C.W. Cook, S.L. LaDeau, R.B. Jackson, and A.C. Finzi. 2010. Re-assessment of plant carbon dynamics at the Duke free-air CO_2 enrichment site: interactions of atmospheric [CO_2] with nitrogen and water availability over stand development. New Phyologist 185:514-528.

McLaughlin, D.L., D.A. Kaplan, and M.J. Cohen. 2013. Managing forests for increased regional water yield in the southeastern US coastal plain. Journal of the American Water Resources Association 49:953-965.

McLaughlin, D.L., D.A. Kaplan, and M.J. Cohen. 2014. A significant nexus: geographically isolated wetlands influence landscape hydrology. Water Resources Research 50:7153-7166.

Menges, E.S., and M.A. Deyrup. 2001. Postfire survival in south Florida slash pine: interacting effects of fire intensity, fire season, vegetation, burn size, and bark beetles. International Journal of Wildland Fire 10:53-63.

Miller, J.H., B.R. Zutter, S.M. Zedaker, M.B. Edwards, and R.A. Newbold. 2003. Growth and yield relative to competition for loblolly pine plantations to midrotation- A southeastern United States regional study. Southern Journal of Applied Forestry 27:237-252.

Mitchell, R.J., J.K. Hiers, J. O'Brien, and G. Starr. 2009. Ecological forestry in the southeast: Understanding the ecology of fuels. Journal of Forestry 107:391-397.

Mitchell, R.J., Y. Liu, J.J. O'Brien, K.J. Elliott, and G. Starr. 2014. Future climate and fire interactions in the southeastern region of the United States. Forest Ecology and Management 327:316-326.

Mixon, M.R., S. Demarais, P.D. Jones, and B.J. Rude. 2009. Deer forage response to herbicide and fir in mid-rotation pine plantations. The Journal of Wildlife Management 73:663-668.

Mulholland, P.J., G.R. Best, C.C. Coutant, G.M. Hornberger, J.L Meyer, P.J. Robinson, J.R. Stenberg, R.E. Turner, F. Vera-Herrera, and R.G. Wetzel. 1997. Effects of climate change on freshwater ecosystems of the south-eastern United States and the Gulf coast of Mexico. Hydrological Processes 11:949-970.

Nowak, J.T., J.R. Meeker, D.R. Coyle, C.A. Steiner, and C. Brownie. 2015. Southern pine beetle infestations in relation to forest stand conditions, previous thinning, and prescribed burning: Evaluation of the southern pine beetle prevention program. Journal of Forestry 113:454-462.

Omernik, J.M. 1976. The influence of landuse on stream nutrient levels. US Environmental Protection Agency, EPA Pub. 600/3-76-014., Seattle, WA.

Oswalt, S.N., W.B. Smith, P.D. Miles, and S.A. Pugh. 2014. Forest resources of the United States, 2012: A technical document supporting the Forest Service Update of the 2010 RPA assessment. General Technical Report WO-91. USDA Forest Service, Washington, D.C. 228 p.

Outcalt K.W. 1984. Influence of Bed Height on the Growth of Slash and Loblolly Pine on a Leon Fine Sand in Northeast Florida. Southern Journal of Applied Forestry 8:29-31.

Pastor, J., and W.M. Post. 1986. Influence of climate, soil moisture and succession on forest carbon and nitrogen cycles. Biogeochemistry 2:3-27.

Porporato, A, E. Daly, and I. Rodriguez-Iturbe. 2004. Soil water balance and ecosystem response to climate change. The American Naturalist 164:625-632.

Porporato, A, G. Vico, and P.A. Fay. 2006. Superstatistics of hydro-climatic fluctuations and interannual ecosystem productivity. Geophysical Research Letters 33: L15402.

Quesada, T., K. Shin, K. Smith, J. Hughes, C. Staub, M. Marsik, and J.A. Smith. 2016. Discovery of biological drivers of pitch canker disease in a changing climate. Abstract. American Phytopathological Society Annual Meeting, July 30-August 3, Tampa, Florida.

Saxe, H., D.S. Ellsworth, and J. Heath. 1998. Tree and forest functioning in an enriched CO_2 atmosphere. New Phytologist 139:395-436.

Schoch, P., and D. Binkley. 1986. Prescribed burning increased nitrogen availability in a mature loblolly pine stand. Forest Ecology and Management 14:13-22.

Scott, R.E., and S.J. Mitchell. 2005. Empirical modelling of windthrow risk in partially harvested stands using tree, neighbourhood, and stand attributes. Forest Ecology and Management 218:193-209.

Sharma, A., K. Bohn, S. Jose, and W.P. Cropper, Jr. 2014. Converting even-aged plantations to uneven-aged stand conditions: A simulation analysis of silvicultural regimes with slash pine (*Pinus elliottii* Engelm.). Forest Science 60: 893-906.

Showalter, D., J. Smith, K. Raffa, R. Sniezko, D.A. Herms, A. Liebhold, and P. Bonello. 2016. Tree resistance as a primary tool to respond to established invasions by cryptic, tree killing forest pathogens and insects. *Proceedings of the 2016 North American Forest Insect Work Conference*, Washington D.C., 31 May – 3 June, pp. 72-73 (http://www.cpe.vt.edu/nafiwc16/NAFIWC_2016_Proceedings.pdf).

Smith, W.B., P.D. Miles, C.H. Perry, and S.A. Pugh. 2009. Forest resources of the United States, 2007: a technical document supporting the forest service 2010 RPA Assessment. General Technical Report-USDA Forest Service WO-78.

Soto, J.S., D.C. Adams, and F.J. Escobedo. 2016a. Landowner Attitudes and Willingness to Accept Compensation from Forest Carbon Offsets: Application of Best-Worst Choice Modeling in Florida USA. Forest Policy and Economics 63:35-42.

Soto, J.S., F.J. Escobedo, D.C. Adams, D.C., and J. Blanco. 2016. A distributional analysis of the socio-ecological and economic determinants of forest carbon stocks. Environmental Science and Policy 60:28-37.

Stainback, G.A., and J.R.R. Alavalapati. 2002. Economic analysis of slash pine forest carbon sequestration in the southern US. Journal of Forest Economics 8:105-117.

Stanturf, J.A., S.L. Goodrick, and K.W. Outcalt. 2007. Disturbance and coastal forests: A strategic approach to forest management in hurricane impact zone. Forest Ecology and Management 250:119-135.

Stewart, J.F., R.E. Will, K.M. Robertson, C.D. Nelson. 2015. Frequent fire protects shortleaf pine (*Pinus echinata*) from introgression by loblolly pine (*P. taeda*). Conservation Genetics 16:491-495.

Stone, J.K., I.A. Hood, M.S. Watt, and J.L. Kerrigan. 2007. Distribution of Swiss needle cast in New Zealand in relation to winter temperature. Australasian Plant Pathol. 36:445-454.

Stone, J.K., L.B. Coop, and D.K. Manter. 2008: Predicting effects of climate change on Swiss needle cast disease severity in Pacific Northwest forests Canadian Journal of Plant Pathology 30:169-176.

Sun, G., D.M. Amatya, S.G. McNulty, R.W. Skaggs, and J.H. Hughes. 2000a. Climate change impacts on the hydrology and productivity of a pine plantation. Journal of the American Water Resources Association 36:367-374.

Sun, G., H. Riekerk, and L.V. Korhnak. 2000b. Ground-water-table rise after forest harvesting on cypress-pine flatwoods in Florida. Wetlands 20:101-112.

Sun, G., S.G. McNulty, J. Lu, D.M. Amatya, Y. Liang, and R.K. Kolka. 2005. Regional annual water yield from forest lands and its response to potential deforestation across the southeastern United States. Journal of Hydrology 308:258-268.

Susaeta, A., D. Adams, D. Carter, and P. Dwivedi. 2016c. Climate change and ecosystem services output efficiency in southern natural loblolly pine forests. Environmental Management 58:417-430.

Susaeta, A., D.C. Adams, D.R. Carter, C. Gonzalez-Benecke, and P. Dwivedi. 2016b. Technical, allocative, and total profit efficiency of loblolly pine forests under changing climatic conditions. Forest Policy and Economics 72:106-114.

Susaeta, A., D.R. Carter, and D.C. Adams. 2014a. Sustainability of forest management under changing climatic conditions in the Southern United States: adaption strategies, economic rents and carbon sequestration. Journal of Environmental Management 139:80-87.

Susaeta, A., D.R. Carter, and D.C. Adams. 2014b. Impacts of climate change on economics of forestry and adaptions strategies in the Southern United States. Journal of Agricultural and Applied Economics 46:257-272.

Susaeta, A., J.R. Soto, D.C. Adams, and D.L. Allen. 2016a. Economic Sustainability of Payments for Water Yield in Slash Pine Plantations in Florida. Water 8:382-398.

Susaeta, A., J.R. Soto, D.C. Adams, and J. Hulcr. 2016d. Pre-invasion economic assessment of invasive species prevention: A putative ambrosia beetle in Southeastern loblolly pine forests. Journal of Environmental Management 183:875-881.

Taylor, K.E., R.J. Stouffer, and G.A. Meehl. 2012. An overview of CMIP5 and the experiment design. Bulletin of the American Meteorological Society 93:485-498.

Teskey, R. 2014. Developing scenarios to use in model simulations. PINEMAP: Year 3 Annual Report. March 2013 – February 2014.

Tian, H., G. Chen, M. Liu, C. Zhang, G. Sun, C. Lu, X. Xu, W. Ren, S. Pan, and A. Chappelka. 2010. Model estimates of net primary productivity, evapotranspiration and water use efficiency in the terrestrial ecosystems of the southern United States during 1895-2007. Forest Ecology and Management 259:1311-1327.

Turner, D.P., G.J. Koerper, M.E. Harmon, and J.J. Lee. 1995. A carbon budget for forests of the conterminous United States. Ecological Applications 5:421-436.

USEPA, 1995. National water quality inventory, 1994. Report to Congress. EPA841-R-95-005. Office of Water, USEPA, Washington, DC.

van Vuuren, D.P., J. Edmonds, M. Kainuma, K. Riahi, A. Thomson, K. Hibbard, G.C. Hurtt, T. Kram, V. Krey, J.F. Lamarque, T. Masui, M. Meinshausen, N. Nakicenovic, S.J. Smith, and S.K. Rose. 2011. The representative concentration pathways: an overview. Climatic Change 109:5-31.

Varner, J.M., D.R. Gordon, F.E. Putz, and J.K. Hiers. 2005. Restoring fire to long-unburned Pinus palustris ecosystems: novel fire effects and consequences for long-unburned ecosystems. Restoration Ecology 13:536-544.

Vorosmarty, C.J., P. Gree, J. Salisbury, and R.B. Lammers. 2000. Global water resources: Vulnerability from climate change and population growth. Science 289:284-288.

Wade, D.D., and R.W. Johansen. 1986. Effects of fire on southern pine: observations and recommendations. Gen. Tech. Rep. SE-41. Asheville, NC: US Department of Agriculture, Forest Service, Southeastern Forest Experiment Station. 14 p.

Wang, D., S.C. Hagen, and K. Alizad. 2013. Climate change impact and uncertainty analysis of extreme rainfall events in the Apalachicola River basin, Florida. Journal of Hydrology 480:125-135.

Wear, D.N., and J.G. Greis. 2012. The southern forest futures project: summary report. USDA Forest Service Southern Research Station General Technical Report SRS-168. Asheville, NC. 54 p.

Wear, D.N., and J.G. Greis. 2012. The southern forest futures project: summary report.

Wertin, T.M., M.A. McGuire, and R.O. Teskey. 2012. Effects of predicted future and current atmospheric temperature and [CO2] and high and low soil moisture on gas exchange and growth of Pinus taeda seedlings at cool and warm sites in the species range. Tree Physiology 32:847-858.

Williams, K., K.C. Ewel, R.P. Stumpf, and F.E. Putz. 1999. Sea-level rise and coastal forest retreat on the west coast of Florida, USA. Ecology 80:2045-2063.

Woods, A. J., K.D. Coates, and A. Hamann. 2005. Is an unprecedented Dothistroma needle blight epidemic related to climate change? Bioscience 55:761-769.

Woods, A.J., J. Martın-Garcıa, L. Bulman, M.W. Vasconcelos, J. Boberg, N. La Porta, H. Peredo, G. Vergara, R. Ahumada, A. Brown, and J.J. Diez. 2016. Dothistroma needle blight, weather and possible climatic triggers for the disease's recent emergence Forest Pathology 46:443-452.

Zhai, L., Jokela E.J., Gezan S., and Vogel J.G. 2015. Family, environment and silviculture effects in pure- and mixed-family stands of loblolly (*Pinus taeda* L.) and slash (*P. elliottii Engelm* var. *ellitotttii*) pine. Forest Ecology and Management. 337:28-40.

Zutter, B.R., G.R. Glover, and D.H. Gjerstand. 1986. Effects of herbaceous weed control using herbicides on a young loblolly pine plantation. Forest Science 32:882-899.

CHAPTER 10

Florida Tourism

Julie Harrington[1], Hongmei Chi[2], and Lori Pennington Gray[3]

[1]*Center for Economic Forecasting and Analysis, Florida State University, Tallahassee, FL;* [2]*Department of Computer and Information Sciences, Florida Agricultural and Mechanical University, Tallahassee, FL;* [3]*Tourism Crisis Management Initiative, University of Florida, Gainesville, FL*

Tourism is one of the largest economic industries in Florida. In 2015, a record 106.3 million tourists visited Florida (about five visitors per resident), with an economic impact of about $90 billion. Tourism also provides additional benefits for federal, state, and local governments in the form of taxes (e.g., excise, sales, income, and property taxes). In Florida, tourism accounts for over one million direct jobs and an additional 1.5 million indirect and supply chain jobs. The three industries or business sectors most impacted by tourism and currently experiencing substantial growth in the state, include: leisure and hospitality (e.g., hotels, restaurants, museums, amusement parks, entertainment), transportation (e.g., cruise ships, taxis, airports), and retail trade (e.g., gas stations, retail stores). The 106.3 million tourists comprise approximately 91.2 million out-of-state visitors, 3.9 million Canadian visitors, and 11.2 million overseas visitors. The domestic visitors are anticipated to grow by 20% in 2018. Tourism and the associated industries in Florida are highly vulnerable to climate change over time. The state population and real estate markets continue to grow in the coastal areas, with corresponding increases in property values at risk. In addition, there are losses associated with the properties used to mitigate the effects of climate change. In summary, indicators of climate change, such as higher sea levels and more frequent and powerful hurricanes and other extreme weather events, have the potential to severely impact the tourism industry in Florida.

Key Messages

- In 2015, a record 106.3 million tourists visited Florida, with an economic impact of ~$90 billion. Over the last five years, tourism has averaged about 6% growth annually.
- The 106.3 million tourists are comprised of 91.2 million out-of-state (or domestic) visitors, 3.9 million Canadian visitors, and 11.2 million overseas visitors.
- Tourism accounts for more than one million direct jobs and an additional 1.5 million indirect jobs.
- The three business sectors most impacted by tourism in the state are leisure and hospitality, transportation, and retail trade.
- Climate change presents significant uncertainties in future Florida tourism and economics.
- The state's population and real estate markets continue to grow in the coastal areas, along with corresponding increases in property values at risk to sea level rise and inundation, storm surge, land subsidence, and wind damage among other things.
- The authors discuss the relationship between tourism's impacts of climate change, vulnerability and adaptation.
- The vulnerability of Florida tourism will decrease if we improve our adaptive capacity with respect to climate change.

Keywords

Tourism and economic impacts; Visitor spending; Climate change; Adaptation; Uncertainty; Vulnerability; Risk and sensitivity analysis; Big data analysis; Rare/extreme event simulation

Introduction

Globally, it is estimated that about 100 million people live within three feet of mean high tide, and another 100 million live within six feet of it. Florida's 1,350 miles of natural coastline, including 825 miles of sandy beaches, translates to a very large tourism-driven economy. For 2015, the number of visitors to Florida totaled 106.3 million people, making it the topmost travel destination in the world. According to Visit Florida, the goal for 2016 was for at least 115 million tourists to visit Florida, and 120 million by 2020. Over the last five years, tourism has averaged about 6% growth on an annual basis. However, when comparing 2014-15 and 2015-16, the number of tourists in Florida increased by more than 10%. In addition, according to a recent University of Florida study, it is estimated that another 14.9 million people will move to Florida by year 2070 (i.e., an additional 680 people will be moving to Florida daily).

Visitor Demographics in Florida

There were 106.3 million visitors to Florida in 2015. As shown in Fig. 10.1, 91.2 million of these visitors were domestic and 15.1 million were from overseas (including Canada). The overall annual growth in tourists to Florida (since 1927) has been about 4.4%.

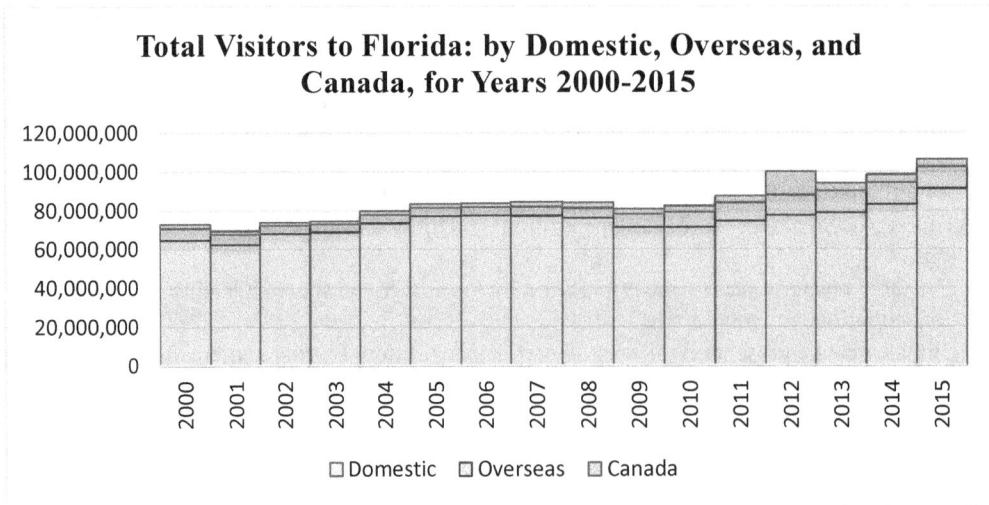

Figure 10.1. Visitors to Florida: Percentages by quarter from 2000 to 2015.

We have forecasted the number of total visitors to Florida to years 2035 and 2060 using three different estimations, ranging from pessimistic to optimistic, as depicted in Fig. 10.2. The figure shown reflects the expected visitor growth to year 2035, with projections ranging from 148 million to 191 million visitors, for pessimistic and optimistic forecasts, respectively.

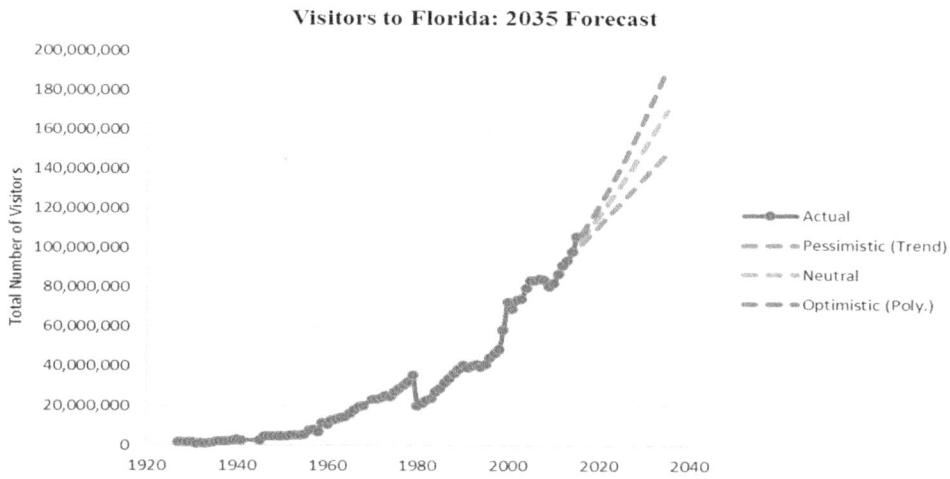

Figure 10.2. Visitors to Florida: Forecasted numbers to years, 2035.

The top five origin states for domestic visitors are expected to be: New York (10.3%), Georgia (8.2%), Texas (6.3%), New Jersey (5.3%), and Illinois (4.9%). The preferred travel seasons for domestic visitors are during summer and spring months, which is when about 56% of visitors to Florida came during 2014.

About one-fifth of all international visitors who come to the U.S visit Florida. In 2014, these visitors came from 190 countries; with the top five countries being Canada, the United Kingdom, Brazil, Argentina, and Columbia. In 2015, international tourists to Florida represented 14.2% of all tourists. Fig. 10.3 provides the quarterly numbers and percentages from years 2000 to 2015.

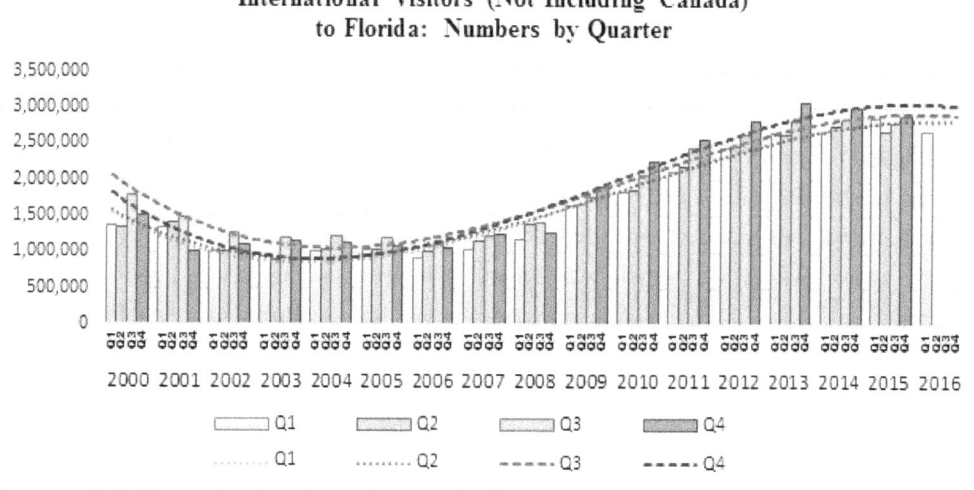

Figure 10.3. International visitors to Florida: Percentages by quarter, from 2000 to 2016.

It should be noted that visitors from Canada, as shown in Fig. 10.4, exhibit strong seasonal effects; their preference is to visit Florida primarily in the winter quarter (Q1) of each year.

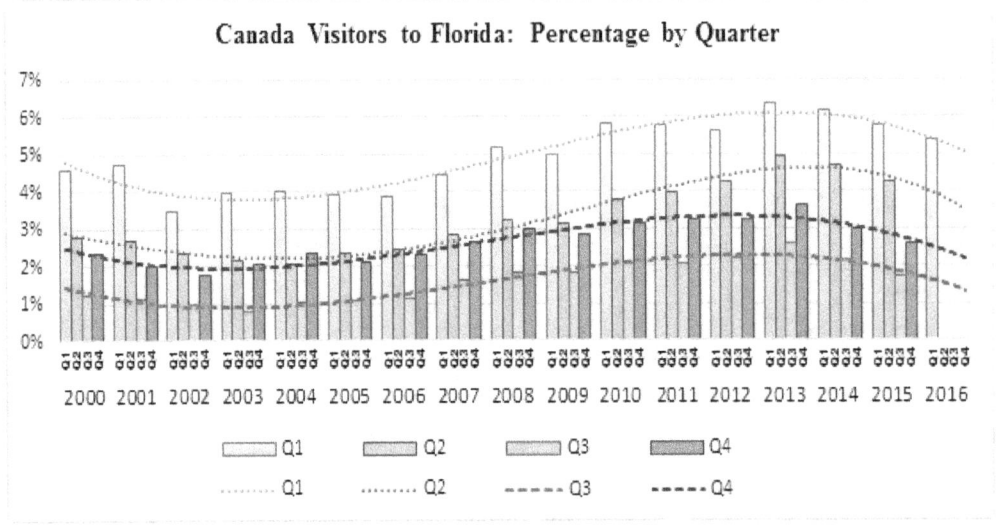

Figure 10.4. Canadian visitors to Florida: Percentages by quarter, from 2000 to 2016.

Tourism Spending in Florida

Visit Florida, a not-for-profit corporation formed in 1996 (as a public-private partnership), is the state's official tourism marketing corporation. Visit Florida's total tourism spending budget for fiscal year (FY) 2014-15 was $212.5 million, with about $74 million and $138.5 million in public and private investment, respectively. In FY 2013-14, the total budget amount was $183.6 million; thus FY 2014-15 represented about a 16% increase in tourism funding for Visit Florida over the previous year. In addition to fundraising to support matching funds, Visit Florida is involved in assisting more than 12,000 tourism industry businesses; participating in travel-related trade shows and missions; working with travel agents, tour operators, meeting and event planners; and operating the five state welcome centers. Florida tourism and travel-related direct and indirect spending reached $90 billion in FY 2014-15, representing an increase of 8.3% over FY 2013-14 spending. In 2014, visitor direct and indirect spending was about 10% of Florida's gross domestic product; approximately $21 billion was generated in total lodging revenue, $36 billion in restaurants/dining, $14 billion in admissions, and $11 billion in other visitor-related spending. The average spending by domestic visitors was $162.40 per day[1], with the majority spent on lodging ($55.30) and transportation ($51.40), 30% and 28%, respectively (Fig. 10.5). As shown in Fig. 10.6, about 51% of total visitors came by air, and 49% by non-air travel.

---

[1] Representing a 4.2% increase from the previous year, 2013.

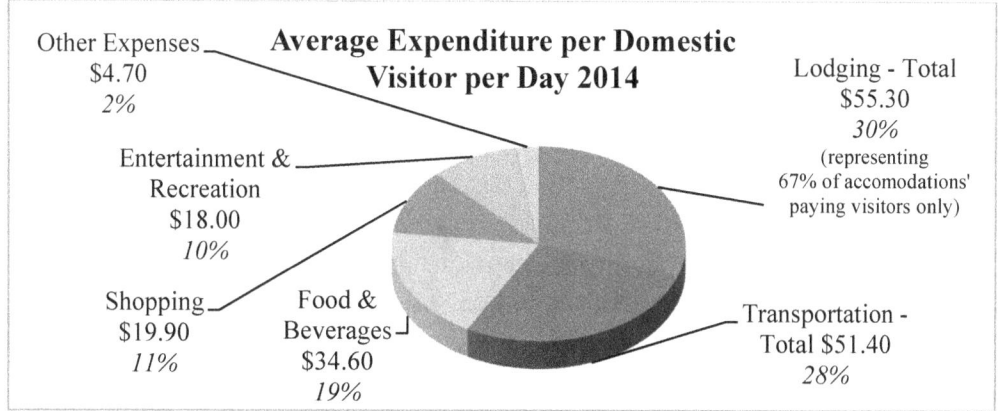

Figure 10.5. Domestic visitor average spending per day, 2014.

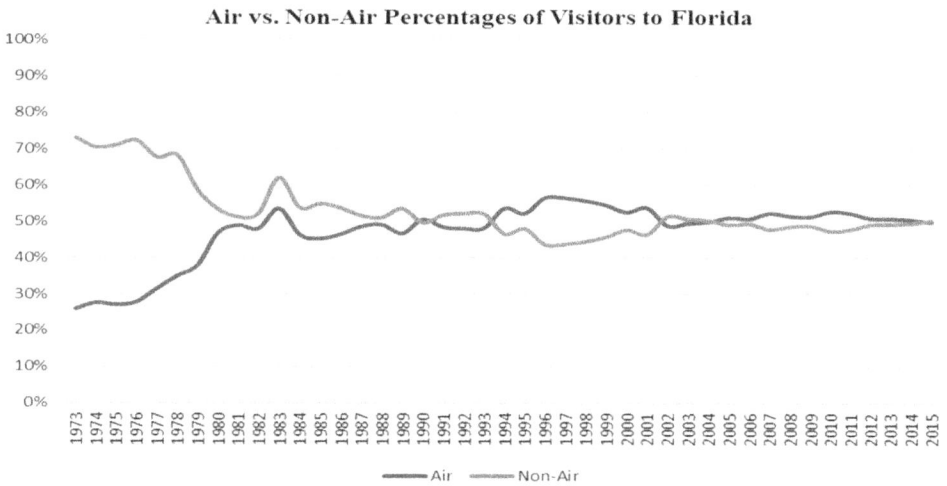

Figure 10.6. Air vs. non-air percentage of visitors to Florida from 1973 to 2015.

The Impact of Climate on Tourism, by Industry or Business Sector

Florida's warmer climate translates to a mecca for spending on recreation and leisure activities. Out of the five Gulf of Mexico states, Florida had the highest rate (52 million people) participating in outdoor leisure activities. In addition, Florida beaches had approximately 810 million beach day visits in 2012, the most of any state or country in the world. Also in 2012, the beach-oriented visitors to Florida totaled 38.4 million or 42% of the total visitor population to Florida.

In 2014, about 90% of domestic visitors traveled to Florida for leisure activities; the primary reasons were for vacation, visiting family and friends, and for a weekend getaway. In Florida, domestic visitors' top five preferences were beach/waterfront activities (41%), dining (34%), shopping (33%), visiting friends/relatives (31%), and theme/amusement parks (19%). Domestic visitors' average length of stay was 4.3 days.

As mentioned earlier, the three top industries or business sectors impacted by tourism in Florida are the leisure and hospitality, transportation, and retail sectors. In the last five years, leisure and hospitality has experienced growth four times that of the rest of Florida's economy, while the transportation and retail sectors have managed a doubling of growth for the same time period. Other secondary sectors highly impacted by tourism (based on non-resident spending) include state and local government (i.e., taxes), real estate (including commercial and vacation homes), transportation investment (in construction and manufacturing), and financial services.

Economic Impact of Tourism in Florida

Tourism is the top industry in Florida. However, the "tourism" sector is not measured as a separate sector, but rather as components of multiple sectors. Thus, it is more difficult to measure economic impacts on tourism with the same precision as other individual sectors. That said, it has been reported by Visit Florida that tourism is "the state's No. 1 industry, and generates 23% of the state's sales tax revenue and employs about 1.2 million Floridians." This corresponds to about one job per every 89 domestic or international visitors.

For 2015, the economic impacts associated with tourism were the highest on record, with an economic impact of about $90 billion, and accounting for more than one million direct jobs and an additional 1.5 million indirect and supply chain jobs. Tourism also provides benefits for federal, state, and local governments in the form of taxes (e.g., excise, sales, income, and property taxes). In 2015, $5.3 billion in state sales tax was collected, representing 23% of taxes collected in Florida that year.

Ecological Impacts on Florida Tourism

From an ecological perspective, Florida is highly vulnerable to the effects of tourism. As a peninsula, Florida is already experiencing an estimated 400 miles of critically eroded coastline. As mentioned earlier, a positive 2.3% annual population growth estimated to year 2060,[2] will place tremendous negative load on the ecosystem, particularly in the coastal areas. In addition, the effects of climate change will serve to further amplify the negative impacts associated with increasing population densities. Over time, it is expected that a warming climate (as a result of global climate change from increases in greenhouse gas concentrations) will manifest in greater

---

[2] Based on a forecast to years 2035, and 2060, conducted by the FSU Center for Economic Forecasting and Analysis, Sept. 2016

sea level rise[3] and warming of the coastal oceans in the Gulf of Mexico that could increase the likelihood of the storm intensity of land falling tropical cyclones (Liu et al. 2012) and extreme weather events (e.g., tornados, flooding, droughts, and other unstable weather patterns; Wuebbles et al. 2014).

In addition, the real estate markets continue to expand in the coastal areas, along with corresponding increases in property values at risk for sea level rise and inundation, storm surge, land subsidence, and wind damage among other things. It is also unlikely, with the prospect of higher coastal property taxes, that coastal development will be reduced over time. In addition, there are losses associated with the properties used to mitigate the effects of climate change. One study conducted in 2008 examined the properties at risk based on varying sea level rise scenarios (for years 2030 and 2080) in six counties in Florida. The authors found that in Dade County the properties at risk ranged from $1.1 billion to $12.3 billion at sea level rise increments ranging from 0.16 feet to 2.13 feet, respectively. It should be noted that these can now be viewed as conservative estimates compared to the most recent Intergovernmental Panel on Climate Change (IPCC) Fifth Assessment Report estimates.

State and local governments have or are developing their adaptation and mitigation planning processes addressing the climate change impacts on tourism. The planning efforts, especially in South Florida, range from moving and elevating infrastructure (e.g., commercial and residential buildings, roads, airports, schools, sidewalks) to making massive investments in water utilities, such as additional pumps, desalinization plants, and other technologies. There are also other adaptation-oriented, longer-term planning operations underway involving the development of "water" or floating cities.

Beach restoration is a mitigation operation shown to have significant benefits with regards to tourism revenue generation and property value protection. It is estimated that each state dollar spent on the protection of Florida's public access beaches prevents the loss of $8 in state taxes paid by out-of-state tourists and resident users of those beaches. For example, in the mid-1970s Miami Beach's sandy beach was almost non-existent and its surrounding infrastructure was deteriorating. Following one year of beach nourishment[4], tourism earnings increased 56% and the project demonstrated a greater than 5 to 1 ratio, benefit (tourism income) to cost (of nourishment). As of 2015, visitor spending had increased to $24.4 billion in the Greater Miami economy, with beaches being the top draw for visitors to Miami Beach. In another study completed following the 2004 and 2005 hurricane seasons, of 28,000 properties in eight Florida counties, the findings showed that restored beaches prevented a loss of $1.8 billion in property values.

---

[3] See http://www.southeastfloridaclimatecompact.org/wp-content/uploads/2015/10/2015-Compact-Unified-Sea-Level-Rise-Projection.pdf

[4] According to James Houston (2013), the Army Corps of Engineers Miami Beach estimated $51 million spent on beach nourishment cost, and $1.6 million annual costs.

Climate Change Adaptations and Uncertainties on Tourism in Florida

Climate change presents significant uncertainties in future Florida tourism and economics. For example, the uncertainty of climate projections associated with different greenhouse gas emission scenarios call for different likelihoods of extreme events (USGCRP 2014). How to minimize those uncertainties and build capacity for climate change adaptation are challenging tasks. Those small-probability (rare or extreme) events could have big impacts for tourism on a global scale.

Tourism is highly susceptible to small-probability events, including terrorism, war, epidemics, and natural disasters. For instance, as shown in Fig. 10.7, global tourism was adversely impacted immediately after the September 11, 2001 terrorist attacks and during the financial crisis during the years 2007–2009. But standard statistical models used to predict future outcomes have their weaknesses (McDowell et al. 2016). Major weaknesses of the current models include predictability power associated with travel flows (Gössling and Hall 2006) as a result of the following uncertainties: (1) the role of weather extremes is unknown; (2) the existence of fuzzy-variables (terrorism, war, epidemics, natural disasters) is problematic; (3) the future costs of transportation is unpredictable; and; (4) a global financial crisis is intermittent.

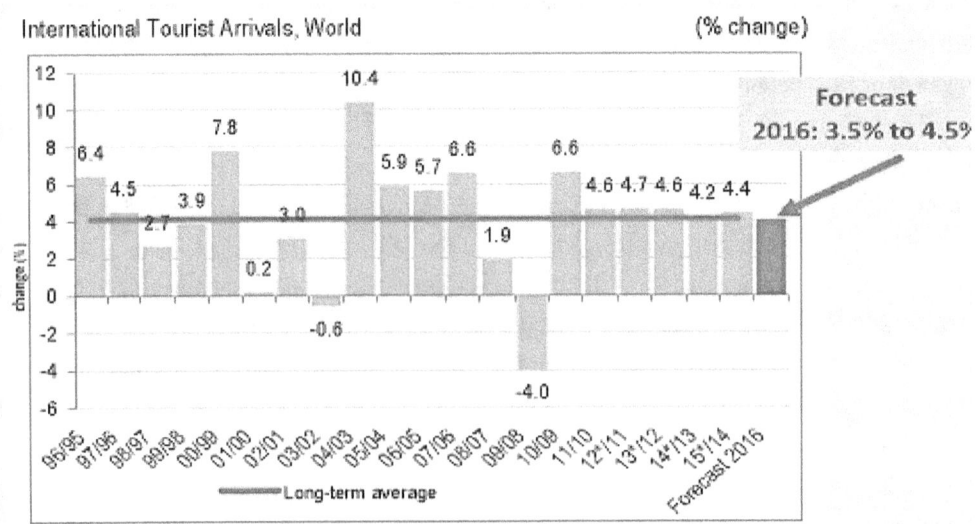

Figure 10.7. Tourism change each year (Source: World Tourism Organization (UNWTO))

All the aforementioned events are rare and unpredictable events, and standard statistical models are based on predictable patterns. In a probabilistic model, a rare event is an event with a very small probability of occurrence. The forecasting of rare events is important in the area of tourism prediction. Monte Carlo simulation (a computerized mathematical technique), as well as other methods, can be used to estimate probability of rare events. Also, in Fig. 10.7, one can easily see how tourism flows were adversely impacted by the terrorist attacks of 9/11 and subsequent global spillover events. Compared with the year 2000, tourism flows are increasing at a negative rate. Meanwhile, the global financial and economic downturn that affected tourism from years 2007 through 2010, and beyond, has cast substantial attention on the role that crisis events play in tourism (Ritchie et al. 2010). In Fig. 10.8 we see how Florida demonstrated corresponding impacts to global crisis events. However, local or statewide events may also have significant impacts on tourism. An oppressive, long-term drought from late March 2006 until late August 2008 impacted the entire state of Florida, as shown in Fig. 10.8; the increasing rate of tourism approached zero for those years.

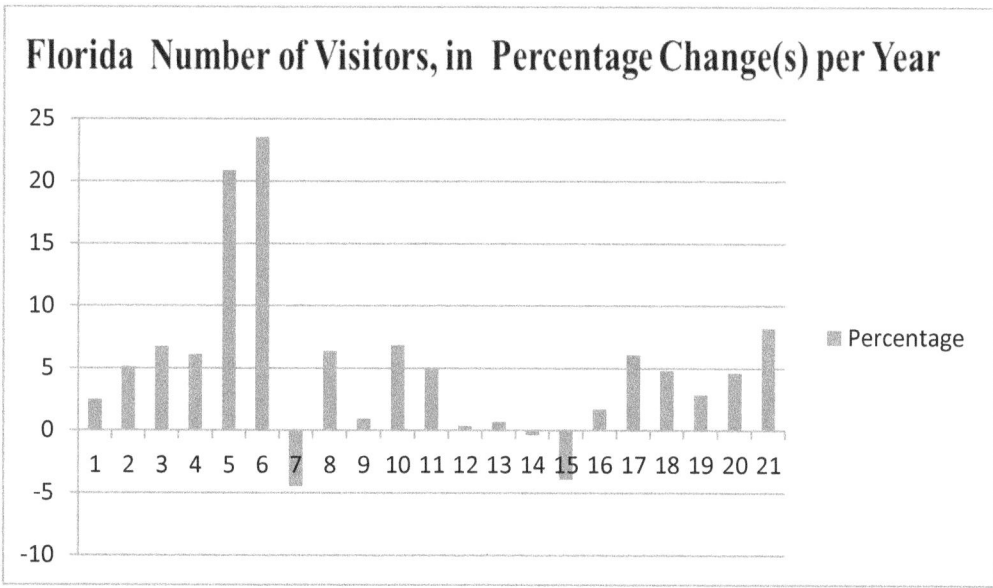

Figure 10.8. Florida tourism annual change(s) (percentage) from years 1995 to 2015.

Strategies for Adapting to Climate Change

Perspectives from stakeholders concerning land use planning, extreme weather events, and sea level rise scenarios provide valuable information to tourism planners who balance short-term growth with longer term sustainability. Stakeholder perspectives can identify competing opportunities and constraints to mitigation and adaptation to current and future planning for hazards. Their perspectives assist community planners in balancing decision-making processes and better equating sustainability regarding resource use, as well as building resiliency in terms of natural hazards.

Understanding different adaptive capacities is a prerequisite for targeting interventions to reduce the adverse impacts of climate change. Fig. 10.9 (Binita, et. al. 2015) provides a broader perspective on vulnerability from a historical examination of climate change, as well as a determination of risk based on the potential of future climate change in Florida. The results also help planning agencies to develop strategies for adaptation to climate change in vulnerable counties or areas in Florida. According to a Chinese Proverb: "One prospers in worries and hardship, and perishes in ease and comfort. We will have to be mindful of possible adversities and be prepared for the worst." With the assistance of big data, it is possible to organize experts and collect more readily available information in order to make more accurate predictions of the effects of climate changes. Fig. 10.9 presents a framework for further understanding the relationship between tourism's impacts of climate change, vulnerability, and adaptation. When we have greater adaptation ability, vulnerability will decrease. Therefore, prediction models and simulations help us to reduce vulnerability and develop a better ability to prepare uncertainty in the future.

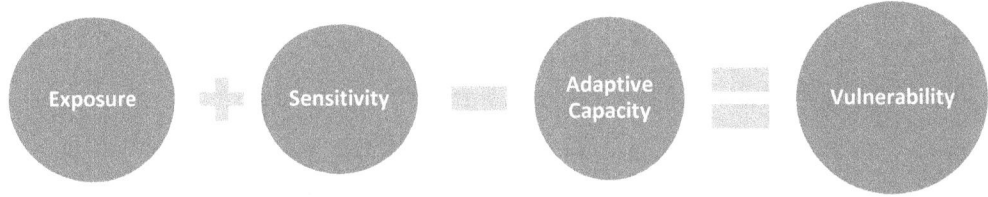

Figure 10.9. Relationship between climate change impacts, vulnerability, and adaptation.

Exposure specifies the projected change of climate that is affecting the system, such as temperature change and climate extremes. **Sensitivity** describes the degree to which a system is affected, either adversely or beneficially. The impacts of climate change may be direct or indirect. **Adaptive capacity** describes the ability of a system to adapt to changes in climate. **Vulnerability** can be defined as the degree to which a system is susceptible to being affected by adverse effects of climate change (Lindner et al. 2008).

The vulnerability of Florida's tourism sector will decrease if we improve our adaptive capacity with respect to climate change. As adaptive capacity improvements are made throughout communities in Florida, the impacts should be monitored and measured, and decisions reached among stakeholders concerning effective planning strategies: e.g., adopting more stringent carrying capacity at the beaches, investing in marketing programs that educate tourists on their impacts, increasing funding to make the experience exclusive (with a corresponding reduction in numbers), or developing programs that focus on boosting tourism numbers, among others. Big data analysis can be integrated into rare-event simulations since all extreme climate change will be categorized as rare-events. With the assistance of big data tools, such as the R programming language or Hadoop technology, Monte Carlo methods can carry out extensive stochastic simulations. Although these types of randomly determined simulations bear immense computational costs, the ability to effectively compute the probability of a rare event will help us to prepare for extreme climate changes, thereby increasing our adaptive capacities. These results will further assist in shaping future long-term state policies on building capacity and more effectively integrate climate change issues into the state strategic planning processes.

Conclusion

In 2015, a record 106.3 million tourists visited Florida (about five visitors per resident), with an economic impact of about $90 billion, including more than one million direct jobs, and an additional 1.5 million indirect and supply chain jobs. The three industries most impacted by tourism, and currently experiencing substantial growth in the state, include: leisure and hospitality (e.g., hotels, restaurants, museums, amusement parks, entertainment), transportation (e.g., cruise ships, taxis, airports), and retail trade (e.g., gas stations, retail stores). In 2015, there were approximately 91.2 million out-of-state visitors, 3.9 million Canadian visitors, and 11.2 million overseas visitors. Tourism and the associated industries in Florida are highly vulnerable to climate changes over time.

The state's population and real estate markets continue to grow in the coastal areas, with corresponding increases in property values at risk. In addition, there are losses associated with the properties used to mitigate the effects of climate change. Indicators of climate change, such as higher sea levels more frequent and powerful hurricanes, and other extreme weather events, have the potential to severely impact the tourism industry in Florida. These climate change indicators, can significantly impact travel decision-making by tourists.

In short, to ensure that Florida remains one of the top tourist attractions and destinations in the U.S. and worldwide, the state should continue making improvements in building capacity towards adaptation, specifically in the tourism industry, and better model the uncertainty associated with a changing climate and its associated impacts concerning the tourism economy. With more accurate predictions and effective communication and collaboration with research professionals, Florida will be well-positioned to continue improvements in building adaptive capacity and increasing economic opportunities. Although it is difficult to reverse the impacts of climate change, building better adaptive capacity will slow the effects of climate change in the tourism industry in Florida.

Acknowledgements

The authors would like to thank Ms. Glencora Haskins, FSU CEFA, for literature review assistance on the chapter.

References

Becketti, S., and B. Lacy. 2016. *"Life's a Beach."* In *Economic and Housing Research Insight*, April 2016. Freddie Mac Housing Corporation, 7 pp.

Berrang-Ford, L., T. Pearce, & J.D. Ford. 2015. "Systematic Review Approaches for Climate Change." In *Adaptation Research. Regional Environmental Change*, 15(5): 755-769.

Binita, K. C., J.M Shepherd, & C.J. Gaither. 2015. "Climate Change Vulnerability Assessment in Georgia." In *Applied Geography*, 62: 62-74.

Bonn, M., and J. Harrington. 2008. "A Comparison of Three Economic Models for Applied Hospitality and Tourism Research". In *Journal of Tourism Economics*, 14 (4).

Catanese Center at Florida Atlantic University. 2005. "Economics of Beach Tourism in Florida." and "Economics of Beaches Literature Review." In a Report to *Florida Department of Environmental Protection Bureau of Beaches and Coastal Systems.*

Catanese Center at Florida Atlantic University. 2005. "The Economic and Fiscal Benefits from Florida's Beaches: a Proposed Methodology."

Copperman, R., J. Lemp, D. Kurth, B. Lipkin, M. Henley, & N. Berry. 2016. "Development of a Risk Analysis Methodology for Quantifying the Uncertainty of Travel Demand Forecasts."

Dolega, Michael. 2015. "Domestic Tourism Will Support Florida Economy Despite International Weakness." In: *Toronto Dominion (TD) Bank Economics.* https://www.td.com/document/PDF/economics/special/Florida_Driven_by_Tourism.pdf

Dow, K., L. Carter, A. Brosius, E. Diaz, R. Durbrow. 2013. "Chapter 13: Climate Adaptations in the Southeast USA." In U.S. Environmental Protection Agency Papers 214. See: http://digitalcommons.unl.edu/usepapapers/214

Florida Department of Environmental Protection. 2016. "Beaches and Coastal Systems". See: http://www.dep.state.fl.us/beaches/publications/pdf/fl_beach.pdf

Florida Shore and Beach Preservation Association. 2015. "Healthy Beaches Drive Florida's Economy."

Florida Trend. 2016. Various Articles Concerning Tourism and the Economy. See: http://www.floridatrend.com/

Frazier, T., N. Wood, and B. Yarnel. 2010. "Stakeholder Perspectives on Land-use Strategies for Adapting to Climate-change-enhanced Coastal Hazards: Sarasota, Florida." In *Applied Geography* 30 (2010) 506-517.

García-Cueto, O. R., & N. Santillán-Soto. 2012. "Modeling Extreme Climate Events: Two Case Studies in México." In *Climate Models*, 137-160.

Gossling, S. and M. Hall. 2006. "Uncertainties in Predicting Tourist Flows under Scenarios of Climate Change." In *Climatic Change*, 79 (3-4), 163-173.

Hall, Colin Michael. 2008. "Tourism and Climate Change: Knowledge Gaps and Issues." In *Tourism Recreation Research* 33(3): 339-350.

Harrington, J. and T. Walton. 2008. "Climate Change in Coastal Areas in Florida: Sea Level Rise Estimation and Economic Analysis to Year 2080." Final Report to *The Bipartisan Policy Center*, and In *Ocean Engineering* 34: 1832-1840, Elsevier Press.

Houston, James R. 2003. "The Economic Value of Beaches – a 2013 Update." In Shore and Beach, 81 (1): 3-11.

Intergovernmental Panel on Climate Change (IPCC). 2013. "The IPCC Fifth Assessment Report on Climate Change 2013." See: https://www.ipcc.ch/publications_and_data/publications_and_data_reports.shtml

Kaján, E., & J. Saarinen. 2013. "Tourism, Climate Change and Adaptation: A Review." In *Current Issues in Tourism*, 16 (2): 167-195.

Kolbert, Elizabeth. 2015. "The Siege of Miami." In *The New Yorker*, Dec. 21 & 28, pp. 42-50.

Lindner, M., et al. 2008. "Impacts of Climate Change on European Forests and Options for Adaptation." Report to: *The European Commission Directorate-General for Agriculture and Rural Development*: 1-173.

Lisle, D. 2007. "Defending Voyeurism: Dark Tourism and the Problem of Global Security." In *Tourism and Politics. Global Frameworks and Local Realities*, 333-345.

Liu, Y., Lee, S.-K., Muhling, B.A., Lamkin, J.T., Enfield, D.B., 2012b. Significant reduction of the Loop Current in the 21st century and its impact on the Gulf of Mexico. J. Geophys. Res., 117, C05039, doi:10.1029/2011JC007555.

Mitchell, Kenneth L., et al. 2013. "Chapter 4: Energy Production, Use, and Vulnerability to Climate Change in the Southeast USA." In *Climate of the Southeast United States: Variability, Change, Impacts, and Vulnerability*. K.T. Ingram, K. Dow, L. Carter, J. Anderson (Eds.). National Climate Assessment (NCA) series, 62-85.

Mozumber, P., E. Flugman, T. Randhir. 2010. "Adaption Behavior in the Face of Global Climate Change: Survey Responses from Experts and Decision Makers Serving the Florida Keys." In *Ocean and Coastal Management* 54 (2011) 37-44.

Parker, L., and G. Steinmetz. 2015. "Treading Water." In *National Geographic*, February 2015, 107-127.

Repetto, Robert. 2011. "Economic and Environmental Impacts of Climate Change in Florida." In *Demos: State-Based Climate Change Series:* http://www.demos.org/sites/default/files/publications/FL_ClimateChangeInTheStates_Demos.pdf

Ritchie, J. B., C. M. Molinar, & D.C. Frechtling. 2010. "Impacts of the World Recession and Economic Crisis on Tourism: North America." In *Journal of Travel Research*, 49(1): 5-15.

Scott, D., & C. Lemieux. 2010. "Weather and Climate Information for Tourism." In *Procedia Environmental Sciences*, 1: 146-183.

Shifflet, D.K. and Associates. 2014. "Profile of Domestic Visitors to Florida." In *Visit Florida*. See: http://www.visitflorida.org/resources/research/

State of Florida. 2016. *Quick Facts*. See: http://www.visitflorida.org/about-us/what-we-do/tourism-fast-facts/

Stronge, William B. 2013. "Economic Impact of Beach Tourism: Florida and Palm Beach County." Presentation at the annual conference of the *Florida Beach and Shore Preservation Association*.

University of Florida Geoplan Center, Florida Department of Agriculture and Consumer Services (FDACS), and 1000 Friends of Florida. 2016. "Mapping Florida's Future – Alternative Patterns of Development in 2070." See: http://www.1000friendsofflorida.org/Florida2070

UNWTO. "International Tourism Results and Prospects for 2016." http://cf.cdn.unwto.org/sites/all/files/pdf/unwto_fitur_2016_hq_jk.pdf

U.S. Global Change Research Program. 2014. "Climate Change Impacts in the United States: The Third National Climate Assessment."Melillo, J. M., T.C. Richmond, and G.W. Yohe (Eds.). See: http://admin.globalchange.gov/sites/globalchange/files/Ch_0a_FrontMatter_ThirdNCA_GovtReview Draft_Nov_22_2013_clean.pdf

Visit Florida. 2016. *"Five Years of Tourism Industry Growth."* In *Visit Florida*. See: http://www.visitflorida.org/resources/research/

Wuebbles, D. J, and co-authors. 2014. "CMIP5 Climate Model Analyses: Climate Extremes in the United States." In *Bull. Amer. Soc.*, 573-581. See: http://journals.ametsoc.org/doi/abs/10.1175/BAMS-D-12-00172.1

Zhao, Y., & K. M. Kockelman. 2002. "The Propagation of Uncertainty Through Travel Demand Models: an Exploratory Analysis." In *The Annals of Regional Science*, 36(1): 145-163.

CHAPTER 11

Adaptation of Florida's Urban Infrastructure to Climate Change

Frederick Bloetscher[1], Serena Hoermann[2], and Leonard Berry[3,4]

[1]Department of Civil, Environmental and Geomatics Engineering, Florida Atlantic University, Boca Raton, FL; [2]Center for Urban and Environmental Solutions, Florida Atlantic University, Boca Raton, FL; [3]Florida Center for Environmental Studies, Florida Atlantic University, Davie, FL; [4]Coastal Risk Consulting, Plantation, FL

This chapter looks at how the impacts of climate change affect different parts of Florida. With more than 1500 miles of coastline that contains numerus differences in character between the state's southern-most point in the Florida Keys to the northwest Florida Panhandle and northeast Florida in Jacksonville, it is easy to see why areas across the state are not all the same; temperature, rainfall rates, and even the potential for sea level rise can vary significantly depending on what part of the state one is in. For example, southeast Florida and the Tampa Bay area are already dealing with sea level rise issues, but there is much work to be done in order to assess the risks and help identify potential solutions. Efforts to adapt to rising seas will need to draw upon prior research and current work to develop tool box strategies that involve the hard and soft components. A background of impacts to water resources (less rainfall has been detected) will be discussed.

Key Messages

- Climate changes, along with population continuing to increase, makes the water supply planning and management a critical challenge for the state.
- Climate change factors in Florida impact different areas differentially, making it unfeasible to develop one-size-fits-all policies. This makes it essential to tailor climate adaptation management strategies to each community's unique needs.
- Storm surge in coastal areas will increase flooding and property damage.
- Flooding is not just a coastal issue, but an inland issue, leading to lower capacity for soil to absorb precipitation, thereby increasing the risk of flooding because aquifers are full and groundwater has no place to go.
- Climate change, especially sea level rise, will have adverse impacts on water, sewer, transportation and stormwater infrastructure. The risk of failure from these systems put private property and economic prosperity at risk.
- The development of a framework to evaluate the impacts of climate change on infrastructure and urban development (as they are intrinsically intertwined) requires (1) identification of vulnerable areas, and (2) the development of successful flood mitigation scenarios to address community vulnerability and cost effectiveness.
- A set of strategies to combat or mitigate climate impacts on a community will be community-specific and usually require significant engineering and planning to determine the best mix.
- Longer-term development policies will need to include the 50- and 100-year vision for development. This vision will address hard and green infrastructure, policy and development objectives and funding needs.

Keywords

Sea level rise adaptation; Groundwater; Aquifer drainage; Toolbox of strategies

A Long Water History

The history of Florida has always been defined by water—either too much or too little, depending on the vision of the decision-makers at the time. One hundred and fifty years ago, much of Florida was swamp; too much water to live and settle in, except for small bands of Native Americans. But at the turn of the 20th century and starting with the creation of the Florida East Coast Railroad by Henry Flagler, the age of infrastructure construction began in Florida. Then-Governor Napoleon Bonaparte Broward proposed draining the wetlands to permit more settlement of people and farming throughout the state, and drainage on the East Coast began in earnest, spurring the accompanying migrations of winter residents. Hurricanes in 1926 and 1928 disrupted this expansion, requiring another round of diking (the Hoover Dike – see Fig. 11.1) and canal building that kicked off 20 more years of construction, this time around Lake Okeechobee and including development of the Everglades Agricultural Area on the lake's south side. Post-World War II, hurricanes prompted the Army Corps of Engineers' next round of canals, dikes, and pumping stations. This effort included completion the largest drainage project of its time—the Central and South Florida Flood Control project, which was authorized by the Flood Control Act of 1948. The project drained the southern half of the State of Florida, altering drainage and recharge patterns, and permanently changing aquifer levels and conditions. The work continued into the late 1960s. In other words, for more than 70 years, decision-makers in Florida had acted upon the idea that there was too much water in the state.

Figure 11.1. Hoover Dike from US 27.

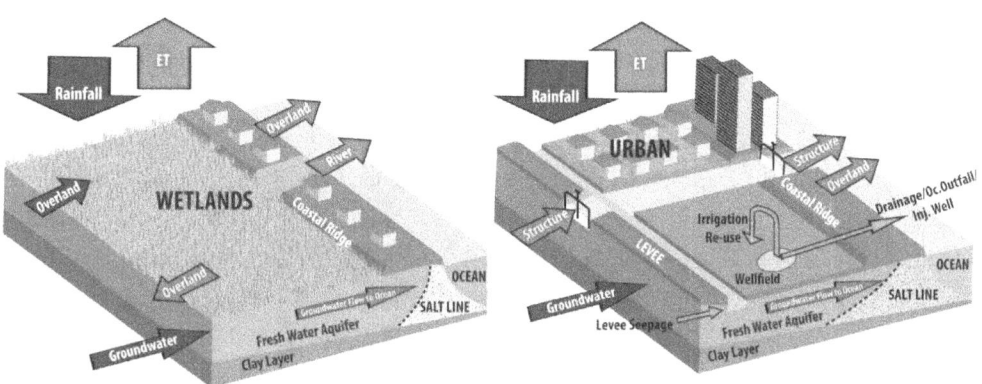

Figure 11.2a. (Left) Pre-1900 South Florida. **Figure 11.2b.** (Right) Post 1970-South Florida.

Concurrent with this continuum of drainage efforts was the state's population explosion, from under 1 million in 1910 to over 10 million by the 1970s; Florida's population exceeded 20 million in 2010. This growth brought demands on drinking and irrigation water that conflicted with 70 years of politically driven drainage programs and the state's annual wet-dry cycle of rainfall whereby the greatest demands of the population occurred in the dry months (December to April). With its dry and warm winters, Florida was a desirable destination for retirees and others seeking to escape cold winters. But the state's dry season also meant that water supplies drained to the ocean during the summer and were no longer available for winter use. By the mid-1970s, it was obvious that water shortages would plague certain parts of the state if action was not taken. The Florida Legislature responded by approving Chapter 373 FS, which created five water management districts (originally called the "flood control districts"). The tenor of the statute was distinctly different than previous efforts, which focused on draining the natural systems and moving water away from people. Instead, intense water supply planning began in the early 1970s and over the ensuing 40 years partial success was realized when the total water demands throughout the state were held relatively constant from 1974 onward, a testament to aggressive planning, regulation and education, although the locations of the changing demands changed over that time.

Since their establishment in the 1970s, the responsibilities of water management districts have expanded to include oversight of the local water supply, water quality, natural systems management, and flood protection. The *Save Our Rivers* program was established in 1981 to help water management districts manage, protect, and conserve the state's water resources. The program allowed the districts to acquire environmentally sensitive land. In response to a growing concern about water quality, the Surface Water Improvement and Management Act of 1987 required the water management districts to develop plans to clean up and preserve the state's rivers, lakes, estuaries, and bays.

Because periodic years of low rainfall are part of Florida's precipitation cycle (they occur every 6-8 years), failure to plan for these low rainfall years creates conditions akin to a drought—water supplies may be limited and competition increases. Under such conditions, the water management districts are responsible for mediating between the interested parties. But meeting all user needs can be difficult, if not impossible, which has led to legislation that addresses water supply planning. As a result, all five water districts in Florida now have water supply plans that include all spectrums of water use. These plans were initiated in the early 1990s and are updated every five years. Several of the districts have even split their plans into regional units to better address local conditions.

For example, in southeast Florida the 2005-2006 Lower East Coast Water Supply Plan Update was adopted right before the area's last major "drought," which occurred in 2007. During that year, rainfall in the upper and central Everglades was about half the norm, while coastal areas saw average rainfall. Consequently, the South Florida Water Management District passed the Regional Water Supply Availability Rule, which limited surficial aquifer water withdrawals to the 2006 levels. Yet the planned alternatives—water reuse, aquifer storage, and deeper brackish sources—have all proven to be a greater challenge to implement than anticipated. As a result, some utilities have tried indirect potable use as a possible future option, but the water supply permitting authority is lacking since recycling wastewater for potable use is not contemplated in any district's rules. Concurrently, coastal wellfields, sanitary sewer systems that are the source of the wastewater to be reclaimed, and storm water systems that replenish water supplies via infiltration have all become more vulnerable to the impacts of drought.

All of this makes clear how the history of Florida, specifically the development of infrastructure, has been deeply impacted by water and attempts to manage it; and it is logical to expect the same moving forward. Florida's continued development and projected climate impacts will help shape the state's future. Florida's continued development and projected water supply trends will be impacted by water supply variations, more frequent extreme weather events, a trend noted by Marshall et al. (2003) toward less total summer rainfall but higher intensity storms, an uncertain rainfall expectations for agricultural irrigation, and higher heat which will increase demands for irrigation, as well as domestic and industrial demands.

Selected Coastal Effects of Relative Sea-Level Rise

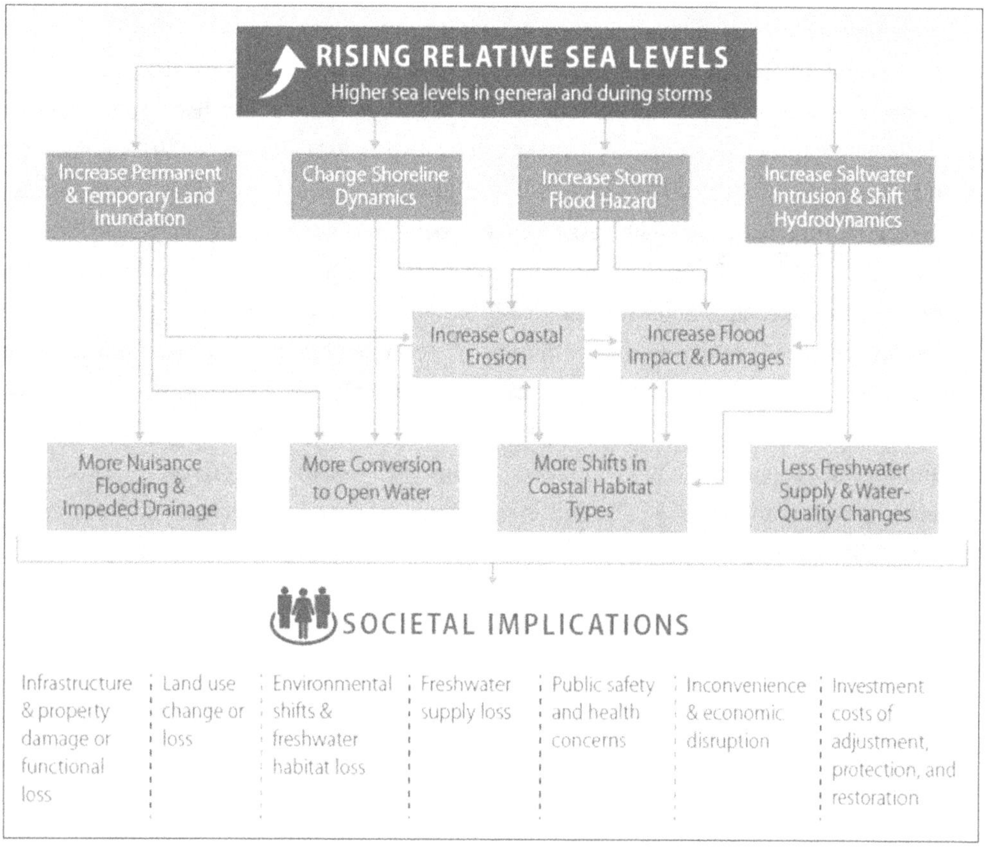

Source: Congressional Research Service, September 12, 2016.

Notes: This illustration depicts how rising higher sea levels may influence a number of coastal processes and affect coastal terrestrial and estuarine ecosystems and society. Each process and impact shown also is affected by other factors in addition to sea-level rise. For example, coastal shoreline dynamics are influenced not only by sea-level rise but also by climate and weather patterns; nearshore hydrodynamics and geomorphology; coastal land development and use; infrastructure projects such as seawalls and dredging; and other factors.

Figure 11.3. Selected coastal effects of sea level rise.

Addressing the Challenges

Much of Florida's economy (tourism, housing, and agriculture) is based on having adequate water supplies. Climate change is already impacting water availability in the state through three key drivers: 1) changes in precipitation patterns, 2) temperature increases that alter evaporation rates while increasing irrigation needs, and 3) sea level rise. As discussed in depth in prior chapters, these three factors may create the perfect storm, resulting in widespread disruption to the state's long-term economic growth and development. Bloetscher (2012) designated regions across the state and showed how each of the regions will be differentially affected by changes in climate. He discussed how water issues will significantly impact the various regions' economic, natural and built environmental systems, and detailed how there are differing perceptions of risk depending on what area of the state is being looked at.

For example, the increase in sea level observed along the Atlantic Coast (Bloetscher et al. 2012) combined with consistent population growth makes it essential for that area to focus on and continue to improve flood management strategies (NFIP 2011; Parkinson 2010; Schmidt et al. 2011; Warner et al. 2012; Zhang et al. 2011, 2011a). Bloetscher (2012) identified southeast Florida, the Florida Keys and southwest Florida as being far more vulnerable to sea level rise than the rest of the state. These three areas have dense populations, low topography, and are at high risk for climate change impacts (specifically sea level rise and surges).

On the other hand, in the northwest, northeast, north central ridge and Kissimmee River Valley regions, the risks associated with sea level rise due to climate change are not viewed with as much concern because those areas are at a higher elevations and less densely populated; that said, future water supplies due to precipitation changes may be the key driver that those areas must contend with (Bloetscher, 2012). This is an example of how different areas of the state will experience distinct impacts and risks; one-size-fits-all policies may turn out to be undesirable and, in some cases, ineffective. Many solutions must be tailored to the local needs and available resources while other challenges will require cooperative efforts between different government entities and even the private sector.

Water Supply

Florida has five primary sources of potable water that are aerially limited: 1) the Biscayne aquifer located in southeast Florida, 2) the Floridan aquifer system located in North and Central Florida as fresh water and in South Florida at brackish water 100 ft below the surface, 3) a series of sand aquifers that have limited production, 4) a few surface water bodies, and 5) the ocean. As a result, Florida has greater abundance and more sustainable water supplies than most other states in the country. Each of Florida's potable water sources is unique, particularly in terms of their challenges and vulnerabilities.

The *Biscayne aquifer* is a phreatic or water table system (not under pressure or protected by an overlying layer of rock) located in southeast Florida. In fact, the Biscayne aquifer is that area's only source of freshwater. But since it is a karst formation, the aquifer's flow channels make it susceptible to influxes of saltwater. Interestingly, it is anticipated that as sea level rises Florida's groundwater level will also rise, which might create an additional water supply through dewatering along roads and developed area to keep groundwater levels low to protect infrastructure (much like dewatering used during construction, only permanent). Higher future water tables could improve the availability of potable water in South Florida, higher water tables can also complicate strategies such as reclaimed water irrigation (higher water tables increase the potential for flooding), and increasing infiltration into sanitary sewer systems with salt water that requires reverse osmosis treatment to remove the salt prior to use for irrigation.

Located south of Lake Okeechobee is *the Floridan aquifer*, which, because there is no local recharge, is confined, brackish, and unsustainable. Over time, utilities using this source have experienced a degradation of water quality making its potential of large scale utilization unlikely. Drilling deeper is not South Florida's future.

In addition, many areas north of Interstate 4 already appear to be at the limit of their sustainable water yields. This means that water supply may be a barrier to future economic development in the northern half of the state. North Florida has a few, mostly small surface water systems. Many of the streams in the Florida Panhandle originate from Georgia and Alabama, posing actual and potential conflict points and a major barrier to future development. The possibility of lower total rainfall in the state may also create significant adverse effects on local agriculture and development.

Many alternative water supply strategies have been proposed as solutions to Florida's water supply issues, but most are expensive, involve large power draws, and are unlikely to be pursued by regional utilities (except in urbanized areas). For example, desalinization (removing salt from water) is touted as a potential solution, and certainly coastal areas could desalinize the nearby ocean water. But the desalinization process is expensive and many of these coastal areas currently have sufficient freshwater so there is no incentive to invest in establishing desalinization capabilities. Meanwhile, the inland areas in need of additional water supplies do not have access to the ocean and cannot easily adopt desalinization as a way to meet their potable water needs. That said, some municipalities in northeast Florida and the Tampa Bay area have seriously looked at or invested in desalinization capacity in a limited manner.

Another alternative solution, aquifer storage and recovery, has met with some success along Florida's West Coast; but it has not been as successful in southeast Florida or North Florida, likely due to confinement/formation, recovery, and metals recovery issues (Bloetscher et al. 2014). Also, while aquifer storage and recovery may be useful for utilities, agriculture is unlikely to pursue it as a water management strategy due to its higher cost that current supplies and the recovery uncertainty. Case in point, the South Florida Water Management District's regional aquifer storage and recovery program, which was originally designed to store billions of gallons of water, has been significantly scaled back due to cost and recovery challenges. Two wells have 10 years of testing. One recovers only 20% of the injected water. .

Surface storage is another alternative strategy sometimes proposed, but areas with large storage potential are virtually non-existent in much of Florida. Discussions are ongoing about using the C51 Reservoir (currently under construction) to store billions of gallons of stormwater in western Palm Beach County as a way of supplementing water supplies for urban utilities; but despite its decade long history this multi-billion dollar project is several years away from being completed. Hence, the potential for increases in conflicts over water resources will no doubt persist.

And finally, water reuse already serves as an effective solution for irrigation demands in many areas of the state (although most agricultural users are not in proximity to wastewater utilities and balk at the cost). However, water quality is a barrier for a variety of reuse options. For example, recharging water conservation areas with reclaimed water treated to higher standards might be possible in southeast Florida, but it can be cost prohibitive in other parts of the state that lack such catchment areas. While indirect potable recharge may work in several regions of the state (e.g., southeast Florida), the high cost and public perception make it unattractive at present. That said, as an example of desperate times supporting more costly solutions, Texas' Wichita Falls and Big Spring both created direct potable water treatment systems for their communities when reservoir levels dropped below 5% of capacity.

Mitigating Risks Associated with Flooding and Sea Level Rise

Bloetscher et al. (2016) made it clear that flooding is not just a coastal issue, but an issue inland as well. Higher groundwater levels equate to reduced soil storage capacity, which means lower capacity for soil to absorb precipitation, thereby increasing the risk of groundwater flooding (Romah 2012). Chang et al. (2011) described an overall "lifting process" by which there is a 1:1 ratio in water table elevation that can be correlated to sea level rise, while Bloetscher and Romah (2015) and Romah (2012) noted that groundwater levels in southeast Florida are intrinsically linked to the sea level. Thus, while coastal populations are particularly at risk for flooding due to erosion, inundation, and storm surge, interior populations are susceptible to rising water tables and extended periods of inundation. It is this limited soil storage that leads to flooding, necessitating the extensive drainage works facilities that discharge large volumes of water during the wet season. Sea level rise, combined with the end of the wet season in September, the king tides in late September and October, and southeast Florida's flat topography puts much of the region, whether coastal or inland, at risk.

Due to the associated loss of soil storage capacity caused by sea level rise, an increase in the intensity of storms due to climate change is expected to overwhelm the current storm water infrastructure. Projections indicate the potential for severe damage to southeast Florida's energy systems, transportation infrastructure, water infrastructure, agricultural lands, and the Everglades ecosystem (Zhang 2011; Karl et al. 2009). Mapping models indicate that infrastructure, primarily roadways (and the associated infrastructure laid underneath them), will be impacted first, followed by property. Higher groundwater elevations created by sea level rise will compromise stormwater and transportation infrastructure in low-lying areas, impacting access to roads, bridges, rail and rail transit (Bloetscher and Romah 2015). Roadway beds can be damaged by inundation and higher water table levels (FDOT 2012). Road bases can become saturated, causing premature base failure. In addition, since soil storage capacity is diminished, the potential for frequently flooded roadways will likely damage pavements (FDOT 2012). Fig. 11.4

illustrates roadway bases before and after sea level rise. Sea level rise will affect FDOT roadways, which are main arteries for transportation as well as emergency evacuation routes, by wetting the base. Many local roads that do not meet stringent FDOT standards are far more vulnerable to failure. As a result, billions of dollars in infrastructure investment would be needed just to maintain the status quo.

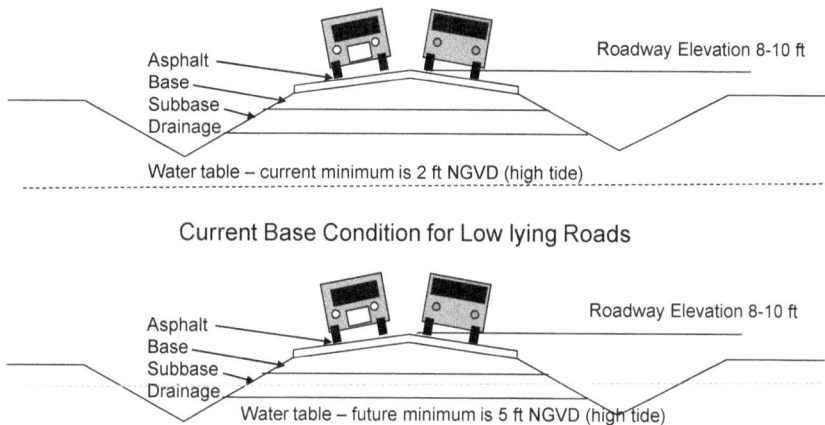

Figure 11.4. Current and future roadway conditions.

Flooding and sea level rise can pose serious threats when it comes to infrastructure systems below the road base. Existing sanitary sewer systems in Florida are constructed at depths below the groundwater level. Unfortunately, submerged pipes increase the potential for infiltration through cracked pipes or poorly constructed pipes systems, thereby consuming capacity in treatment plants. Inflow during rainstorms or other inundation can leach into the sanitary sewer systems through unsealed manholes, open cleanouts, and problem surface connections. Inflow can lead directly to sanitary sewer overflows and result in fines assessed against the utility and property damage. Water mains are typically shallower than sewer systems, but they too may also periodically be submerged in groundwater. Near the coast, pipelines may suffer brackish or freshwater conditions, depending on the aquifer levels. This is particularly damaging to cast or ductile iron pipes, both of which may eventually leak and both are prone to movement as groundwater can alter the pipe beds.

Beach erosion is another impact of the combination of sea level rise and coastal storms. As beaches erode, coastal developments face greater risks. Sea walls fail, building foundations can be disrupted, and concrete is more prone to damage from the lower pH, high salinity seawater. This means that effective methods for protecting the coastal infrastructure must be found. And finally, there are indications that the frequency of certain vector and waterborne illnesses due to climate changes (Bloetscher et al. 2016). Therefore, a better understanding of future trends in mosquito-spread diseases (e.g., Zika, dengue fever, or chikungunya) or waterborne diseases (e.g., giardia and cryptosporidium) is needed to adequately address the challenges posed by climate change.

In short, the future Florida condition will be characterized by a warmer climate, greater variability in storm intensity, greater extremes in temperature, (Marshall et al 2003) as well as uncertainty about rainfall and timing of same. Floridians should anticipate sea level rise that will threaten infrastructure reliability, economic activity, property values, public health and put population risks (Bloetscher et al. 2012, 2014, 2016), all of which will be tied to Florida's changing climate.

Level of Service in Vulnerable Areas

An important consideration prior to any infrastructure investment by local communities should begin by defining the acceptable "level of service" for areas identified as vulnerable. For example, king tides occur annually (in September/October) in southeast Florida often resulting in coastal flooding. Thus, defining the level of service will involve determining how often it is acceptable for flooding to occur in that community on an annual basis. The effects of sea level rise on the level of service should be used to update the mapping in terms of demonstrating changes in vulnerability and increased flooding frequency. For example, a 1% flooding frequency translates to four flood days per year.

Fig. 11.5 shows the vulnerable and potentially vulnerable areas in Miami-Dade and Broward counties, illustrating that the impacts of sea level rise and groundwater changes are significantly higher than the bathtub models project. Modeling by Romah (2012) showed that the areas in dark gray may be vulnerable at the 99 percentile condition (four days per year), the white indicates potentially vulnerable, and the light gray indicates areas not currently vulnerable. This figure shows the current, 1, 2, and 3 ft sea level rise conditions. Of interest is that the inland areas are at far greater risk than perceived by residents, public officials and government agency staffs, given that the groundwater continues to rise and the topography slopes to the west from the coastal ridge (Romah 2012; Bloetscher and Romah 2015).

Failure to identify potentially vulnerable areas and establish an acceptable level of service prior to an extreme event is often the cause of a loss of confidence in public officials because of increases in flooding and/or property losses, and a perceived weakening level of service by residents.

ADAPTATION OF FLORIDA'S URBAN INFRASTRUCTURE TO CLIMATE CHANGE • 321

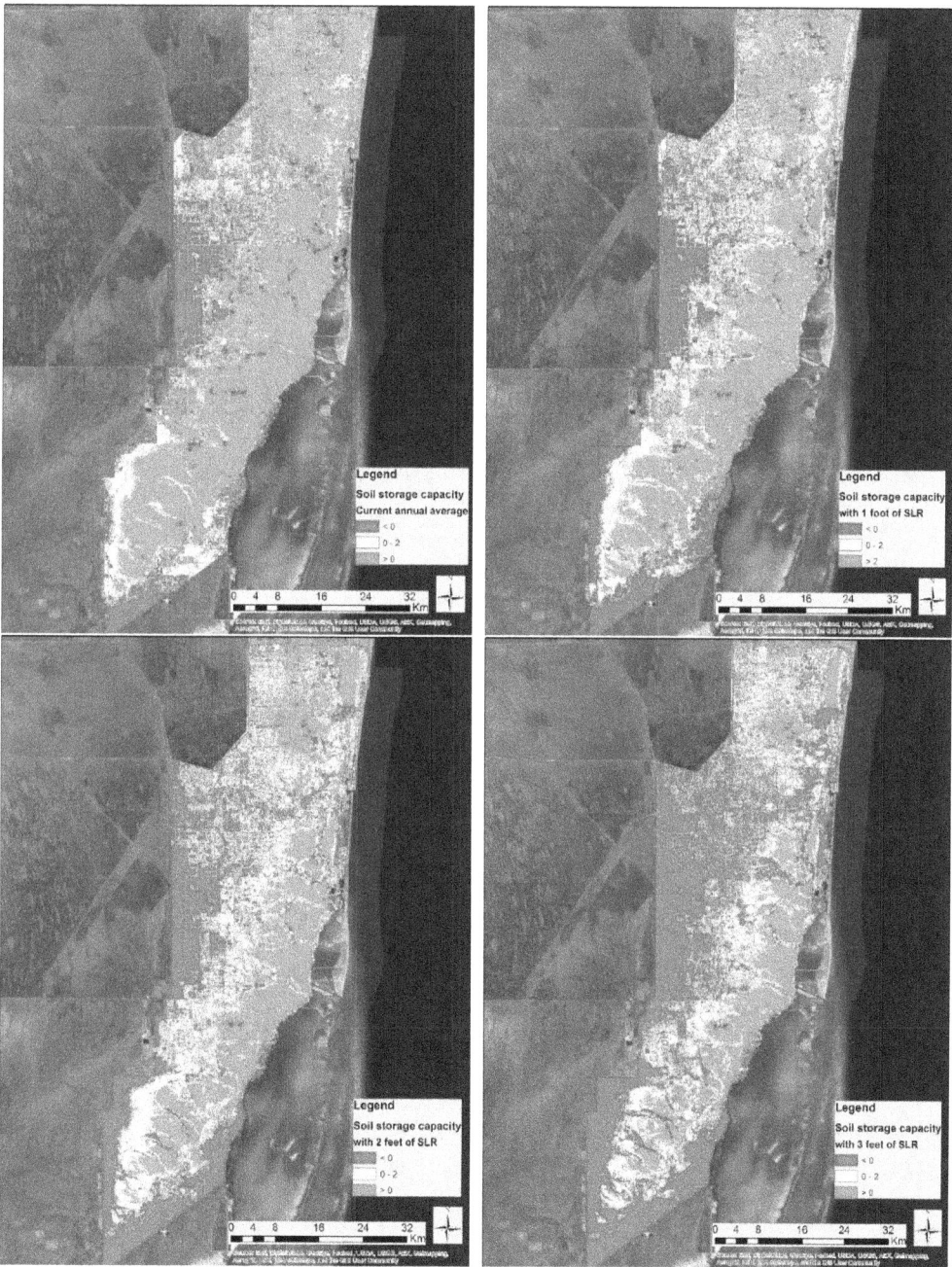

Figure 11.5. Miami-Dade and Broward counties—vulnerability at 0, 1, 2, 3 ft sea level rise at 99 percentile groundwater/tidal elevations (ignoring current infrastructure).

Proposed Frameworks

Once vulnerable areas are identified, various scenarios can be developed and options can be considered to address a community's potential needs. One approach is to identify successful mitigation strategies used by other cities that face similar problems based on identified vulnerabilities and cost effectiveness. These two issues can then be combined to develop a framework that can evaluate the impacts of climate change on infrastructure and urban development (as they are intrinsically intertwined). Fig. 11.6 outlines a simplified flow chart that can used as a basis for this type of evaluation.

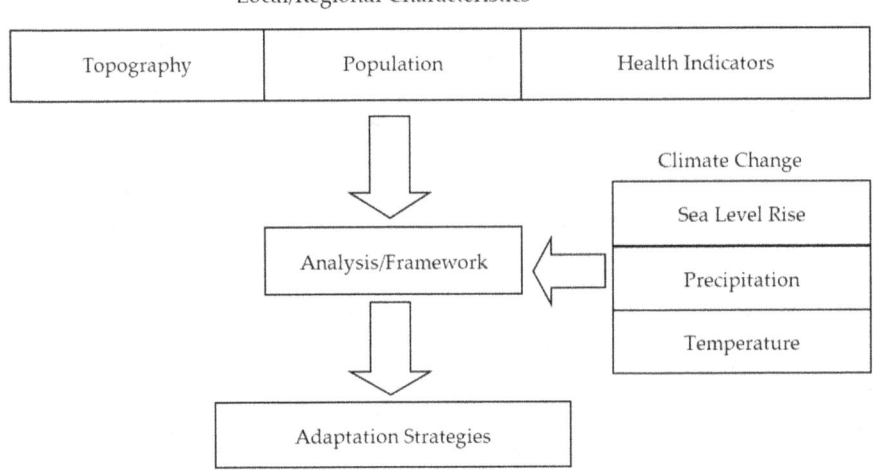

Figure 11.6. Analytical framework for toolbox development.

The strength of this framework lies in the proposed holistic and incremental approach to addressing climate change impacts, which includes understanding the combined social and health vulnerabilities in the context of higher exposure of the physical infrastructure to hazards. It combines physical vulnerability with health indicators and social evaluation criteria, and conveys the notion that a plan is not a fixed document, but rather a process that evolves with the changing conditions. Therefore, the approach requires that a mechanism be established to reassess the plan at regular intervals at the community level so that adjustments can incorporate improvements to various infrastructure systems.

Developing a Toolbox

Solving the infrastructure problems discussed in this chapter may be difficult, but not at all impossible. Much of the current work on adaptation to sea level rise focuses on understanding the physical and economic vulnerability of infrastructure, as well as on developing adaptation strategies for the natural and built environments using new infrastructure systems (Zhang 2011; Hansen et al. 2009; Parkinson 2009; SFRCC 2011; Tebaldi et al. 2012; Titus and Richman 2012; Weiss et al. 2011). Far more statewide coordination will be needed to balance Florida's urban, natural, water, wastewater and groundwater level needs. Below are some of the primary concerns that will need to be addressed and the steps that can be taken to meet the challenges.

Communities will need to develop a toolbox (i.e., a series of solutions) that can be used to address their specific vulnerabilities. Identifying viable adaptation solutions requires (1) the prior analysis of vulnerability, and (2) overlaying development priorities with expected climate change on GIS maps to identify hotspots where adaptation activities should be focused. This approach will allow for more precise identification of at-risk infrastructure and prediction of impacts on physical infrastructure and the community, as well as local-scale evaluation of conditions. A long-term view is required, since roadways and other infrastructure are normally designed for a 50- to 100-year service life, and they are rarely abandoned. Long-term planning for the effects of climate may also consider possible temporal impacts that can damage infrastructure and limit accessibility to certain services (e.g. health services).

Risk Assessments by Utilities

In order for a community to develop effective mitigation and adaptation strategies, local utilities must judge the vulnerability of their infrastructure and operating protocols (Wallis et al. 2008). Utilities are encouraged to use risk assessments to account for the uncertainties associated with stormwater and any impacts of changes in climate patterns. Strategies should include adaptive management techniques employed during planning and operations, as well as infrastructure improvements. Determining the appropriate actions or strategies to adapt to climate change involves using a vulnerability assessment approach to evaluate the need for operational changes and/or to install new infrastructure or "harden" existing infrastructure. Risk assessment requires evaluation of the likelihood of the climate change impact occurring. For Florida, the changes in rainfall patterns and sea level rise are the most pressing concerns for stormwater master planning; sea level rise will impact flooding and precipitation patterns have changed. However, there is uncertainty about how much and how fast sea level will rise and how precipitation patterns will change the perception of stormwater infrastructure needs.

Deyle et al. (2007) outlined the need for planning for adaptations, protection and potential retreat scenarios given the competition for scarce public dollars over the next 20 to 100 years to deal with sea level rise. However, Frederick and Gleick (1999) noted that socioeconomic implications may restrict the ability of managers and planners to act on required plans or force development changes in the near term. To resolve this, Freas et al. (2008) suggested using risk assessments as a way to determine system vulnerabilities for infrastructure and supply sources using a dual analytical approach known as the threshold–scenario risk assessment framework. The *threshold assessment* is a qualitative approach that relies on the experience and judgment of professionals to define vulnerabilities and adaptive strategies. Freas et al. (2008) suggested that in order to set proper thresholds for water systems, experienced water managers' experience and judgment of meteorological and natural systems is required. Threshold impacts can be assigned based on the infrastructure most susceptible to climate impacts. The method can be applied to any infrastructure system. For example, the most at-risk infrastructure in Iowa during floods in June 2008 were wellheads that were not above the flood stage. Freas et al. (2008) said that setting these thresholds for utilities involves four steps:

1. Define the performance criteria of the system infrastructure
2. Establish variables of importance
3. Define infrastructure component responses to climate change variables
4. Develop adaptation strategies that will reduce or eliminate the impact based on vulnerability assessment and performance risk.

Threshold analyses should identify the weak areas of the system that need hardening through either adaptive strategies or infrastructure changes that will reduce or eliminate the vulnerability. For example, in the case of water systems identified as weak, a diversified approach to water supplies is the best method to minimize future risks.

A more quantitative approach is the *scenario risk assessment,* which identifies the likelihood of failure in a system. The purpose of this approach is to quantify risks of a utility's current system in the event of climate change impacts. Freas et al. (2008) suggested that a scenario risk assessments for utilities involves seven steps. Two additional steps are added here to identify the hardening measures and decision-making steps.

1. Select a range of climate change scenarios based on commonly accepted models
2. Translate these to local scenarios
3. Identify climate change variables of importance (e.g. rainfall frequency, rainfall volume, temperature, sea level)
4. Determine system responses locally (e.g., incorporate rainfall changes into surface flow models)
5. Develop adaptation strategies
6. Evaluate robustness of the adaptive strategies
7. Identify hardening measures.
8. Evaluate overall system performance
9. Make the decisions needed to adapt to the change

Adaptations to Address Infrastructure Issues

The next task for communities develop effective mitigation and adaptation strategies involves imagining scenarios whereby toolbox options can be utilized to address flooding in the community. The goal is identifying successful mitigation strategies used by other cities facing similar challenges (for example, for flooding mitigation strategies, a community would look at cities faced with similar drainage and construction problems).

In order to complete risk assessments, a utility (regional or local) will need sufficient knowledge of the local hydrology. This includes development of surface water flow forecasting models and an integrated surface water–groundwater hydrologic model, all of which incorporate rainfall and temperature variables as driving forces. These models can use the output of downscaled climate models that provide different rainfall and temperature time series and make assessments of the effects of changes in these parameters (i.e., how they affect the service area).

Developing a toolbox of strategies can improve the regional resiliency in response to sea-level rise. It is important to note that any proposed/planned solutions must be site- or community-specific, and most require significant engineering and planning to determine the best set of strategies to pursue. Communities may need to spend millions of dollars to identify their solutions. Hard infrastructure systems and roadways are usually the first systems to be impacted because they are typically positioned lower than buildings and are critical for maintaining transportation. Additionally, most infrastructure systems (water, sewer, stormwater, power) are located within the roadways. As a result, adequate resources for transportation infrastructure and other major investments may need to be given the highest priority. So, for example, catastrophic flooding should be expected during heavy rain events and reducing the vulnerability of a community's transportation infrastructure will require the design of more resistant and adaptive infrastructure and network systems. This will, in turn, require the development of new performance measures to assess the ability of transportation infrastructure (e.g., roadways, bridges, rail, sea ports, airports) to withstand sea level rise and/or rainfall and to enhance resilience standards and guidelines for design and construction of transportation facilities. Considerations must include retrofitting, material protective measures, rehabilitation and, in some cases, the relocation of facilities to accommodate sea level rise impacts. As they are related, groundwater is also expected to have a significant impact on flooding in low-lying areas as a result of the loss of soil storage capacity. So, groundwater needs to be an important consideration of planning efforts.

Other Measures

In addition to the specific measures outlined above, it makes sense that long-term development policies should include the 50- and 100-year vision of development and require developers to include hardening within ordinances. Additional development in flood prone areas should not be permitted without local solutions or mitigation plans in place. But state and local agencies have been averse to setting such regulations due to private property rights arguments. The challenge is to balance property rights with the practicality of developing or rebuilding projects in areas that are flood prone now or in the future. Lending agencies will ultimately seek relief from the state if properties lack long-term viability or fall into default due to climate change impacts. Therefore, codes and design guides will need to be modified and standards of practice will also need to change.

Reducing Risk vs. Creating Resiliency
While uncertainties about the scale, timing, and location of climate change impacts can make decision-making difficult, response strategies can be effective if planning is initiated early on. Because vulnerability can never be estimated with 100% accuracy, the conventional *anticipation* approach should be replaced or supplemented with one that recognizes the importance of building *resiliency*. The resiliency approach focuses on adaptive capacity, which can be seen as the ability of a system (natural or social) to handle surprises, respond, learn from mistakes, and recover. Policies that aim to strengthen adaptive capacity/resiliency work to not only avoid potential impacts, but also to create benefits, regardless of the actual occurrence of any or all impacts.

The literature on climate change focuses on two plausible strategic responses: 1) "*mitigation of climate change,*" which for the most part relies on policies to reduce greenhouse gas emissions, and 2) "*adaptation to climate change,*" which consists of actions aimed at meaningfully addressing the vulnerability to climate change impacts (IPCC 2007; Füssel and Klein 2006). Unfortunately, most infrastructure planning is done on timeframes too short to adequately consider climate change, and day-to-day resource management is often crisis driven, subject to political and budgetary constraints. Adaptation planning must merge scientific understanding with political and institutional capacity on an appropriate scale and horizon. This is the challenge.

What Communities Are Doing—Case Studies

Southeast Florida, the Tampa Bay area and Charlotte County have long been among the most active areas in the state looking at climate change adaptation largely because sea level rise poses a real threat to those communities' valuable real estate in close proximity to the ocean. Several Miami-Dade and Broward county communities have developed climate plans (e.g., Delray Beach, Dania Beach, Coconut Creek, and Surfside) and others have well-developed resiliency programs (for example, Miami Beach has a $500 million program to raise roads and sidewalks and to install pumps and pipes to address ongoing flooding).

The cities of Miami Beach, West Palm Beach, and Fort Lauderdale all have active stormwater projects that model seasonal flooding to identify hot spots. The City of Davie has a similar model, albeit on a broader scale. Similarly, the City of Hollywood is examining flooding in eastern communities. These efforts will lead to capital projects much similar to those already well underway in Miami Beach. But first, the costs must be calculated.

A major catalyst for community action began in 2009 when the Southeast Florida Regional Climate Compact was created after elected officials had come together at the Southeast Florida

Climate Leadership Summit, as way to discuss challenges and strategies for responding to the impacts of climate change. Since then, climate issues, particularly southeast Florida's susceptibility to sea level rise and the need for green infrastructure and building techniques— have been hot topics at the annual Climate Compact event. However, the Regional Climate Compact is a forum for discussion, not a forum for activity to address problems— that responsibility falls to the local governments.

Below, are three summaries of three plans that demonstrate how local governments are turning discussions into action plans and preparing their communities for current and future climate impacts. These examples provide a flavor of what actions can be taken and the anticipated impacts of sea level rise on these communities. Note that only two of the case study communities are coastal, and neither is actually on the ocean.

CASE STUDY 1 – City of Dania Beach
Dania Beach, a city in Broward County, completed their climate action plan in 2010 with a focus on hard infrastructure and future planning. The plan began by noting that coastal areas should begin preparing for the impacts of climate change in order to safeguard their communities' social, cultural, environmental, and economic resources. Policies needed to focus on both mitigation and adaptation strategies. Policy formulation had to be informed by sound science, and administration of both policy and science was ideally to be conducted at the local level to better engage the community and formulate local decisions.

Thinking of the social, political, and natural systems as layers, it is easy to envision the world as one where all things share space, but move at different speeds. Change is constant, but not consistent. The natural environment changes relatively slowly, and people have less ability to affect this layer in the short-term. The built environment changes relatively quickly, with new stock replacing the old on a cycle of about 30 years. Infrastructure, such as transportation networks, water treatment systems, and energy providers, can take even longer—sometimes up to 50 years to retrofit. This means that infrastructure is the layer that takes the longest to develop, has the most impact on our future ability to cope with impacts and adapt to change; thus, it is the layer that should be the focus of planning efforts now.

Dania Beach's Risk
The City of Dania Beach is a community that includes mangroves, 660 ft of sandy beaches and a pier, and 30,000 residents who reside in a low-lying area. Over a third of the city is below 5 ft of sea level (NAVD88)the expected high tide in 2100; some streets are currently below 3 ft NAVD88. Adaptation measures will most certainly be required to protect the city from the impacts of climate change and groundwater rise. The city has more than $100 million in infrastructure at risk—roads, sewer systems, stormwater lines, along with buried water mains. And as protection of existing public infrastructure and coastal private property becomes less feasible, and the need to relocate population centers more intense, the city's climate strategies

will have to go beyond just mitigation. In order to reduce the impact of sea level rise, three types of *adaptive* responses should be undertaken (Nicholls *et al*. 1999, Klein *et al*. 2001, Deyle et al. 2007) by the City of Dania Beach:
- Planned retreat (zoning away from coastal areas)
- Accommodation (retrofitting buildings)
- Protection (levees and seawalls)

Fig. 11.7 shows the vulnerability of Dania Beach to sea level rise—dark gray areas are currently vulnerable using the 99 percentile event; white areas are potentially vulnerable, and light gray indicates areas not currently vulnerable. Note that theses maps focus on the vulnerability of Dania Beach's roadway system to flooding.

Reducing Economic Vulnerability

Dania Beach's economic vitality is at risk as well. The Federal Emergency Management Agency's National Flood Insurance Program ranked Florida third in the nation for repetitive-loss properties (those having two or more flood insurance claims within a 10-year period) (FCOC 2008). To reduce the costs associated with maintaining, insuring, and rebuilding flood-prone properties, a number of strategies are being considered by Dania Beach, such as: rolling easements, targeted coastal land acquisition, tax incentives for landward relocation, stricter setbacks, reclassifying hazard zones, conservation easements, restrictions on rebuilding after storm destruction, and improved comprehensive planning (Deyle *et al*. 2007, Poulter *et al*. 2008, FCOC 2008).

The Role of Local Ecosystems

Dania Beach's relationship between natural systems and the ability to protect and adapt to anticipated impacts of climate change is also important to recognize. Coastal wetlands provide habitat for wildlife, such as nurseries for fish and rookeries for birds, and critical functions for humans, such as the natural cleaning of ground and surface water, as well as flood and storm protection (Stumpf and Haines 1998; Titus 1991). Wetlands are under enormous pressure already, largely due to human density, infrastructure, and water management practices. So as sea level rises, all elements in a coastal wetland (salt marshes, mangrove forests, cypress swamps, etc.) can get caught in the middle of rising tides and development and, effectively, bee "squeezed" out (Titus 1991; Poulter et al. 2008). While the extent of coastal wetlands within Dania Beach's city limits is limited, there are mangrove forests in West Lake (between the city and the ocean) that will migrate westward as sea level rises. Migrating mangroves will impact private property owners' ability to develop. However, those same mangroves will reduce the force of surges, helping low-lying eastern areas during storms. Protecting remnant wetlands in and near the city is an important strategy to that the city is pursuing through continuation of good coastal management, as well as careful future infrastructure and land use planning.

According to Mukheibir and Ziervogel, 2007, there are 10 steps to consider when creating an adaptation strategy on the municipal level. The City of Dania Beach's climate plan specifically addressed the first two, and provided a framework for the remaining eight:

1. Assess current climate trends and future projections for the region (defining the science).
2. Undertake a preliminary vulnerability assessment of the city and communicate results through vulnerability maps (using GIS and other tools).
3. Analyze vulnerability spatially, by overlaying development priorities with expected climate change on GIS maps to identify hotspots where adaptation activities should be focused. Supplement analysis with participatory and quantitative methods.
4. Survey current strategic plans and development priorities to reduce redundancy and understand institutional capacity.
5. Develop an adaptation strategy that focuses on highly vulnerable areas. Make sure the strategy offers a range of adaptation actions that are appropriate to the local context.
6. Prioritize adaptation actions using tools such as multi-criteria analysis, cost-benefit analysis, and/or social accounting matrices.
7. Develop a document which covers the scope, design and budget of such actions (a Municipal Adaptation Plan).
8. Engage stakeholders and decision-makers to build political support. Implement the interventions prioritized in the Municipal Adaptation Plan.
9. Monitor and evaluate the interventions on an ongoing basis.
10. Regularly review and modify the plans at predefined intervals.

The strength of this framework for developing a local adaptation plan lies in the initial focus on location-specific science, the use of both monetary and non-monetary evaluation criteria, and the notion that the plan is not a fixed document, but rather a process by which people learn to change in harmony with a changing environment.

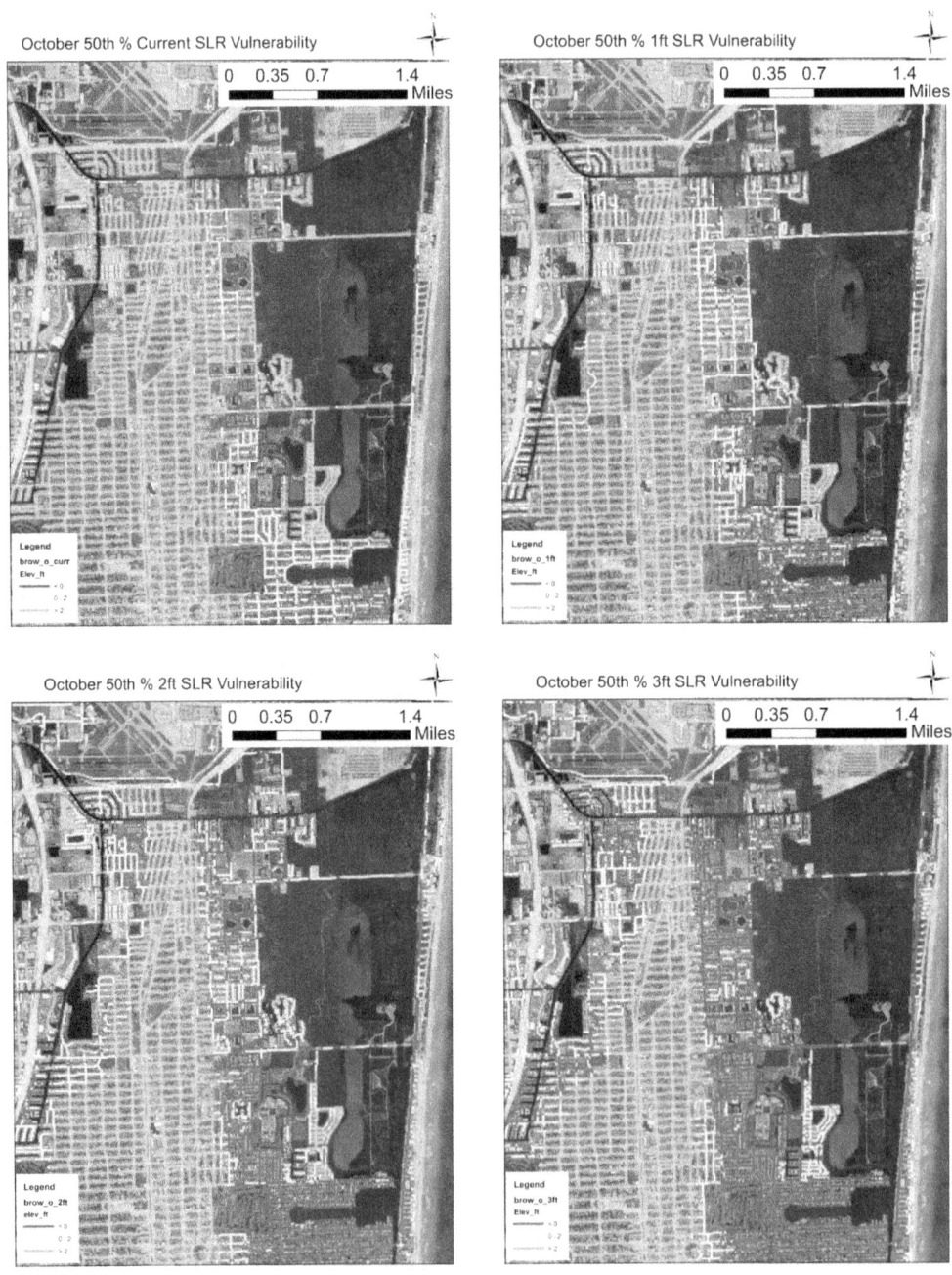

Figure 11.7. Dania Beach risk of roadway flooding as sea level rises.

CASE STUDY 2 – Davie

While the Dania Beach case study is an example of a city in the early planning stages of developing their climate plan, the inland Broward community of Davie (population: 91,000) is further along and actively looking at project funding options to begin implementing their plans. City leaders agree it is in the community's best interest to develop a stormwater planning framework to adapt to sea level rise and protect vulnerable infrastructure through a long-term plan. The stormwater conveyance system serving Davie residents includes primary, secondary, and tertiary canals operated by others; a canal operated by the South Florida Water Management District (SFWMD), and still others operated by local drainage districts. The canal systems in Davie pump the stormwater west to the Everglades and east to the Atlantic Ocean.

The community's stormwater system consists of structures (including catch basins, curb inlets, culverts, canals, swales, pump stations, ditches, and manholes) that help channel stormwater to the canals. Davie's stormwater system must comply with Broward County's Municipal Separate Storm Sewer System (MS4) Program, a publicly-owned system that collects or transports stormwater and discharges it to the state's surface waters . Like most South Florida communities, the city's current stormwater management program has significant financial constraints on the Town's General Fund.

To determine capital needs, an understanding of the existing assets and a means to evaluate potential flooding areas was required. The impacts of sea level rise on groundwater were considered; note the Town is 10 miles from the coast, stretches over 10 miles east-west and is generally sloped to the west, which means that western areas may flood faster than the eastern areas due to groundwater levels and elevation. Discussion with the Town's staff was held regarding the storm scenario for modeling. The decision was 1:10 storm event (FDOT), but hydrographs were created for the 1:10 event, 5-inch rain event (Florida Building Code) and 25 year 3 day (12 inch) event (SFWMD) for comparative purposes. were run at the 99 percentile groundwater and tidal dates and levels. Modeling at the current condition, 1, 2 and 3 ft sea level rise scenarios were run that incorporate the loss of soil storage capacity created by rising groundwater levels. Figure 11.8 shows the modeling process. An example basin is shown in Figure 11.9 for the current, and future scenarios.

For this southeast Florida community, a number of potential options are available to deal with sea level rise. The perception that they were less vulnerable was found to be incorrect given that the coastal ridge appears to control groundwater levels in the community and surrounding areas. Table 11.1 outlines the estimated cost infrastructure needs, from a long-range planning perspective, needed for each basin under each scenario, with costs. Specific sizing of improvements must be designed to achieve water quantity and quality requirements (quality is not part of the modeling software). The community needs to define which event they are planning to address and the timelines as the costs vary from an initial need of $30 million to nearly $335 million might be needed to address stormwater issues arising from sea level rise before 2100 depending on the level of protection community leaders wish to pursue for residents.

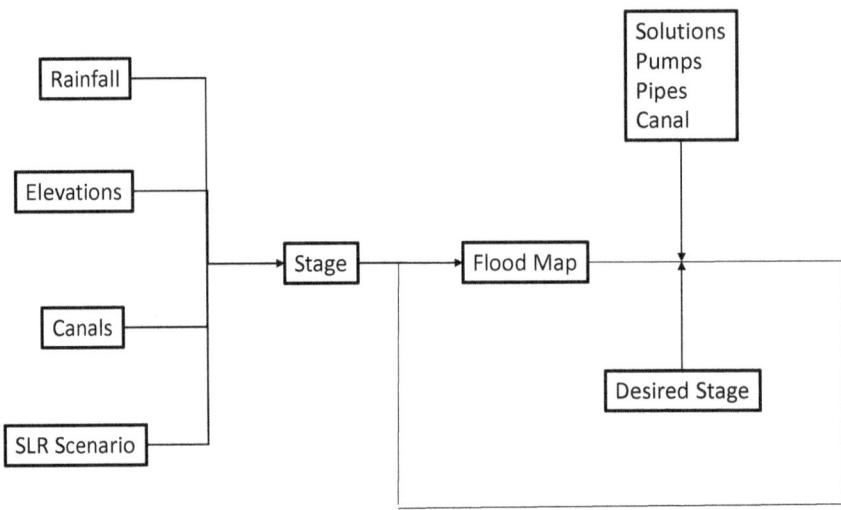

Figure 11.8. Modeling process for Davie risk modeling.

Table 11.1. Cost Estimate Current and Future Needs (Millions of Dollars).

Current	
Min Need	$ 38
Max Need	$ 148
1 ft SLR	
Min Need	$ 78
Max Need	$ 159
2 ft SLR	
Min Need	$ 123
Max Need	$ 255
3 ft SLR	
Min Need	$ 178
Max Need	$ 335

ADAPTATION OF FLORIDA'S URBAN INFRASTRUCTURE TO CLIMATE CHANGE • 333

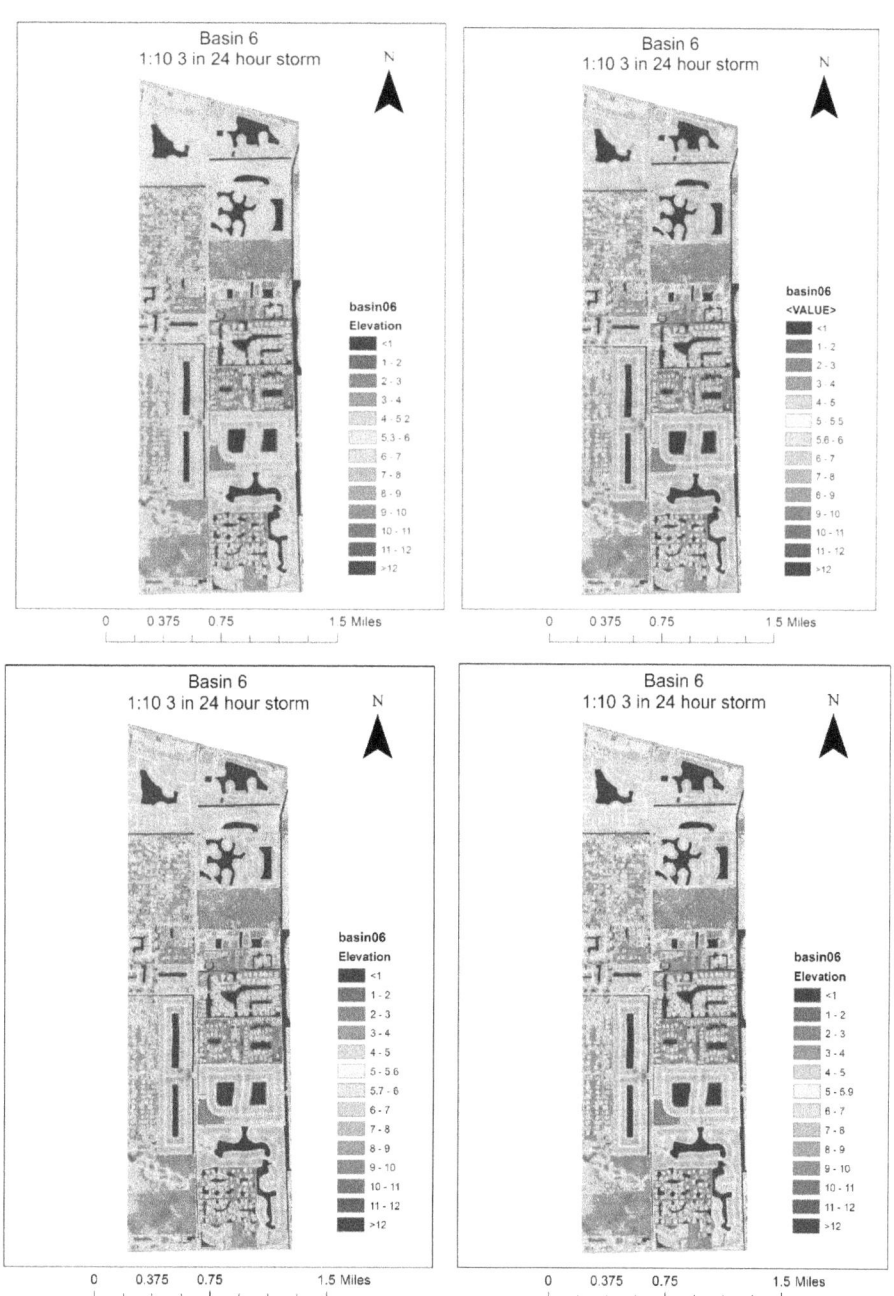

Figure 11.9. Change in flooding from current, to 3 ft SLR for the 1:10 storm event- dark gray areas indicate flooding, black means deeper water.

CASE STUDY 3 – Punta Gorda

The City of Punta Gorda is located in Charlotte Harbor of Florida's southwestern coast. The area is very low, and while the community is inland, it is directly connected to the Gulf of Mexico. Efforts by several universities and the regional planning council have evaluated the understanding of impacts of sea level rise in Punta Gorda. A Florida Atlantic University model of downtown Punta Gorda is shown in Figure 11.10. The blue area have more flooding that the adjacent areas. Of interest is that certain residential areas adjacent to downtown may have less risk, but newer areas may have been more due to elevation and exposure to Charlotte Harbor. Topography drives risk. Storm surge remains a major concern as there is little to slow surge. The regional planning council has taken the lead in evaluating the appropriate responses to sea level rise. The major focus has been community engagement and education given the residents' relatively low level of concern about the long-term effects of sea level rise. This is an area received a direct hit from hit by Hurricane Charley and its 145mph winds in August 2004. Heavy destruction in Charlotte and Desoto counties was sustained from wind and associated, limited storm surge. 33 were killed with $14 billion in damage. Surge was less than 7ft on the Gulf and only 1.5 ft in coastal waters. Storms remain a focus as well as sea level rise (Beever et al 2009).

The City's plan was adopted in 2009 (Beever, et al 2009). The plan indicated that "Severe tropical storms and hurricanes with increased wind speeds and storm surges have already severely damaged the community. Significant losses of mature mangrove forest, water quality degradation, and barrier island geomorphic changes have already occurred in the adjacent Charlotte Harbor" (Beever et al 2009). Results have included language to specifically address sea level rise in development codes. Goals included those identified by Deyle, et al (2007), goals to enhance green building initiatives, redevelopment of downtown areas prone to flooding, and the use of pavers instead of asphalt, among other efforts. It should be noted that the actions receiving the highest agreement included protection of seagrasses, xeriscaping, identifying area to protect natural shorelines, restricting fertilizer use and constraining locations for certain infrastructure. Communication with stakeholders was a large part of the plan, including a series of meetings that were held as a part of the plan preparation stage. There were many ideas proposed by the public, but cost was a limiting factor for residents. The risk scenarios focused on adaptation and acceptance of climate impacts.

The city administration has remained attentive to sustainability issues that emerged from the plan through ongoing community boards, comprehensive plan, and infrastructure efforts that meet the plan goals —building infrastructure on higher, safer ground to creating an energy-independent city. The land development codes require all foundations built in Punta Gorda to be higher than state and federal regulations and the city has also purchased shoreline property lined with mangroves to resist sea level rise and storm surge.

Figure 11.10. Downtown Punta Gorda Flooding.

Conclusion

Climate change will impact Florida's infrastructure primarily through sea level rise and changes in precipitation patterns-making it critical that current planning efforts by state and local governments, as well as utilities incorporate climate change issues. Impacts are expected to vary regionally, but climate change will likely result in increased demands on Florida's infrastructure systems statewide, both in terms of operations and maintenance costs and the need for capital expenditures. Climate changes will impact the state's supply of drinking water, wastewater, and stormwater systems; impacts will include not only physical property damage, but also increases in the cost of water treatment and treatment infrastructure.

A number of tools and outreach efforts can be adapted or created to foster the inclusion of climate change when considering and planning for infrastructure sustainability. These include advanced asset management strategies, ongoing modeling and monitoring of climate impacts, requiring utilities to examine their environmental footprints, and constantly working towards improvements in their infrastructure systems.

It is worth noting that preventive measures such as pumping, piping, hardening and power grid reinforcement are less costly than reactive measures. And while many local utilities have invested millions to harden their systems, but water supplies, stormwater infrastructure and wastewater systems remain vulnerable. Climate solutions are local, and funding is likely to be local, but the state will need to allocate disaster funding to help those utilities who make investments but still suffer losses. Giving those utilities that take proactive steps first priority may serve as an incentive for others in the system to make similar proactive investments.

It is essential that state and local government, as well as businesses, use a systems approach that takes into account population, economics, and environmental conditions when making long-term decisions. They will need to examine long-term viability, particularly with regard to investment decisions. Property values are also dependent on the maintenance and proper operation of transportation and utilities, especially storm water, wastewater treatment, and water supply.

Of greater concern is the future practices of the insurance industry, which has traditionally focused on a one-year vision of loss risk. As the insurance industry begins to discuss long-term risks of losses, there will be an accompanying impact on lending practices. Where properties are at risk, lending options may be reduced by insurance limitations—i.e., if the insurance industry sees the potential for significant losses from sea level rise within 30 years, the mortgage industry will limit the length of loans and increase interest rates due to insurance risk, thereby increasing costs to buyers and reducing the attractiveness of the purchase for sellers. The result may be declining property values and slower sales. Therefore, it is certainly in the community's interest to develop a planning framework to adapt to sea level rise and protect vulnerable infrastructure through a long-term plan.

References

Beever, J.W.; Gray, W.; Trescott, D.;Cobb, D.; Utley, J.; Hutchinson, D.; Gibbons, J.; Walker, T.; Abimbola, M.; Beever, L.B.; and Ott, J. 2009, *City of Punta Gorda Adaptation Plan*, Southwest Florida Regional Planning Council, Technical Report 09-4, SFRPC, 1926 Victoria Avenue Fort Myers FL 33901.

Bloetscher, Frederick; Polsky, C.; Bolter, K.; Mitsova, D.; Pablicke Garces, K.; King, R.; Cosio Carballo, I.; and Hamilton, K. 2016, Assessing Potential Impacts of Sea Level Rise on Public Health and Vulnerable Populations in Southeast Florida and Providing a Framework to Improve Outcomes, *Sustainability* **2016**, 8, 315; doi:10.3390/su8040315

Bloetscher, F.; Romah, T. 2015. Tools for Assessing Sea Level Rise Vulnerability. *Journal of Water and Climate Change* Vol 6 No 2 pp 181–190 © IWA Publishing 2015 doi:10.2166/wcc.2014.045.

Bloetscher, F.; Sham, C.H.; Danko, J.J.; and Ratick, S. 2014. Lessons Learned from Aquifer Storage and Recovery (ASR) Systems in the United States, *Journal of Water Resource and Protection*, 2014, 6, 1603-1629. http://dx.doi.org/10.4236/jwarp.2014.617146

Bloetscher, F. 2012. Protecting People, Infrastructure, Economies, and Ecosystem Assets: Water Management and Adaptation In The Face Of Climate Change, *Journal of Water*. Water 2012, 4; doi:10.3390/w40x000x

Bloetscher, F.; Heimlich, B.N. and Romah, T. 2011. Counteracting the Effects of Sea Level Rise in Southeast Florida, *Journal of Environmental Science and Engineering*, 5:11, 1507-1525, November 2011.

Chang, S. W., Clement, T. P., Simpson, M. J., and Lee, K. K. 2011. Does sea-level rise have an impact on saltwater intrusion? *Advances in Water Resources*, 34(10), 1283-1291.

Congressional Research Service, September 12, 2016.

Deyle, R.E.; Bailey, K.C.; and Matheny, A. (2007), Adaptive Response Planning to Sea Level Rise in Florida and Implications for Comprehensive and Public Facilities Planning, Florida State University, Tallahassee, Fla.

FDOT. 2012. BDK79 977-01 Development of a methodology for the assessment of sea level rise impacts on Florida's transportation modes and infrastructure. Summary [PDF - 470 KB], Final Report [PDF - 13,326 KB].

FCOC. 2009. *Florida Coastal and Ocean Coalition Policy Report*, FCOC, Tampa, FL.

Freas, K.; Bailey, R.; Muneavar, A.; Butler, S. Incorporating climate change in water planning. J. Am. Water Works Ass. 2008, 100, 93–99.

Frederick, K.D. and P.H. Gleick. 1999. Water and Global Climate Change: Potential Impacts on U.S. Water Resources. Pew Center on Global Climate Change. Washington, D.C.

Füssel, HM. & Klein, R.J.T. 2006. Climate Change Vulnerability Assessments: An Evolution of Conceptual Thinking, *Climatic Change,* 75: 301. doi:10.1007/s10584-006-0329-3

Global Insight. The role of metro areas in the US Economy; Reported for The United States Conference of Mayors, Washington, DC, USA, 1 March 2006. Available online: http://usmayors.org/metroeconomies/Top100_2006.pdf (accessed on 14 February 2012).

Hanson, S.; Nicholls, R.; Ranger, N.; Hallegatte, S.; Corfee-Morlot, J.; Herweijer, C.; Chateau, J. A global ranking of port cities with high exposure to climate extremes. Clim. Chang. **2011**, 104, 89–111.

Johns, G.; Leeworthy, V.R.; Bell, F.W.; Bonn, M.A. Socioeconomic Study of Reefs in Southeast Florida: Final Report; Hazen and Sawyer: Hollywood, FL, USA, 2003.

Karl, T.; Melillo, J.; Peterson, T. (Eds.) Global Climate Change Impacts in the United State; Cambridge University Press: Cambridge, UK, 2009.

Keith, D.W., Giardina, J.A., Morgan, M.G. and Wilson, E.J. (2005) Regulating the Underground Injection of CO2. *Environmental Science and Technology*, 39, 499A-505A. http://dx.doi.org/10.1021/es0534203Lettenmaier et al 2008

Kundzewicz,Z. W.; Mata, L. J.; Arnell, N. W.; Döll, P ; Jimenez, .;Miller, **K.**; Oki, T.; Şen, Z. and Shiklomanov, I (2007), "The implications of projected climate change for freshwater resources and their management," *Hydrological Sciences Journal*, Volume: 53, Issue: 1, Page(s): 3-10.

Mukheibir, P. and Ziervogel, G. (2007). Developing a Municipal Adaptation Plan (MAP) for climate change: the city of Cape Town. Environment & Urbanization. 19(1).

NFIP, 2011. N*ational Flood Insurance Program Loss Statistics from January 1, 1978* Through Report as of 07/31/2011. /http://bsa.nfipstat.com/reports/1040.html, (cited October 2015).

Nicholls, R.J. *Ranking Port Cities with High Exposure and Vulnerability to Climate Extremes*; OECD Environment Working Papers: Paris, France, 2008-Nicholls, R.J.; Hoozemans, F.M.J. and Marchand, M. 1999. Increasing flood risk and wetland losses due to global sea-level rise: regional and global analyses, *Global Environmental Change*, V. 9: S69-S87. doi: 10.1016/S0959-3780(99)00019-9.

Parkinson, R.W. *Adapting to Rising Sea Level: A Florida Perspective. In Sustainability 2009: The Next Horizon*; In Proceedings of the AIP Conference, Melbourne, FL, USA, 3–4 March 2009; Available online: https://411.fit.edu/sustainability/documents/FINAL%20-%20With%20Cover%2010-1-09.pdf (accessed on 28 March 2016).

Parkinson Randall W. 2010. *Municipal adaptation to sea-level rise: City of Satellite Beach, July 30, 2010.* R. W. Parkinson Consulting, Inc. Melbourne, Florida.

Poulter, B. and Halpin, P.N. (2008) Raster modeling of coastal flooding from sea level rise. *International Journal of Geographical Information Sciences* 22:167–82

Reilly, T.E.; Dennehy, K.F.; Alley, W.M.; William L.C. *Ground Water Availability in the United States*; USGS: Reston, VA, USA, 2009.

Romah T. 2012. *Advanced Methods In Sea Level Rise Vulnerability Assessment*, master thesis. Florida Atlantic University, Boca Raton, FL.

Schmidt, K.A.., Hadley, B.C. and Wijekoon, M. 2011. Vertical Accuracy and Use of Topographic LIDAR Data in Coastal Marshes. *J.Coastal Res.* 27 6A 116–132 West Palm Beach, Florida November 2011.

Southeast Florida Regional Climate Compact (SFRCCC). Analysis of the Vulnerability of Southeast Florida to Sea-Level Rise. 2012. Available online: http://www.southeastfloridaclimatecompact.org/wp-content/uploads/2014/09/regional-climate-action-plan-final-ada-compliant.pdf (accessed on 29 March 2016).

Stumpf, R. P. and Haines, J. W., 1998, Variations in tidal level in the Gulf of Mexico and implications for tidal wetlands: Estuarine, *Coastal and Marine Science*, 46: 165-173.

Tebaldi, C.; Strauss, B.H.; Zervas, C.E. Modelling sea level rise impacts on storm surges along US coasts. *Environ. Res. Lett.* **2012**, 7. Article 1.

Titus, J.G. and Wang, J. 2008 *Maps of lands close to sea level along the middle Atlantic coast of the United States: an elevation data set to use while waiting for LIDAR. Section 1.1 In: Background Documents Supporting Climate Change Science Program Synthesis and Assessment Product 4.1* (J. G. Titus & E. M. Strange, eds). EPA 430R07004, US EPA, Washington, DC, USA. http://papers.risingsea.net/federal_reports/Titus_and_Strange_EPA_section1_1_Titus_and_Wang_may2008.pdf.

Titus, J.G.; Richman, C. Maps of lands vulnerable to sea level rise: Modeled elevations along the US Atlantic and Gulf coasts. *Clim. Res.* **2001**, 18, 205–228.

Wallis, M.J.; Ambrose, M.kR. and Chan, CC. 2008 Climate Change: Charting a Water Course in an Uncertain Future, *JAWWA* v100:6. Pp 70-79.

Warner, N. N. and Tissot, P.E. 2012. Storm flooding sensitivity to sea level rise for Galveston Bay, Texas. *Ocean Engineering* 44 (2012) 23–32.

Weiss, J.L.; Overpeck, J.T.; Strauss, B. Implications of recent sea level rise science for low-elevation areas in coastal cities of the conterminous USA. *Clim. Chang.* **2011**, 105, 635–645.

Zhang, K. (2011). Analysis of non-linear inundation from sea-level rise using LIDAR data: a case study for South Florida. *Climatic Change.* 106, 537-565.

Zhang, K., Dittmar, J., Ross, M., and Bergh, C. (2011). Assessment of sea level rise impacts on human population and real property in the Florida Keys. *Climatic Change*, 107(1-2), 129-146.

CHAPTER 12

Climate Change Impacts on Florida's Biodiversity and Ecology

Beth Stys[1], Tammy Foster[2], Mariana M.P.B. Fuentes[3], Bob Glazer[4], Kimberly Karish[5], Natalie Montero[3], and Joshua S. Reece[6]

[1]*Florida Fish and Wildlife Conservation Commission, Tallahassee, FL;* [2]*Ecological Program, Kennedy Space Center, FL;* [3]*Department of Earth, Ocean and Atmospheric Science, Florida State University, Tallahassee, FL;* [4]*Florida Fish and Wildlife Conservation Commission and Fish and Wildlife Research Institute, Marathon, FL;* [5]*GeoAdaptive, Boston, MA;* [6]*Department of Biology, California State University, Fresno, CA*

Florida's rich biodiversity is the product of climatic conditions, geographic position, and underlying geology. Interactions of these factors over time have led to the state's unique biota, with Florida ranking fourth in the nation for total number of endemic species. The ability of Florida's ecosystems to support plants and animals is intimately tied to its geographic location, climatic and hydrologic variables, including timing and amount of precipitation, the frequency and intensity of storms, the range and duration of temperature extremes, and water chemistry. The ecosystems and species of Florida have adapted to past periods of climatic change. However, these ecosystems are now under stress and less resilient due to past and existing human-caused alterations and impacts, affecting their ability to withstand and adapt to additional stressors such as climate change. The overall vulnerability of some systems and species is primarily driven by the severity and extent of these non-climate stressors. Florida's biodiversity may be very different in the future, with some species and ecosystems affected to a greater extent than others. Community-level changes will occur as plant and animal species move and adapt at different rates. There are tools available to assist in determining relative vulnerability (vulnerability assessments) and potential impacts (scenario planning) that can aid in developing adaptation strategies. Awareness that change is likely to happen is critical to planning for the future and allowing for adaptation in management practices that will maximize Florida's biodiversity for future generations.

Key Messages

Climate Change Impacts on Biodiversity and Ecology
- Climate change has differential impacts on: coastal ecosystems, freshwater wetlands, and upland ecosystems. Coastal ecosystems, in particular, are subject to the "squeeze" of human impacts, changing climate, and rising sea levels.
- The Florida Keys and the Everglades are particularly vulnerable to sea level rise over the next 50 to 100 years due to their low elevation (typically less than 1 m).
- Out of 1,200 species tracked by the Florida Natural Areas Inventory, 25% are likely to lose at least half of their current habitat due to sea level rise alone.
- Florida's species have migrated and adapted to climate change in the past, but that ability is severely compromised now due largely to human modification of the landscape. Up to 76% of 236 surveyed species were deemed unlikely to be able to relocate inland in response to rising sea level.
- Several keystone species are particularly vulnerable to the impacts of climate change and the loss of these species can have cascading impacts on natural communities.
- Sea turtles are likely to respond to climate change through altered sex ratios of hatchlings, northward movements of rookeries, decreased reproductive output due to storm events, and potential shifts in foraging ground locations.

- Phenology, or the timing of life history events, are likely to change in response to climate shifts, both as the climate becomes warmer but also as it becomes more variable. This is particularly true for plants and can cause major disruptions to co-evolutionary relationships, such as those between pollinators and the plants they pollinate.

Existing Stressors and Climate Change
- Habitat loss and degradation are the leading causes of extinctions in Florida and globally. The impacts of climate change on species and natural communities are greatly magnified by decreased adaptive capacity due to habitat loss and degradation.
- Many invasive species are projected to have enhanced fitness under future climate change scenarios, potentially causing greater disruption to natural communities.
- Climate change is projected to increase the vulnerability of native species to foreign and domestic pathogens and parasites.
- Overexploited species have diminished capacity to adapt to climate change, making them especially vulnerable.

Preserving Biodiversity for the Future
- Planning for climate change involves impact assessments, adaptation scenario planning, and research and monitoring.
- While many of the ways in which species and natural communities respond to climate change are gradual, other changes can be abrupt and non-linear. These so-called thresholds, trigger points, or paradigm shifts are harder to predict, but are often more consequential than linear patterns of change through time.

Keywords

Ecosystem; Habitat; Species; Phenology; Biodiversity; Adaptation; Vulnerability Assessments; Scenario Planning

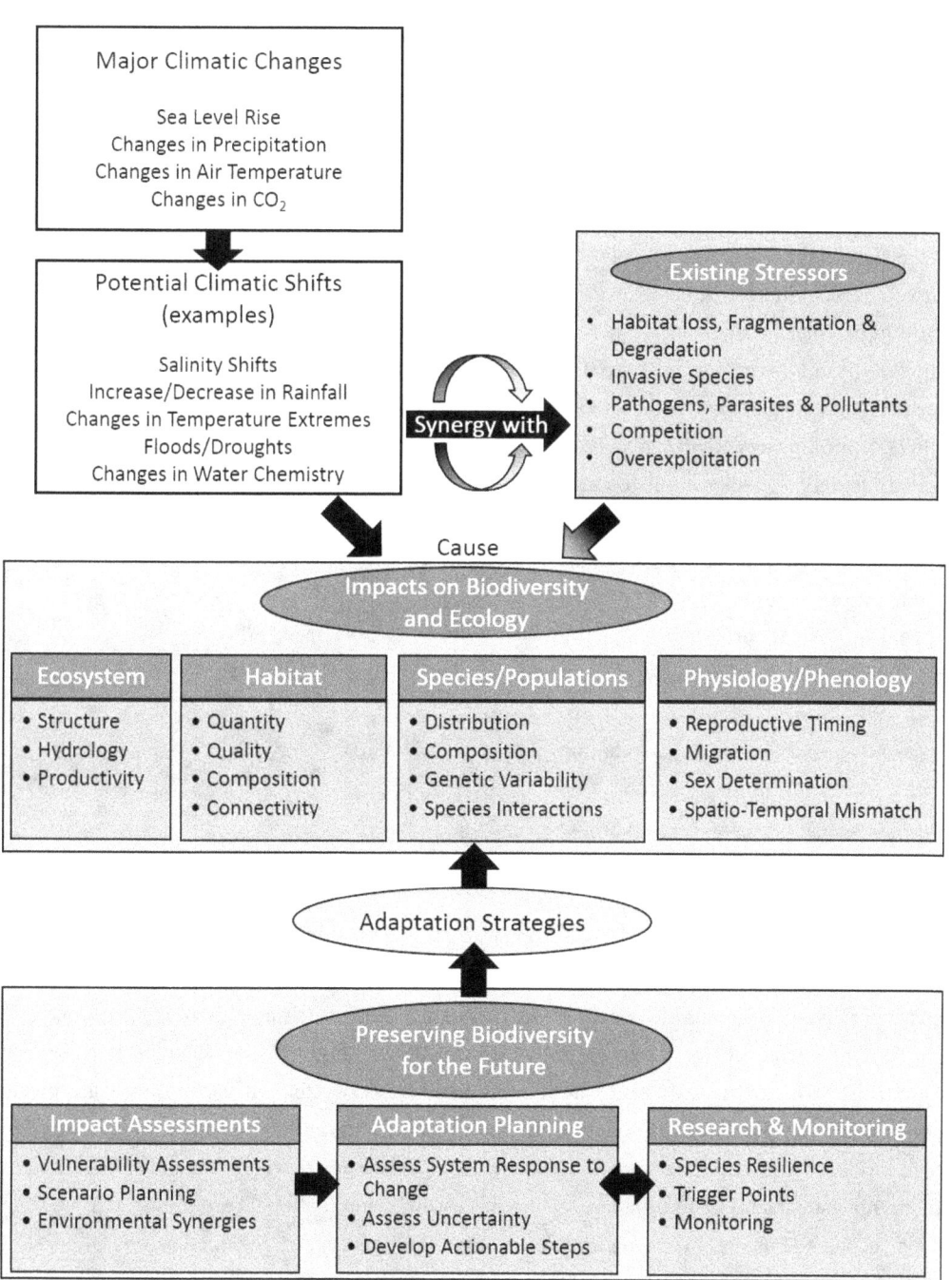

Figure 12.1. Diagrammatic illustration of this chapter showing simplified connections between major components (chapter sections). The grey boxes include overview of the chapter content, the white boxes are described in other chapters, but form the foundation of content in this chapter.

Introduction

Florida's biodiversity is extremely rich and contains a multitude of unique systems; it is identified as a "hotspot" of rare and imperiled species (Noss et al. 2015). Florida's geographic position and latitudinal range mean that the state is situated such that it encompasses both temperate and sub-tropical climate regimes, contributing to Florida's systems, communities, species diversity and distribution. Florida has the highest number of plant families, is the sixth highest in native plants species richness, has the highest number of fern species in the United States, and has the highest diversity of orchid flora and the densest concentration of carnivorous plant species in all of North America (Knight et al. 2011). Florida has more than 16,000 species of native fish, wildlife, and invertebrates, including 147 endemic vertebrate species and approximately 400 terrestrial and freshwater endemic invertebrates (Muller et al. 1989). There are currently 82 species designated as federally endangered or threatened in Florida. An additional 59 species are listed as endangered or threatened by the state, including 21 birds, eight mammals, 13 reptiles, four amphibians, nine fish, and four invertebrates (FWC 2016b). The unique scrub systems of Florida's dry, sandy ridges have the highest level of endemism for terrestrial habitats in the Southeastern United States, with more than 95 species of plants, lichens, arthropods, and vertebrates, including the iconic Florida scrub jay (*Aphelocoma coerulescens*). Coastal areas provide critical habitat for many of Florida's threatened species, including seaside sparrows, beach mice, sea turtles, beach nesting birds, and many endemic plant species. Many of Florida's rarest and most diverse communities occur as small isolated areas, such as pine rocklands, rockland hammocks, upland glades, seepage slopes, cutthroat seeps and springs (Knight et al. 2011). Florida has an extremely diverse estuarine and marine ecosystem; it is the only state in the continental U.S. with an extensive shallow reef system. The mild tropical-maritime climate of the Florida Keys provides habitats for a number of terrestrial and marine plant and animal species found nowhere else. The Florida Everglades has received international recognition and has long been recognized as one of our nation's most imperiled landscapes, included as one of 44 sites globally and one of two sites in the United States on the UNESCO List of World Heritage in Danger (Mitchell and Krueger 2011, Aumen et al. 2015). The Everglades is home to 68 threatened and endangered species (USFWS 1999). The Lake Okeechobee ecosystem is unique

Figure 12.2. Florida scrub jay. Photo by Alex Kropp

within North America, due to its large size, shallowness (the average depth is nine feet), and habitat diversity.

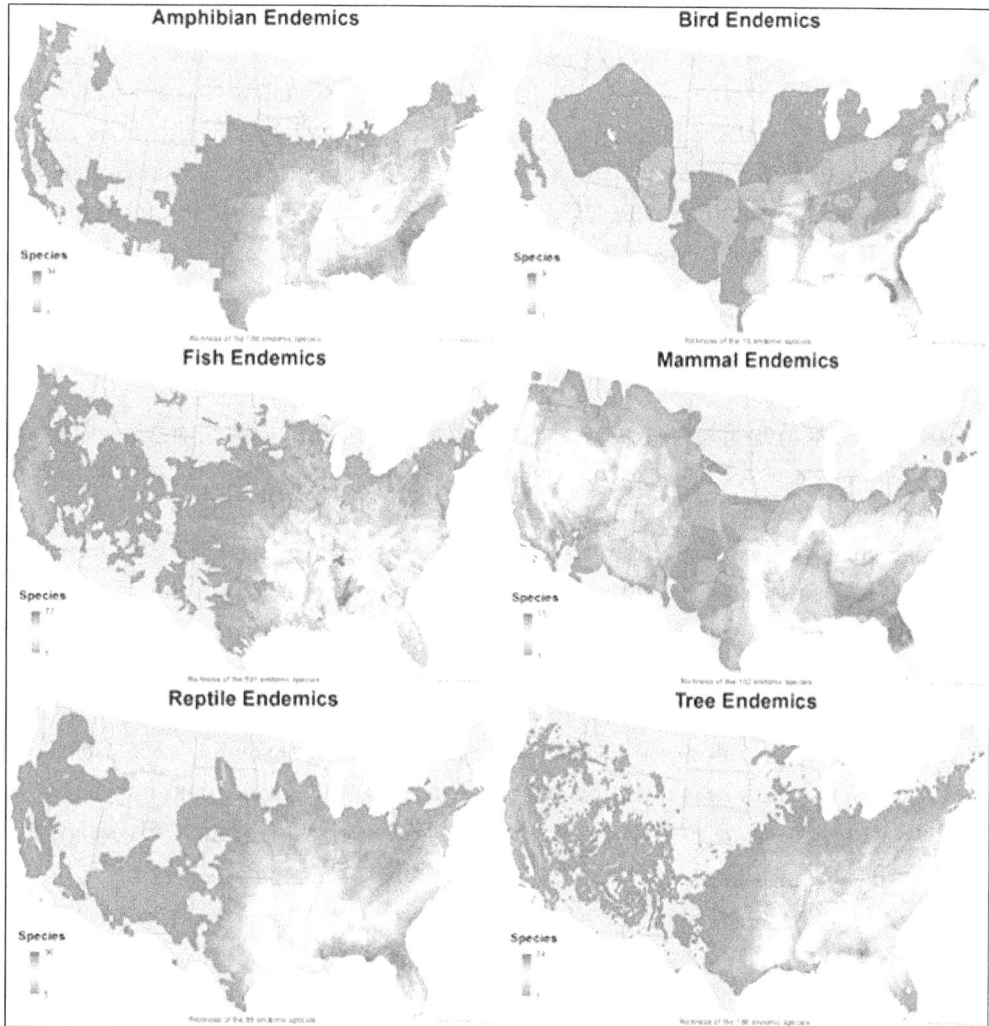

Figure 12.3. Maps of endemic species. Source: Jenkins et al. 2015.

Climate Change Impacts on Biodiversity and Ecology

Climate change is one of the most important determinants of changes in biodiversity (Sala et al. 2000; Schweiger et al. 2008). It will have impacts on biodiversity that operate at the individual, population, community, ecosystem and biome scales (Ackerly et al. 2010, Bellard et al. 2012), altering species distributions, life histories, community composition, and ecosystem function (Graham and Grimm 1990; Gates 1993; Kappelle et al.1999; Hughes 2000; McCarty 2001). The rate of climate change may become the most important feature in terms of consequences for biodiversity, potentially leading to escalating extinctions and widespread reorganizations of ecosystems, particularly where the rate of change is too fast and overwhelms the capacity of current ecosystems to adapt (Steffen et al. 2009). Those species, populations, and communities that cannot keep pace with the rate of change will be most adversely impacted (Thomas et al. 2004; Visser 2008; Ackerly et al. 2010; Bellard et al. 2012). Potential impacts on biodiversity, even under the most modest climate change scenario, will increase through most of this century (Steffen et al. 2009). Distributions of species have already been affected by climate change (Hill et al. 2001; Parmesan and Yohe 2003; Hickling et al. 2006) and it is expected that future climatic changes will have even more severe impacts (Sala et al. 2000; Thuiller et al. 2005; Araujo et al. 2006).

Climate Change Impacts on Ecosystems

The local physical geography (e.g., elevation, soil type, hydrology, climate) largely determines the type and extent of the natural communities. Florida's elevation range is extremely small, ranging from sea level to a high of approximately 107 meters. However, the subtle changes in elevation; in combination with variations in the physical geography; have led to an incredible range of ecosystems within the state. The relationships between characteristics of individual species and the surrounding environment, the role of individual species in the communities and ecosystems, the structure and function of ecosystems, and the phenomena associated with changes at all levels (genetic to biome) are important for dealing with the climate change threat to ecosystems (Steffen et al. 2009). A variety of ways exist to delineate Florida's ecosystems; for this discussion they have been divided into three groups following the divisions used by Myers and Ewel (1990): Coastal Ecosystems, Freshwater Wetlands and Aquatic Ecosystems, and Upland Ecosystems.

Coastal Ecosystems

The community structure of coastal ecosystems is governed by the tolerances of species to environmental conditions, such as light availability, temperature, moisture, disturbance, tides, water depth, salinity, and nutrient availability (Burkett et al. 2008). These systems have the natural ability to adapt to the dynamic conditions that formed and maintains them; however, these capacities are being overwhelmed by sea level rise, particularly in areas that have already been damaged by development, coastal armoring, and other activities (Anderson et al. 2016). Depending on the relative rates of sea level rise and barrier island retreat, the lagoonal area between the barrier island and the mainland may remain constant, expand, or shrink (Michener et al. 1997). Changes in wind circulation patterns and increases in wave actions due to storms will impact the interactions of sand with the pioneer grasses that build dunes. Loss of pioneer grass species and other dune vegetation likely will increase dune erosion and degradation, especially given the predicted increase in storm events. Much of the large swaths of salt marsh in Florida's Big Bend region will likely convert to open water as sea levels rise, but predicted transition of inland habitats to salt marsh will likely offset major changes in salt marsh extent. While this capacity is high in many of the undeveloped regions of the Central/North Florida Gulf Coast, that capacity is severely compromised in the heavily urbanized areas of South Florida and Florida's East Coast.

Estuarine productivity will be impacted by changes in the timing and amount of freshwater, nutrient, and sediment delivery (NFWPCAS 2012). Seagrass supports many ecological processes, including: regulation of water column dissolved oxygen; modification of the physical and chemical environments; reduction of suspended sediments, chlorophyll, and nutrients; stabilization of bottom sediments; and filtration of suspended matter (Nixon and Oviatt 1972; Short and Short 1984; Ward et al. 1984; Stevenson 1988; Koch 1996; Komatsu 1996). Changes in sea level, salinity, temperature, atmospheric carbon dioxide (CO_2), and ultra violet (UV) radiation can affect seagrass. One of the primary effects of increased temperature on seagrass will be the alteration of growth rates and other physiological functions of the plants (Short and Neckles 1999). Sea level rise and associated increases in water depth will decrease light availability and impact seagrass distribution, productivity, and structure.

The increase in sea surface temperatures (SST) associated with climate change is perhaps one of the best-documented impacts to marine ecosystems, especially in tropical coral reefs. The steady increase of global SSTs over the past century (Solomon et al. 2007) may increase susceptibility of corals to disease (Bruno et al. 2007) and often exceeds a critical threshold beyond which 'coral bleaching' occurs (Hoegh-Guldberg 1999). This phenomenon is the result of corals expelling their symbiotic zooxanthellae (algae) leading to a ghostly condition in which the coral turns white. Depending on the severity of the 'bleaching' event, corals may recover to a weakened condition, or may die altogether. Over recent years, modest SST increases have resulted in catastrophic impacts to the world's coral reefs at the global (Brown and Dunne 2016), regional (Wilkinson and Souter 2008), and Florida-wide (Manzello 2015) scales.

Freshwater Wetlands and Aquatic Ecosystems
A significant portion of Florida's landscape is covered by wetlands, ranging from expansive systems (e.g., Everglades, Big Cypress, Paynes Prairie) to isolated features located in a mosaic of upland communities (e.g., ephemeral wetlands, pitcher plant bogs). Regardless of size, wetland systems are expected to be impacted through changes in precipitation, temperature, sea level rise, and the synergisms among these factors. Annual length of soil saturation, amount of organic matter, source of water, and fire frequency all contribute to determining the major characteristics of forested wetlands in Florida. Decreased precipitation coupled with increased temperature will likely alter plant composition—allowing for encroachment of upland woody species and increased fragmentation of larger systems through reduced flow and connectivity.

The Everglades ecosystem forms the interface between temperate and subtropical biomes—creating habitats unique to Florida (Pearlstine et al. 2010). With approximately 40% of the Everglades National Park at elevations below 1 m, a potential sea level rise of more than 1 m combined with predicted temperature increases poses a significant challenge for the future ecological integrity of the park, especially in light of the disruption of the region's natural hydrology over the past century (Mitchell and Krueger 2011; Catano et al. 2014). Decreases in water quantity and quality will continue to stress the system and cause degradation; however, if the region receives more rainfall, the habitat suitability could be enhanced for aquatic prey productivity and apex predators (Catano et al. 2014). Within the Everglades ecosystem and other freshwater marshes, fire is used as a management tool to prevent mangrove and other woody vegetation encroachment into marshes, and to eliminate invasive exotics that frequently occur at the upland–marsh interface (Smith et al. 2013). Climate change effects that reduce the ability to conduct prescribed burns will contribute to shifts within the ecosystem.

The large river systems in northern Florida have the highest diversity of freshwater fish species in the state, with some watersheds having up to 100 species. The highest diversity of aquatic invertebrates is also found in northern Florida due to the higher gradient of rivers and streams, proximity to the continental landmass, and the presence of karst features such as sinkholes and caves (Knight et al. 2011). Warming water temperatures, altered stream flow patterns, and increasing storm events will impact freshwater systems (Poff et al. 2002). Additionally, sea level rise will lead to saltwater inundation of freshwater areas, groundwater contamination, and higher tidal/storm surges. Florida's karst system of sinkholes, submerged caves and springs depend upon the connection between the surface and the underground, with even slight changes in soil moisture, elevation, and temperature causing profound effects.

Upland Ecosystems
Upland ecosystems in Florida range from systems similar to those found in the Southeastern Coastal Plain to systems more commonly found in sub-tropical and Caribbean areas. The species composition of forest systems and their location and ranges are influenced by winter temperatures and other climatic factors, as well as by local factors such as fire, substrate, elevation, and species interactions. Increased temperatures will lead to increases in forest pest damage, changing fire patterns, longer growing seasons, higher evapotranspiration/drought stress, and the spread of non-native species. Crumpacker et al. (2001) found that even moderate increases in temperature (1 °C) may cause serious effects on temperate hardwood forests of northern Florida. Some of the most severe impacts indicated potential shifting of the ecosystem from forested to open woodland, scrub and savanna. Some tree species already at their southern range boundaries are predicted to have range reductions, such as southern red oak (*Quercus falcate*) and American beech (*Fagus grandiflora*) in the panhandle, and range contraction of longleaf pine (*Pinus palustris*), with the southern boundary moving northward up the Florida Peninsula. Loss of key woody species could affect forest suitability for nesting, roosting, or foraging. The majority of

Florida's upland ecosystems are dependent upon fire, with the frequency, intensity, and seasonality of fire varying between communities. The ability to maintain these systems through the use of prescribed fire will become more challenging with increased temperatures and changes in precipitation. Altered patterns of precipitation could lead to changes in the seasonality of prescribed burns, potentially altering the effectiveness of the burn for some species and systems. An increased number and intensity of extreme storm events can cause a build-up of debris leading to increases in wildfires, hotter prescribed fires, and even the inability to use prescribed fire as a management tool. Additionally, drought can alter the decomposition rates of forest floor organic materials, impacting fire regimes and nutrient cycles (Hanson and Weltzin 2000).

Habitats

The degree to which habitat conversion will favor some communities over others and how those conversions differ among areas is a major unknown factor in assessing the vulnerability of natural communities (Noss et al. 2014). Noss et al. (2014) applied the Standardized Index of Vulnerability and Value (SIVVA) framework to 30 natural communities in Florida. This assessment included quantitative model overlays of projections from sea level rise and land use change. On average, these 30 communities will lose 12% of their extent to sea level rise, as projected by high resolution statewide Sea Levels Affecting Marshes Model (SLAMM) overlays of 1 m of sea level rise by 2100. Some natural communities, such as maritime hammocks, coastal interdunal swales, and saltwater marsh will lose nearly 50% of their current extent to sea level alone. Some rare natural communities in extreme South Florida will suffer greater losses when projected changes due to land use conversion are coupled with losses from sea level rise; cactus barrens and tidal rock barrens in the Florida Keys are likely to lose 85% and 75% of their extent, respectively. These direct losses of habitat will have significant impacts on the species dependent upon them.

Coastal Habitats

An increase in storm surge associated with hurricanes could affect the sustainability of some natural coastal systems and the species that depend upon them. Loss of beaches would affect species such as sea turtles, terns, American oystercatcher (*Haematopus palliates*), and black skimmers (*Rynchops niger*), as well as critical habitat for wintering shorebirds and migrating neo-tropical migrants. Some aquatic and terrestrial species limited to coastal areas (e.g., beach mouse, Okaloosa Darter (*Etheostoma okaloosae*)) may be threatened throughout their range (Burkett 2008). Salt marshes are expected to move upslope with sea level rise (Brinson et al. 1995), but human development is likely to limit retreat and migration (Donnelly and Bertness 2001; Feagin et al. 2005; Desantis et al. 2007). The most severe loss will likely occur at sites where the coastline is unable to move inland because of steep topography or seawalls (Galbraith et al. 2002). These conditions can result in the crowding of foraging and beach-nesting birds, as well as loss of crucial coastal habitat for species such as diamondback terrapin (*Malaclemys terrapin rhizophorarum*), which requires both marsh and beach habitats (Shellenbarger Jones et al. 2009). Inundation of coastal habitats will increase fragmentation as patches are divided by areas of open sea water. Sea level rise threatens small and low-lying islands with erosion or inundation (Baker et al. 2006; Church et al. 2006), many of which support high concentrations of rare, threatened, and endemic species (Baker et al. 2006). Of

Figure 12.4. Florida Key Deer. Photo credit: USFWS Digital Library.

40 species identified as being vulnerable to sea level rise, the mangrove diamondback terrapin, Key deer (*Odocoileus virginianus claivum*), Peninsula ribbon snake (*Thamnophis sauritus sackenii*), Lower Keys marsh rabbit (*Sylvilagus palustris heneri*), mangrove cuckoo (*Coccyzus minor*), Florida panther (*Puma concolor coryi*), loggerhead sea turtle (*Caretta caretta*), Florida brown snake (*Storeria dekayi*), and Florida bonneted bat (*Eumops floridanus*) had the highest relative vulnerability ranks when assessed using the SIVVA framework (Reece and Noss 2014).

Mangroves are one of the most productive habitats, providing integral nursery habitats for fish species; shorelines fringed by mangrove prop roots harbor diverse fish assemblages in high densities (Thayer et al. 1987). Many species of birds use the mangrove canopy as nesting sites, including wading birds, mangrove cuckoos (*Coccyzus minor*), and white-crowned pigeons (*Patagioenas leucocephala*). Relatively small changes in winter climate can result in dramatic mangrove range expansion at the expense of salt marsh; salt marsh could be reduced by 60% in Florida with only a 2-4 °C increase in annual mean minimum temperature (Osland et al. 2013). Saltmarsh-dependent bird species such as seaside sparrows (*Ammospiza maritima*) may be forced to leave the area if suitable habitat no longer exists.

Increased soil salinity in coastal uplands will lead to changes in species composition as salt-intolerant plants decline and plants with higher salt tolerances increase. Cabbage palm (Sabal palmetto) mortality on coastal islands and along the marsh/upland transition zone has already impacted coastal areas along the Big Bend region of Florida. Cabbage palm seedling mortality is correlated with tidal flooding, suggesting that salinity, flooding or the combination may be responsible for the regeneration failure of cabbage palms in low-lying coastal areas (Perry and Williams 1996).

Fire-maintained Habitats

As previously discussed, many of Florida's systems are dependent upon fire. Altered fire regimes or the absence of fire, along with other climatic changes, could lead to compositional changes to these habitats, potentially altering their suitability to the current suite of species. The absence of fire in the longleaf pine sandhill community can lead to an increase in woody vegetation, creating a dense mid-story. Species such as the red-cockaded woodpecker (*Picoides borealis*) rely on the openness of the sandhill for foraging. Florida scrub jays depend on fire to keep scrub oak habitats short and maintain plenty of open sandy areas in which to store acorns. Dry prairie provides habitat for multiple distinctive species including the crested caracara (*Polyborus plancus*), burrowing owl (*Athene cunicularia*), the Florida sandhill crane (*Grus Canadensis pratensis*), and the federally endangered Florida grasshopper sparrow (*Ammodramus savannarum floridanus*). Without appropriate fire regimes, trees and other woody vegetation move into dry prairie, creating unsuitable conditions for these and other species. In the absence of periodic fires, broadleaf plant species invade the pine rockland communities that sustain a rich diversity of plants and animals and, if left unchecked, transition to a broadleaf "hammock" (Burg 2010).

Florida Keys and Coral Habitats

Even small changes in water temperatures can have profound effects, especially in the marine environment where coral reefs, seagrasses, and mangroves predominate. In addition to the previously discussed effects of temperature on corals, reduction of ocean carbonate ion concentrations due to ocean acidification impacts their ability to build skeletons (Cooper et al. 2008). The net result of these temperature-induced impacts is the eventual loss of coral structure and a shifting community structure (Ruzicka et al. 2013). Since corals play a pivotal role in supporting biodiversity (Connell 1978), harboring the highest diversity of marine species (Carpenter 2008), impacts to their long-term survival can have devastating effects on reef-associated biodiversity. Most of Florida's sport fish species and many other marine animals spend significant parts of their lives (particularly early development stages) on or around coral reef habitats.

How changes in temperature will impact the other dominant habitat types in the Florida Keys is less well-known (Koch et al. 2015). Sea level rise will alter the landscape of the Florida Keys where elevations, with few exceptions, are between 3 and 6 ft (1–1.9 m). The Florida Keys contain approximately 75% of the state's rockland hammocks, which provide habitat for many endemic species, including 10 mammals and five reptiles (Snyder et al. 1990). Adjacent freshwater wetlands provide breeding habitat for amphibians and sources of prey for reptiles. These wetlands, as well as other important sources of fresh and brackish water, are expected to become more saline with rising sea levels and increased tidal/storm surges. In addition to these impacts, altered soil salinities will alter plant composition of the terrestrial habitats. There have already been adaptation efforts to "buy time" for the Key tree-cactus (*Pilosocereus robinii*). A project in 2015-16 relocated Key tree cactuses to higher elevations due to the species limited tolerance for saline soils (S. Traxler, USFWS, Pers. Commun.).

Freshwater Wetland and Aquatic Habitats

Herbaceous wetlands provide the foraging and nesting habitat for many species, including waterfowl, Florida sandhill crane (*Grus canadensis pratensis*), snail kite (*Rostrhamus sociabilis plumbeus*), limpkin (*Aramus guarauna),* mink (*Mustella vison*), river otter (*Lutra Canadensis lataxina*), Florida gopher frog (Rana capito), tiger salamander (*Ambystoma tigrinum*), and flatwoods salamanders (*Ambystoma cingulatum*). These wetland-dependent species will be impacted through loss and degradation of habitat when water levels and the timing of water inputs become incompatible with their foraging, nesting, or roosting requirements. Ephemeral wetland-dependent species will be affected by changes in precipitation, regardless of direction of change. Due to the typical shallow structure of ephemeral wetlands, they will be more susceptible to increased evapotranspiration rates, leading to a shorter wet period. This could lead to interrupted or terminated life stage development, as well as the replacement of herbaceous species by woody species. Increased precipitation could permanently connect these isolated wetlands to other water bodies, introducing predators. Palis (1997) found that the timing of salamanders' breeding migration is tied to precipitation and temperature, both of which could be impacted by climate change. Wading birds' nesting success is tied to appropriate nesting and foraging habitats and their proximity to one another. Nesting success is reduced when nesting sites become dry, allowing terrestrial predators easier access to the nests, and when foraging sites are located at distances too far away, beyond their physiological ability to survive and rear their young.

Figure 12.5. Flatwoods salamander. Photo by Pierson Hill.

The suitability of riverine habitats is based on variations in flow, substrate, temperature, dissolved oxygen, and other water chemistry factors. The riverine systems in the Florida Panhandle are unique in that they provide habitat for many rare fish species that are at their various range limits, either at the eastern range limit of the Mississippi River Valley system or the western range limit of the Atlantic Coastal Plain system (Bailey et al. 1954). Northwest Florida also contains 17 of the 27 first magnitude artesian springs and spring groups (Rosenau et al. 1977). Florida's fish species may be impacted by increased water temperatures, with projected decreases in precipitation and increases in temperature. Even if higher water temperatures don't cause direct mortality, they can increase the stress on the fish, leading to declines in health and increases in vulnerability to parasites and disease. Many aquatic species will be affected by bank erosion, increased siltation, and run-off caused by increased precipitation and storm events. Sea level rise will result in the inland movement of seawater, shifting the tidal influence zone of streams and rivers upstream and permanently inundating downstream riparian/coastal habitats with brackish water. Tidal and storm surges can degrade aquatic habitats through oxygen depletion, changes in salinity, and increased siltation and turbidity.

Climate Change Impacts on Species

A recent vulnerability assessment of 300 species in Florida presents some opportunities for generalizing the unique and synergistic threats among a variety of taxonomic groups from across the state (Reece et al. 2013a). As predicted by Pilkey and Young (2009), sea level rise is a major threat to many of Florida's rare and endangered species. Nearly one quarter of the approximately 1,200 species tracked by the Florida Natural Areas Inventory are projected to have at least 50% of their range lost to a sea level rise of 1 m by the year 2100. The greatest threat to species is anthropogenic habitat fragmentation (Benscoter et al. 2013), but synergisms with threats from climate change are especially dangerous for many species. Lessons learned from this assessment (Reece et al. 2013a) include: there is good data demonstrating species' risk of extinction, but insufficient data to make meaningful conservation interventions; the adaptive capacity of many species is compromised by human alterations to the landscape; and planning horizons for climate change vulnerabilities are extremely important. Reece et al. (2013a) documented a complete lack of published records or models of predicted responses to climate change or sea level rise in 88% of 300 species surveyed. Across all 300 species assessed, 30% were scored as having strong anthropogenic geographic barriers limiting their ability to shift distributions in response to threats. Sea turtles are a good example of a species with the potential capacity to shift their nesting location away from areas with inadequate incubating environment (e.g., eroded); however, shifts in nesting location may result in exposure to other threats and the use of areas that are not protected (Reece et al. 2013b). For many species, protecting existing habitats from land use change is highly likely to prevent extinction over the next 50 years; however, for nearly 25% of species assessed, that same strategy is unlikely to prevent extinction over 100-year timescales.

Climate change impacts on species will be driven by one or multiple climate-related factors acting in concert or synergistically (NFWPCAS 2012). Impacts of climate change on species can lead to changes in geographic range, species composition, risk of extinction, and species interactions (predator/prey, competition). Species with poor dispersal ability, long generation times, long time to sexual maturity, low reproductive rates, low genetic variability, narrow environmental tolerances, specialized requirements or relationships with other species, specialized habitat and/or microhabitat requirements, a narrow geographic range, or a dependence on specific triggers or cues likely to be disrupted by climate change will be the most vulnerable to climate change (Foden et al. 2008;, Steffen et al. 2009; NFWPCAS 2012). Many generalist species; such as white-tailed deer *(Odocoileus virginianus)* or feral hogs (*Sus scrofa*); are likely to continue to thrive in a changing climate (Johnston and Schmitz 2003; Campbell and Long 2009). Species, both native and exotic, with traits that assist in invading or colonizing disturbed areas will have an advantage in a rapidly changing climate (Steffen et al. 2009). Mechanisms of species' adaptation include shifting their climatic niche by adjusting their range, phenology, and physiology (Bellard 2012).

Changes in Geographic Ranges
Species distributions are influenced through species-specific temperature and precipitation thresholds and, as these thresholds are crossed, species will need to change their movement patterns, shift their ranges, or disperse further distances to reach suitable habitat as they are forced to move away from unsuitable habitat conditions (NFWPCAS 2012). Some species will be unable to relocate due to lack of suitable habitat or anthropogenic barriers obstructing their movement, Noss et al. (2014) found that 76% of 236 species threatened by sea level rise would be unable to relocate further inland. While climatic changes will lead to contraction of the range of some species, these same changes could lead to the range expansion of other species, particularly non-native invasive species (as discussed later in this chapter).

Migratory species are likely to be strongly affected by climate change. Migratory species may be impacted at multiple geographic scales, possibly experiencing alterations of habitat in their wintering grounds, breeding grounds, and along their migratory routes (Ahola et al. 2004). Mechanisms that aid in migrations, such as wind and water currents, may have positive or negative consequences depending on whether changes increase or decrease required energy expenditures to complete their migration. Altered directions of winds/currents can impact species' ability to navigate to the desired location, even delivering individuals to the wrong location. However, the ability to move and utilize multiple habitats and resources may make some migratory species relatively less vulnerable.

Due to their vulnerability to reductions in water flows and water quality, and their limited capacity to migrate to new waterways, climate change may have a strong influence on fish species distributions and abundance (Brander et al. 2003; Reid 2003). Fish species composition may change as species with lower tolerances move or suffer the impacts of rising water temperatures, as previously discussed. Some aquatic species may be able to expand their range due to increasing winter temperatures. The distribution of coastal species is closely linked with soil and water salinity (Burkett 2008). Additionally, changes in freshwater flow inputs into the estuaries may affect the distribution of suspension feeders, such as mussels, clams, and oysters (Wildish and Kristmanson 1997). There are a multitude of factors that can individually and synergistically impact marine species distributions, including changes in sea level, ocean stratification, oxygen availability, patterns of ocean circulations, storms, precipitation and freshwater input, and ocean physical and chemical conditions (NFWPCAS 2012).

Changes in Species Composition

In response to climate change, many native and non-native species may increase in abundance to such an extent that they have a transformative, and often negative, impact on other species and ecosystems (Steffen et al. 2009). As species respond to changing habitat conditions, shifts in composition are likely to alter important competitive and predator–prey relationships, which can reduce local or regional biodiversity (Parmesan and Galbraith 2004). Factors in aquatic systems, such as changes in thermal regimes, flow regimes, or salinity could alter the competitive interactions or predator–prey relations among species in ways that are detrimental to species of conservation concern (Rahel et al. 2008). The structure and function of coastal systems may change as species with a greater tolerance of increased salinity outcompete those with lower tolerance; these changes in community structure can be episodic, potentially leading to elimination of some ecosystems if thresholds are exceeded (Burkett et al. 2005).

In marine systems, climate-induced changes in community composition and food web structure resulting from the shifts in ecological niches for individual species are likely to be significant (Harley et al. 2006). Changes in temperature may influence key species interactions through which small changes in climate could generate large changes in natural communities, such as a decrease in key predator populations. Seasonal changes in freshwater inflow will be a contributing factor that may induce changes in species composition of mangrove fishes along estuarine gradients (Ley 1999).

Risk of Extinction

A review of various models predicting future biodiversity found that the majority of the models indicated significant consequences for biodiversity, with the worst case scenario models leading to extinction rates that would qualify as the Earth's sixth mass extinction (Bellard et al. 2012). Many of the species with the highest vulnerability or predicted extinction rates are species with limited or isolated populations. However, endangered species with large home range sizes and greater dispersal limitations are also associated with greater risk of extinction, possibly indicative of higher resource requirements or lower habitat quality (Benscoter et al. 2013). Both characteristics may affect their ability to adapt to rapid environmental change. Those species with low adaptive capacity will have low likelihood of finding distant habitats to colonize, ultimately resulting in increased extinction rates (Walther et al. 2002). Additionally, species with narrow geographic ranges and specific habitat requirements will be at even greater risk due to interactions of climate change and existing and future habitat fragmentation (NFWPCAS 2012).

Sea level rise will likely have a significant negative effect on species persistence, impacting the size and quality of habitat patches for coastal species through changes in the coastline and transitions among coastal habitats. Sea level rise will cause a decline in suitable habitat and carrying capacity for the Snowy Plover (*Charadrius alexandrinus*) and increase its risk of extinction (Aiello-Lammens et al. 2011). A recent vulnerability assessment (Reece et al. 2013a), evaluating sea level rise of up to 2 m and synergistic effects of climate change and anthropogenic factors identified several species as highly likely to be extinct by 2100, including the Florida grasshopper sparrow, Miami blue butterfly (*Cyclargus thomasi bethunebakeri*), Florida duskywing (*Ephyriades brunnea floridensis*), Gulf Coast solitary bee (*Hesperapis oraria*), Key deer, Florida Keys tree snail (*Orthalicus reses nesodryas*), Key tree cactus, Bartram's scrub-hairstrak (*Strymon acis bartrami*), Lower Keys marsh rabbit, and Key ringneck snake (*Diadophis punctatus acricus*). The primary threats for these species were identified as sea level rise, barriers to dispersal, storm surge, lack of freshwater, habitat loss, invasive species, disease, collection, small range size, and habitat degradation.

Keystone Species

A keystone species is a species that has a significant effect on its environment relative to its abundance and plays a critical role in maintaining the structure of an ecological community, affecting a suite of other species in an ecosystem. Examples of keystone species in Florida include the gopher tortoise, reef building species and the American alligator (*Alligator mississippiensis*). The gopher tortoise is considered a keystone species for the sandhill

Figure 12.6. Gopher tortoise. Photo by Jay Exum.

community in that it "engineers" the habitat of many other species. Species that have been reported using gopher tortoise burrows include at least 36 amphibians and reptiles, 19 mammals, seven birds, and more than 300 species of invertebrates (Jackson and Milstrey 1989; Diemer 1992; Brandt et al. 1993; Kent and Snell 1994). Climatic changes that impact gopher tortoise abundance or survival, such as alterations to fire regimes, will impact a large suite of associated species.

The species that create worm reefs and coral reefs are also considered to be keystone species. Coral reef systems composed of species such as *Oculina* provide habitat for many recreational and commercially important species, such as scallop, shrimp, grouper, snapper, and amberjack. The sedentary polychaete worms (*Sabellaria vulgaris* Verrill) build tubes from sand and shells, forming colonies that attract fish, birds, and algae. Changes in circulation patterns, wave actions, SSTs, and ocean acidification may impact the coral and worm reef species and, in turn, the species that depend upon their structure as habitat.

The American alligator is considered to be a keystone species of the Everglades ecosystem and wetlands systems, creating important habitat for other species and aiding in ecological processes. The deep holes that they create in the wetland systems retain water during the dry season, providing habitat for a variety of other species. The nesting activity of the females is important for the creation of peat, as well as providing a nesting substrate for several species of turtle. Although alligators seem to be quite resilient to habitat alterations, if climate changes, particularly those changes impacting hydrological processes can cause changes in the alligator's range or nesting; multiple other species would also be impacted.

Species Highlight – Sea Turtles and Climate Change

Figure 12.7. Green Sea Turtle. FWC Digital Library.

Although sea turtles have been in existence for millions of years and have adapted to past changes in climate, it is still unsure whether they will be able to adapt to present day climate change; it is occurring more rapidly than has been observed in the past and it is accompanied by a variety of anthropogenic threats (Poloczanska et al. 2009; Fuentes et al. 2013). The cumulative impact of the various non-climatic threats sea turtles face makes their populations more vulnerable to climate threats and decreases their resilience (Fuentes et al. 2013). Sea turtles, being ectotherms, have specific thermal requirements, with their distribution often constrained by the 15 °C isotherm (Hamann et al. 2012). Therefore, shifts in ocean temperatures will likely result in distribution shifts (Weishampel et al. 2004). Indeed many populations of sea turtles worldwide have started to shift their range as a response to alterations in temperature (Witt et al. 2010). For instance, Kemps Ridley turtles (Lepidochelys kempii) have always nested on a 1000 km stretch of beach in Mexico. But over recent years, these turtles have expanded their nesting range to various beaches along the Gulf Coast of Florida, potentially as a result of changes in temperature (Pike 2013). Species distribution models for these turtles predict further expansion within the Gulf of Mexico and even along the Northern Atlantic Ocean (Pike 2013). Shifts may also occur at a more regional scale. There has been a northward shift in loggerhead sea turtle (Caretta caretta) nests along Melbourne Beach, Florida, the largest loggerhead turtle rookery in the Atlantic Ocean, likely due to warming temperatures (Reece et al. 2013b). Range shifts may be accompanied by increased exposure to other threats or, more optimistically, to areas where fewer threats exist. Availability of suitable habitat will be crucial for sea turtle adaptation in the future; however, several models predict that changes in climate and sea level rise may reduce the availability of suitable sea turtle nesting areas and the locations of where turtles nest (Fuentes et al. 2010; Katselidis et al. 2014; Pike et al. 2015). For example, it is projected that Melbourne Beach will decrease in area by 43% from 1986 to a future with 0.5 m of sea level rise; this will restrict nesting to narrow beaches, increasing risk of erosion and crowding resulting in nests overlapping with each other (Reece et al. 2013b). As temperatures and sea level rise continue to increase, protecting future suitable habitat for nesting sea turtles will greatly increase their chances of adapting to both climate change and anthropogenic impacts. However, the heavily developed coasts in the United States may hinder adaptation (Pike 2013; Fuentes et al. 2016).

Sea turtles play important ecological roles in both oceanic and terrestrial habitats (Hawkes et al. 2009). They help maintain sea grass meadows and coral reefs by grazing on sea grass plots and sponges, respectively (Bjorndal 1980), they provide transportation for epibionts, and their egg clutches and dead hatchlings provide nutrients to beach and dune vegetation (Bouchard and Bjorndal 2000; Hannan et al. 2007). Sea turtles also have important cultural, social, and economic significance (Campbell 2002; Campbell and Smith 2006). In Florida for example, residents are willing to pay $42–$57 per year for five years to protect sea turtle habitats from sea level rise (Hamed et al. 2016). As emblematic species, sea turtles can help promote awareness to the threats that climate change poses to marine species (Hamann et al. 2012; Fuentes and Saba 2016).

A sea turtle's life history, behavior, and physiology are strongly influenced by environmental temperature, which makes sea turtles particularly vulnerable to environmental changes (Fuentes et al. 2011; Hamann et al. 2012; Dudley and Porter 2014). Besides the thermal limitations that sea turtles face in their oceanic habitat, sea turtles have other strict thermal thresholds on land as well. Embryo development, hatchling sex ratio, and hatching success are all influenced by temperature and rainfall at nesting beaches (Standora and Spotila 1985; Janzen 1994; Wyneken and Lolavar 2015). Successful egg incubation typically occurs when sand temperatures are between 25 and 34 °C, with variability (Miller 1985; Howard et al. 2014) between different species; for instance, loggerheads, flatbacks (Natator depressus), hawksbills (Eretmochelys imbricata), and greens (Chelonia mydas) have been shown to tolerate nest temperatures as high as 35 °C (Howard et al. 2014). Incubation outside this range results in lower hatching success and higher morphological abnormalities in hatchlings (Miller 1985). Sea turtles also have temperature-dependent sex determination, where the sex of the hatchlings is determined by the nest temperature (Mrosovsky 1980; Yntema and Mrosovsky 1980). Temperatures above the pivotal temperature, where the result is a 1:1 sex ratio, produces more females, while temperatures below the pivotal temperature produces more males (Yntema and Mrosovsky 1980). For example, in Florida, similar to other nesting grounds worldwide, there is evidence of a bias in the production of female hatchlings (Mrosovsky and Provancha 1989; Hanson et al. 1998; Wibbels et al. 1991; Blanvillain 2007; Wibbels 2012a, b, c). Knowledge of the primary sex ratio of nestlings on nesting grounds is crucial to accurately understand the projected impacts of climate change on the reproductive output of sea turtles (Fuller et al. 2013; Marcovaldi et al. 2016). Although some knowledge does currently exist on the general sex ratio of hatchlings on Florida beaches, a systematic long-term monitoring program is still necessary to obtain data at the appropriate temporal and spatial scale.

Changes in temperature will also impact the phenology of sea turtles, including the frequency and timing of nesting (Limpus and Nicholls 1988; Saba et al. 2007; Fuentes and Saba 2016). Some populations, such as the leatherbacks (Dermochelys coriaceain) located in Costa Rica and the US Virgin Islands, have shown a delay in nesting due to increased temperatures in their foraging areas (Neeman et al. 2015). In comparison, other populations, such as the loggerheads along Florida's Atlantic Coast, have started to nest earlier due to warmer temperatures prior to the typical start of the nesting season (Weishampel et al. 2004; Pike et al. 2006). The nesting season may differ between species and populations worldwide such that earlier nesting may result in a shorter or, more optimistically, a longer nesting season. For example, SSTs resulted in loggerheads experiencing an earlier and shorter nesting season on Florida's Canaveral National Seashore, whereas loggerheads on Cape San Blas, Florida experienced a longer season as a result of earlier nesting, which is similar to more northerly rookeries (Wieshampel et al. 2004; Lamont and Fujisaki 2014). Shorter nesting seasons may cause females to lay fewer clutches in a season (Pike et al. 2006). Extended seasons may allow more individual females to nest within the season (Lamont and Fujisaki 2014); however, there is still uncertainty about the implications of nesting season lengths for sea turtle population stability.

Phenology/Physiology

Phenology is the timing of seasonal activities of an organism, which is typically highly adapted to the climatic seasonality of the environment in which it evolved (IPCC 2007). Species can cope with climate change through phenotypic plasticity and microevolution, in addition to shifting their range (Hulin et al. 2009; McGuire et al. 2016). Phenotypic plasticity is the ability of an organism to change its characteristics or traits, including morphological, physiological, and behavioral. Species with phenotypic plasticity can quickly compensate for a moderate change in environmental conditions (Jump and Penuelas 2005). Microevolution is the changes in the gene pool of a population over time that result in relatively small changes to the organism.

Microevolution can be observed over short periods of time, even between one generation and the next, and it can occur via mutation, gene flow, genetic drift, or natural selection.

The annual phenology of many species has changed in the past few years in response to modified environmental conditions (Walther et al. 2002; Parmesan 2006; Pertoldi and Bach 2007). Spring activities, such as breeding or first singing of birds, arrival of migrant birds, appearance of butterflies, choruses and spawning in amphibians, and shooting and flowering of plants, have been occurring progressively earlier since the 1960s (Walther et al. 2002).

Although migratory species are adapted to adjust their behavior with annual changes in the weather, shifts in climatic variables are beginning to result in mistimed migration (Robinson et al. 2009), with some species abandoning migration altogether and others shifting their migratory pattern (Foden et al. 2008). Changes in cues (e.g., temperatures, precipitation) for migration initiation or pathways could lead to mismatched availability of resources required for successful completion of migration or reproductive success and survival upon arriving at the spring or winter destination.

Synchrony of phenological changes in species that interact with one another, such as competitors, food species, and pollinators, will be extremely important to many species. If these timing shifts are synchronous across species that normally interact then the system is likely to remain healthy; however, if responses to change (e.g., temperature increases) vary across species then species' interactions may become out of synchrony and could lead to population declines (Parmesan and Galbraith 2004). For example, if the arrival of a migrating bird to its breeding ground and the insect it depends on for food both occur two weeks earlier, they remain in synchrony and may persist; however, if the bird arrives before the insect's hatch/emergence they become out of synchrony and the bird may experience population declines. Schweiger et al. (2008) found a pronounced spatial mismatch in future niche spaces of a butterfly and its larval host plant under three global change scenarios, suggesting that climate change has the potential to disrupt trophic interactions because co-occurring species do not necessarily react in a similar manner to global change.

There is particular interest in the effects of climate change on the population dynamics of species with temperature-dependent sex determination (Walther et al. 2002). All crocodilians (Deeming and Ferguson 1989; Lang and Andrews 1994), many turtles (Ewert et al. 1994), and several lizards (Viets et al. 1994) have temperature-dependent sex determination. Two parameters—the pivotal temperature and the transitional range of temperature—control sex determination; species with a larger transitional range of temperature are expected to be at a lower risk to climate change (Hulin et al. 2009). Ewert et al. (2005) determined that the sex ratio of the American snapping turtle (*Chelydra serpentine*) is female-biased at cool temperatures, male-biased at moderate temperatures, and only females are produced at warm temperatures. Climate change-induced shifts in thermal regimes of incubation may lead to a bias in sex ratios in populations of temperature-dependent sex determination species (Janzen 1994, Walther et al. 2002).

Plant phenology studies have generally shown earlier onset of leafing out, flowering, and fruiting as temperatures have increased (Menzel et al. 2006). As yet, few plant phenology studies have been published for Florida. Von Holle et al. (2010) used herbarium specimens and long-term climate data to assess whether the phenologies of 70 plant species varied with a changing climate; these species included 29 invasive species and 41 native species related closely to each of the invasive species. Only three species sampled were found to have flowering times that differed significantly with climate changes: two flowered later (*Albizia lebbeck* and *Sassafras albidum*) and one flowered earlier in the year (*Morus rubra*). Von Holle et al. (2010) did not find a difference in phenological response between invasive and native species. Both exhibited a trend of delayed flowering in years where minimum temperatures fluctuated.

Plant Physiology – Case Study

Plants grown under elevated concentrations of CO_2 use resources more efficiently than plants growing at ambient CO_2 (Drake et al. 1997). Photosynthesis is often stimulated while stomatal conductance and leaf nitrogen are reduced resulting in greater water-use and nitrogen-use efficiency (Drake et al. 1997; Ainsworth and Long 2005). Growth and biomass production are also often stimulated by CO_2 (Ainsworth and Long 2005). An open top chamber study at the Kennedy Space Center evaluated the impacts of elevated CO_2 (ambient +350 µmol mol-1 CO_2) in Florida scrub over an 11-year period (Hungate et al. 2013). This is the only long-term study of the effects of CO_2 on native Florida vegetation to date. Exposure to elevated CO_2 stimulated aboveground biomass accumulation in Florida scrub over the duration of the study. The biomass stimulation response was species specific: elevated CO_2 stimulated Myrtle oak (Quercus myrtifolia) and Chapman Oak (Quercus chapmanii) but had no impact on the aboveground biomass of sand live oak (Quercus geminata) (Seiler et al. 2009; Dijkstra et al. 2002). Elevated CO_2 stimulated fine root biomass following disturbance; but the effect was temporary (Day et al. 2013). Net primary production (aboveground and below ground) was stimulated by elevated CO_2 following disturbance, peaking with high availability of soil nutrients (Hungate et al. 2013). Belowground biomass was the main driver of the net primary production response. Species-specific net primary production responses were the same as for biomass; productivity of Q. geminata did not respond to elevated CO_2 (Hungate et al. 2013).

Photosynthesis was stimulated for the scrub oaks and stomatal conductance reduced with growth in elevated CO_2 (Lodge et al. 2001; Ainsworth et al. 2002; Li et al. 2003). Q. geminata was the only oak that showed consistent evidence of photosynthetic acclimation to elevated CO_2 (Ainsworth et al. 2002; Hymus et al. 2002). Q. geminata and Q. myrtifolia grown in elevated CO_2 were more nitrogen use efficient than plants grown under ambient conditions (Ainsworth et al. 2002).

Plants grown under elevated CO_2 had decreased leaf foliar nitrogen concentrations and increased C:N ratios; there was less damage from herbivores on these lower quality leaves (Stiling et al. 2003, Hall et al. 2005). Few legacy effects of elevated CO_2 were found to persist one year after exposure to elevated CO_2 concentrations was terminated; no differences remained in leaf nitrogen concentration or in herbivore densities (Stiling et al. 2013).

Long-term stomatal adaptation to increased atmospheric CO_2 may occur, which decreases water loss while maximizing carbon gain (Drake et al. 1997). There was no evidence of stomatal adaptation for oaks in the open top chamber experiments at the Kennedy Space Center: stomatal densities were similar between ambient and elevated treatments (Lodge et al. 2001). Changes in stomatal density and dimensions with increasing atmospheric CO_2 have been documented for several common Florida species by studying specimens preserved in peat and herbaria (Wagner et al. 2005; Wagner et al. 2007; Lammertsma et al. 2011). Decreases in the stomatal index of five species—water oak (Q. nigra), red maple (Acer rubrum), wax myrtle (Myrica cerifera), dahoon holly (Ilex cassine), and royal fern (Osmunda regalis)— occurred *as atmospheric CO_2 increased from 310 to 370 ppm over 60 years (Wagner et al. 2005). Lammertsma et al. (2011) identified changes in stomatal density and pore size in nine common Florida species (red maple, wax myrtle, dahoon holly, laurel oak, water oak, slash pine, longleaf pine, bald cypress, royal fern), which led to an average 34% decrease in maximum stomatal conductance per 100 ppm rise in CO_2.*

The latest IPCC assessment (2014) reported that there is high confidence (much evidence, medium agreement) that climate change-induced phenological shifts will continue to alter the interactions between species in regions with a marked seasonal cycle. Phenological changes may be the simplest process to track ecological changes of species in response to climate change (Walther et al 2002).

Existing Stressors and Climate Change

The biodiversity and ecology of Florida are already suffering from a number of existing stressors, including habitat loss, fragmentation and degradation, invasive plants and animals, altered hydrologic regimes, overexploitation, and pathogens, parasites and pollutants. In a study conducted by Wilcove et al. (1998), habitat degradation was identified as the top threat, contributing to the endangerment of 85% of the listed species analyzed; competition with or predation by alien species was the second-ranked threat, with the exception of aquatic vertebrate and invertebrate species, where pollution (including siltation) was the second-ranked threat. The ability of species and systems to adapt to climate change will be further challenged when considering the effects of these other stressors (Parmesan and Galbraith 2004). It is expected that the overall vulnerability of some ecosystems may be primarily driven by the severity of these non-climate stressors and by how they interact with climate change (NFWPCAS 2012). The synergistic effects of climate and non-climate stressors, leading to range reductions and population declines, may be severe enough to threaten some species with extinction or extirpation from significant portions of their ranges (NFWPCAS 2012). Parmesan and Galbraith (2004) found that there is a growing consensus that climate change will compound existing threats and lead to an increased rate of biodiversity loss Three key drivers of biodiversity loss include existing threats, direct effects of climate change, and the interaction between the existing threats and climate change (Driscoll et al. 2012). As described in other chapters, climate change is expected to vary regionally across Florida. This will make it even more challenging to predict how the interactions of climate change with other stressors will affect species and population responses (Noss 2011). The reduction of existing stressors is a key strategy in natural resource adaptation planning in response to climate change.

Habitat Loss, Fragmentation and Degradation

Habitat loss, destruction, and degradation are the most pervasive threats to biodiversity (Wilcove et al. 1998), with anthropogenic habitat fragmentation the greatest threat to species (Benscoter et al. 2013). Habitat loss has been identified as the most significant challenge Florida's biodiversity has faced over the past century (Knight et al. 2011). These threats are expected to continue as human populations are predicted to continue to increase and lead to additional land use changes. Fragmented habitats and human land uses will hinder movement of species, further reducing their ability to shift their distributions in response to climate change (Lawler et al. 2009; Marini et al. 2009; McGuire et al. 2016).

Figure 12.8. Habitat fragmentation. USFWS National Digital Library.

Shifting patterns of human habitation, either into new locations to accommodate new residents or away from existing locations as areas, particularly along the coast, become uninhabitable, will lead to loss and degradation of habitats and ecological processes. Additionally, as people withdraw from coastal areas impacted by sea level rise, pollution from abandoned infrastructure, such as septic tanks and underground gasoline tanks, will be a major obstacle to the maintenance of communities in terms of ecological structure and function (FWC 2012). The ability of plant and animal species to retreat in response to rising waters (both sea level rise and flood events) will be affected by barriers preventing their retreat, including human-made structures, such as buildings, bulkheads, roadways, and other obstructions. Additionally, manmade ecosystem alterations, either those already existing or those put in place in response to effects from climate change, may lead to increased habitat loss, degradation, and fragmentation. For example, the use of hardened shoreline stabilization measures coupled with more intense storms could lead barrier islands (and their habitats) to fragment and disappear. As previously mentioned, climate change is expected to impact the use of prescribed fire, an important tool for the management of Florida's pyrogenic communities. Encroachment of development into and adjacent to natural systems will further reduce the ability to use fire as a management tool.

Transportation and associated infrastructure affects the structure, function, and composition of ecosystems, causing cumulative ecological effects on landscapes, with fragmentation of the landscape being the most obvious impact since roads bisect large patches of a contiguous land cover (Coffin 2007). Trombulak and Frissell (2000) developed a framework for assessing ecological effects of roads, categorizing the impacts into seven general ways that roads affect terrestrial and aquatic ecosystems. The framework included: increased mortality during road construction, increased mortality from collision with vehicles, modification of animal behavior, alteration of the physical environment, alteration of the chemical environment, spread of exotic species, and increased alteration and use of habitats by humans (increased accessibility).

Collision with vehicles has been documented to be a significant cause of mortality for several Florida species, including the Florida black bear (Neal et al. 2003) and the Florida panther (Kautz et al. 2006, Coffin 2007), and as a limiting factor in the recovery of the endangered American crocodile in Florida (Kushlan 1988). In the case of the Florida panther, the incidence of roadkill mortality has been reduced with the installation of underpasses along Interstate 75 and State Road 29 (Kautz et al. 2006). The interactions of climate change and human population growth will increase the impacts of roads and associated infrastructure on fish and wildlife species, ecosystems, and ecological processes. The effects of roads as barriers altering natural hydrology will be exacerbated by changes in the amount of precipitation and large storm events. If precipitation patterns shift to fewer rainfall events but with larger amounts of rainfall, existing transportation infrastructure, such as bridges and culverts, may not be sufficient to accommodate the increased flow. Additionally, if the number and duration of flood events increase, the number of roadkill mortality events may increase. In low-lying areas that are frequently flooded, roads built on raised beds/berms often serve as travel corridors for wildlife, with the road shoulders frequently used as the only dry foraging sites for species such as deer. During increased storm events and floods, wildlife using the roads to escape flooded habitats will be more exposed to collision with vehicles.

In coral reef systems, local anthropogenic impacts can reduce the resilience of corals to withstand threats (including climate change), resulting in a deterioration of reef structure and the ability to sustain their complex ecological interactions (Hodgson 1999; Knowlton 2001; Gardner et al. 2003; Hughes et al. 2003; Pandolfi et al. 2003; Wilkinson 2004; Bruno and Selig 2007; Hoegh-Guldberg et al. 2007).

Invasive Species

Dispersal of species to regions beyond their normal range of dispersal (i.e. introduced or alien organisms) has been a major force shaping biodiversity (Wilson et al. 2009). Climate change is expected to exacerbate impacts from non-native invasive species (Dukes and Mooney 1999; Mooney and Hobbs 2000; McNeely et al. 2001) by facilitating the introduction of invasive species (Rahel et al. 2008) and by increasing the invasiveness (rate of spread, competitiveness) of species (Clout and Williams 2009). Biological invasion is recognized as a significant threat to biodiversity, with climate and land use changes leading to drastic range shifts of invasive species (Bellard et al. 2013). Invasive species can have significant impacts on fundamental biological processes and these will most likely increase as climate change affects the distribution, spread, abundance, and impact of the invasive species (Gritti et al. 2006). Invasive species affect native populations via competition, predation, and disease (Rahel et al. 2008), as well as by alterations of habitat structure and the food web dynamics, such as replacing natives that serve as a food source (e.g., plants providing fruits, seeds, nectar, pollen). Climate change has already enabled range expansion of some invasive species and will likely create welcoming conditions for new invaders (Dukes and Mooney 1999; NFWPCAS 2012). As noted by Hellman et al. (2008), there are five possible consequences of climate change for invasive species: (1) altered mechanisms of transport and introduction, (2) altered climatic constraints on invasive species, (3) altered distribution of existing invasive species, (4) altered impact of existing invasive species, and (5) altered effectiveness of management strategies for invasive species.

Invasions of new species are assisted by land use changes, the alteration of nutrient cycles, and climate change (Vitousek et al. 1996; Dukes and Mooney 1999; Mooney and Hobbs 2000; Hellman et al. 2008). Climate changes, including extreme climatic events (i.e., storms, floods), can enhance invasion processes from initial introduction through establishment and spread (Dukes and Mooney 1999; Walther et al. 2009; Diez et al. 2012). Human-aided transport of invasive species occurs through purposeful introductions for a variety of reasons (e.g., biocontrol, sport fishing, horticulture, agriculture, aquaculture) and through accidental introductions during the course of other economic activities. Climate change could alter these patterns of human transport (Hellman et al. 2008). Changes in the amount and timing of precipitation can alter the pathways of species introductions as new or increased flow routes transport invasive species, including animals, plants and plant propagules. Changes in precipitation may also allow for additional areas to be invaded by existing species, such as the Brazilian pepper (*Schinus terebinthifolius*), where it is restricted by inundation periods longer than three to six months (Ferriter 1997).

Figure 12.9. Brazilian pepper. USFWS National Digital Library.

Climate change can lead to the establishment of new invasive species via three mechanisms: removal/reduction of climatic constraints, tolerance of climate leading to persistent populations, and increased competitive ability or rate of spread (Hellman et al. 2008). The competitive resistance of native species may be reduced as climate change causes native species to shift out of the conditions to which they are adapted (Byers 2002). It is expected that, on average, mechanisms (e.g., dispersal) enabling invasion will allow existing invasive species to expand their ranges into newly suitable habitat more quickly than native species. Therefore, those species that have the ability to shift ranges quickly would have a competitive advantage if native populations become progressively poorer competitors for resources in a changing climate (Hellman et al. 2008). For example, some invasive species such as kudzu (*Pueraria lobataor*) may benefit when CO_2 concentrations increase or historical fire regimes are disturbed (Dukes and Mooney 1999).

In addition to facilitating the colonization of new invasive species, climate change could exacerbate the effects of existing invasive species, including selective mortality of native versus invasive species, reversals in competitive dominance, increased consumption by predators, or increased virulence of disease organisms (Rahel et al. 2008). Florida has a well-documented list of invasive plants and animals—a list that is expected to increase as temperatures warm, number of frost/freeze nights decrease, intensity and/or frequency of storm events increase, and Florida's human population increases and responds to climate change. More than 170 species of ferns and flowering plants are naturalized in southeastern Florida and hundreds of exotic plants have been introduced into the region (Austin 1978). Some of these species are not currently invasive or have not spread beyond South Florida; however, with climate change, these species may become invasive in the future or expand their current range into other regions of the state. Category I plants, defined as invasive exotics that are altering native plant communities by displacing native species, changing community structures or ecological functions, or hybridizing with natives include species such as Melaleuca (*M. quinquenervia*), Australian Pine (*Casuarina equisetifolia*), Water-hyacinth (*Eichhornia crassipes*), and old world climbing fern (*Lygodium microphyllum*). These species are invading native habitats and decreasing diversity, in some cases becoming so abundant that they interfere with species use of the area (e.g., nesting sea turtles and crocodiles) and contribute to the degradation of the habitat (e.g., erosion, clogging water bodies) (Austin 1978). There are more than 400 documented non-native animals in Florida, although not all are currently considered invasive. The Gambian pouch rat (*Cricetomys gambianus*), Burmese python (*Python bivattatus*), green iguana (*Iguana iguana*), giant toad (*Bufo marinus*), walking catfish (*Clarias batrachus*), Cuban tree frog (*Osteopilus septentrionalis*) and lionfish (*Pterois volitans*) are examples of invasive animals found in Florida. These species are known to prey upon and compete with native species. The Burmese python, native to Asia, is now found throughout much of southern Florida and has been the focus of several recent studies on impacts to native species (Dorcas et al. 2011; Dove et al. 2011; Holbrook and Chesnes 2011; Mazzotti et al. 2011, Willson et al. 2011, Dorcas et al. 2012). Many of the invasive plant and animal species found in Florida are constrained to their current extent by temperature. As temperatures increase and the number of frost/freeze nights is reduced or eliminated, many of these species will be able to migrate northward, expanding their range and potentially increasing the density of their infestations/populations.

How species respond "naturally" to the impacts of climate change may make it necessary to re-evaluate the definitions of "non-native" and "invasive" species, , as some species not currently considered native, but instead transient or occasional, expand or shift their range more permanently into Florida. For example, there is speculation that climate change is a contributing factor in the natural invasion and recent establishment in Florida of two species of tropical dragonflies from Cuba and the Bahamas (Paulson 2001). Climate change could also facilitate the movement of native species into a new area of habitat or increase its abundance in an area, and in doing so it may harm other native species in ways we typically associate with invasive species

(Rahel et al. 2008), possibly leading to localized mass extinctions, speciation, and the formation of new ecosystems (Wilson et al. 2009). Climate change impacts on the population size and scarcity of native species will influence the significance of the impact from invasive species (Hellman et al. 2008). In aquatic systems, climate change could exacerbate the effects of invasive species through selective mortality of native versus invasive species, reversals in competitive dominance, increased consumption by predators, or increased virulence of disease organisms due to increased water temperatures (Rahel et al. 2008).

Management techniques to prevent invasive plant establishment and spread include mechanical, chemical, and biological methods. Changes in temperature, precipitation, growth rates and patterns, and overall health of a particular species will affect the feasibility of these management techniques as well as the response of the species to the applied treatment. The total impact of an invasive species on a community, ecosystem, or resource includes the size of the range occupied by the invasive species (its spatial extent), its average abundance within that range, and its per capita (or per unit biomass) impact (Parker et al. 1999). Anticipating future distributions of invasive species is essential to facilitate preemptive and effective management (Bellard et al. 2013). The ability to manage invasive species due to changes in temperature, precipitation, and sea level rise will require new research, more monitoring, and a coordinated response system.

Pathogens, Parasites, and Pollutants

The climate change may result in increasing pathogen development and survival rates, disease transmission, and host susceptibility (NFWPCAS 2012). Warmer temperatures allow disease organisms to complete their life cycle more rapidly and attain higher population densities (Marcogliese 2001). Additionally, native diseases that currently only have minor effects on host organisms could have more devastating impacts under future climatic conditions (Rahel et al. 2008). Many marine and terrestrial species' pathogens are sensitive to shifts in temperature, rainfall, and humidity. As temperatures increase, many diseases are expected to become more lethal and to spread more readily (Epstein 2001; Harvell et al. 2002). There are climate-linked predictions that amphibians will decline in unusually warm years due the influence of temperature on disease dynamics (Epstein 2001; Harvell et al. 2002). A study of frogs and a pathogenic chytrid fungus (*Batrachochytrium dendrobatidis*) in Central and South America concluded that climate-driven epidemics are an immediate threat to biodiversity (Pounds et al. 2006). Changes in temperature, precipitation, soil moisture, and relative humidity can also affect the dispersal and colonization success of forest pathogens, which may impact forest ecosystem biodiversity among other important indicators of forest health (Brasier 1996; Lonsdale and Gibbs 1996; Chakraborty 1997; Houston 1998).

Increased temperatures impact parasites by increasing their growth rates, sexual maturation, mortality, and number of generations per year; higher temperatures can also promote earlier maturation, transmission, and potential maintenance of transmission year-round (Marcogliese 2001). The predicted changes in temperature and precipitation will have a serious impact on almost all environmental conditions in aquatic systems, affecting the distribution and abundance of free-living organisms, and thus by extrapolation, their parasites (Marcogliese 2001).

Alterations to temperature, pH, dilution rates, salinity, and other environmental conditions due to climate change can affect the impacts of pollutants on species and systems. The effects from these climatic changes can modify the availability of pollutants, the exposure and sensitivity of species to pollutants, transport patterns, and the uptake and toxicity of pollutants (Noyes et al. 2009). Altered transport patterns of environmental pollutants may lead to accumulations in new places thereby exposing biota in different habitats. Increased coastal flooding and inundation may result in release of contaminants from coastal soils, sediments, and infrastructure and increase exposure of fish, wildlife, and plants to these pollutants (NFWPCAS 2012). Climatic changes may lead to increased sensitivity to pollutants due to metabolic stress or inhibition of physiological processes that govern detoxification (NFWPCAS 2012).

Climate change is one of the cited causes of harmful algal blooms (Moore et al. 2008; Hallegraeff 2010), with warmer temperatures boosting the growth of harmful algae (Jöhnk et al. 2008; Paerl and Huisman 2008). The amount of runoff of phosphorus and other nutrients from farms and other landscapes currently contributes to harmful algal blooms and is expected to worsen with predicted increases in floods and other extreme precipitation events (NFWPCAS 2012).

Competition for Resources/Overexploitation

Florida's natural systems, in addition to their role in supporting biodiversity, provide a multitude of public services—supporting working landscapes, commercial and recreational activities. When well-maintained and well-managed, Florida's ecosystems can support these activities; however, overexploitation, misuse, and illegal activities can cause harm to the systems, communities, and species. Climate change can heighten species' vulnerability to overexploitation and, inversely, exploitation has made species particularly vulnerable to changes in climate (Harley and Rogers-Bennett 2004).

Activities such as hunting, fishing, wildlife viewing, hiking, biking, swimming, boating, and kayaking are popular recreational activities in Florida. Under future climatic conditions, current harvesting regimes may no longer be sustainable; furthermore, the indirect consequences of species harvest on non-target species may require special attention where a common resource base is likely to alter under climate change (Hulme 2005). Species and populations already stressed by the effects of climate change, could be pushed beyond their ability to adapt and survive—even when the natural areas they inhabit remain intact. Their environment could be degraded simply by the presence of humans as well as impacts through removal (collecting, pet trade), handling by recreationists, increased number of predators attracted by food waste, and disturbance by dogs (Gibbons et al. 2000). Due to concern about the increasing popularity of turtles for over-harvest, the Florida Fish and Wildlife Conservation Commission passed stronger rules to protect turtle species, prohibiting the taking or possession of six species of freshwater turtles from the wild (FWC 2016c). Species and populations impacted by climate change may be more vulnerable to over-harvesting if they become easier to harvest due to altered and increased movements as they react to loss or degradation of habitat, if they are forced to find alternative food sources, if their behavior is altered, or if they become stressed or diseased.

Additive and synergistic interactions between climate change and exploitation are becoming increasingly important to the dynamics of marine ecosystems and the sustainability of marine fisheries, such that stress and reduction in population size from existing fishing pressure in combination with the effects of intense events such as extreme temperature or storms may lead to increased risk of extinction of local populations (Harley and Rogers-Bennett 2004). In oceanic fish populations, human exploitation may further exacerbate the effects of oceanic warming (Walther et al. 2002). Florida is the number one destination in the US for saltwater anglers (Anderson et al. 2016). As the climate continues to change, the subtropical and tropical flats upon which species such as bonefish, tarpon, and permit depend upon, will be threatened by sea level rise. As sea level rise impacts these systems and habitats, fish stock could decrease and fishing regulations may have to change, reducing or eliminating harvest of these species.

The timber industry, cattle ranching, fishery and aquaculture industries are examples of compatible commercial use of the landscape. Climate change may impact the ability of the land to support existing levels of commercial use. For example, decreases in precipitation coupled with increases in temperature may reduce the landscape's ability to grow the same number of trees or cattle, or affect their growth rate or health. Modifications in stand density or cattle stocking rates, or the expansion of these systems to maintain existing yields, may impact the ability of the landscape to continue to support compatible populations of fish, wildlife, and plants. Other chapters of this book contain more information on climate impacts to forestry, land use and land cover, and fisheries.

Changes in groundwater and surface water, both in the amount and quality, have significant effects on biodiversity (Knight et al. 2011). Water resources may have the highest demand for competitive uses. Reduced water availability as a result of climate change is expected to affect the greatest number of species (Robinson et al. 2009). Increases to ground and surface water withdrawal to accommodate current and increased human populations as well as potential shifts in land use could further degrade systems that are stressed by decreased precipitation and droughts. Extraction of fresh water can significantly alter natural water flows, leading to impacts on habitats and the populations and species dependent upon them. Reduced precipitation will act in concert with water extractive activities, leading to decreases in water availability and flows causing potential alterations in food/prey abundance and availability, misalignment of reproductive cycles of aquatic organisms, increased rates of disease/parasite transmission as species are crowded into fewer remaining suitable areas, and direct loss of habitat. Increased water demands for domestic, industrial, and agricultural use along with rising temperatures will lower water tables, severely impacting wetlands (Sala et al. 2000).

Preserving Biodiversity for the Future

Biodiversity conservation in a changing climate requires a re-evaluation of management goals and objectives. Compared to the rate of past environmental change, the rate of future change within natural systems could be very swift and the magnitude of change could be large. Management approaches will need to be forward-looking and focus on maintaining a diversity of well-functioning ecosystems, as it could be very difficult to maintain the current spatial arrangements and composition of systems, communities, and species under a changing climate (Steffen et al. 2009). The Climate Smart Conservation framework outlines seven major steps for climate adaptation (Stein et al. 2014). The process begins with defining the planning purpose and scope, and an assessment of climate impacts and vulnerabilities to be used to review and possibly revise the conservation goals and objectives defined in the planning purpose and scope. The next steps include identifying possible adaptation options, evaluating and selecting adaptation actions, and then implementing priority adaptation actions. Implementation is followed by monitoring to track the effectiveness and ecological response of the adaptation actions. The entire process is iterative, with each step potentially looping back to the previous one, as needed. There are tools available to assist in determining potential impacts and relative vulnerability that can aid in developing adaptation strategies. Three main practices have been proposed to identify priority areas to protect biodiversity, including: 1) focus on areas where species are predicted to have the highest loss or the highest stability (areas to serve as refugia) (Lawler et al. 2009), 2) provide connectivity to allow species to move and shift their ranges (Heller and Zavaleta 2009), and 3) maintain landscape features that control species richness (geophysical variables) (Anderson and Ferree 2010). Identification of knowledge gaps can motivate research and monitoring efforts to improve future adaptation strategies development and implementation.

Impact Assessments

Species and the natural communities that they comprise have evolved through episodes of climate change more severe and more rapid than even the post-industrial age of anthropogenic global warming (Balsillie and Donoghue 2004; Donoghue 2011). However, the adaptive capacity of species and natural communities to respond to that change has been severely compromised by human modification of the landscape (Hughes 2000; Brooks et al. 2002; Thomas et al. 2004). As such, a critical step in analyzing the potential impacts of current and future climate change is the assessment of the vulnerabilities of species and natural communities (Miller et al. 2006). These vulnerabilities may be mediated by other threats and impacts, such as land-use change.

Vulnerability assessments are a suite of tools that reflect the relative and cumulative vulnerability of populations, species, or groups of species comprising a natural community, to stressors. Vulnerabilities are typically partitioned into two components: exposure and sensitivity; often, the adaptive capacity of species or natural communities is also assessed. Inclusion of at least these three factors is important because the data typically used in the assessment process includes geographic overlays of projected conditions. In such cases, exposure can be precisely calculated through geospatial model overlays (e.g., inundation from sea level rise, conversion of natural areas to agricultural uses, etc.), while other factors such as sensitivity and adaptive capacity are determined by less empirical and more qualitative methods of assessment.

> Vulnerability assessments are a suite of tools that reflect the relative and cumulative vulnerability of populations, species, or groups of species comprising a natural community, to stressors.

Land use change represents the strongest stressor on natural communities and on most species globally (Pimm and Raven 2000) and in Florida (Reece et al. 2013a). Anthropogenic climate change may exacerbate this stress through altered temperatures, precipitation, seasonality, and most importantly, sea level rise. Uncertainty of various types is an important factor to consider when implementing the results of a vulnerability assessment. For example, a high vulnerability to a particular threat, such as altered precipitation patterns, should be modulated by the relatively high uncertainty in precipitation projections relative to the more predictable change in temperature and sea level rise.

Several vulnerability assessment tools are available and have been implemented widely throughout Florida and surrounding regions. Each of the following tools focuses on a different aspect of vulnerability. Many of these tools are not equivalent and, importantly, the assessment of the same species often varies wildly depending on the choice of vulnerability assessment (Reece and Noss 2014). The Climate Change Vulnerability Index (CCVI; Young et al. 2009; Dubois et al. 2011) is widely-used and very easy to implement for rapid assessments of species. The scorer chooses a number value corresponding to a degree of exposure and sensitivity for threats, and then these are summarized into an overall assessment score. The relative importance of each of the criteria assessed is unclear, as not all vulnerabilities weigh equally on the overall assessment. The CCVI includes temperature and precipitation change as well as sea level rise and other land use-related threats exacerbated by climate change. However, the focus of this assessment is on future vulnerability to climate change, ignoring other types of threats and the current status of the species. The Conservation Status Assessment (CSA; Faber-Langendoen et al. 2012) is a widely-used system instituted by NatureServe for both species and natural communities. Similar to the CCVI, it uses a quantitative assessment system to produce a numerical score. The CSA uses a statewide and global assessment for spatial scale, and it focuses on past and present threats more than future threats. Global assessment tools such as the US

Endangered Species Act prioritization system (ESA 1973) and the International Union for the Conservation of Nature Red List system (IUCN 2010, 2015) both include quantitative assessments for species and natural communities, but do not differentiate by state or regional lines consistently enough for broad-scale prioritization or assessment efforts. The Standardized Index of Vulnerability and Value Assessment (SIVVA; Reece and Noss 2014) is a relatively recent addition to the list of vulnerability assessment tools for both species and natural communities. The SIVVA framework includes a mix of quantitative and qualitative criteria and a capacity for the user to set and manipulate the relative importance of criteria. In addition to assessing exposure and sensitivity (vulnerability) and adaptive capacity, SIVVA also includes criteria on conservation value and information availability. When the purpose of a vulnerability assessment is not only to calculate extinction risk but also to prioritize conservation efforts, the relative value of a species and the amount of information available to properly manage that species are extremely important to include (Reece et al. 2013a). This assessment has been used throughout Florida (Benscoter et al. 2013; Reece et al. 2013a), Georgia (Lowery 2016), and more broadly in the Gulf Coast (Watson et al. 2015).

There are a variety of vulnerability assessments available depending on the goals of the user. In each case, the user should, at a minimum, assess the exposure, sensitivity, and adaptive capacity of the species or natural community. It is also important to examine synergisms among threats, the spatial scale of the assessment, and the focus on past, present, and/or future threats.

Awareness that change is likely to happen is critical to planning for the future. Visualization of that awareness can be achieved through a method of study called scenario planning. Scenario planning is a discipline for developing a visual or narrative description of plausible future outcomes based on the combination of a range of complex and often intertwined factors that are projected into the future (Steinitz and Rogers 1970). It can be used to investigate the variables involved in the low controllability and high uncertainty of future states of the environment, and to determine the feasibility or likelihood of long-term biodiversity conservation in the face of climate change and resource consumption (Peterson et al. 2003). Scenario planning facilitates the comparative measurements of the rates and types of changes in biodiversity response variables, such as habitats, indices, and land cover (Gude et al. 2007). This information can then be used in support of various vulnerability assessments of natural systems or species persistence across a modeled range of future climate perturbations. Predetermined variables are applied to a series of modeled decisions over specified time steps to visualize how events or environments could look in the future. When the variables and decisions are altered, the outcomes of the scenarios change to reflect how these differences are represented or expressed, and how they

> **Scenario planning is a discipline for developing a visual or narrative description of plausible future outcomes.**

interact with each other. By comparing multiple future scenarios, and identifying the strategic issues and causes that led to each of those particular outcomes, the potential impacts of individual decisions or modeled events can be directly visualized, evaluated, and contrasted between scenarios, thereby reducing uncertainty.

Inputs to biodiversity-oriented scenario planning models are dependent on the impacts being examined and the types of outcome information sought. Many scenario planning projects use spatial data-manipulating programs such as geographic information systems (GIS) to visually depict the locations and extent of impacts. In other studies, scenario planning can be a narrative that describes a series of decisions and policy directions and the resulting environment, including descriptive visions of the future ecosystem states. In all cases, the inputs to the planning process are directly related to the construction of the scenario framework. The framework is the set of bounding data, assumptions, and desired analysis questions that constrain the extent of the study.

There is a high degree of uncertainty as to how the extent and speed of both current and future anthropogenic alterations that will influence climate change, putting pressure on species and ecosystems to adapt, possibly rapidly and in unknown directions. By applying scenario planning to these issues, employing credible information, informed questions, robust models, and a willingness to explore a range of alternatives, the percentage of what is unknown can be reduced to a manageable range of plausible futures. This process allows for informed decision-making to the benefit of biodiversity.

There are a broad range of scenario planning models for the environmental realm that can be grouped into three subcategories: (a) "exploratory scenarios," which represent different plausible futures; (b) "target-seeking scenarios," also termed "normative scenarios," which represent an agreed-upon future target and the scenarios that provide alternative pathways for reaching this target; and (c) "policy-screening scenarios," also known as "ex-ante scenarios," which represent the outcomes of various policy options under consideration (IPBES 2016). A recent example of an "exploratory" scenario model for the state of Florida was the 2014 project: Landscape Conservation and Climate Change Scenarios for the state of Florida: A Decision Support System for Strategic Conservation. This project used scenario modeling to predict future conservation opportunities to maximize protection of biodiversity, as well as potential areas of conflict, for locations of high ecological importance that overlapped with predicted urban growth or climate change impacted areas (Vargas et al. 2014). These models used were GIS-based spatial analyses that incorporated spatial infrastructure data, population growth projections, land use categories, financial conservation allocation strategies, and sea level rise predictions to create a range of future scenarios for Florida. This information could then be reflected back to management and funding policies to inform decision-making for biodiversity conservation objectives or goals, including mandates or conservation target development. A similar "exploratory" assessment of future climate change impacts was analyzed using the same set of Florida scenarios, specifically to determine where and what would be affected by simulated sea level rise projections.

Scenario planning results are used as inputs for a range of supplementary investigations. Additional analyses that incorporate the comparative aspect of multiple futures can include the evaluation of uncertainty, degree of impacts to constructed and natural systems, risk analysis and strategic forecasting, and cost-benefit analyses, among others. In the case of the Florida scenarios example, the future scenario outputs were subsequently used in a spatial comparison with imperiled species' habitat areas. This was done to determine where critical habitat would potentially be prime targets for urban development under different growth drivers and to highlight the amounts and locations of critical lands that could be lost to inundation due to sea level rise projections. In these analyses, spatial calculations quantified the amounts of direct impacts to habitats for each scenario. This information could then be used to estimate the exposure that ecological systems might incur due to the amounts, types, and locations of loss based on each scenario.

Adaptation Planning

Adaptation, as defined by the Intergovernmental Panel on Climate Change (IPCC 2007), is an "adjustment in natural or human systems in response to actual or expected climatic stimuli or their effects, which moderates harm or exploits beneficial opportunities." Tools such as scenario planning give scientists and managers indications of the array of potential environmental impacts due to climate change, and vulnerability assessments identify the realistic range of changes that could be withstood before detrimental effects on biodiversity are observed. Similar to scenario planning, adaptation planning is not necessarily constrained to biodiversity conservation, but it is most often applied to impacts to natural resources and the environment. Adaptation planning is a tool that assesses observed and forecasted impacts, acknowledges the uncertainty of future states of the environment, and develops actionable steps and a flexible implementation strategy in order to prepare for changes to the environment. It is a process that manages change, versus maintenance of existing conditions, and adjusts management techniques and goals as needed (Stein et al. 2013). Adaptation planning includes two initial processes: determining the scope of the system being planned for, and developing the strategies employed to encompass the uncertainty of climate change impacts to that system.

> **Adaptation planning is a process that assesses the future change of current conditions, and adjusts management techniques and goals as needed to support positive outcomes.**

Conservation planning strategies related to climate change adaption can be grouped into three general categories: (a) the continuation and support of "best practice" strategies, (b) "building off of or "extending 'best practice' principles," and (c) "integrating assessments on species vulnerability to climate change into a conservation planning framework" (Watson et al. 2012).

"Best practice" strategies are those that continue current accepted or implemented conservation actions. These strategies include the identification and protection of key habitats and critical populations, habitat conservation replication and extent minimums to reduce vulnerability from stochastic events or other variables, and efficient design and management of conservation networks to maximize effectiveness and minimize existing and potential threats (Watson et al. 2012). "Best practice" strategies on their own are no longer considered sufficient to preserve biodiversity over the long term due to the static nature of their conservation techniques (Cameron-Devitt et al. 2012).

Extensions of "best practice" principles take the methods of achieving current conservation objectives and apply forward-looking assessments of how the environment will change over time to affect the existing conservation system and the diversity it supports. These extensions include strategies such as expansion and connectivity of current conservation lands to maximize the potential for populations to adapt to change within their environment, the inclusion of species refugia in conservation objectives, and management strategies that prioritize protection and maintenance of entire ecosystems versus individual species (Watson et al. 2012). These techniques incorporate potential climate change impacts into conservation planning, but are not truly adaptive because they do not set forth a system to reassess goals over time and alter or change strategies as needed.

The third category of conservation planning strategies has developed into what most scientists and managers currently consider adaptation planning (Stein et al. 2013). This form of planning incorporates species or ecosystem vulnerability assessments into the planning framework and alters conservation strategies to fit the needs of the environment. The vulnerability assessments will be altered by changing conditions over time, both climate-induced and direct anthropogenic impacts. To address future climate uncertainties, conservation planning strategies require management support in order to rerun vulnerability assessments to reassess the potential for natural resources to persist through a range of environmental changes. The incorporation of these assessments broadens the scope of the process to address both current and potential future needs under an array of scenarios. These types of supporting assessments provide data and information for the development of plausible, flexible, and effective management approaches to improve resilience, reduce vulnerability, and adapt to changing conditions.

Adaptation planning strategies developed specifically for biodiversity or ecological diversity preservation can be grouped into categories related to their area of application, such as habitat protection or management, species management, planning and monitoring, or law and policy (Mawdsley et al. 2009). However, effective adaptation planning should consider an

> **Adaptation strategies are specific steps enacted at pre-determined times or levels to reduce risk from, or increase resilience to, detrimental climate change impacts.**

array of applicable strategies, as climates and ecosystem-dependent conservation goals shift over time. Examples of adaptation planning in Florida range from a suite of all-encompassing policy actions in the Florida Energy and Climate Change Action Plan (Center for Climate Strategies 2008) and other sector-wide policy frameworks (e.g. Murley et al. 2008; Beever et al. 2010, Southeast Florida Regional Climate Change Compact Counties 2012), to specific biodiversity conservation techniques for inclusion in state wildlife action plans (Association of Fish & Wildlife Agencies 2009) or for other natural resource agencies (Cameron-Devitt et al. 2012).

The Florida Fish and Wildlife Conservation Commission recently completed their planning document, *A Guide to Climate Change Adaptation for Conservation: Resources and Tools for Climate Smart Management of Florida's Fish and Wildlife Species and Their Habitats* (FWC 2016a). The report identifies climate-related threats to species and natural communities based on impact and vulnerability assessment tools, and proposes priority conservation strategies. It also offers an approach to identifying adaptation strategies by grouping ecosystems with shared vulnerabilities and similar sets of ecological impacts, while tailoring strategies to the climate change stressors for each group. This leads to the determination of specific adaptation strategies to reduce risk that address the consequences of climate change for each natural community. These strategies form the basis of the climate change adaptation plan for Florida's natural resources. For many of the ecosystems analyzed in the document, collecting ecological values and feedback through the use of monitoring (described in detail in the next section) is specified. This element of the planning guide contributes to its function as an adaptive management plan for climate change. This information is used to alter goals, strategies, or implementation of actions and adaptation strategies as climate changes occur which shift biodiversity or natural resource targets and potential vulnerabilities.

Watson et al. (2012) described the elements of a "good adaptation strategy" as one that includes the following characteristics:
- Incorporate clear planning principles (flexibility, efficiency)
- Account for uncertainty
- Understand trade-offs
- Manage for both climate variability and long-term climate change
- Integrate human response
- Clarity of adaptation goal: resilience vs. resistance

As one of a series of climate change planning tools, adaptation planning incorporates alternative scenarios, impact and vulnerability assessments, conservation prioritization and "best practice" methodologies, and climate change predictions to produce a plan to preserve biodiversity in the face of major environmental changes. However, the step in the process that is integral to adaptation planning effectiveness is the inclusion of an element of flexibility and acceptance of uncertainty of climate variability and the resulting impacts to the environment. This adaptability can be acknowledged through updated vulnerability assessments, policy and management alterations, and modified strategies, but the ability to revise underlying conservation goals and implement adjustments will be critical to the long-term preservation of biodiversity and natural resources.

Research and Monitoring

Species Resilience

In previous sections of this chapter, vulnerability assessments were mentioned as tools to evaluate the degree to which species or ecosystems can withstand perturbations and alterations to the habitats relied upon for continued existence or functioning. An assessment of species resiliency is a similar tool used in adaptation planning to predict impacts from potential shifts in future states of the environment. Resiliency is the ability of species or ecosystems to endure direct or indirect changes to their environment, and either recover or adapt to those changes. Resiliency is an attribute that allows for biological flexibility and continued existence in highly variable natural systems, and species or ecosystems with higher resiliency have the ability to 'weather the storm' more successfully. Species or systems with low resiliency may reach disturbance or variation thresholds sooner or at lower impact levels, from which there may be either no recovery or the occurrence of a regime shift (Folke et al. 2004). Regime shifts can often be to a condition or state of existence that is less productive, less stable, or threatens the survival of species. Climate change will cause unknown impacts to ecosystems and dependent species; but by employing the planning tools previously discussed, potential impacts and magnitudes of change may be possible to describe and predict. Enhanced by the inclusion of resiliency assessments, adaptive resource management and planning can be improved by taking into account the points in time or state of the environment at which species or ecosystems would be irrevocably affected.

> **Resiliency is the ability of species or ecosystems to endure direct or indirect changes to their environment, and either recover or adapt to those changes.**

Trigger Points

Adaptive management's 'early warning system' for conservation is called a trigger point: an event, change in status, or measureable level that indicates the system or object being monitored has reached a crucial state in advance of a critical threshold. A trigger point provides a preventative warning or alarm that indicates

> **A trigger point is an event, change in status, or measureable level that indicates the system or object being monitored has reached a crucial state in advance of a critical threshold.**

some type of action needs to be taken to prevent the state from deteriorating further and reaching a critical threshold. Trigger points are developed from potential threats deduced from scenario planning, the vulnerability of species or ecosystems from vulnerability assessments, and the determination of detrimental regime shifts or significant events from species resiliency assessments, incorporated with an understanding of the time lag needed to reassess plans and activate a management response. Trigger points enable adjustments to adaptation plans and strategies in response to new or updated information and changing circumstances (Moss and Martin 2012). It is important to note that a trigger point is not a tipping point or a critical threshold; it is a status or level identified during the planning stage that indicates a critical threshold may be imminent if actions are not taken to prevent it.

The CoastAdapt tool developed by the National Climate Change Adaptation Research Facility is an example of how trigger points can be used in adaptive management for biodiversity conservation (NCCARF 2016). This online program, partially funded by the Australian Department of the Environment and Energy, seeks to provide a tool and guidance to managers and scientists to approach climate change and sea level rise issues with an adaptive management process. In step six of their Coastal Climate Adaptation Decision Support (C-CADS) methodology, it is recommended to develop trigger points that are robust and inclusive of the range of potential climate variability (NCCARF 2016). Trigger points can be physical, environmental, social, or economic depending on the scope of the study, and should be observable, measureable, and comprehensible to all stakeholders involved (NCCARF 2016).

Monitoring

Plans that continually incorporate updated information on the status of the resources, and adapt their policies and strategies accordingly, will be more robust and responsive in the long term. Monitoring programs specific to the threatened species and ecosystems in question are essential but often overlooked components of adaptation planning. Conservation monitoring programs will be most effective when they are embedded in and inform management plans, including the necessary ability to detect spatial and temporal changes early on (Beever 2006; Lindenmayer et al. 2013). Without monitoring systems in place, system variables cannot be evaluated on a continual basis to recognize when trigger points are reached, environmental subtleties can go unnoticed, the amount of reaction time available to alter or adapt policies to extreme events is reduced or eliminated, and key indicator data for species and ecosystems is not consistent or available to other research endeavors. Equally as important, if not more so, is the necessity of using monitoring programs to assess the performance of adaptation plan efforts to determine whether strategies are effective, relevant, and efficient (NCCARF 2016).

Adaptation management monitoring programs can be scaled from local resource levels to entire countries, and the usefulness of the program depends on the objective of the study and the information it contributes to evaluation, planning, and management processes. An ongoing monitoring program established in Everglades National Park in Florida collects an array of variables associated with hydrology, climate, and salinity (Mitchell and Krueger 2011). This data contributes to evaluations of risk and resiliency of species and habitats to climate change impacts, such as sea level rise (Mitchell and Krueger 2011). The collected information is also used in conjunction with additional research and modeling efforts to understand hydrological, species, carbon, and ecosystem dynamics in and around the park, and is critical to efforts to evaluate the potential magnitude of climate change effects on the natural resources in the park (Mitchell and Krueger 2011). Without long-term monitoring programs, all research programs that inform adaptive management strategies for the park, such as habitat suitability modeling, mangrove carbon dynamics, or Florida Bay restoration planning, would not be supported.

The tools described in this section build upon each other and are integral to effective climate change adaptation planning for the long-term conservation of biodiversity. Natural resource managers and policy makers can take advantage of this wealth of information to make informed decisions when they have awareness of the range of potential impending changes to the environment, acknowledgement of the uncertainty to be faced, and access to a flexible adaptation plan with supportive monitoring programs and relevant trigger points. "The future is not predictable and as a result, adaptation depends on learning and responding effectively to lessons learnt, as well as experience, changing circumstances and new knowledge" (South West Climate Change Portal 2016).

References

Ackerly, D.D., S.R. Loarie, W.K. Cornwell, S.B. Weiss, H. Hamilton, R. Branciforte, and N.J.B. Kraft. 2010. "The Geography of Climate Change: Implications for Conservation Biogeography." *Diversity and Distributions* 16:476-487.

Ahola, Markus, Toni Laaksonen, Katja Sippola, Tapio Eeva, Kalle Rainio, and Esa Lehikoinen. 2004. "Variation in Climate Warming Along the Migration Route Uncouples Arrival and Breeding Dates." *Global Change Biology* 10 (9):1610-1617.

Aiello-Lammens, Matthew E., Ma Chu-Agor, Matteo Convertino, Richard A. Fischer, Igor Linkov, and H. Resit Akcakaya. 2011. "The Impact Of Sea-Level Rise on Snowy Plovers in Florida: Integrating Geomorphological, Habitat, and Metapopulation Models." *Global Change Biology* 17 (12):3644-3654.

Ainsworth, Elizabeth A., Phillip A. Davey, Graham J. Hymus, Bert G. Drake, and Stephen P. Long. 2002. "Long-Term Response of Photosynthesis to Elevated Carbon Dioxide in a Florida Scrub-Oak Ecosystem." *Ecological Applications* 12 (5):1267-1275.

Ainsworth, Elizabeth A., and Stephen P. Long. 2005. "What Have We Learned From 15 Years of Free-Air CO2 Enrichment (FACE)? A Meta-Analytic Review of the Responses of Photosynthesis, Canopy Properties and Plant Production to Rising CO2." *New Phytologist* 165:351-372.

Anderson, L., P. Glick, S. Heyck-Williams, and J. Murphy. 2016. "Changing Tides: How Sea-Level Rise Harms Wildlife and Recreation Economies along the U.S. Eastern Seaboard." National Wildlife Federation: Washington, DC.

Anderson, Mark G., and Charles E. Ferree. 2010. "Conserving the Stage: Climate Change and the Geophysical Underpinnings of Species Diversity." *PLoS ONE* 5 (7): e115.doi:10.1371/journal.pone.0011554

Araujo, M.B., W. Thuiller, and R.G. Pearson. 2006. "Climate Warming and the Decline of Amphibians and Reptiles in Europe." *Journal of Biogeography* 33:1712-1728. doi:10.1111/j.1365-2699.2006.01482.x

Association of Fish & Wildlife Agencies. 2009. "Voluntary Guidance for States to Incorporate Climate Change into State Wildlife Action Plans & Other Management Plans." *Climate Change Wildlife Action Plan Guidance Document.* Climate Change Wildlife Action Plan Work Group, 50pp.

Aumen, Nicholas G., Karl E. Havens, G. Ronnie Best, and Leonard Berry. 2015. "Predicting Ecological Responses of the Florida Everglades to Possible Future Climate Scenarios: Introduction." *Environmental Management*. DOI 10.1007/s00267-014-0439-z

Austin, Daniel F. 1978. "Exotic Plants and Their Effects in Southeastern Florida." *Environmental Conservation* 5 (1):25-34.

Bailey, R. M., H. E. Winn, and C. L. Smith. 1954. "Fishes from the Escambia River, Alabama and Florida, with Ecologic and Taxonomic Notes." *Proceedings of the Academy of Natural Sciences of Philadelphia* 106:109–164.

Balsillie, James H., and Joseph F. Donoghue. 2004. High Resolution Sea-level History for the Gulf of Mexico Since the last Glacial Maximum. In Florida Geological Survey Report of Investigation.

Baker, J.D., C.L. Littnan, and D.W. Johnston. 2006. "Potential Effects of Sea Level Rise on the Terrestrial Habitats of Endangered and Endemic Megafauna in the Northwestern Hawaiian Islands." *Endangered Species Research* 4:1-10.

Beever, E.A. 2006. "Monitoring Biological Diversity: Strategies, Tools, Limitations, and Challenges." *Northwestern Naturalist* 87:66-97.

Beever, J. W., W. Gray, J. Utley, D. Hutchinson, T. Walker, and D. Cobb. 2010. "Lee County Climate Change Resiliency Strategy (CCRS)." Southwest Florida Regional Planning Council, 163pp.

Bellard, Celine, Cleo Bertelsmeier, Paul Leadley, Wilfried Thuiller, Franck Courchamp. 2012. "Impacts of Climate Change on the Future of Biodiversity." *Ecol Lett* 15:365–377.

Bellard, C., W. Thuiller, B. Leroy, P. Genovesi, M. Bakkenes, and F. Courchamp. 2013. "Will Climate Change Promote Future Invasions?" *Global Change Biology* 19 (12):3740-3748.

Benscoter, Allison M., Joshua S. Reece, Reed F. Noss, Laura A. Brandt, Frank J. Mazzotti, Stephanie S. Romañach, and James I. Watling. 2013. "Threatened and Endangered Subspecies with Vulnerable Ecological Traits Also Have High Susceptibility to Sea Level Rise and Habitat Fragmentation." *PLoS ONE* 8:e70647.

Bjorndal, Karen A. 1980. "Nutrition and Grazing Behavior of the Green Turtle Chelonia Mydas." *Marine Biology* 56 (2):147-154.

Blanvillain, Gaelle. 2007. "Sex Ratio Prediction of Juvenile Hawksbill Sea Turtles (Eretmochelys imbricata) from South Florida, USA." *Herpetological Conservation and Biology* 3 (1):21-27.

Bouchard, S. S., and K. A. Bjorndal. 2000. "Sea Turtles as Biological Transporters of Nutrients and Energy from Marine to Terrestrial Ecosystems." *Ecology* 81 (8):2305-2313.

Brander, K. M., G. Blom, M.F. Borges, K. Erzini, G. Henderson, B.R. MacKenzie, H. Mendes, J. Ribeiro, A.M.P. Santos, and R. Toresen. 2003. "Changes in Fish Distribution in the Eastern North Atlantic: Are We Seeing a Coherent Response to Changing Temperature?" ICES Marine Science Symposia 219:261-270.

Brandt, L. A., K. L. Montgomery, A. W. Saunders, and F. J. Mazzotti. 1993. "Life History Notes: Gopherus polyphemus (gopher tortoise) Burrows." *Herpetological Review* 24:149.

Brasier, C.M. 1996. "Phytophthora cinnamomi and Oak Decline in Southern Europe: Environmental Constraints Including Climate Change." *Annales des Sciences Forestières* 53:347-358.

Brinson, M.M., R.R. Christian and L.K. Blum. 1995. "Multiple States in the Sea-Level Induced Transition from Terrestrial Forest to Estuary." *Estuaries* 18 (4):648-659.

Brooks, Thomas M., Russell A Mittermeier, Cristina G. Mittermeier, Gustavo A. B. Da Fonseca, Anthony B. Rylands, William R. Konstant, Penny Flick, John Pilgrim, Sara Oldfield, Georgina Magin, and Craig Hilton-Taylor. 2002. "Habitat Loss and Extinction in the Hotspots of Biodiversity." *Conservation Biology* 16:909-923. doi: 10.1046/j.1523-1739.2002.00530.x.

Brown, Barbara E. and Richard P. Dunne. 2016. "Coral Bleaching: The Roles of Sea Temperature and Solar Radiation." pp. 266-283 In: Cheryl M. Woodley, Craig A. Downs, Andrew W. Bruckner, James W. Porter and Sylvia B. Galloway (eds). Chapter 18. Diseases of coral. 582 p.

Bruno, John F., and Elizabeth R. Selig. 2007. "Regional Decline of Coral Cover in the Indo-Pacific: Timing, Extent, and Subregional Comparisons." *PLoS One* 2:1.

Bruno, John F., Elizabeth R. Selig, Kenneth S. Casey, Cathie A. Page, Bette L. Willis, C.Drew Harvell, Hugh Sweatman, and Amy M. Melendy. 2007. "Thermal Stress and Coral Cover as Drivers of Coral Disease Outbreaks". *PLoS Biol* 5(6): e124. doi:10.1371/journal.pbio.0050124.

Burg, C. 2010. "Initial Estimates of the Ecological and Economic Consequences of Sea Level Rise on the Florida Keys Through the Year 2100." The Nature Conservancy. Unpublished report 36.

Burkett, Virginia A., Douglas A. Wilcox, Robert Stottlemyer, Wylie Barrow, Dan Fagre, Jill Baron, Jeff Price, Jennifer L. Nielsen, Craig D. Allen, David L. Peterson, Greg Ruggerone, and Thomas Doyle. 2005. "Nonlinear Dynamics in Ecosystem Response to Climatic Change: Case Studies and Policy Implications." *Ecological Complexity* 2:357-394.

Burkett, V. R., Robert J. Nicholls, Leandro Fernandez, and Colin D. Woodroffe. 2008. "Climate Change Impacts on Coastal Biodiversity." University of Wollongong - Research Online: 167-193.

Byers, James. E. 2002. "Impact of Non-Indigenous Species on Natives Enhanced by Anthropogenic Alteration of Selection Regimes." *Oikos* 97:449–458.

Cameron-Devitt, Susan E., Jennifer R. Seavey, Sieara Claytor, Tom Hoctor, Martin Main, Odemari Mbuya, Reed Noss, and Rainyn Corrie. 2012. "Florida Biodiversity under a Changing Climate." *Florida Climate Task Force*. 128pp

Campbell, Lisa M. 2002. "Contemporary Culture, Use, and Conservation of Sea Turtles." In The Biology of Sea Turtles, 301-331.

Campbell, Lisa M., and Christy Smith. 2006. "What Makes Them Pay? Values of Volunteer Tourists Working for Sea Turtle Conservation." *Environmental Management* 38 (1):84-98. doi: 10.1007/s00267-005-0188-0.

Campbell, Tyler A. and David B. Long. 2009. "Feral Swine Damage and Damage Management in Forested Ecosystems." *Forest Ecology and Management* 257:2319-2326.

Carpenter, Kent E., Muhammad Abrar, Greta Aeby, Richard B. Aronson, Stuart Banks, Andrew Bruckner, Angel Chiriboga et al. 2008. "One-Third of Reef-Building Corals Face Elevated Extinction Risk from Climate Change and Local Impacts." *Science* 321 (5888):560-563.

Catano, Christopher P., Stephanie S. Romanach, James M. Beerens, Leonard G. Pearlstine, Laura A. Brandt, Kristen M. Hart, Frank J. Mazzotti, and Joel C. Trexler. 2014. "Using Scenario Planning to Evaluate the Impacts of Climate Change on Wildlife Populations and Communities in the Florida Everglades." *Environmental Management* 55 (4):807-823.

Center for Climate Strategies. 2008. "Florida's Energy & Climate Change Action Plan." Contribution of Governor's Action Team on Energy & Climate Change to the State of Florida, Michael Sole, Chairman, 609pp.

Chakraborty, S. 1997. "Recent Advances in Studies of Anthracnose of Stylosanthes. V. Advances in Research on Stylosanthes Anthracnose Epidemiology in Australia." *Tropical Grasslands* 31:445-453.

Church, John A., Neil J. White, and John R. Hunter. 2006. "Sea-Level Rise at Tropical Pacific and Indian Ocean Islands." *Global and Planetary Change* 53 (3):155-168

Clout, Mick N., and Peter A. Williams. 2009. Invasive Species Management: A Handbook of Principles and Techniques. Oxford University Press.

Coffin, Alisa W. 2007. "From Roadkill to Road Ecology: A Review of the Ecological Effects of Roads." *Journal of Transport Geography* 15 (5):396-406.

Connell, Joseph H. 1978. "Diversity in Tropical Rain Forests and Coral Reefs." *Science*, New Series. 199:1302-1310.

Cooper, Timothy F., Glenn De'Ath, Katharina E. Fabricius, and Janice M. Lough. 2008. "Declining Coral Calcification in Massive Porites in Two Nearshore Regions of the Northern Great Barrier Reef." *Global Change Biology* 14 (3):529-538.

Crumpacker, David W., Elgene O. Box, and E. Dennis Hardin. 2001. "Implications of Climatic Warming for Conservation of Native Trees and Shrubs in Florida." *Conservation Biology* 15 (4): 1008-1020.

Day, Frank P., Rachel E. Schroeder, Daniel B. Stover, Alisha L.P. Brown, John R. Butnor, John J. Dilustro, Bruce A. Hungate, P. Dijkstra, B. D. Duval, T.J. Seiler, B. Drake, and C. R. Hinkle. 2013. "The Effects of 11 Yr of CO_2 Enrichment on Roots in a Florida Scrub-Oak Ecosystem." *New Phytologist* 200 (3):778-787. doi: 10.1111/nph.12246.

Deeming, Denis C., and Mark W.J. Ferguson. 1989. "The Mechanism of Temperature Dependent Sex Determination in Crocodilians: A Hypothesis." *Amer. Zool.* 29:973-985.

Desantis, Larisa R.G., Smriti Bhotika, Kimberlyn Williams, and Francis E. Putz. 2007. "Sea-Level Rise and Drought Interactions Accelerate Forest Decline on the Gulf Coast of Florida, USA." *Global Change Biology* 13 (11):2349-2360.

Diemer, J. E. 1992. "Gopher tortoise, Gopherus Polyphemus (Daudin)." Pages 123–127 in P. E. Moler, editor. Rare and endangered biota of Florida. Volume III. Amphibians and reptiles. University Press of Florida, Gainesville, Florida, USA

Diez, Jeffrey M., Carla M. D'Antonio, Jeffrey S. Dukes, Edwin D. Grosholz, Julian D. Olden, Cascade J.B. Sorte, Dana M. Bluemnthal, Bethany A. Bradley, Regan Early, Ines Ibanez, Sierra J. Jones, Joshua J. Lawler, and Luke P. Miller. 2012. "Will Extreme Climatic Events Facilitate Biological Invasions?" *Frontiers in Ecology and the Environment* 10:249–257.

Dijkstra, Paul, Graham Hymus, Debra Colavito, David Vieglais, Christina M. Cundari, David P. Johnson, Bruce A. Hungate, C. R. Hinkle, and Bert G. Drake. 2002. "Elevated Atmospheric CO_2 Stimulates Aboveground Biomass in a Fire-Regenerated Scrub-Oak Ecosystem." *Global Change Biology* 8:90-103.

Donnelly, Jeffrey P., and Mark.D. Bertness. 2001. "Rapid Encroachment of Salt Marsh Cordgrass in Response to Accelerated Sea-Level Rise." *PNAS* 98 (25):14218-14223.

Donoghue, Joseph. 2011. "Sea Level History of the Northern Gulf Of Mexico Coast and Sea Level Rise Scenarios for the Near Future." *Climatic Change* 107:17-33. doi: 10.1007/s10584-011-0077-x.

Dorcas Michael E., John D. Willson, and J. Whitfield Gibbons. 2011. "Can Invasive Burmese Pythons Inhabit Temperate Regions of the Southeastern United States?" *Biol Invasions* 13:793–802.

Dorcas, Michael E., John D. Willson, Robert N. Reed, Ray W. Snow, Michael R. Rochford, Melissa A. Miller, Walter E. Meshaka, Jr., Paul T. Andreadis, Frank J. Mazzotti, Christina M. Romagosa, and Kristen M. Hart. 2012. "Severe Mammal Declines Coincide with Proliferation of Invasive Burmese Pythons in Everglades National Park." *PNAS* 109 (7):1-5.

Dove, Carla J., Ray W. Snow, Michael R. Rochford, and Frank J. Mazzotti. 2011. "Birds consumed by the Invasive Burmese Python (Python molurus bivittatus) in Everglades National Park, Florida, USA." *The Wilson Journal of Ornithology* 123:126–131.

Drake, Bert G., Miquel A. Gonzalez-Meler, and Steve P. Long. 1997. "More Efficient Plants: A Consequence of Rising Atmospheric CO_2." *Annual Review of Plant Physiology and Plant Molecular Biology* 48:609-39.

Driscoll, Don A., Adam Felton, Philip Gibbons, Annika M. Felton, Nicola T. Munro, and David B. Lindenmayer. 2012. "Priorities in Policy and Management When Existing Biodiversity Stressors Interact with Climate Change." *Climatic Change* 111 (3-4): 533-557.

Dubois, N., A. Caldas, J. Boshoven, and A. Delach. 2011. "Integrating Climate Change Vulnerability Assessments Into Adaptation Planning: A Case Study Using the Natureserve Climate Change Vulnerability Index to Inform Conservation Planning for Species in Florida." A Report Prepared for the Florida Fish and Wildlife Conservation Commission.

Dudley, Peter N., and Warren P. Porter. 2014. "Using Empirical and Mechanistic Models to Assess Global Warming Threats to Leatherback Sea Turtles." *Marine Ecology Progress* Series 501:265-278. doi: 10.3354/meps10665.

Dukes, Jeffrey S., and Harold A. Mooney. 1999. "Does Global Change Increase the Success of Biological Invaders?" *Trends in Ecology and Evolution* 14 (4):135-139.

Epstein, Paul R. 2001. "Climate Change and Emerging Infectious Diseases." *Microbes and Infection* 3 (9):747-754.

ESA. 1973. "Endangered Species Act of 1973." *Title 16 United States Code*, Sections 1531-1544.

Ewert, Michael. A., Dale R. Jackson, Craig E. Nelson. 1994. "Patterns of Temperature-Dependent Sex Determination in Turtles." *J. exp. Zool*. 270:3–15.

Ewert, Michael A., Jeffrey W. Lang, and Craig E. Nelson. 2005. "Geographic Variation in the Pattern of Temperature-Dependent Sex Determination in the American Snapping Turtle (Chelydra serpentina). *J. Zool., Lond*. 265:81-95.

Faber-Langendoen, D., J. Nochols, L. Master, K. Snow, A. Tomaino, R. Bittman, G. Hammerson, B. Heidel, L. Ramsay, A. Teucher, and B. Young. 2012. "NatureServe Conservation Status Assessments: Methodology for Assigning Ranks." Arlington, VA: NatureServe.

Feagin, Rusty A., Douglas J. Sherman, and William E. Grant. 2005. "Coastal Erosion, Global Sea-Level Rise, and the Loss of Sand Dune Plant Habitats." *Frontiers in Ecology and the Environment* 3 (7):359-364.

Ferriter, Amy. (ed). 1997. "Brazilian Pepper Management Plan for Florida." A Report from the Florida Exotic Pest Plant Council's Brazilian Pepper Task Force. 31pp.

Florida Fish and Wildlife Conservation Commission - FWC. 2012. "Florida's Wildlife Legacy Initiative: Florida's State Wildlife Action Plan." Tallahassee, Florida.

Florida Fish and Wildlife Conservation Commission. - FWC 2016a. "A Guide to Climate Change Adaptation for Conservation." Version 1. Tallahassee, Florida.

Florida Fish and Wildlife Conservation Commission - FWC. 2016b. "Florida's Endangered and Threatened Species." Tallahassee, FL 14 pp.

Florida Fish and Wildlife Conservation Commission - FWC. 2016c. Website – "Freshwater Turtles", http://myfwc.com/wildlifehabitats/managed/freshwater-turtles/. Accessed September 27, 2016.

Foden, Wendy., Georgia Mace, Jean-Christophe Vié, Ariadne Angulo, Stuart Butchart, Lyndon DeVantier, Holly Dublin, Alexander Gutsche, Simon Stuart, and Emre Turak. 2008. "Species Susceptibility to Climate Change Impacts." in The 2008 Review of the IUCN Red List of Threatened Species, edited by Jean-Christophe Vié, Craig Hilton-Taylor and Simon N. Stuart. IUCN Gland, Switzerland.

Folke, C., S. Carpenter, B. Walker, M. Scheffer, T. Elmqvist, L. Gunderson, and C. S. Holling, 2004. "Regime Shifts, Resilience, and Biodiversity in Ecosystem Management." *Annual Review of Ecology, Evolution, and Systematics* 35:557-58.

Fuentes, Mariana M.P.B., Christian Gredzens, Brooke L. Bateman, Ruth Boettcher, Simona A. Ceriani, Matthew H. Godfrey, David Helmers, Dianne K. Ingram, Ruth L. Kamrowski, Michelle Pate, Robert L. Pressey, and Volker C. Radeloff. 2016. "Conservation Hotspots for Marine Turtle Nesting in the United States Based on Coastal Development." *Ecological Applications*. doi: 10.1002/eap.1386.

Fuentes, M. M. P. B., C. J. Limpus, and M. Hamann. 2011. "Vulnerability of Sea Turtle Nesting Grounds to Climate Change." *Global Change Biology* 17 (1):140-153. doi: 10.1111/j.1365-2486.2010.02192.x.

Fuentes, M.M.P.B., C. J. Limpus, M. Hamann, and J. Dawson. 2010. "Potential Impacts of Projected Sea-Level Rise on Sea Turtle Rookeries." *Aquatic Conservation: Marine and Freshwater Ecosystems* 20 (2):132-139. doi: 10.1002/aqc.1088

Fuentes, M. M., D. A. Pike, A. Dimatteo, and B. P. Wallace. 2013. "Resilience of Marine Turtle Regional Management Units to Climate Change." *Global Change Biology* 19 (5):1399-406. doi: 10.1111/gcb.12138.

Fuentes, M. M. P. B., and V Saba. 2016. "Impacts and Effects of Ocean Warming on Marine Turtles." in Explaining Ocean Warming: Causes, Scale, Effects, and Consequences, edited.

Fuller, W.J., B.J. Godley, D.J. Hodgson, S.E. Reece, M.J. Witt, and A.C. Broderick. 2013. "Importance of Spatio-Temporal Data for Predicting the Effects of Climate Change on Marine Turtle Sex Ratios." *Marine Ecology Progress Series* 488:267-274. doi: 10.3354/meps10419

Galbraith, H., R. Jones, R. Park, J. Clough, S. Herrod-Julius, B. Harrington, and G. Page. 2002. "Global Climate Change and Sea Level Rise: Potential Losses of Intertidal Habitat for Shorebirds." *Waterbirds* 25 (2):173-183.

Gardner, Toby A., Isabelle M. Côte, Jennifer A. Gill, Alastair Grant, and Andrew R. Watkinson. 2003. "Long-Term Region-Wide Declines in Caribbean Corals." *Science* 301:958-960.

Gates, David Murray. 1993. Climate Change and Its Biological Consequences. Sinauer Associates, Inc., Sunderland, MA.

Gibbons, J. Whitfield, David E. Scott, Travis J. Ryan, Kurt A. Buhlmann, Tracey D. Tuberville, Brian S. Metts, Judith L. Greene, Tony Mills, Yale Leiden, Sean Poppy, and Christopher T. Winne. 2000. "The Global Decline of Reptiles, Déjà Vu Amphibians." *BioScience* 50 (8): 653-666.

Graham, Russell W., and Eric C. Grimm. 1990. "Effects of Global Climate Change on the Patterns of Terrestrial Biological Communities." *Trends in Ecology & Evolution* 5 (9):289-292.

Gritti, E. S., B. Smith, and M. T. Sykes. 2006. "Vulnerability of Mediterranean Basin Ecosystems to Climate Change and Invasion by Exotic Plant Species." *Journal of Biogeography* 33:145–157.

Gude, Patricia H., Andrew J. Hansen, and Danielle A. Jones. 2007. "Biodiversity Consequences of Alternative Future Land Use Scenarios in Greater Yellowstone." *Ecological Applications* 17 (4):1004-1018.

Hall, Mary, C., Peter Stiling, Daniel C. Moon, Bert Drake, and Mark D. Hunter. 2005. "Effects of Elevated CO_2 on Foliar Quality and Herbivore Damage in a Scrub Oak Ecosystem." *Journal of Chemical Ecology* 31 (2):267-286.

Hallegraeff, Gustaaf M. 2010. "Ocean Climate Change, Phytoplankton Community Responses, and Harmful Algal Blooms: A Formidable Predictive Challenge." *Journal of Phycology* 46:220-235.

Hamann, M., M. M. P. B. Fuentes, N. C. Ban, and V. J. L. Mocellin. 2012. "Climate Change and Marine Turtles." In Biology of Sea Turtles, 353-376.

Hamed, Ahmed, Kaveh Madani, Betsy Von Holle, James Wright, J. Walter Milon, and Matthew Bossick. 2016. "How Much are Floridians Willing to Pay for Protecting Sea Turtles from Sea Level Rise?" *Environmental Management* 57:176-188.

Hannan, Larua B., James D. Roth, Llewellyn Ehrhart, and John F. Weishampel. 2007. "Dune Vegetation Fertilization by Nesting Sea Turtles." *Ecology* 83 (4):1053-1058.

Hanson, JoAnne, Thane Wibbels, and R. Erik Martin. 1998. "Predicted Female Bias in Sex Ratios of Hatchling Loggerhead Sea Turtles From a Florida Nesting Beach." *Canadian Journal of Zoology* 76 (10):1850-1861.

Hanson, Paul J., and Jake F. Weltzin. 2000. "Drought Disturbance from Climate Change: Response of United States Forests." *The Science of the Total Environment* 262 (3):205-220.

Harley, Christopher D.G., and Laura Rogers-Bennett. 2004. "The Potential Synergistic Effects of Climate Change and Fishing Pressure on Exploited Invertebrates on Rocky Intertidal Shores." California Cooperative Oceanic Fisheries Investigations Report 45:98-110.

Harley, Christopher D.G., A. Randall Hughes, Kristin M. Hultgren, Benjamin G. Miner, Cascade J.B. Sorte, Carol S. Thornber, Laura F. Rodriguez, Lars Tomanek, and Susan L. Williams. 2006. "The Impacts of Climate Change in Coastal Marine Systems." *Ecology Letters* 9 (2):228-241.

Harvell, C. Drew, Charles E. Mitchell, Jessica R. Ward, Sonia Altizer, Andrew P. Dobson, Richard S. Ostfeld, and Michael D. Samuel. 2002. "Climate Warming and Disease Risks for Terrestrial and Marine Biota." *Science* 296 (5576):2158-2162.

Hawkes, L. A., A. C. Broderick, M. H. Godfrey, B. J. Godley, and M. J. Witt. 2009. "The Impacts of Climate Change on Marine Turtle Reproductive Success." In The Impacts of Climate Change, 287-310.

Heller, Nicole E., and Erika S. Zavaleta. 2009. "Biodiversity Management in the Face of Climate Change: A Review of 22 Years of Recommendations." *Biological Conservation* 142:14–32.

Hellmann, Jessica J., James E. Byers, Britta G. Bierwagen, and Jeffrey S. Dukes. 2008. "Five Potential Consequences of Climate Change for Invasive Species." *Conservation Biology* 22 (3):534-543.

Hickling, Rachel, David B. Roy, Jane K. Hill, Richard Fox, and Chris D. Thomas. 2006. "The Distributions of a Wide Range of Taxonomic Groups are Expanding Polewards." *Global Change Biology* 12:450-455.

Hill, J. K., Y. C. Collingham, C D. Thomas, D. S. Blakeley, R. Fox, D. Moss, and B. Huntley. 2001. "Impacts of Landscape Structure on Butterfly Range Expansion." *Ecology Letters* 4: 313-321.

Hodgson, G. 1999. "A Global Assessment of Human Effects on Coral Reefs." *Marine Pollution Bulletin* 38 (5):345-355.

Hoegh-Guldberg, Ove. 1999. "Climate Change, Coral Bleaching and the Future of the World's Coral Reefs." *Marine and Freshwater Research* 50:839-866.

Hoegh-Guldberg, Ove, P.J. Mumby, A.J. Hooten, R.S. Steneck, P. Greenfield, E. Gomez, C. D. Harvell, P.F. Sale, A.J. Edwards, K. Caldeira, N. Knowlton, C.M. Eakin, R. Iglesias-Prieto, N. Muthiga, R.H. Bradbury, A. Dubi, and M.E. Hatziolos. 2007. "Coral Reefs under Rapid Climate Change and Ocean Acidification." *Science* 318 (5857):1737-1742.

Holbrook, Joshua, and Thomas Chesnes. 2011. "An effect of Burmese pythons (Python molurus bivittatus) on mammal populations in southern Florida." *Fla Sci* 74:17–24.

Houston, D.R. 1998. "Beech Bark Disease." in K. Britton (ed.), Exotic pests of eastern forests. USDA Forest Service 29-41.

Howard, Robert, Ian Bell, and David A. Pike. 2014. "Thermal Tolerances of Sea Turtle Embryos: Current Understanding and Future Directions." *Endangered Species Research* 26 (1):75-86. doi: 10.3354/esr00636.

Hughes, Lesley. 2000. "Biological Consequences of Global Warming: Is the Signal Already Apparent?" *Trends in Ecology & Evolution* 15 (2):56-61.

Hughes, T.P., A.H. Baird, D.R. Bellwood, M. Card, S.R. Connolly, C. Folke, R. Grossberg, O. Hoegh-Guldberg, J.B.C. Jackson, J. Kleypas, J.M. Lough, P. Marshall, M. Nystrom, S.R. Palumbi, J.M. Pandolfi, B. Rosen, and J. Roughgarden. 2003. "Climate Change, Human Impacts, and the Resilience of Coral Reefs." *Science* 301:929-933.

Hughes, Lesley. 2000. "Biological Consequences of Global Warming: Is the Signal Already Apparent?" *Trends in Ecology & Evolution* 15:56-61.

Hulin, Vincent, Virginie Delmas, Marc Girondot, Matthew H. Godfrey, Jean-Michel Guillon. 2009. "Temperature-Dependent Sex Determination and Global Change: Are Some Species at Greater Risk?" *Oecologia* 160:493-506. doi 10.1007/s00442-009-1313-1

Hulme, Philip E. 2005. "Adapting to Climate Change: Is There Scope for Ecological Management in the Face of a Global Threat?" *Journal of Applied Ecology* 42 (5): 784-794.

Hungate, Bruce A., Frank P. Day, Paul Dijkstra, Benjamin D. Duval, C. Ross Hinkle, J. Adam Langley, J. Patrick Megonigal, Peter Stiling, Dale W. Johnson, and Bert G. Drake. 2013. "Fire, Hurricane and Carbon Dioxide: Effects on Net Primary Production of a Subtropical Woodland." *New Phytologist* 200:767-777.

Hymus, Graham J., Tom G. Snead, David P. Johnson, Bruce A. Hungate, and Bert G. Drake. 2002. "Acclimation of Photosynthesis and Respiration to Elevated Atmospheric CO_2 in Two Scrub Oaks." *Global Change Biology* 8:317-328.

IPBES. 2016. "Summary for Policymakers of the Methodological Assessment of Scenarios and Models of Biodiversity and Ecosystem Services of the Intergovernmental Science-Policy Platform on Biodiversity and Ecosystem Services." Eds. S. Ferrier, K. N. Ninan, P. Leadley, R. Alkemade, L.A. Acosta, H. R. Akçakaya, L. Brotons, W. Cheung, V. Christensen, K. A. Harhash, J. Kabubo-Mariara, C. Lundquist, M. Obersteiner, H. Pereira, G. Peterson, R. Pichs-Madruga, N. H. Ravindranath, C. Rondinini, B. Wintle. Secretariat of the Intergovernmental Science-Policy Platform on Biodiversity and Ecosystem Services, Bonn, Germany. 32 pages.

IPCC. 2007. "Climate Change 2007: Impacts, Adaptation and Vulnerability." Contribution of Working Group II to the Fourth Assessment Report of the Intergovernmental Panel on Climate Change, Parry, M. L. et al. Eds., Cambridge University Press, 996pp.

IPCC. 2014. "Climate Change 2014: Impacts, Adaptation, and Vulnerability." Contribution of Working Group II to the Fifth Assessment Report of the Intergovernmental Panel on Climate Change, Field, C.B. et al., Eds., Cambridge University Press, Cambridge, United Kingdom and New York, NY, USA, 1132 pp.

IUCN. 2010. "Guidelines for using the IUCN Red List Categories and Criteria Version 8.1."

IUCN. 2015. Guidelines for the application of IUCN Red List of Ecosystems Categories and Criteria, Version 1.0. ed. L.M. Bland, D.A. Keith, N.J. Murray and J.P. Rodríguez. Gland, Switzerland: International Union for the Conservation of Nature.

Jackson, D. R., and E. G. Milstrey. 1989. "The Fauna of Gopher Tortoise Burrows." Pages 86–98 in J. E. Diemer et al., editors. Proceedings gopher tortoise relocation symposium. Florida Game and Fresh Water Fish Commission Nongame Wildlife Program Technical Report No. 5, Tallahassee, Florida, USA.

Janzen, Frederic J. 1994. "Climate Change and Temperature-Dependent Sex Determination in Reptiles." *Population Biology* 91:7487-7490.

Jenkins, Clinton N., Kyle S. Van Houtan, Stuart L. Pimm, and Joseph O. Sexton. 2015. "US Protected Lands Mismatch Biodiversity Priorities." PNAS 112 (16) doi/10.1073/pnas.1418034112.

Jöhnk, K.D., J. Huisman, J. Sharples, B. Sommeijer, P.M. Visser, and J.M. Stroom. 2008. "Summer Heatwaves Promote Blooms of Harmful Cyanobacteria." *Global Change Biology* 14 (3):495-512.

Johnston, K., and O. Schmitz. 2003. "Wildlife and Climate Change: Assessing the Sensitivity of Selected Species to Simulated Doubling of Atmospheric CO_2." *Global Change Biology* 3 (6):531-544.

Jump, Alistair S., and Josep Penuelas. 2005. "Running to Stand Still: Adaptation and the Response of Plants to Rapid Climate Change." *Ecol Lett* 8:1010–1020. doi: 10.1111/j.1461-0248.2005.00796.x

Kappelle, Maarten, Margret MI Van Vuuren, and Pieter Baas. 1999. "Effects of Climate Change on Biodiversity: A Review and Identification of Key Research Issues." *Biodiversity and Conservation* 8 (10):1383-1397.

Katselidis, Kostas A., Gail Schofield, Georgios Stamou, Panayotis Dimopoulos, and John D. Pantis. 2014. "Employing Sea-Level Rise Scenarios to Strategically Select Sea Turtle Nesting Habitat Important for Long-Term Management at a Temperate Breeding Area." *Journal of Experimental Marine Biology and Ecology* 450:47-54. doi: 10.1016/j.jembe.2013.10.017.

Kautz, Randy, Robert Kawula, Thomas Hoctor, Jane Comiskey, Deborah Jansen, Dawn Jennings, John Kasbohm, Frank Mazzotti, Roy McBride, Larry Risharsdson, and Karen Root. 2006. "How Much is Enough? Landscape-scale Conservation for the Florida Panther." *Biological Conservation* 130 (1) 118-133.

Kent, D. M., and E. Snell. 1994. 'Vertebrates Associated with Gopher Tortoise Burrows in Orange County, Florida." *Florida Field Naturalist* 22:8–10.

Knight, G.R., J.B. Oetting, and L. Cross. 2011. "Atlas of Florida's Natural Heritage – Biodiversity, Landscapes, Stewardship, and Opportunities." Tallahassee, FL: Institute of Science and Public Affairs, Florida State University.

Knowlton, Nancy. 2001. "The Future of Corals." *Proc. Natl. Acad. Sci.* U.S.A. 98 (10):5419-5425.

Koch, Eva Maria W. 1996. "Hydrodynamics of Shallow Thalassia testudinum Beds in Florida, USA." In: Kuo, J., R.C. Phillips, D.I. Walker, and H. Kirkman (Eds.). Seagrass Biology: Proceedings of an International Workshop, Rottnest Island, Western Australia, pp.25-29

Koch, M. S., C. Coronado, M. W. Miller, D. T. Rudnick, E. Stabenau, R. B. Halley, and F. H. Sklar. 2015. "Climate Change Projected Effects on Coastal Foundation Communities of the Greater Everglades Using a 2060 Scenario: Need for a New Management Paradigm." *Environmental Management* 55:857–875

Komatsu, T. 1996. "Influence of a Zostera Bed on the Spatial Distribution of Water Flow Over a Broad Geographic Area. In: Kuo, J., R.C. Phillips, D.I. Walker, and H. Kirkman (Eds.). Seagrass Biology: Proceedings of an International Workshop, Rottnest Island, Western Australia, pp. 111-116.

Kushlan, J.A. 1988. "Conservation and Management of the American Crocodile." *Environmental Management* 12:777-790.

Lammertsma, Emmy I., Hugo Jan de Boer, Stefan C. Dekker, David L. Dilcher, Andre F. Lotter, and Friederike Wagner-Cremer. 2011. "Global CO2 Rise Leads to Reduced Maximum Stomatal Conductance in Florida Vegetation." *Proceedings of the National Academy of Sciences, USA* 108 (10):4035-4040.

Lamont, Margaret M., and Ikuko Fujisaki. 2014. "Effects of Ocean Temperature on Nesting Phenology and Fecundity of the Loggerhead Sea Turtle (Caretta caretta)." *Journal of Herpetology* 48 (1):98-102. doi: 10.1670/12-217.

Lang, Jeffrey W., and Harry V. Andrews. 1994. "Temperature-Dependent Sex Determination in Crocodilians." *J. Exp. Zool.* 270:28–44.

Lawler, Joshua J., Sarah L. Shafer, Denis White, Peter Kareiva, Edwin P. Maurer, Andrew R. Blaustein, and Patrick J. Bartlein. 2009. "Projected Climate-Induced Faunal Change in the Western Hemisphere." *Ecology* 90:588–597.

Ley, J. A., C. C. McIvor, and C. L. Montague. 1999. "Fishes in Mangrove Prop-Root Habitats of Northeastern Florida Bay: Distinct Assemblages Across an Estuarine Gradient." *Estuarine, Coastal and Shelf Science* 48 (6):701-723.

Li, J.-H., W.A. Dugas, G.J. Hymus, D.P. Johnson, C.R. Hinkle, B.G. Drake, and B.A. Hungate. 2003. "Direct and Indirect Effects of Elevated CO_2 on Transpiration from Quercus Myrtifolia in a Scrub-Oak Ecosystem." *Global Change Biology* 9:96-105.

Limpus, Colin J., and N. Nicholls. 1988. "The Southern Oscillation Regulates the Annual Numbers of Green Turtles (Chelonia mydas) Breeding Around Northern Australia." *Australian Journal of Wildlife Research* 15:157-61

Lindenmayer, D.B., M. P. Piggott, and B. A. Wintle. 2013. "Counting the Books While the Library Burns: Why Conservation Monitoring Programs Need a Plan for Action." *Frontiers in Ecology and the Environment* 11 (10):549-555.

Lodge, F. J., Paul Dijkstra, B. G. Drake, and James IL Morison. 2001. "Stomatal Acclimation to Increased CO2 Concentration in a Florida Scrub Oak Species Quercus myrtifolia Willd." *Plant, Cell and Environment* 24:77-88.

Lonsdale, D. and J.N. Gibbs. 1996. "Effects of Climate Change on Fungal Diseases of Trees." British Mycological Society Symposium; Fungi and Environmental Change 20:1-19.

Lowery, M.B. 2016. "A Climate Change and Sea-Level Rise Assessment and Prioritization Protocol of 20 Natural Communities in Georgia's Lower Coastal Plain." M.S., Biology, Valdosta State University.

Manzello, Derek P. 2015. "Rapid Recent Warming of Coral Reefs in the Florida Keys." *Scientific Reports*, 5, 16762. http://doi.org/10.1038/srep16762

Marcogliese, David J. 2001. "Implications of Climate Change for Parasitism of Animals in the Aquatic Environment." *Canadian Journal of Zoology* 79 (8):1331-1352.

Marcovaldi, Maria A.G., Milagros López-Mendilaharsu, Alexsandro S. Santos, Gustave G. Lopez, Matthew H. Godfrey, Frederico Tognin, Cecília Baptistotte, Joao C. Thomé, Augusto C. C. Dias, Jaqueline C. de Castilhos, and Mariana M. P. B. Fuentes. 2016. "Identification of Loggerhead Male Producing Beaches in the South Atlantic: Implications for Conservation." *Journal of Experimental Marine Biology and Ecology* 477:14-22. doi: http://dx.doi.org/10.1016/j.jembe.2016.01.001.

Marini, Miguel Angelo, Morgane Barbet-Massin, Leonardo Lopes, and Frederic Jiguet. 2009. "Predicted Climate-Driven Bird Distribution Changes and Forecasted Conservation Conflicts in a Neotropical Savanna." *Conservation Biology* 23:1558-1567.

Mawdsley, Jonathan R., Robin O'Malley, and Dennis S. Ojima. 2009. "A Review of Climate-Change Adaptation Strategies for Wildlife Management and Biodiversity Conservation." *Conservation Biology* 23 (5):1080-1089.

Mazzotti, Frank J., Michael S. Cherkiss, Kristen M. Hart, Ray W. Snow, Michael R. Rochford, Michael E. Dorcas, and Robert N. Reed. 2011. "Cold-induced Mortality of Invasive Burmese Pythons in South Florida." *Biological Invasions* 13:143–151.

McCarty, John P. 2001. "Ecological Consequences of Recent Climate Change." *Conservation Biology* 15 (2):320-331.

McGuire, Jenny L., Joshua J. Lawler, Brad H. McRae, Tristan A. Nuñez, and David M. Theobald. 2016. "Achieving Climate Connectivity in a Fragmented Landscape." *Proceedings of the National Academy of Sciences* doi/10.1073/pnas.1602817113.

McNeely, Jeffrey, A., Mooney HA, Neville L, Schei PJ and Waage JK. 2001. "A Global strategy on Invasive Alien Species." IUCN Gland, Switzerland, and Cambridge, U

Menzel, Annette, Tim H. Sparks, Nicole Estrella, Elisabeth Koch, Anto Aasa, Rein Ahas, Kerstin Alm-Kubler, Peter Bissolli, Ol'Ga Braslavska, Agrita Briede, et al. 2006. "European Phenological Response to Climate Change Matches the Warming Pattern." *Global Change Biology* 12 (10):1969-1976. doi: 10.1111/j.1365-2486.2006.01193.x.

Michener, William K., Elizabeth R. Blood, Keith L. Bildstein, Mark M. Brinson, and Leonard R. Gardner. 1997. "Climate Change, Hurricanes and Tropical Storms, and Rising Sea Level in Coastal Wetlands." *Ecological Applications* 7 (3):770-801.

Miller, J. D. 1985. "Embryology of Marine Turtles." In Biology of the Reptilia Vol.14, edited by C. Gans, F. Billett and P. F. A. Maderson, 271-328. New York: Wiley Interscience.

Miller, Rebecca M, Jon Paul Rodrí-guez, Theresa Aniskowicz-Fowler, Channa Bambaradeniya, Ruben Boles, Mark A Eaton, Ulf Gardenfors, Verena Keller, Sanjay Molur, Sally Walker, and Caroline Pollock. 2006. "Extinction Risk and Conservation Priorities." *Science* 313:441. doi: 10.1126/science.313.5786.441a.

Mitchell, Carol. L., and Jerome A. Krueger. 2011. "Climate Change Science in the Everglades National Park." *ParkScience* 28 (2):1-10.

Mooney, H.A. and R.J. Hobbs (eds.). 2000. "Invasive Species in a Changing World." Island Press, Washington, DC.

Moore, Stephanie K., Vera L. Trainer, Nathan J. Mantua, Micaela S. Parker, Edward A. Laws, Lorraine E. Fleming, and Lora C. Backer. 2008. "Impacts of Climate Variability and Future Climate Change on Harmful Algal Blooms and Human Health". *Environmental Hea*lth 7 (Suppl 2).

Moss, A. and S. Martin. 2012. "Flexible Adaptation Pathways." ClimateXChange, Edinburgh Centre for Carbon Innovation. Accessed September 2016. http://www.climatexchange.org.uk/adapting-to-climate-change/flexible-adaptation-pathways/.

Mrosovsky, N. 1980. "Thermal Biology of Sea Turtles." *American Zoologist* 20:531-547.

Mrosovsky, N., and J. Provancha. 1989. "Sex Ratio of Loggerhead Sea Turtles Hatching on a Florida Beach." *Canadian Journal of Zoology* 67 (10):2533-2539.

Muller, J. W., E.D. Hardin, D.R. Jackson, S.F. Gatewood and N. Caire. 1989. "Summary Report on the Vascular Plants, Animals and Plant Communities Endemic to Florida." Florida Game and Freshwater Fish Commission, Nongame Wildlife Program Technical Report No. 7. Tallahassee, Florida, USA.

Murley, James, Barry N. Heimlich, and Nicholas Bollman. 2008. "Florida's Resilient Coasts: A State Policy Framework for Adaptation to Climate Change." Florida Atlantic University 72pp.

Myers, R. L. and J. J. Ewel. eds. 1990. "Ecosystems of Florida." University of Central Florida Press, Orlando, Florida, USA.

National Climate Change Adaptation Research Facility - NCCARF. 2016. CoastAdapt (beta). Griffith University, Australia. Accessed September 2016. https://coastadapt.com.au/.

National Fish, Wildlife and Plants Climate Adaptation Strategy Partnership - NFWPCAS. 2012. "National Fish, Wildlife and Plants Climate Adaptation Strategy." Association of Fish and Wildlife Agencies, Council on Environmental Quality, Great Lakes Indian Fish and Wildlife Commission, National Oceanic and Atmospheric Administration, and U.S. Fish and Wildlife Service. Washington, DC.

Neal, Letitia, Terry Gilbert, Thomas Eason, Lisa Grant, and Tom Roberts. 2003. "Resolving Landscape Level Highway Impacts on the Florida Black Bear and Other Listed Wildlife Species." Road Ecology Center. UC Davis.

Neeman, Noga, Nathan J. Robinson, Frank V. Paladino, James R. Spotila, and Michael P. O'Connor. 2015. "Phenology Shifts in Leatherback Turtles (Dermochelys coriacea) Due to Changes in Sea Surface Temperature." *Journal of Experimental Marine Biology and Ecology* 462:113-120. doi: 10.1016/j.jembe.2014.10.019.

Nixon, Scott. W., Candace A. Oviatt. 1972. "Preliminary Measurements of Midsummer Metabolism in Beds of Eelgrass, Zostera marina." *Ecology* 53:150-153.

Noss, Reed F. 2011. "Between the Devil and the Deep Blue Sea: Florida's Unenviable Position with Respect to Sea Level Rise." *Climate Change* 107:1-16.

Noss, Reed F., Joshua S. Reece, Thomas Hoctor, and Jon Oetting. 2014. "Adaptation to Sea-level Rise in Florida: Biological Conservation Priorities". Final Report to the Kresge Foundation Grant Request Number 244146

Noss, R. F., W.J. Platt, B.A. Sorrie, A.S. Weakley, D.B. Means, J. Costanza, and R.K. Peet. 2015. "How Global Biodiversity Hotspots May Go Unrecognized: Lessons from the North American Coastal Plain." *Diversity and Distributions* 21:236-244. 10.1111/ddi.12278

Noyes, Pamela D., Matthew K. McElwee, Hilary Miller, Bryan Clark, Lindsey A. Van Tiem, Kia C. Walcott, Kyle Erwin, and Edward D. Levin. 2009. "The Toxicology of Climate Change: Environmental Contaminants in a Warming World." *Environmental International* 35 (6):971-986.

Osland, Michael J., Nicholas Enwright, Richard H. Day, and Thomas W. Doyle. 2013. "Winter Climate Change and Coastal Wetland Foundation Species: Salt Marshes vs. Mangrove Forests in the Southeastern United States" *Global Change Biology* 19:1482-1494, doi: 10.1111/gcb.12126.

Paerl, Hans W., and Jef Huisman. 2008. "Blooms Like it Hot." *Science* 320:57-58.

Palis, John. G. 1997. "Breeding Migration of Ambystoma Cingulatum in Florida." *Journal of Herpetology* 31 (1):71-78.

Pandolfi, John M., Roger H. Bradbury, Enric Sala, Terence P. Hughes, Karen A. Bjorndal, Richard G. Cooke, Deborah McArdle, Loren McClenachan, Marah J.H. Newman, Gustavo Paredes, Robert R. Warner, and Jeremy B.C. Jackson. 2003. "Global Trajectories of the Long-Term Decline of Coral Reef Ecosystems." *Science* 301:955.

Parker, I. M., D. Simberloff, W.M. Lonsdale, K. Goodell, M. Wonham, P.M. Kareiva, M.H. Williamson, B. Von Holle, P.B. Moyle, J.E. Byers, and L. Goldwasser. 1999. "Impact: Toward a Framework for Understanding the Ecological Effects of Invaders." *Biological Invasions* 1:3–19.

Parmesan, C. 2006. "Ecological and Evolutionary Responses to Recent Climate Change." *Annu Rev Ecol Evol Syst* 37:637–669

Parmesan, Camille, and Hector Galbraith. 2004. "Observed Impacts of Global Climate Change in the US." Pew Center on Global Climate Change, Vol. 12. Arlington, VA, USA.

Parmesan, Camille and Gary Yohe. 2003. "A Globally Coherent Fingerprint of Climate Change Impacts Across Natural Systems." *Nature* 42 (2):37-42.

Paulson, Dennis R. 2001. "Recent Odonata Records from Southern Florida - Effects of Global Warming?" *International Journal of Odonatology* 4 (1):57-69.

Pearlstine, Leonard G., E.V. Pearlstine, Nicholas G. Aumen. 2010. "A Review of the Ecological Consequence and Management Implications of Climate Change on the Everglades." *J N Am Benthol Soc* 29:1510-1526.

Perry, L., and K. Williams. 1996. "Effects of Salinity and Flooding on Seedlings of Cabbage Palm (Sabal palmetto)." *Oecologia* 105 (4):428-434.

Pertoldi C., and L.A. Bach. 2007. "Evolutionary Aspects of Climate-Induced Changes and the Need for Multidisciplinarity." *J Therm Biol* 32:118–124

Peterson, Garry D., Graeme S. Cumming, and Stephen R. Carpenter. 2003. "Scenario Planning: A Tool for Conservation in an Uncertain World." *Conservation Biology* 17 (2):358-366.

Pike, David A. 2013. "Forecasting Range Expansion into Ecological Traps: Climate-Mediated Shifts in Sea Turtle Nesting Beaches and Human Development." *Global Change Biology* 19 (10):3082-92. doi: 10.1111/gcb.12282.

Pike, David A., Elizabeth A. Roznik, and Ian Bell. 2015. "Nest Inundation from Sea-Level Rise Threatens Sea Turtle Population Viability." *Royal Society Open Science* 2 (7):150127. doi: 10.1098/rsos.150127.

Pike, David A., Rebecca L. Antworth, and John C. Stiner. 2006. "Earlier Nesting Contributes to Shorter Nesting Seasons for the Loggerhead Seaturtle, Caretta caretta." *Journal of Herpetology* 40 (1):91-94. doi: 10.1670/100-05n.1.

Pilkey, Orrin H., and Rob Young. 2009. The Rising Sea. Washington DC: Island Press.

Pimm, Stuart L., and Peter Raven. 2000. "Biodiversity: Extinction by Numbers." *Nature* 403 (6772):843-845.

Poff, N. LeRoy., Mark M. Brinson, and John W. Day, Jr. 2002. "Aquatic Ecosystems and Global Climate Change: Potential Impacts on Inland Freshwater and Coastal Wetland Ecosystems in the United States." Pew Center on Global Climate Chang, Arlington, VA 44pp.

Poloczanska, Elvira S., Colin J. Limpus, and Graeme C. Hays. 2009. "Vulnerability of Marine Turtles to Climate Change." In Advances in Marine Biology, 151-211. Academic Press.

Pounds, J. Alan, Martı́n R. Bustamante, Luis A. Coloma, Jamie A. Consuegra, Michael P. L. Fogden, Pru N. Foster, Enrique La Marca, Karen L. Masters, Andre´s Merino-Viteri, Robert Puschendorf, Santiago R. Ron, G. Arturo Sanchez-Azofeifa, Christopher J. Still, and Bruce E. Young. 2006. "Widespread Amphibian Extinctions from Epidemic Disease Driven by Global Warming." *Nature* 439 (12): doi: 10.1038/nature04246

Rahel, Frank J., Britta Bierwagen, and Yoshinori Taniguchi. 2008. "Managing Aquatic Species of Conservation Concern in the Face of Climate Change and Invasive Species." *Conservation Biology* 22 (3):551-561.

Reece, Joshua Steven, Reed F. Noss, Jon Oetting, Tom Hoctor, and Michael Volk. 2013a. "A Vulnerability Assessment of 300 Species in Florida: Threats from Sea Level Rise, Land Use, and Climate Change." *PLoS ONE* 8:e80658. doi: 10.1371/journal.pone.0080658.

Reece, Joshua S., Davina Passeri, Llewellyn Ehrhart, Scott C. Hagen, Allison Hays, Christopher Long, Reed F. Noss, Matthew Bilskie, Cheryl Sanchez, Monette V. Schwoerer, Betsy Von Holle, John Weishampel, and Shaye Wolf. 2013b. "Sea Level Rise, Land Use, and Climate Change Influence the Distribution of Loggerhead Turtle Nests at the Largest USA Rookery (Melbourne Beach, Florida)." *Marine Ecology Progress Series* 493:259-274. doi: 10.3354/meps10531.

Reece, Joshua S., and Reed F. Noss. 2014. "Prioritizing Species by Conservation Value and Vulnerability: A New Index Applied to Species Threatened by Sea-Level Rise and Other Risks in Florida." *Natural Areas Journal* 34 (1):31-45. doi:10.3375/043.034.0105. doi: 10.3375/043.034.0105.

Reid, P. C. 2003. "Climate Change, Ocean Processes and Plankton Regime Shifts." In: Global Climate Change and Biodiversity eds. R. E. Green, M. Harley, L. Miles, J. Scharlemann, A. Watkinson and O. Watts. Pp.13-14. The RSPB, Sandy, UK, University of East Anglia, Norwich, UK

Robinson, Robert A., Humphrey QP Crick, Jennifer A. Learmonth, Ilya Maclean, Chris D. Thomas, Franz Bairlein, Mads C. Forchhammer, Charles M. Francis, Jennifer A. Gill, Brendan J. Godley, John Harwood, Graeme C. Hays, Brian Huntley, Anthony M. Hutson, Graham J. Pierce, Mark M. Rehfisch, David W. Sims, M. Begona Santos, Timothy H. Sparks, David A. Stroud, and Marcel E. Visser. 2009. "Travelling Through a Warming World: Climate Change and Migratory Species." *Endangered Species Research* 7 (2):87-99.

Rosenau, J. C., G. L. Faulkner, C. W. Hendry, Jr., and R. W. Hull. 1977. "Springs of Florida." *Florida Bureau of Geology Bulletin* No. 31.

Ruzicka, R.R., M.A. Colella, J.W. Porter, J.M. Morrison, J.A. Kidney, V. Brinkhuis, K.S. Lunz, K.A. Macaulay, L.A. Bartlett, M.K. Meyers, and J. Colee. 2013. "Temporal Changes in Benthic Assemblages on Florida Keys Reefs 11 Years After the 1997/1998 El Niño." *Marine Ecology Progress Series* 489:125–141.

Saba, Vincent S., Pilar Santidrian-Tomillo, Richard D. Reina, James R. Spotila, John A. Musick, David A. Evans, and Frank V. Paladino. 2007. "The Effect of the El Niño Southern Oscillation on the Reproductive Frequency of Eastern Pacific Leatherback Turtles." *Journal of Applied Ecology* 44 (2):395-404. doi: 10.1111/j.1365-2664.2007.01276.x.

Sala, Osvaldo E., F. Stuart Chapin, Juan J. Armesto, Eric Berlow, Janine Bloomfield, Rodolfo Dirzo, Elisabeth Huber-Sanwald, Laura F. Huenneke, Robert B. Jackson, Ann Kinzig, Rik Leemans, David M. Lodge, Harold A. Mooney, Martin Oesterheld, N. LeRoy Poff, Martin T. Sykes, Brian H. Walker, Marilyn Walker, and Diana H. Wall. 2000. "Global Biodiversity Scenarios for the Year 2100." *Science* 287 (5459):1770-1774. Sanford, Eric. 1999. "Regulation of Keystone Predation by Small Changes in Ocean Temperature." *Science* 283:2095-2097.

Schweiger, Oliver, Josef Settele, Otakar Kudrna, Stefan Klotz, and Ingolf Kühn. 2008. "Climate Change Can Cause Spatial Mismatch of Trophically Interacting Species." *Ecology* 89 (12):3472-3479.

Seiler, Troy J., Daniel P. Rasse, Jiahong Li, Paul Dijkstra, Hans P. Anderson, David P. Johnson, Thomas L. Powell, Bruce A. Hungate, C. Hinkle, and Bert G. Drake. 2009. "Disturbance, Rainfall and

Contrasting Species Responses Mediated Aboveground Biomass Response to 11 Years of CO2 Enrichment in a Florida Scrub-Oak Ecosystem." *Global Change Biology* 15:356-367.

Shellenbarger Jones, A., C. Bosch, and E. Strange. 2009. "Vulnerable Species: the Effects of Sea-level Rise on Coastal Habitats." in Coastal Sensitivity to Sea-Level Rise: A Focus on the Mid-Atlantic Region. A Report by the U.S. Climate Change Science Program and the Subcommittee on Global Change Research. J.G. Titus (coordinating lead author), K.E. Anderson, D.R. Cahoon, D.B. Gesch, S.K. Gill, B.T. Gutierrez, E.R. Thieler, and S.J. Williams (lead authors). U.S. environmental Protection agency, Washington, DC. 43-56 pp.

Short, Frederick T., and Hilary A. Neckles. 1999. "The Effects of Global Climate Change on Seagrasses." *Aquatic Botany* 63:169-196.

Short, F.T., and C.A. Short. 1984. "The Seagrass Filter: Purification of Estuarine and Coastal Water." In: Kennedy, V.S. (Ed.), The Estuary as a Filter. Academic Press, Orlando, pp. 395±413.

Smith, Thomas J. III, Ann M. Foster, Ginger Tiling-Range, and John W. Jones. 2013. "Dynamics of Mangrove-Marsh Ecotones in Subtropical Coastal Wetlands: Fire, Sea-Level Rise, and Water Levels." *Fire Ecology* 9:66-77.

Snyder, J. R., A. Herndon, and W. B. Robertson, Jr. 1990. "South Florida Rockland." Pages 230–277 in R. L. Myers and J. J. Ewel, editors. Ecosystems of Florida. University of Central Florida Press, Orlando, Florida, USA.

Solomon, Susan, Daha Qin, Martin Manning, Z. Chen, M. Marquis, K.B. Averyt, M. Tignor, and H.L. Miller, eds. 2007. Contribution of Working Group I to the Fourth Assessment Report of the Intergovernmental Panel on Climate Change. Cambridge University Press, Cambridge, United Kingdom and New York, NY, USA

South West Climate Change Portal. 2016. "Adaptation Pathways, 5. Monitoring, Evaluation, Reporting, Improvement and Learning." The Climate Resilient Communities of the Barwon South West Project. Accessed September 2016. http://www.swclimatechange.com.au/cb_pages/adaptation_pathways.php

Southeast Florida Regional Climate Change Compact Counties. 2012. "A Region Responds to a Changing Climate." Regional Climate Action Plan, 84pp.

Standora, Edward, and James R. Spotila. 1985. "Temperature Dependent Sex Determination in Sea Turtles." *Copeia* 1985 (3):711-722.

Steffen, Will., Andrew Burbridge, Lesley Hughes, Roger Kitching, David Lindenmayer, Warren Musgrave, Mark Stafford Smith, and Patricia Werner. 2009. "Australia's Biodiversity and Climate change: Summary for Policy Makers 2009." Australia's biodiversity and climate change: summary for policy makers.

Stein, Bruce A., Patty Glick, Naomi Edelson, and Amanda Staudt. 2014. "Climate-Smart Conservation: Putting Adaption Principles into Practice." National Wildlife Federation.

Stein, Bruce A., Amanda Staudt, Molly S. Cross, Natalie S. Dubois, Carolyn Enquist, Roger Griffis, Lara J. Hansen, Jessica J. Hellmann, Joshua J. Lawler, Erik J. Nelson and Amber Pairis. 2013. "Preparing for and Managing Change: Climate Adaptation for Biodiversity and Ecosystems." *Frontiers in Ecology and the Environment* 11 (9):502-510. doi: 10.1890/120277

Steinitz, Carl, and Peter Rogers. 1970. A Systems Analysis Model of Urbanization and Change: An Experiment in Inter-disciplinary Education. Cambridge, MA: MIT Press.

Stevenson, J. 1988. "Comparative Ecology of Submersed Grass Beds in Freshwater, Estuarine, and Marine Environments. *Limnol. Oceanogr.* 33:867-893.

Stiling, Peter, Daniel C. Moon, Mark D. Hunter, Jamie Colson, Anthony M. Rossi, Graham J. Hymus, and Bert G. Drake. 2003. "Elevated CO_2 Lowers Relative and Absolute Herbivore Density Across All Species of a Scrub-Oak Forest." *Oecologia* 134:82-87.

Stiling, P., D. C. Moon, A. Rossi, R. Forkner, B. A. Hungate, F. P. Day, R. E. Schroeder, and B. Drake. 2013. "Direct and Legacy Effects of Long-Term Elevated CO2 on Fine Root Growth and Plant-Insect Interactions." *New Phytologist* 200 (3):788-795. doi: 10.1111/nph.12295.

Thayer, Gordon W., David R. Colby, and William F. Hettler, Jr. 1987. "Utilization of the Red Mangrove Prop Root Habitat by Fishes in South Florida." *Marine Ecology Progress Series* 35: 25–38.

Thomas, Chris D, Alison Cameron, Rhys E Green, Michel Bakkenes, Linda J Beaumont, Yvonne C Collingham, Barend F N Erasmus, Marinez Ferreira de Siqueira, Alan Grainger, Lee Hannah, Lesley Hughes, Brian Huntley, Albert S van Jaarsveld, Guy F Midgley, Lera Miles, Miguel A Ortega-Huerta, A Townsend Peterson, Oliver L Phillips, and Stephen E Williams. 2004. "Extinction Risk from Climate Change." *Nature* 427:145-148.

Thuiller, Wilfried, Sandra Lavorel, Miguel B. Araujo, Martin T. Sykes, and I. Colin Prentice. 2005. "Climate Change Threats to Plant Diversity in Europe." *Proceedings of the National Academy of Sciences* (USA) 102:8245-8250.

Trombulak, Stephen C., and Christopher A. Frissell. 2000. "Review of Ecological Effects of Roads on Terrestrial and Aquatic Communities." *Conservation Biology* 14 (1):18-30.

United States Fish and Wildlife Service - USFWS. 1999. "South Florida Multi-species Recovery Plan." South Florida Ecological Services Field Office. Vero Beach, Florida.

Vargas, J.C., M. Flaxman, and B. Fradkin. 2014. "Landscape Conservation and Climate Change Scenarios for the State of Florida: A Decision Support System for Strategic Conservation." Summary for Decision Makers. GeoAdaptive LLC, Boston, MA and Geodesign Technologies Inc., San Francisco CA.

Viets, Brian E., Michael A. Ewert, Larry G.Talent, and Craig E. Nelson. 1994. "Sex-Determining Mechanisms in Squamate Reptiles." *J. Exp. Zool.* 270:45–56.

Visser, Marcel E. 2008. "Keeping Up With a Warming World; Assessing the Rate of Adaptation to Climate Change." *Proc R Soc B* 275:649–659.

Vitousek, Peter M., Carla M. D'Antonio, Lloyd L. Loope, and Randy Westbrooks. 1996. "Biological Invasions as Global Environmental Change." *American Scientist* 84 (5):468-478.

Von Holle, B., Y. Wei, and D. Nickerson. 2010. "Climatic Variability Leads to Later Seasonal Flowering of Floridian Plants." *PLoS ONE* 5 (7):1-9: e11500. doi:10.1371/journal.pone.0011500.

Wagner, Friederike, David L. Dilcher, and Henk Visscher. 2005. "Stomatal Frequency Responses in Hardwood-Swamp Vegetation from Florida During A 60-Year Continuous CO_2 Increase." *American Journal of Botany* 92 (4):690-695.

Wagner, Friederike, Henk Visscher, Wolfram M. Kurschner, and David L. Dilcher. 2007. "Influence of Ontogeny and Atmospheric CO_2 on Stomata Parameters of Osmunda regalis." Cour. Forsch.-Inst. Senckenberg 258:183-189

Walther, Gian-Reto, Eric Post, Peter Convey, Annette Menzel, Camille Parmesan, Trevor J.C. Beebee, Jean-Marc Fromentin, Ove Hoegh-Guldberg, Franz Bairlein. 2002. "Ecological Responses to Recent Climate Change." *Nature* 416:389–395.

Walther, Gian-Reto, Alain Roques, Philip E. Hulme, Martin T. Sykes, Petr Pyšek, Ingolf Kühn, Martin Zobel, et al. 2009. "Alien Species in a Warmer World: Risks and Opportunities." *Trends in Ecology & Evolution* 24 (12):686-693.

Ward, Larry G., W. Michael Kemp, and Walter Boynton. 1984. "The Influence of Waves and Seagrass Communities on Suspended Particulates in an Estuarine Embayment." *Mar. Geol.* 59:85-103.

Watson, Amanda, Joshua S. Reece, Blair E. Tirpak, Cynthia Kallio Edwards, Laura Geselbracht, Mark Woodrey, Megan LaPeyre, and Patricia Soupy Dalyander. 2015. The Gulf Coast Vulnerability Assessment: Mangrove, Tidal Emergent Marsh, Barrier Islands, and Oyster Reef. http://gulfcoastprairielcc.org/science/science-projects/gulf-coast-vulnerability-assessment/.

Watson, James. E., Madhu. Rao, Kang Ai-Li, and Xie Yan. 2012. "Climate Change Adaptation Planning for Biodiversity Conservation: A Review." *Advances in Climate Change Research* 3 (1):1-11.

Weishampel, John F., Dean A. Bagley, and Llewellyn M. Ehrhart. 2004. "Earlier Nesting by Loggerhead Sea Turtles Following Sea Surface Warming." *Global Change Biology* 10: 1424-1427. doi: 10.1111/j.1365-2486.2004.00817.x.

Wibbels, T. 2012a. "Evaluating the Effects of Global Warming, Beach Nourishment, and Beach Location on Temperature-Dependent Sea Determination in Loggerhead Sea Turtles in Florida." University of Alabama, Birmingham.

Wibbels, T. 2012b. "Implications of Temperature-Dependent Sea Determination for the Conservation & Biology of Loggerhead Sea Turtles in Florida." University of Alabama, Birmingham.

Wibbels, T. 2012c. "Simultaneous Evaluation of Loggerhead Nesting Beach Temperatures Throughout Florida: Implications for Population Sex Ratio". University of Alabama, Birmingham.

Wibbels, Thane, R. Erik Martin, David W. Owens, and Max S. Amoss Jr. 1991. "Female-Biased Sex Ratio of Immature Loggerhead Sea Turtles Inhabiting the Atlantic Coastal Waters of Florida." *Canadian Journal of Zoology* 69 (12):2973-2977. doi: 10.1139/z91-419.

Wilcove, David S., David Rothstein, Jason Dubow, Ali Phillips, and Elizabeth Losos. 1998. "Quantifying Threats to Imperiled Species in the United States." *BioScience* 48 (8):607-615.

Wildish, D. and D. Kristmanson. 1997. "Benthic Suspension Feeders and Flow." Cambridge: Cambridge University Press. 409 pp.

Wilkinson, C. 2004. "Status of Coral Reefs of the World." Australian Institute of Marine Studies (AIMS), Townsville, Australia.

Wilkinson, Clive and David Souter, eds. 2008. "Status of Caribbean Coral Reefs After Bleaching and Hurricanes in 2005." Global Coral Reef Monitoring Network.

Willson, John D., Michael E. Dorcas, and Raymond W. Snow. 2011. "Identifying Plausible Scenarios for the Establishment of Invasive Burmese Pythons (Python molurus) in Southern Florida." *Biological Invasions* 13:1493–1504.

Wilson, John R.U., Eleanor E. Dormontt, Peter J. Prentis, Andrew J. Lowe, and David M. Richardson. 2009. "Something in the Way You Move: Dispersal Pathways Affect Invasion Success." *Trends in Ecology & Evolution* 24 (3): 136-144.

Witt, Matthew J., Lucy A. Hawkes, M. H. Godfrey, B. J. Godley, and A. C. Broderick. 2010. "Predicting The Impacts of Climate Change on a Globally Distributed Species: The Case of the Loggerhead Turtle." *Journal of Experimental Biology* 213 (6):901-11. doi: 10.1242/jeb.038133.

Wyneken, Jeanette, and Alexandra Lolavar. 2015. "Loggerhead Sea Turtle Environmental Sex Determination: Implications of Moisture and Temperature for Climate Change Based Predictions for Species Survival." *Journal of Experimental Zoology* 324 (3):295-314. doi: 10.1002/jez.b.22620.

Yntema, C. L., and Nicholas Mrosovsky. 1980. "Sexual Differentiation in Hatchling Loggerheads (Caretta caretta) Incubated at Different Controlled Temperatures." *Herpetologica* 36:33-36.

Young, B E, E Byers, K Gravuer, K R Hall, G Hammerson, A Redder, K Szabo, and J E Newmark. 2009. "Using the NatureServe Climate Change Vulnerability Index: A Nevada Case Study." Arlington, Virginia.

CHAPTER 13

Florida's Oceans and Marine Habitats in a Changing Climate

Steven Morey[1], Marguerite Koch[2], Yanyun Liu[3], and Sang-Ki Lee[4]

[1]Center for Ocean-Atmospheric Prediction Studies, Florida State University, Tallahassee, FL;
[2]Department of Biological Sciences, Florida Atlantic University, Boca Raton, FL;
[3]Cooperative Institute for Marine and Atmospheric Studies, University of Miami, Miami, FL; [4]Atlantic Oceanographic and Meteorological Laboratory, National Oceanic and Atmospheric Administration, Miami, FL

Florida's peninsula extending ~700 km north-to-south, extensive shoreline (2,100 km), and broad carbonate platform create a diversity of marine habitats (estuaries, lagoons, bays, beach, reef, shelf, pelagic) along the coast, shelf, and deep ocean that are influenced by continental, oceanographic, and atmospheric processes all predicted to shift with a rapidly changing climate. Future changes of the global ocean circulation could result in a 25% reduction in the Atlantic Meridional Overturning Circulation (AMOC), leading to a subsequent slowing of Florida's regional/local current systems (Yucatan, Loop, Florida and Gulf Stream) and eddies. While downscaled climate models suggest that slowing of the Loop Current by 20-25% during the 21st century will moderate the increase in surface temperatures in the Gulf of Mexico to 1.4ºC - 2.8ºC, this warming is predicted to have wide-ranging consequences for Florida's marine habitats (e.g., enhanced coral bleaching, lower O_2 in surface waters, increased harmful algal blooms, reduced phytoplankton and fisheries production, and lower sea turtle reproduction). The reduction in the AMOC is also predicted to reduce hurricane frequency, albeit with increased intensity (2-11%) due to ocean warming. Climate projections affecting Florida's oceans include rises in sea level, changes in coastal circulation impacting larval and nutrient transport, changes in marine biogeochemistry including ocean acidification, and loss of coastal wetlands that protect Florida's coastline. Understanding the consequences of these projected climate impacts and gaining a more complete understanding of complex changes in atmospheric processes (e.g., ENSO, AMO, convection, wind shear), air-sea interaction, currents, and stratification under a changing climate is critical over the next few decades to prepare and protect the state of Florida.

Key Messages

- Florida has a unique peninsular geography that creates an extensive shoreline with a diversity of marine habitats along the coast, shelf, and deep ocean influenced by continental, oceanographic, and atmospheric processes all predicted to shift with a rapidly changing climate.
- Climate projections affecting Florida's oceans include rise in sea level, warmer sea surface temperatures, changes in coastal circulation impacting larval and nutrient transport, changes in marine biogeochemistry including ocean acidification, and loss of coastal wetlands and reefs that protect Florida's coastline.
- Downscaled ocean models have proven successful for understanding future changes for the region given climate projections, and their continued revision and improvement will result in a more complete understanding of complex changes in air-sea interaction, large-scale currents, and the rates of climate change impacts, a critical research need over the next few decades to prepare and protect the state of Florida.

Keywords

Ocean climate; Sea level rise; Florida climate; Gulf of Mexico; AMOC; Caribbean climate; Florida hydrology; Florida reefs; Global warming

Introduction

Geomorphology

Florida's unique peninsular shape, with over 2,100 km (1,304 mi) of shoreline extending approximately 700 km (435 mi) north-to-south and a broad subsurface carbonate platform, creates a diversity of marine environments along the coast, shelf, and deep basins that are subject to influence by regional and global climate and ocean circulation patterns (Fig. 13.1). The geomorphology of the shelf that encircles Florida influences coastal connectivity to deep basins in the Gulf of Mexico and Atlantic Ocean (Fig. 13.1). For example, a narrow shelf at the head of the De Soto Canyon along the western extent of the state connects shelf and coastal waters with deep oceanic basins within the Gulf of Mexico (Fig. 13.1). The West Florida Shelf (WFS) – a long, broad, flat carbonate platform – buffers the west coast of the peninsula from the offshore waters of the Gulf. At the southern tip of the peninsula, the Florida Keys are proximate to the shelf edge with the shallow Florida Bay to the north and the deeper Straits of Florida to the south (Fig. 13.1). Along Florida's East Coast, the shelf is a narrow strip at the southern part of the peninsula that widens northward as part of the South Atlantic Bight, which extends offshore of North Carolina.

Coastal Hydrography

Coastal regions are buffered from the open ocean by wide continental shelves along much of Florida's coastline. The hydrography (physical properties of seawater) of these coastal waters is more heavily influenced by local atmospheric forcing and rivers than by the deep ocean, an example of how geomorphology is linked to hydrography. The northern parts of Florida, including the Panhandle, Big Bend, and northeastern regions, experience distinct seasonal climate regimes during the year, with a strong continental influence in the cool months and coastal humid subtropical conditions during the warm months. Numerous rivers along the coast discharge fresh water that mixes with ocean waters before being exported to deeper water. Many of Florida's rivers, which discharge into estuaries and coastal lagoons, have relatively local watersheds. Notable exceptions include rivers in the Apalachicola Basin, which encompasses the Chattahoochee and Flint rivers in eastern Alabama and western Georgia. These coastal, shallow water systems are adjacent to larger regional and global current systems that have significant influences on Florida's climate, coastal hazards (storm surges), seawater temperature regimes, and productivity of marine habitats.

Current Systems

The major regional and global ocean currents that are proximate to the Florida Peninsula are illustrated in Fig. 13.1. Along the eastern coastline of Florida, the subtropical North Atlantic Ocean has an intensified western boundary current linked to the Florida and Antilles currents, generally referred to as the Gulf Stream system, which flows northward (Fig. 13.1). The Gulf Stream system is a component of the global thermohaline circulation or Atlantic Meridional Overturning Circulation (AMOC; Fig. 13.1). The proximity of the Gulf Stream system to Florida provides a strong connection between Florida's marine environments and the global ocean circulation. A branch of this western Atlantic boundary current that enters the Gulf of Mexico from the Caribbean Sea through the Yucatan Channel forms the Loop Current, which turns eastward and exits the Gulf through the Straits of Florida. This current then flows northward along the east Florida coast as the Florida Current (Fig. 13.1).

The Florida Current/Gulf Stream system is globally significant as it is responsible for poleward transport of approximately 32 Sv (Sverdrups) of seawater (Barringer and Larsen 2001), equating to 32 million cubic meters of water per second (10^6 m^3 s^{-1}) and 1.3 PW (Petawatts), or 1.3 quadrillion watts (10^{15} W) of thermal energy (Larsen 1992). Thus, Florida currents are an integral part of the global current system (AMOC) and contribute to the planetary energy balance between the tropics and poles. Because these global currents are so inextricably linked to Florida's regional currents, the state and its coastline experience climate and hydrological reverberations as these global currents change in response to large-scale global shifts in flow (Ezer 2015). This local-global current interaction and the low elevation of Florida make the state vulnerable to even modest changes in global ocean circulation patterns.

Natural Variability of Florida's Oceans

Florida's marine areas experience large natural variability at seasonal, interannual, and longer time scales, largely due to atmospheric forcing. Certain climate modes (repeatable, naturally occurring patterns in the dynamic climate system) have very pronounced impacts on Florida's oceans. Shallow nearshore regions generally exhibit greater variability as a response to the overlying atmosphere than do the deeper offshore regions, which are influenced both by local atmospheric forcing and remote atmospheric forcing acting on large-scale ocean circulation patterns. It is important to understand the variability of oceanic properties that responds to natural fluctuations in climate or seasonal forcing in order to evaluate projections of influences of future climate change on Florida's oceans in the proper context.

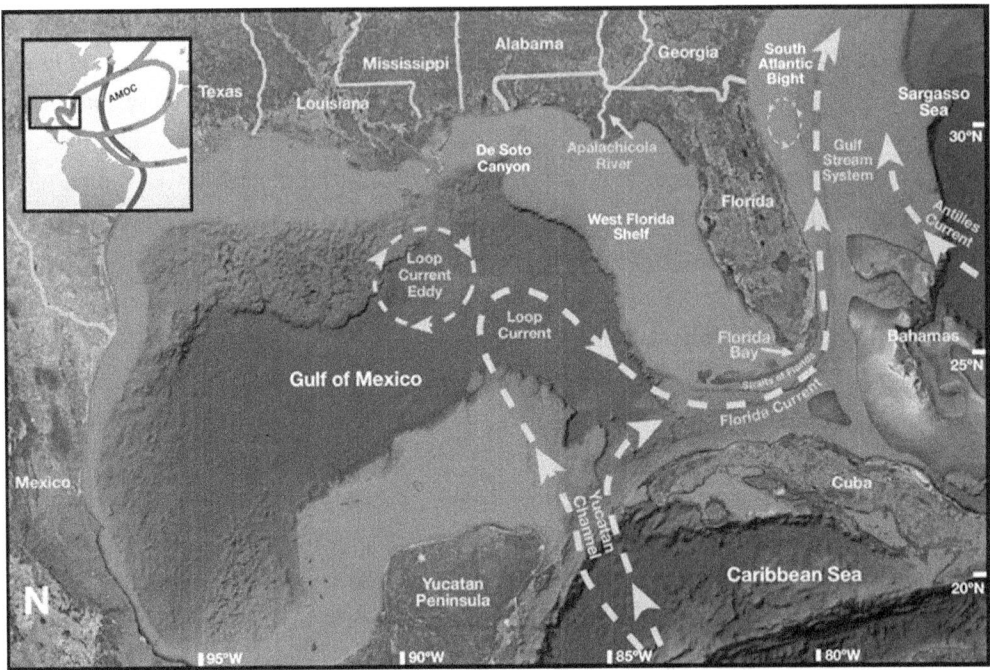

Figure 13.1. Map depicting the peninsular state of Florida. The West Florida Shelf (WFS), eastern shelf, as the southern extent of the South Atlantic Bight, De Soto Canyon, Gulf of Mexico, and western subtropical Atlantic Ocean (Sargasso Sea) are shown with a schematic representation of large-scale currents influencing waters around the state. The variable Loop Current is represented in its retracted and extended states emerging from the Yucatan Channel. A single representation of a Loop Current eddy is shown, though these generally westward migrating features (cyclonic and anticyclonic) are nearly ubiquitous throughout the deep western Gulf. The insert in upper left corner illustrates the connectivity of Florida's local and regional currents (Loop/Florida/Gulf Stream System) with the Atlantic Meridional Overturning Circulation (AMOC); arrows indicate dominant flow of red warm surface currents and blue arrows cold subsurface currents within the AMOC. *Illustration credit: Chris Johnson. Image credit: Google Maps.*

Seasonal Climatology

Currents

The deep water currents offshore of Florida's coasts are generally either persistent (such as the Florida Current) or have large-amplitude variability due to their stochastic (random, but behaving in a statistically meaningful fashion) nature (for instance, the Loop Current). However, there is evidence of some seasonality in the Florida Current. Variations in its transport are roughly ±10% of the mean, and the annual cycle is strongly affected by longer climate modes, such as the North Atlantic Oscillation (NAO) (Baringer and Larsen 2001; Peng et al. 2009).

Over the shallow shelf regions, currents vary more strongly due to direct atmospheric forcing. Low frequency variations in the coastal and shelf currents are along isobaths (lines of constant depth) and are a response to the along-shelf component of local wind stress. Over the Florida Gulf coastal waters, winter cold front passages are characterized by rotation of local winds from the southeast to the northwest at typically three- to ten-day intervals. These winds force alternating flow direction along much of the northern WFS and Panhandle region, with asymmetry in the strength of the northwesterly winds driving eastward and southeastward seasonal flow (Todd et al., 2014). The local wind climatology shows that dominantly northerly wintertime winds shift to easterly and southerly through the spring and summer, weakening during the summer months before strengthening and rotating back easterly and northeasterly during the fall (Zavala-Hidalgo et al., 2014). The effect of the seasonal shift between light southerly winds and stronger northeasterly winds is evident in the trajectories of surface drifters that measure currents over the WFS and Panhandle coastal waters, with movement strongly westward during the fall-winter and eastward during the summer (Morey et al., 2003).

A similar seasonal shift in the wind regime also affects currents inshore of the Florida Current along the eastern (particularly northeastern) coast of Florida. Northerly winds during the winter prevail, forcing a southward flowing coastal counter current. However, as winds shift to a more southerly direction, this countercurrent ceases. The southerly winds create upwelling along the narrow shelf off Florida's East Coast. This upwelling forces cool and more nutrient-rich water from depth onto the shelf in the spring and summer.

Changes in seawater density structure can also induce seasonally varying buoyancy-driven currents, particularly near river plumes. Thermal gradients due to differential cooling over the sloping shelf can also induce buoyancy-driven currents. For example, a springtime mid-shelf cold tongue along the northern WFS impacts circulation over the shelf at seasonal time scales (He and Weisberg 2002).

Temperature

Coastal ocean temperature is strongly affected by the distinct warm and cool seasons through surface heat fluxes. Net surface heat flux is the sum of the shortwave and longwave radiative energy that is absorbed or emitted by the ocean, and sensible (conduction) and latent (evaporative) heat fluxes. The radiative fluxes are largely dependent upon the season (amount of incoming solar radiation at the top of the atmosphere), cloudiness, ocean temperature, and ocean turbidity. The sensible and latent heat fluxes depend on the atmospheric state and ocean surface temperature. During much of the year net heat flux acts to warm the ocean around Florida. As the amount of incoming solar radiation decreases going into the winter season, and cool dry air masses with strong winds associated with cold fronts pass over the ocean, surface waters cool. This cooling is stronger in North Florida than in South Florida due to the change in solar radiation, frequency and strength of the cold fronts (Fig. 13.2). However, southern seasonal cooling can occur when cold fronts occasionally reach as far south as the Florida Keys, as well as when the Loop Current stretches far into the northern Gulf and this cooled water is subsequently advected south into the Florida Straits (Rudzin et al, 2013).

The transition between cooling and warming regimes over coastal waters (largely isolated from strong temperature advection by offshore currents) occurs when the net surface heat flux shifts between negative and positive (using the convention of positive heat fluxes being directed into the ocean). The timing of these seasonal transitions varies from year to year and with latitude, but climatologically the spring transition occurs around late February to early March and the fall transition occurs during September in Florida waters.

Salinity

Another important factor influencing variability of hydrographic conditions in Florida's coastal waters is the variation of discharge from rivers, particularly those with large watersheds such as the Apalachicola River. This river exhibits strong seasonal variability (Fig. 13.3), with important changes at interannual and longer time scales, mimicking the hydrological conditions over the watershed with a lag of several weeks (Morey et al., 2009). The spring maximum discharge by this and other regional rivers not only affects the coastal salinity, but also ocean optical properties, including those associated with phytoplankton pigments, over a broad region of the northern WFS extending out to 200 km from the coast through input of nutrients and organic matter. In contrast, Florida's East Coast has little freshwater input by rivers compared to the Gulf Coast, but these may substantially impact coastal water quality due primarily to dense human development and agricultural activities along their watersheds.

FLORIDA'S OCEANS AND MARINE HABITATS IN A CHANGING CLIMATE • 397

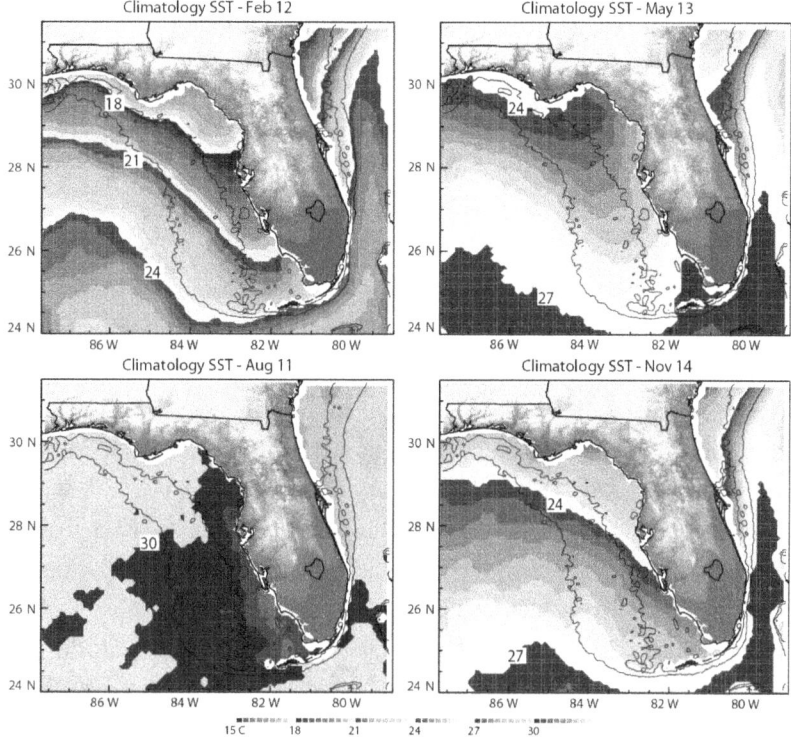

Figure 13.2. Sea surface temperature (SST) climatology maps from the NASA/JPL AVHRR 9 km 5-Day Climatology (1985-1999) (Casey and Cornillon, 1999) shown near the peaks of the cold and warm seasons (February 12 and August 11) and from spring and fall transition periods (May 13 and November 14).

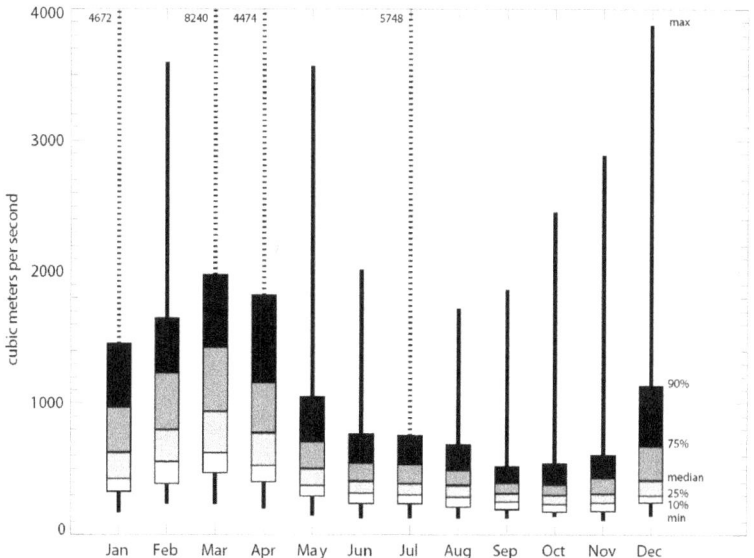

Figure 13.3. Distributions of the Apalachicola River daily flow rates (measured at Chattahoochee, Florida from 1929 through 2007) by month. The 10th, 25th, 50th (median), 75th, and 90th percentiles are shown with the recorded maximum and minimum. Dotted maxima lines extend beyond the plotting limits and the true maxima values are indicated. From Morey et al. (2009).

Interannual and Interdecadal Ocean Climate Variability

Interannual variability in the region is strongly linked with the El Niño/Southern Oscillation (ENSO). Though ENSO is an equatorial Pacific phenomenon, atmospheric teleconnections cause impacts across the southeastern United States, Gulf of Mexico and subtropical Atlantic. In particular, a significant precipitation signal over the southeastern US is connected to ENSO, with higher rainfall during the warm (El Niño) phase (Ropelewski and Halpert 1986) and reduced precipitation over the region during the cold (La Niña) phase (Smith et al. 1998). This leads to interannual variability in the coastal ocean salinity and optical properties through changes in river discharge rates (Morey et al. 2009). These ENSO impacts on precipitation and discharge of Florida's rivers are primarily observed in the late winter (Schmidt et al. 2001).

ENSO also modulates the occurrence of hurricane landfalls in the US, with fewer strikes during the El Niño phase (Bove et al. 1998), but the long-term variability of the impacts of such episodic events on Florida's oceans is difficult to ascertain given their relative infrequency. However, ENSO also modulates the occurrence and tracks of extratropical cyclones in the Gulf of Mexico (Eichler and Higgins 2006), which in turn impacts the likelihood of extreme coastal sea level fluctuations associated with these storms with greater extrema during the El Niño phase along the Gulf Coast (Kennedy et al. 2007).

Though the direct observational record remains too short to clearly show strong links between variability in the Gulf of Mexico and interdecadal climate modes such as the Atlantic Multidecadal Oscillation (AMO, with period of roughly 70 years), there is compelling evidence to suggest that the longer modes of variability can have important consequences on Florida's coastal waters. These impacts may include modulation of interannual climate signals. For example, Enfield et al. (2001) examined the interdecadal modulation of the ENSO teleconnections, suggesting a greater correlation between rainfall over the southeastern United States and the Southern Oscillation during the cold phase of the AMO. Using proxy records from tree ring analysis and chemical analysis of sediment box cores, Poore et al (2009) showed that multidecadal-scale variability occurs in the northern Gulf of Mexico sea surface temperature (SST) and may be linked with the AMO.

Atmospheric Forcing in a Future Climate

As discussed in the previous sections, atmospheric variability is a strong driver of variability in ocean currents and hydrographic properties around Florida, particularly in shallow shelf and coastal waters. Thus, in order to understand how Florida's waters may change under a future climate, it is necessary to understand the potential atmospheric changes over various spatial and temporal scales that are likely to occur. The Intergovernmental Panel for Climate Change's Fifth Assessment Report (IPCC-AR5 2013) showed that the most pronounced atmospheric changes in the ocean regions surrounding Florida during the 21st century are expected to occur in the boreal summer season (IPCC, 2013). They include the reductions of summer rainfall in the Caribbean Sea (Rauscher et al. 2008; 2011), a reduction in Atlantic hurricane activity (Vecchi and Soden 2007), and the intensification of the North Atlantic subtropical high pressure system (Li et al. 2012). The following sections briefly review recent works on these projected changes and the associated physical mechanisms.

Warming and Drying of the Caribbean Sea

Under the different IPCC-AR5 climate projections based on different atmospheric greenhouse gas concentration scenarios, the global ocean is generally expected to warm. However, there are substantial regional differences in warming rates. This regional variability in warming has important consequences for Florida's future oceanic climate. The Caribbean Sea is projected to warm less than other tropical (specifically, Indo-Pacific) waters. Cooler North Atlantic SSTs are associated with a decrease in the AMOC, which transports warm water northward (e.g., Delworth and Mann 2000; Knight et al. 2006). IPCC-AR5 projects a significantly weakened (by about 25%) AMOC in the 21st century that could be responsible for suppressed warming of the Caribbean. Another potential contribution to the suppression of warming of the Caribbean Sea is that a uniform increase of global SSTs may result in a greater evaporative cooling response in a region of high mean surface wind speed, such as in the Caribbean Sea (Leloup and Clement 2009; Xie et al. 2010).

According to the second lowest greenhouse gas concentration trajectory adopted by IPCC-AR5 (IPCC 2013), the Representative Concentration Pathway (RCP) 4.5 scenario, the Caribbean Sea is likely to experience a reduction in rainfall of ~10% in the boreal summer by 2100. As shown in Rauscher et al. (2008), the projected drying in the Caribbean Sea can be described as an extension and intensification of the Meso-American mid-summer drought (e.g., Mapes et al. 2005). Lee et al. (2011) and Rauscher et al. (2011) demonstrated that the reduced warming of the tropical North Atlantic compared to surrounding ocean decreases convection, promoting lower relative humidity and reduced precipitation in the Caribbean Sea. These changes have implications for hurricanes impacting Florida's oceans and coastal regions, as well as basin-scale circulation patterns affecting Florida's offshore regions.

Projected Reduction of Atlantic Hurricane Activity

Hurricanes, though episodic events, have profound impacts on Florida's oceans and coastal regions and thus serve as a major link between Caribbean climate and Florida's climate. The reduced convection over the Caribbean Sea is anticipated to result in an increase in wind shear that would reduce the frequency of hurricanes in the region (Lee et al. 2011). According to the IPCC Fourth Assessment Report (IPCC-AR4), global climate model simulations under various scenarios predict an overall increase in the vertical wind shear in the Main Development Region (MDR) for Atlantic hurricanes (10°–20°N, 85°–15°W), with relatively large multi-decadal variation in the 20th and 21st centuries. This occurs along with significantly reduced convection in the MDR from 1900 to 2100 (Lee et al. 2011), another condition not favorable for hurricanes.

Future projections based on theory and high-resolution dynamical models consistently indicate that, due to the increased vertical wind shear and decreased convective instability in the MDR, Atlantic cyclone activity could be significantly reduced in the 21st century despite a large increase in the SSTs (Vecchi and Soden 2007). However, existing dynamic models also project that greenhouse warming could cause the globally averaged intensity of tropical cyclones to shift towards stronger storms, with intensity increases of 2–11% by 2100. (Knutson et al. 2010).

Intensification of the North Atlantic Subtropical High Pressure System

Li et al. (2012) projected that drying in the Caribbean Sea in the 21st century increases the regional sea level pressure, and is thus linked to the westward expansion and intensification of the North Atlantic subtropical high pressure system. This high pressure system not only is a major driver of Florida's climate, but also forces the large-scale anticyclonic subtropical gyre circulation in the North Atlantic.

Intensification of the North Atlantic subtropical high under future climate scenarios will lead to enhanced easterly trade winds in the tropics and mid-latitude westerlies in the North Atlantic. This strengthening of the trades and westerlies is expected to force a stronger circulation of the North Atlantic subtropical gyre that is the major oceanic circulation feature of the Atlantic located east of Florida.

Temperature and Salinity Changes in Florida's Oceans in a Future Climate

Due to increasing greenhouse gas emissions, climate model simulations project a greater than 2°C increase in upper ocean temperatures in the Gulf of Mexico by the end of this century (Coupled Model Inter-comparison Project phase-3 – CMIP3 – and phase-5 – CMIP5; Liu et al. 2012, 2015). Further, a 20-25% slowing of the AMOC is also predicted by 2100 (Liu et al. 2012, 2015; Cheng et al. 2013). These changes could substantially affect the hydrographic and biogeochemical properties of the seawater, with important consequences for marine ecosystems in the state. However, the global climate models (CMIP3/CMIP5) have a typical spatial resolution (grid spacing) of about 1° of latitude/longitude, which is too coarse to properly resolve the strength, position, and eddy shedding characteristics of the Western Boundary Current systems such as the Caribbean Current, Yucatan Current, and Loop Current (Oey et al. 2005). Thus, the global climate models cannot be used by themselves to address future changes in these currents.

Liu et al. (2012) addressed the issues of using coarse-resolution global models for studies of the region by using higher resolution (0.1° grid spacing) models nested within the global climate models (CMIP3) to dynamically downscale the global model projections for the Gulf of Mexico region. The downscaled simulations predict that the Loop Current transport will be reduced by up to 20-25% during the 21st century. These simulations further show that the projected Loop Current reduction and associated weakening of warm Loop Current eddies could suppress the surface warming in the Gulf of Mexico, particularly in the northern deep basin. These results are in contrast to the low-resolution global climate models, which underestimate the projected reduction of the Loop Current and predict greater warming in the Gulf, partly due to the inability of these models to accurately simulate oceanic eddies. These results clearly show the utility of using high-resolution models to downscale global climate projections to a regional level.

SST Changes in the Gulf of Mexico during the 21st Century

A similar downscaling approach using the CMIP5 as forcing for historical and two future climate change scenarios (RCP4.5 for medium-low emission scenario and RCP8.5 for high emission scenario, Taylor et al. 2012) was used to further understand the warming and natural climate variability in the Gulf of Mexico (Liu et al. 2015). This model reproduced basin-averaged SST variability in the Gulf of Mexico during the 20th century reasonably well compared to analysis of historic observations (Fig. 13.4a), which supports the use of the downscaled modeling approach for climate studies. Under the RCP8.5 scenario, the downscaled model projects the annual average SSTs in the Gulf of Mexico to increase from 26°C in the late 20th century to slightly above 29°C by 2100 (Fig. 13.4b). It is important to note the uncertainty of future projections due to natural climate variability, given by the standard deviation (STD) of SST anomalies within the Gulf (STD = 0.21° C, Fig. 13.4b). Under RCP4.5 (RCP8.5), a trend of SST in the Gulf of Mexico shorter than 26 years (13 years) cannot be used to distinguish the greenhouse gas effect from natural variability.

The Gulf of Mexico warms by 1.2 ~ 2°C (3°C or more) under RCP4.5 (RCP8.5) by 2100, based on global low resolution CMIP5 models. In the downscaled simulations (Liu et al. 2015), the Gulf of Mexico also shows extensive warming (Fig. 13.5), but with significant differences in the spatial pattern of the warming, especially during the boreal spring months of April, May, and June (AMJ). During AMJ, the downscaled model-simulated SST increase under RCP4.5 (RCP8.5) in the northern deep Gulf of Mexico is only about 1.4°C (2.8°C), much less than the low resolution model SST increase of 1.8°C (3.4°C). In fact, the northern deep Gulf of Mexico is characterized as the region of minimum warming, whereas it is the region of maximum warming in the low-resolution model projections. The SST increases in the western Gulf of Mexico and the Straits of Florida region are also greatly reduced in the downscaled model compared to the low-resolution global model.

A potential cause for this difference resulting from model resolution may be the weakening of the Loop Current and the associated reduction in the warm water transport through the Yucatan Channel in future scenarios, which are not well simulated in low-resolution global models (e.g.,

Lee et al. 2007; Liu et al. 2012, 2015). The effect of the Loop Current in the present climate is to warm the Gulf of Mexico. Therefore, a reduction in the Loop Current and the associated weakening of the warm transient Loop Current eddies can result in less warming of the Gulf.

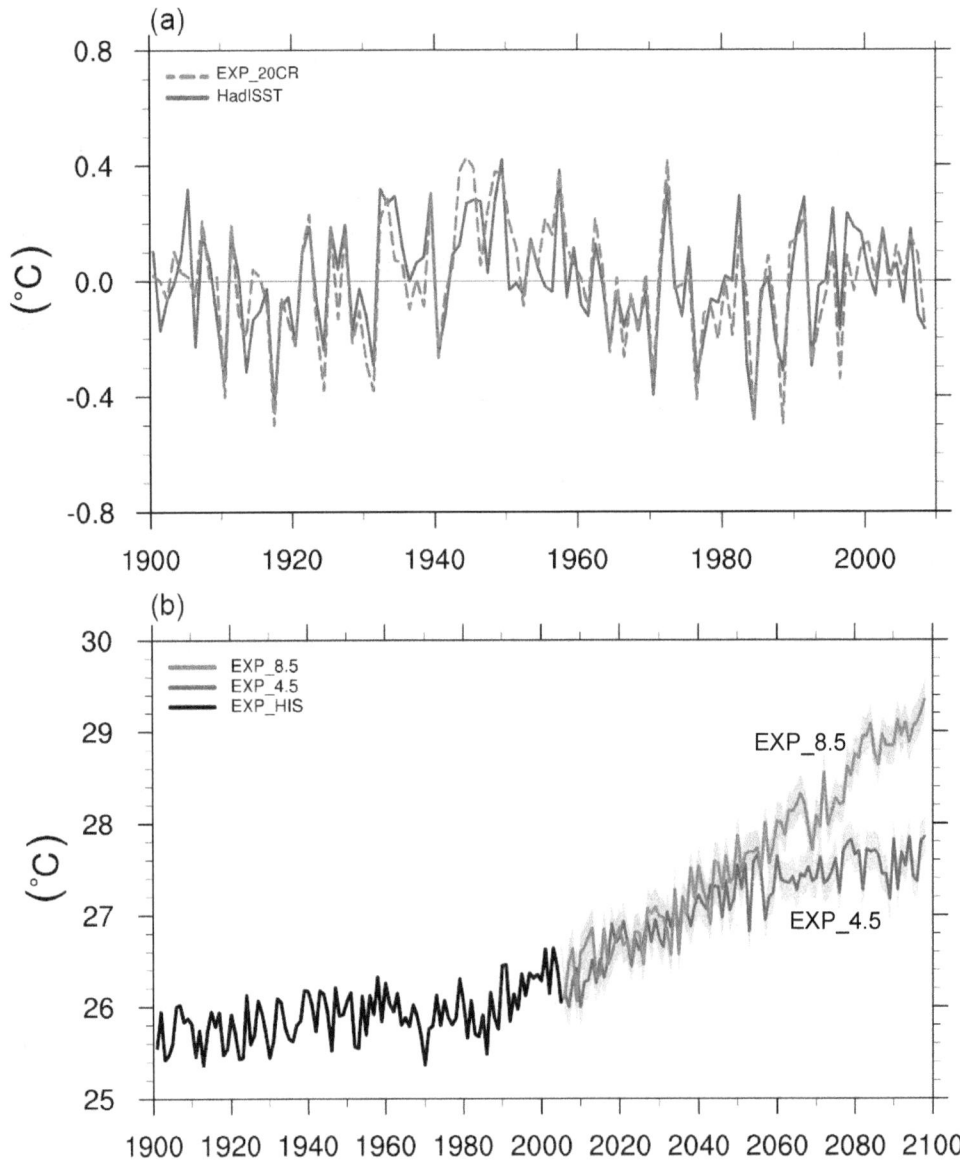

Figure 13.4. (a) Time series of annual mean sea surface temperature (SST) anomalies averaged over the Gulf of Mexico (100°W-82°W, 21°N-30°N) during 1900-2008 obtained from a downscaled model (EXP_20CR, red) and HadISST (blue). (b) Time series of the annual mean SSTs averaged over the Gulf of Mexico during 1900-2098 obtained from downscaled MOM4.1 simulations (20th century simulation [black], RCP4.5 forcing [blue] and RCP8.5 forcing [red]). The standard deviation (STD) of the SST anomalies in the Gulf of Mexico for the period of 1900-2008 is calculated (STD = 0.21°C) and the ± 0.21°C is added to each time point of the future SST projections (light color regions). From Liu et al. (2015).

In contrast to the low-resolution models showing reduced warming in the northern deep Gulf of Mexico, the downscaled model predicts an enhanced warming over the shallow shelf region (<200 m) of the northeastern Gulf, especially during the boreal summer and early autumn months of August, September, and October (ASO) (Liu et al. 2015). As shown in Figs. 13.5c and 13.5d, the projected SST increase in the northeastern Gulf Coast for ASO is about 4.0°C in the downscaled model under RCP8.5, while the global low-resolution model predicts a SST increase about 3.5°C. In the shallow northeastern Gulf Coast region, the surface ocean circulation is quite weak and dynamically detached from the Loop Current in the deep Gulf of Mexico, and the shelf supresses mixing with cooler deeper waters. Therefore, there is no mechanism to counter the increased surface heating over the shallow northeastern Gulf Coast region. The enhanced summertime warming over the northeastern Gulf Coast could greatly increase the chance for rapid intensification of hurricanes making landfall across the northeastern Gulf Coast in the 21st century and cause greater stratification of surface waters over the shelf.

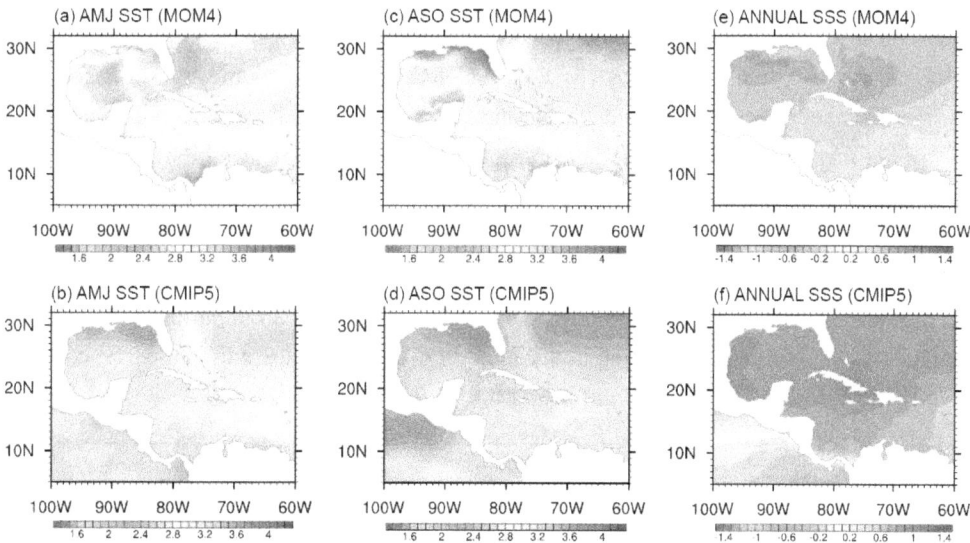

Figure 13.5. Sea surface temperature (SST) differences (°C) in the Gulf of Mexico between the late 21st century (2090 ~ 2098) and late 20th century (1990 ~ 1998) during the boreal spring months of April, May, and June (AMJ) obtained from (a) the downscaled simulation (indicated by "MOM4" in the title) and (b) the weighted ensemble of 18 CMIP5 low- resolution model simulations. Fig (c) and (d) are same as (a) and (b), except for the boreal summer-late autumn months of August, September, and October (ASO). Annual mean sea surface salinity difference in the Gulf of Mexico between the late 21st century and late 20th century obtained from (e) the downscaled model simulation and (f) the weighted ensemble of 18 CMIP5 models simulations. Modified from Liu et al. (2015).

Sea Surface Salinity Changes in the Gulf of Mexico during the 21st Century

As shown in Fig. 13.5e, the sea surface salinity (SSS) is greatly increased almost everywhere in the Gulf of Mexico during the 21st century (up to 1 part-per-thousand [PSU] by 2100 under RCP8.5), consistent with the CMIP5 projected SSS changes as shown in Fig. 13.5f (Terray et al. 2012). This is largely due to the decrease in net surface freshwater flux to the ocean (precipitation minus evaporation) in the Gulf of Mexico during the 21st century. Additionally, in the North Atlantic, the slowing down of AMOC and associated reduced warming of the tropical North Atlantic could also contribute to the projected reduced rainfall in the Gulf of Mexico (Lee et al. 2011).

Projected Reduction of the AMOC and Impact on Gulf of Mexico in the 21st Century

The downscaled climate model simulations project that reductions of the Loop and Caribbean currents in the 21st century play important roles in determining regional warming patterns in the Gulf of Mexico. Therefore, it is important to understand what processes are responsible for the reductions in the strengths of these currents. Fig. 13.6a shows the time series of the simulated annual mean volume transport across the Yucatan Channel for the period 1900-2098 under the historical and two future scenarios (RCP4.5 and RCP8.5). The volume transport across the Yucatan Channel is considerably reduced during the 21st century. The reduction is about 25% of the present mean under RCP8.5. The Caribbean Current is also reduced during the late 21st century (Fig. 13.6b). As shown in Fig. 13.6c, the AMOC at 30°N is reduced during the 21st century under both scenarios. Figs. 13.6d-f further show that the AMOC is highly reduced at all latitudes by the late 21st century (Liu et al. 2012; Cheng et al. 2013). Since the western boundary current system, including the Loop and Caribbean currents, forms an important pathway of the AMOC, it is likely that the reduction in the strength of these currents during the 21st century is driven by the projected deceleration of the AMOC (Liu et al. 2012, 2015).

Consequences of Climate Change for Florida's Marine Habitats

Florida Marine Habitats

Predicted changes in Florida's climate, ocean temperature, tropical storm (frequency and intensity), sea level rise, and current systems will significantly affect marine habitats in Florida (Fig. 13.7), which include: (1) Coastal Estuaries, Bays and Lagoons, (2) Coral Reefs, (3) Beaches, (4) Pelagic Zone, and (5) Shelf Zone. These five broad marine habitats of Florida are described below in the context of the drivers that lead to their function as a habitat, distribution, and ecological services. In the subsequent section, the primary climate change threats to these habitats are discussed.

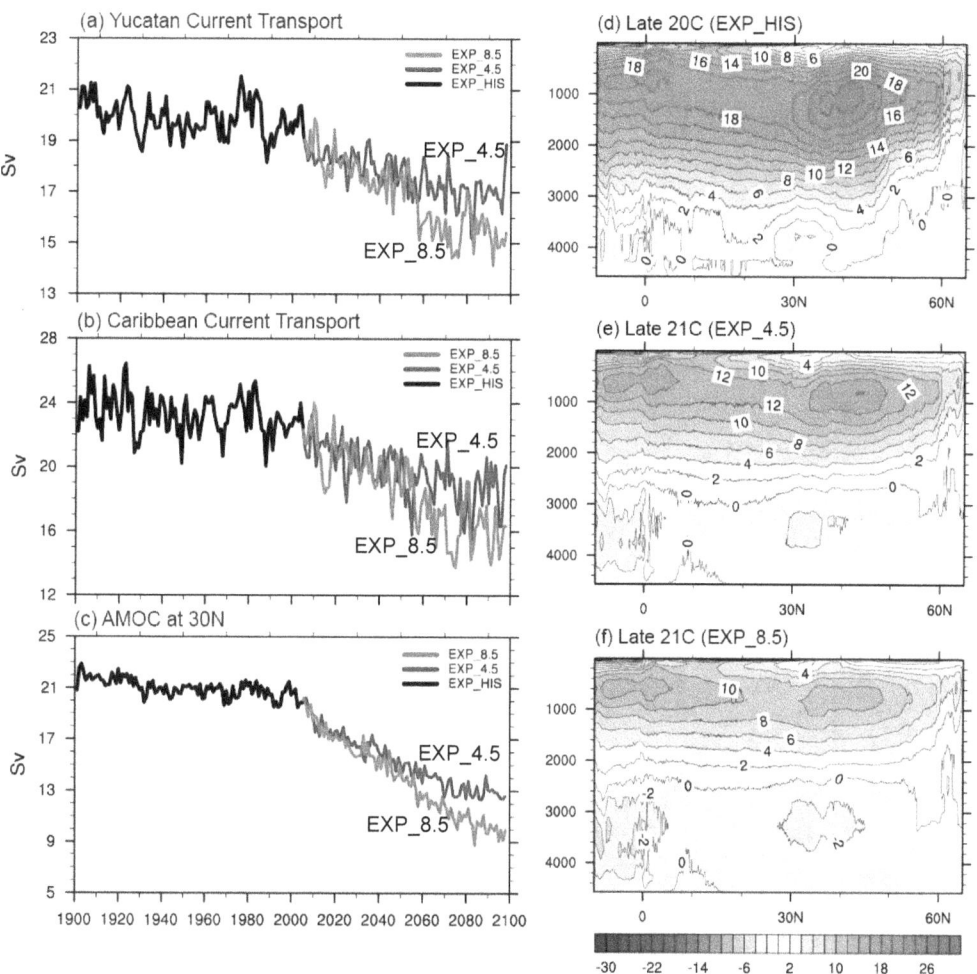

Figure 13.6. Time series of the simulated annual mean volume transport (Sverdrup; 10^6 m^3 s^{-1}) (a) across the Yucatan Channel, (b) in the Caribbean Current, and (c) the Atlantic Meridional Overturning Circulation (AMOC; see Fig. 13.1) at 30°N for the period 1900-2098 obtained from downscaled model simulations under the historical, RCP4.5 and RCP8.5 scenarios (EXP_HIS, EXP_4.5 and EXP_8.5). Time-averaged AMOC (Sv) in (d) the late 20th century and (e) the late 21st century under RCP4.5 and (f) RCP8.5 scenarios obtained from downscaled model simulations. Depth (1000-4000) is in meters (d-f). From Liu et al (2015). The AMOC plots can be interpreted as depicting circulation along contour lines with higher contour values to the right of the flow when looking at the image.

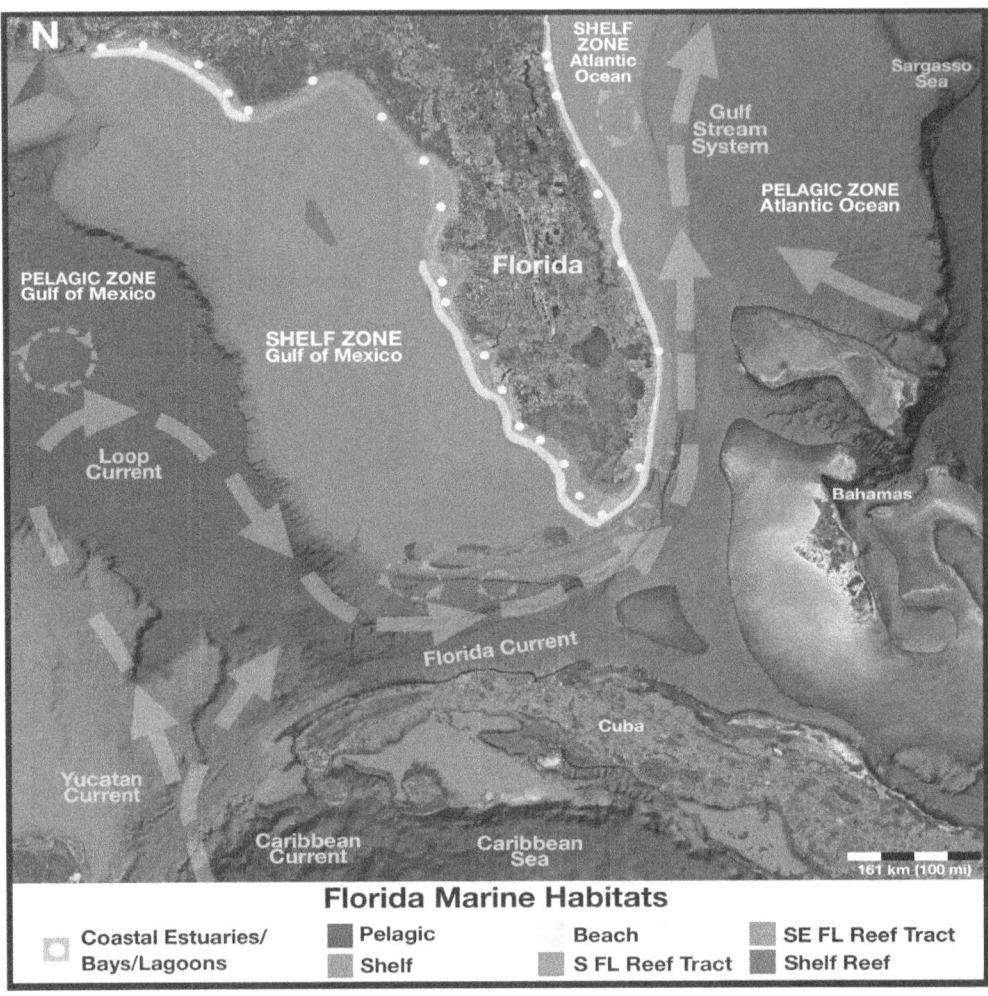

Figure 13.7. Map depicting five broad Florida marine habitats across the landscape and seascape including Coastal Estuaries, Bays and Lagoons, Pelagic and Shelf Zones, Beaches and South and Southeast Florida Reef tracts and western shelf reefs (e.g., Florida Middle Ground [northwest shelf], Pulley Ridge [southwest shelf]). Regional current systems that are important drivers of marine habitats in Florida are also depicted (defined in Fig. 13.1). Northwest to northeast the following Estuaries, Bays and Lagoons are identified by yellow circles: Pensacola Bay, Choctawhatchee Bay, Grand Lagoon, St Joseph Bay, Apalachicola Bay, Apalachee Bay, Deadman Bay, Waccasassa Bay, Homosassa Bay, Tampa Bay, Sarasota Bay, Charlotte Harbor, Estero Bay, Rookery Bay, Fakahatchee Bay, Chokoloskee Bay, White Water Bay, Florida Bay, Biscayne Bay, Lake Worth Lagoon, Indian River Lagoon, Banana River, Mosquito Lagoon, Matanzas River, Guana River, Saint Johns River. *Illustration credit: Chris Johnson. Image credit: Google Maps.*

Coastal Estuarine, Bay and Lagoon Habitats

Florida's peninsula geography and low-lying coastline create extensive networks of shallow estuaries, bays, and lagoons (Fig. 13.7). These include Pensacola Bay, Apalachicola Bay and Grand Lagoon along the panhandle in the northern Gulf of Mexico; Tampa Bay, Charlotte Harbor, and Rookery Bay along the western peninsula; White Water Bay, Florida Bay and Biscayne Bay at the southern tip of the peninsula; and long linear lagoon systems interior of barrier islands, Lake Worth Lagoon, Indian River Lagoon and Mosquito Lagoon along the eastern seaboard, representative of the hundreds of coastal systems encircling the state. Maintenance of the foundation communities within these habitats, salt marshes, in upper reaches of the state influenced by continental temperatures, mangroves in the southern subtropical regions, and oyster reefs in mesohaline (intermediate salinity) estuaries, is critical in order to preserve the highly productive marine habitats of Florida's coastline. Coastal estuaries, bays, and lagoons provide ecosystem services for human populations, including primary productivity that supports fisheries production, three-dimensional habitat structure, and improved water quality through sediment deposition and nutrient uptake. Organic matter and nutrient recycling support microbial- or algal-based food webs, both within the coastal ecosystems and adjacent pelagic ecosystems through transient marine species of ecological and economic importance, for example tarpon, grouper, mackerel, snapper, pink shrimp and lobster (Ault et al. 2005; Lellis-Dibble et al. 2008; Ault et al. 2014). Mangroves and saltmarshes also attenuate storm surge and lessen wave and flooding impacts on coastal built infrastructure along Florida's coastline (Zhang et al. 2012). Estuaries and lagoons provide recreational opportunities, e.g., boating and fishing, that add to Florida's economy (Ault et al. 2005; FOA 2013). Seagrass ecosystems are recognized for their nursery role in the development of juvenile fish and shellfish, and for provide forage grounds for endangered, threatened and at risk marine species, including manatees, queen conch, sea turtles and sharks (Heck et al. 2003; Ault et al. 2014).

Reef Habitats

In contrast to soft-bottom sedimentary environments, reef habitats are formed on hard-bottom substrate, primarily carbonate platforms and matrices. Florida reefs form along 580 km of coast from the Dry Tortugas on the shelf, around the southern peninsula, and half-way up the southeast coast to 27° 10'N (Fig. 13.7). Although geographically contiguous, the origin and formations of these reefs are quite distinct. The South Florida Reef Tract is the second longest (240 kilometers) offshore bank-barrier reef system in the world, extending from Biscayne Bay to the Tortugas Banks and dominated by reef-building (hermatypic) corals (Fig. 13.8). Florida's barrier reef is generally composed of an inner and outer ridge parallel to the Florida Keys (Lidz et al. 2003 2006). Patch reefs and seagrass meadows develop on the interior banks underlain by carbonate sedimentary features of the Holocene (~5,000 bp). Outer reefs are composed of narrow coral ridge-and-swale structures established on Pleistocene coral bedrock. The hermatypic reef-building stony corals on the barrier reef are slow growing, but have kept pace with sea level rise during the last 6,000 years, shifting community composition in response to wave energy (Lidz et al. 2006).

The southeast Florida continental reef tract (~1450 km^2) along the eastern peninsula of Florida (Fig. 13.8) sustains a high biodiversity of corals, sponges, and other marine benthic organisms. While not reef building, due to a minimal presence reef-building corals and other physiographic conditions, contemporary southeast reefs create important habitats. For example, they are within the *Acropora* sp. critical zone, an important endangered reef-building coral in the Caribbean (Wirt et al. 2013), and provide habitat for the highly endangered hawksbill turtle (Wood et al. 2013) in addition to a wide range of fish, shellfish, sharks, rays, and marine mammal species. The southeastern Florida reef system is composed of outer, middle, and inner relict reef terraces (Precht and Aronson 2004), with only the outer reef extending northward to Palm Beach County. The southern reefs indicate distinct spur and groove characteristics (Dade and Broward counties) and may have been continuous with the southern Florida Reef Tract. However, southeast reefs appear to have formed on dune ridges (Stathakopoulos and Riegl 2015) parallel to the Florida shoreline (Banks et al. 2007) in contrast to the southern Florida Reef Tract that underlies Pleistocene coral reefs. Southeastern corals also established on platforms built by Sabellariid marine worms or polychaetes through the consolidation of sands and carbonate shells. In addition to their ecological value, the Florida continental reef tract supports an active diving and sports fishery industry in the coastal counties of southeast Florida (FOA 2013).

Along Florida's West Coast discontinuous reefs are found on ledges and outcrops on the Florida shelf, one of the largest carbonate platforms on earth (~225,000 km^2). A recent review of benthic data from the shelf by Jaap (2015) indicates that, while not as speciose and continuous as southern platform reefs, the benthic communities of the western shelf represent a complex mosaic of corals and other benthic organisms (e.g., sponges, hydrozoans, macroalgae, molluscs). These reef systems develop on pinnacles and outcroppings of carbonate constructed by vermetid gastropods over sand ridges ~8,000 years ago (Reich et al. 2013; Jaap 2015). Some of these sites, such as the Florida Middle Ground reef (28°35'N) and Pulley Ridge, have high biodiversity and a complex structure that contributes to high secondary fisheries production in a relatively low productivity shelf environment (Jaap 2015), thus also having a high ecological and economic value.

Pelagic and Shelf Habitats

Florida's pelagic and shelf habitats in the Gulf of Mexico and Atlantic Ocean are coupled through regional currents, including the Caribbean, Yucatan, Loop, Florida and Gulf Stream complex. Regional currents affect the transport of nutrients and plankton that support secondary production and larvae for finfish, shellfish and invertebrates. Currents are also critical in establishing larval recruitment into coastal estuarine, bay, lagoon, reef, and shelf habitats. Upwelling of deep nutrient-rich water onto the Florida shelf from pelagic zones (<150 m) is influenced by the Loop Current and its eddies. This depends on current proximity and strength along the Florida shelf slope (Weisberg et al. 2016) and upwelling associated with mixing of deep layers under cooler surface water temperatures. The inter-annual mixing of new nutrients with isolated oligotrophic (low nutrient) shelf water (Dixon et al. 2014) enhances primary production above the modest production rates driven by local wind-driven mixing (Weisberg et al. 2016) and entrainment of riverine nutrient flux, particularly from the Mississippi River plume, which also interacts with Loop Current eddies (Jones and Wiggert 2015). High fisheries production on the oligotrophic Florida shelf habitat depends on nutrient subsidy primarily driven by the Loop Current across the shelf slope. Cooler temperatures during the winter months are associated with deeper mixed layers that mix high nutrient deep water with surface waters leading to enhanced surface chlorophyll concentrations (Muller-Karger et al. 2015).

Along the Atlantic Shelf, the southern extent of the South Atlantic Bight (Fig. 13.1), nutrients are upwelled along the shelf break. Nutrient transport is primarily driven by eddies, meanders and subsurface intrusions of the Gulf Stream complex toward the shelf (Fig. 13.8; Lee et al. 1991). These short, but significant, nutrient pulses support primary production and account for high secondary production along the eastern Florida shelf (Fiechter and Mooers 2007). The South Atlantic Bight is considered a low-nutrient shelf where phytoplankton rely on nutrients from rivers and the high-magnitude nutrient pulses propelled by Gulf Stream dynamics (Yoder et al. 1983; Miles and He 2010). Thus, on both the east and west Florida shelves, primary productivity is linked to regional currents and eddies upwelling nutrient-enriched subsurface water.

The deeper pelagic zones of the Gulf of Mexico and Atlantic Ocean not only provide nutrients in support of shelf primary and secondary productivity, but represent critical spawning grounds for large pelagic fish with high commercial value, for example tuna (Muhling et al. 2013). The Gulf of Mexico contains spawning grounds for tuna, mackerels, billfishes and other important commercial and sport fisheries. Studies of larval stages of these fish from the Florida Current indicate they are close to 100% satiated (no empty guts) compared to other regions of the world (Gulf of California; Northwest Australia) (Llopiz and Hobday 2014). The Loop-Florida-Gulf Stream Current system, with its complex oceanographic eddies, waves and meanders, represents fundamental recruitment mechanisms that sustain Florida marine habitats' high biodiversity and productivity (Lee et al. 1991). Loop Current eddies transport tropical invertebrate larvae, including corals, into the continentally-influenced cooler waters in the northern Gulf of Mexico, for example the Florida Middle Ground Reefs (Fig. 13.8; Reich et al. 2013). Frontal eddies of the Florida Current also coincide with multi-taxa (29 fish families) larval coral reef fish recruitment in the upper Keys (Sponaugle et al. 2005) from spawning sites in the Dry Tortugas (Lee and Williams 1999). Reef fish larvae entrained into eddies along the Florida Current exhibit higher growth rates due to greater resource availability, which ultimately correspond to high survival rates and recruitment onto the reef (Shulzitski et al. 2016). There is new evidence that even large game fish (permit) are recruited from local spawning sites in the southern Keys (Dry Tortugas region) rather than from spawning locations within the Caribbean (Bryan et al. 2015). In Florida, spawning aggregation sites (Tortuga and Pulley Ridge reefs) appear to be critical for reef fish self-recruitment (Cowen and Sponaugle 2009; Ault et al. 2014), thus population sustainability is dependent on current entrainment eddies (Bryan et al. 2015) rather than long-distance dispersal.

Similar to Loop and Florida currents, the Gulf Stream intrusions and frontal zone eddies along the west Florida Atlantic shelf support high phytoplankton, zooplankton, and larval fish abundances with enhanced growth rates (Yoder et al. 1983; Govoni et al. 2009). Eddies with associated upwelling of deep water provide nutrients to support rapid development of larval fish and time for development through retention that increase survival and retain populations proximate to juvenile and adult foraging grounds (Yoder et al. 1983; Govoni et al. 2009). Therefore, while the pelagic zones themselves (for example, the central Gulf of Mexico or Gulf Stream) are low in nutrients and primary productivity, eddies generated from the Loop Current in the Gulf of Mexico and the Florida Current and Gulf Stream circumnavigating Florida promote highly productive fisheries habitats along Florida's coastal zones.

Pelagic currents also assist in the transport of larval and juvenile tropical organisms throughout the subtropics and temperate regions, including sea turtle hatchlings that utilize the Sargasso Sea gyre (Figs. 13.1, 13.8) for trans-Atlantic transport during development (Putman et al. 2010). Hatchlings are entrained into wracks of floating *Sargasssum* seaweed, which provide them with food and protection as they are transported to western Atlantic foraging sites (e.g., Azores).

Beach Habitats

Once adults, sea turtles migrate back to the beach where they hatched. Loggerhead, green, and leatherback turtles nest on Florida beaches across the entire coastal zone of the state, with the exception of the Big Bend area (Fig. 13.7). Florida beaches provide 80% of the nesting sites of loggerhead sea turtles in the US. Sustaining sea turtle populations, including the rare and endangered green, Kemp's ridley, and hawksbill turtle, is a state and national priority. Preserving beaches in Florida is critical not only for endangered sea turtles, but also to protect one of the most important economic sectors of Florida's economy, tourism (FOC 2013). The value of beaches is enhanced by the presence of sea turtle nesting and residents are committed to their conservation (Hamed et al. 2016).

Consequences of Climate Change for Florida's Marine Habitats

Major Drivers of Change

The highly diverse and productive marine habitats of Florida, and their associated ecosystem services, are threatened by climate change impacts (Fig. 13.8). In this section, dominant drivers of climate change are discussed along with their impacts on Florida habitats. The major direct climate change drivers that presently and will continue to degrade Florida's marine habitats are (1) accelerated rates of sea level rise, (2) increasing ocean temperatures, and (3) ocean acidification. These primary drivers lead to secondary changes that further degrade habitats. Increasing ocean temperatures lead to (4) low oxygen levels in surface waters or hypoxia. Higher evaporation rates with greater atmospheric warming raises the (5) sea surface salinity, which stresses organisms because they must utilize metabolic energy for osmoregulation rather than for growth or reproduction. Rapid sea level rise results in (6) coastal erosion with primary effects on coastlines and their associated habitats, such as beaches, lagoons and estuaries. Under coastal erosion, sediment and nutrient loads increase, leading to turbidity and (7) low light transmittance to benthic (bottom dwelling) communities that require high light (e.g., seagrasses and coral reefs). At the broader scale, (8) changes in current patterns and flow rates will significantly modify Florida marine habitats and their primary and secondary productivity.

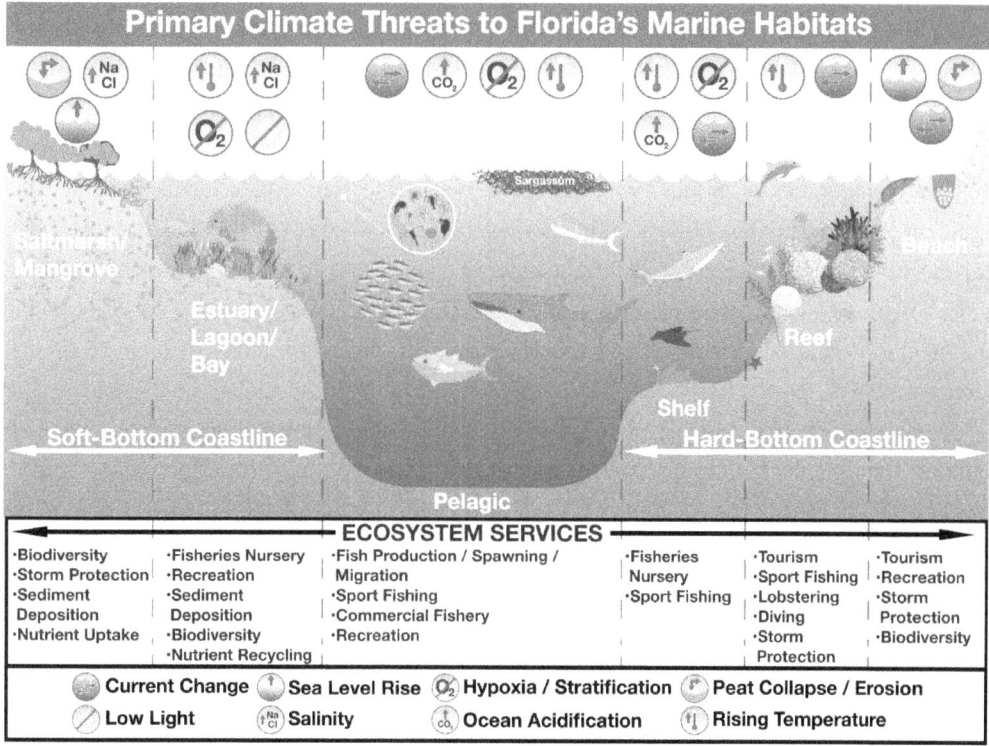

Figure 13.8. Illustration of the large-scale marine habitats of Florida and the primary drivers that threaten their sustainability and ecosystem services under climate change. *Illustration credit: Chris Johnson.*

Coastal Wetlands and Beach Shorelines

Mangrove forests and salt marshes represent the transition zone between terrestrial and aquatic systems where they buffer waves and storm surges for upland areas and filter land sediment and nutrients to adjacent estuaries, lagoons, and bays. These ecosystem services, along with high biodiversity, are primarily threatened by high rates of sea level rise, particularly in locations around Florida where migration inland is not an option because of human development.

Developments and infrastructure will also be subject to greater coastal hazards without coastal wetlands (FOCC 2010; Geselbracht et al. 2015). The predicted rates of sea level rise globally and in Florida are based on different scenarios of global greenhouse gas mitigation measures with the low range assuming strong mitigation measures and the high range reflecting the "business as usual" scenario. The current unified sea level rise predictions for Florida are 6-10 inches (15.2-25.4 cm; 4-7 mm y^{-1}) by 2030, 14-26 inches (35.6-66.0 cm; 5-10 mm y^{-1}) by 2060, and 31-61 inches (78.7-154.9 cm; 7-14 mm y^{-1}) by 2100, with less certainty as the rates are projected further in time (SFRCCC 2015).

Models of coastal wetland loss using vegetation and elevation maps for Florida were used to forecast sea level rise effects on six major Florida estuaries along the Gulf of Mexico: Pensacola Bay, St. Andrews/Choctawhatchee Bays, Apalachicola Bay, Southern Big Bend, Tampa Bay, and Charlotte Harbor (Geselbracht et al. 2015). This assessment showed that urban areas (Tampa Bay and Charlotte Harbor) lost a higher percentage of wetlands (~15K hectares; 15-20%) and almost a complete loss of tidal flats and freshwater marsh, under 1 m (100 cm) of sea level rise. Where wetlands had the opportunity to move upslope, less total wetlands were lost but shifts in wetland communities occurred. For example, in Pensacola Bay, Apalachicola Bay, and the Southern Big Bend region, 44-80% of tidal swamp (0.3-3K ha) and 18-49% of coastal forest (10-25K hectares) were replaced by tidal flats, brackish/salt marsh, and estuarine beach shorelines under 1 m of sea level rise (Geselbracht et al. 2015). This analysis suggests that if area is available around the Florida coastline, wetland species may shift, but retain many of their ecosystem services.

In coastal areas with low elevation gradients, for example South Florida with less than 4.5 cm km^{-1} from the mangroves through the Everglades' marsh to Lake Okeechobee (~100 miles; 160 km), coastal mangroves and marshes are likely to be overwhelmed by a rapid rate of sea level rise (> 5 mm y^{-1}; Koch et al. 2015). They can, however, keep pace with modest rates of sea level rise. Florida's recent (last ~100 years) sea level rise rates of 2.14 ± 0.03 mm y^{-1}, estimated using 26 gauges across Florida's coastline (Bâki et al. 2012), are similar to earlier estimates of 1.5 to 2.4 mm y^{-1} using tide gauge data (Maul and Martin 1993) and compare closely with regional altimetry-based estimates of 1.5 to 1.7 ± 0.3-0.6 mm y^{-1} (Bâki et al. 2012; Palanisamy et al. 2012). At these relatively slow rates of rise, belowground biomass of saltmarsh and mangrove vegetation contributes to inorganic sediment deposition and keeps pace with sea level changes, illustrating the importance of a biological feedback in this habitat (McKee et al. 2007; McKee 2011; Alizad et al. 2016). Promoting organic matter accumulation is enhanced by minimizing other stressors, such as hypersalinity with freshwater enhancement. Peat loss of mangrove sediments in the Caribbean and Florida can also be subject to non-linear "collapse" or erosion following tropical storms when vegetation is compromised (Wanless et al. 1994). Thus, sustaining vegetation is critical to promote positive elevation change along Florida's wetland coastline. Although accretion and erosional processes occur, mangroves in Florida and the wider Caribbean have kept pace with sea level rise rates of 2-4 mm y^{-1} over thousands of years, but not

5 mm y^{-1} (Koch et al. 2015), and modeling studies of salt marshes in northeast Florida indicate that overall marsh biomass density increases at 3.1 mm yr^{-1} (11 cm by 2050) but declines at 13.7 mm y^{-1} (48 cm by 2050, Alizad et al. 2016). These data highlight the importance of global mitigation of climate warming to maintain Florida's sea level rise at the lowest projections and relatively stable present shoreline. This point is critical, as even the upper unified Florida projections for 2030 equate to 7 mm y^{-1}, assuming linearity of rise, which would likely drown coastal wetlands with no potential for shoreward expansion or shallow elevation gradients. Further, many of the wetlands in Florida developed in quiescent waters interior of coastal carbonate reef or sand barrier systems that are also subject to erosion as sea levels rise.

Beach Shorelines

Beach mainland and barrier islands are subject to sand migration in response to local hydrodynamics, prevailing currents, counter currents, and wave profiles (Finkl and Makowski 2013). Along Florida's East and West coasts as well as the Panhandle, sand-dominated shorelines are dynamic, requiring input of sands to maintain a static position of Florida's coastline (Dean and Houston 2016). Stronger tropical cyclones and a rise in sea levels predicted with increasing sea surface temperatures have the potential to destabilize shorelines. The dynamic nature of Florida's beach shorelines that reside on bedrock or migrating sands results in a high sea level rise vulnerability of beach habitats on which sea turtles and many coastal human residents depend. Based on an analysis of the dominant sea turtle nesting in Florida, the loggerhead turtle, ~43% of beach nesting is predicted to be lost with a 50 cm increase in sea level by 2050 (Reece et al. 2013). This sea turtle species and others are predicted to shift northward to accommodate high temperatures that negatively influence a balance of sex ratios and successful reproduction (Witt et al. 2010). Hard armoring of dynamic shorelines, contraction of land available for human settlement, and erosion of beaches may constrain sea turtle nesting in alternative suitable locations to adapt to climate change, as they have done over evolutionary (100 million years) time scales (Reece et al. 2005; Witt et al. 2010). The rates of change are likely to limit sea turtles' ability to respond because they are a long-lived species (80-100 years) and have site fidelity to beach nesting sites adjacent to urban development throughout Florida.

Coastal Estuaries, Bays and Lagoons – Open Water Systems

Estuaries, bays, and lagoons encompass Florida's coastline (Fig. 13.7) and sustain a sport fishing, boating, and tourist economy worth more than one billion dollars. This economy depends on "healthy" seagrass ecosystems. The three-dimensional structure and high primary productivity of seagrasses transform a depauperate bare bottom estuary, lagoon or bay into a highly biodiverse habitat for invertebrates (e.g., lobsters, conch, crabs, oysters) and larval/juvenile fish (e.g., seatrout, mullet, bonefish, sheepshead, red drum, tarpon), and provide foraging grounds for sea turtles, manatees, dolphin, and elasmobranchs (rays, sharks). Globally, estuarine and lagoon seagrass ecosystems are threatened by poor water quality that reduces the light reaching the benthos (Orth et al. 2006), principally from upland non-point sources of nutrients and sediments. Seagrasses stabilize the sediment and provide organic matter to infauna (clams, shrimp, worms) with their large proportion of belowground biomass; but this non-photosynthetic tissue has a high metabolic demand, necessitating a high light environment (4-30% of surface light levels; Duarte 1991 ; Dennison et al. 1993). Thus, as sea levels rise and coastal erosion and nutrient fluxes increase, seagrasses across the state will be exposed to lower light, degrading ecosystem services presently provided by these ecosystems (Fig. 13.9).

Given Florida's broad coastal ocean temperature range (seasonal average range = 10-30 °C; 50-87 °F), increases of 1-4 °C above maximum average temperatures to ~31-34 °C are within the physiological limits of tropical seagrasses (Koch et al. 2013). Under these conditions, seagrass species in North Florida may shift to tropical species, but direct effects will likely be tolerable. However, ocean warming also results in high salinity (evaporation) and lower oxygen solubility, which causes physiological stress to vegetation, increases root exposure to sulfide, a known phytotoxin (Koch et al. 2007), and limits the ability of estuaries and lagoons to serve as viable fish habitats, particularly in lagoons and estuaries with restricted circulation and long water residence times. Florida Bay, one of the largest (1,000 square miles; 2,600 km^2) and most productive estuaries in Florida that supports important fisheries (e.g., pink shrimp, stone crab, spiny lobster, bone fish) and associated industries (sport fishing, tourism), has succumbed to large-scale seagrass die-off (~10,000 ha; 25K acres) and fish mortality events over the last few decades (Hall et al. 2016). These mortality events have been attributed to hypoxia and sulfide production driven by warm, hypersaline waters in this shallow seagrass-dominated ecosystem (Koch et al. 2007; Hall et al. 2016). The Florida Bay case study represents a precursor for estuaries around the state that are building up biomass (seagrass, algae and phytoplankton) under coastal nutrient enrichment leading to a high oxygen demand. Increased thermal loads and erosion with climate change will further exacerbate this problem statewide. Based on downscaled climate models, coastal estuaries and bays along Florida's northwest coast may encounter the greatest rise in temperature (3-4 °C by 2100). This localized warming along the West Florida coast is a consequence of stratification and reduced upwelling as the Loop Current and AMOC weaken (Liu et al. 2015). High ocean temperatures and enhanced nutrient enrichment synergistically promote hypoxia and foster harmful algal blooms (primarily cyanobacteria and dinoflagellates) in coastal systems causing fish kills and human health concerns (Paerl and Paul 2012). Harmful algal blooms are already an issue for human health and fisheries along Florida's West Coast and in Florida Bay (Weisberg et al. 2016; Berry et al. 2015).

Hard-armoring of shorelines in Florida to protect human infrastructure from sea level rise is also likely to restrict the connectivity between wetlands and estuaries, bays, and lagoons. Without wetland buffers, pulsed nutrients and sediment loads to coastal water bodies are likely to increase. Loss of wetlands also and eliminate an important larval to juvenile fish and shellfish habitat, further reducing fisheries production potential in these coastal ecosystems (Gittman et al. 2016).

Coral Reefs

Ocean Warming

Coral reefs worldwide are negatively affected by stressors at the local and global scale, ushering in the current age of corals: "Conservation or Extinction" (Hixon 2011). Widespread local stressors (siltation, pollution, eutrophication, disease, exotics and overfishing) that have been driving reef decline over the last few decades globally (Jackson et al. 2014) and in Florida (Ault et al. 2005; Kruczynski and Fletcher 2012) are now compounded by coral thermal bleaching in response to warming oceans under climate change (Jackson et al. 2014). A massive coral reef die-off (80% bleaching; 40% mortality) in response to a modest rise in temperature (+1.2 °C) throughout the wider Caribbean during the ENSO event of 2005 (Eakin et al. 2010) and the present 81% coral beaching on the northern Great Barrier Reef in 2016 (CCOA 2016) illustrate the thermal sensitivity of corals residing at their upper thermal limits. Degree heating weeks (DHWs), an index used for regional assessments of thermally-induced coral bleaching (loss of symbiotic algae), has been used to assess bleaching probabilities and reef decline. At eight DHWs, corals reached a threshold of annual severe bleaching (ASB) that results in mortality (Frieler et al. 2012) unless acclimation or adaptation ensues. Based on SST predictions (RCP 8.5) applying low resolution, climate models (33 CMIP 5 ensemble) and dynamically downscaled (MOM4.1, ~11 km) or statistical models were used to define the year that ASB would be reached across the wider Caribbean (3781 reef locations), including Florida (van Hooidonk et al. 2015). Regardless of models used or the downscaling approach, ASB was predicted to be reached by 2040 to 2043 ± 10 years at the broad regional scale and all reef locations by 2070. Dynamically downscaled models that incorporate local-scale hydrodynamics and regional current systems exhibit high temporal variability. For example, on the southern Florida Reef Tract, western reefs are predicted to experience ASB by late 2020 compared to the early 2050s for eastern reefs in response to slowing of the Gulf Stream System and Loop Current. Downscaled models also identify regions for transient coral refugia against severe bleaching, which may allow some corals to survive and more slowly adapt to rising temperatures. Coral refugia from thermal stress would also occur along Florida's southeastern Reef Tract and on solitary shelf reefs in deeper waters, but larval transport to the southern Florida Reef Tract may be limited by prevailing currents eastward and reduced excursions of the Loop Current into the Gulf of Mexico, respectively.

Elevated temperatures and stressed corals from thermal bleaching may also promote coral diseases that have already decimated reef-building corals in Florida and the wider Caribbean region over the last few decades (Weil and Rogers 2011). Specifically, a virulent outbreak of white band disease caused mass mortality of the *Acroporas* (*palmata* and *cervicornis*; Gladfelter 1982), which have not recovered within the southern Florida Reef Tract, although individuals are found amongst the three reef habitats discussed herein. As microbial activity increases with temperature, particularly when organisms are stressed, ocean warming could drive greater pathogenic activity on Florida's reefs.

Ocean Acidification

In addition to causing ocean warming, CO_2 in the atmosphere is dissolving into and reacting with seawater leading to a lower ocean pH called "ocean acidification" (OA). The effect of OA on coral calcification has been the focus of intense research over the last decade. Based on this research, some coral species and locations have been shown to be more resilient to OA than others (Shamberger et al. 2014; Barkley et al. 2015) making generalizations on the effects of OA on coral calcification challenging. Some corals acquire the energetics to calcify even at a low pH predicted for 2100 (pH ~7.8). However, these corals are subject to greater bioerosion (Manzello et al. 2008; Barkley et al. 2015). In Florida's southern Reef Tract, seagrass sequestration of CO_2 in reef lagoons has provided seasonal resilience to OA for interior patch reef corals (Manzello et al. 2012). At a broad scale, high net ecosystem metabolism in coastal ecosystems of Florida, particularly with high seagrass biomass and under nutrient enrichment, exhibit highly seasonal and daily changes in pH (Millero et al. 2001; Yates and Halley 2006; Manzello et al. 2012) that dwarf open ocean estimates of pH decline from OA (Bates et al. 2012). Determining the effect of OA on net calcification at the reef scale provides the ability to ascertain reef ecosystem potential to build structure and keep pace with sea level rise (Andersson and Gledhill 2013). Reef building is currently uncertain across the southern Florida Reef Tract based on evidence of erosional or modest rates of calcium carbonate deposition (Muehllehner et al. 2016) and significant declines in reef building corals over the last few decades (Ruzicka et al. 2013) from 1996-2009 (CREMP, 2012). Further, there is evidence that corals that build carbonates are being replaced in part by bioeroders of the reef framework (e.g., sponges, urchins, mollusks; Muehllehner et al. 2016).

Reef Fisheries

As coral and other marine organisms possess planktonic early life stages, and recruitment is local or self-recruiting, population sustainability will be reliant on local current systems (Cowen et al. 2000). Recruitment is essential in the southern Florida Reef Tract given target fisheries, such as snapper, grouper, grunts, jacks progies, and hogfish, have all been exhaustively exploited for 85 years (Ault et al. 2005). Further, reef fish are sensitive to extraction and loss of reef structure and adjacent habitats, including the coral-seagrass-mangrove complex. In Florida, eddies associated with coastal currents transport larval invertebrates, including coral, shrimp, and fish from local spawning grounds (e.g., Tortugas, Pulley Ridge Reefs) to Florida Reef Tracts (Lee and Williams 1999; Sponaugle et al. 2005; Ault et al. 2014; Shulzitski et al. 2016; Vaz et al. 2016) and local estuaries (e.g., Florida Bay pink shrimp; Criales et al. 2007). Thus, predictions of a slowing AMOC and Gulf Stream system under climate change that reduce the Florida and Loop currents by 25% and weaken accompanying gyres (Liu et al. 2015) are likely to have wide-ranging consequences for population dynamics, fisheries management, and overall secondary productivity of Florida's coral reefs and associated estuary, bay, and lagoon ecosystems. While Florida reef recruitment from the wider Caribbean may be limited for some species, albeit still significant for others (e.g., spiny lobster; Kough et al. 2013), limited larval entrainment from Caribbean and Yucatan current systems could constrain genetic diversity and resilience in populations over longer time scales.

Pelagic and Shelf
Phytoplankton Production and Harmful Algal Blooms
Florida's shelf and pelagic habitats have been subject to multiple stressors over the last few decades, shifting how these ecosystems function and modifying their conceivable response to climate change. Primary amongst these is overfishing of target piscivore fish species (e.g., red snapper, grouper and mackerel) on Florida's western shelf (Walsh et al. 2011). The systematic removal of predatory pressure on zooplankton-grazing fish (clupeoid sardine, herring, anchovy, and menhaden), shrimp, jelly fish and other grazers depressed zooplankton ten-fold (1973-1993) on the shelf and likely contributed to a reduction in zooplankton grazers (pink shrimp) as their prey was diminished (Walsh et al. 2011; Muhling et al. 2012). In the absence of zooplankton, phytoplankton increased along the West Florida Shelf, including harmful algal blooms that negatively affect marine organisms and human health and a food web perturbation observed globally (Landsberg et al. 2009; Walsh et al. 2012).

Under ocean warming and a reduction of nutrient flux across the Florida shelves, and as the Loop Current excursions into the Gulf of Mexico subside, harmful algal blooms will continue to be favored. Harmful algal species competitively dominate the phytoplankton community under low-nutrient stratified conditions promoted by increased warming, which are predicted to be greatest along the west Florida Shelf (Liu et al. 2015) where harmful algal blooms dominate (Landsberg et al. 2009). Further, lack of upwelling of Loop or Florida (Gulf Stream) currents via eddies and other transgressions along the Florida shelves will diminish deep nutrient sources to the shelf (Weisburg et al. 2014) selecting for low-nutrient adapted harmful algal bloom species. Finally, harmful algal bloom development with greater toxin potency, including saxitoxin dangerous for humans, has been linked to low nutrient conditions (Walsh et al. 2011). If nutrient limitation on the Florida shelves accompanies climate change due to a reduction of the Loop and Florida currents, the Florida shelves and South Atlantic Bight are likely to depend on wind-induced upwelling and riverine sources, becoming a less productive fisheries habitat.

Pelagic Fish Spawning
In the open pelagic realm of the Gulf of Mexico, modest ocean warming predictions from downscaled models (MOM4) indicate changes in distributions and reproduction of commercially important pelagic fish species such as Atlantic bluefin, yellowfin, and skipjack tuna (Muhling et al. 2015). The temperate Atlantic bluefin tuna is most at risk from climate change because of its low adult and larval thermal thresholds (28-29 and 26°C, respectively) and restricted spawning habitat in the Gulf of Mexico, which is predicted to be unsuitable for adults or larval stages by 2090 (RCP 8.5; Muhling et al. 2015). The wider-ranging tropical tuna species, yellowfin and

skipjack, are predicted to replace Atlantic bluefin and expand their range (Muhling et al. 2015) into the Gulf of Mexico as temperate waters warm to 30-32 °C (>16 °C), temperature ranges where they are currently found (Boyce et al. 2008). These model scenarios predict a loss of bluefin tuna and a shift to tropical tuna species in the Gulf of Mexico and Florida Current; however, bluefin tuna accommodate surface water hypoxia and high temperature by residing in deeper waters (Muhling et al. 2015), which may provide a short-term strategy for adapting to rapid temperature increases in the Gulf of Mexico, particularly the northern Gulf, which may warm more slowly as the Loop Current incursions subside.

Ocean Acidification

The scale of atmospheric CO_2 invasion into oceans is staggering at a current rate of 1-3.2 billion (10^9) metric tons C y^{-1} (1-3.2 petagrams [10^{15} g] C y^{-1}), equating to 155 cumulative billion tons of C over the last 250 years (1750-2010) with another ~400 billion tons of C to be added by 2100 (2012-2100) without atmospheric CO_2 mitigation (RCP 8.5; IPCC 2013). Elevated CO_2 concentrations in seawater can constrain highly mobile marine organisms with high metabolic rates thorough hypercapnia acidosis (Pörtner et al. 2004) or directly affect development, as was recently found for yellowfin tuna (Frommel et al. 2016). Other calcifiers in the plankton including calcified molluscs (pteropods) and phytoplankton (coccolithophores) widely distributed in the Gulf of Mexico and Atlantic have shown sensitivity to OA in temperate oceans (Fabry et al. 2008; Guinotte and Fabry 2008). Changes in ocean chemistry with respect to atmospheric CO_2 sequestration, and OA were examined in a synoptic cruise along the Gulf of Mexico shelf and eastern seaboard of the United States in 2007 and again in 2012 (Wang et al. 2013; Wanninkhof et al. 2015). Cruise data indicated that the highly buffered water (high alkalinity that resists OA) from the Loop Current and Gulf Stream presently maintains a high saturation state for calcium carbonate (Ω) in surface waters of the Gulf of Mexico and Florida Shelf (Wang et al. 2013). All measurements indicated that shelf waters were supersaturated with respect to the aragonite carbonate mineral ($\Omega_A > 1$), inferring mineral stability rather than dissolution. Thus, today the subtropical surface waters around Florida's shelf and open water are relatively well buffered from OA in contrast to temperate and polar oceans, and undersaturated deep waters (Wang et al. 2013; IPCC-AR5 2013).

In a future without significant mitigation of atmospheric CO_2 emissions (RCP8.5), however, carbonate saturation states will be undersaturated (<1 Ω) by 2050 in the Arctic and Southern Ocean and 2150 in the tropics (IPCC-AR5 2013). Thus, tropical/subtropical oceans will buffer Florida's shelf and pelagic habitats in the short term from corrosive waters. While this is relatively positive, low pH and Ω synergistically interact with rising temperature can affect tropical organisms before Ω becomes undersaturated. Further, because of the dominance of marine calcifiers and carbonate sediment environments in Florida habitats, the consequences of carbonate undersaturation will be stark, including dissolution of calcified organisms and release of a large pool of nutrients presently stored in carbonate sediments.

Conclusion

Florida's marine resources, already under cumulative stress, and coastal human populations are highly vulnerable to a changing climate because of the strong linkage between the state's ocean current systems (Loop, Florida, and Gulf Stream Complex), climate to global ocean circulation (AMOC) and atmospheric processes (ENSO, AMO, convection, wind shear). Predicted ocean warming will significantly affect Florida's coastline via sea level rise, as well as marine species and ecosystems due to their susceptibility to high temperatures. A 25% reduction in the Loop Current is predicted based on downscaled models to restrict warming in the Gulf of Mexico by 2100 to between 1.4°C to 2.8°C based on different scenarios of CO_2 mitigation, as the current will transport less warm tropical water into the Gulf than at present. However, even modest ocean warming or more extreme warming without CO_2 mitigation is predicted to have wide-ranging consequences for Florida's marine habitats (e.g., enhance coral bleaching, lower O_2 in surface waters, promote harmful algal blooms, reduce phytoplankton and fisheries production, and lower sea turtle reproduction). Further, as current systems around the peninsula of Florida entrain and transport marine organisms (e.g., fish, coral, lobster, crab, shrimp larvae and juveniles) the loss of current-driven connectivity and/or gyre systems will significantly lower the potential to sustain marine fisheries and ecosystems throughout the state. Without CO_2 mitigation at the global level, sea level rise will overwhelm wetlands along Florida's coast and lead to hard armament or retreat of human populations. Continuing to revise downscaled models and gain a more complete understanding of complex changes in air-sea interaction, large-scale currents, and the rates of climate change impacts will be critical over the next few decades to prepare and protect the state of Florida.

References

Alizad, K. et al., 2016. "A coupled, two-dimensional hydrodynamic-marsh model with biological feedback." *Ecological Modelling* 327:29–43.

Andersson, A.J. & Gledhill, D., 2011. "Ocean Acidification and Coral Reefs: Effects on Breakdown, Dissolution, and Net Ecosystem Calcification." *Annual Review of Marine Science* 5 no. 1:120717164858000.

Ault, J.S. et al., 2005. "Towards sustainable multispecies fisheries in the Florida, USE, coral reef ecosystem." *Bulletin of Marine Science* 76 no 2:28.

Ault, J.S. et al., 2014. "Indicators for assessing the ecological dynamics and sustainability of southern Florida's coral reef and coastal fisheries." *Ecological Indicators* 44:164–172.

Bâki Iz, H., Berry, L. & Koch, M., 2012. "Modeling regional sea level rise using local tide gauge data." *Journal of Geodetic Science* 2 no. 3:188–199.

Banks, K.W. et al., 2007. "Geomorphology of the Southeast Florida continental reef tract (Miami-Dade, Broward, and Palm Beach Counties, USA)." *Coral Reef* 26 no. 3:617–633.

Barkley, H.C. et al., 2015. "Changes in coral reef communities across a natural gradient in seawater pH." *Science advances*, 1 no 5: e1500328.

Baringer, M.O., and J.C. Larsen. 2001. "Sixteen years of Florida Current transport at 27 N." *Geophysical Research Letters* 28 no. 16: 3179-3182.

Bates, N.R. et al., 2012. "Detecting anthropogenic carbon dioxide uptake and ocean acidification in the North Atlantic Ocean." *Biogeosciences* 9 no. 7:2509–2522.

Berry, D.L. et al., 2015. "Shifts in Cyanobacterial Strain Dominance during the Onset of Harmful Algal Blooms in Florida Bay, USA." *Microbial Ecology* 70 no. 2:361–371.

Bove, M.C., J.J. O'Brien, J.B. Elsner, C.W. Landsea, and X. Niu. 1998. "Effect of El Niño on U.S. Landfalling Hurricanes, Revisited." *Bull. Am. Met. Soc.* 79 no. 11: 2477-2482.

Boyce, D.G., Tittensor, D.P.,Worm, B., 2008. Effects of temperature on global patterns of tuna and billfish richness.Mar. Ecol. Prog. Ser. 355, 267–276.

Bryan, D.R. et al., 2015. "Transport and connectivity modeling of larval permit from an observed spawning aggregation in the Dry Tortugas, Florida." *Environmental Biology of Fishes*, 98 no. 11:2263–2276.

Casey, K.S. and P. Cornillon. 1999. "A Comparison of Satellite and In Situ based Sea Surface Temperature Climatologies." *Journal of Climate*, 12, no. 6: 1848-1863.

CCOA. 2016. Australia's Coral Reefs Under Threat form Climate Change. Climate Council of Australia Limited ISBN: 978-0-9945973-0-4 (print) 978-0-9944926-9-2 (web) © Climate Council of Australia Ltd 2016

Cheng, W., J. C. H. Chiang, and D. Zhang, 2013. "Atlantic Meridional Overturning Circulation (AMOC) in CMIP5 models: RCP and historical simulations." *J. Climate* 26: 7187–7197.

Cowen, R.K. and S. Sponaugle, 2009. "Larval Dispersal and Marine Population Connectivity." *Annu. Rev. Mar. Sci.* 1:443–66.

Cowen, R.K. et al., 2000. "Connectivity Populations: Open or Closed?" *Science* 287 no.5454: 857–859.

Criales, M.M. et al., 2007. "Cross-shelf transport of pink shrimp larvae: Interactions of tidal currents, larval vertical migrations and internal tides." *Marine Ecology Progress Series*, 345:167–184.

Dennison, W.C., R.J. Orth, K.A. Moore, C.J. Stevenson, V. Carter, S. Kollar, P.W. Bergstrom, and R.A. Batiuk, 1993. "Assessing water quality with submersed aquatic vegetation." *Bioscience* 43 no. 2:86–89.

Dixon, L.K. et al., 2014. "Nitrogen, phosphorus and silica on the West Florida Shelf: Patterns and relationships with *Karenia* spp. occurrence." *Harmful Algae*, 38:8–19.

Duarte, C.M., 1991. "Seagrass depth limits." *Aquatic Botany* 40: 363–377.

Eakin, C.M. et al., 2010. "Caribbean corals in crisis: Record thermal stress, bleaching, and mortality in 2005." *PLoS ONE* 5 no. 11: e13969.

Enfield, D.B., A.M. Mestas-Nuñez, and P.J. Trimble, 2001. "The Atlantic multidecadal oscillation and its relation to rainfall and river flows in the continental U.S." *Geophys. Res. Lett.* 28: 2077-2080.

Eichler, T., and W. Higgins. 2006. "Climatology and ENSO-related variability of North American extratropical cyclone activity." *Journal of Climate* 19: 2076-2093.

Ezer, T., 2015. "Detecting changes in the transport of the Gulf Stream and the Atlantic overturning circulation from coastal sea level data: The extreme decline in 2009-2010 and estimated variations for 1935-2012." *Global and Planetary Change* 129:23–36.

Fabry, V. et al., 2008. "Impacts of ocean acidification on marine fauna and ecosystem processes." *ICES Journal of Marine Science* 65:414–432.

Fiechter, J. and C.N.K. Mooers, 2007. "Primary production associated with the Florida Current along the East Florida Shelf: Weekly to seasonal variability from mesoscale-resolution biophysical simulations." *Journal of Geophysical Research: Oceans* 112:1–21.

Finkl, C. and C. Makowski, 2013. "The Southeast Florida Coastal Zone (SFCZ): A Cascade of Natural, Biological, and Human-Induced Hazards." In: C.W. Finkl (ed.), Coastal Hazards, Coastal Research Library 6, DOI 10.1007/978-94-007-5234-4_1,#Springer Science+Business Media Dordrecht 2013 3.

FOA. 2013. "Florida's oceans and coasts: An economic and cluster analysis. Florida Ocean Alliance Report". www.floridaoceanalliance.org pp. 56.

FOCC. 2010. Climate Change and Sea-Level Rise in Florida: An Update of "The Effects of Climate Change on Florida's Ocean and Coastal Resources." Tallahassee, Florida. vi + 26 p. www.floridaoceanscouncil.org.

Frieler, K. et al., 2012. "Limiting global warming to 2°C is unlikely to save most coral reefs." *Nature Climate Change*, 3 no.2:165–170.

Frommel, A.Y. et al., 2016. "Ocean acidification has lethal and sub-lethal effects on larval development of yellowfin tuna, *Thunnus albacares*." *Journal of Experimental Marine Biology and Ecology* 482:18–24.

Geselbracht, L.L. et al., 2015. "Modeled sea level rise impacts on coastal ecosystems at six major estuaries on Florida's gulf coast: Implications for adaptation planning." *PLoS ONE* 10 no. 7:e0132079.

Gill, A.E., 1980. " Some simple solutions for heat-induced tropical circulation." *Quart. J. Roy. Met. Soc.*, 106: 447-462.

Gittman, R.K. et al., 2016. "Ecological Consequences of Shoreline Hardening: A Meta-Analysis." *BioScience* p.biw091.

Gladfelter W.B., 1992. "White band disease in *Acropora palmata*: impli-cations for the structure and growth of shallow reefs." *Bull Mar Sci* 32:639–643.

Gnanadesikan, A., and Coauthors, 2006. "GFDL's CM2 global coupled climate models. Part II: the baseline ocean simulation." *J. Climate*, 19:675–697.

Govoni, J.J. et al., 2009. "Mesoscale, cyclonic eddies as larval fish habitat along the southeast United States shelf: A Lagrangian description of the zooplankton community." *ICES Journal of Marine Science*, 67 no. 3:403–411.

Griffies, S. M., M. . Harrison, R. C. Pacanowski, and A. Rosati, 2004. "A technical guide to MOM4." GFDL ocean group technical report No. 5, Princeton, NJ: NOAA/Geophysical Fluid Dynamics Laboratory, 342 pp.

Guinotte, J.M. and V.J. Fabry, 2008. "Ocean acidification and its potential effects on marine ecosystems." *Annals of the New York Academy of Sciences*, 1134:320–342.

Hall, M., B.T. Furman, M. Merello, and M.J. Durako, 2016. "Recurrence of *Thalassia testudinum* seagrass die-off in Florida Bay: initial observations." *Marine Ecology Progress Series* 560:243–249.

Hamed, A., K. Madani, B. Von Holle, J. Wright, J.W. Milon, and M. Bossick, 2016. "How Much Are Floridians Willing to Pay for Protecting Sea Turtles from Sea Level Rise?" *Environmental Management* 57 no. 1:76–188.

He, R., and R.H. Weisberg. 2002. "West Florida shelf circulation and temperature budget for the 1999 spring transition." *Continental Shelf Res.* 22, no. 5: 719-748.

Heck, K.L., Hays, G. and Orth, R.J., 2003. "Critical evaluation of the nursery role hypothesis for seagrass meadows." *Marine Ecology Progress Series* 253:123–136.

Hixon, M.A., 2011. "60 Years of Coral Reef Fish Ecology: Past, Present, Future." *Bulletin of Marine Science* 87 no. 4:727–765.

IPCC, 2013. "Climate Change 2013: The Physical Science Basis. Contribution of Working Group I to the Fifth Assessment Report of the Intergovernmental Panel on Climate Change." Stocker, T.F., D. Qin, G.-K. Plattner, M. Tignor, S.K. Allen, J. Boschung, A. Nauels, Y. Xia, V. Bex and P.M. Midgley (eds.), Cambridge University Press, Cambridge, United Kingdom and New York, NY, USA, 1535 pp, doi:10.1017/CBO9781107415324.

Jaap, W.C., 2015. "Stony coral (Milleporidae and Scleractinia) communities in the eastern Gulf of Mexico: A synopsis with insights from the Hourglass collections." *Bulletin of Marine Science* 91 no. 2:207–253.

Jackson J.B.C., M.K. Donovan, K.L. Cramer, and V.V. Lam (editors). 2014. "Status and Trends of Caribbean Coral Reefs: 1970-2012." Global Coral Reef Monitoring Network, IUCN, Gland, Switzerland.

Jones, E.B. and J.D. Wiggert, 2015. "Characterization of a high chlorophyll plume in the northeastern Gulf of Mexico." *Remote Sensing of Environment* 159:152–166.

Kennedy, A.J., M.L. Griffin, S.L. Morey, S.R. Smith, and J.J. O'Brien. 2007. "Effects of El Niño – Southern Oscillation on sea level anomalies along the Gulf of Mexico coast." *J. Geophys. Res.* 112: doi:10.1029/2006JC003904.

Knutson, T.R., McBride, J.L., Chan, J., Emanuel, K., Holland, G., Landsea, C., Held, I., Kossin, J.P., Srivastava, A.K. and Sugi, M., 2010. "Tropical cyclones and climate change." *Nature Geosci.* 3 no. 3:157-163.

Koch, M.S. et al., 2014. "Climate Change Projected Effects on Coastal Foundation Communities of the Greater Everglades Using a 2060 Scenario: Need for a New Management Paradigm." *Environmental Management* 55 no. 4:857–875.

Koch M.S., S.A. Schopmeyer, O.I. Nielsen, C. Kyhn-Hansen, and C.J. Madden, 2007. "Conceptual model of seagrass die-off in Florida Bay: links to biogeochemical processes." *J. Exp. Mar. Biol. Ecol.* 350:73–88.

Kough, A.S., C.B. Paris, and M.J. Butler IV, 2013. "Larval Connectivity and the International Management of Fisheries." *PLoS ONE* 8 no. 6: e64970.

Kruczynski, W.L. and Fletcher, P.J. 2012. "Tropical Connections: South Florida's Marine Environments." IAN Press, University of Maryland Center for Environmental Science.

Landsberg, J.H., L.J. Flewelling, and J. Naar, 2009. "*Karenia brevis* red tides, brevetoxins in the food web, and impacts on natural resources: Decadal advancements." *Harmful Algae* 8 no. 4: 598–607.

Larsen, J. C. 1992. "Transport and heat flux of the Florida current at 27 degrees N derived from cross-stream voltages and profiling data: Theory and observations." *Philosophical Transactions of the Royal Society of London A: Mathematical, Physical and Engineering Sciences* 338, no. 1650: 169-236.

Lee, S.-K., D. B. Enfield, and C. Wang, 2007. "What drives seasonal onset and decay of the Western Hemisphere warm pool?" *J. Climate* 20: 2133-2146.

Lee, S.-K., D. B. Enfield, and C. Wang, 2011. "Future impact of differential inter-basin ocean warming on Atlantic hurricanes." *J. Climate* 24: 1264-1275.

Lee, T.N. and E. Williams, 1999. "Mean distribution and seasonal variability of coastal currents and temperature in the Florida Keys with implications for larval recruitment." *Bulletin of Marine Science* 64 no. 1:35–56.

Lee, T.N., J.A. Yoder, J.A., and L.P. Atkinson, 1991. "Gulf Stream frontal eddy influence on productivity of the southeast U.S. continental shelf." *Journal of Geophysical Research: Oceans* 96 no. C12:22191–22205.

Lellis-Dibble, K. A., K.E. McGlynn, and T.E. Bigford, 2008. "Estuarine fish and shellfish species in U.S. commercial and recreational fisheries : economic value as an incentive to protect and restore Habitat." NOAA NMFS, Office of Habitat Conservation, Habitat Protection Division.

Li, W., L. Li, M. Ting, and Y. Liu, 2012. "Intensification of Northern Hemisphere subtropical highs in a warming climate." *Nat. Geosci.*, 5: 830-834.

Lidz, B.H., 2006. "Pleistocene Corals of the Florida Keys: Architects of Imposing Reefs—Why?" *Journal of Coastal Research* 224:750–759.

Lidz, B.H., C.D. Reich, and E.A. Shinn, 2003. "Regional quaternary submarine geomorphology in the Florida Keys." *Bulletin of the Geological Society of America*, 115 no. 7:845–866.

Liu, Y., S.-K. Lee, B. A. Muhling, J. T. Lamkin, and D. B. Enfield, 2012. "Significant reduction of the Loop Current in the 21st century and its impact on the Gulf of Mexico." *J. Geophys. Res.*, 117: C05039.

Liu, Y., S.-K. Lee, D. B. Enfield, B. A. Muhling, J. T. Lamkin, F. E. Muller-Karger, and M. A. Roffer, 2015. "Potential impact of climate change on the Intra-Americas Sea: Part-1. A dynamic downscaling of the CMIP5 model projections." *J. Mar. Syst.* 148:56-59.

Llopiz, J.K. and A.J. Hobday, 2015. "A global comparative analysis of the feeding dynamics and environmental conditions of larval tunas, mackerels, and billfishes." *Deep-Sea Research Part II: Topical Studies in Oceanography* 113:113–124.

Manzello D.P., I.C. Enochs, N. Melo, D.K. Gledhill, and E.M. Johns, 2012. "Ocean acidification refugia of the Florida Reef Tract." *PLoS ONE* 7 no. 7:e41715.

Mapes, B. E., P. Liu, and N. Buenning, 2005. "Indian monsoon onset and the Americas midsummer drought: out-of-equilibrium responses to smooth seasonal forcing." *J. Climate*, 18:1109–1115.

Maul, G.A., and D.M. Martin,1993. "Sea-Level Rise at Key-West, Florida, 1846-1992 - America Longest Instrument Record." *Geophysical Research Letters*, 20 no. 18:1955–1958.

McKee, K.L., D.R. Cahoon, and I.C. Feller, 2007. "Caribbean mangroves adjust to rising sea level through biotic controls on change in soil elevation." *Global Ecology and Biogeography*, 16 no. 5:545–556.

McKee, K.L., 2011. "Biophysical controls on accretion and elevation change in Caribbean mangrove ecosystems." *Estuarine, Coastal and Shelf Science*, 91 no. 4:475–483.

Millero F.J., W.T. Hiscock, F. Huang, M. Roche, and J.Z. Zhang, 2001. Seasonal variation of the carbonate system in Florida Bay." *Bull. Mar. Sci.* 68:101–123.

Miles, T.N. and R. He, 2010. "Temporal and spatial variability of Chl-a and SST on the South Atlantic Bight: Revisiting with cloud-free reconstructions of MODIS satellite imagery." *Continental Shelf Research*, 30 no. 18:1951–1962.

Morey, S.L., P.J. Martin, J.J. O'Brien, A.A. Wallcraft, and J. Zavala-Hidalgo. 2003. "Export pathways for river discharged fresh water in the northern Gulf of Mexico." *J. Geophys Res.* 108. Doi: 10.1029/2002JC001674.

Morey, S.L., D.S. Dukhovskoy, and M.A. Bourassa. 2009. "Connectivity of the Apalachicola River flow variability and bio-optical oceanic properties of the northern West Florida shelf." *Cont. Shelf. Res.* 29: 1264-1275.

Muehllehner, N., C. Langdon, A. Venti, and D. Kadko, 2016. "Dynamics of carbonate chemistry, production, and calcification of the Florida Reef Tract (2009–2010): Evidence for seasonal dissolution." *Global Biogeochemical Cycles* 30 no. 5:661–688.

Muhling, B.A., J.T. Lamkin, and W.J. Richards, 2012. "Decadal-scale responses of larval fish assemblages to multiple ecosystem processes in the northern Gulf of Mexico." *Marine Ecology Progress Series* 450:37–53.

Muhling, B.A., P. Reglero, L. Ciannelli, D. Alvarez-Berastegui, F. Alemany, J.T. Lamkin, and M.A. Roffer, 2013. "Comparison between environmental characteristics of larval bluefin tuna *Thunnus thynnus* habitat in the Gulf of Mexico and western Mediterranean Sea." *Marine Ecology Progress Series*, 486:257–276.

Muller-Karger, F.E., J.P. Smith, S. Werner, R. Chen, M. Roffer, Y. Liu, B. Muhling, D. Lindo-Atichati, J. Lamkin, S. Cerdeira-Estrada, and D. Enfield. 2015. "Natural variability of surface oceanographic conditions in the offshore Gulf of Mexico." *Prog. Oceanogr.* 134: 54-76.

NOAA Office for Coastal Management, 2016. "General Coastline and Shoreline Mileage of the United States. https://coast.noaa.gov/data/docs/states/shorelines.pdf. Accessed 20 July 2016.

Oey, L.-Y., T. Ezer, and H. C. Lee, 2005. "Loop Current, rings and related circulation in the Gulf of Mexico: A review of numerical models and future challenges," in *Circulation in the Gulf of Mexico: Observations and Models*, W. Sturges and A. Lugo-Fernandez, Eds., Amer. Geophys. Union, pp. 31-56

Orth, R.J., T.J. Carruthers, W.C. Dennison, C.M. Duarte, J.W. Fourqurean, K.L. Heck, A.R. Hughes, G.A. Kendrick, W.J. Kenworthy, S. Olyarnik, and F.T. short, 2006. "A Global Crisis for Seagrass Ecosystems." *Bioscience* 56 no. 12:987–996.

Paerl, H.W., and V.J. Paul, 2012. "Climate change: links to global expansion of harmful cyanobacteria" *Water Res* 46 no.5:1349-63.

Palanisamy, H., M. Becker, B. Meyssignac, O. Henry, and A. Cazenave, 2012. "Regional sea level change and variability in the Caribbean sea since 1950." *Journal of Geodetic Science* 2 no. 2:125–133.

Peng, G., Z. Garraffo, G.R. Halliwell, O.M. Smedstad, C.S. Meinen, V. Kourafalou, and P.Hogan, 2009. "Temporal variability of the Florida Current transport at 27°N." in *Ocean Circulation and El Nino: New Research*, J.A. Long and D.S. Wells, eds. 119-137.

Pörtner, H.O., M. Langenbuch, and A. Reipschläger, 2004. "Biological Impact of Elevated Ocean CO_2 Concentrations: Lessons from Animal Physiology and Earth History." *Journal of Oceanography* 60:705–718.

Precht, W.F. and R.B. Aronson, 2004. "Climate flickers and range shifts of reef corals." *Frontiers in Ecology and the Environment* 2 no. 6:307–314.

Putman, N.F., J.M. Bane, and K.J. Lohmann, 2010. "Sea turtle nesting distributions and oceanographic constraints on hatchling migration." *Proceedings of the Royal Society B: Biological Sciences* 277 no. 1700: 3631–3637.

Rauscher, S. A., F. Giorgi, N. S. Diffenbaugh, and A. Seth, 2008. "Extension and intensification of the Meso-American mid-summer drought in the twenty-first century." *Clim. Dynam.* 31: 551-571.

Rauscher, S. A., F. Kucharski, and D. B. Enfield, 2011. "The role of regional SST warming variations in the drying of Meso-America in future climate projections." *J. Climate* 24:2003-2016.

Reece, J.S., D. Passeri, L. Ehrhart, S.C. Hagen, A. Hays, C. Long, R.F. Noss, M. Bilskie, C. Sanchez, M.V. Schwoerer, and B. Von Holle, 2013. "Sea level rise, land use, and climate change influence the distribution of loggerhead turtle nests at the largest USA rookery (Melbourne Beach, Florida)." *Marine Ecology Progress Series* 493:259-274.

Reich, C.D., R.Z. Poore, and T.D. Hickey, 2013. "The Role of Vermetid Gastropods in the Development of the Florida Middle Ground, Northeast Gulf of Mexico." *Journal of Coastal Research* 63:46–57.

Ropelewski, C.F., and M.S. Halpert. 1986. "North American precipitation and temperature patterns associated with the El Niño/Southern Oscillation (ENSO)." *Mon. Wea. Rev.* 114: 2352-2362.

Rudzin, J.E., S.L. Morey, M.A. Bourassa, S.R. Smith. 2013. "The influence of Loop Current position on winter sea surface temperatures in the Florida Straits." *Earth Interactions* 17: 16, doi:10.1175/2013EI000521.1.

Ruzicka, R.R., M.A. Colella, J.W. Porter, J.M. Morrison, J.A. Kidney, V. Brinkhuis, K.S. Lunz, K.A. Macaulay, L.A. Bartlett, M.K. Meyers, and J. Colee, 2013. "Temporal changes in benthic assemblages on Florida Keys reefs 11 years after the 1997/1998 El Niño." *Marine Ecology Progress Series* 489:125–141.

Schmidt, N., E.K. Lipp, L.B. Rose, and M.E. Luther. 2001. "ENSO influences on seasonal rainfall and river discharge in Florida." *J. Climate*, 14: 615-628.

Shulzitski, K., S. Sponaugle, M. Hauff, K.D. Walter, and R.K. Cowen, 2016. "Encounter with mesoscale eddies enhances survival to settlement in larval coral reef fishes." *Proceedings of the National Academy of Sciences* 113 no. 25:6928–6933.

Smith, S.R., P.M. Green, A.P. Leonardi, and J.J. O'Brien. 1998. "Role of multiple-level tropospheric circulations in forcing ENSO winter precipitation anomalies." *Mon. Wea. Rev.* 126: 3102-3116.

SFRCCC, 2015. "Unified Sea Level Rise Projection for Southeast Florida." Southeast Florida Regional Climate Change Compact Steering Committee. 35 p.

Shamberger, K.E.F., A.L. Cohen, Y. Golbuu, D.C. McCorkle, S.J. Lentz, and H.C. Barkley, 2014. "Diverse coral communities in naturally acidified waters of a western Pacific reef." *Geophysical Research Letters*, 41 no. 2:1–6.

Sponaugle, S., T. Lee, V. Kourafalou, and D. Pinkard, 2005. "Florida Current frontal eddies and the settlement of coral reef fishes." *Limnology and Oceanography*, 50 no. 4:1033–1048.

Stathakopoulos, A. and B.M. Riegl, 2015. "Accretion history of mid-Holocene coral reefs from the southeast Florida continental reef tract, USA." *Coral Reefs*, 34 no. 1:173–187.

Taylor, K. E., R. J. Stouffer, and G. A. Meehl, 2012. "An overview of CMIP5 and the experiment design." *Bull. Amer. Meteor. Soc.*, 93:485–498.

Terray, L., L. Corre, S. Cravatte, T. Delcroix, G. Reverdin, and A. Ribes, 2012. "Near-surface salinity as nature's rain gauge to detect human influence on the tropical water cycle." *J. Climate*, 25:958–977.

Todd, A.C., S.L. Morey, and E.P. Chassignet. 2014. "Circulation and cross-shelf transport in the Florida Big Bend." *J. Marine Res.* 72: 445-475.

van Hooidonk, R., J.A. Mayard, Y. Liu, and S.K. Lee., 2015. "Downscaled projections of Caribbean coral bleaching that can inform conservation planning." *Global Change Biology*, 21 no. 9:3389–3401.

Vaz, A.C., C.B. Parins, M.J. Olascoaga, V.H. Kourafalou, H. Kang, and J.K. Reed, 2016. "The perfect storm: Match-mismatch of bio-physical events drives larval reef fish connectivity between Pulley Ridge mesophotic reef and the Florida Keys." *Continental Shelf Research*, 125:136–146.

Walsh, J.J., C.R. Tomas, K.A. Steidinger, J.M. Lenes, F.R. Chen, R.H. Weisberg, L. Zheng, J.H. Landsberg, G.A. Vargo, and C.A. Heil, 2011. "Imprudent fishing harvests and consequent trophic cascades on the West Florida shelf over the last half century: A harbinger of increased human deaths from paralytic shellfish poisoning along the southeastern United States, in response to oligotrophication." *Continental Shelf Research*, 31 no. 9:891–911.

Wang, Z.A., R. Wanninkhof, W.J. Cai, R.H. Byrne, X. Hu, T.H. Peng, and W.J. Huang, 2013. "The marine inorganic carbon system along the Gulf of Mexico and Atlantic coasts of the United States : Insights from a transregional coastal carbon study." *Limnology and Oceanography*, 58 no. 1:325–342.

Wanless H, R.W. Parkinson, and L.P. Tedesco, 1994. "Sea level control on stability of Everglades wetlands." In: Davis SM, Ogdon JC (eds) *Everglades: the ecosystem and its restoration*. St Lucie Press, FL.

Wanninkhof, R., L. Barero, R. Byrne, W.J. Cai, W.J. Huang, J.Z. Zhang, M. Baringer, and C. Langdon, 2015. "Ocean acidification along the Gulf Coast and East Coast of the USA." *Continental Shelf Research*, 98:54–71.

Weil, E. and C.S. Rogers 2011. "Coral Reef Diseases in the Atlantic-Caribbean." In: Z. Dubinsky and N. Stambler (eds.), *Coral Reefs: An Ecosystem in Transition*, Springer Science+Business Media B.V. 201.

Weisberg, R.H., L. Zheng, L.and Y. Liu, Y., 2016. "West Florida shelf upwelling: Origins and pathways." *Journal of Geophysical Research: Oceans*.

Weisberg, R.H, L. Zheng, Y. Liu, C. Lembke, J.M. Lenes, and J.J. 2014. "Why no red tide was observed on the West Florida Continental Shelf in 2010." *Harmful Algae*, 38 no. C:119–126.

Witt, M.J., L.A. Hawkes, M.H. Godfrey, B.J. Godley, and A.C. Broderick, 2010. "Predicting the impacts of climate change on a globally distributed species: the case of the loggerhead turtle." *The Journal of Experimental Biology*, 213:901–911.

Wirt, K.E., P. Hallock, D. Palandro, and K.L. Daly, 2013. "Potential habitat of *Acropora* spp. on Florida reefs." *Applied Geography*, 39:118–127.

Wood, L.D., R. Hardy, P. Meylan, and A. Meylan 2013. "Characterization of a hawksbill turtle (*Eretmochelys imbricata*) foraging aggregation in a high-latitude reef community in southeastern Florida, USA." *Herpetological Conservation and Biology*, 8 no. 1:258–275.

Yates K.K., R.B. Halley, 2006. "Diurnal variation in rates of calcification and carbonate sediment dissolution in Florida Bay." *Estuar Coast* 29 no. 1:24–39.

Yoder, J.A., L.P. Atkinson, S.S. Bishop, E.E. Hofmann, and T.N. Lee, 1983. "Effect of upwelling on phytoplankton productivity of the outer southeastern United States continental shelf." *Continental Shelf Research*, 1 no. 4:385–404.

Zavala-Hidalgo, J., R. Romero-Centeno, A. Mateos-Jasso, S.L. Morey, and B. Martinez-Lopez. 2014. "The response of the Gulf of Mexico to wind and heat flux forcing: What has been learned in recent years?" *Atmosfera* 27, No. 3: 317-334.

Zhang, K., H. Liu, Y. Li, H. Xu, J. Shen, J. Rhome, and T.J. Smith, 2012. "The role of mangroves in attenuating storm surges." *Estuarine, Coastal and Shelf Science*, 102:11–23.

CHAPTER 14

Climate Change Impacts on Florida's Fisheries and Aquaculture Sectors and Options for Adaptation

Kai Lorenzen[1], Cameron H. Ainsworth[2], Shirley M. Baker[1], Luiz R. Barbieri[3],
Edward V. Camp[1], Jason R. Dotson[4], and Sarah E. Lester[5]

[1]*Fisheries and Aquatic Sciences Program, School of Forest Resources & Conservation, University of Florida, FL;* [2]*College of Marine Science, University of South Florida, St. Petersburg, FL;* [3]*Fish and Wildlife Research Institute, Florida Fish and Wildlife Conservation Commission, St. Petersburg, FL;* [4]*Fish and Wildlife Research Institute, Florida Fish and Wildlife Conservation Commission, Gainesville, FL;* [5]*Department of Geography, Florida State University, Tallahassee, FL*

Florida supports diverse marine and freshwater fisheries and a significant aquaculture industry with a combined economic impact of approximately 15 billion US$. We begin by describing the characteristics of the different fisheries and aquaculture sectors. This is followed by a description of the relevant climate change and confounding drivers. We then present an integrated social-ecological systems framework for analyzing climate change impacts and apply this framework to the different fisheries and aquaculture sectors. We highlight how the characteristics of each sector gives rise to distinct expected climate change impacts and potential adaptation measures. We conclude with general considerations for monitoring and adaptation.

Key Messages

- Sea level rise, more frequent severe storms, coastal habitat loss associated with both factors, changes in nutrient dynamics, and ocean acidification are likely to impact the productivity of Florida's marine fisheries. Some of these factors will also affect fisheries access.
- Florida's freshwater fisheries will be impacted by increased hydrological variability, increased temperatures, and more frequent severe storms. Shallow lakes may respond by switching from a clear to a turbid, phytoplankton-dominated state that provides poor sport fishing. Greater hydrological variability will also exacerbate fishing access issues.
- Among the aquaculture sectors, shellfish aquaculture is particularly sensitive to multiple drivers including sea level rise, coastal habitat loss, increased frequency of harmful algal blooms and ocean acidification. Ornamental fish culture and other forms of intensive aquaculture under controlled conditions will be relatively insensitive to climate change.
- Key adaptation options for marine fisheries include switching of species, locations and fishing methods, while adapting catch limits to changes in productivity. In freshwater fisheries, on the other hand, water and habitat management will be key to adaptation. Change in farming methods will be important in aquaculture, along with species and location changes, particularly in the shellfish industry. Aquaculture for fisheries enhancement and ecological restoration can aid adaptation in both marine and freshwater fisheries. Adaptation will benefit from awareness of drivers and impact pathways, monitoring of a broad suite of impact indicators, and adaptive decision-making.

Keywords

Fisheries; Aquaculture; Sea level rise; Coastal habitat; Social-ecological system; Fisheries enhancement; Restoration aquaculture

Introduction

Florida's fisheries and aquaculture sectors are of major economic, social, and cultural importance to the state, generating some $15 billion in economic activity, supporting over 150,000 jobs, and attracting more than 2.4 million visiting anglers to the 'fishing capital of the world' (Table 14.1). Both fisheries and aquaculture use aquatic biological resources for human ends, but they do so in very different ways (FAO 1990; Anderson 2002). Fisheries involve the harvesting of wild aquatic organisms held in some form of common ownership. Aquaculture, on the other hand, entails the active husbandry of aquatic organisms that are privately-owned (Bostock et al. 2010). Fisheries and aquaculture, therefore, differ in both the technology used and the way in which resources are owned and regulated. Not all fisheries and aquaculture enterprises fall neatly into these two categories. For example, in some fisheries, catch shares confer private use rights to wild aquatic organisms (Fujita & Brozon 2005). Also, some fisheries are enhanced by using technologies akin to active husbandry, such as the deployment of artificial reefs or the release of hatchery-reared fish (Bortone et al. 2011; Lorenzen et al. 2013).

Fishing may be for commercial purposes (income), for recreation, or for subsistence (meeting nutritional needs). All three types of fishing exist in Florida, but recreational and commercial fisheries predominate. Commercial and subsistence fishers are motivated by a desire to harvest fish; they get satisfaction primarily from income earned or nutrition gained. Recreational fishers, on the other hand, are motivated by a variety of factors, including enjoyment of the outdoors, relaxation, escape, and self-actualization, as well as or instead of a desire to harvest fish (Fedler & Ditton 1994; Cooke et al. 2017; Garlock & Lorenzen 2017). Little is known about subsistence-oriented fishing in Florida, which exists in both the marine and freshwater realms. It is often carried out from shore and involves a broad range of fish species, including many that are not targeted in the commercial or recreational fisheries.

Being reliant on sensitive ecosystems and support infrastructure (e.g., docks, boat ramps), fisheries and aquaculture are likely to be among the sectors of Florida's economy most vulnerable to climate change. Here, we briefly discuss the importance and diversity of Florida's fisheries and aquaculture sectors before outlining key considerations for understanding potential climate change impacts and adaptation options. We then apply these considerations to select fisheries and aquaculture sectors, and point out similarities and differences in vulnerabilities and adaptation options.

Overview of Florida's Fisheries and Aquaculture Sectors

Florida's fisheries and aquaculture sectors are exceptionally diverse and of great economic, social, and cultural importance to the state (Table 14.1). Marine fisheries—which consists of recreational, commercial, and commercial marine life fishing—is the largest subsector overall, with an economic impact of over $13 billion annually. Marine recreational fishing is responsible for more than $12 of that $13 billion, producing 70,000 jobs and attracting 6.5 million participants each year (Southwick Associates 2012; NMFS 2016). In fact, Florida has the largest and most valuable marine recreational fisheries of any state in the U.S., with its economic impact generated through expenditures by recreational anglers and by jobs in the fishing equipment, hospitality, guide services, and other support areas. Out-of-state visitors are responsible for about 18% of Florida recreational fisheries' economic impact (USFWS & USCB 2013), and Florida is a net gainer in the movement of recreational fishers; in other words, more fishing trips are 'imported' to the state than are 'exported' from it (Ditton et al. 2002). Much of Florida's marine recreational fishing takes place on private boats or from shore, but about 15% of fishing trips involve charter vessels ranging from smaller personalized guide services to large 'head boats.' Charter operators are not allowed to sell fish caught during charter trips (i.e., they are not commercial fishermen); they exist to provide a service to anglers. Marine commercial fishing adds another $1 billion to the state's marine fisheries and produces some 12,000 jobs (NMFS 2016). The economic impact of commercial fishing in Florida is generated through sales of harvested seafood. Seafood is a widely-traded commodity, and a substantial share of the seafood produced in Florida is exported out of state and out of the country. On the other hand, much of the seafood consumed in Florida is imported, with locally-produced seafood accounting for only about 15% of consumption. As a result of such trade patterns, seafood production and consumption in the state are only weakly linked. In addition to the commercial fishery for seafood, a smaller and very specialized 'marine life' fishery provides live organisms for the aquarium trade and for research (Larkin et al. 2001).

Freshwater fishing in Florida is also predominantly recreational, but it does have subsistence-oriented and commercial components. The state's freshwater recreational fishing generates about $1.7 billion annually and is responsible for some 14,000 jobs, the highest impact of freshwater recreational fishing in any state in the U.S. (Southwick Associates 2012). And while the freshwater commercial fishing industry in Florida is of less economic importance than recreational, it still contributes several million dollars to the state's bottom line each year (FWC 2011).

Finally, Florida's aquaculture industries (also known as aquafarming) generate about $69 million in sales volume annually and has an economic impact that is double, possibly triple that generated by 686 aquaculture operators across the state and supporting an estimated 2000 jobs (USDA 2013). The largest aquaculture subsector in terms of economic impact is the ornamental fish industry, which produces a variety of tropical species for the aquarium trade. Second in economic importance is the shellfish industry, which harvests mainly clams and oysters. In addition, there are several smaller but profitable industries, including sturgeon farming for caviar and culture of baitfish for recreational anglers. Emerging aquaculture subsectors include open ocean aquaculture, which is being considered for the Gulf of Mexico. Restoration and fisheries enhancement aquaculture (i.e., producing organisms for release into natural ecosystems) is currently carried out only at the research-level but it may expand into a more significant industry over the long term. In its totality, Florida's aquaculture industry presents a diversity of production systems, ranging from the extensive and environmentally-open systems used in shellfish farming to the intensive and highly controlled indoor systems used in sturgeon farming or in the marine ornamental industry.

Table 14.1. Economic impact, employment, and participation in Florida's fisheries and aquaculture sectors and subsectors.

Sector	Economic Impact or Value ($ Million)	Employment	Participation (Thousands)	Harvest
Marine fishing				
Recreational	12,249[1]	70,109[1]	6500[1]	
Commercial	1,060[1]	12,241[1]	12[1]	99 million lbs[1]
Commercial (marine life)	7[2]			12 million individuals
Freshwater fishing				
Recreational	1,689[3]	14,040[3]	3100[3]	
Commercial	5[4]			10 million lbs[4]
Aquaculture	69[5]	2000[5]	2[5]	
Ornamental	27[5]	400[5]	< 1[5]	
Shellfish	12[5]	400[5]	< 1[5]	
Other	30[5]	1000[5]	1[5]	

Sources: (1) NMFS 2016; (2) Larkin et al. 2001; FWC 2017 (3) Southwick Associates 2012; (4) FWC 2011; (5) USDA 2013.

Key Considerations for Understanding Climate Change Impacts on Fisheries and Aquaculture

Climate change will impact fisheries and aquaculture through multiple drivers and pathways in ways that will be strongly dependent on specific characteristics of the different systems. It is, therefore, important to adopt an integrated, social-ecological systems approach to assessing potential impacts and adaptation measures. (Hollowed et al. 2013; Bush et al. 2016; Hunt et al. 2016). Here, we conceptualize climate change impacts on these systems by applying the considerations outlined in Table 14.2.

Table 14.2. Key considerations for understanding climate change impacts and adaptation options relating to Florida's fisheries and aquaculture sectors.

Climate Change and Confounding Drivers	Fishery or Aquaculture System Attributes	Impact Pathways	Outcomes	Adaptation Options
Climate Change Temperature Rainfall Altered circulation Altered hydrology Storm frequency and severity Sea level rise Geomorphic changes Acidification Habitat Mitigation and adaptation policies **Confounding Factors** Demography Land use change Water demand	Environment Resource Technology Users/producers Governance	Resource pathway Resource user/producer pathway	Resource conservation Resource use/production Economic Social Governance	Technical change Behavioral change Governance change

Climate change and confounding drivers outside the control of fisheries/aquaculture stakeholders or governance systems constitute external drivers. These drivers include changes in temperature, rainfall, circulation, hydrology, frequency and severity of storms, sea level, geomorphology, ocean acidity, habitats and infrastructure, and even mitigation policies. However, the impact of climate change-related drivers may be confounded by other anthropogenic drivers (e.g., changes in human population demography, land use, and markets for inputs and outputs of fisheries and aquaculture). Attributes of a fishery or aquaculture system influence its exposure to climate change drivers and, ultimately, the likely outcomes and potential adaptation options available to it. These attributes include characteristics of the environment where the fishery or aquaculture system is situated; the resource/cultured species (e.g. temperature preferences, life history, and habitat use); the degree of technical control over the environment and biological production (typically low in fisheries but high in certain aquaculture systems); motivations, socio-economic status, and adaptive capacity of resource users/producers; and governance arrangements (e.g., rules and regulations in place, compliance and effectiveness, adaptive capacity). The pathways through which impacts of climate change and confounding drivers on the fisheries/aquaculture system occur fall into one of two broad categories: (1) A resource pathway that includes all impacts on the exploited/cultured resources (and thus eventually, on users). For example, this could include a temperature-induced range or productivity shift in a fish stock. (2) A resource user/producer pathway that includes all impacts that act directly on the resource users/producers and on the governance system. Examples would include changes in resource access due to the destruction of boats or landing facilities by a storm, or an increase in operating costs due to a carbon tax on fuel.

Climate change impacts can be characterized and measured in multiple ways: effects on resource conservation status, effects on resource productivity, economic impacts (e.g. overall economic activity, viability of businesses), social impacts (e.g. exclusion of poorer sections of the population from fishing), and performance or sustainability of governance systems. Adaptation to climate change, those actions generally aimed at reducing negative impacts, comes in different forms as well, including technical changes (e.g., modifications made to fishing gear), behavioral changes by resource users/producers, or changes in the governance system.

Drivers of Climate Change Impacts and Confounding Factors in Florida

Key drivers of climate change impacts on Florida's fisheries and aquaculture sectors include (Carter et al. 2014):

- Temperature increases (Carter et al. 2014)
- Moderate increases in average rainfall and increases in variability (Carter et al. 2014; Moser et al. 2014)
- Altered hydrology with an increase in average and variability of river flows, lake water levels, groundwater recharge, and freshwater outflow into coastal systems (Georgakakos et al. 2014; Obeysekera et al. 2015)
- Changes in large- and meso-scale circulation features in the Gulf of Mexico (Liu et al. 2012)
- Changes in ocean stratification (Doney et al. 2014)
- Changes in the frequency and intensity of harmful algal blooms (Moore et al. 2008)
- Greater frequency and severity of storms (Carter 2014)
- Sea level rise; Florida is highly vulnerable and this is perhaps the single most important driver of climate change impacts in the state (Carter et al. 2014)
- Salt water intrusion (FWC 2009; Barlow & Reichard 2010)
- Ocean acidification , although higher latitudes tend to face a greater challenge (Ekstrom et al. 2015)
- Changes in coastal and riparian geomorphology (Glick 2006; FWC 2008; Moser et al. 2014)
- Changes in infrastructure (e.g. boat ramps, docks, roads; Moser et al. 2014)
- Mitigation policies (e.g., a carbon tax on fuel or carbon credits for sequestration in shellfish farming)

In addition, several confounding factors have the potential to affect Florida's fisheries and aquaculture sectors, including:

- Human population growth (FWC 2008 predicted a doubling of over the next 50 years) and increased resource utilization
- Conversion of natural and agricultural land to urban land use, leading to an overall reduction in freshwater and coastal wetland area and more intensive use and modification of remaining areas (FWC 2008)
- Increased demand for fresh water (FWC 2008), including surface and groundwater.
- Increased introduction of nutrients from land-based sources
- Invasive species (e.g. lionfish)
- Economic factors (e.g., changing operating costs for fisheries, changing demand)

Impacts on Marine Fisheries and Adaptation Options

Marine fisheries operate in large natural ecosystems where human influence and control over environmental conditions is comparatively limited. Fishing itself has been the predominant human influence on many marine fish stocks. Florida's marine fisheries are characterized by a high diversity of resources and fishers (Lowther 2011). Key fisheries include offshore commercial shrimp fisheries, offshore reef (mostly snapper/grouper) and pelagic (mackerels, mahi mahi, tunas and billfishes) fisheries that are shared by commercial and recreational fishers, predominantly nearshore lobster and crab fisheries that are likewise shared between sectors, and inshore finfish fisheries (red drum, snook, spotted seatrout, and others) that are almost exclusively recreational. Many marine recreational fisheries in Florida are harvest-oriented (fishers tend to harvest fish they can legally keep), but in some fisheries such as those offshore for billfishes or inshore for snook, voluntary release of legally harvestable fish has become common. Voluntary release can help maintain the fishing quality (e.g., catch rates and size structure) under high recreational fishing pressure (Arlinghaus et al. 2007).

Florida's marine fisheries are managed by the Florida Fish and Wildlife Conservation (FWC) Commission in coastal waters (nine miles from the coast in the Gulf of Mexico and three miles in the Atlantic). In federal waters outside the state limits, fisheries are managed by the Gulf of Mexico and South Atlantic Fisheries Management Councils (GMFMC and SAFMC), respectively, with administrative oversight from the National Marine Fisheries Service (NMFS) and the U.S. Department of Commerce. Coordination routinely occurs between these entities with respect to stocks and issues that straddle management boundaries. Management of tuna and billfish is complicated by the wide-ranging migratory habits of these fish that take them through the exclusive economic zones of various eastern and western Atlantic countries. The International Commission for the Conservation of Atlantic Tunas sets catch limits for about 15 pelagic species landed in Florida waters. The mainstay of marine fisheries management is to regulate harvest, often to a level that will allow the stock to produce the greatest average catch in the long term ('Maximum Sustainable Yield'). Harvest regulations are informed by regular, scientific stock assessments for the major fisheries (Cooper 2004; Methot 2009). Stock assessments use data collected from the fisheries and from fisheries-independent monitoring programs to track the abundance of stocks and their responses to fishing and environmental variation. Hence, fisheries management systems have many features that make them well suited to track and respond to change. Most fisheries have experienced major natural and/or fishing-induced variation in stock abundance and historically, many have been overfished. Today, effective management has largely rectified this situation, with most major marine fish stocks around Florida exploited near their sustainable limit, while some remain overfished and a few are underexploited (NOAA 2017a). Demand for marine fisheries products and recreational fishing remains such that any relaxation of management efforts or failure to account for changes in stock productivity could easily lead to overfishing in many stocks.

In addition to harvest regulations, deployment of artificial reefs that aggregate fish (and fishers) at known locations is a common fisheries management measure in Florida, often conducted by coastal counties with the aim of enhancing local recreational fisheries and associated economic activity (Bortone et al. 2011; Lindberg & Seaman 2011). Stocking of hatchery fish is another way of enhancing fisheries, conducted experimentally by the FWC and Mote Marine Laboratory (Tringali et al., 2008; Camp et al. 2014).

Although Florida's offshore habitats (with the exception of coral reefs) have been relatively removed from human impacts other than fishing, inshore habitats have been heavily impacted by coastal development, with substantial losses of saltmarsh, mangrove, and seagrass habitats over the past century. Such inshore habitat changes potentially affect not only inshore fish and fisheries but also many offshore stocks, which rely on inshore habitats as juveniles. Nonetheless, even though the importance of habitat in maintaining fisheries has been widely acknowledged, relationships between habitat and fish stock dynamics are complex and clear quantitative links have proved elusive (Rose 2000). Efforts have been made by fisheries management agencies to identify essential fish habitat and, to a lesser extent, to conserve and restore such habitat (Rosenberg et al. 2000). However, these efforts have remained limited in scope and they are somewhat separate from the fisheries management process. Marine fisheries management systems are set up primarily to regulate fishing rather than manage environmental conditions or habitats. Marine protected areas may include zones in which harvest is largely or entirely restricted, and these zones may be associated with enhanced stock and community resilience.

Climate Change Impacts on Marine Fisheries Resources

The physical environment of the ocean has a major influence on the productivity and distribution of organisms at all trophic levels (Karnauskas et al. 2015). Climate change alters the amount of salt and heat in different parts of the ocean, leading to changes in the major currents of the ocean (the thermohaline circulation, Schlesinger et al. 2006). Wind and rainfall patterns are being changed as well (Mann and Emanuel 2006). These changes affect species abundance, biodiversity, and fisheries catch composition in the Gulf of Mexico and the Atlantic Ocean coasts of Florida. The scope and nature of the impact on the ecosystem and on fisheries are difficult to predict (Brander 2010). Some effects may be harmful and others may be beneficial to exploited and functionally important marine species.

Changes in Oceanic Flows and Winds

The Gulf Stream and the Loop Current are the dominant circulation features on the east and west coasts of Florida, respectively (Schmitz et al. 2005). On the Atlantic continental shelf, upwelling of the Gulf Stream caused by tides and winds provides nutrients and stimulates production (Mann and Lazier 2013). Additional nutrients are delivered to the ocean by rivers and coastal wetlands, including the Everglades. In the Gulf of Mexico, anticyclonic (warm-core) and cyclonic (cold-core) eddies pinched off from the Loop Current can show some upwelling of nutrients around their periphery and in their centers, respectively (Mann and Lazier 2013, Chérubin et al. 2006). Although the Loop Current does not impinge on the West Florida Shelf (WFS) directly, it can establish a cross-shelf pressure gradient that intensifies upwelling onto the shelf (Hetland et al 1999). Upwelling is further strengthened by seasonal southeasterly winds that act via Ekman transport of surface water offshore, and bottom water toward the inshore (Weisberg at al 2005). A change in strength, location, or variability of oceanic flows and winds implies a change in overall productivity of the shelf. Some have suggested a mechanism by which ocean warming globally could intensify alongshore wind stress and accelerate upwelling (Bakun 1990), potentially leading to increased rates of primary production. The WFS is upwelling-favorable on long-term average (Weisberg et al., 2009), particularly around the spring transition (Liu and Weisberg 2012). This is a critical period in larval feeding for many exploited species (e.g., the shallow-water grouper complex, Farmer et al. 2017). However, any change in seasonal forcing would change the phenology of species and favor a different set of species and fisheries. For example, a 1–2 month delay in peak upwelling on the WFS could favor summer spawners, such as red snapper. So too, temporal changes in stratification or water mass convergence will benefit species that can take advantage of concentrated food during critical life stage periods. Besides providing nutrients for photosynthesis, cross-shelf movement of water plays a role in larval dispersal and retention on both the east and west coasts of Florida (Weisberg et al. 2014, Werner et al. 1997). Gag grouper spawning aggregation sites are well positioned to take advantage of seasonal cross-shore currents (Todd 2013). Thus, the adaptability of populations will depend on the plasticity of animal behavior and stock demographics. Ubiquitous and year-round spawners may be better able to mitigate the effects of more variable productivity and current flows.

Changes in Marine Productivity, the Food Web, and Habitats

Marine primary productivity forms the basis of the food web on which all other productivity depends. Substantial uncertainty surrounds likely changes in marine primary productivity. Globally, different modeling approaches have yielded predictions ranging from a 20% decline to an 8% increase (Sumaila et al. 2011).

A large, seasonal anoxic zone in the northern Gulf of Mexico has well-documented temporary effects on the distribution of fisheries resources. This affects some of the fisheries assisted by Florida-based boats. The future spatial extent and severity of this anoxic zone will depend on river flow and agricultural practices in the Mississippi River Basin, and the degree to which this and other anoxic zones cause an overall reduction in fisheries yields is subject to ongoing scientific debate (Rabalais et al. 2002; Diaz and Rosenberg 2008; Breitburg et al. 2009).

Ocean acidification is predicted to proceed at a relatively moderate pace around Florida, but has the potential to greatly affect coral reefs, the resources directly associated with live corals, exploited epibenthic invertebrates (shrimp, crabs, bivalves), and indirectly, benthic fish preying upon macrobenthos. These impacts will be felt most strongly in the nearshore fisheries. Another issue connected with ocean acidification is the potential change in plankton community structure and secondary production rates; this indirectly affects predatory fish species. A detailed ecosystem and fisheries modeling study on the effects of ocean acidification has recently been completed for the California Current, where acidification is projected to proceed more rapidly than in Florida (Marshall et al. 2017). Impacts around Florida can be expected to follow similar patterns.

Many of Florida's marine fish stocks rely on coastal habitats for at least part of their lifecycle. That is true even for the snapper and grouper species harvested in deeper offshore waters, many of which rely on saltmarsh or mangrove areas as juveniles. Juvenile stages are critical to the population dynamics of fish, and availability of juvenile habitat often limits the overall abundance of stocks. Therefore, changes in costal wetland habitats due to sea level rise and changes in rainfall and freshwater flow patterns may well be among the most important drivers of climate change impact on Florida's marine fisheries (Glick 2006). At a statewide level, it is predicted that sea level rise will cause the area of saltmarsh habitat and tidal flats to decline substantially, whereas the area of mangroves is set to increase (Saintilan et al. 2014), as is the area of brackish marsh. These habitats may not be functionally redundant as nursery areas for many coastal fish and shellfish species, leading to altered faunal assemblages. Additionally, most habitat considered important for juvenile fish are produced by organisms such as sea grasses, salt marsh grasses, mangroves, and oysters, all of which have their own population dynamics that may be affected differently by sea level rise or climate change. While some (oysters, mangroves) may, under some conditions, rapidly respond to colonize newly inundated areas (Saintilan et al. 2014; Rodriguez et al. 2014), it is not clear if others (seagrasses, marsh grasses) can do so (Morris et al. 2002; Orth et al. 2006). Changes in these critical juvenile habitats can reasonably be expected to have impacts on fish stocks and fisheries yields, but such impacts are likely to be complex and sometimes counterintuitive (Zimmerman et al. 2002). For example, research on saltmarsh loss in Louisiana showed that local shrimp production actually increased as the area of the marsh declined, most likely because the remaining saltmarsh area became more accessible to the shrimp. In concert, it is reasonable to expect habitat-mediated changes in fish populations and communities, but the intensity and even the directionality of such changes will depend on the speed at which habitat forming organisms and fish populations respond to altered environmental conditions, and this is not yet well described.

Climatic warming is expected to result in poleward distributional shifts of species and assemblages. Such shifts are well-documented for marine fish stocks, particularly in temperate latitudes (Perry et al. 2005; Pinsky et al. 2013, Doney et al. 2014). So far, studies in the Gulf provide some evidence of such shifts occurring (Tolan & Fisher 2009; Fodrie et al. 2010). For Atlantic marine waters, temperature increases are expected to be less pronounced around Florida than along the U.S. mid- and North Atlantic coasts, which will warm rapidly due to a northerly shift in the Gulf Stream (Saba et al. 2016). Distributions in the South Atlantic appear to be responsive to short-term climate fluctuations but, so far, have not shown long-term directional changes (Morley et al. 2016). Even if the magnitude and speed of distribution shifts around Florida are uncertain, it is useful to consider how such shifts might affect the stocks currently fished. Florida's marine fisheries target a range of south-temperate and tropical stocks, the majority of which have distributions that extend somewhat south of Florida or are centered in the tropics (Robertson & Van Tassell 2015; Froese & Pauly 2017). This is true for the crustacean stocks (shrimps, spiny lobster, stone and blue crab), for offshore reef fish (groupers, snappers, and others), offshore pelagics (mackerels, tuna, billfishes) and for inshore fish (red drum, snook, spotted seatrout). Therefore, these stocks can be expected to maintain their distributions or expand further into Florida. Red drum has its southern limit around South Florida, so the species may become increasingly rare in the southern parts of the state. It is not, however, the mainstay of the fishery in those areas even at present. A detailed analysis of potential changes in habitat suitability for a range of juvenile fish and lobster in Florida Bay concluded that changes varied between scenarios but were on average small (Kearney et al. 2015).

Impacts of distributional shifts on stock abundance and fisheries yields have been explored on a global level using dynamic bioclimate envelope models, which suggests that average abundance and yield potential will decline in the tropics, increase in polar oceans, and remain largely unchanged in the temperate zone (Cheung et al. 2010). However, species interactions may have a moderate dampening effect on distribution and yield changes (Fernandes et al. 2013). No specific assessments have been conducted for the stocks around Florida.

Climate Change Impacts on Marine Fisheries Users

Changes in the abundance and distribution of fisheries resources will affect the benefits that fishers attain from their activities and the costs incurred. Climate change and confounding factors may also impact fishing activities through, for example, changes in boating access or costs of fishing inputs such as fuel. Sea level rise, associated habitat alterations, and coastal defense responses may affect boating access to marinas and boat ramps, as well as the spatial extent of habitats normally sought out by fishers (FWC 2009). As discussed in more detail below, fishers can adapt to such changes in a variety of ways including switching target species, changing fishing locations, modifying the overall effort they expend on fishing, and changing fishing methods and/or motivations (Colburn et al. 2017). Overall reductions in individual and collective fishing efforts may occur where neither species switching nor location change are viable adaptation strategies.

Given their proximity to the coast, fishers, support industries and related infrastructure are highly vulnerable to direct impacts of storms, weather, and sea level rise (Colburn et al. 2017). A vulnerability assessment of commercial fishing communities identified those in South Florida, including the Florida Keys, as particularly vulnerable in this respect (Colburn et al. 2017). The dramatic impacts of hurricanes on the marine fisheries sector are well-documented, and it is clear that an increase in frequency and severity of such events will have major economic and social consequences (Tilmant et al. 1994; Buck 2005; Solis et al. 2013).Recreational fishing effort in Florida is substantially influenced by migration to the state and by tourism, both of which may decline somewhat with climatic warming due to increasing attractiveness of currently temperate regions. Impacts of sea level rise and extreme weather on Florida's tourism infrastructure may further reduce recreational fishing in the state (Weatherdon et al. 2016). Climate in and of itself is likely to affect the level of recreational fishing effort (Carter & Letson 2009; Whitehad & Willard 2016).

Adaptation Options

Marine fisheries may adapt to climate change through changes in fishers' targeting and spatial behavior, changes in governance, and use of certain fisheries enhancement and restoration measures. Switching of target species is a common feature of fisheries sub-sectors in which a number of different species can be caught with broadly similar means and at similar locations (for example, within the recreational inshore or recreational and commercial reef fisheries). Marine recreational fishers in Florida tend to target multiple species and switch between them in response to changes in abundance, even in the inshore fisheries that rely on only 2-3 major species (Camp et al. 2016). Such behaviors may, however, be constrained by species-specific regulations. Switching between fisheries sub-sectors (e.g. from inshore to reef fisheries) is less common and more costly since it typically requires investment in new gear and a steep learning curve with respect to fishing practices, locations, etc. Changes in fishing locations can be an alternative to switching species, i.e., fishers may choose to follow changing spatial abundance patterns of their traditional targets. Both strategies are found in fisheries but for reasons of cost, switching species within the sub-sector is likely more common unless the market or recreational value of alternative species differs greatly. In Florida, many fishers are familiar with and value both tropical and temperate species within their sub-sectors. This, combined with the fact that a majority of recreational fishers appear to conduct the majority of trips within a limited 'home range' of less than 50 miles (Camp et al. 2017), suggests that switching species will be the predominant adaptation strategy among Florida's fishers.

Overall reductions in individual and collective fishing effort may occur where neither species switching nor location change are viable adaptation strategies. This may be the case in response to changes in overall resource availability, access, or costs and prices of inputs and outputs. Effort changes in response to such factors are well-documented, for example, in the Gulf of Mexico shrimp fisheries (Nance et al. 2008).

Since catching fish is one of a wide range of motivations behind recreational fishing, and since anglers can attain satisfaction even if no or few fish are being caught, there is scope for new and different fishing approaches to compensate for certain negative impacts on traditional fisheries (Radomski et al. 2001; Hunt et al. 2016). Thus, recreational fishers are likely to show the greatest adaptive capacity. This is illustrated, for example, in the switch to predominantly catch-and-release fishing in the snook fishery, which has allowed for improvements in fishing quality despite increasing fishing effort and habitat changes that would otherwise have resulted in declines in fish abundance and fishing quality. Commercial fishers are often more specialized, more constrained by economic factors, and less inclined and able to switch to other ways of making a living than are charter boat captains (Seara et al. 2016).

No major changes in geographical boundaries of governance structures (FWC, GMFMC, and SAFMC) will be required for climate adaptation of Florida's marine fisheries, since distribution shifts are likely to be small on average and the northward boundary of tropical stocks occurs within current governance boundaries.

The focus of climate adaptation in Florida's marine fisheries management will be on adapting catch limits and fishing regulations to changes in stock distributions and productivity. Failure to adapt catch limits would result in overfishing of stocks that are declining within the management area and forego the potential for higher sustainable catches from stocks that are expanding in range and productivity. As discussed above, fisheries management systems are well set up to track and respond to changes in stock abundance and fishing pressure (Melnychuk et al. 2014). However, stock assessments that inform the setting of catch limits typically are based on the assumption that stock productivity will not undergo long-term changes. Explicitly incorporating long-term changes in stock productivity into stock assessment is, therefore, an important priority for adapting assessment and management systems, even though it may not be easy to discern such changes from available data (Punt et al. 2014). Several approaches can increase the ability anticipate climate-related changes and help inform monitoring and management strategies. A methodology for assessing the climate change vulnerability of individual fish stocks, which is based on combining existing information on the exposure of a stock to climate stressors and its sensitivity to the stressors, has been devised by NOAA and completed for the northeast U.S.

Continental Shelf (Hare et al. 2016). Application of this vulnerability assessment method to the Gulf and South Atlantic is planned under NOAA's Climate Change Strategy and Regional Action Plans (Bush et al. 2016; Lovett 2016). More quantitative assessments of changes in stock productivity are likely to remain elusive until quantitative relationships can be established between key climate change drivers and stock dynamics. In the meantime, it may be most appropriate to consider the implications of broad, plausible forecasts related to how biological parameters may change in the future as a way to assess the robustness of management strategies rather than attempting specific predictions per se (Punt et al. 2014).

Climate change effects acting on habitats and different species may combine in synergistic ways and lead to unintuitive consequences (Ainsworth et al. 2011). Likewise, human responses to ecological changes at all levels will be important drivers of fisheries outcomes (Haynie & Pfeiffer 2012). It is, therefore, important to complement analysis of changes in stock dynamics with ecosystem-scale and socio-economic assessments. This will involve identifying and monitoring relevant ecological or socio-economic indicators to establish trends and provide early warning of climate change impacts. Moreover, thresholds could be set to demark a qualitative change in fishery performance and trigger adaptation actions (Bush et al. 2016). Advancing place-based and cooperative management of fisheries by promoting more locally-adapted and stakeholder-involved management strategies may enhance adaptation to changing environmental conditions and stakeholder needs (Lorenzen et al. 2010; Camp et al. 2017).

Marine fisheries, both recreational and commercial, are highly dependent on coastal access infrastructure (boat ramps, docks) and on support industries (marinas, fish houses). It is, therefore, crucial to maintain such infrastructure in the face of sea level rise, impacts from storms and coastal erosion, and confounding factors such as increasing coastal population density and property values.

Fisheries enhancements, technical interventions aimed at enhancing or restoring fisheries such as the provision of artificial habitat and the release of hatchery-reared fish, may have some scope to aid climate change adaptation (Sale et al. 2014). Artificial reefs are already widely deployed in Florida and have the effect of creating reef fish habitat in areas where it is naturally scarce. Artificial reefs attract reef fish (e.g. snappers and groupers) and fishers to known locations. Their benefits, from a fisheries management perspective, are primarily the result of the aggregation of fish at known and often easily accessible locations; but artificial reefs may also support overall increases in fish production (Bortone et al. 2011). Hatchery programs raise early life stages and juveniles of fish under controlled conditions and can help sustain fisheries under conditions where natural habitats for these sensitive life stages are reduced in extent or quality (Camp et al. 2014). In practice, the effectiveness of such approaches has been found to be variable but often low. Moreover, such approaches are expensive to develop and maintain; they will, therefore, be an option only for certain high-value species (Lorenzen et al. 2010, 2013). In addition to resource enhancements, there may be way to improve fisheries through infrastructure, for example by enhancing boating access (FWC 2009).

The expected increase in the frequency and severity of storms, combined with the extreme vulnerability of the fisheries sector to such events, may call for greater attention to disaster response as an explicit function of the fisheries governance system. Previous experience has indicated the value of systematic attention to recovery planning for the fisheries sector (Dyer & McGoodwin 1999; Land 2015).

Impacts on Freshwater Fisheries and Adaptation Options

Freshwater fisheries operate in a large number of water bodies that are ecologically separated and confined to various degrees, and often are strongly influenced by human water and land use (Arlinghaus et al. 2016). Compared to marine fisheries, freshwater fisheries in Florida are also characterized by a lower diversity of exploited species and of fishers. Largemouth bass is the single most important resource, primarily targeted by some 40% of freshwater anglers, followed by sunfishes (23%), crappie (16%), catfish (12%), and striped bass (4%) (Morales 2016). Freshwater fishing is almost exclusively recreational, and strongly catch-and-release-oriented in the bass fishery, but much less so for the other species. Freshwater fisheries operate in ecosystems that are heavily influenced by multiple anthropogenic pressures. In various combinations, eutrophication due to accelerated nutrient loading from agricultural and domestic sources, hydrological alterations for water supply and flood control, and spread of invasive aquatic plants have affected a majority of Florida's lakes and rivers (Williams 1985). Despite a broad suite of pollution control and water management measures, these stressors remain relevant and continue to have adverse impacts on fish stocks and fisheries (Dotson et al. 2015). In addition, inter-annual variation in rainfall causes substantial variation in water levels, which affects both fish stocks and fisheries access and use. Fishing has a relatively minor impact on freshwater fish stocks in Florida. Owing to the widespread use of voluntary catch and release, largemouth bass fisheries are lightly exploited and harvest regulations have only a small impact on fisheries outcomes (Myers at al. 2008; Kerns et al. 2015). Fisheries for the other freshwater species are more harvest-oriented, but overall less intensive than the marine sector.

Florida's freshwater fisheries are managed by the FWC. Freshwater fisheries management in Florida, as elsewhere, involves habitat management for fish and fisheries enhancement measures, such as placement of fish attractors and stocking of hatchery fish, in addition to harvest regulations (Arlinghaus et al. 2016). Stock assessments and other scientific approaches used to inform freshwater fisheries management need to account for the large number of freshwater systems, limited resources for sampling, and the wide range of anthropogenic factors affecting freshwater fisheries (Lorenzen et al. 2016). Harvest regulations are widely used, but due to the generally low rates of exploitation of largemouth bass stocks, the scope for improving stock abundance using such regulations is limited (Myers et al. 2008). Nonetheless, restrictive harvest regulations have the potential to improve trophy bass opportunities (Dotson et al. 2013). Recent initiatives aim to conserve very large 'trophy-sized' fish while simultaneously engaging anglers in scientific data collection through a program called TrophyCatch that encourages non-lethal documentation and release of trophy fish (Dutterer et al. 2014). Overfishing of crappie stocks is of some concern (Dotson et al. 2009) and harvest regulations can be used to improve stock abundance (Allen et al. 2013).

Freshwater habitats and their environmental quality are managed by multiple organizations, including the U.S. Army Corps of Engineers, the Florida Water Management Districts, the Florida Department of Environmental Protection, local counties, and the FWC. Of these, the FWC is the only agency focused primarily on fisheries and wildlife management. In response to the fish habitat issues outlined above, the FWC has initiated more than 50 major restoration projects in the past 40 years (Dotson et al. 2015). These include measures such as extreme drawdown, muck removal, tussock removal, control of nuisance plants, and planting of native plants. However, due to their high costs, such projects have become increasingly rare. In addition, the FWC conducts stocking to restore fish populations after natural or intentional drawdowns and to supplement weak largemouth bass year-classes in systems with limited recruitment owing to poor habitat quality (Porak et al. 2002; Mesing et al. 2008).

Climate Change Impacts on Freshwater Fisheries Resources

Overall freshwater availability is predicted to increase moderately in most of Florida, but to decline in the Panhandle and the southwest (Carter et al. 2014). However, variability in rainfall is predicted to increase along with the frequency and severity of storms, which means that hydrological variability is likely to be an equally or more important driver of fish habitat quality and population dynamics under climate change. Since most of Florida's lakes are shallow and well-mixed, no major changes in dissolved oxygen are expected as a result of rising temperatures. However, increased temperature and severity of storms could exacerbate eutrophication, degrading habitat availability and quality, thereby creating an alternate stable state that is less desirable for sport fisheries (Scheffer 1990; Ficke et al. 2007). Wind-driven wave action and high water levels from hurricanes can uproot plants, suspend nutrients, and increase turbidity, which can have deleterious effects on shallow lakes with aquatic macrophyte communities dominated by submersed aquatic vegetation. Clear lakes with expansive submersed aquatic vegetation and premier sport fisheries can quickly transition to a turbid, phytoplankton-dominated system with poor sport fisheries. These effects in Florida lakes are well-documented (Bachmann et al. 1999; Havens et al. 2001; Havens 2005; Rogers and Allen 2008; Johnson et al. 2014). Increased frequency of storms, along with increased temperature and other confounding anthropogenic influences, will make it exceedingly difficult to slow or reverse eutrophication. Some freshwater habitats near the coast may suffer saltwater intrusion as a consequence of sea level rise, and this may severely impact local freshwater fisheries if it results in substantial increases in salinity (FWC 2009; Barlow & Reichard 2010; Herbert et al. 2015). Freshwater mussels are of particular concern, as there are currently 15 federally-protected endangered and threatened mussels that are found in major Gulf Coast basins in Florida between the Escambia and Hillsborough rivers; and all but one have designated critical habitats in state coastal rivers and streams (Williams et al. 2014).

Climate-related distributional shifts are expected for freshwater fish, but may occur at a slower rate compared to marine fish due to lower connectivity between freshwater systems and in particular watersheds. Of the commonly fished species, most are relatively flexible and have distribution ranges that extend throughout Florida and to the north of the state (Hocutt & Wiley 1986). Striped bass (Atlantic and Gulf strain) and American shad, as well as many non-game imperiled fishes (e.g., gulf sturgeon, Atlantic sturgeon, alligator gar, Alabama shad, crystal darter, harlequin darter, tessellated darter, saltmarsh topminnow, blackmouth shiner, bluenose shiner, blackbanded sunfish, spotted bullhead, snail bullhead) that are limited to temperate Florida, are more sensitive to environmental fluctuation and may reduce their range in the state or disappear entirely. At least 34 exotic freshwater fishes are already naturally reproducing in Florida waters, a phenomena occurring more in Florida than any other state (Fuller et al. 1999; Shafland et al. 2007). Most of these species are currently restricted to South Florida (south of State Road 70), but rising temperatures may allow for range expansions northwards and establishment of additional exotic fishes currently thermally restricted. The potential ecological impacts of exotic species on Florida's freshwater ecosystems is not well understood, but displacement or suppression of native fish populations is of serious concern.

Several interrelated confounding factors are likely to have major impacts on inland fisheries habitats and resources: population growth, conversion of natural and agricultural land to urban land use, and increase in demand for fresh water (FWC 2008). Although the precise magnitude of these changes is uncertain, they could be substantial (FWC 2008 predicted a doubling of all three over a period of 50 years). Consequences will include an overall reduction in freshwater habitat area and more intensive use and modification of remaining areas.

Climate Change Impacts on Freshwater Fisheries Users

In addition to impacts mediated through the resources, climate change and confounding factors can affect anglers directly. Most importantly in the case of freshwater fisheries, changes in the spatial distribution and hydrology of freshwater systems will impact accessibility, both in terms of travel distances and boating access. Low water levels severely restrict boat and shore access to freshwater fisheries. The occurrence and magnitude of extremely low water levels preventing access and use to freshwater lakes in Florida has substantially increased over the last two decades.

Overall, it is likely that climate change and confounding factors will increase costs for many freshwater anglers and result in an overall reduction in fishing satisfaction and fishing-related economic activity. This reduction is likely to disproportionally affect the poorer sections of the angling public, who face greater challenges in adapting.

Adaptation Options

Freshwater recreational anglers can adapt to the expected changes in habitats, abundance and distribution of species, and access by changing fishing locations or switching target species. Both short-term (inter-annual) variation and long-term changes are likely to be regionally differentiated, and changes in fishing location are therefore likely to be a major adaptation option (see Ward et al. 2016). However, anglers are likely to vary greatly in their propensity for location change, which typically involves increased costs, time, and inconvenience. The majority of freshwater anglers conduct most of their fishing within 50 miles of their residence, which is a good indication of the limits to costs and time that they are willing to invest (Morales 2016). Such anglers, many of whom are poor, are likely to reduce their fishing activities. A much smaller portion of freshwater anglers, mostly those who fish tournaments at least occasionally, are highly mobile and likely to adapt easily to changing opportunities. There is potential for recreational anglers to shift their focus from freshwater to marine, and vice versa, as fisheries respond to climate change. The hydrological, habitat, and population impacts expected from climate change and confounding factors are likely to affect all exploited resources in a broadly similar manner so that in freshwater systems, switching species is unlikely to be a major adaptation option. In Florida, the establishment and possible expansion of exotic freshwater fish present the greatest opportunity for expanding traditional freshwater fisheries. Popular sport fisheries for peacock bass, mayan cichlids, oscars, among others, already exist in the Miami area.

Due to the widespread adoption of voluntary catch-and-release, the bass fisheries are lightly exploited and harvest regulations have only a small impact on fisheries outcomes (Myers at al. 2008). This implies that adapting fishing regulations to climate-driven changes in stock productivity is not a major concern. Substantial increases in fishing intensity due to population growth, shifting effort from marine to freshwater fisheries, or a reduction in available freshwater habitat could affect this conclusion, but increasing catch-and-release orientation and stable or declining per capita participation in freshwater fishing make this unlikely.

The most important fisheries management responses are likely to be active habitat restoration and stock enhancement measures such as those already in use. The difference is that such measures may be more widely and frequently required. Protection of freshwater habitats is a key issue that largely extends beyond the fisheries management real, but to which the importance of fisheries and the expertise and engagement of anglers and fisheries professionals can make vital contributions (Lynch et al. 2017). Maintenance and enhancement of access facilities, such as boat ramps, will also be important in order to maintain fishery access under conditions of great hydrological variability.

Since freshwater fishing is a recreational activity, there are many ways in which participation, satisfaction, and economic impact can be enhanced (Radomski et al. 2001). This includes the provision of fishing opportunities in urban and other modified habitats, or development of programs that incentivize participation through organized competitions and rewards. Such approaches are, perhaps, most developed in the largemouth bass fisheries, which are already the mainstay of Florida's fisheries.

Impacts on Marine Shellfish Aquaculture and Adaptation Options

Shellfish aquaculture in Florida involves growing hard clams in mesh bags on the bottom or oysters either on planted shell or in suspended cages (UF 2011). These activities take place on state-owned submerged land leases. With the exception of spawning and rearing of early life stages, which are carried out under more controlled conditions, shellfish aquaculture is reliant on suitable natural conditions and primary productivity for production, and is, therefore, vulnerable to environmental stressors and public health threats. Moreover, although the cultured shellfish are privately-owned, they are grown in public waters that are subject to a wide range of other uses. For effective production, shellfish require suitable environmental conditions (substrate, depth, tidal range, salinity, primary productivity) at the culture site. To protect the health of consumers, shellfish culture is permitted only in designated areas with low levels of waterborne human pathogens. Harvesting is temporarily suspended in response to heavy rainfall in the watershed, which increases the risk of illness, or in response to high cell counts of "red tide" organisms that could expose consumers to neurotoxins. Added to these constraints is a policy to limit environmental impacts by not permitting shellfish culture in sensitive habitats such as seagrass beds or in areas of potential use conflicts. As a result, the extent of suitable lease areas for shellfish farming in Florida is very limited (FDACS 2013).

Climate Change Impacts on Production and Producers

Clearly, many of the general conditions required for shellfish cultivation are sensitive to climate change drivers, from temperature and rainfall patterns to sea level rise. It is, therefore, likely that the quality of current shellfish growing areas will change and the distributions of optimal areas shift (Allison et al. 2011; Anderson et al. 2013). In addition to these general changes, ocean acidification poses a fundamental threat to shellfish culture because it affects the ability of mollusks, particularly their larval stages, to build shells (Ekstrom et al. 2015). Recognized as a major threat to shellfish culture at the national and international level, ocean acidification is, however, expected to progress comparatively slowly in the southeastern U.S., including the marine waters around Florida (Ekstrom et al. 2015). Furthermore, ocean acidification is affected by local conditions, including freshwater inflow (reduces acidification) and eutrophication (enhances acidification), and is therefore likely to vary spatially (Clements & Chopin 2016; Ekstrom et al. 2015). Rising water temperatures may increase the prevalence of human pathogens and the frequency of harmful algal blooms (Rose et al. 2001; Moore et al. 2008), particularly in North Florida where both are currently less of a problem than in the south and where the state's major shellfish industry is located.

Adaptation Options

The principal adaptation options for the shellfish aquaculture industry include technology changes and relocation of farms, which may be facilitated by some changes in governance. Production systems can be adapted to adverse environmental conditions by raising the most vulnerable juvenile stages in closed systems with controlled water temperature and chemistry. Development of water treatment systems, principally buffering, is a key aspect of adapting shellfish hatcheries to the threat of ocean acidification (Barton et al. 2015). In the longer term, selective breeding of shellfish for greater tolerance to higher temperatures or acidic conditions may further strengthen adaptive capacity (Barton et al. 2015). Relocation of operations in response to environmental changes will be a key aspect of adaption. This is likely to be a gradual process except when distinct events, such as hurricanes, cause major changes in geomorphology. Growers may also adapt by shifting to other species or developing new strains.

The existing management system by which leases are granted based on site surveys and various criteria lends itself, in principle, to adaptation when conditions change and culture operations may seek to relocate. Regular monitoring of conditions in existing and potential lease areas may support adaptation planning. With respect to ocean acidification, monitoring of acidification in the environment and in culture facilities is an important step toward identifying impacts and developing management responses (Barton et al. 2015). In addition to the adaptation measures outlines above, which can be taken by individual producers, curtailing eutrophication will be a key management priority for shellfish growing areas, as it can reduce both ocean acidification and the risk of harmful algal blooms (Ekstrom et al. 2015).

Overall, adaptive capacity in the shellfish industry is likely to be limited due its strong reliance on environmental conditions (Ekstrom et al. 2015). Moreover, shellfish culture tends to be concentrated in regions of the state that are characterized by low adaptive capacity (Colburn et al. 2017).

Impacts on Freshwater Ornamental Aquaculture and Adaptation Options

Freshwater aquaculture is predominantly carried out in earthen ponds, combined with indoor/tank-based hatcheries for reproduction and rearing of early life stages. Florida's ornamental aquaculture industry (production for the aquarium trade) produces over 30 species of tropical freshwater fish (Hill & Yanong 2010). While pond systems are more environmentally open than indoor tanks, producers can exercise a high degree of control over environmental conditions by means of water management, aeration, provision of cover, treatment of pond water, feeding, etc. (Watson & Shireman 1996). The species currently produced differ widely in their environmental requirements and tolerances, and ornamental aquaculture producers are therefore well-positioned to manage these conditions. Ornamental aquaculture in Florida is regulated by the Florida Department of Agriculture and Consumer Services (FDACS), in particular with respect to containment of non-native species and effluent control (Tuckett et al. 2016).

Since ornamental production is focused on tropical species, a change in climate towards more tropical conditions is not, in general, problematic for the industry. Indeed, the length of the growing season for certain sensitive species may well increase in the industry's focal area around Tampa. However, production may become more challenging due to greater variability in rainfall and higher temperatures that would influence oxygen saturation and other water quality parameters. Producers have the technical means to address these issues, but production costs could increase moderately as a consequence.

No major management/regulatory changes are likely required to help producers adapt. However, changing climate may alter the survivability of escapees from ornamental farms. This may necessitate additional policy measures to guard against the inadvertent introduction of potentially invasive species.

Aquaculture in Support of Restoration and Fisheries Enhancement

Aquaculture can be used to maintain or restore populations of aquatic organisms or to enhance fisheries. Demand for restoration aquaculture and fisheries enhancements is likely to increase since these adaptation approaches can make climate change impacts more manageable (Lorenzen et al. 2013; Barton et al. 2017). Restoration and enhancement aquaculture uses some of the same technologies that are used in commercial aquaculture, but it often requires different husbandry and genetic management approaches in order to produce organisms that maintain wild-like characteristics and survive well upon release (Lorenzen et al. 2012). Furthermore, aquaculture-based enhancement or restoration initiatives need to be integrated into overarching fisheries management or restoration programs using a planning framework such as the 'responsible approach' (Lorenzen et al. 2010). Overall, use of aquaculture offers some potential for climate change adaptation, but this is likely to be effective and economically viable only in certain cases.

Conclusion

Complex interactions among climate change and confounding drivers and the characteristics of Florida's diverse fisheries and aquaculture industries make it difficult to predict the magnitude and sometimes even the directionality of climate change impacts. Nonetheless, even a qualitative, conceptual assessment such as presented here is valuable because it helps identify impact pathways and adaptation options that are likely to be most relevant under the specific conditions found in Florida.

While colloquial debate about climate change impacts often focuses on increases in temperature and associated impacts such as species range changes, the assessment presented here points to the likely importance of additional and more complex drivers. For example, sea level rise and increased frequency and intensity of storm events can be expected to exert major impacts on coastal habitats and fishing-related infrastructure, which in turn will impact on the productivity and accessibility of marine fisheries. Likewise, in fresh waters, increased variability in rainfall is likely to be a major driver, possibly combined with impacts of frequent storm events. Anticipating such impacts helps in designing indicators and monitoring programs to track climate impacts, and in identifying possible adaptation options.

Acknowledgements

The lead author's (K.L.) work on this chapter was partially supported by a data synthesis grant from the Gulf Research Program of the National Academies of Sciences, Engineering, and Medicine under the Grant Agreement number 2000006433. The content is solely the responsibility of the authors and does not necessarily represent the official views of the Gulf Research Program or the National Academies of Sciences, Engineering, and Medicine.

References

Ainsworth, C. H., Samhouri, J., Busch, D. S., Cheung, W., Dunne, J. & Okey, T. (2011) Potential impacts of climate change on Northeast Pacific marine foodwebs and fisheries. ICES Journal of Marine Science 68: 1217-1229.

Allen, M.S., Ahrens, R.N.M., Hansen, M.J. & Arlinghaus, R. (2013). Dynamic angling effort influences the value of minimum-length limits to prevent recruitment overfishing. Fisheries Management and Ecology 20: 247-257.

Allison, E.H., M.C. Badjeck, K. Meinhold. 2011. The implications of global climate change for molluscan aquaculture. In Shellfish aquaculture and the environment, ed. S.E. Shumway, 461-490. West Sussex, UK: John Wiley & Sons, Inc.

Anderson, J.L. (2002). Aquaculture and the future: why fisheries economists should care. Marine Resource Economics 17: 133-151.

Anderson, J.A., Baker, S.M., Graham, G.L., Harby, M.G., Hall, S.G., Swann, L., Walton, W.C. & Wilson, C.A. (2013) Effects of climate change on fisheries and aquaculture in the Southeast USA. In Climate of the Southeastern United States: Variability, Change, Impacts, and Vulnerability, ed. K.T. Ingram, K. Dow, L. Carter, J. Anderson, 190-209. Washington DC, USA: Island Press.

Arlinghaus, R., Cooke, S. J., Lyman, J., Policansky, D., Schwab, A., Suski, C., Sutton, S.G. & Thorstad, E. B. (2007). Understanding the complexity of catch-and-release in recreational fishing: an integrative synthesis of global knowledge from historical, ethical, social, and biological perspectives. Reviews in Fisheries Science 15: 75-167.

Arlinghaus, R., Lorenzen, K., Johnson, B.M., Cooke, S.J. & Cowx, I.G. (2016) Management of freshwater fisheries: addressing habitat, people and fishes. In: Craig, J.F. Freshwater Fisheries Ecology. pp. 557-579. Oxford, UK: Wiley.

Bachmann, R.W., Hoyer, M.V. & Canfield, D.E. (1999). The restoration of Lake Apopka in relation to alternative stable states. Hydrobiologia 394: 219-232.

Bakun, A., 1990. Global climate change and intensification of coastal ocean upwelling. Science 247(4939): 198-201.

Barlow, P. M., & Reichard, E. G. (2010). Saltwater intrusion in coastal regions of North America. Hydrogeology Journal 18: 247-260.

Barton, A., Waldbusser, G. G., Feely, R. A., Weisberg, S. B., Newton, J. A., Hales, B., Cudd, S., Eudeline, B., Langdon, C.J., Jefferds, I., King, T., Suhrbier, A. & McLaughlin, K. (2015). Impacts of coastal acidification on the Pacific Northwest shellfish industry and adaptation strategies implemented in response. Oceanography 28: 146-159.

Barton, J.A., Willis, B.L. & Hutson, K. S. (2017). Coral propagation: a review of techniques for ornamental trade and reef restoration. Reviews in Aquaculture 9: 238-256.

Bortone, S. A., Brandini, F. P., Fabi, G., & Otake, S. (Eds.). (2011). Artificial Reefs in Fisheries Management. CRC Press.

Bostock, J., McAndrew, B., Richards, R., Jauncey, K., Telfer, T., Lorenzen, K., Little, D., Ross, L., Handisyde, N. & Gatward, I. (2010) Aquaculture: global status and trends. Philosophical Transactions of the Royal Society B 365: 2897-2912.

Brander, K. (2010). Impacts of climate change on fisheries. Journal of Marine Systems 79: 389-402.

Breitburg, D.L., Hondorp, D.W., Davias, L.A. & Diaz, R.J. (2009). Hypoxia, nitrogen and fisheries: Integrating effects across local and global landscapes. Annual Reviews in Marine Science 1: 329-349.

Buck, E.H. (2005). Hurricanes Katrina and Rita: Fishing and Aquaculture Industries--Damage and Recovery. Congressional Research Service, Library of Congress. http://nationalaglawcenter.org/wp-content/uploads/assets/crs/RS22241.pdf

Busch, D. S., Griffis, R., Link, J., Abrams, K., Baker, J., Brainard, R. E., Ford, M., Hare, J.A., Himes-Cornell, A., Hollowed, A., Mantura, N.J., McClatchie, S., McClure, M., Nelson, M.W., Osgood, K., Peterson, J.O., Rust, M., Saba, V., Sigler, M.F., Sykora-Bodie, S., Toole, C., Thunberg, E., Waples, R.E. & Merrick, R. (2016). Climate science strategy of the US National Marine Fisheries Service. Marine Policy 74: 58-67.

Camp, E.V., Lorenzen, K., Ahrens, R.N.M. & Allen, M.S. (2014). Stock enhancement to address multiple recreational fisheries objectives: an integrated model applied to red drum Sciaenops occelatus in Florida. Journal of Fish Biology 85: 1868-1889.

Camp, E.V., Ahrens, R.N.M., Allen, M.S. & Lorenzen, K. (2016) Relationships between angler effort and fish abundance in recreational marine fisheries. Fisheries Management and Ecology 23: 264-275.

Camp, E.V., Ahrens, R.N.M., Crandall, C.A. & Lorenzen, K. (2017). Angler travel distances: implications for spatially explicit governance of recreational fisheries. Marine Policy.

Carter, D. W., & Letson, D. (2009). Structural vector error correction modeling of integrated sportfishery data. Marine Resource Economics 24: 19-41.

Carter, L. M., Jones, J.W., Berry, L., Burkett, V., Murley, J. F., Obeysekera, J., Schramm, P. J., & Wear, D. (2014). Ch. 17: Southeast and the Caribbean. Climate Change Impacts in the United States: The Third National Climate Assessment, J. M. Melillo, Terese (T.C.) Richmond, and G. W. Yohe, Eds., U.S. Global Change Research Program, 396-417. doi: 10.7930/J0NP22CB.

Chérubin, L.M., Morel, Y. & Chassignet, E.P., 2006. Loop Current Ring Shedding: The Formation of Cyclones and the Effect of Topography. Journal of Physical Oceanography 36: 569–591.

Cheung, W. W., Lam, V. W., Sarmiento, J. L., Kearney, K., Watson, R. E. G., Zeller, D., & Pauly, D. (2010). Large-scale redistribution of maximum fisheries catch potential in the global ocean under climate change. Global Change Biology 16: 24-35.

Clements, J. C., & Chopin, T. (2016). Ocean acidification and marine aquaculture in North America: potential impacts and mitigation strategies. Reviews in Aquaculture.

Colburn, L. L., Jepson, M., Weng, C., Seara, T., Weiss, J., & Hare, J. A. (2016). Indicators of climate change and social vulnerability in fishing dependent communities along the Eastern and Gulf Coasts of the United States. Marine Policy 74: 323-333.

Cooke, S. J., Twardek, W. M., Lennox, R. J., Zolderdo, A. J., Bower, S. D., Gutowsky, L. F., Danylchuk, A.J. Arlinghaus, R. & Beard, D. (2017) The nexus of fun and nutrition: Recreational fishing is also about food. Fish and Fisheries.(early view).

Cooper, A. (2006) Guide to Fisheries Stock Assessment: from Data to Recommendations. University of New Hampshire/NH Sea Grant. https://seagrant.unh.edu/sites/seagrant.unh.edu/files/media/pdfs/stockassessmentguide.pdf

Diaz, R. J., & Rosenberg, R. (2008). Spreading dead zones and consequences for marine ecosystems. Science 321: 926-929.

Ditton, R.B., Holland, S.M., & Anderson, D.K. (2002). Recreational fishing as tourism. Fisheries 27(3): 17-24.

Doney, S., Rosenberg, A.A., Alexander, M., Chavez, F., Harvell, C.D., Hofmann, G., Orbach, M. & Ruckelshaus, M. (2014). Ch. 24: Oceans and Marine Resources. Climate Change Impacts in the United States: The Third National Climate Assessment, J. M. Melillo, Terese (T.C.) Richmond, and G. W. Yohe, Eds., U.S. Global Change Research Program, 557-578. doi: 10.7930/J0RF5RZW.

Dotson, J.R., Allen, M.S., Johnson, W.E. & Benton, J. (2009). Impacts of commercial gill net bycatch and recreational fishing on a Florida Black Crappie population. North American Journal of Fisheries Management 29: 1454-1465.

Dotson, J.R., Allen, M.S., Kerns, J.A. & Pouder W.F. (2013). Utility of restrictive harvest regulations for trophy Largemouth Bass management. North American Journal of Fisheries Management 33: 499-507.

Dotson, J.R., Bonvechio, K.I., Thompson, B.C., Johnson, W.E., Trippel, N.A., Furse, J.B., Gornack, S., McDaniel, C.K., Pouder, W.F. & Leone, E.H. (2015). Effects of large-scale habitat enhancement strategies on Florida bass fisheries. American Fisheries Society Symposium 82: 387-404.

Dutterer, A.C., Wiley, C., Wattendorf, B., Dotson, J.R. & Pouder, W.F. (2014). TrophyCatch: A conservation program for trophy bass in Florida. Florida Scientist 77, 167-183.

Dyer, C.L. & McGoodwin, J.R. (1999). 'Tell them we're hurting': Hurricane Andrew, the culture of response, and the fishing peoples of South Florida and Louisiana. In: The Angry Earth: Disaster in Anthropological Perspective (Eds. Anthony Oliver-Smith & Susanna Hoffman). pp. 213-231. New York: Routledge.

Ekstrom, J. A., Suatoni, L., Cooley, S. R., Pendleton, L. H., Waldbusser, G. G., Cinner, J. E., Ritter, J., Langdon, C., van Hooidonk, R., Gledhill, D., Wellman, K., Beck, M.W., Brander, L.M., Rittschof, D., Doherty, C., Edwards, P.E.T. & Portela, R. (2015). Vulnerability and adaptation of US shellfisheries to ocean acidification. Nature Climate Change 5: 207-214.

Farmer, N.A., Heyman, W.D., Karnauskas, M., Kobara, S., Smart, T.I., Ballenger, J.C., Reichert, M.J.M., Wyanski, D.M., Tishler, M.S., Lindeman, K.C., Lowerre-Barbieri, S.K., Switzer, T.S., Solomon, J.J., McCain, K., Marhefka, M., Sedberry, G.R. (2017). Timing and locations of reef fish spawning off the southeastern United States. PLoS ONE 12(3): e0172968.

FAO (Food and Agriculture Organization of the United Nations) (1990). The definition of aquaculture and collection of statistics. Aquaculture Minutes No. 7. Inland Fisheries and Aquaculture Service, Fisheries Department. Rome: FAO.

FDACS (Florida Department of Agriculture and Consumer Services) (2013). Florida's aquaculture lease program. DACS-P-00070. https://shellfish.ifas.ufl.edu/wp-content/uploads/Floridas-Aquaculture-Lease-Program_UPDATED.pdf

Fedler, A. J., & Ditton, R. B. (1994). Understanding angler motivations in fisheries management. Fisheries 19: 6-13.

Fernandes, J. A., Cheung, W.W., Jennings, S., Butenschön, M., Mora, L., Frölicher, T.L., Barange, M. & Grant, A. (2013). Modelling the effects of climate change on the distribution and production of marine fishes: accounting for trophic interactions in a dynamic bioclimate envelope model. Global Change Biology 19: 2596-2607.

Ficke, A.D., Myrick, C.A. & Hansen, L.J. (2007). Potential impacts of global climate change on freshwater fisheries. Reviews in Fish Biology and Fisheries 17: 581-613.

Froese, R. & Pauly, D. (Editors) (2017). FishBase. World Wide Web electronic publication. www.fishbase.org, version (06/2017).

FWC (Florida Fish and Wildlife Conservation Commission). 2008. Wildlife 2016: What's at stake for Florida? http://myfwc.com/media/129053/FWC2060.pdf

FWC (Florida Fish and Wildlife Conservation Commission). 2009. Florida's wildlife: On the frontline of climate change. http://myfwc.com/media/135483/ClimateChange_SummitRept.pdf

FWC (Florida Fish and Wildlife Conservation Commission). 2011. The economic impact of freshwater fishing in Florida. http://myfwc.com/conservation/value/freshwater-fishing/.

FWC (Florida Fish and Wildlife Conservation Commission). 2017. The economic impact of saltwater fishing in Florida. http://myfwc.com/conservation/value/saltwater-fishing/

Fodrie, F., Heck, K. L., Powers, S. P., Graham, W. M., & Robinson, K. L. (2010). Climate-related, decadal-scale assemblage changes of seagrass-associated fishes in the northern Gulf of Mexico. Global Change Biology 16: 48-59.

Fujita, R., & Bonzon, K. (2005). Rights-based fisheries management: an environmentalist perspective. Reviews in Fish Biology and Fisheries 15: 309-312.

Fuller, P.L., Nico, L. & Williams, J. D. (1999) Nonindigenous fishes introduced into inland waters of the United States. American Fisheries Society Special Publication. Bethesda, MD: American Fisheries Society.

Georgakakos, A., Fleming, P., Dettinger, M., Peters-Lidard, C., Richmond, T.C., Reckhow, K. White, K. & Yates, D. (2014). Ch. 3: Water Resources. Climate Change Impacts in the United States: The Third National Climate Assessment, J. M. Melillo, Terese (T.C.) Richmond, and G. W. Yohe, Eds., U.S. Global Change Research Program, 69-112. doi: 10.7930/J0G44N6T.

Glick, P. (2006). An unfavorable tide: Global warming, coastal habitats and sportfishing in Florida. National Wildlife Federation.

Hare, J. A., Morrison, W. E., Nelson, M. W., Stachura, M. M., Teeters, E. J., Griffis, R. B., Alexander, M.A., Scott, J.D., Alade, L., Bell, R.J., Chute, A.S., Curti, K.L., Curtis, T.H., Kircheis, D., Kocik, J.F., Lucey, S.M., McCandless, C.T., Milke, L.M., Richardson, D.E., Robillard, E., Walsh, H.E., McManus, M.C., Marancik, K.E., Griswold, C.A. (2016). A vulnerability assessment of fish and invertebrates to climate change on the Northeast US Continental Shelf. PloS one, 11(2), e0146756.

Havens, K.E., Jin, K., Rodusky, A.J., Sharfstein, B., Brady, M.A. & East, T.L. (2001). Hurricane effects on a shallow lake ecosystem and its response to a controlled manipulation of water level. Sci. World 1: 44-70.

Havens, K.E. (2005). Lake Okeechobee: hurricanes and fisheries. LakeLine 25: 25-28.

Haynie, A. C., & Pfeiffer, L. (2012). Why economics matters for understanding the effects of climate change on fisheries. ICES Journal of Marine Science 69: 1160-1167.

Herbert, E. R., Boon, P., Burgin, A. J., Neubauer, S. C., Franklin, R. B., Ardón, M., Hopfensprenger, K.N., Lamers, L.P.M & Gell, P. (2015). A global perspective on wetland salinization: ecological consequences of a growing threat to freshwater wetlands. Ecosphere 6: 1-43.

Hetland, R.D., Hsueh, Y., Leben, R. & Niller, P. (1999). A Loop Current-induced jet along the edge of the West Florida Shelf Geophysical Research Letters 26: 2239–2242

Hill, J.E. & Yanong, R.P.E. (2010). Freshwater ornamental fish commonly cultured in Florida. Ruskin, FL: Institute of Food and Agricultural Sciences, University of Florida.

Hocutt, C.H. & Wiley, E.O. 1986. The zoogeography of North American freshwater fishes. New York, NY: John Wiley & Sons, Inc.

Hollowed, A. B., Barange, M., Beamish, R., Brander, K., Cochrane, K., Drinkwater, K., Foreman, M., Hare, J., Holt, J., Ito, S-I., Kim, S., King, J., Loeng, H., MacKenzie, B., Mueter, F., Okey, T., Peck, M. A., Radchenko, V., Rice, J., Schirripa, M., Yatsu, A., and Yamanaka, Y. (2013). Projected impacts of climate change on marine fish and fisheries. – ICES Journal of Marine Science, 70: 1023–1037.

Hunt, L.M., Fenichel, E.P., Fulton, D.C., Mendelsohn, R.J., Smith, W., Tunney, T.D., Lynch, A.J., Paukert, C.P., & Whitney, J.E. (2016). Identifying Alternate Pathways for Climate Change to Impact Inland Recreational Fishers. Fisheries 41: 362-373.

IPCC (Intergovernmental Panel on Climate Change) (2014) Climate Change 2014: Synthesis Report. Contribution of Working Groups I, II and III to the Fifth Assessment Report of the Intergovernmental Panel on Climate Change. Core Writing Team, R.K. Pachauri and L.A. Meyer (eds.). IPCC, Geneva, Switzerland, 151 pp.

Johnson, K.G., Dotson, J.R., Pouder, W.F., Trippel, N.A. & Eisenhauer, R.L. (2014). Effects of hurricane-induced hydrilla reduction on the Largemouth Bass fishery at two central Florida lakes. Lake and Reservoir Management 30: 217-225.

Karnauskas, M., Schirripa, M. J., Craig, J. K., Cook, G. S., Kelble, C. R., Agar, J. J., Black, B.A., Enfield, D.B., Lindo-Atichati, D., Muhlig, B.A., Purcell, K. M., Richards, P.M. & Wang, C. (2015). Evidence of climate-driven ecosystem reorganization in the Gulf of Mexico. Global Change Biology 21: 2554-2568.

Kearney, K. A., Butler, M., Glazer, R., Kelble, C. R., Serafy, J. E., & Stabenau, E. (2015). Quantifying Florida Bay habitat suitability for fishes and invertebrates under climate change scenarios. Environmental Management 55: 836-856.

Kerns, J.A., Allen, M.S., Dotson, J.R. & Hightower, J.E. (2015). Estimating regional fishing mortality for freshwater systems: a Florida Largemouth Bass example. North American Journal of Fisheries Management 35: 681-689.

Land, L. (2015). The old model still works: face-to-face university extension after disaster. Current Opinion in Environmental Sustainability 17: 57-65.

Larkin, S.L., Adams, C.M., Degner, R.L., Lee, D.J., & Milon, J. W. (2001). Economic profile of Florida's marine life industry. Technical Paper 111. Gainesville, FL: Florida Sea Grant.

Lindberg, W.J. & Seaman, W. (editors). 2011. Guidelines and Management Practices for Artificial Reef Siting, Use, Construction, and Anchoring in Southeast Florida. Florida Department of Environmental Protection. Miami, FL. xi and 150 pages. https://www.dep.state.fl.us/coastal/programs/coral/reports/MICCI/MICCI_18_19.pdf

Liu, Y. & Weisberg, R.H. (2012). Seasonal variability on the West Florida Shelf. Progress in Oceanography 104: 80–98.

Liu, Y., Lee, S.-K., Muhling, B.A., Lamkin, J.T. & Enfield, D.B. (2012). Significant reduction of the Loop Current in the 21st century and its impact on the Gulf of Mexico. Journal of Geophysical Research 117: C5039. doi: 10.1029/2011JC007555.

Lorenzen, K. (2014). Understanding and managing enhancements: why fisheries scientists should care. Journal of Fish Biology 85: 1807-1829.

Lorenzen, K., Steneck, R.S., Warner R.R., Parma, A.M., Coleman, F.C. & Leber, K.M. (2010a) The spatial dimensions of fisheries: putting it all in place. Bulletin of Marine Science 86: 169-177.

Lorenzen, K., Leber, K.M. & Blankenship, H.L. (2010b) Responsible approach to marine stock enhancement: an update. Reviews in Fisheries Science 18: 189-210.

Lorenzen, K., Beveridge, M.C.M. & Mangel, M. (2012) Cultured fish: integrative biology and management of domestication and interactions with wild fish. Biological Reviews 87: 639-660.

Lorenzen, K., Agnalt, A.L. Blankenship, H.L. Hines, A.H., Leber, L.M., Loneragan, N.R. & Taylor, M.D. (2013) Evolving context and maturing science: aquaculture-based enhancement and restoration enter the marine fisheries management toolbox. Reviews in Fisheries Science 21: 213-221.

Lorenzen, K., Cowx, I.G., Entsua-Mensah, R.E.M., Lester, N.P., Koehn, J.D., Randall, R.G., Nam, S., Bonar, S.A., Bunnell, D.B., Venturelli, P., Bower, S.D. & Cooke, S. (2016). Stock assessment in inland fisheries: a foundation for sustainable use and conservation. Reviews in Fish Biology and Fisheries 26: 405-440.

Lovett, H.B., S.B. Snider, K.R. Gore, R.C. Muñoz (Editors) 2016. Gulf of Mexico Regional Action Plan to Implement the NOAA Fisheries Climate Science Strategy. NOAA Technical Memorandum NMFS-SEFSC-699. https://www.sefsc.noaa.gov/gmrap/NMFS-SEFSC-699_Final.pdf

Lowther, A. 2011. Fisheries of the United States 2010. Silver Spring, MD: National Marine Fisheries Service http://www.st.nmfs.noaa.gov/st1/fus/fus10/FUS_2010.pdf.

Lynch, A.J., Myers, B.J.E., Chu, C., Eby, L., Falke, J.A., Kovach, R.P., Krabbenhoft, T.J., Kwak, J., Lyons, T.J., Paukert, C.P. & Whitney, J. E. (2016). Climate Change Effects on North American Inland Fish Populations and Assemblages. Fisheries 41: 346-361.

Lynch, A.J., Cooke, S.J., Beard, T.D., Kao, Y.-C., Lorenzen, K., Song, A.M., Allen, M.S. Basher, Z., Bunnell, D.B., Camp, E.V., Cowx, I.G. Freedman, J.A., Nguyen, V.M., Nohner, J. K., Rogers, M.W. Siders, Z.A., Taylor, W.W. & Youn. S. (2017). Grand challenges in the management and conservation of North American inland fish and fisheries. Fisheries 42: 115-124.

Mann, M.E. & Emanuel, K.A. (2006). Atlantic hurricane trends linked to climate change. Earth and Space Science News 87: 233-241.Mann, K.H. & Lazier, J.R.N. (2013). Dynamics of Marine Ecosystems: Biological-Physical Interactions in the Oceans: Third Edition,

Marshall, K. N., Kaplan, I. C., Hodgson, E. E., Hermann, A., Busch, D. S., McElhany, P., Essington, T.E., Harvey, C.J. & Fulton, E. A. (2017). Risks of ocean acidification in the California Current food web and fisheries: ecosystem model projections. Global Change Biology.

Melnychuk, M. C., Banobi, J. A., & Hilborn, R. (2014). The adaptive capacity of fishery management systems for confronting climate change impacts on marine populations. Reviews in Fish Biology and Fisheries 24: 561-575.

Mesing, C.L., Cailteux, R.L, Strickland, P.A., Long, E.A. & Rogers, M.W. (2008). Stocking of advanced-fingerling Largemouth Bass to supplement year-classes in Lake Talquin, Florida. North American Journal of Fisheries Management 28: 1762-1774.

Methot, R.D. (2009). Stock assessment: operational models in support of fisheries management. In The Future of Fisheries Science in North America (Eds. R.J. Beamish & B.J. Rothschild), pp. 137-165. New York: Springer.

Moore, S.K., Trainer, V.L., Mantua, N.J., Parker, M.S., Laws, E.A., Backer, L.C. & Fleming, L.E. (2008). Impacts of climate variability and future climate change on harmful algal blooms and human health. Environmental Health 7: S4.

Morales, N. (2016) A snapshot of Florida's freshwater anglers: Angler composition, behaviors and attitudes. Annual Project Report, Florida Fish and Wildlife Research Institute - Freshwater Fisheries Research. 17 pp.

Morris, J. T., Sundareshwar, P. V., Nietch, C. T., Kjerfve, B., & Cahoon, D. R. (2002). Responses of coastal wetlands to rising sea level. Ecology 83: 2869-2877.

Morley, J. W., Batt, R. D., & Pinsky, M. L. (2016). Marine assemblages respond rapidly to winter climate variability. Global Change Biology 23: 2590-2601.

Moser, S.C., Davidson, M.A., Kirshen, P., Mulvaney, P., Murley, J.F., Neumann, J.E. Petes, L. & Reed, D. (2014). Ch. 25: Coastal Zone Development and Ecosystems. Climate Change Impacts in the United States: The Third National Climate Assessment, J. M. Melillo, Terese (T.C.) Richmond, and G. W. Yohe, Eds., U.S. Global Change Research Program, , 579-618. doi: 10.7930/J0MS3QNW.

Myers, R., Taylor, J., Allen, M. & Bonvechio, T. F. (2008). Temporal trends in voluntary release of largemouth bass. North American Journal of Fisheries Management 28: 428-433.

Nance, J., Keithly, W., Caillouet, C., Cole, J., Gaidry, W., Gallaway, B., Griffin, W., Hart, R., Travis, M. (2008). Estimation of effort, maximum sustainable yield, and maximum economic yield in the shrimp fishery of the Gulf of Mexico. NOAA Technical Memorandum NMFS-SEFSC-570, 71pp.

NMFS (National Marine Fisheries Service) (2016). Fisheries Economics of the United States, 2014. U.S. Dept. of Commerce, NOAA Tech. Memo. NMFS-F/SPO-163, 237p. https://www.st.nmfs.noaa.gov/Assets/economics/publications/FEUS/FEUS-2014/Report-and-chapters/FEUS-2014-FINAL-v5.pdf

NOAA (National Oceanic and Atmospheric Administration). (2017a). Status of U.S. Fisheries. http://www.nmfs.noaa.gov/sfa/fisheries_eco/status_of_fisheries/

Obeysekera, J., Barnes, J., & Nungesser, M. (2015). Climate sensitivity runs and regional hydrologic modeling for predicting the response of the greater Florida Everglades ecosystem to climate change. Environmental Management 55: 749-762.

Orth, R.J., Carruthers, T.J.B., Dennison, W.C., Duarte, C.M., Fourqurean, J.W., Heck, K.L. & Hughes A.R. (2006) A global crisis for seagrass ecosystems. Bioscience 56: 987-996.

Perry, A.L., Low, P.J., Ellis, J.R. & Reynolds J.D. (2005). Climate change and distribution shifts in marine fishes. Science 308: 1912-1915.

Pinsky, M. L., Worm, B., Fogarty, M. J., Sarmiento, J. L., & Levin, S. A. (2013). Marine taxa track local climate velocities. Science 341: 1239-1242.

Porak, W.E., Johnson, W.E., Crawford, S., Renfro, D.J., Schoeb, T.R., Stout, R.B., Krause, R.A. & DeMauro. R.A. (2002). Factors affecting survival of Largemouth Bass raised on artificial diets and stocked into Florida lakes. American Fisheries Society Symposium 31: 649-665.

Punt, A.E., A'mar, T., Bond, N. A., Butterworth, D. S., de Moor, C. L., De Oliveira, J. A., ... & Szuwalski, C. (2013). Fisheries management under climate and environmental uncertainty: control rules and performance simulation. ICES Journal of Marine Science 71: 2208-2220.

Rabalais, N. N., Turner, R. E., & Wiseman Jr, W. J. (2002). Gulf of Mexico hypoxia, aka "The dead zone". Annual Review of Ecology and Systematics 33: 235-263.

Radomski, P. J., Grant, G. C., Jacobson, P. C. & Cook, M. F. (2001). Visions for recreational fishing regulations. Fisheries 26: 7-18.

Robertson, D.R. & Van Tassell, J. (2015). Shorefishes of the Greater Caribbean: online information system. Version 1.0 Smithsonian Tropical Research Institute, Balboa, Panamá. http://biogeodb.stri.si.edu/caribbean/en/pages

Rodriguez, A. B., Fodrie, F. J., Ridge, J. T., Lindquist, N. L., Theuerkauf, E. J., Coleman, S. E., Grabowski, J.H., Brodeur, M.C., Gittman, R.K., Keller, D.A. & Kenworthy, M. D. (2014). Oyster reefs can outpace sea-level rise. Nature Climate Change 4: 493-497.

Rogers, M.W., & Allen, M.S. (2008). Hurricane impacts to Lake Okeechobee: Altered hydrology creates difficult management trade-offs. Fisheries 33: 11-17.

Rose, K.A. (2000). Why are quantitative relationships between environmental quality and fish populations so elusive? Ecological Applications 10: 367-385.

Rose, J.B., Epstein, P.R. Lipp, E.K., Sherman, B.H., Bernard, S.M., & Patz, J.M. (2001). Climate variability and change in the United States: Potential impacts on water- and foodborne diseases caused by microbiologic agents. Environmental Health Perspectives 109: 211-220.

Rosenberg, A., Bigford, T. E., Leathery, S., Hill, R. L., & Bickers, K. (2000). Ecosystem approaches to fishery management through essential fish habitat. Bulletin of Marine Science 66: 535-542.

Saba, V.S., Griffies, S.M., Anderson, W.G., Winton, M., Alexander, M.A., Delworth, T.L., Hare, J.A., Harrison, M.J., Rosati, A., Vecchi, G.A., Zhang, R. (2016). Enhanced warming of the Northwest Atlantic Ocean under climate change. Journal of Geophysical Research: Oceans 121: 118-132

Saintilan, N., Wilson, N. C., Rogers, K., Rajkaran, A., & Krauss, K.W. (2014). Mangrove expansion and salt marsh decline at mangrove poleward limits. Global change biology 20: 147-157.

Sale, P.F. Agardy, T., Ainsworth, C.H., Feist B.E., Bell, J.D., Christie, P., Hoegh-Guldberg., O., Mumby, P.J., Feary, D.H., Saunders, M.I., Daw, T.M., Foale, S.J., Levin, P.S., Lindeman, K.C., Lorenzen, K.,

Pomeroy, R.S. Allison, E.H., Bradbury, R.H., Corrin, J., Edwards, A.J., Oburat, D.O., Sadovy de Mitcheson, Y., Samoilys, M. A. & Sheppard, C.R.C. (2014) Transforming management of tropical coastal seas to cope with challenges of the 21st century. Marine Pollution Bulletin 85: 8-23.

Schlesinger, M.E., Yin, J., Yohe, G., Andronova, N.G., Malyshev, S. & Li, B. (2006). Assessing the risk of a collapse of the Atlantic thermohaline circulation. Pages 37-49 in Shellnhuber, H.J., Cramer, W., Nakicenovic, N., Wigley, T., Yohe, G. (Eds.), Avoiding Dangerous Climate Change. Cambridge University Press.

Schmitz, W.J., Biggs, D.C., Lugo-Fernandez, A., Oey, L.-Y., & Sturges, W. (2005). In Sturges, W., Lugo-Fernandex, A. (Eds.) Circulation in the Gulf of Mexico: observations and models. American Geophysical Unition. Geophysical Monograph 161. Washington, DC.

Scheffer, M. (1990). Multiplicity of stable states in freshwater systems. Biomanipulation Tool for Water Management, 475-486.

Seara, T., Clay, P. M., & Colburn, L. L. (2016). Perceived adaptive capacity and natural disasters: A fisheries case study. Global Environmental Change, 38, 49-57.

Shafland, P.L., Gestring, K.B. & Stanford, M.S. (2007). Florida's exotic freshwater fisheries. Florida Scientist 71: 220-245.

Solís, D., Perruso, L., del Corral, J., Stoffle, B., & Letson, D. (2013). Measuring the initial economic effects of hurricanes on commercial fish production: the US Gulf of Mexico grouper (Serranidae) fishery. Natural Hazards 66: 271-289.

Southwick Associates. 2012. Sportfishing in America: An Economic Force for Conservation. Produced for the American Sportfishing Association (ASA) under a U.S. Fish and Wildlife Service (USFWS) Sport Fish Restoration grant (F12AP00137, VA M-26-R) awarded by the Association of Fish and Wildlife Agencies (AFWA), 2012.

Sumaila, U. R., Cheung, W. W., Lam, V. W., Pauly, D., & Herrick, S. (2011). Climate change impacts on the biophysics and economics of world fisheries. Nature Climate Change 1: 449-456.

Tilmant, J.T., Curry, R.W., Jones, R., Szmant, A., Zieman, J.C., Flora, M., Roblee, M.B., Smith, D., Snow, R.W., & Wanless, H. (1994). Hurricane Andrew's effects on marine resources. BioScience 44: 230-237.

Todd, A.C. (2013). Circulation Dynamics and Larval Transport Mechanisms in the Florida Big Bend. PhD Diss. Department of Earth, Ocean and Atmospheric Science. Florida State University. 90 pp.

Tolan, J. M., & Fisher, M. (2009). Biological response to changes in climate patterns: population increases of gray snapper (Lutjanus griseus) in Texas bays and estuaries. Fishery Bulletin, 107: 36-45.

Tringali, M. D., Leber, K. M., Halstead, W. G., McMichael, R., O'hop, J., Winner, B. Cody, R., Young, C., Neidig, C., Wolfe, H. & Forstchen, A. (2008). Marine stock enhancement in Florida: a multi-disciplinary, stakeholder-supported, accountability-based approach. Reviews in Fisheries Science 16: 51-57.

Tuckett, Q.M., Ritch, J.L., Lawson, K.M. & Hill, J.E. (2016). Implementation and enforcement of best management practices for Florida ornamental aquaculture with an emphasis on nonnative species. North American Journal of Aquaculture 78: 113-124.

UF (University of Florida). (2011). Online resource guide for Florida shellfish aquaculture. Cedar Key, FL: UF Shellfish Extension Office. http://shellfish.ifas.ufl.edu/industry/

USDA (U.S. Department of Agriculture). (2013). Aquaculture. National Agricultural Statistics Service. https://www.nass.usda.gov/Statistics_by_State/Florida/Publications/Aquaculture/Aquaculture2013-FDA.pdf

USFWS & USCB (U.S. Fish and Wildlife Service & U.S. Census Bureau). (2013) 2011 National Survey of Fishing, Hunting and Wildlife-Associated Recreation. U.S. Census Bureau.

Ward, H.G.M., Allen, M.S., Camp, E.V., Cole, N., Hunt, L.M., Matthias, B.J., Post, R., Wilson, K. & Arlinghaus, R. (2016). Understanding and managing social-ecological feedbacks in spatially structured recreational fisheries: The overlooked behavioral dimension. Fisheries, 41: 524-535.

Watson, C. A., & Shireman, J. V. (1996). Production of ornamental aquarium fish. University of Florida Cooperative Extension Service, Institute of Food and Agriculture Sciences, EDIS.

Weatherdon, L. V., Magnan, A. K., Rogers, A. D., Sumaila, U. R., & Cheung, W. W. (2016). Observed and projected impacts of climate change on marine fisheries, aquaculture, coastal tourism, and human health: an update. Frontiers in Marine Science, 3, 48.

Weisberg, R.H., Liu, Y. & Mayer, D. (2009). West Florida Shelf mean circulation observed with long-term moorings Geophysical Research Letters 36: L19610

Weisberg, R.H., Zheng L. & Peebles, E. (2014). Gag grouper larvae pathways on the West Florida Shelf. Continental Shelf Research 88: 11-23

Weisberg, R., He, R., Liu, Y. & Virmani, J. (2005). West Florida shelf circulation on synoptic, seasonal, and interannual time scales. Circulation in the Gulf of Mexico: Observations and Models. Geophysical Monographs 161: 325–347.

Werner, F.E., Quinlan, J.A., Blanton, B.O. & Luettich, R.A. (1997). The role of hydrodynamics in explaining variability in fish populations. Journal of Sea Research 37: 195-212.

Whitehead, J., & Willard, D. (2016). The Impact of Climate Change on Marine Recreational Fishing with Implications for the Social Cost of Carbon. Journal of Ocean and Coastal Economics 3: 7.

Williams, V. P., Canfield Jr, D. E., Hale, M. M., Johnson, W. E., Kautz, R. S., Krummrich, J. T., Langford, F.H., Langland, K., McKinney, S.P., Powell, D.M. & Shafland, P. L. (1985). Lake habitat and fishery resources of Florida. WE Seaman, Jr. Florida aquatic habitat and fishery resources. American Fisheries Society, Florida Chapter, Bethesda, Maryland, 43-119.

Williams, J.D., Butler, R.S., Warren, G.L. & Johnson, N.A. (2014). Freshwater Mussels of Florida, University of Alabama Press.

Zimmerman, R. J., Minello, T. J., & Rozas, L. P. (2002). Salt marsh linkages to productivity of penaeid shrimps and blue crabs in the northern Gulf of Mexico. In Concepts and controversies in tidal marsh ecology (pp. 293-314). Springer Netherlands.

CHAPTER 15

Paleoclimate of Florida

Albert C. Hine[1], Ellen E. Martin[2], John M. Jaeger[2], and Mark Brenner[2,3]

[1]*College of Marine Science, University of South Florida, Saint Petersburg, FL;* [2]*Department of Geological Sciences, University of Florida, Gainesville, FL;* [3]*Land Use and Environmental Change Institute, University of Florida, Gainesville, FL*

We present our understanding of Florida's paleoclimate for the past ~50 million years (Myr). The paleoclimate of the Florida Platform is closely linked to global paleoclimate. Global climate change over the past 50 Myr is a record of declining atmospheric carbon dioxide, decreasing temperature, and progressive addition of ice sheets. The overall global climate narrative is one of transition from a greenhouse Earth (warm temperatures with higher sea levels) to an icehouse Earth (colder temperatures with lower sea levels). The early 21st century has been a period of extreme climate conditions in Florida, in that we have already seen very low lake levels, including complete drying of some water bodies for the first time in recorded history. Such complete drying was never reported previously and suggests that we have entered a new climate regime in this millennium.

Key Messages

- The peninsular morphology of Florida, created during the near-simultaneous tectonic opening of the Gulf of Mexico, Caribbean Sea, and western North Atlantic Ocean starting ~200 million years ago has always played a fundamental role in Florida's climate. When a large fraction of the peninsular land mass was exposed during sea level lowstands, huge thunderstorms formed thus defining a unique component of Florida's climate.
- The topographically low and flat morphology of the Florida Platform has also allowed climate-driven sea level changes to leave a robust stratigraphic record. From these rocks and sediments the paleoceanography and, to a lesser extent, the paleoclimate of the Florida Platform has been reconstructed.
- Over the past 50 Myr the climate of the Florida Platform followed the global climate change of declining atmospheric carbon dioxide (CO_2) and cooling, i.e., a transition from a greenhouse (warm) to an icehouse (cooler with cyclical glaciations and deglaciations) Earth. There were three major warming events that occurred during this prolonged cooling that impacted Florida's paleoclimate.
- Pleistocene data reveal a terrestrial climate comparable to the modern climate, with evidence of cool climate episodes that may have been influenced by regional upwelling of cold marine waters. As climate in Florida warmed after the Last Glacial Maximum and early Holocene (~18-~11.7 ka), there were profound consequences for Florida's terrestrial environment, as vast areas that had served as habitat for Pleistocene land plants and animals, some now extinct (e.g., mammoths, horses, giant sloths, tapirs), were inundated by rising seawater.
- Shortly after the onset of the Holocene Epoch (11.7 ka), rainfall increased contributing to rising groundwater tables and initial filling of Florida's more than 8,000 shallow lakes.

Keywords

Paleoclimate; Florida Platform; Sea level; Basement rocks; Carbonate rocks; Greenhouse Earth; Icehouse Earth; Milankovitch cycles; Quartz; Phosphate; Pollen; Geologic age; Biotic assemblages

Introduction

Inferring Paleoclimate from the Geologic Record

Climate can be broadly characterized by temperature and precipitation, and fluctuations in these two variables throughout Earth history are referred to as paleoclimate change. Although factors that influence climate are enormously complex, geoscientists know that Earth's climate has displayed extremes in temperature and precipitation over many time scales. Geology is the discipline that studies changes on Earth through time. Therefore, understanding paleoclimate variations relies on interpretation of the rocks and sediments left behind. That is what we have to work with to decipher the past.

One challenge for geologists is determining the age of sediments or rocks, which in turn enables us to calculate the timing and rate of key events in the past. The age and sequence of events helps elucidate Earth processes such as sea level change, mountain building, biological extinctions, continental breakup, etc. A fundamental tool in geological studies is the Geologic Time Scale. Specific time intervals (Eons, Eras, Periods, Epochs) are defined by terms such as Precambrian, Paleozoic, Jurassic, Eocene, and represent durations of billions, millions, or thousands of years. Discussion of the Earth's geologic past requires use of these terms even when "absolute" time, e.g. 50 million years ago, is presented. We provide The Geological Society of America Geologic Time Scale in Figure 15.1 for the readers' use.

Rather than referring to Eons, Eras, Periods, or Epochs, a variety of quantitative dating methods has been developed based on sound scientific principles to determine absolute geologic age (number of years in the past). They include use of the natural radioactivity in rocks as a clock (radiometric age dating), the magnetic properties of rocks (paleomagnetic properties such as polarity reversals), the isotopic composition of rocks (isotope geochemistry), and biological evolution of microfossils in rocks (biostratigraphy). As new techniques are developed and tested, the absolute age boundaries of the Eons, Eras, Periods and Epochs are refined, and events such as the massive meteor impact that created Yucatan's Chicxulub Crater in southeast Mexico can be dated more accurately and precisely. It is now determined to have struck Earth close to 66 Ma, perhaps within 100 kyr of that time. More accurate and precise age dating of Earth's past events are an essential, ongoing line of research.

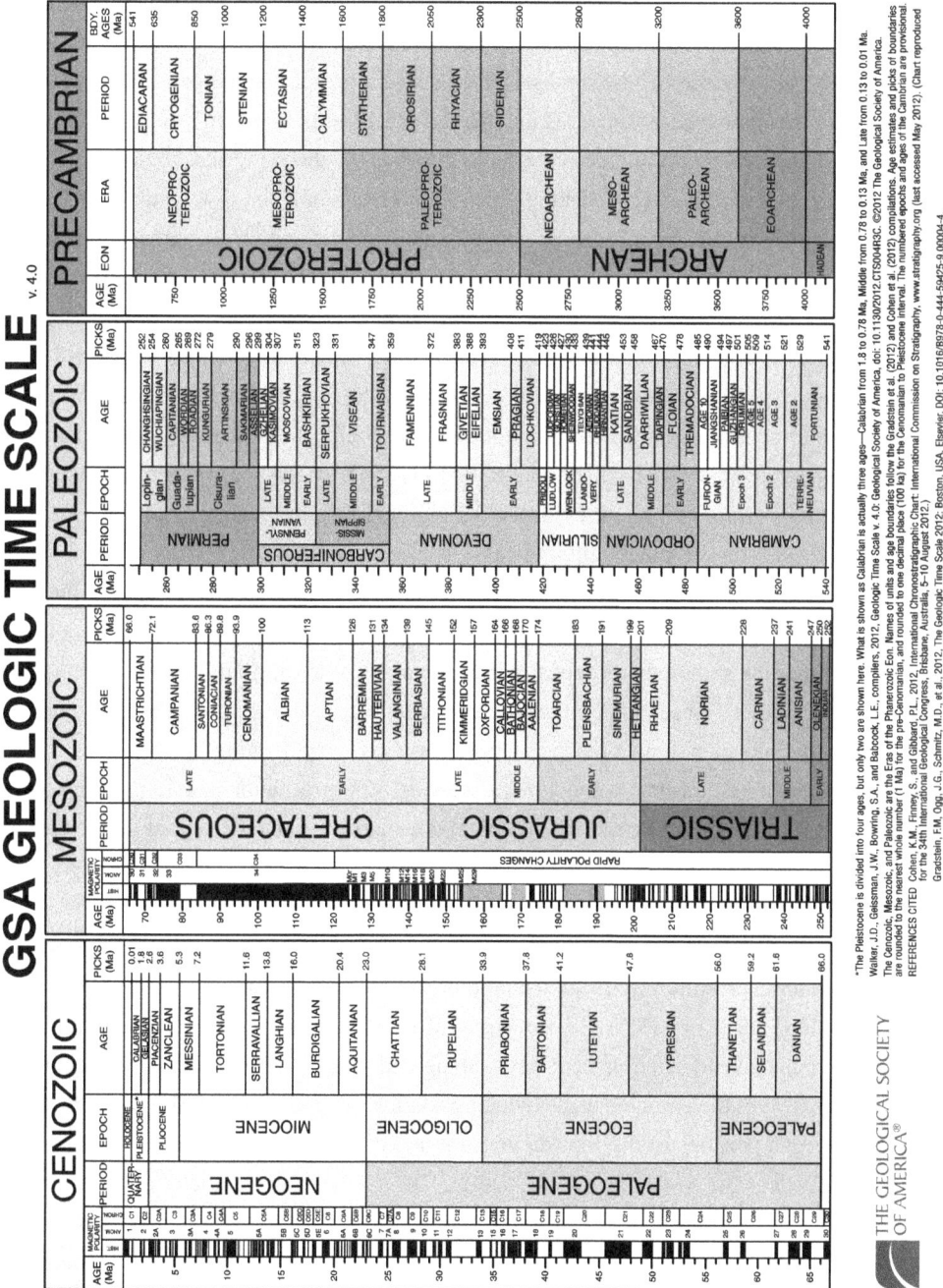

Figure 15.1. Geological Society of America Geologic Time Scale (2012, most recent version from GSA; Walker et al. 2013). Herein, **Ga** means a billion (1,000,000,000) years ago. **Ma** means a million (1,000,000) years ago (i.e., 100 **Ma** refers to something that happened 100,000,000 years ago; **ka** means a thousand (1,000) years ago; Additionally, **Gyr, Myr,** and **kyr**, express *geological duration*, i.e., an event lasted 10 **Myr** or the event's duration was 10 million years (10,000,000 years) long (Christie-Blick 2012).

A third challenge in reconstructing Florida's paleoclimate after age determination and gaps in the rock record is that Florida's sedimentary geological formations, which date back to 50 Ma, formed largely underwater. These rocks reflect ocean behavior and not rainfall or paleo-temperature of the atmosphere per se. Thus, most of the Florida Platform's rock and sediment formations only generally and indirectly preserve information about Florida's paleoclimate. The rock and sediment formations (called lithologic units) were emplaced as a result of global rise and fall of sea level. In particular, during warm climate intervals, which are associated with high sea level, carbonate sediments derived from corals, shell-bearing organisms such as mollusks, and mud from algae, were deposited in shallow seas that covered the Florida Platform. Because Florida's land elevation is so close to sea level, vertical fluctuations cause widespread land inundation (sea level highstands) or exposure of land to the atmosphere (sea level lowstands). Even though marine sedimentary rocks are not direct paleoclimate indicators, they can serve as important indirect indicators of paleoclimate. For instance, warm sea surface temperature is linked to warm air temperature and evidence for regional salinity decreases can be caused by increased freshwater runoff, potentially indicating higher rainfall.

During sea level lowstands, direct paleoclimate indicators such as tree pollen are buried in inland lake sediments and a stratigraphic study of pollen grains can be used to infer paleoclimate. But these non-marine sedimentary deposits in Florida are often thin and discontinuous. Also, oxidation and erosion of these sediment units during exposure to the atmosphere can remove and alter them, rendering paleoclimate reconstructions even more challenging.

Before we examine Florida's paleoclimate, we must answer a fundamental question—What is Florida? Florida is much more than the political boundaries and coastline that define the state of Florida. The entire Florida Platform extends well beyond the state of Florida's borders (Figs. 15.2 and 15.3) and is structurally bounded by faults created during the near-simultaneous tectonic opening of the Gulf of Mexico, Caribbean Sea, and western North Atlantic Ocean. The creation of these ocean basins resulted from the tectonic breakup of the mega-continent Pangea, which began ~200 Ma (Figs. 15.4, 15.5). So any study of Florida's geological history, including its paleoclimate, must consider the present-day, submerged portions (modern continental shelves and slopes), as well as the emerged areas, which we know of as the state of Florida.

The submerged portions of the Florida Platform are difficult to reach because of their large area and the depth of the overlying coastal ocean. Therefore, we are restricted to the rocks and sediments that are easily accessed, i.e., those rock and sediment formations exposed on land, which extend back ~50 Myr. These 50-Ma and younger sedimentary rocks and sediments can be studied more comprehensively than small rock samples retrieved from a few deep drill sites. This accessible rock record is revealed on the geologic map of Florida (See Florida Geological Survey). For these reasons we begin our narrative of Florida's paleoclimate at that point in time, 50 Ma.

We will follow a simple strategy of first presenting the global climate record and then discussing how the Florida Platform has responded to these global climate changes. As with practically all geologic studies that cover vast amounts of time, the more recent events (hundreds to tens of thousands of years ago) are better represented in the rock record and provide much more detail than events that occurred hundreds or even tens of millions of years ago.

First, however, we present a brief look at the Florida Platform's geologic history in deep time (hundreds of millions of years ago). We do this to set the stage for our examination of the last 50 Myr of the Florida Platform's paleoclimate.

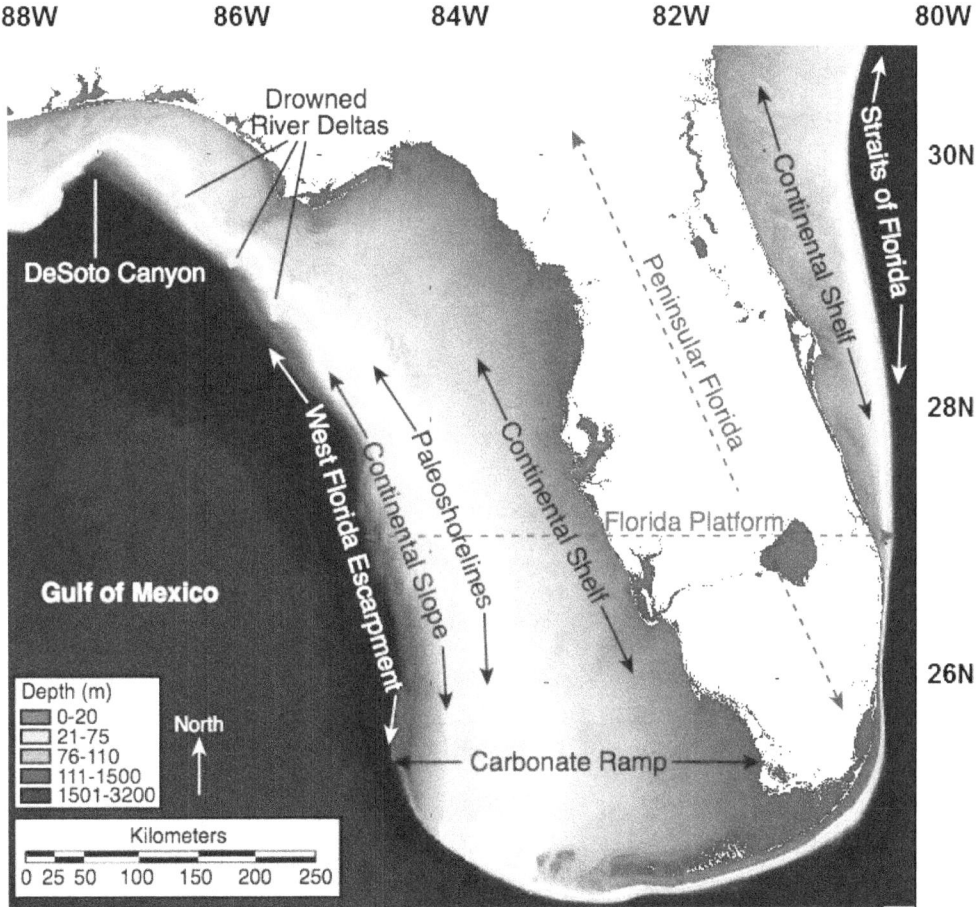

Figure 15.2. Exposed and submerged portions of the Florida Platform. Geologists consider the Florida Platform to be a single entity that includes the emerged state of Florida, as well as the vast area that today lies under water. The size of the exposed portion of the Platform changed dramatically over geologic time as sea level rose and fell (from Hine 2013; *Geologic History of Florida: Major Events That Formed the Sunshine State* by Albert C. Hine. Gainesville: University Press of Florida, 2013. Reprinted by permission; modified from USGS Open File Report 2007-1397; courtesy of Dr. L. Robbins).

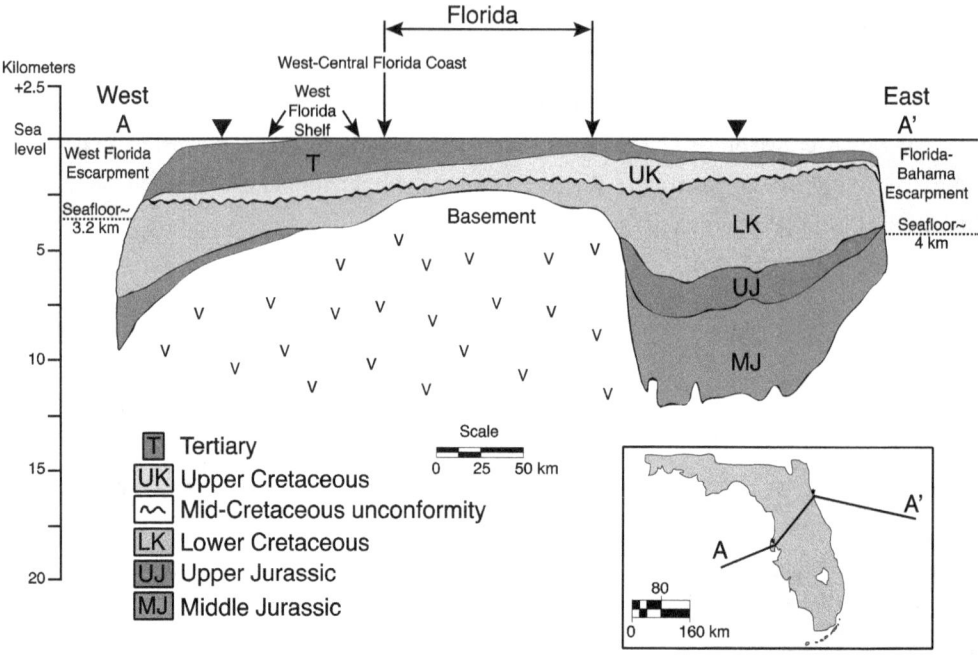

Figure 15.3. West-to-east depth section across the central Florida Platform—part of the even larger Florida-Bahamas Platform. This diagram shows two distinct rock types that underlie the Florida Platform: (1) >500-Ma-old basement rocks (white component marked with "v") overlain by (2) much younger, mostly limestone (marked with colors; MJ, UJ, LK, UK, T). There is a third geologic unit, mostly quartz-rich sand, which overlies the limestone. The sand covers most of the exposed Florida Platform and forms most of our beaches. But this youngest geologic unit is so thin that it cannot be represented in the above figure. It is thinner than the thickness of the black line that represents the top of the Florida Platform. This chapter focuses on the climate inferred from the rock record extending back 50 Ma, i.e., the brown unit labeled "T" (Tertiary and Quaternary; modified from Hine et.al 2003; Hine 2013; *Geologic History of Florida: Major Events That Formed the Sunshine State* by Albert C. Hine. Gainesville: University Press of Florida, 2013. Reprinted by permission.), which is exposed at the surface today.

Long-Term History of the Florida Platform

Seven hundred million years ago, the basement rocks that underlie the Florida Platform were located near the South Pole and were part of a larger continental landmass called Gondwana, which eventually collided over millions of years with another large land mass called Laurasia, forming the megacontinent Pangea in the Paleozoic, from ~350 Ma to ~250 Ma (Hine 2013). Thus, the > 500-Ma-old basement rocks that underlie the present Florida Platform (Fig. 15.3) once formed part of the African and South American continents. Basement rocks are generally crystalline igneous and metamorphic rocks. They form the basic crust of continents and possess very little paleoclimate information that geologists can exploit. The climate at the South Pole ~500 Ma was freezing and the basement rocks were probably covered by a thick ice sheet. Over the next ~500 Myr, these basement rocks migrated northward about 13,000 km (8,150 miles) via plate tectonic processes, traveled through the tropics, crossed the Equator and arrived at their present location ~200 Ma (Fig. 15.5).

The Paleozoic sedimentary rock on top of the Platform's basement rocks has been eroded or is deeply buried, and has been rarely sampled (Klitgord et al 1984; Sheridan et al. 1988; Pindell and Barrett 1990; DeBalko and Buffler 1992; Randazzo and Jones 1997; Redfern 2001). Before the Pangean megacontinent breakup, starting around 200 Ma, Florida's basement rocks were at a mid-continent location, probably thousands of kilometers (1,000 km = 621 miles) from the oceans that existed at that time (Fig. 15.4). Although there is no rock record from that time for study of past climate, we surmise that tropical rain forests probably dominated because the area was at low latitudes.

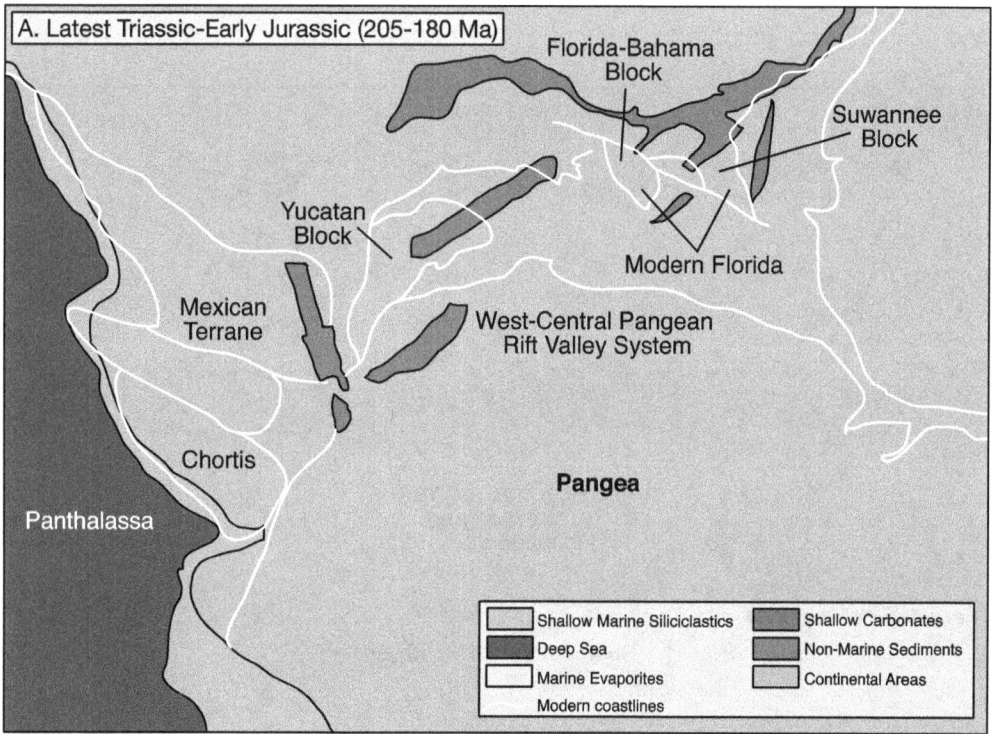

Figure 15.4. Paleogeographic reconstruction of mega-continent Pangea, from 205 to 180 Myr. The Equator runs across the center of the diagram. Note the components of Florida's basement rocks—Florida-Bahama Block and Suwannee Block, which together form the basement rock beneath the modern Florida Platform. The location was far inland, probably >1,000s of km from the global oceans at that time. The terrestrial environment must have been something like the equatorial rain forests we see today in South America and Africa (used by permission from Iturralde-Vinent 2003).

Florida's paleoclimate is unique because the platform became a peninsula during the early breakup of the Pangean megacontinent around 200 Ma and was almost completely surrounded by seawater by 160 Ma (Fig. 15.5). The peninsular morphology of the Florida Platform has fundamentally controlled Florida's climate at times when large areas of land mass were exposed

464 • ALBERT C. HINE ET AL.

to the atmosphere. Because of its relatively low heat capacity compared to ocean water, land heats up quickly on hot days, forcing air to rise, which brings in warm moist air from above the ocean to both Florida's east and west coasts. These local winds are known today as sea breezes. Rising air over the heated land produces, and has produced for millions of years, thunderstorms that reach great altitudes (>18,000 m ~>59,000 ft). Such storms, which extend high into the atmosphere, capture dust and aerosols that originated on other continents and deposit them on Florida. These seasonal thunderstorms are some of the largest and most intense in the world, thus earning Florida the nickname of the "thunderstorm and lightning capital of the United States." The Florida Platform has probably been influenced by substantial thunderstorm activity since the western North Atlantic Ocean to the east, the Gulf of Mexico to the west, and the Caribbean Sea to the south first formed.

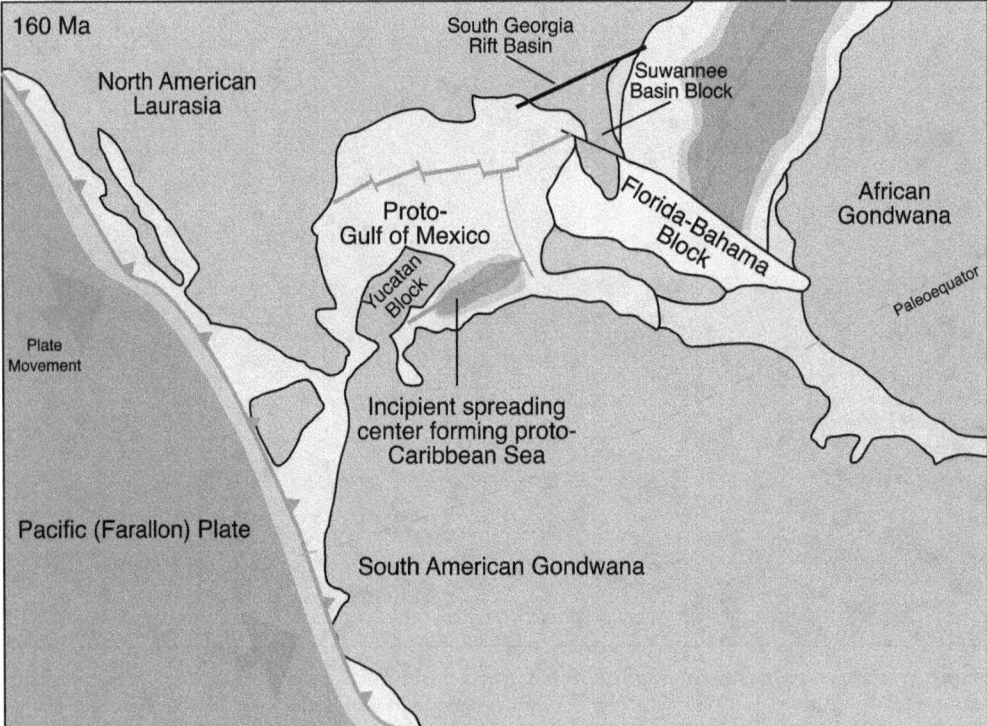

Figure 15.5. Paleogeographic map showing the late-stage breakup of the Pangean megacontinent. Note position of the Equator at this time relative to the land masses. Darker blue indicates deep ocean water, and light blue indicates shallower water. Darker brown indicates continental rocks. Light brown indicates non-marine sediments erode from continental rocks. Florida-Bahama Block is highlighted in yellow. The map illustrates the newly opened Gulf of Mexico and the newly opened western North Atlantic Ocean that formed by seafloor spreading. At that time (160 Ma), the Caribbean Sea was starting to open as well. Note that the Florida Platform was already north of the Equator, roughly at the same latitude as it is today. The basement rocks beneath the Florida Platform were surrounded by seawater and formed the beginning of the peninsular Florida Platform (modified from Redfern 2001; published in Hine 2013).

From about 200 to 150 Ma, the Florida Platform's basement rocks were exposed to the atmosphere, with non-marine sediments deposited only sporadically. A Late Jurassic (164-157 Ma), non-marine formation superimposed on the igneous and metamorphic basement rocks indicates that an arid climate must have dominated, given the presence of sand dunes (Norphlet Formation) of Upper Jurassic (UJ) age (the purple formation in Fig. 15.3). Inferring paleoclimate during the very early development of the Florida Platform is problematic and quite speculative.

Marine carbonate rocks (limestones, dolomites) and other sedimentary rocks (evaporites) covered the Florida Platform's basement rocks and early non-marine rocks, and extend back to the Late Jurassic/Early Cretaceous (~150 - 140 Ma). These formations are nearly six km thick. The climate signals contained within this deeply buried rock likely reflect the global warm temperate/tropical climate of the Earth at that time.

For the next 100 Myr (150-50 Ma), the Florida Platform was mostly under water, with occasional periods of exposure when sea level fell (Fig. 15.6, 7B). It was during that time that much of the sedimentary carbonate cover on the Florida Platform accumulated. Not until the late Oligocene (~ 25 Ma) did widespread carbonate sedimentation cease and become largely restricted to the extreme southern portion of peninsular Florida. This carbonate rock, which covers the deeper basement rocks, contains the Floridan aquifer and features the famous springs and sinkholes that dominate much of today's exposed Florida Platform.

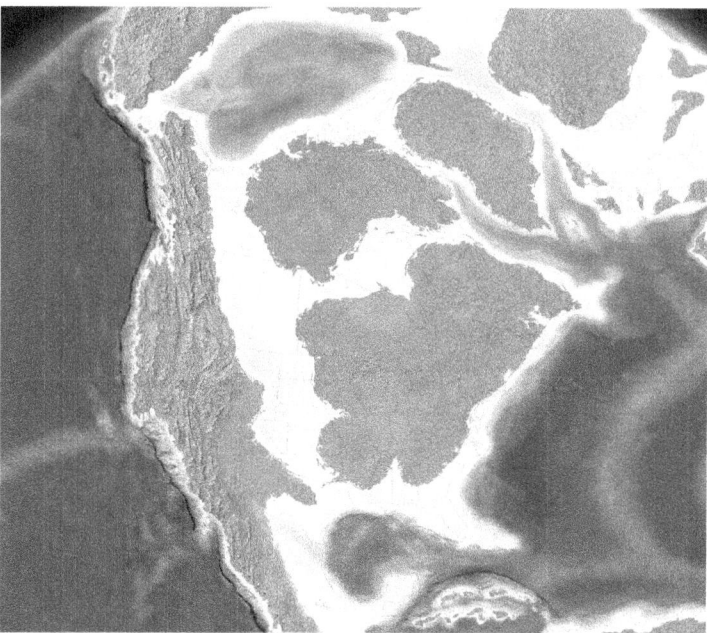

Figure 15.6. A sea level highstand flooded all of Florida, much of the area surrounding the modern Gulf of Mexico, and the seaway that connects the Gulf of Mexico to the Arctic Ocean. This is the paleogeography of North America at ~100 Ma, when warm, shallow seas covered much of the region. During that time, the rocks labeled "LK, UK" and the lower part of "T" (Fig. 15.3) were deposited. Although this rendering is for 100 Ma, sea levels were similarly high at 50 Ma (from Ron Blakey, Colorado Plateau Geosystems with permission; Hine 2013, Hine et al. 2016).

Climate from the Eocene Greenhouse to the Pleistocene Icehouse

Global Climate

The history of global climate change over the past 50 Myr is a story of declining atmospheric carbon dioxide (CO_2), decreasing temperature, and progressive addition of ice sheets, first on Antarctica and later in the Northern Hemisphere. This period also includes an overall decline of global sea level (Fig. 15.7A, B, C). Global cooling did not occur gradually and continuously through time. Instead, the declining temperature trend was interrupted by intervals of rapid cooling and ice growth, and warmer intervals of ice retreat that are referred to as "climatic optima."

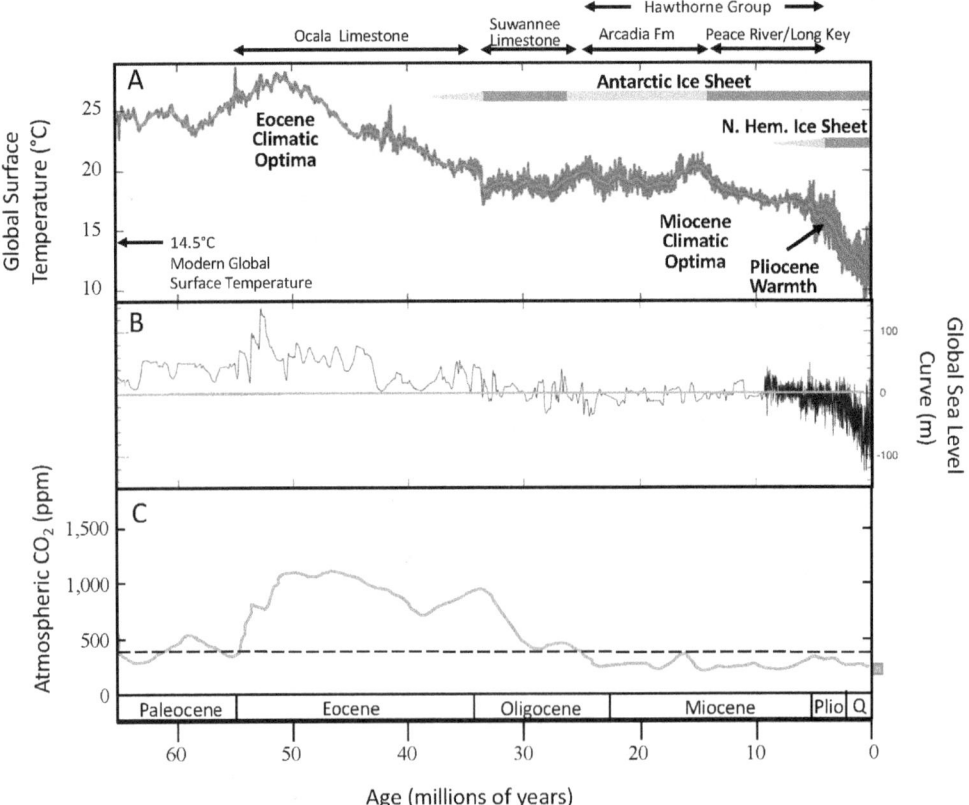

Figure 15.7 A, B, C. Cenozoic climate conditions. A. Global surface temperature derived from stable isotope measurements ($\delta^{18}O$) on carbonate shells of deep-sea benthic foraminifera (Zachos et al. 2001, 2009). Blue bars represent a qualitative estimate of ice volume in each hemisphere. Darker blue coincides with maximum extent of the ice sheets. B. Variations in global sea level relative to modern sea level (0 m) are from Miller et al. (2005). C. Earth's atmospheric CO_2 history is from Beerling and Royer (2011). Line connects data points from terrestrial and marine proxies for atmospheric CO_2. Error bars for individual data points are typically several hundred ppm (parts per million) CO_2. Horizontal dashed line indicates modern CO_2 value of ~400 ppm. Blue box at bottom right represents range of Pleistocene glacial to interglacial CO_2 values from ice cores.

There is no evidence for development of major ice sheets in either hemisphere during the early Eocene Climatic Optimum, at ~50 Ma. Estimates indicate that CO_2 was ~1,000 ppm, roughly three times pre-industrial atmospheric concentrations (Beerling and Royer 2011), and that the global mean temperature, averaged across land and sea at all latitudes, was ~28 °C (80 °F), i.e., about 14 °C (20 °F) warmer than today! Although the Antarctic continent sat over the South Pole at that time, it was covered by subtropical vegetation (ferns, cypress and beech trees), rather than ice, because of warm global temperatures. Determining sea level so far back in time is difficult, but estimates range from 20-100 m (66-328 ft) higher than today, with the most recent estimate ~75 m (246 ft) (Müller et al. 2008). Under such conditions, almost the entire Florida Peninsula would have been under water.

Over the next 16 Myr (ending ~34 Ma), global surface temperatures gradually cooled by about 4 °C (~8 °F). At that point, the Earth's climate system appears to have crossed a critical threshold, possibly caused by lower atmospheric CO_2 levels (DeConto and Pollard 2016), and ice grew rapidly on Antarctica. Within ~ 300 kyr, an ice sheet approximately the size of the modern Antarctic Ice Sheet that extended to the coastline covered this southernmost continent. As water was removed from the ocean to form the ice, global sea level dropped by ~70 m (230 ft).

For the next 18 Myr (ending ~16 Ma), the Antarctic Ice Sheet retreated (shrank) and advanced (grew), varying between about 33% and 100% of its current size. Yet no major ice sheets formed in the Northern Hemisphere. Atmospheric CO_2 concentrations declined from ~800 ppm, when the ice sheet first formed, to ~500 ppm, only about 100 ppm higher than today.

This waxing and waning of the Antarctic Ice Sheet was interrupted at ~16 Ma by dramatic warming and ice melt during the period referred to as the Miocene Climatic Optimum (Miocene Epoch; 23.0-5.3 Ma). This interval is puzzling because there is clear evidence of warming and ice sheet retreat, but the change in atmospheric CO_2 associated with this warming appears to have been small. This observation has led scientists to suggest that the climate system is more sensitive to changes in atmospheric carbon dioxide concentrations when CO_2 values are relatively low. The Miocene Climatic Optimum ended at ~14 Ma with renewed cooling and ice growth on Antarctica, which continued until another climate threshold was crossed—the transition to global icehouse conditions.

A period of renewed warmth at ~3 Ma in the Pliocene is particularly interesting because this is the last time period for which estimates of atmospheric CO_2 concentrations were as high as they are today (~400 ppm). It is somewhat surprising then that the global mean temperature was believed to be 2-3 °C warmer than pre-industrial times, or 1-2 °C warmer than today. This magnitude of warming is similar to Intergovernmental Panel on Climate Change (IPCC 2013) estimates for global temperatures for the 21st century and represents the warmest temperature experienced by Earth in the past 3 million years. Estimates of sea level during warm intervals of the Pliocene range from 6 to 25 m higher than today, with likely contributions from ice melt on both Greenland and Antarctica. There is also evidence that the Arctic Ocean was seasonally ice-

468 • ALBERT C. HINE ET AL.

free. One possible explanation for warmer conditions in the Pliocene, despite atmospheric CO_2 concentrations similar to those today, is that the modern climate system has not yet reached equilibrium with respect to CO_2 concentrations driven largely by human activity, which rose very rapidly compared to past rates of increase, driven solely by natural additions to atmospheric CO_2. Another idea is that there may have been additional feedbacks within the climate system, such as changes in ocean circulation, which enhanced heat transport by the ocean and contributed additional warming.

Major growth of ice in the Northern Hemisphere began ~2.6 Ma, when the Earth's climate system began to alternate between colder glacial and warmer interglacial intervals. During glacial periods, large ice sheets covered Greenland, much of North America, and Scandinavia (Fig. 15.8).

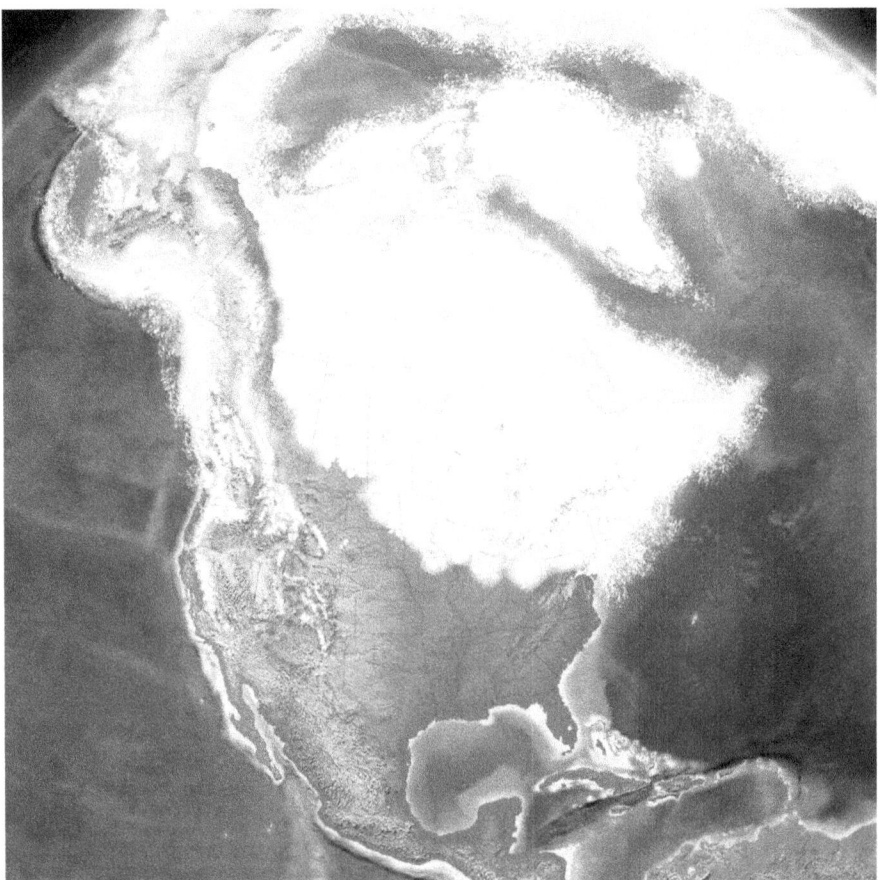

Figure 15.8. Map of North America during the Last Glacial Maximum (~20 ka). Light blue areas were exposed by a drop in sea level of ~130 m (427 ft) associated with ice growth at high latitudes. The Florida Platform was much wider at that time (from Ron Blakey, Colorado Plateau Geosystems with permission; Hine 2013, Hine et al. 2016).

During interglacial intervals, conditions were similar to today, with continental ice sheets persisting at high latitudes; for example, Antarctica, Greenland, and smaller glaciers present atop high mountains at lower latitudes (e.g., in the Andes). Fluctuations between these glacial and interglacial states are believed to have been a consequence of subtle changes in the Earth's orbit around the Sun, i.e., Milankovitch Orbital Cycles (Fig. 15.9). The Earth's tilt on its axis and the shape of its orbit around the Sun are affected by interactions with other planets and result in predictable changes in the way the Sun's incoming energy is distributed across the Earth surface. These variations were small, but under the generally cool conditions that developed over the past 50 Myr, such small changes were enhanced (amplified) by fluctuations in atmospheric CO_2 and ocean circulation, which led the Earth in and out of glacial intervals.

During the last major glacial advance around 20 ka, known as the Last Glacial Maximum or end of the last Ice Age, global temperatures were ~5 °C (9 °F) cooler than today, atmospheric CO_2 concentrations decreased from ~280 ppm to ~200 ppm, and global sea level was an astonishing 130 m (427 ft) lower than today (Fig. 15.10). See Ruddiman (2008) for a detailed explanation of the Earth's global climate history.

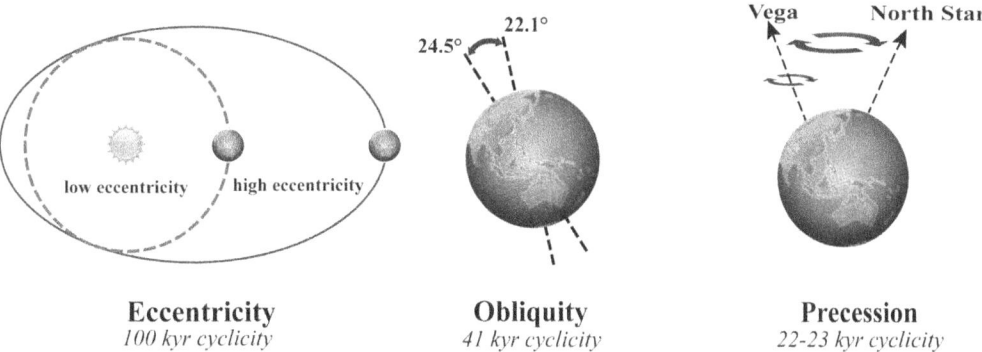

Figure 15.9. Milankovitch cycles of 100 kyr, 41 kyr, and 22-23 kyr, caused by changes in the Earth's orbit around the Sun (eccentricity), the tilt of the Earth's axis of rotation (obliquity), and the wobble (precession) (modified from Hine et al. 2016; *Sea Level Rise in Florida; Science, Impacts, and Options* by Albert C. Hine, Don P. Chambers, Tonya D. Clayton, Mark R. Hafen, and Gary T. Mitchum, Gainesville: University Press of Florida, 2016. Reprinted by permission).

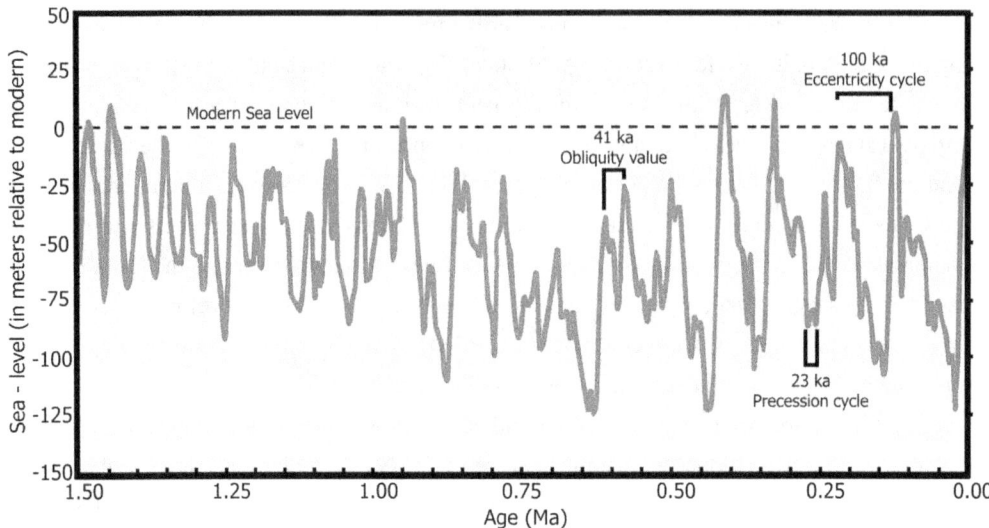

Figure 15.10. Global sea level fluctuations over the past 1.5 Mya related to Milankovitch orbital cycles. These cycles have periodicities of 22-23 ka (precession), 41 ka (obliquity), and 100 ka (eccentricity). Note that the last sea level lowstand at approximately 20 ka (0.02 Ma) was about 130 m (427 ft) below present-day sea level. These cycles are driven by variations in Earth's orbital and rotational characteristics (modified from Hine 2013; *Geologic History of Florida: Major Events That Formed the Sunshine State* by Albert C. Hine. Gainesville: University Press of Florida, 2013. Reprinted by permission.).

Florida's Climate History

We follow the same time line over the past 50 Myr to examine the effect of the global climate variations (Eocene Climatic Optimum, Miocene Climatic Optimum, Pliocene Warm Period) on the Florida Platform glacial/interglacial events). The response of the Florida Platform to glacial/interglacial events is presented separately in a later section (Florida's Climate in the Global Glacial/Interglacial World—The Icehouse).

The Marine Narrative (Eocene Climatic Optimum, Miocene Climatic Optimum, Pliocene Warm Period)

During the Eocene Climatic Optimum, the Florida Platform was mostly covered by warm tropical/subtropical seas during an extended sea level highstand (Fig 15.7B). Carbonate sedimentation dominated in these shallow seas. The accumulating fossiliferous sediments formed the upper limestone portion of Florida's carbonate platform and generated the rocks seen today in mines, pits, sinkholes, and sometimes at the land surface. These sediments and rocks formed the Ocala Limestone (Fig. 15.7A).

The Miocene Climatic Optimum was also a period of relatively high sea level, during which Florida's abundant phosphate deposits accumulated. These sediments constitute part of the Hawthorn Group of formations (Fig. 15.7A), which yield ~30% of the world's phosphate, mined primarily to produce agricultural fertilizer. The mining industry contributes to the economy of

Central Florida where phosphate-rich minerals, originally formed on the shallow seafloor during periods of elevated sea level, are extracted. From ~23 to ~10 Ma (i.e., during much of the Miocene Epoch), strong ocean currents crossed the central Florida Peninsula, providing the ideal environment for accumulation of the economically important phosphate deposits. Even though these minerals formed on the seafloor, it can be argued that the global climate change that flooded the central Florida Platform was ultimately responsible for their deposition (Popenoe 1990; Riggs 1979; Riggs 1984; Riggs et al. 2000; Hine 2013).

During the late Pliocene Warm Period, river deltas, starting in the Caloosahatchee River/Lake Okeechobee area, migrated at least 200 km (125 miles) southward, forming the Peace River Formation and the Long Key Formation. These quartz-rich sediments underlie younger limestone that are exposed in the Florida Everglades and form the Florida Keys (Pleistocene-age Miami Limestone and Key Largo Formations). These deltaic sediments are characterized by quartz-rich sands and gravels, and prograding deltaic slopes that display up to 100 m (328 ft) relief, indicating that these river delta deposits accumulated in water at least that deep (Warseski et al. 1996; Cunningham et al. 1998; Missimer 1976; Missimer 1999; Guertin et al. 1999; Cunningham et al. 2003). No modern rivers in Florida carry that much sediment. Water and sediment discharged in these large Pliocene rivers and streams were much greater than anything we see today and was probably caused by much heavier rainfall (Hine et al. 2009; Hine 2013).

The Terrestrial Narrative (Eocene Climatic Optimum, Miocene Climatic Optimum, Pliocene Warm Period)

Florida's terrestrial paleoclimate record has been inferred using preserved plant macrofossils and microfossils (e.g., leaves and pollen). One challenge associated with using fossil pollen and plant records in Florida for paleoclimate inference is that floral diversity is low in the southeastern US for the past 50 Myr (Rich 1995). This means that there were relatively few plants to serve as unique bio-indicators of specific climate regimes. Since Florida's mean elevation was close to sea level, the potential for terrestrial or marginal-marine sediments to preserve plant fossils likely varied with relative sea level. Also, the oldest terrestrial or marginal-marine sediments that did accumulate would have had a greater probability of erosion. Consequently, sediments that are available for study provide the most complete record of Florida's climate mainly for the past ~3 Ma, and we have only spotty coverage of terrestrial plants extending back to the Eocene (56.0-33.9 Ma).

Florida's terrestrial climate from the Eocene to the early Pliocene (56-5 Ma) changed as the Earth transitioned from globally warm to cooler temperatures (Fig. 15.7A). The oldest known terrestrial plant remains in Florida are from the Eocene and indicate a climate similar to modern-day South Florida and the circum-Caribbean. These fossils are geographically scattered and fragmentary, and mostly represented by seagrass macrofossils (Ivany et al. 1990). Fossil fern spores and pollen grains in Eocene deposits from central and southern Florida locales indicate a terrestrial landscape that ranged from a tropical to subtropical coastal climate, as observed on modern beaches in Central America and islands of the Caribbean. Inland south-central Florida likely had a somewhat more temperate climate in exposed, inland, forested areas, as indicated by pollen derived from pine, oak, hickory, mallow, and perhaps sycamore (Jarzen and Dilcher 2006; Jarzen and Klug 2010; Gradstein et al. 2012).

By the time of the Miocene (~23.0-5.3 Ma), pollen and other plant fossils indicate that north-central Florida had transitioned to a climate similar to that of the present. Leaves of hickory, elm, and buckthorn, as well as a large number of temperate taxa represented by pollen, fruits, and seeds, suggest a warm-temperate climate in North Florida (Corbett, 2004). The plant community type was similar to that in the modern, northern Gulf Coast of Florida, where elm-hickory-cabbage palm forests occur near oak- and pine-dominated landscapes.

The culmination in the late Pliocene (~3.6-2.6 Ma) of the transition to colder global temperatures is represented in Florida by a climate similar to that of today, but with cooler conditions and evidence for fluctuations between dry and wet. There were abundant cypress, arrowhead, black gum, sweetgum, and elm in Central Florida, all suggestive of a freshwater swamp landscape (Emslie et al. 1996), but one that became drier at times, as indicated by abundant pine and occasional oak pollen. Occasionally, there were also colder conditions in South Florida, similar to modern coastal mid-Atlantic states, which may have been related to coastal upwelling of cooler, deeper water along the coastal Gulf of Mexico (Willard et al. 1993).

Florida's Climate in the Global Glacial/Interglacial World—The Icehouse

The Pleistocene (2.6 Ma to 11.7 ka) climate of Florida under global icehouse conditions can be reconstructed using fossils from the same central and northern Florida locations that were used for Pliocene reconstructions. During glacial maxima, extensive, thick ice sheets covered North America and Western Europe, and there was significant enlargement of mountain, alpine-type glaciers worldwide.

During Pleistocene, sea level lowstands, strong, easterly trade winds transported iron-rich dust from the Sahara to the Caribbean, to the southeast US including Florida, and even northward up to Bermuda. Such dust events still occur today, causing limited visibility and pulmonary stress for some people. But during cooler periods, when the Northern Hemisphere supported huge glaciers, Saharan dust events were more common. The result in Florida was the formation of extensive reddish soils (caliche/duricrust) that accumulated on exposed limestone—particularly in the Florida Keys. During these sea level lowstands, prolonged exposure of the Pleistocene limestone to slightly acidic rainwater etched and dissolved these rocks, forming the karst topography we see today (Multer et al. 2002). In addition, the exposed portion of the Florida Platform was cooler, windier, and drier than today.

Similar to the Pliocene, Pleistocene terrestrial climate was comparable to the modern climate, with evidence of episodic climate transitions that may have been influenced by regional upwelling of cold marine waters. Baseline Pleistocene climate was also generally similar to the modern climate, with a range of mesic (moist) to xeric (dry) woodlands, scattered marshes and wetlands, inferred from geologic studies in North Florida's Trail Ridge, Central Florida's Leisey Shell Pits (Hillsborough County) and Peace Creek sinkhole (Polk County), and southwest Florida (Sarasota County) (Rich 1985; Rich and Newsom 1995; Hansen et al. 2001; Emslie et al. 1996; Willard et al. 1993).

In North Florida, climate at the Plio-Pleistocene transition (~3 Ma to 2.6 Ma) was similar to that of the Holocene (11.7 ka to present), with evidence for cypress forests that possessed shrubby undergrowth and standing water (Rich 1985). The transition then, from the Pliocene into the Pleistocene, was accompanied by generally drier conditions, dominated by shrubs and herbs, with periods of drought and fire, the latter indicated by abundant charcoal.

This drier episode was followed by an interval at ~2.2 Ma, when shrub-dominated swamps indicate wetter conditions. Alternations between wetter and drier conditions are also observed in Polk County, where three pine-oak pollen cycles are described for samples from ~2.8 Ma (Hansen et al. 2001). The longer pine phases are interpreted as reflecting drier climate conditions, whereas the shorter oak phases are thought to represent wetter conditions. These wet-dry variations may have been driven by the same orbital cycles (Milankovitch orbital cycles) that drove glacial-interglacial cycles. Hansen et al. (2001) noted an almost complete absence of pollen from tropical plant taxa that are now common in southern-most Florida. This suggests the presence of a long-term climate barrier that confined tropical species to areas south of Central Florida, and may point to occasional Pleistocene incursions into Florida of Arctic air and frost, as occur today.

In Southwest Florida, the climate of the early-middle Pleistocene was similar to that of today, but with notable episodes of cooler conditions in the earliest Pleistocene (Willard et al. 1993). In South Central Florida, pollen associations argue for a climate similar to that of the modern Florida Panhandle, with a transition in the earliest Pleistocene to conditions like those of modern coastal Florida. Invertebrate and vertebrate marine fossils from the later Pleistocene Epoch (2.6 Ma to 11.7 ka) showed this interval was less tropical, with slight cooling of coastal waters (15–25 °C; 59–77 °F), which may be explained by upwelling of cold water in the Gulf of Mexico resulting from changes in the circulation of the Gulf of Mexico due to tectonic movements in the lower Caribbean Sea, the absence of El Nino events, or the absence of red tides (Willard et al. 1993; Emslie et al. 1996).

Coming Out of the Last Ice Age—Up to the Arrival of Humans

The Last Glacial Maximum (26 to 18 ka) spans the period during the most recent ice age when ice sheets achieved their greatest spatial coverage. Because so much water was stored in ice at that time, and ocean temperatures were much colder than today, sea level was approximately 130 m (427 ft) lower than present. Extensive areas off the modern coast of Florida, especially in the shallow Gulf of Mexico, were exposed at that time, i.e. there was much more "real estate" on the peninsula than there is today (Fig. 15.11).

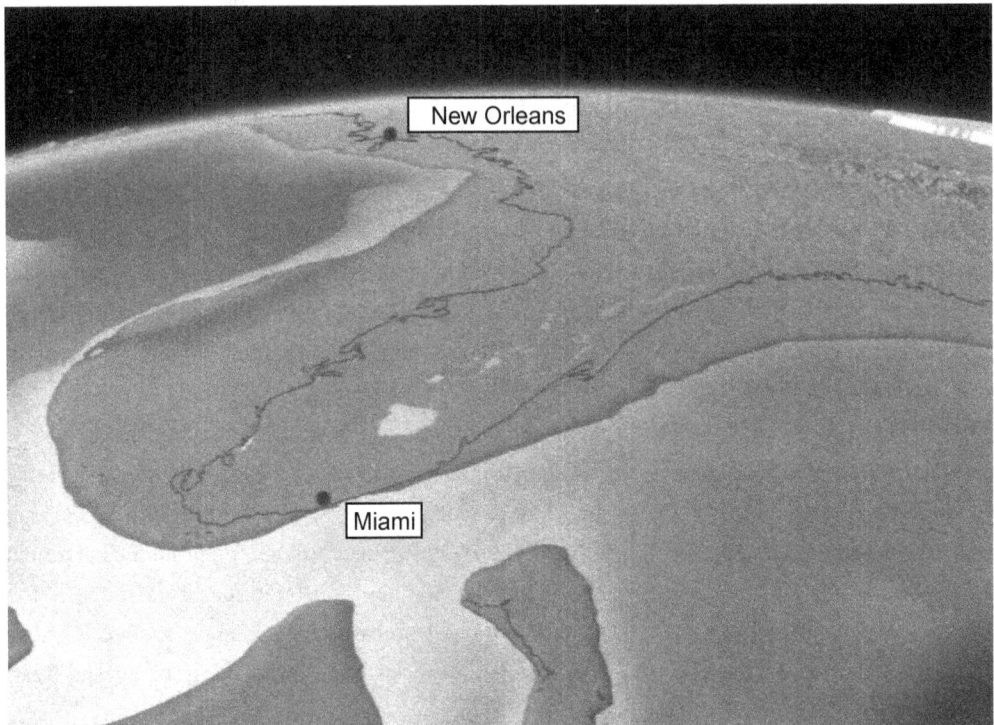

Figure 15.11. Most of the Florida Platform was exposed during the Last Glacial Maximum (~26 to 18 ka), when sea level was 130 m (427 ft) below present. This land area was approximately twice as large as the present state of Florida (modified by Hine, 2013; *Geologic History of Florida: Major Events That Formed the Sunshine State* by Albert C. Hine. Gainesville: University Press of Florida, 2013. Reprinted by permission; originally from Geotimes, now Earth Magazine; used by permission).

Periods of sea level lowstands were cooler, windier and drier, and parabolic dunes dating back some 20 ka, to the Last Glacial Maximum must have been very common, as they are so abundant along the modern west – central Florida coastline and coastal interior (Wright et al 2005). For example, Cedar Key and Seahorse Key in Levy County are the modern remains of these sand dunes. These dunes formed from a persistent wind (from the southwest) and limited sand supply, and were anchored by sparse vegetation. As sea level rose, the dunes became isolated sandy beaches along the shoreline, temporary offshore sandy islands, and eventually sand shoals when they were completely submerged or eroded by waves. Further inland, they retained their parabolic shape and became vegetated hills as humidity increased, allowing trees and shrubs to become more common. These elevated areas, surrounded by flat topography, were very useful to pre-Columbian, Native Americans (Sassaman et al., 2017 in press).

A period of global warming commenced about ~18 ka, signaling the end of the Last Glacial Maximum. Melting of the glaciers (deglaciation) at higher latitudes began, associated with a dramatic, rapid rise in sea level, especially during the period from ~18 to ~8 ka (Fig. 15.12). By about ~8 ka, sea level was within ~10–20 m (33–67 ft) of its current position. Thereafter, it continued to rise, albeit at a much slower rate.

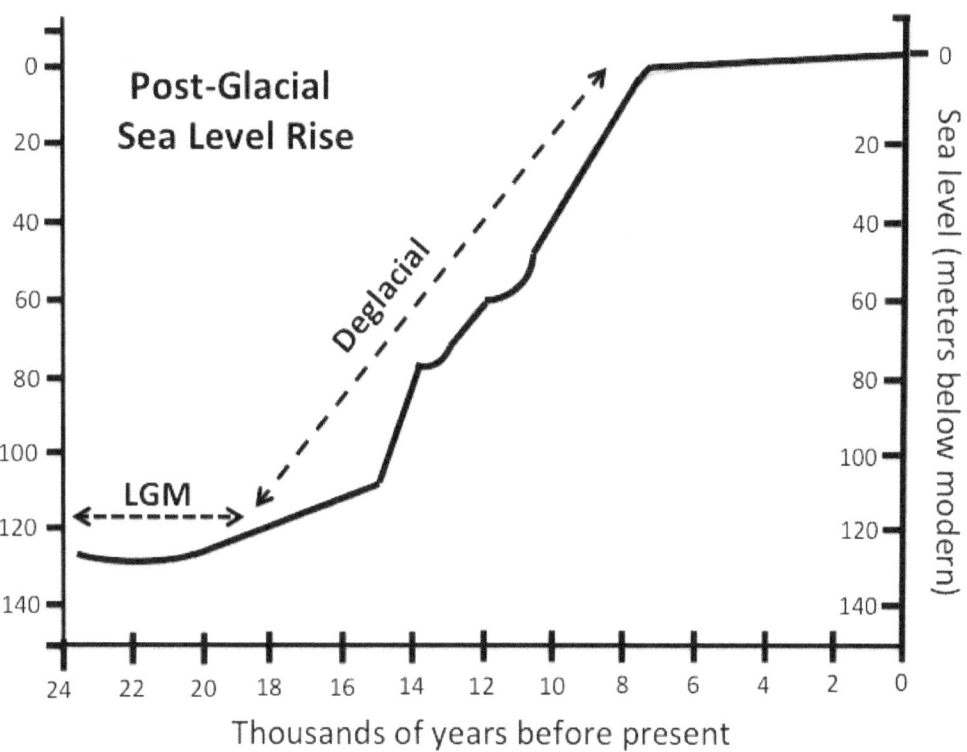

Figure 15.12. Sea level history for the past 24 kyr. Note the irregular character of the sea level curve since the Last Glacial Maximum (LGM) (modified from Lambeck et al. 2014).

Some of what we know about late glacial, deglacial, and recent climate conditions in Florida comes from study of sediment cores collected in Florida's lakes. These sediment profiles provide a window into past climate and environment on the peninsula. Lake sediments accumulate in an orderly fashion, with younger sediments continuously accruing atop older deposits. Lake sediments typically accumulate at rates ranging from <1 to several mm/yr (0.04 to ~0.1 inch), enabling fairly high-resolution reconstructions of past conditions if cores are sampled at close intervals. Sediment cores are relatively easy to obtain and age–depth relations for such profiles, extending back in time >40 ka or more, can be established by radiocarbon (^{14}C) dating. The paleoclimate and paleoenvironmental information preserved in these cores can be gleaned from stratigraphic study of pollen grains (which are informative about past vegetation), geochemical measurements on carbonate shells of snails or ostracodes (which reflect changing regional evaporation to precipitation ratios), diatom (siliceous algae) assemblages, and other physical, chemical and biological characteristics of the sediments.

Some lakes located along the upland, central "spine" of the state are relatively deep, with maximum water depths on the order of 20–30 m (67–98 ft). Studies of sediment cores from several such lakes show that some held water through even the driest episodes of the late glacial (Watts 1975, 1980; Watts and Stuiver 1980; Watts et al. 1992). At least one, Lake Tulane in Highlands County, held water continuously on the landscape for some 60 ka (Grimm et al. 1993, 2006). High amounts of *Ambrosia* (ragweed) pollen in the Pleistocene sediments of Lake Tulane indicate that Florida climate at the end of the Last Glacial Maximum was generally much cooler than present, but displayed fluctuations between times of somewhat warmer and wetter conditions as were revealed by high relative abundance of pine pollen and elevated lake levels, and times of relatively colder and drier conditions, dominated by oak pollen and lower lake levels.

As climate in Florida warmed during the "deglacial" and early Holocene (~11.7 ka), there were profound consequences for Florida's terrestrial environment. Vast areas that had served as habitat for Pleistocene land plants and animals, some now extinct (e.g., mammoths, horses, giant sloths, tapirs), were inundated by rising seawater. And as sea level rose, the freshwater aquifers that underlie the region were forced upwards, ultimately causing Florida's famous artesian springs to begin to flow.

Archaeological excavations at Salt Springs near Lake Kerr revealed wood beneath anthropogenic deposits, and a radiocarbon date on that wood suggests spring flow began about 9.45 to 9.25 ka (O'Donoughue et al. 2011). Similarly, at Silver Glen Run on the west side of Lake George a radiocarbon assay on organic deposits that sit atop mineral sediment, and therefore reflect rising water, yielded a date of 8.59 to 8.45 ka (O'Donoughue 2017).

Shortly after the onset of the Holocene Epoch (11.7 ka), rainfall in the region increased. The combination of greater precipitation and rising groundwater tables probably contributed to initial filling of Florida's more than 8,000 shallow lakes, many of which have maximum depths less than 5 m (~<16 ft) (Brenner et al. 1990). Similar to evidence from the springs, filling of the shallow lakes in the state commenced about 9.0 to 6.0 ka. The onset of filling has been established at a number of water bodies by collecting sediment cores and radiocarbon dating the organic material that lies directly atop the clay seal or sand that lines the basin bottoms.

Sediment cores from upland lakes along the Lake Wales Ridge and Trail Ridge (Fig. 15.13) show that early Holocene pollen assemblages were dominated by oaks; but by about 5.0 ka, there was a transition to pine dominance (Watts 1969; 1971; 1980; 1983; Watts and Hansen 1988).

The flora, as we know it today, was largely in place by about 5 ka ago. Since that time, however, there have been some notable shifts, such as the spread of bald cypress (*Taxodium*), beginning a little more than 2.0 ka. The spread of these trees, which typically occupy saturated soils, may reflect a rise of the shallow groundwater table or simply the colonization of new sites as the plants moved across the landscape and occupied appropriate habitats. Geochemical measurements on snail shells in a sediment core from Lake Panasoffkee, north–central Florida, suggest that the climate became substantially wetter in the earliest Holocene, in general agreement with pollen records, with a more gradual increase in moisture thereafter.

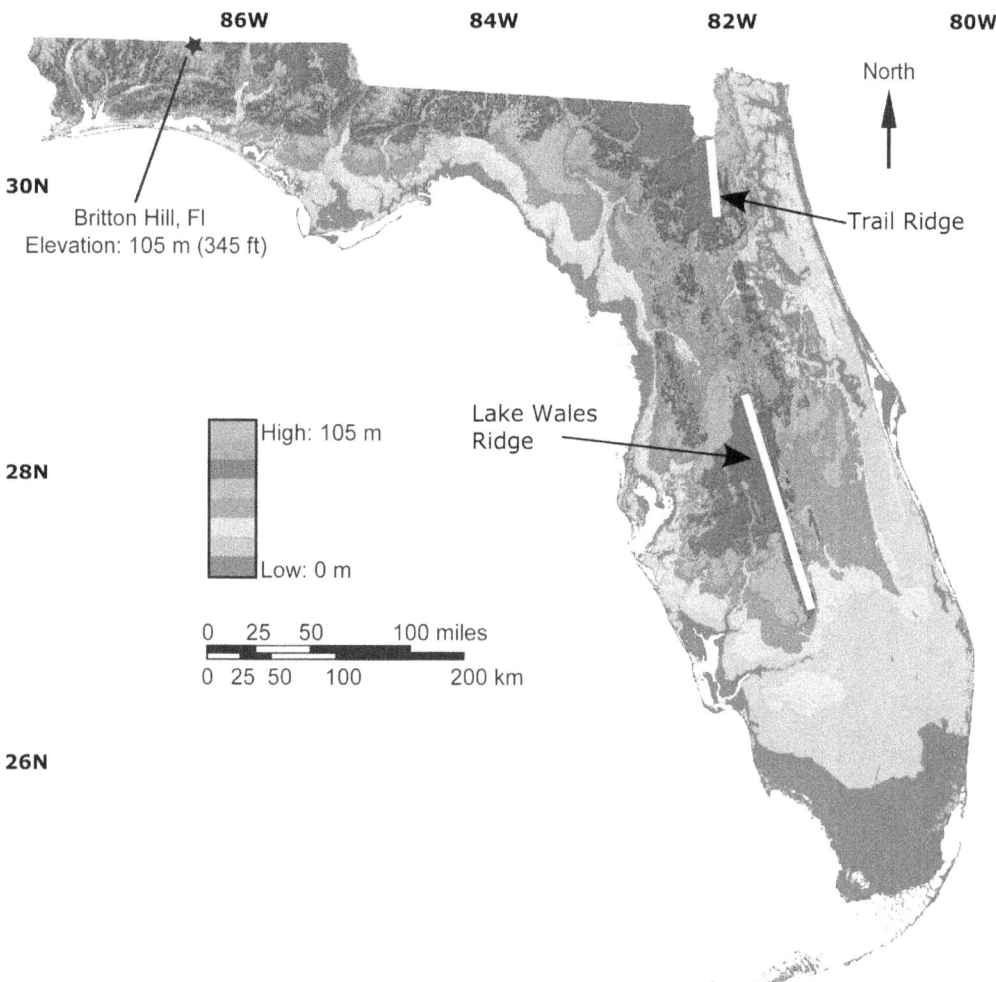

Figure 15.13. Location of Lake Wales Ridge and Trail Ridge (base map from the Florida Geological Survey, Tallahassee, FL; modified from Hine 2013; *Geologic History of Florida: Major Events That Formed the Sunshine State* by Albert C. Hine. Gainesville: University Press of Florida, 2013. Reprinted by permission.).

Most of the long paleoclimate records from Florida come from upland lakes in well-drained terrains. The Newnans and Lochloosa lakes near Gainesville, however, occupy poorly-drained areas. A recent study of sediment cores from these two water bodies found that pine dominated the pollen assemblages over the past >8.0 ka at both sites, and that charcoal concentrations and accumulation rates were high only in the early parts of the records, before about 7.0 ka (Larios 2015). In contrast to records from lakes in the well-drained uplands, the evidence for abundant fires around Newnans and Lochloosa might suggest overall drier conditions in the early

Holocene. Alternatively, the data from the Newnans and Lochloosa cores may indicate greater seasonality in the early Holocene, with wetter wet seasons and drier dry seasons at low latitudes in the Northern Hemisphere. Such a pattern would have enabled a build-up of "fuel" during the rainy season, drying of that organic material during the intense dry season, and ignition by lightning at the onset of each new wet season.

We are able to obtain a "brushstroke" picture of climate change in Florida since the Last Glacial Maximum using lake cores (Fig. 15.14). There were pronounced changes in both temperature and rainfall during the Pleistocene–Holocene transition (11.7 ka), marked by a shift from generally colder and drier to warmer and wetter conditions. Studies of recent shifts in instrumentally-measured rainfall, lake levels, and groundwater tables illustrate how dynamic the state's climate can be, even on short timescales (Deevey 1988). Regional lakes rise and fall synchronously, with little to no lag, in response to fluctuations in monthly rainfall. And longer-term "ups and downs" in lake levels are linked to the height of the deep, Eocene-age Florida aquifer, which exerts pressure on the lakes and surficial water table through thick, overlying, Miocene-to-Pliocene deposits. Droughts lasting just a few years (e.g. those of the middle 1950s, early 1980s, and early 2000s) are characterized by low lake stands and declines in the level of the Floridan aquifer.

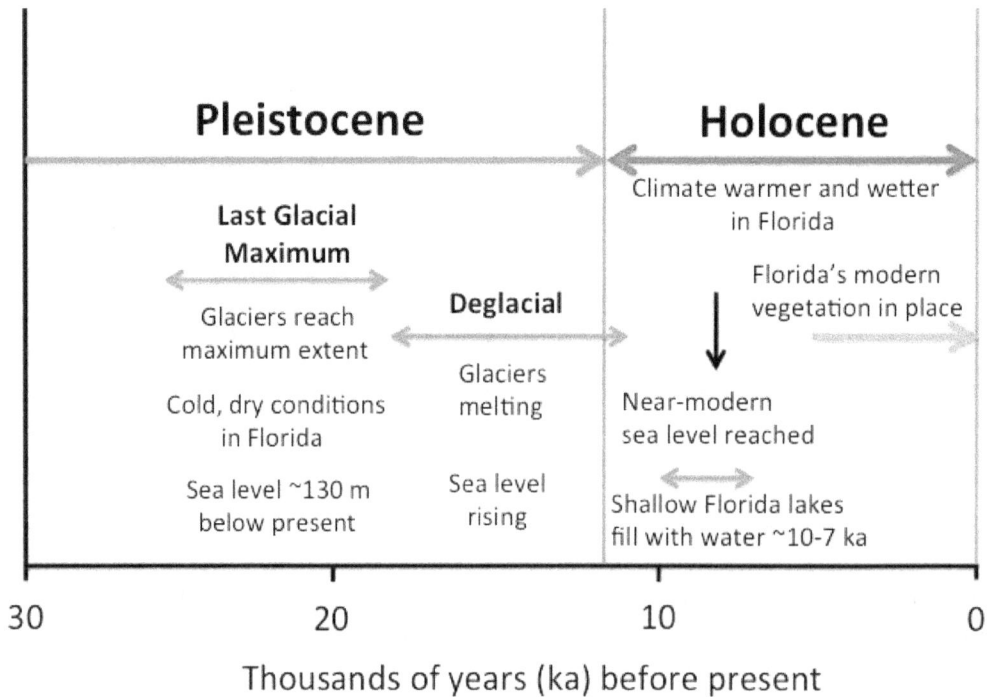

Figure 15.14. Summary of climate change in Florida derived from lake cores.

Modern Day

The early 21st century marked a period of what appears to be extreme climate conditions in Florida, in that we have already seen very low lake levels on two occasions and complete drying of some Florida lakes for the first time in recorded history. Rainfall deficits in the first years of the new millennium led to a pronounced decline in stage at Newnans Lake. More than 100 prehistoric canoes were discovered along the exposed shoreline in 2000, and radiocarbon dates on these vessels spanned a range from about 5000 to 500 years ago (Wheeler et al. 2003). The fact that the discovery was made so recently suggests low lake stands of such magnitude had probably not occurred at any other time in the past 500 years.

Dry conditions were evident into the second decade of the new millennium, and Little Lake Johnson, in Mike Roess Gold Head Branch State Park, north of Keystone Heights, dried completely in 2012 (Fig. 15.15 A, B), also revealing a prehistoric canoe.

Anthropogenic water withdrawals have been invoked by some to explain the lake desiccation, but there is ample evidence that persistent rain deficits were the true culprit. Again, the fact that complete drying was never reported prior to this time suggests that we have moved into a new climate regime.

Figure 15.15 A, B. Little Lake Johnson is a recreational water body in Mike Roess Gold Head Branch State Park, north of Keystone Heights, Florida. In 2012 (A), the lake dried completely, an unprecedented event in historic times and one that may signal a pronounced change in our local climate. By 2014 (B), the lake again held water, but had not returned to previously measured higher levels.

Conclusion

We focused on the past ~50 Myr of geologic time because rocks of this age and younger are accessible near or at the surface on the Florida Platform, thus allowing geoscientists to obtain a more complete understanding of past climate. We tied global climate changes over the past 50

Myr to specific climate responses on the Florida Platform. In general, the climate of the Florida Platform tracked global climatic events.

A key response to past climate change was sea level fluctuation, which covered the continental margins with seawater or exposed them to the atmosphere. Sedimentary rocks, initially of non-marine origin, followed by marine limestone (carbonate), began to record past sea level fluctuations after the Florida Platform was formed during tectonic, mega-continent breakup at ~200 Ma. The Platform has been topographically flat and low, thereby promoting deposition of carbonate sediments derived from corals, shell-bearing organisms such as mollusks, and mud-producing algae in shallow, surrounding seas. During sea level lowstands, the flat, topographically low Florida Platform was exposed to the atmosphere, subjecting terrestrial plants and animals to temperature and humidity variations. For at least 160 Myr, the surrounding oceans created a unique climate on the peninsular Florida Platform. Powerful seasonal thunderstorms developed, particularly when a significant portion of the platform was exposed to the atmosphere.

The history of global climate change over the past 50 Myr is a story of declining atmospheric carbon dioxide (CO_2) and cooling, i.e., a transition from a greenhouse (warm) to an icehouse (cooler with cyclical glaciations and deglaciations) Earth. This global cooling did not occur gradually and continuously through time. Instead, the declining temperature trend was interrupted by intervals of rapid cooling and ice growth, and warmer intervals of ice retreat that are referred to as "climatic optima."

Three global climatic optima occurred during this 50-Myr timeframe, which resulted in distinct geologic responses in Florida. First was the early Eocene Climatic Optimum at ~50 Ma, during which CO_2 was estimated to be ~1,000 ppm, roughly three times pre-industrial concentrations. Warm, shallow seas covered the Florida Platform, depositing widespread carbonate sediments, which are called the Ocala Limestone.

Second, the waxing and waning of the Antarctic Ice Sheet was interrupted at about 16 Ma by dramatic warming and ice melt during a period referred to as the Miocene Climatic Optimum. In Florida, the resulting higher sea level led to the deposition of economically valuable phosphate deposits, emplaced on the west–central portion of the peninsula. These are some of the richest phosphate deposits in the world and form a key component of Florida's Hawthorn Group of rocks and sediments.

Third, the global Pliocene Warm Period around 3 Ma resulted in the migration of river deltas with 100 m (327 ft) relief at least 200 km (125 miles) southward, to positions beneath what are now the modern Florida Everglades and the Florida Keys. These sediments were quartz-rich sands and gravels and formed the Peace River and Long Key formations. These large water- and sediment-laden Pliocene rivers and streams must have been fed by very large amounts of rainfall.

The major growth of ice in the Northern Hemisphere, initiating the icehouse Earth, began ~2.6 Ma when the Earth's climate system started to alternate between colder glacial and warmer interglacial intervals, driven by Milankovitch cycles. Numerous glacial and interglacial periods occurred since that time, and Earth today is in an interglacial interval. Florida climate responded to both sea level lowstands, and highstands.

Florida's Pleistocene terrestrial paleoclimate record has been reconstructed using plant macrofossils and microfossils (e.g., leaves and pollen). Similar to data from the Pliocene, Pleistocene data reveal a terrestrial climate comparable to the modern, with evidence of cool climate episodes that may have been influenced by regional upwelling of cold marine waters. As climate in Florida warmed after the Last Glacial Maximum, which ended at ~18 ka, and in the early Holocene (~11.7 ka), there were profound consequences for Florida's terrestrial environment as much of the Florida Platform was exposed prior to that time. Vast areas that had served as habitat for Pleistocene land plants and animals, some now extinct (e.g., mammoths, horses, giant sloths, tapirs), were inundated by rising seawater.

Shortly after the onset of the Holocene Epoch (11.7 ka), rainfall in the region increased; it was probably the combination of greater precipitation and rising groundwater tables that contributed to initial filling of Florida's more than 8,000 shallow lakes.

The early 21st century has marked a period of what appears to be extreme climate conditions in Florida, in that we have already seen very low lake levels on two occasions and complete drying of some Florida lakes for the first time in recorded history. Such complete drying was never reported previously and suggests that we have entered a new climate regime in this millennium.

Acknowledgements

We thank Tom Missimer, Tom Scott, Bob Hatcher, and Kevin Cunningham for their perspective on various aspects of the lithostratigraphy of the Florida Platform and the tectonic behavior of the southern Appalachian Mountains. We thank Dr. Eric Chassignet for a very helpful early review of the manuscript. Albert Hine thanks Dr. Shane Dunn for his assistance with some of the figures.

References

Beerling, David J. and Royer, Dana L. "Convergent Cenozoic CO2 history." *Nature Geoscience* 4 (2011): 418-420.
Brenner, Mark, Michael W. Binford, and Edward S. Deevey. "Lakes." In *Ecosystems of Florida*, edited by Ron L. Myers and John J. Ewel, 364-391. Orlando: University of Central Florida Press, 1990.
Christie-Blick, N. "Geological Time Conventions and Symbols." *Geological Society of America Today* 22, no. 2 (2012): 28-29. doi: 10.1130/G132GW.1.
Corbett S.L. 2004. The middle Miocene Alum Bluff flora, Liberty County, Florida [M.S. thesis]. [Gainesville (FL)]: University of Florida.
Cunningham, K. J., D. F. McNeill, L. A. Guertin, P. F. Ciesielski, T. M. Scott, and L. de Verteuil. "New Tertiary Stratigraphy for the Florida Keys and Southern Peninsula of Florida." *Geological Society of America Bulletin* 110, no. 2 (1998): 231-58. doi: 10.1130/0016-7606(1998).
Cunningham, K. J., S. D. Locker, A. C. Hine, D. Bukry, J. A. Barron, and L. A. Guertin. "Interplay of Late Cenozoic Siliciclastic Supply and Carbonate Response on the Southeast Florida Platform." *Journal of Sedimentary Research* 73, no. 1 (2003): 31-46. doi: 10.1306/062402730031.
DeBalko, D. A., and R. T. Buffler. "Seismic Stratigraphy and Geologic History of Middle Jurassic through Lower Cretaceous Rocks, Deep Eastern Gulf of Mexico." *Gulf Coast Association of Geological Societies Transactions* 42 (1992): 89-105.
doi: 10.1306/A1ADDD5A-0DFE-11D7-8641000102C1865D.

DeConto, Robert M., and Pollard, David. "Contributions of Antarctic to past and future sea-level rise. *Nature* 531 (2016): 591-597. doi: 10.1038/nature17145.

Deevey, Edward S. Jr. "Estimation of Downward Leakage from Florida Lakes." *Limnology and Oceanography* 33, no. 6 part 1 (1988). 1308-1320.

Emslie, Steven D., W. D. Allmon, F. J. Rich, J. H. Wrenn, and S. D. DeFrance. 1996. Integrated taphonomy of an avian death assemblage in marine sediments from the late Pliocene of Florida. Palaeogeography, Palaeoclimatology, Palaeoecology 124:107–136.

Evans, M. W., and A. C. Hine. "Late Neogene Sequence Stratigraphy of a Carbonate-Siliciclastic Transition: Southwest Florida." *Geological Society of America Bulletin* 103, no. 5 (1991): 679-99. doi: 10.1130/0016-7606(1991)103<0679:lnssoa>2.3.co;2.

Gradstein, F.M., Ogg, J.G., Schmitz, M., Ogg, G., 2012. The Geologic Time Scale 2012. Elsevier, Amsterdam (1176 pp.).

Grimm, Eric C., G. L. Jacobson Jr., W. A. Watts, B. C. S. Hansen, and K. A. Maasch. "A 50,000-Year Record of Climate Oscillations from Florida and Its Temporal Correlation with the Heinrich Events." *Science* 261, no. 5118 (1993). 198-200.

Grimm, Eric .C., W.A. Watts, G. L. Jacobson Jr., B.C.S. Hansen, H. R. Almquist, and Ann C. Dieffenbacher-Krall. "Evidence for Warm Wet Heinrich Events in Florida." *Quaternary Science Reviews* 25, No. 17-18 (2006). 2197–2211.

Guertin, L.A., D.F. McNeill, B.H. Lidz, and K.J. Cunningham. Chronologic Model and Transgressive-Regressive Signatures in the Late Neogene Siliciclastic Foundation Long Key Formation) of the Florida Keys. Journal of Sedimentary Research 69, no. 3 (1999): 653-666.

Hansen, Barbara C.S., Eric C. Grimm, William A. Watts. 2001. Palynology of the Peace Creek Site, Polk County, Florida. Geological Society of America Bulletin 113: 682–692.

Hine, A. C., B. Suthard, S. D. Locker, C. K.J., D. D. Duncan, M. W. Evans, and R. A. Morton. "Karst Subbasins and Their Relation to the Transport of Tertiary Siliciclastic Sediments on the Florida Platform." *International Association of Sedimentologists Special Publication* 41, no. Perspectives in Sedimentary Geology: A Tribute to the Career of Robert N. Ginsburg (2009): 179-97.

Hine, A.C., 2013, Geologic History of Florida; Major Events that Formed the Sunshine State; Gainesville, FL: University Press of Florida, 229p.

Hine, A.C., D. P. Chambers, T.D. Clayton, M.R Hafen, and G. T. Mitchum, 2016, Sea Level Rise in Florida; Science, Impacts, and Options, Gainesville, FL: University Press of Florida, 179p.

IPCC, 2013: *Climate Change 2013: The Physical Science Basis. Contribution of Working Group I to the Fifth Assessment Report of the Intergovernmental Panel on Climate Change* [Stocker, T.F., D. Qin, G.-K. Plattner, M. Tignor, S.K. Allen, J. Boschung, A. Nauels, Y. Xia, V. Bex and P.M. Midgley (eds.)]. Cambridge University Press, Cambridge, United Kingdom and New York, NY, USA, 1535 pp, doi: 10.1017/CBO9781107415324.

Iturralde-Vinent, M.A., The conflicting paleontologic versus stratigraphic record of the formation of the Caribbean Seaway. In Circum-Gulf of Mexico and the Caribbean Hydrocarbon Habitats, Basin Formation and Plate Tectonics, ed. C. Bartolini, R. T. Buffler, and J.T. Bilickwede, 75-88. Tulsa, American Association of Petroleum Geologists, 2003.

Ivany, L. C., R. W. Portell, and D. S. Jones. 1990. Animal-plant relationships and paleobiogeography of an Eocene seagrass community from Florida. Palaios 5:244–258.

Jarzen, D.M. and D.L. Dilcher (2006): Mid-Eocene terrestrial palynmorphs from the Doline Minerals and Gulf Hammock Quarries, Florida; Palynology, 30: 89-110.

Jarzen, D.M., and C. Klug. 2010. A preliminary investigation of a lower to middle Eocene palynoflora from Pine Island, Florida, USA. Palynology Vol. 34 (2): 164-179.

Klitgord, K. D., P. Popenoe, and H. Schouten. "Florida: A Jurassic Transform Plate Boundary." *Journal of Geophysical Research* 89, no. B9 (1984): 7753-72. doi: 10.1029/JB089iB09p07753.

Lambeck, K., H. Rouby, A. Purcell, Y. Sun, M. Sambridge, 2014, Sea level and global ice volumes from the Last Glacial Maximum to the Holocene, PNAS, v. 14, no. 43, p. 15296-15303; www.pnas.org/cgi/doi/10.1073/pnas.1411762111.

Larios, Kalindhi, "Florida Wildfires During the Holocene Climatic Optimum (9,000-5,000 cal yr BP)" MS Thesis, University of Florida, Gainesville, 2015.

Miller, K.G., Kominz, M.A., Browning, J.V., Wright, J.D., Mountain, G.S., Katz, M.E., Sugarman, P.J., Cramer, B.S., Christie-Blick, N., Pekar, S.F., 2005. The Phanerozoic record of global sea-level change. Science, 310, 1293-1298. doi: 10.1126/science.116412.

Missimer, T. M., and R. A. Gardner. "High-Resolution Seismic Reflection Profiling for Mapping Shallow Water Aquifers in Lee County, Fl." In *Water-Resources Investigations Report* 30. Tallahassee: U.S. Geological Survey, 1976.

Missimer, T. M. "Sequence Stratigraphy of the Late Miocene—Early Pliocene Peace River Formation, Southwestern Florida." *Gulf Coast Association of Geological Societies Transactions* 49 (1999): 358-68. doi: 10.1306/2DC40C53-0E47-11D7-8643000102C1865D.

Müller, R.D., M. Sdrolias, C. Gaina, B. Steinberger, and C. Heine, Long-term sea-level fluctuations driven by ocean basin dynamics. *Science* 319 (2008): 1356-1362.

Multer, H.G., E. Gischler, J.Lundberg, K.R. Simmons, and E.A. Shinn. "Key Largo Limestone Revisited: Pleistocene Shelf Edge Facies, Florida Keys, USA". *Facies* 46, no. 1 (2002): 229-72 (doi: 10.1007/BF02668083).

O'Donoughue, J.M. 2017. Water from stone: archeology and conservation at Florida's springs. University of Florida Press, Gainesville. 245 p.

O'Donoughue, J.M., K.E. Sassaman, M.E. Blessing, J. B. Talcott, and J.C. Byrd. 2011. Archaeological investigations at Salt Springs (8MR2322), Marion County, Florida. Technical Report 11, Laboratory of Southeastern Archaeology, Department of Anthropology, University of Florida Gainesville. 151 p.

Pindell, J. L., and S. F. Barrett. Geological Evolution of the Caribbean Region; a Plate Tectonic Perspective. In *The Caribbean Region*, edited by G. Dengo and J. E. Case. The Geology of North Florida, 405-32. Boulder, CO: The Geological Society of America, 1990.

Popenoe, P. "Paleoceonography and Paleogeography of the Miocene of the Southeastern United States." In *Phosphate Deposits of the World*, edited by W. C. Burnett and S. R. Riggs. Cambridge Earth Science Series, 352-80. New York: Cambridge University Press, 1990.

Randazzo, A. F., and D. S. Jones, eds. *The Geology of Florida*. Gainesville, FL: University Press of Florida, 1997.

Redfern, R. *Origins: The Evolution of Continents, Oceans and Life.* Norman, OK: University of Oklahoma Press by special arrangement with Cassell and Co, UK, 2001.

Rich, F. J. 1985. Palynology and paleoecology of a lignitic peat from Trail Ridge, Florida. Florida Bureau of Geology Information Circular 100.

Rich, F. J., and L. A. Newsom. 1995. Preliminary palynological and macrobotanical report for the Leisey Shell Pits, Hillsborough County, Florida. Bulletin of the Florida Museum of Natural History 37 Pt. I(4):117-126.

Riggs, S.R.Phosphorite Sedimentation in Florida; a Model Phosphogenic System. *Economic Geology* 74, no. 2 (1979): 285-314. doi: 10.2113/gsecongeo.74.2.285.

Riggs, S. R. "Paleoceanographic Model of Neogene Phosphorite Deposition, U.S. Atlantic Continental Margin." *Science* 223, no. 4632 (1984): 123-31. doi: 10.1126/science.223.4632.123.

Riggs, S.R., S. Snyder, D. Ames, and P. Stille. Chronostratigraphy of the Upper Cenozoic Phosphorites on the North Carolina Continental Margin and the Oceanographic Implications for Phosphogenesis. Marine Authigenesis: From Global to Microbial 66, SEPM Special Publication (2000), 369-385. doi: 10.2110/pec.00.66.0369.

Ruddiman, W. F. *Earth's Climate: Past and Future.* 2nd ed. New York: W.H. Freeman and Company, 2008, 388p.

Sassaman, K.E., and 13 others, Keeping pace with rising sea: The first six years of the lower Suwannee archeological survey, Gulf coastal Florida: Journal of Island and Coastal Archeology.

Sheridan, R. E., H. T. Mullins, J. A. Austin, M. M. Ball Jr., and J. W. Ladd. "Geology and Geophysics of the Bahamas." In *The Atlantic Continental Margin: U.S.*, edited by R. E. Sheridan and J. A. Grow. Geology of North America, 329-64. Boulder, CO: The Geological Society of America, 1988.

Walker, J.D., J. W. Geissman, S. A. Browning, and L. E. Babcock, 2013, The Geological Society of America Time Scale: Geological Society of America Bulletin 125, (3/4) p. 259-274.

Warzeski, E. R., K. J. Cunningham, R. N. Ginsburg, J. B. Anderson, and Z.-D. Ding. "A Neogene Mixed Siliciclastic and Carbonate Foundation for the Quaternary Carbonate Shelf, Florida Keys." *Journal of Sedimentary Research* 66, no. 4 (1996): 788-800. doi: 10.1306/d426840a-2b26-11d7-8648000102c1865d.

Watts, W. A., A Pollen Diagram from Mud Lake, Marion County, North Central Florida. *Geological Society of America Bulletin* 80, no. 4 (1969). 631-642.

Watts, W. A., Postglacial and Interglacial Vegetation History of Southern Georgia and Central Florida. *Ecology* 52, no. 4 (1971). 676-690.

Watts, W. A., A Late Quaternary Record of Vegetation from Lake Annie, South-Central Florida. *Geology* 3, June (1975). 344-346.

Watts, W. A. The Late Quaternary Vegetation History of the Southeastern United States. *Annual Review of Ecology and Systematics* 11 (1980). 387-409.

Watts, W. A., Vegetational History of the Eastern United States 25,000 to 10,000 Years Ago. In *Late Quaternary Environments of the United States*. Volume 1, edited by Herbert E. Wright, 294-310. Minneapolis: University of Minnesota Press, 1983.

Watts, W.A. and M. Stuiver. 1980. Late Wisconsin climate of northern Florida and the origin of species-rich deciduous forest. Science 210: 325-327.

Watts, W. A., and B.C.S. Hansen, Environments of Florida in the Late Wisconsin and Holocene. In *Wetsite Archaeology*, edited by Barbara A. Purdy, 307-323. Caldwell: The Telford Press, 1988.

Watts, W.A., B. C.S. Hansen, and E. C. Grimm. Camel Lake: A 40,000-Yr Record of Vegetational and Forest History from Northwest Florida. *Ecology* 73, no. 3 (1992). 1056-1066.

Wheeler, Ryan J., James J. Miller, Ray M. McGee, Donna Ruhl, Brenda Swann, and Melissa Memory. Archaic Period Canoes from Newnans Lake, Florida. *American Antiquity* 68, no. 3 (2003).

Willard, D.A., Cronin, T.M. Ishman, S.E., and Litwin, R.J., 1993, Terrestrial and marine records of climatic and environmental changes during the Pliocene in subtropical Florida: Geology, 21: 679-682.

Wright, E.E., Hine, A.C., Goodbred, S.L., and Locker, S.D., 2005, The effect of sea-level and climate change on the development of a mixed siliciclastic –carbonate deltaic coastline; Suwannee River, FL; Journal of Sedimentary Research, v. 75. p. 621-635; DOI: 10.2110/jsr.2005.051

Zachos, James, Pagani, Mark, Sloan, Lisa, Thomas, Ellen, Billups, Katharina. "Trends, rhythms, and aberrations in global climate." *Science* 292 (2001): 686-693.

Zachos, J. C., G. R. Dickens, and R. E. Zeebee, An early Cenozoic perspective on greenhouse warming and carbon-cycle dynamics. *Nature* 451 (2009): 279-283. doi:10.1038/nature06588.

CHAPTER 16

Terrestrial and Ocean Climate of the 20th Century

Vasubandhu Misra[1,2,3], Christopher Selman[4], Amanda J. Waite[5,6], Satish Bastola[7], and Akhilesh Mishra[3,8]

[1]*Department of Earth, Ocean and Atmospheric Science, Florida State University, Tallahassee, FL;* [2]*Florida Climate Institute, Florida State University, Tallahassee, FL;* [3]*Center for Ocean-Atmospheric Prediction Studies, Florida State University, Tallahassee, FL;* [4]*Center for Ocean-Land-Atmosphere Studies, George Mason University, Fairfax, VA;* [5]*Department of Geological Sciences, University of Florida, Gainesville, FL;* [6]*ANGARI Foundation, West Palm Beach, Florida;* [7]*School of Civil and Environmental Engineering, Georgia Institute of Technology, Atlanta, GA;* [8]*Amity Center for Ocean-Atmosphere Science and Technology, Amity University, Rajasthan, Jaipur, India*

The Florida peninsula, with its close proximity to the equator surrounded by robust surface and deep water ocean currents, has a unique climate. Generally, its climate is mild with variations on numerous time scales, punctuated by periodic extreme weather events. In this chapter, we review the mechanisms by which some well-known natural variations impact the regional climate and modulate the occurrence of extreme weather over Florida and its neighboring oceans. In addition, we explore the role of land cover and land use changes on the regional climate over the same area. It is made apparent from the review that remote variations of climate have an equally important impact on the regional climate of Florida as the local changes to land cover and land use.

Key Messages

- Florida is a unique region to the east of the Rocky Mountains with a very distinct monsoonal type of wet season in the summer that distinguishes it from the rest of the seasons.
- Florida's climate is as much affected by remote climate variations as local variability over land and its neighboring water bodies. Florida's climate is affected by more global scale natural variations like ENSO, AMO, PDO. Similarly, there is a discernible impact of local land cover and land use change on surface temperatures in Florida.
- There are important interactions of the observed climate across time and spatial scales to consider. For example, the sea breeze over the Florida Panhandle is shown to be affected by the subtle variations of the Bermuda High. Similarly, ENSO forcing on Florida's winter climate is affected by decadal variations such as the PDO and the AMO.

Keywords

Seasonal cycle; Diurnal variations; Sea breeze; ENSO; Tropical cyclones; Hurricanes; AWP; AMO; PDO; PIZA

Introduction

The atypical peninsular geography and relatively close proximity to the equator give Florida, a land of flowers in a latitude of deserts, a unique and desirable climate. This is a leading factor for the state's growing population and tourism. Florida has a relatively mild climate throughout the year, significant freshwater resources that are replenished naturally by seasonal rains, and picturesque coastlines that have led to rapid growth in human settlement and urbanization along the coast. However, amidst its rather serene climate, Florida often experiences a considerable number of weather and climate anomalies that pose a threat to the region and those living in it. This chapter will focus on this variability around the mean climate in the region, where variations span a range of time scales, from days to decades to centuries. Our understanding of these variations is largely limited by the availability of reliable observations of certain critical meteorological and oceanographic variables. We have, therefore, restricted discussions in this chapter to variations that could be reasonably resolved in the past 100 years. However, it is a challenge to isolate natural variability of climate when artificially-introduced variations from changes in location of observation, instrumentation, and method of measurement are not accounted for (Misra and Michael 2012). In addition, complex interactions in natural climate variation across space and time, as well as anthropogenic climate change, make understanding the manifestation of specific climate anomalies and weather events difficult. Some of these issues will be closely examined in this chapter.

The Seasonal Cycle

The distinct seasonality of Florida manifests in its hydroclimate, with a clear and apparent rainy season coinciding with the June-July-August (JJA) season (Fig. 16.1). In fact, Fig. 16.1 suggests that by volume of rain (mean seasonal rain multiplied by the surface area over which it falls), the state of Florida receives the highest in the Continental United States (CONUS). Much of this rainfall seasonality is sustained by diurnal variations (Fig. 16.2; wherein rain maximizes in certain times of the day). This diurnal variability is typical of marine environments in lower latitudes, where day and night temperature differences between land and ocean cause sufficient density variations in the atmospheric layer to drive an atmospheric circulation (e.g., sea breeze) that would spawn thunderstorm activity during certain times of the day. But more fundamentally, with a warmer atmosphere in the summer season there is a higher moisture-holding capacity that gives rise to more diurnal variations in precipitation if there is a sufficient moisture supply to the atmospheric column to condense. This is partly because of the Clausius-Clayperon equation, which relates the saturation vapor pressure (a measure of the maximum moisture-holding capacity in a given volume of air to temperature) and implies water-holding capacity of the atmosphere increases by about 7% for every Kelvin rise in temperature (Trenberth et al. 2003).

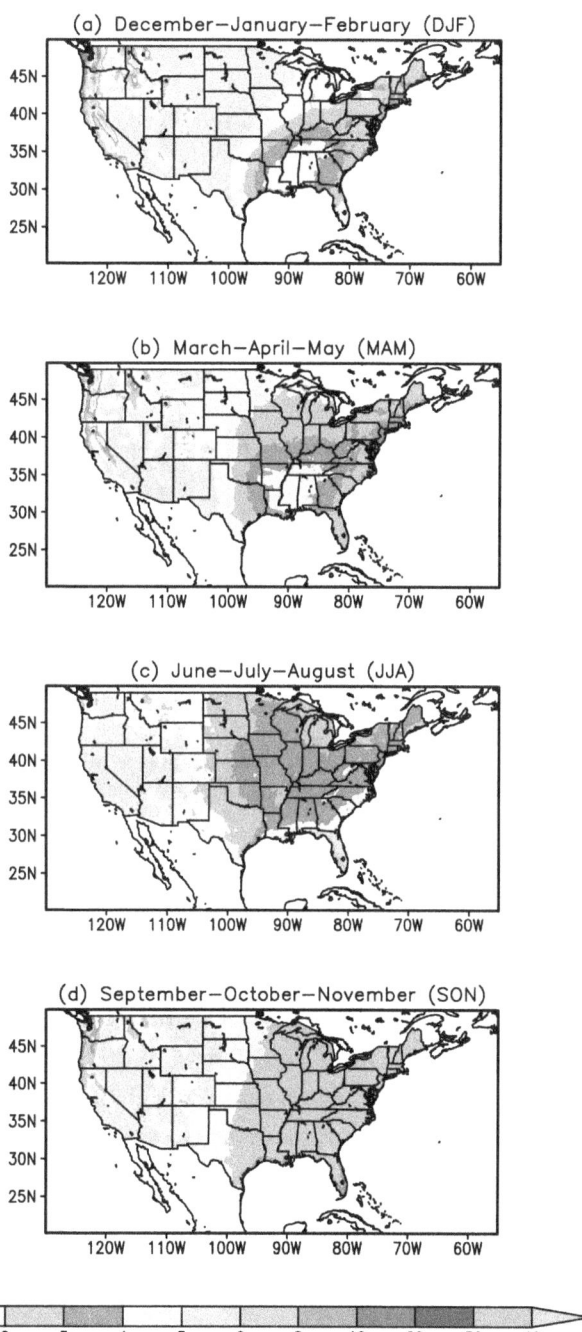

Figure 16.1. Climatological seasonal mean precipitation over the Continental United States (CONUS) for a) December-January-February (DJF), b) March-April-May (MAM), c) June-July-August (JJA), and d) September-October-November (SON) seasons from the National Centers for Environmental Prediction (NCEP) Climate Prediction Center (Higgins et al. 2000). The units are in mm/day.

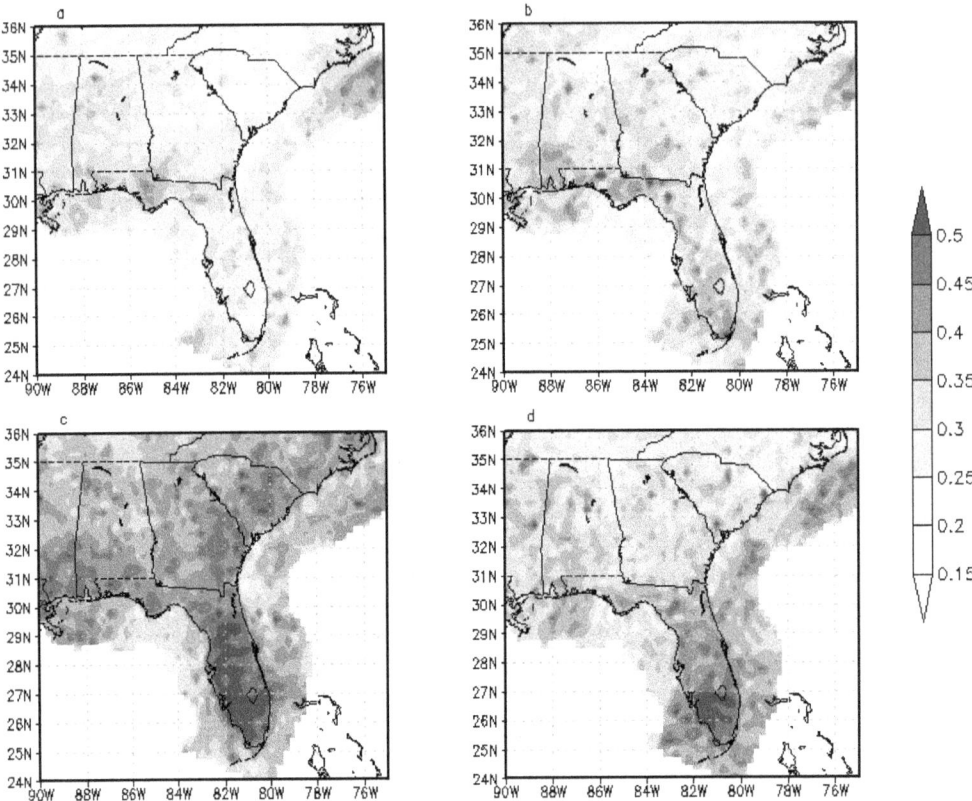

Figure 16.2. The fraction of diurnal variability that explains the total seasonal variability in a) DJF, b) MAM, c) JJA, and d) SON seasons from NCEP Stage IV hourly data (Lin and Mitchell, 2005). From Bastola and Misra (2013).

Seasonality in the temporal correlations between precipitation and surface temperature over Florida reveal that they change from positive correlations in the winter to negative correlations in the summer (Trenberth and Shea 2005; Misra and Dirmeyer 2009). In the winter time, surface evaporation that leads to cooling of the surface temperature is not an important contributor of moisture for local precipitation (Misra and Dirmeyer 2009). Therefore, positive correlations between surface precipitation and temperature are a likely indicator of warm moist advection ahead of the cold front that favors precipitation and warms the surface (Trenberth and Shea 2005). However, in the summer season, surface evaporation becomes an important source of moisture for local precipitation that leads to negative correlations with precipitation, as evaporation tends to cool the surface temperature.

Some of this seasonality in the hydroclimate of Florida is also caused by the seasonal vacillation of the North Atlantic Subtropical High (NASH; Davies et al. 1996; Misra et al. 2011; Li et al. 2013). The NASH, also known as the Azores or Bermuda High, displays a distinct high pressure anticyclone in the lower troposphere during the summer over the subtropical Atlantic basin that changes to two distinct high pressure systems over eastern North America and northwestern Africa by the winter season. These seasonal migrations of the NASH modulate the large-scale moisture advection (Li et al. 2013; Chan and Misra 2010), low-level divergence (Misra et al., 2011), and atmospheric stability (Selman et al. 2013), which impacts seasonal precipitation over Florida.

Diurnal Variability

In this section we delve further into the drivers of diurnal variability of precipitation and temperature in Florida, and discuss their delicate interplay. In order to analyze these variables, we extracted the diurnal harmonic from the NCEP's STAGE-IV radar-based precipitation data (STAGEIV; Lin and Mitchell 2005) and surface temperature from the North American Land Data Assimilation System-1 (NLDAS-1; Cosgrove et al. 2003).

We see (Fig. 16.3) that in the winter and fall months the amplitude of diurnal variability of precipitation is much lower than in the spring and summer months. In the winter and fall seasons, it is the passage of transient (synoptic) frontal systems that dictates precipitation variability. This is also reflected by the spatially varied timing of diurnal maximum precipitation in these seasons. However, as spring and summer set in, the timing of diurnal maximum precipitation becomes quite uniform, centering around 2000 UTC. This is, in fact, slightly offset from the timing of the diurnal maximum temperature, depicted as the black vectors in Fig. 16.4.

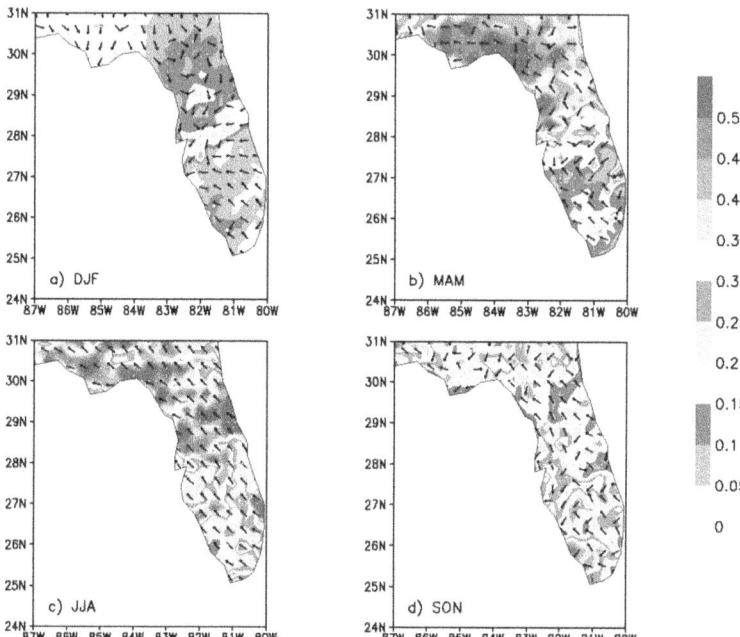

Figure 16.3. The seasonal average of diurnal amplitude (mm/day) and phase or timing (overlaid vectors) of precipitation (shaded) from NCEP STAGEIV (Len and Mitchell 2005). Vectors indicate the UTC time of maximum diurnal precipitation and are arranged as if on a 24-hour clock face (e.g., vectors pointing north represent 0000 UTC, east 0600 UTC, south 1200 UTC and west 1800 UTC).

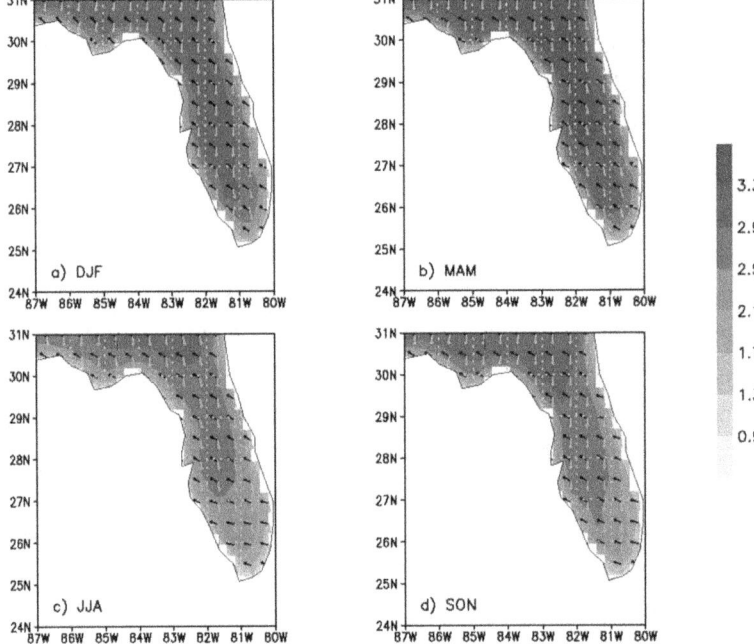

Figure 16.4. Same as Fig. 16.3 but for surface temperature from the North American Land Data Assimilation System-1 (NLDAS-1; Cosgrove et al. 2003). The seasonal average of diurnal amplitude (shaded) and the phase or timing (overlaid vectors) of T_{max} and T_{min} in black and pink vectors respectively.

Shown in Fig. 16.4 is the diurnal amplitude of surface temperatures with the timing of maximum temperature (black) and minimum temperature (pink) overlaid. In general, we find that the diurnal amplitude of daily temperatures is quite uniform in Florida — generally on the order of 2 °C in the interior of the state and 1 °C near the coastlines. The relatively smaller amplitude of diurnal variations of surface temperature in South Florida compared to the rest of the state during the summer season is a result of the cooling induced by precipitation from persistent sea breeze circulations (Case et al. 2005). Likewise, the larger diurnal amplitude of surface temperature near the coasts in the winter months relative to other seasons is a result of the absence of sea breeze circulations.

One of the most influential features of Florida's delicate ecosystem is related to the presence of daytime sea breezes and nocturnal land breezes that most commonly occur in the summer season. This relatively shallow atmospheric circulation is a direct consequence of the differential heating of air over land and ocean, which builds up a strong temperature contrast especially in the summer season. This differential heating between land and ocean is a result of the differential in the heat capacity (the amount of heat energy required to raise the temperature by a °C per unit mass) of land, which is typically far less than that of ocean. This causes a horizontal pressure gradient leading to the so-called sea breeze circulation. Cloud and thunderstorm development usually occurs in the ascending part of the circulation and typically matures in the afternoon after the downwelling shortwave flux results in maximum heating of land. At nighttime, this circulation is reversed, establishing the land breeze, again owing to the differential rate of cooling between land and ocean that leads to a thermal contrast opposite to that of sea breeze. Numerous sub-regional features, such as local land use (Pielke et al. 1999), coastal orientation (Baker et al. 2001), and even changes in the local maritime environment (Van der Molen et al. 2006) dictate the inland propagation of the sea breeze.

There is usually a convergence of a double sea breeze front over peninsular Florida, one translating westward from the Atlantic Coast and the other moving eastward from the Gulf Coast (Byers and Rodebush 1948; Blanchard and Lopez 1985; Gibson and Vonder Haar 1990). The sea breeze circulation, being relatively shallow in depth, is sensitive to the prevailing wind speed and direction (Nicholls et al. 1991). For example, Misra et al. (2011) showed that the sea breeze over the Florida Panhandle is associated with the variability of the Bermuda High pressure system. They showed that by way of the Sverdrup vorticity balance, the large-scale meridional flow of the high pressure system along the Florida Panhandle will cause modulation of the large-scale divergence in the region, which then modulates the strength of the late afternoon sea breeze. In seasons when the Bermuda High is stronger, it promotes stronger upper level divergence over the Florida Panhandle coast (Fig. 16.5a), which makes the sea breeze along this coast stronger than usual. Similarly, in summer seasons when the Bermuda High is weaker and retracted further east, the large-scale conditions become less favorable for sea breeze along the Florida Panhandle (Fig. 16.5b).

ENSO Variability

El Niño-Southern Oscillation (ENSO) variations in the equatorial Pacific have a global impact on seasonal anomalies and extreme weather through atmospheric teleconnections. This remote teleconnectivity of ENSO variations is a result of atmospheric (stationary Rossby) waves emanating from the anomalous release of diabatic heating with associated changes in large-scale atmospheric circulation and upper ocean heat content redistribution in the equatorial Pacific. One such teleconnection that affects the winter and early spring climate of the Southeastern US (SEUS), including Florida, is the shift in the subtropical jet stream related to upper air geopotential height in response to the shift in the atmospheric convection from the western to central equatorial Pacific (Fig. 16.6). This 200 hPa height pattern (Fig. 16.6) consists of alternating high and low pressure centers along a great circle route featuring stronger (weaker) Aleutian low, high (low) pressure over Canada, and low (high) pressure over the SEUS. Typically, this results in cold and wet (warm and dry) winters in warm (cold) eastern equatorial Pacific (ENSO) years over the SEUS (Ropelewski and Halpert 1986, 1987). The episodic weather events during these anomalous winters can be inferred from such upper air circulation patterns (Fig. 16.6). For example, the warm ENSO associated with wet and cold winters in the

SEUS are related to increased frequency of frontal and cyclone activity steered along the southern part of the US by the zonally-oriented and equatorward displaced subtropical jet stream (Eichler and Higgins 2006). In contrast, during La Niña winters the subtropical jet stream is more meridionally-oriented and shifted poleward, bringing a wetter winter to Canada (Smith et al. 1998).

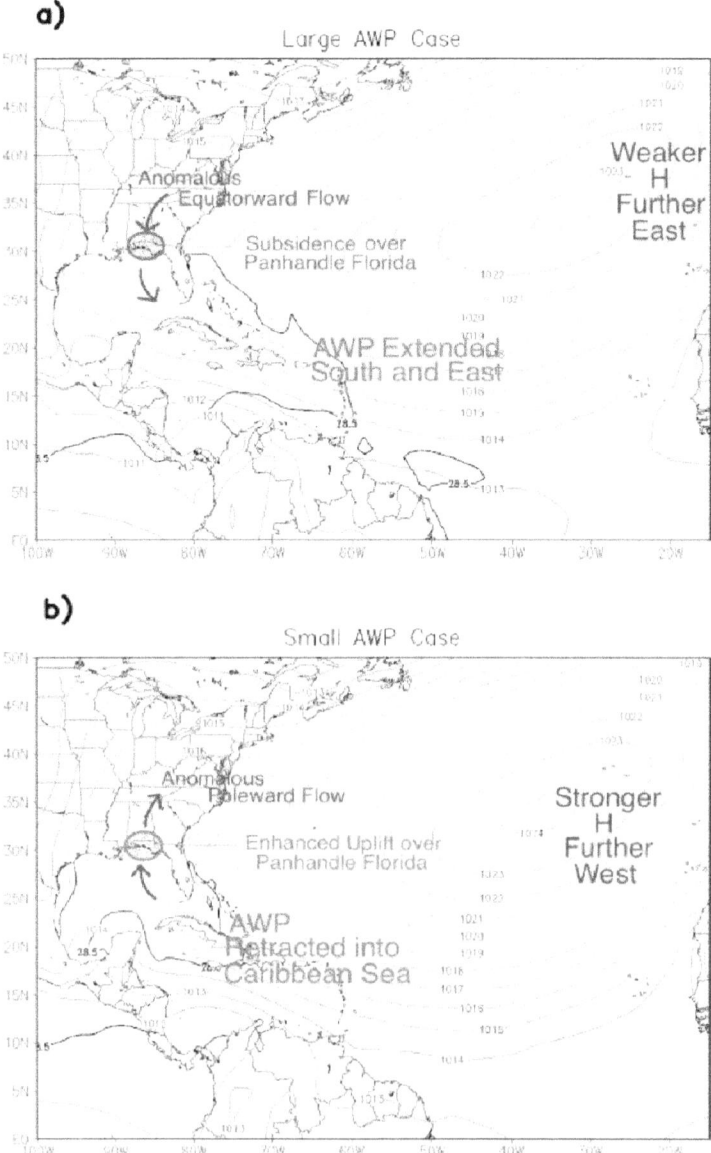

Figure 16.5. Schematic of the anomalous conditions over the Florida Panhandle generated by the modulation of the North Atlantic Subtropical High (NASH) in the a) large and b) small Atlantic Warm Pool (AWP) years. The composite mean sea level pressure (hPa) from NCEP-R2 reanalysis (Kanamitsu et al. 2002) is contoured for the five a) large and the b) small AWP years. From Misra et al. (2011).

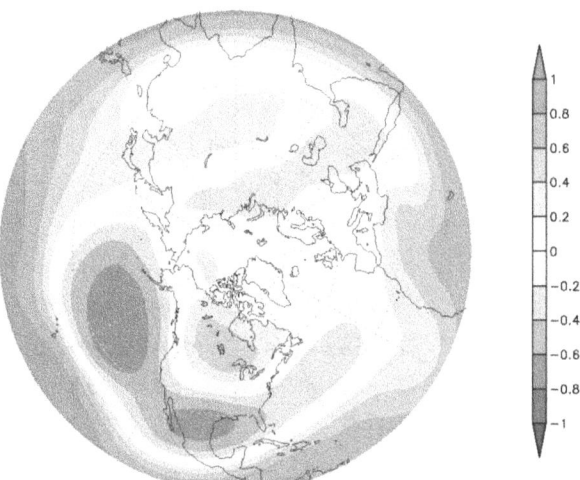

Figure 16.6. The contemporaneous correlation of the mean December-January-February Niño3.4 SST index with corresponding seasonal mean 500 hPa geopotential heights from NCEP-DOE reanalysis (Kanamitsu et al. 2002).

The winter weather in the SEUS, especially the surface temperatures, are also strongly influenced by other variations such as the Arctic Oscillation (AO; Thompson and Wallace 1998), the North Atlantic Oscillation (NAO; Stephenson et al. 2003) and the Pacific North American pattern (Higgins et al. 2002; Hagemeyer 2006). These variations can occasionally mask the ENSO influence. For example, the winter of 2009-2010 was an El Niño event in the equatorial Pacific that set record low temperatures across Florida. This was a result of both the canonical El Niño forcing from the Pacific and a very strong negative NAO. However, the winter of 2010-2011, despite being a La Niña event in the equatorial Pacific, continued to be dominated by the negative NAO influence resulting in very cold surface temperatures in the SEUS, contrary to the ENSO forcing. These extremely cold temperature episodes during the winter in Florida are devastating to the local agriculture, especially citrus (Miller 1991; Attaway 1997). In a related study, Stefanova et al. (2013) showed that both ENSO and the AO have a significant influence on the skewness and kurtosis of the surface temperature in the SEUS. The authors indicated that negative skewness of the surface temperature over Florida is exacerbated in La Niña or positive AO winters. But the converse of the El Niño or the negative AO forcing on the reduction of negative skewness of surface temperature is found to be less impactful.

However, it is important to recognize that the influence of ENSO variations are not restricted to the boreal winter and spring seasons. They are also found to influence the seasonal Atlantic tropical cyclone activity during the boreal summer and fall seasons (Gray 1984; Shapiro 1987). It is widely recognized that tropical cyclogenesis and development occur when these necessary conditions are satisfied: a) warm ocean waters (≥ 26 °C), b) an atmosphere that is potentially unstable for convection to occur, c) relatively moist lower troposphere, d) a minimum distance of about 500 km away from the equator for Coriolis force to be effective, and e) relatively low values of vertical shear between 850 and 200 hPa. That said, these conditions alone are not sufficient to lead to tropical cyclogenesis, as many disturbances that occur in such favorable conditions do not develop into tropical cyclones (Velasco and Fritsch 1987; Emanuel 1993). The Atlantic Basin feels the impact of ENSO remotely through changes to its vertical shear in the so-called main development region (MDR) of Atlantic tropical cyclones (10 °N to 20 °N and from northwest Africa to Central America; Gray 1984; Shapiro 1987; Goldenberg and Shapiro 1996). During El Niño (La Niña) years, the vertical shear increases (decreases) in the MDR owing primarily to an increase (decrease) in the upper level westerlies (Landsea 2000). In a related observational study, Bove et al. (1998) indicated the probability of two or more landfalling hurricanes in the US to be 28%, 48%, and 66% during El Niño, neutral, and La Niña years, respectively (Fig. 16.7a).

Atlantic tropical cyclones are critical to the hydroclimate of Florida (Knight and Davis 2009; Maxwell et al. 2012, 2013; Prat and Nelson 2013 a, b). Knight and Davis (2009) ascertained from observations that peninsular Florida, especially south peninsular Florida, has the highest density of landfalling tropical cyclones in the continental US besides the Carolina coasts. Using the Tropical Rainfall Measuring Mission (TRMM) 3B42 rainfall analysis product, Prat and Nelson (2013a, b) found that tropical cyclones contributed 10-15% of the annual rainfall in Florida. More importantly, they found there is an increase in the probability of landfalling hurricanes (especially category 1 and 2) in Florida during neutral and cold ENSO years relative to warm ENSO years (Fig. 16.7b).

There is, however, a temporal variability of the ENSO teleconnection as a result of other low frequency variations such as the Pacific Decadal Oscillation (PDO; Manatua et al. 1997), the Atlantic Multi-decadal Oscillation (AMO; Kerr 2000), and the non-stationarity of ENSO itself (Trenberth and Shea 1987; Allan et al. 1996). Since the strength of these atmospheric teleconnections are linearly dependent on the equatorial Pacific SST anomalies (Kumar and Hoerling 1998), it is reasonable to expect these teleconnections to be modulated by variations in the amplitude of ENSO SST anomalies (Diaz et al. 2001). Using observations, several studies have shown that the positive (negative) phase of AMO weakens (enhances) the ENSO teleconnection with the SEUS rainfall (Enfield et al. 2001; Mo 2010; Misra et al. 2012; Nag et al. 2015). Similarly, Diaz et al. (2001) indicated ENSO teleconnections with North America have intensified post-1976. Several things might explain such a change. The first is that the character of ENSO variations changed in the mid-1970s due to the stochasticity of the coupled ocean-atmosphere system of the equatorial Pacific; ENSO began occurring later in spring and extending far less to the west from the eastern equatorial Pacific in comparison to earlier decades (Diaz et al. 2001). Second, the PDO SST variations in the North Pacific could be influencing the ENSO teleconnection, with cold (warm) SST anomalies in the North Pacific enhancing (weakening) the teleconnection (Gershunov and Barnett 1998). Third, the multi-decadal variation of the equatorial Pacific SST could also result in such modulation of the teleconnection (Diaz et al. 2001).

In a related study, Mo (2010) found contrasting ENSO teleconnections for the early period of 1915-1960 and the more recent period of 1962-2006. The study further finds that in recent decades the ENSO teleconnection in the SEUS has become weaker and attributed this to two different kinds of ENSO identified by their characteristic evolution and manifestation of SST anomalies (Ashok et al. 2007; Yu and Kao 2007). These two types of ENSO are: the central Pacific ENSO and the eastern Pacific ENSO. The central (eastern) Pacific ENSO has a maximum SST anomaly in the central (eastern) Pacific, which Mo (2010) determined to have different teleconnection patterns with surface meteorology over North America. The frequency of central Pacific ENSOs has increased in recent decades, which produces a wave train that is consistent with the west-east contrast in surface temperature, as opposed to the wave train created by eastern Pacific ENSOs in past decades that produced north-south contrast over North America (Mo 2010).

Atlantic Warm Pool Variability

The Atlantic Warm Pool (AWP) is a seasonal feature that is defined by the appearance of SSTs warmer than or equal to 28.5 °C in the Gulf of Mexico, Caribbean Sea, and the parts of the subtropical northwestern Atlantic Ocean (Wang and Enfield 2001). The AWP constitutes the dominant part of the Western Hemisphere Warm Pool (WHWP), which is preceded by a similar appearance of warm SSTs $\geq 28.5°C$ in the northeast tropical Pacific Ocean. However, Misra et al. (2016) found that the AWP variations are unrelated to variations of the warm pool in the tropical northeast Pacific. The AWP has a seasonal peak in terms of its areal coverage in the August-September-October season that coincides with the seasonal peak of the Atlantic tropical cyclone activity (Wang et al. 2007). Surface heat budget studies of AWP have indicated that the radiative fluxes dominate in the Gulf of Mexico, while in the Caribbean Sea upwelling and advective cooling also play a significant role in regulating SSTs (Lee et al. 2007; Misra et al. 2013).

The seasonal peak of the AWP-induced heating forces a Gill-type atmospheric response (to off-equator diabatic heating) in the form of extratropical stationary waves (Lee et al. 2007). These waves produce rainfall variability over the continental US (Fig. 16.8a), while modulating the subtropical highs in the North Atlantic Ocean (Fig. 16.8b) and in the southeastern Pacific (cf Fig. 8 in Wang et al. 2010). In other words, Fig. 16.8a suggests that in large (small) AWP years the June-October seasonal rainfall over the Mississippi and Ohio valleys is reduced (increased) accompanied by a reduced (increased) strength in the NASH (Fig. 16.8b). Furthermore, AWP variations and their teleconnections to North American hydroclimate are also observed to be largely independent of the ENSO variations in the equatorial Pacific (Wang et al. 2006, 2008; Misra et al. 2013).

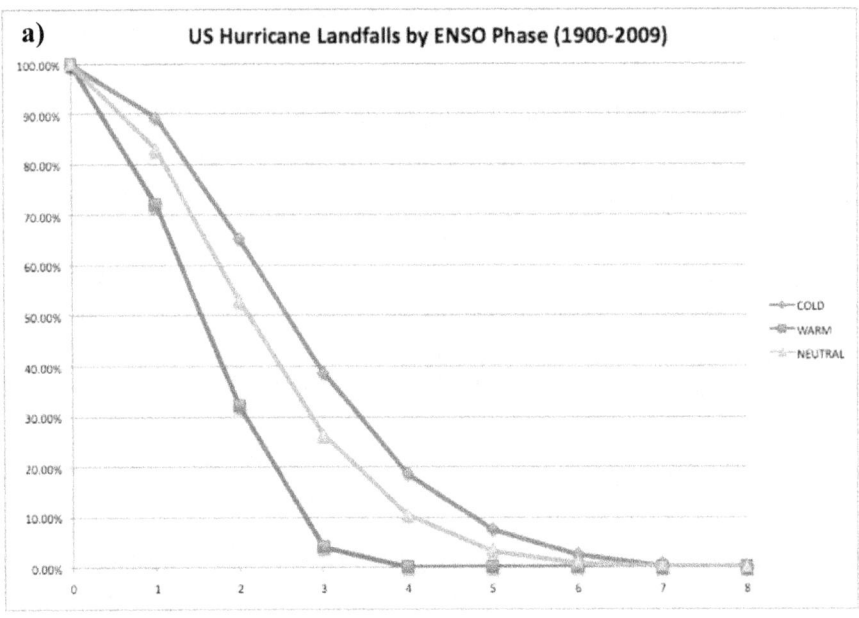

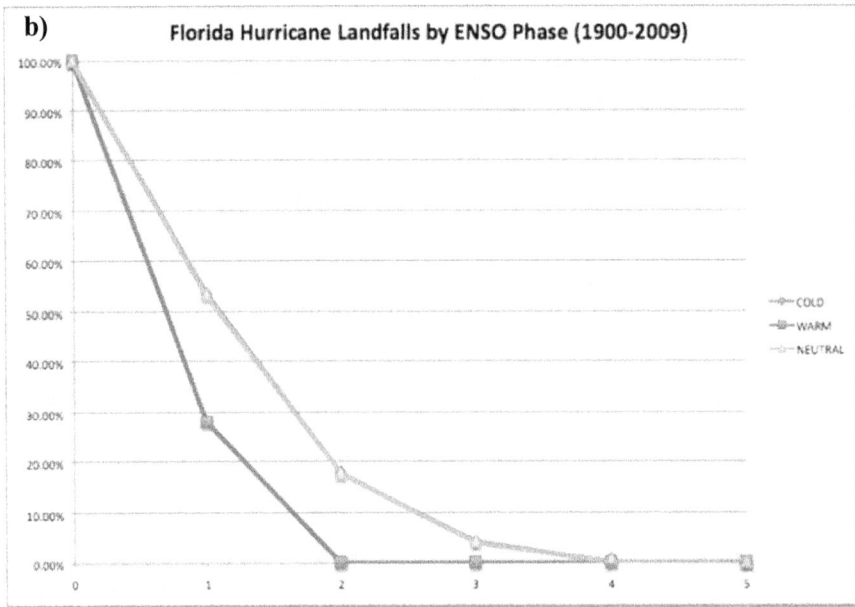

Fig. 16.7. Inverse cumulative frequency distribution of a) US and b) Florida landfalling hurricanes for the period 1900-2009 using HURDAT (http://www.aoml.noaa.gov/hrd/hurdat/Data_Storm.html).

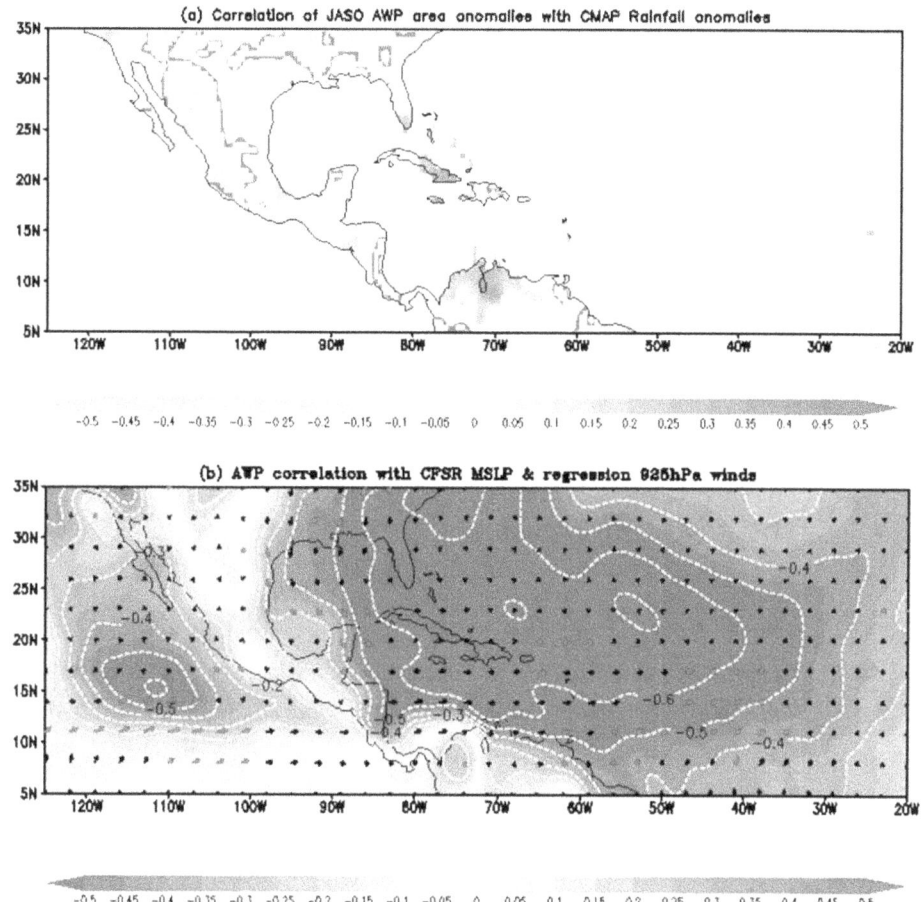

Figure 16.8. Correlation of the June-October AWP area anomalies (SST from Reynolds et al. 2007) with corresponding a) rainfall anomalies (shaded; rainfall from Chen et al. 2002) and b) mean sea level pressure (MSLP) anomalies (shaded; MSLP from Saha et al. 2010) and regression of June-September AWP area anomalies on corresponding 925 hPa wind anomalies (winds from Saha et al. 2010). The significant values at 95% confidence interval according to t-test are contoured and vectors are shown in red.

Misra and DiNapoli (2013) observed that Florida has a distinct seasonal cycle of rainfall that gives rise to a monsoonal type of climate, and they objectively defined the onset and demise of the Florida wet season for a particular grid point over peninsular Florida as the day after the cumulative rainfall anomaly ($C'_m(i)$) reaches a minimum and maximum for the year, respectively. The cumulative rainfall anomaly is given by:

$$C'_m(i) = \sum_{n=1}^{i}[D_m(n) - \bar{\bar{C}}], \qquad (1)$$

where

$$\bar{\bar{C}} = \frac{1}{MN}\sum_{m=1}^{M}\sum_{n=1}^{N}D(m,n). \qquad (2)$$

$D_m(n)$ is the daily rainfall for day n of year m, and $\overline{\overline{C}}$ is the climatology of the annual mean rainfall over N (=365/366) days for M years. This definition breaks down when the unique rainy season begins to disappear moving north of Jacksonville (Misra and DiNapoli 2013). Misra and DiNapoli (2013) showed that the onset and demise of the wet season are profoundly modulated by ENSO and AWP variations. They demonstrated that warm (cold) ENSO events in the winter are followed by early (later) onset of the rainy season in peninsular Florida. Similarly, Misra and DiNapoli (2013) indicate that large (small) AWP events associated with weakening (strengthening) of the NASH, leads to later (early) demise of the wet season thereby resulting in longer (shorter) length of the wet season in parts of peninsular Florida. This teleconnection is consistent with the positive correlation of June-October seasonal rainfall with AWP area variations observed over South Florida in Fig. 16.8a.

The Atlantic Multi-decadal Oscillation (AMO)

A roughly 60- to 80-year periodicity is observed in SSTs in and around the Atlantic Ocean basin after removal of the warming signature of the last 150 years and this is known as the AMO (Figure 16.9). The AMO is characterized by alternating multi-decade phases of persistently above and below average SSTs in the Atlantic between 0 and 70 °N. The range in SSTs between the warm (+) and cool (-) phases is ~0.4 °C (Delworth and Mann 2000; Enfield et al. 2001; Kerr 2000; Schlesinger and Ramankutty 1994). Historically, warm phases have been documented from 1860-1900 and 1925-1965, and cool phases from 1905-1925 and 1965-1995. Analysis of North Atlantic climate time series from tree rings further suggest that an "organizational phase" exists between warm and cool phase shifts during which SST variability is dampened on multi-annual to decadal time scales (Gray et al. 2004). Persistence of multi-decadal variability consistent with the AMO in paleo proxy records prior to the anthropogenic period suggests that the oscillation is natural in origin (Gray et al. 2004; Saenger et al. 2009; Waite 2011).

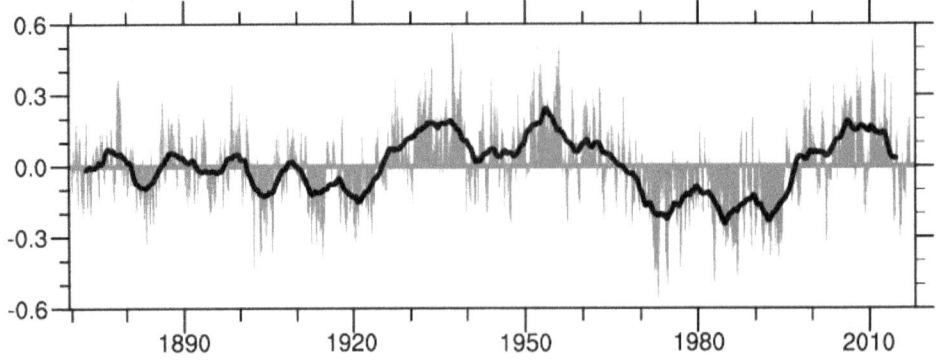

Figure 16.9. Figure from K. Trenberth (personal communication, November 2017) depicting the Atlantic Multi-decadal Oscillation (AMO) Index from HadISST data over the period from 1871 to 2016, updated from Trenberth and Shea (2006). The AMO Index is derived by subtracting the global-mean SSTA time series from the North Atlantic average time series (Trenberth and Shea 2006).

While the impact of the AMO on SSTs is implicit in the nature of the oscillation itself, the AMO has also been shown to affect many aspects of the climate system throughout the Atlantic sector. Of particular relevance to Florida are linkages between the AMO and changes in air temperature, precipitation, and storm occurrences that have significant implications for hazard mitigation, resource management, and sustainability across the state.

Multi-decadal changes in air temperatures have also been associated with variability in evaporation and precipitation budgets of the continents bordering the Atlantic. During warm phases of the AMO, the majority of the United States receives below average rainfall; this is particularly true in the midwest and southwest where droughts tend to be more frequent and severe. One notable example is the 1930s Dust Bowl event, which occurred during a warm phase of the AMO. On the other hand, cool phases of the AMO tend to result in these same areas experiencing above average rainfall. While North Florida conforms to this relationship, Central and South Florida are marked exceptions to this rule as they receive above average rainfall during AMO (+) phases and below average rainfall, resulting in increased drought and wildfires, during AMO (-) phases. Evidence of this relationship can be observed in Lake Okeechobee where inflow can vary up to 40% between phases, resulting in significant implications for water management practices (Enfield et al. 2001; Kelly and Gore 2008). As previously stated, inter-annual winter rainfall associated with ENSO variations is also impacted by the AMO phase. Studies have shown that water transparency in Florida lakes, which has been linked to precipitation and the resulting runoff of dissolved organic material, is also sensitive to the AMO. Gaiser et al. (2009) found that observations and hindcasts of lake transparency demonstrate multi-decadal trends, with transparency being greatest in AMO cool phases associated with below average rainfall, and vice versa. Given the increasing frequency and severity of algae bloom events in Florida waterways and coastal communities, these multi-decadal drivers should be considered in future mitigation measures.

Compilations of AMO indices, proxy reconstructions, and observations indicate that the number of landfalls of major hurricanes in the Gulf of Mexico and along the eastern seaboard of the US is significantly higher during warm, or positive, phases of the AMO than during cool phases (Figure 16.10). These phenomena are most likely connected through the AMO's impact on SST and vertical wind shear ($|Vz|$) in the MDR of the Atlantic Ocean. Studies suggest that circulation anomalies in the Atlantic Ocean's MDR reduce $|Vz|$ during warm phases of the AMO, allowing storms to grow and intensify, while increased shear in this region during cool AMO phases inhibits growth and organization (Bell and Chelliah 2006; Nyberg et al. 2007).

Research also suggests that the AMO is linked to Atlantic tropical cyclone activity through the AWP. Multi-decadal trends are evident in the warm pool region, where AMO warm (cool) phases are associated with repeated large (small) AWPs (Wang et al. 2008). The AWP region is comprised of the Gulf of Mexico, the Caribbean Sea, and the western tropical North Atlantic, and lies in the path of the development region for Atlantic hurricanes. Thus, it may also play a role in connecting them to the AMO. The growth of anomalously large AWPs in association with warm phases of the AMO would reduce $|V_Z|$ in the region of cyclogenesis and increase the moist static instability of the troposphere, favoring the formation of hurricanes (Wang et al. 2008). Recently, there has been a relative increase in storm intensity and landfall. Goldenberg et al. (2001) provided a well-supported synthesis of the mechanisms argued to be contributing to this increase, proposing a feedback mechanism whereby faster thermohaline circulation (THC) results in warmer North Atlantic SSTs and cooler South Atlantic SSTs that enhances Sahel rainfall and decreases $|V_Z|$ in the MDR. However, it is important to note that anthropogenically-induced warming may also amplify these feedbacks with hurricane development in the Atlantic sector. Currently, it is difficult to discern the relative contributions of this versus phase changes of the AMO.

Despite being one of the dominant climatic controls for the Atlantic sector, the precise forcing and stability of the AMO remains poorly understood. Coupled ocean-atmosphere models and paleo reconstructions suggest that the multi-decadal mode is linked to variability in the exchange of water and heat associated with the Atlantic Meridional Overturning Circulation (AMOC). This driver also offers the best mechanism to explain connections between the AMO and other climatic oscillations on decadal to multi-decadal timescales, as well as associations beyond the Atlantic. However, recent modeling investigations by Clement et al. (2015) suggest that the AMO may be a response to stochastic forcing by mid-latitude atmospheric circulation and thermal coupling in the tropics, where changes in ocean circulation (e.g. AMOC) may be dominantly responding to, rather than forcing, the AMO. Furthermore, both models and observations suggest that changes in heat flux to the surface, as produced by low-level clouds and atmospheric dust, contribute to the tropical manifestation of the AMO (Brown et al. 2016; Yuan et al. 2016).

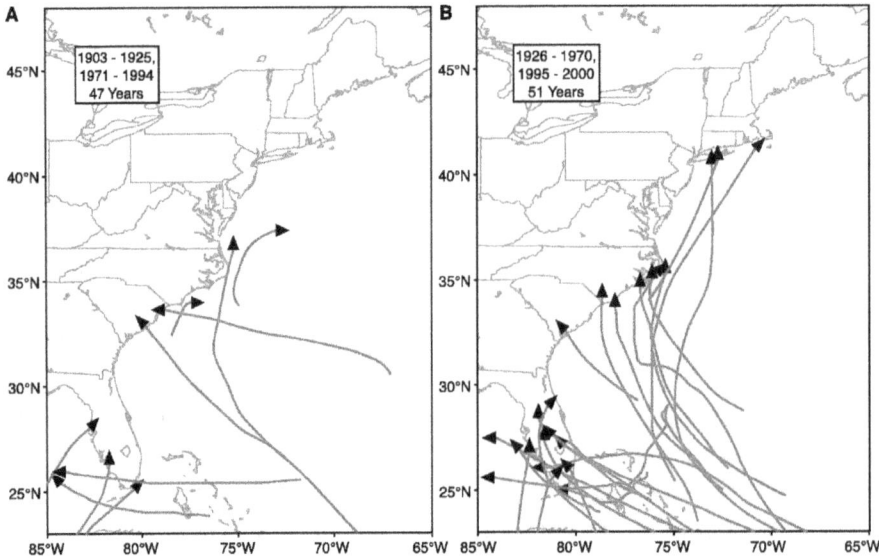

Figure 16.10. Panels from Goldenberg et al. (2001) contrasting US East Coast major hurricane landfalls between (A) cool and (B) warm phases of the AMO. Reprinted with permission from AAAS.

Unfortunately, understanding the mechanisms associated with the AMO is limited in large part by the relatively short duration of the instrumental period, which only covers the last 110 to 150 years. In response, considerable effort is being put into generating long-term paleo proxy reconstructions of multi-decadal change. And the impacts of anthropogenic warming on the long-term stability of the AMO also remain unclear. Modeling simulates long-lived, low frequency persistent variability through time; however, several proxy reconstructions indicate instabilities in the oscillation that, when coupled with anthropogenic uncertainties, hinder predictability. For the time being, the multi-decadal variability in temperature associated with the AMO acts to alternately obscure (during cool phases) and enhance (during warm phases) anthropogenic warming in the North Atlantic region.

Land-Atmosphere Interactions

Land-atmosphere interactions manifest over many different time scales, some of which have already been discussed in relation to diurnal variations. The SEUS is one of the few regions on earth that shows a cooling trend during the 20th century (Trenberth et al. 2007; Ji et al., 2014), often termed the "warming hole." The cooling trend in the SEUS has been attributed to several factors including changes in SST (Robinson et al. 2002), land-atmosphere feedback (Pan et al. 2004), and internal dynamics or chaos (Kunkel et al. 2006). This cooling trend is found to be strongest in the late spring–early summer period, which is coincident with the seasonal increase in rainfall (Portmann et al. 2009). In other words, rainfall tempers the surface warming during the season thus conforming to the well-known near linear relationship of trends in temperature and diurnal temperature ranges to precipitation amounts (Trenberth and Shea 2005; Zhou et al. 2004; Dai et al. 1999; Madden and Williams 1978). However, Portmann et al. (2009) suggested the additional influences, including increasing strength in the direct and indirect impact of rising concentration of aerosols and rapid changes in vegetation could also be causing the region's temperatures to cool. But Ji et al. (2014) showed that the spatial extent and magnitude of the "warming hole" in the SEUS has diminished gradually over recent decades.

Land use/land cover changes can also affect climate through variations in the partitioning of the available energy between sensible and latent heat and the breakup of precipitation between runoff, canopy storage, and evapotranspiration, which can alter the consequent atmospheric feedback (Zhao et al. 2001; Feddema et al. 2005; Kalnay and Cai 2003). Irrigation, a form of land use, raises evaporation during the day and changes the Bowen ratio by way of wetting the soil, which leads to apparent cooling of the surface temperature (Kueppers et al. 2007; Sacks et al. 2009; Misra et al. 2012). On the other hand, irrigation can lead to surface warming of the nighttime minimum temperature (T_{min}) under weak wind conditions (typically at night, when the boundary layer is decoupled from the rest of the atmosphere), as wet soil has a raised heat capacity and conductivity compared to dry soil (Elsner et al. 1996). The urban heat island effect can also have a significant impact on warming trends (Oke 1973; Karl et al. 1988a; Karl and Jones 1989; Misra et al. 2012). The heat capacity and conductivity of building and paving materials allow for more heat to be absorbed during the day in urban areas than in rural areas. The heat then becomes available at night to partially compensate for the radiational cooling from the outgoing long-wave radiation loss. Another cause of increased urban heating is from the trapping of the reflected solar radiation by the narrow arrangement of buildings (often referred as the reduced sky view factor), which is ultimately absorbed by the walls of the buildings in the urban areas. Additional factors such as increased atmospheric pollutants, production of waste heat from air-conditioning, refrigeration systems and industrial processes, and obstruction of rural air flows by the windward face of built-up surfaces can also contribute to the urban heat island effect. As a result of these factors, a higher T_{min} is usually observed in the urban areas relative to the rural areas (Karl et al. 1988b).

Misra et al. (2012) found considerable spatial heterogeneity in the observed linear trends of monthly mean maximum (T_{max}) and minimum temperatures (T_{min}) from station observations in the SEUS (specifically Florida, Alabama, Georgia, South Carolina, and North Carolina). In a majority of these station sites, the warming trends in T_{min} were found to be stronger in urban areas relative to rural areas (Fig. 16.11). This was determined by examining the scatter of the T_{min} with the Population Interaction Zone for Agriculture (PIZA; USDA-ERS 2005), which is an index that represents the residential, commercial, and industrial urban activities affecting agriculture. The PIZA values range from 1 to 4, with lower (higher) values being representative of rural (urban) areas.

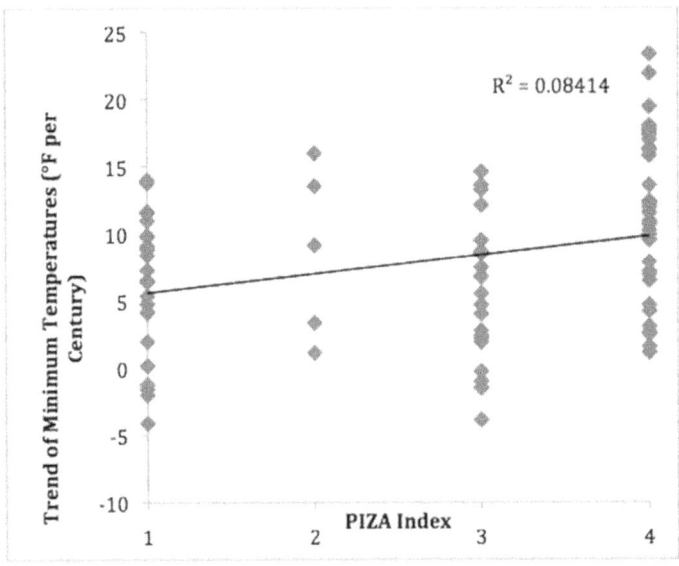

Figure 16.11. The scatter plot of the linear trends (in °F/century) over the southeast US (which includes Florida, Alabama, Georgia, South Carolina, and North Carolina) of T_{min}. Adapted from Misra et al. (2012).

The linear trends of T_{min} in urban areas of the SEUS is approximately 7 °F/century compared to about 5.5 °F/century in rural areas. It should be noted that trends in T_{max} had an insignificant relationship with PIZA (Misra et al. 2012). However, it was shown that during the summer season, T_{max} has weaker warming (or stronger cooling) trends with irrigation, while trends in the summer season T_{min} show stronger warming trends (Fig. 16.12). The corresponding figures for other seasons show an insignificant relationship (Misra et al. 2012). The relationships implied in Fig. 16.12 are consistent with theory, which suggests that by way of evaporation, irrigation cools T_{max} that is measured during the day. On the other hand, theory also suggests that irrigation would raise the heat capacity and conductivity of the otherwise dry soil, which under weak wind conditions (anticipated at night) would lead to warming of the T_{min}. Furthermore, Misra et al. (2012) revealed that linear trends in T_{max} in the boreal summer season show a cooling trend of about 0.5 °F/century with irrigation, while the same observing stations on an average display warming trends in T_{min} of about 3.5 °F/century. The seasonality and the physical consistency of these relationships with urbanity and irrigation would suggest their non-negligible influence on the spatial heterogeneity of the surface temperature trends over the southeastern United States.

Secular Changes

As noted in the previous section, the so-called warming hole of the SEUS has diminished over recent decades, suggesting an overall warming of the surface temperature (Ji et al. 2014). On the other hand, an insignificant linear trend can be observed in the monthly mean precipitation of SEUS in the 20th century (Portmann et al. 2008; NOAA 2012). However, there is a growing body of evidence to suggest that the frequency of extreme precipitation events in the SEUS has increased in recent decades (Kunkel et al. 2013; Wuebbles et al. 2014). While the reason for the observed rising trend in extreme precipitation events is not known for certain, the increasing atmospheric water vapor content in a warming climate is considered to be the most probable cause (Karl and Trenberth 2003; Kunkel et al. 2013). In a related study, Knight and Davis (2009) indicated that extreme precipitation from tropical cyclones has been increasing by approximately 5-10% per decade in the SEUS. They attributed this increase to storm wetness (precipitation per storm) and storm frequency over their period of analysis (1972-2007). However, analysis of longer time periods of tropical cyclone data has revealed conflicting conclusions, with some suggesting increases in Atlantic tropical cyclone frequency over time (Holland and Webster 2007; Mann and Emanuel 2007) while others suggest the opposite (Vecchi and Knutson 2008; Chang and Guo 2007). The former studies, using the strong statistical relationship of tropical Atlantic SST variations to changes in Atlantic tropical cyclone frequency, arrived at the conclusions of rising tropical cyclone frequency from the observation of similar rising trend in Atlantic SSTs attributed to anthropogenic forcing (Holland and Webster 2007; Mann and Emanuel 2007). But when accounting for lower reporting of short-lived tropical cyclones in the pre-satellite era (prior to 1966) leads to a long-term trend of Atlantic tropical cyclones that is statistically insignificant (Vecchi and Knutson 2008; Landsea et al. 2009). Although there may be no significant trend in the Atlantic TC frequency, sea level rise has exacerbated the damage from flooding of modern storms relative to historical storms of comparable strength (Zhang et al. 1997, 1999).

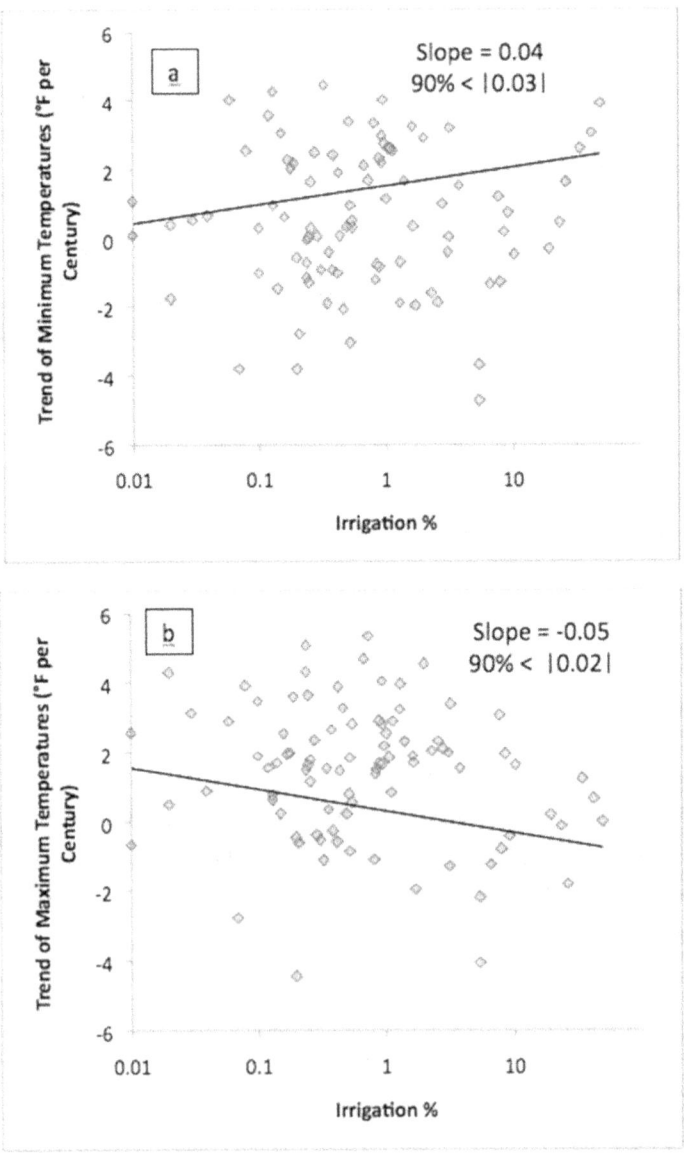

Figure 16.12. The scatter plot of the linear trends (^{0}F/century) in JJA over the southeast US (which includes Florida, Alabama, Georgia, South Carolina, and North Carolina) of a) T_{min} and b) T_{max} with the irrigation data. The slope and its 90% confidence level obtained from a Monte Carlo approach are shown on the top right corner. From Misra et al. (2012).

Conclusion

With its mild climate and its 1,350 miles of picturesque coastline, Florida is one of the most desirable places for populations to live and migrate to. However, Florida's geographic position also leaves it exposed to significant climate variability, including periodic weather extremes. Florida is a unique region east of the Rocky Mountains with a very distinct monsoonal type wet season in the summer that distinguishes it from the other seasons. Similarly, given its proximity to tropical latitudes, Florida also has significant diurnal variations of rainfall especially in late spring and summer season. Finally, Florida has a high density of landfalling Atlantic tropical cyclones in the Continental US that contributes significantly to the hydroclimate of the region.

This chapter establishes that Florida's climate is as much affected by remote climate variations as local variability over land and its neighboring oceans. There is a discernible impact of local land cover and land use change on surface temperature and irrigation across the SEUS, including Florida, with urban areas showing a warming trend comparatively stronger than that in the rural areas. Similarly, SEUS regions with sustained irrigation over time have been shown to have a cooling (warming) trend in the maximum (minimum) temperatures. Likewise, variations of the SST in the neighboring oceans (e.g. variations in the size of the Atlantic Warm Pool [AWP] residing over Gulf of Mexico, Caribbean Sea and parts of northwestern subtropical Atlantic Ocean) affect the local climate, from sea breezes in the Florida Panhandle to variations in the Atlantic tropical cyclone activity.

This chapter's discussion also highlights the scale interactions of the observed climate across time and spatial scales. For example, the Florida Panhandle sea breeze is shown to be affected by the subtle variations of the Bermuda High. Similarly, ENSO forcing on Florida's winter climate is affected by decadal variations such as the PDO and the AMO. The manifestation of decadal-scale variations such as the PDO and AMO at the local scale (e.g., influence on Lake Okeechobee inflow, AWP variations) raises hope for regional decadal predictability for Florida. This attempt at decadal predictability could go a long way toward mitigating the vulnerability of the state to climate anomalies.

Acknowledgements

This work was supported by grants from NOAA (NA12OAR4310078, NA11OAR4310110) and USGS G13AC00408. A. J. Waite acknowledges that portions of the discussion on the Atlantic Multi-decadal Oscillation are adapted from her Ph.D. dissertation from the University of Miami.

References

Allan RJ, Lindesay J, Parker D. 1996. El Niño–Southern Oscillation and Climatic Variability. CSIRO Publishing: Collingwood, Victoria.

Ashok, K., S. K. Beheria, S. A. Rao, H. Weng, and T. Yamagata, 2007: El Niño Modoki and its possible teleconnection. J. Geophys. Res., 112, C11007, doi:10.1029/2006JC003798.

Attaway, J. A., 1997: A History of Florida Citrus Freezes. Florida Science Source, Inc., 368 pp.

Bastola S, Misra V 2013: Sensitivity of hydrological simulations of southeastern United States watersheds to temporal aggregations of rainfalls. J Hydrometeor. doi:10.1175/JHM-D-12-096.1

Bell, G. D., and Chelliah, M., 2006, Leading tropical modes associated with interannual and multidecadal fluctuations in North Atlantic hurricane activity: Journal of Climate, v. 19, no. 4, p. 590-612.

Blanchard, D. O. and R. E. Lopez, 1985: Spatial patterns of convection in South Florida. Mon. Wea. Rev., 113, 1282-1299.

Bove MC, Elsner JB, Landsea CW, Niu X, O'Brien JJ (1998) Effect of El Niño on U. S. Landfalling Hurricanes, revisited. Bull Am Soc 79:2477–2482.

Brown, P.T., Lozier, M.S., Zhang, R. and Li, W., 2016. The necessity of cloud feedback for a basin-scale Atlantic Multidecadal Oscillation. Geophysical Research Letters, 43(8), pp.3955-3963.

Byers, H., and H. Rodebush (1948), Causes of thunderstorms of the Florida Peninsula, *J. Meteor.*, 5, 275-280.

Case, J. L., M. M. Wheeler, J. Manobianco, J. W. Weems, and W. P. Roeder (2005), A 7-yr climatological study of land breezes over the Florida spaceport. *J. Appl. Meteor. Climatol.*, 44, 340-356.

Cosgrove, B., and coauthors (2003), Real-time and retrospective forcing in the North American Land Data Assimilation System (NLDAS) Project, *J. Geophys. Res.*, 108, 8842, doi:10.1029/2002D003118.

Chan, S. and V. Misra, 2010: A diagnosis of the 1979-2005 extreme rainfall events in the southeastern United States with isentropic moisture tracing. Mon. Wea. Rev., 138, 1172-1185.

Chang, E. K. M. & Guo, Y., 2007: Is the number of North Atlantic tropical cyclones significantly underestimated prior to the availability of satellite observations? Geophys. Res. Lett. 34, L14801.

Clement, A., Bellomo, K., Murphy, L.N., Cane, M.A., Mauritsen, T., Rädel, G. and Stevens, B., 2015. The Atlantic Multidecadal Oscillation without a role for ocean circulation. *Science*, 350(6258), pp.320-324.

Davies, R. E., B. P. Hayden, D. A. Gay, W. L. Phillips, and G. V. Jones, 1997: The North Atlantic Anticyclone. J. Climate, 10, 728-744.

Delworth, T. L., and Mann, M. E., 2000, Observed and simulated multidecadal variability in the Northern Hemisphere: Climate Dynamics, v. 16, no. 9, p. 661-676.

Diaz, H. F., M. P. Hoerling, and J. K. Eischeid, 2001: ENSO variability, teleconnections and climate change. Int. J. Climatol., 21, 1845-1862.

Eichler, T., and W. Higgins, 2006: Climatology and ENSO-related variability of North American extratropical cyclone activity. J. Climate, 19, 2076–2093.

Emanuel, K. A., 1993: The physics of tropical cyclogenesis over the Eastern Pacific. Tropical Cyclone Disasters. J. Lighthill, Z. Zhemin, G. J. Holland, K. Emanuel, (Eds.), Peking University Press, Beijing, 136-142.

Enfield DB, Mayer DA (1997) Tropical Atlantic sea surface temperature variability and its relation to El Niño Southern Oscillation. J Geophys Res 102:929–945

Enfield DB, Mestaz-Nunez AM, Trimble PJ (2001) The Atlantic multidecadal oscillation and its relation to rainfall and river flows in the continental US. Geophys Res Lett 28(10):2077–2080

Gaiser, E. E., Deyrup, N. D., Bachmann, R. W., Battoe, L. E., and Swain, H. M., 2009, Multidecadal climate oscillations detected in a transparency record from a subtropical Florida lake: Limnology and Oceanography, v. 54, no. 6, p. 2228-2232.

Gershunov A, Barnett TP (1998) Interdecadal modulation of ENSO teleconnections. Bull Am Meteor Soc 80:2715–2725

Gibson, H., and T. Vonder Haar (1990), Cloud and convection frequencies over the southeast United States as related to small-scale geographic features, *Mon. Weather Rev.*, 118, 2215-2227.

Goldenberg, S. B., Landsea, C. W., Mestas-Nunez, A. M., and Gray, W. M., 2001, The recent increase in Atlantic hurricane activity: Causes and implications: Science, v. 293, no. 5529, p. 474-479.

Gray, W. M., 1984a: Atlantic seasonal hurricane frequency: Part I: El Niño and 30 mb quasi-biennial oscillation influences. Mon. Wea. Rev., 112, 1649-1668.

Gray, S. T., Graumlich, L. J., Betancourt, J. L., and Pederson, G. T., 2004, A tree-ring based reconstruction of the Atlantic Multidecadal Oscillation since 1567 AD: Geophysical Research Letters, v. 31, no. 12, p. -.

Hagemeyer, B. C., 2006: ENSO, PNA and NAO scenarios for extreme storminess, rainfall and temperature variability during the Florida dry season. Preprints, 18th Conf. on Climate Variability and Change, Atlanta, GA, Amer. Meteor. Soc., P2.4. [Available online at http://ams.confex.com/ams/pdfpapers/98077.pdf.]

Higgins, R. W., A. Letmaa, and V. E. Kousky, 2002: Relationships between climate variability and winter temperature extremes in the United States. J. Climate, 15, 1555–1572.

Holland, G. J. & Webster, P. J., 2007: Heightened tropical cyclone activity in the North Atlantic: natural variability or climate trend? Phil. Trans. R. Soc. A 365, 2695–2716.

Ji, F., Z. Wu, J. Huang, and E. P. Chassignet, 2014: Evolution of land surface air temperature trend. Nature Climate Change, doi:10.1038/NCLIMATE2223.

Landsea, C. W., 2000: El Niño-Southern Oscillation and the seasonal predictability of tropical cyclones. In press in El Niño: Impacts of Multiscale Variability on Natural Ecosystems and Society, edited by H. F. Diaz and V. Markgraf.

Landsea, C., Vecchi, G. A., Bengtsson, L. & Knutson, T. R., 2009: Impact of duration thresholds on Atlantic tropical cyclone counts. J. Clim. doi:10.1175/2009JCLI3034.1.

Kanamitsu, M., and Coauthors, 2002: NCEP-DOE AMIP-II Reanalysis (R-2). Bull. Amer. Meteor. Soc., **83**, 1631-1643.

Karl, T. R. and K. E. Trenberth, 2003: Modern climate change. Science, 302, 1719–1723, doi:10.1126/science.1090228

Kelly, M. H., and Gore, J. A., 2008, Florida river flow patterns and the Atlantic multidecadal oscillation: River Research and Applications, v. 24, no. 5, p. 598-616.

Kerr, R. A., 2000, A North Atlantic climate pacemaker for the centuries: Science, v. 288, no. 5473, p. 1984-1986.

Knight DB, Davis RE (2009) Contribution of tropical cyclones to extreme rainfall events in the Southeastern United States. J Geophys Res 114:D23102. doi:10.1029/2009JD012511

Kunkel, K. E., X. Z. Liang, J. Zhu, and Y. Lin, 2006: Can CGCM's simulate the twentieth- century "warming hole" in the central United States. J. Climate, 19, 4137-4153.

Kunkel, K. E. and coauthors, 2013: Monitoring and understanding trends in extreme storms. Bull. Amer. Soc., 94, 499-514.

Li, L., W. Li and A. P. Barros, 2013: Atmospheric moisture budget and its regulation of the summer precipitation variability over the Southeastern United States. Clim. Dyn., 41, 613-631.

Lin, Y., and K. E. Mitchell, 2005: The NCEP stage II/IV hourly precipitation analyses: Development and applications. 19th Conf. on Hydrology, San Diego, CA, Amer. Meteor. Soc., 1.2. [Available online at https://ams.confex.com/ams/pdfpapers/ 83847.pdf].

Madden, R. A., and J. Williams, 1978: The correlation between temperature and precipitation in the United States and Europe. Mon. Wea. Rev., 106, 142-147.

Mann, M. & Emanuel, K., 2006: Atlantic hurricane trends linked to climate change. Eos 87, 233–241.

Mantua NJ, Hare SR, Zhang Y, Wallace JM, Francis RC. 1997. A Pacific interdecadal climate oscillation with impacts on salmon production. Bulletin of the American Meteorological Society 78: 1069–1079.

Maxwell JT, Soulé PT, Ortegren JT, Knapp PA (2012) Drought-busting tropical cyclones in the Southeastern Atlantic United States: 1950–2008. Ann Assoc Am Geogr 102(2):259–275.

Maxwell JT, Ortegren JT, Knapp PA, Soulé PT (2013) Tropical cyclones and drought amelioration in the gulf and Southeastern coastal United States. J Clim 26:8440–8452

McPherson, R. D. (1968), A three dimensional numerical study of the Texas Coast sea breeze, Report No. 15, Atmospheric Science Group, University of Texas at Austin, 252 pp.

Miller, K. A., 1991: Response of Florida citrus growers to the freezes of the 1980s. Climate Res., 1, 133–144.

Misra, V. and P. A. Dirmeyer, 2009: Air, Sea, and Land Interactions of the Continental U. S. Hydroclimate. J. Hydromet., 10, 353-373.

Misra, V., L. Moeller, L. Stefanova, S. Chan, J. J. O'Brien, T. J. SmithIII, and N. Plant, 2011: The influence of Atlantic warm pool on panhandle Florida sea breeze. J. Geophys. Res., 116, doi:10.1029/2010JD015367.

Misra, V. and S. DiNapoli, 2013: Understanding wet season variations over Florida. Clim. Dyn., 40, 1361-1372.

Misra, V., J. -P. Michael, R. Boyles, E. P. Chassignet, M. Griffin, and J. J. O'Brien, 2012: Reconciling the spatial distribution of the surface temperature trends in the Southeastern United States J. Climate, 25(10), 3610-3618, doi:http://dx.doi.org/10.1175/JCLI-D-11-001701.1

Mo KC (2010) Interdecadal modulation of the impact of ENSO on precipitation and temperature over the United States. J Clim 23:3639–3656.

Nag, B., V. Misra, and S. Bastola, 2014: Validating ENSO teleconnections on Southeastern United States Winter Hydrology Earth Interactions. In press. DOI: EI-D-14-0007.1.

Nicholls, M. E., R. A. Pielke, and W. R. Cotton, 1991: A two-dimensional numerical investigation of the interaction between sea breezes and deep convection over the Florida peninsula. Mon. Wea. Rev., 119, 298-323.

Nyberg, J., Malmgren, B. A., Winter, A., Jury, M. R., Kilbourne, K. H., and Quinn, T. M., 2007, Low Atlantic hurricane activity in the 1970s and 1980s compared to the past 270 years: Nature, v. 447, no. 7145, p. 698-U611.

Pan, Z. and co-authors, 2004: Altered hydrologic feedback in a warming climate introduces a "warming hole". Geophys. Res. Lett., 10.1029/2004/GL020528.

Pielke, R. A. (1974), A three-dimensional numerical model of the sea breezes over South Florida, *Mon. Weather Rev., 102,* 115-139.

Pielke, R. A., R. L. Walko, L. T. Steyaert, P. L. Vidale, G. E. Liston, W. A. Lyons, and T. N. Chase (1999), The influence of anthropogenic landscape changes on weather in South Florida, *Mon. Weather Rev., 127,* 1663-1673.

Portmann, R. W., S. Solomon, and G. C. Hegel, 2009: Spatial and seasonal patterns in climate change, temperatures, and precipitation across the United States. Proc. Nat. Acad. Sci., USA, 106: 7324-7329.

Prat OP, Nelson BR (2013a) Precipitation contribution of tropical cyclones in the Southeastern United States from 1998 to 2009 using TRMM satellite data. J Clim 26(3):1047–1062 Prat OP, Nelson BR (2013b) Mapping the world's tropical cyclone rainfall contribution over land using the TRMM multi-satellite precipitation analysis. Water Resour Res 49:7236–7254. doi:10.1002/wrcr.20527.

Ropelewski CF, Halpert MS (1986) North American precipitation and temperature patterns associated with the El Nino/Southern Oscillation (ENSO). Mon Wea Rev 114:2352–2362.

Robinson, W. A., R. Reudy, and J. E. Hansen, 2002: General circulation model simulations of recent cooling in the east-central United States. J. Geophys. Res., 10.1029/2001JD001577.

Saenger, C., Cohen, A. L., Oppo, D. W., Halley, R. B., and Carilli, J. E., 2009. Surface-temperature trends and variability in the low-latitude North Atlantic since 1552: Nature Geoscience, v. 2, no. 7, p. 492-495.

Schlesinger, M. E., and Ramankutty, N., 1994, An oscillation in the global climate system of period 65-70 years: Nature, v. 367, no. 6465, p. 723-726.

Selman, C., V. Misra, L. Stefanova, S. DiNapoli, and T. J. SmithIII, 2013: On the twenty-first century wet season projections over the Southeastern United States. Reg. Env. Change, S153-164, doi:10.1007/s10113-013-0477-8.

Shapiro, L. J., 1987: Month-to-month variability of the Atlantic tropical circulation and its relationship to tropical storm formation. Mon. Wea. Rev., 115, 2598-2614.

Smith, S. R., P. M. Green, A. P. Leonardi, and J. J. O'Brien, 1998: Role of multiple-level tropospheric circulations in forcing ENSO winter precipitation anomalies. Mon. Wea. Rev., 126, 3102–3116.

Stefanova, L., P. Sura, and M. Griffin, 2013: Quantifying the non-gaussianity of wintertime daily maximum and minimum temperatures in the southeast. J. Climate, 26, 838-850.

Stephenson, D. B., H. Wanner, S. Bronnimann, J. Luterbacher, 2003: The history of scientific research on the North Atlantic Oscillation, in the North Atlantic Oscillation: Climatic Significance and Environmental Impact, edited by J. W. Hurrell, Y. Kushnir, G. Ottersen and M. Visbeck, pp 37-50, American Geophysical Union, Washington D. C., doi:10.1029/134GM02.

Thompson, D. W. and J. M. Wallace, 1998: The Arctic oscillation signature in the wintertime geopotential height and temperature fields. Geophys. Res. Lett., 25, 1297-1300.

Tian, Y., C. D. Peters-Lidard, B. J. Choudhury, and M. Garcia (2007), Multitemporal analysis of TRMM-based satellite precipitation products for land data assimilation applications, *J. Hydrometeor., 8,* 1165-1183.

Trenberth KE, Shea DJ. 1987. On the evolution of the Southern Oscillation. Monthly Weather Reiew 115: 3078–3096

Trenberth, K. E., A. Dai, R. M. Rasmussen, and D. B. Parsons, 2003: The changing character of precipitation. Bull. Met. Soc., 84, 1205-1217.

Trenberth, K. E. and D. J. Shea, 2005: Relationships between precipitation and surface temperature. Geophys. Res. Lett., 32, L1703, doi:10.1029/2005GL022760.

Trenberth, K.E. and D.J. Shea (2006): Atlantic hurricanes and natural variability in 2005. Geophysical Research Letters 33, L12704, doi:10.1029/2006GL026894

Trenberth, K. E., and co-authors, 2007: Observations: surface and atmospheric climate change. Climate change 2007: The physical science basis. Contribution of working group I to the Fourth Assessment Report of the Intergovernmental Panel on Climate Change, eds Solomon, S. and co-authors, Cambridge University Press, UK, 235-336.

Vecchi, G. A. & Knutson, T. R., 2008: On estimates of historical North Atlantic tropical cyclone activity. J. Clim. 21, 3580–3600.

Waite, A. J., 2011. Geochemical Insights into Multi-decadal Climate Variability: Proxy Reconstructions from Long-lived Western Atlantic Corals and Sclerosponges: Open Access Dissertations, University of Miami, 693.

Wang, C. Z., Lee, S. K., and Enfield, D. B., 2008, Atlantic Warm Pool acting as a link between Atlantic Multidecadal Oscillation and Atlantic tropical cyclone activity: Geochemistry Geophysics Geosystems, v. 9, p. -.

Wu, Z., N. Huang, J. Wallace, B. Smoliak, and X. Chen (2011), On the time-varying trend in global-mean surface temperature, *Clim. Dyn., 37,* 759-773.

Wuebbles DJ, Meehl G, Hayhoe K, Karl TR, Kunkel K, Santer B, Wehner M, Colle B, Fischer EM, Fu R, Goodman A, Janssen E, Kharin V, Lee H, Li W, Long LN, Olsen S, Seth A, Sheffield J, Tao Z, Sun L (2014) CMIP5 climate model analyses: Climate extremes in the United States. Bullet. Am. Meteorol. Soc., doi: http://dx.doi.org/10.1175/BAMS-D-12-00172.1.

Yu, J.-Y. and H.-Y. Kao, 2007: Decadal Changes of ENSO Persistence Barrier in SST and Ocean Heat Content Indices: 1958-2001. *Journal of Geophysical Research,* 112, D13106, doi: 10.1029/2006JD007654.

Yuan, T., Oreopoulos, L., Zelinka, M., Yu, H., Norris, J.R., Chin, M., Platnick, S. and Meyer, K., 2016. Positive low cloud and dust feedbacks amplify tropical North Atlantic Multidecadal Oscillation. *Geophysical Research Letters, 43*(3), pp.1349-1356.

Velasco, I., and J. M. Fritsch, 1987: Mesoscale convective complexes in the Americas. J. Geophys. Res., 92, 9561-9613.

Zhang, K., B. C. Douglas, and S. P. Leatherman, 1997: East coast storm surges provide unique climate record. Eos, Trans. Amer. Geophys. Uninon, 78, 395-397.

Zhang, K., B. C. Douglas, S. P. Leatherman, 1999: Twentieth-Century Activity along the U. S. East Coast. J. Climate, 13, 1748-1761.

CHAPTER 17

Florida Climate Variability and Prediction

Ben P. Kirtman[1], Vasubandhu Misra[2], Robert J. Burgman[3], Johnna Infanti[4], and Jayantha Obeysekera[5]

[1]Rosenstiel School of Marine & Atmospheric Science, University of Miami, Miami, FL; [2]Florida Climate Institute/Center for Ocean-Atmospheric Prediction Studies/Department of Earth, Ocean and Atmospheric Science, Florida State University, Tallahassee, FL; [3]Department of Earth and Environment, Florida International University, Miami, FL; [4]University Corporation for Atmospheric Research/Florida Atlantic University/University of Miami/United States Geological Survey; [5]South Florida Water Management District, West Palm Beach, FL

This chapter describes the sources and mechanisms for climate variability in Florida across timescales (i.e., seasonal-to-decadal) and how they are used to make predictions. Current capabilities in terms of prediction quality, with an emphasis on precipitation and land surface temperature on seasonal timescales, are introduced as well as challenges and opportunities for the future. The longer decadal time scales are discussed in the next chapter in conjunction with climate change associated with anthropogenic forcing.

Key Messages

- There is known large-scale climate variability (e.g., El Niño) that affect Florida's local climate.
- While this large-scale climate variability can be predicted several months in advance, correctly capturing the regional impacts remains challenging.

Keywords

Multi-model ensembles; Regional climate prediction; Dynamical downscaling; Statistical downscaling

A Scientific Basis for Regional Climate Prediction – Global Drivers of Regional Florida Climate Variability

Florida's climate, including rainfall and temperature, is influenced by many modes of natural variability. Some of these modes are more significant than others; for example, the El Niño Southern Oscillation (ENSO), the Atlantic Multi-decadal Oscillation (AMO), and the Pacific Decadal Oscillation (PDO) have the strongest influence on Florida's climate variability. Other modes do influence Florida's climate, such as the North Atlantic Oscillation (NAO)/Arctic Oscillation (AO) and the solar cycles, but their impact is not as well known. These naturally occurring modes of variability impact Florida by their modification of global circulation anomalies, in particular, the subtropical and polar jet streams. In this section, we describe these modes of climate variability and their imprints on Florida's climate. The subsequent sections will show how these modes of variability can be used to provide the scientific basis for short-term climate prediction on regional scales. Additional information on Florida's climate and its drivers can be found in Misra et al. (2011), Obeysekera et al. (2011a), and Obeysekera et al. (2011b).

El Niño Southern Oscillation (ENSO)

The El Niño Southern Oscillation (ENSO) is characterized by large-scale atmosphere–ocean interaction in the Tropical Pacific, in which a 2- to 10-year oscillation of warm, neutral, and cool sea surface temperature anomalies (SSTAs) exists (Rasmusson and Carpenter 1982; Philander 1983; Rasmusson and Wallace 1983; Trenberth 1997; Cobb et al. 2003; among many others). These oscillatory swings in SSTAs are referred to as El Niño (warm phase), La Niña (cold phase), and neutral. El Niño and La Niña phases typically persist for 6 to 18 months. Though the center of SSTA activity for an El Niño or La Niña event is in the Tropical Pacific, the imprints of these events can be seen globally. As the tropical sea surface temperatures (SSTs) change, the location of tropical convection shifts, which in turn leads to large-scale, global changes in the atmospheric circulation that can alter temperatures, humidity, winds, clouds, and more (Alexander et al. 2002). For example, during an El Niño event (warm Tropical Pacific), warm SSTAs force evaporative anomalies leading to anomalous rainfall. These rainfall anomalies can be thought of as an atypical tropical heat source (or anomalous tropical forcing) in the middle of the atmosphere, which produces a Rossby wave response that reaches into the extra tropics (Sardeshmukh and Hoskins 1988; Trenberth 1997; Barsugli and Sardeshmukh 2002; Straus and Shukla 2002; among many others). This "teleconnection" of the Tropical Pacific to the North Pacific is sometimes referred to as "The Atmospheric Bridge" (Alexander et al. 2002). In turn, anomalous tropical heating (or forcing) leads to impacts in atmospheric circulation fields across the globe, but particularly influences climate over North America. Anomalous tropical forcing is strongest in winter, as this coincides with the mature stage of events, and teleconnections are often strongest in winter (January-March, or JFM), as well (Trenberth et al. 1998). Though ENSO has a periodicity of 2-10 years, the impacts can be seen on timescales ranging from sub-seasonal (on the order of 10 days; e.g. Hudson et al. 2011; Rasmussen et al. 2015, though these results are not specific to North America) to seasonal (on the order of three months) and longer (Trenberth et al. 1998). ENSO teleconnections are far-reaching; however, the southern US, including Florida, is a key zone of strong association with ENSO forcing (Ropelewski and Halpert 1986).

Numerous references discuss the influence of ENSO on Florida's climate. The influence of El Niño (La Niña) on precipitation manifests as an increase (decrease) in wintertime precipitation (Gershunov and Barnett 1998a; Gershunov 1998; Goly and Teegavarapu 2014; Nag et al. 2014), and the influence of El Niño (La Niña) on wintertime temperature is cooling (warming). However, these impacts vary seasonally and spatially (see Table 17.1 adapted from the Florida

Climate Center and Fig. 17.1). Fig. 17.1 depicts the observed composite precipitation[1] anomaly over the southeastern US during JFM and July-August-September (JAS) El Niño events and La Niña events for the period 1982–2009. The strongest anomalies are seen during winter season El Niño and La Niña events, while the summer season shows a mixed, weak response; hence, we separately consider JFM and JAS. A similar assessment for 2 m temperatures[2] is illustrated in Fig. 17.2 where we see the strongest response is for JFM El Niño events, while the remaining composites show only a weak response to El Niño and La Niña events. Because of the strong teleconnection in the winter season, an El Niño or La Niña event leads to predictability of temperature and precipitation over the southeast US and Florida, although there is less predictability in summer months due to the weak response in the region. This notion is discussed later in this chapter.

The mechanism for the southeastern US response to warm tropical forcing is that changes in Tropical Pacific convection lead to a shift in the subtropical jet stream causing moisture to be advected across the region, and a trough in the midlatitude jet stream across the southeast allows precipitation in the region (Ropelewski and Halpert 1986; Leathers et al. 1991; Zorn and Waylen 1997). In contrast, during a La Niña event, there is poleward displacement of the midlatitude jet leading to zonal airflow and drying over the region (Ropelewski and Halpert 1986). The relevant 500 mb circulation anomalies are depicted in Fig. 17.3 for winter and summer El Niño and La Niña[3]. The strongest circulation response is seen in wintertime El Niño and La Niña, consistent with the strong response in temperature and rainfall patterns depicted in Figs. 17.1 and 17.2. In contrast, circulation anomalies are very weak in the summer season and during neutral years (neutral years not shown). While this pattern may be similar to the Pacific North American (PNA) teleconnection pattern, there is an important distinction. The positive (negative) PNA pattern tends to be associated with El Niño (La Niña) events, however the ENSO-forced and PNA teleconnection may be distinct (for further information, see Straus and Shukla 2002).

Table 17.1. El Niño, La Niña impacts for the four seasons (OND, JFM, AMJ and JAS) in Florida. Adapted from the Florida Climate Center Office of the State Climatologist (https://climatecenter.fsu.edu/topics/climate-variability).

Phase	Region	OND	JFM	AMJ	JAS
El Niño	Peninsular Florida	Wet Cool	Very Wet Cool	Slightly Dry	Slightly Dry or No Impact
	Western Panhandle	No Impact	Wet	Slightly Dry	No Impact
La Niña	Peninsular Florida	Dry Slightly Warm	Very Dry Warm	Slightly Wet	No Impact
	Western Panhandle	Slightly Dry	Dry	Dry	No Impact
Neutral	All Regions	No Impact	No Impact	No Impact	No Impact

[1] C CPC Unified Gauge-Based Analysis of Daily Precipitation over CONUS (Xie et al. 2010)
[2] GHCN-CAMS Gridded 2 m temperature over land (Fan and van den Dool 2008)
[3] NCEP-DOE Reanalysis 2 (Kanamitsu et al. 2002)

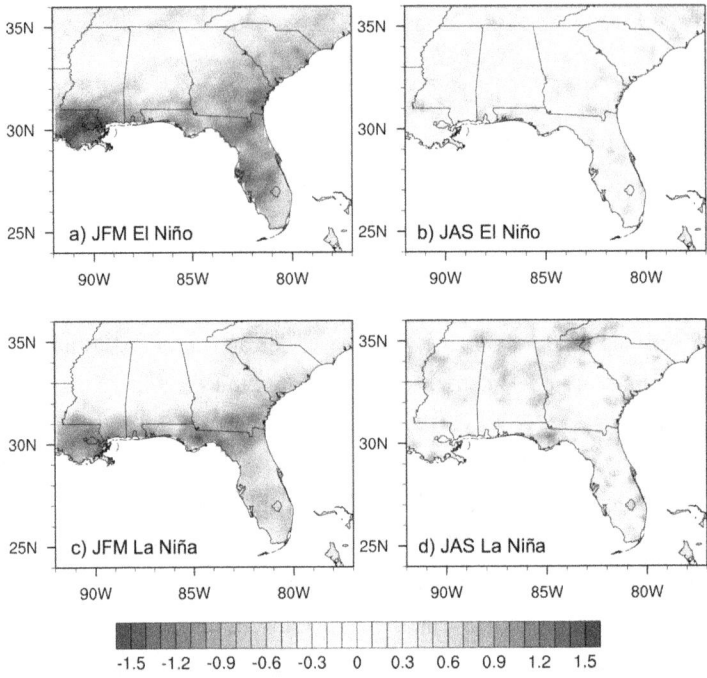

Figure 17.1. 1982–2009 precipitation anomalies (mm/day) during (a) JFM El Niño events, (b) JAS El Niño events, (c) JFM La Niña events, and (d) JAS La Niña events.

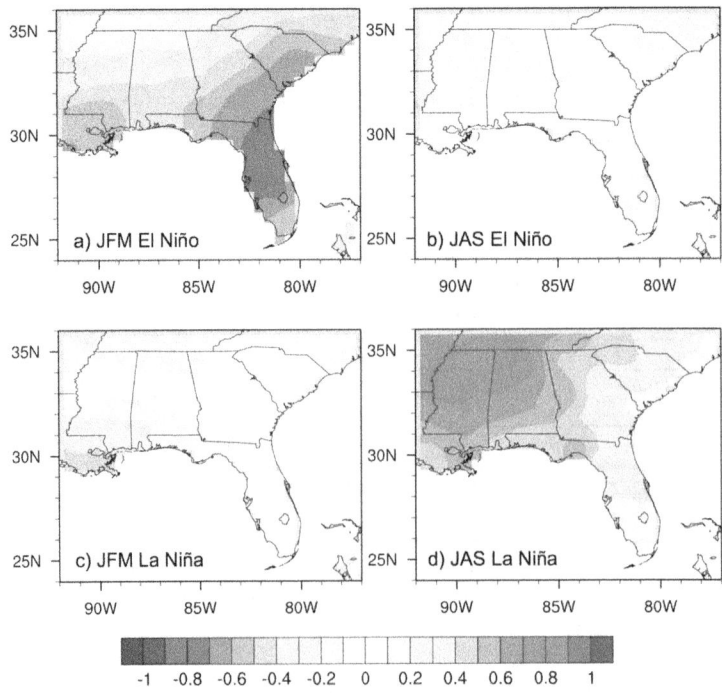

Figure 17.2. As in Fig. 17.1, but for 2 m temperatures.

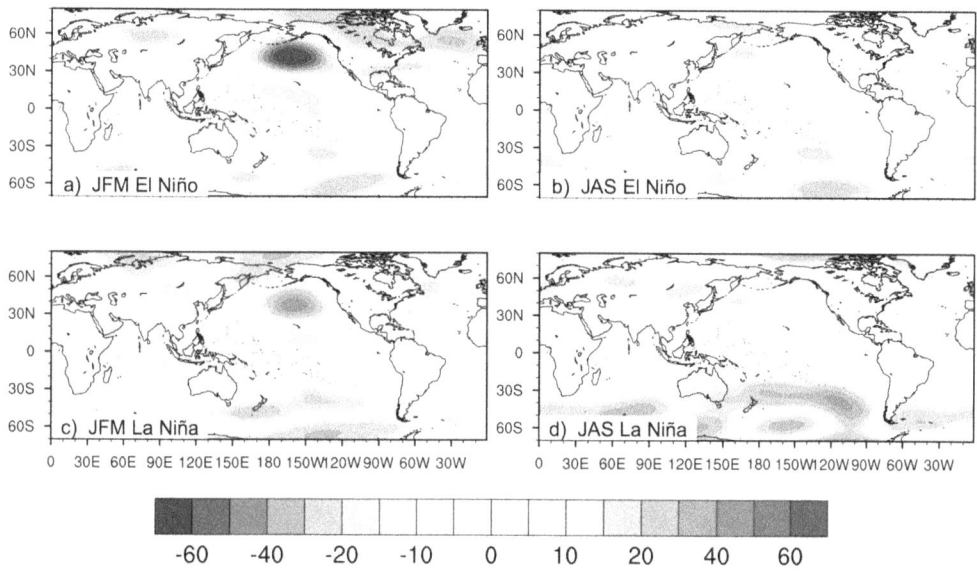

Figure 17.3. As in Fig. 17.1, but for 500 mb heights in meters.

The Pacific Decadal Oscillation

The Pacific Decadal Oscillation (PDO) is sometimes described as a long-lived, "El Niño-like" pattern of Pacific climate variability, and it is a natural mode of variability active on decadal timescales (e.g. Mantua and Hare 2002). Though ENSO and the PDO have similar spatial characteristics, they are very different in their time characteristics; where ENSO events persist for six to 18 months, PDO events persist for 20 to 30 years (but they can be sometimes as short as five years). While the SST signatures for the PDO and ENSO are very similar, the PDO shows strong SSTAs in the North Pacific and a weaker tropical signature, opposite to ENSO. The PDO index can be seen in Fig. 17.4[4]. As the PDO SST signature is very similar to ENSO, the imprints on Florida's climate are also very similar, but they occur across a longer timescale (Misra et al. 2011). Again, this relationship is primarily seen in the winter season, namely November through April (NDJFMA). The correlation of NDJFMA precipitation and temperature over Florida with the PDO index is illustrated in Fig. 17.5., which shows similarities in precipitation and temperature during ENSO events. We also see a strong positive correlation between PDO and precipitation and a weaker negative correlation between PDO and 2 m temperatures.

While the PDO and ENSO are distinct phenomena acting on different timescales, there is evidence of interaction between active PDO and ENSO phases. For example, when the PDO and ENSO are in phase (i.e. both are warm or cold), they constructively interfere, and when they are out of phase (i.e. one is cold and the other is warm, or vice versa), they destructively interfere (Gershunov and Barnett 1998b; Fuentes-Franco et al. 2016). During the destructive phase, ENSO

---

[4] http://ds.data.jma.go.jp/tcc/tcc/products/elnino/decadal/pdo.html

signal over the US is distorted, and it is strengthened during the constructive phase (Gershunov and Barnett 1998b). Overall, while the relationship between Florida precipitation or temperature and the PDO may be weaker than that between Florida precipitation or temperature and ENSO, it still explains about 25% of interannual dry season rainfall variability, making it a key player in Florida's climate variability.

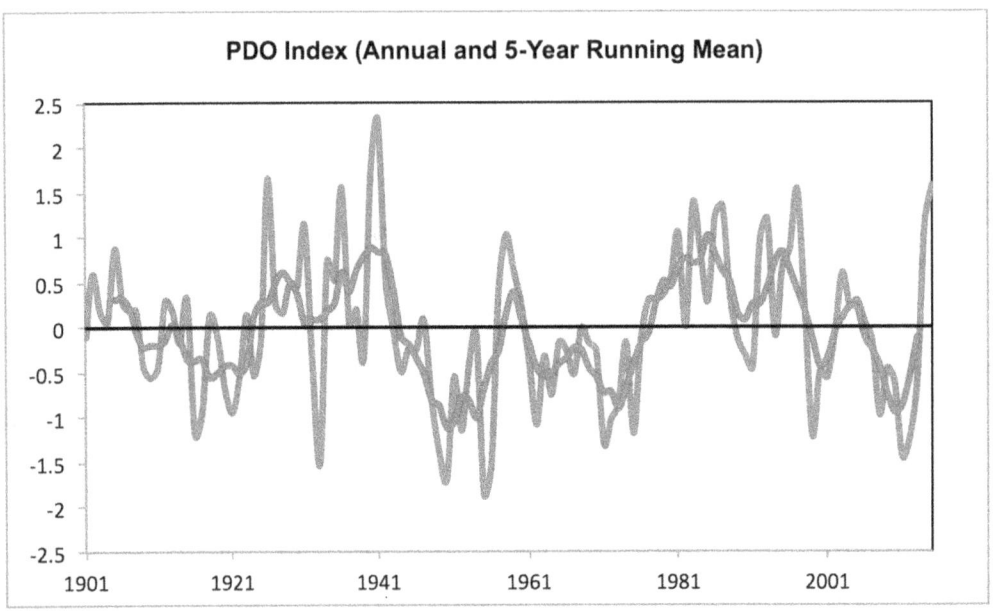

Figure 17.4. The PDO index. Annual means are depicted in blue, and 5-year running mean in red. Averages are defined at the center of each five-year period. This time series uses data from http://ds.data.jma.go.jp/tcc/tcc/products/elnino/decadal/pdo.html

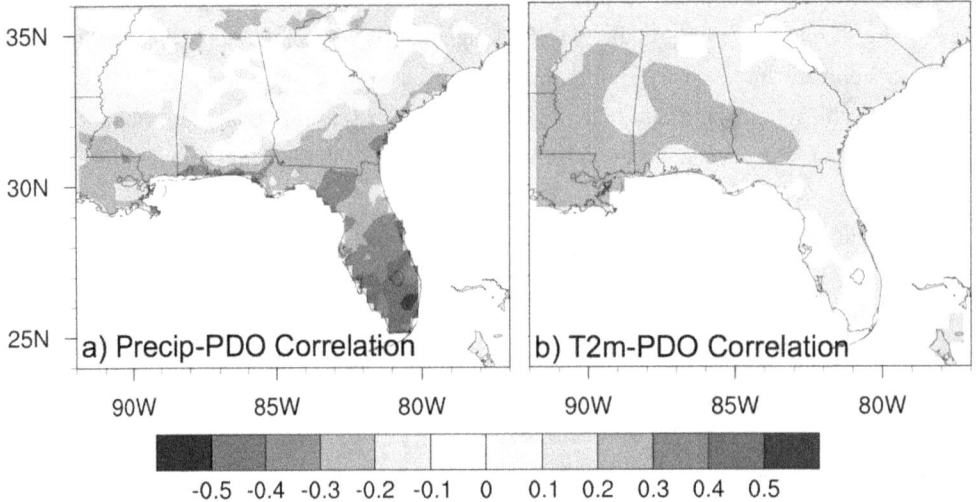

Figure 17.5. Correlation of the PDO index (see Fig. 17.6) with precipitation and temperature over Florida. The time period considered is NDJFMA from 1951 to 2010.

The Atlantic Multi-Decadal Oscillation

The Atlantic Multi-Decadal Oscillation (AMO) is a naturally occurring fluctuation in North Atlantic SSTs, characterized by a 65- to 70-year oscillation (Schlesinger and Ramankutty 1994). Traditionally, the AMO index is based on 10-year average SST anomalies in the North Atlantic, with the northern boundary at 60N (Enfield et al. 2001). The cool and warm phases of the AMO have a difference of about 1 °C. Shifts in the AMO cycle have been occurring for at least the last 1,000 years. While the AMO is an important feature in the long-term climate cycle, predicting it is difficult, as most climate models cannot skillfully predict when the AMO will change and they have difficulties simulating the AMO structure (Ruiz-Barradas et al. 2013). However, the AMO is one of the major contributors to multi-decadal variability over Florida.

While the AMO is a very long-term SST pattern that may not lead to year-to-year shifts in rainfall or temperature over Florida, the influence of this mode of climate variability can be seen across the globe (Knight et al. 2006). AMO SST anomalies most significantly influence summer precipitation over North America (i.e. June through August, or JJA) (Sutton and Hodson 2005), and the influence can be seen on Florida precipitation and temperature (Morss et al. 2005). Although the AMO is difficult to predict, its impacts on decadal-scale variations in summertime North American precipitation are consistent in observations and climate models when climate models represent the AMO (e.g. Schubert et al. 2009). Florida is no exception, and the influence can mainly be seen on summertime precipitation (Enfield et al. 2001) and Atlantic hurricanes (Trenberth and Shea 2006). Studies have shown that wintertime Florida precipitation shows no significant correlation with the AMO (Moses et al. 2013). At least twice as many tropical storms mature into hurricanes during AMO warm phases compared to AMO cold phases (Misra et al. 2011).

Analysis of precipitation extremes has shown that rainfall in South and Central Florida is more plentiful (less plentiful) during warm (cold) AMO phases (Enfield et al. 2001; Curtis 2008; Teegavarapu 2012). Florida precipitation, particularly in Central Florida, is in phase with the AMO. Fig. 17.6 shows the 10-year running mean Central Florida precipitation anomalies from 1905 to 2010[5] (top) and the 10-year running mean AMO index[6] (bottom) (Figure adapted from Enfield et al. 2001). The main signatures for the AMO can be seen for summertime precipitation over Florida; however, Moses et al. (2013) noted that the AMO can also account for some temperature variability in summertime (daytime high temperatures) in Miami, Fort Lauderdale, and Belle Glade (about 15 to 20%).

Hu et al. (2011) discussed the influence of the AMO on large-scale circulation fields, focusing on the summer season impacts on North American Precipitation. During the warm phase, there

---

[5] NOAA National Centers for Environmental information, Climate at a Glance: U.S. Time Series, Precipitation, published October 2016, retrieved on October 13, 2016 from http://www.ncdc.noaa.gov/cag/

[6] AMO smoothed long time series accessed from http://www.esrl.noaa.gov/psd/data/timeseries/AMO/. Timeseries are calculated from the Kaplan SST dataset (Kaplan et al. 1998)

is an anomalous low over North America, the subtropical North Atlantic, and over the eastern subtropical Pacific. During the cold phase, the pattern shows an anomalous low over North America, with a high-pressure anomaly over the eastern subtropical Pacific and subtropical North Atlantic. A schematic diagram (adapted from Hu et al. 2011) is shown in Fig. 17.7. This modification of the pressure pattern due to the AMO is one of the mechanisms leading to climatic shifts in precipitation over Florida on multi-decadal timescales.

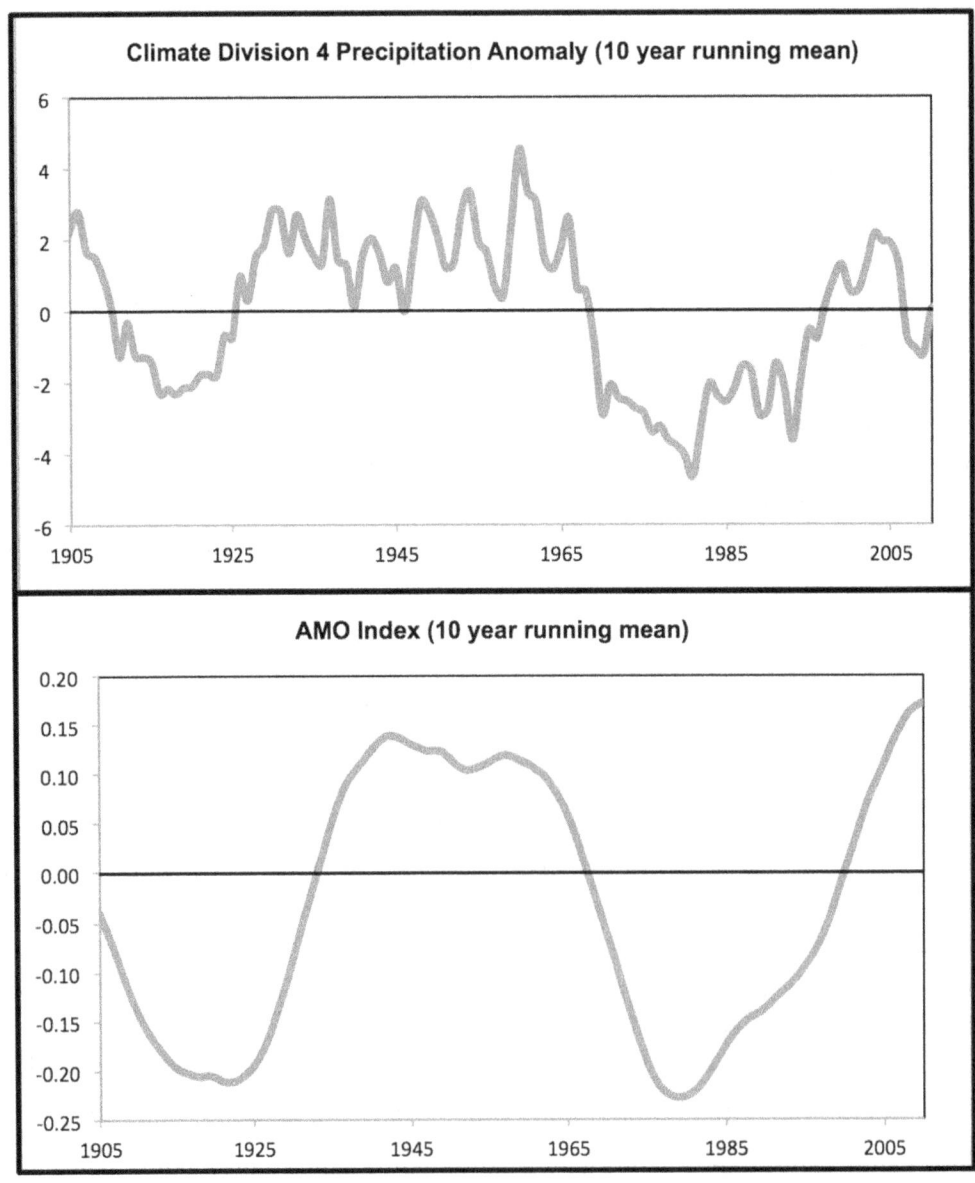

Fig. 17.6. Florida Division 4 rainfall anomalies (10-year running mean) and the AMO index from 1905 to 2010, adapted from (Enfield et al. 2001). The 10-year running mean is defined as the period from 1891–1905 (for example), and this 10-year period is defined at 1905.

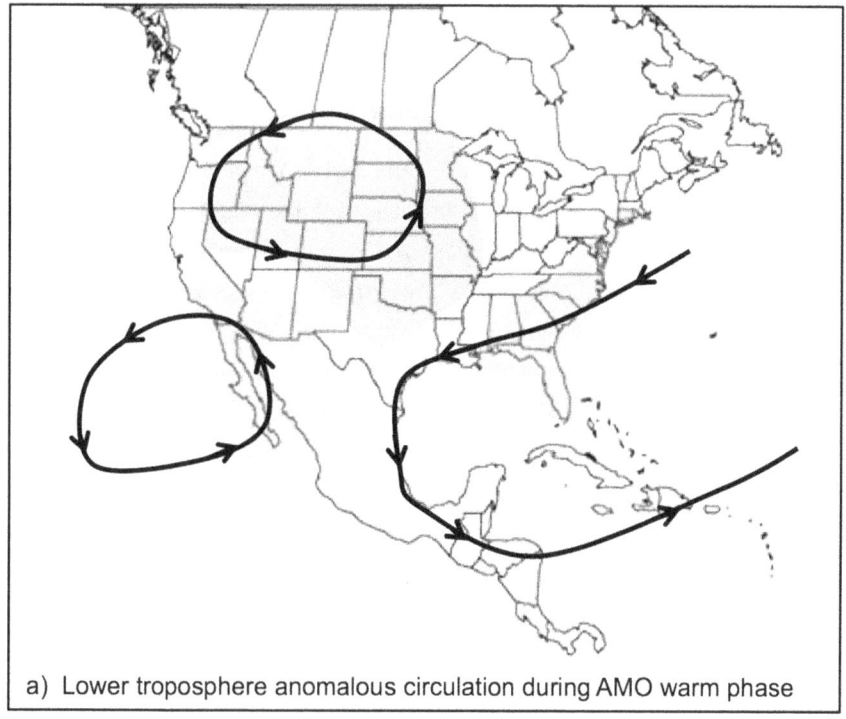

a) Lower troposphere anomalous circulation during AMO warm phase

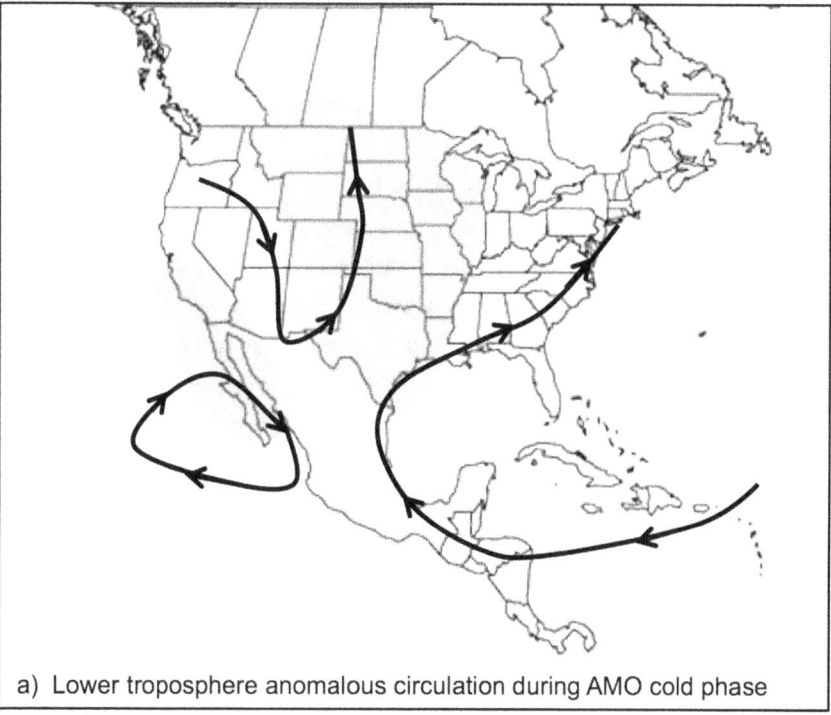

a) Lower troposphere anomalous circulation during AMO cold phase

Figure 17.7. A schematic summary of pressure and flow anomalies in the lower troposphere during AMO phases, as well as the expected summer (JJA) precipitation anomalies. This schematic is adapted from Hu et al. (2011) and is an *approximate* representation.

Sources and Mechanisms for Florida Climate Prediction: Why We Can Predict

The discussion above of ENSO, PDO, and AMO emphasizes the physical phenomena leading to global climate variability and the regional impacts over Florida. Why do these phenomena provide the scientific basis for climate prediction? Certainly, a large part of the answer to this question is persistence, particularly for the lower frequency decadal modes of variability. Persistence can be understood by imagining the following scenario. Suppose there is a strong warm ENSO event today. We know that the SST anomalies associated with this warm event will remain in place or *persist* for several weeks – hence persistence. Similarly, if the PDO is in the anomalously warm state for several years, we expect increased chances of enhanced rainfall over Florida. Essentially, there is substantial predictive information in persistence, and this is quite useful. For the higher frequency modes such as ENSO it is necessary to predict the evolution or life cycle and this requires capturing the complex physical interactions and feedbacks that are the underlying mechanism for their existence. Again, using our example of ENSO, the life cycle of individual events is governed by air–sea interactions in the Tropical Pacific and the thermal inertia associated with temperature anomalies in the upper 400 m of the ocean. As the life cycle of ENSO progresses, SSTs in the Pacific shift, driving associated local shifts in rainfall that, in turn, disrupt atmosphere circulation patterns and storm tracks leading to climate variability in remote regions such as Florida. Predicting this life cycle is complex, and the remote impact is influenced by interactions and interference with other modes of climate variability, such as the PDO as discussed earlier.

The regional prediction is further complicated by local climatic interactions and feedback. For instance, suppose that the soil over a large region of Florida is anomalously wet. Soil moisture or soil water content has significant inertia or memory. This means that the wet conditions would persist for several weeks, leading to enhanced evaporation during this period. This enhanced evaporation leads to cooling of the surface temperatures and potential moistening the atmospheric boundary layer, which then leads to changes in local circulation and rainfall. Similar local interactions can also occur with SSTAs in the nearby oceans. Ultimately, useful predictions will need to include the effects of all large-scale climate drivers, the local impact of these drivers, as well as local climatic feedbacks and interactions. Indeed, the regional prediction challenge is daunting.

Tools for Predicting Near-Time Florida Climate Variability

The fundamental building blocks of any prediction system include: (a) observational data and observing networks, (b) systems for assimilating or filling in gaps in the observational estimates, and (c) statistical or (semi-) empirical models and/or dynamical models (e.g., computer models primarily based on governing physical laws). These building blocks are also intimately connected. For example, the development of statistical models strongly depends on the robustness of our observing networks, and the fidelity of data assimilation systems are often affected by the quality of dynamical computer models. Moreover, the design of observing networks is based on our understanding of the physical phenomenon, which is in part based on our dynamical and statistical models. The intent of the section is to briefly summarize the basic tools for predicting Florida climate variability and describe their current capabilities. A more detailed discussion on the strengths and limitations of prediction systems can be found in NRC 2010 (National Research Council 2010).

Observational estimates are essential as they: form the basis of our physical and dynamical understanding of climate system; are used to derive empirical/statistical models; are the basis of the "initial condition" of any statistical or dynamical prediction system; and are used to assess the quality of our prediction systems. Observations are measurements of climatically relevant variables (e.g., SSTs, rainfall, sea level, …) that are made in situ (rain gauge network or ocean buoy) and are also made using remote platforms such as satellites or weather balloons. The development and implementation of global prediction systems require observational estimates of the entire state of the climate system (ocean, land, atmosphere, and cryosphere). This is different from weather prediction, which largely focuses on just the state of the atmosphere (and to some degree the land surface). For example, we know that ENSO and the PDO (see section above) affect Florida rainfall anomalies. To predict the future evolution of ENSO and the PDO, we need observational estimates of the state of the ocean and the atmosphere in the Pacific. Moreover, observations of Florida land surface temperatures and rainfall are needed to understand, initialize, and verify the predictions of the local manifestation of the global drivers (i.e., ENSO and the PDO). The local land surface state is also likely to be important since, for example, soil moisture anomalies can affect the local recycling of rainfall and the persistence of drought, which can either destructively or constructively interfere with the signal from the global climate drivers. Indeed, regional climate prediction is a daunting observational challenge.

Another challenge is that observational estimates rarely come in a form that can be easily adapted for understanding, verifying, or initializing predictions. For example, there are many spatial-temporal gaps in the observational estimates, and techniques are required to fill in these gaps. There are a variety of these techniques with a wide range of complexity and sophistication that are generically referred to as data assimilation. Again, we do not provide an exhaustive discussion of data assimilation here and refer the reader to NRC 2010. Nevertheless, from the perspective of prediction, data assimilation blends observational data and models (either statistical or dynamical) to estimate the entire state of the climate system, and these state estimates are used to initialize predictions and, in some cases, to verify predictions.

The capstone of any prediction system is the dynamical and/or statistical model that is used to take the state of the climate system today (i.e., the initial condition) and evolve it into some estimate of the state of the climate system in the future. The dynamical or computer model-based approach uses the physical laws from geophysical fluid dynamics and thermodynamics, whereas the statistical or empirical models use relationships derived from observational estimates. Both approaches have strengths and weaknesses, and ultimately prediction data that is used to guide decisions about the future is based on a suite of dynamical and statistical tools. The pragmatic use of multiple prediction tools/systems is typically referred to as the multi-model approach, and is currently viewed as the best practice for prediction across various timescales (see Kirtman et al. 2014 for discussion of the utility of the multi-model approach). The current capability of seasonal prediction is summarized in the section below.

Seasonal Prediction

As discussed above, our ability to predict seasonal Florida climate variability is largely due the large-scale drivers associated with the ENSO. Lower frequency phenomena such as the AMO and PDO can interact and interfere with the ENSO signal over Florida, and are important components of decadal prediction. However, since decadal prediction is also largely influenced by climate change associated with changes in atmospheric composition (e.g., CO_2, methane, aerosols), these decadal modes are discussed in more detail in the next chapter. The interested reader is referred to Kirtman et al. (2013) for a complete discussion of the issues and challenges associated with prediction over timescales from weeks to decades. Here, we focus on current capabilities in seasonal prediction with an emphasis on Florida. We begin with a brief description of the North American Multi-Model (NMME; Kirtman et al. 2014; www.cpc.noaa.gov/products/NMME), which is an official NOAA operational product and an excellent example of the current state-of-the-art in real-time/operational prediction. However, we also use the NMME as an example of one of the grand challenges in regional seasonal prediction, namely, how to make predictions with sufficient regional spatial-temporal resolution as to be useful for decision support.

The North American Multi-Model Ensemble Prediction System

Weather and climate forecasts are necessarily uncertain, and if the forecasts are to be used for effective risk assessment and decision support this uncertainty must be quantified. In terms of global prediction systems, the uncertainty or forecast probability distribution is typically estimated by making ensemble predictions with perturbed initial conditions and multiple models (see Kirtman et al. 2014). The use of multiple models or perturbing the physics in a single model serves to probe uncertainty due structural errors in model formulation. The NMME Project is an integrated research and operations partnership specifically designed to ensure that forecast uncertainty is adequately quantified.

The multi-model ensemble approach has proven extremely effective at quantifying prediction uncertainty due to uncertainty in model formulation, and it has proven to produce better prediction quality (on average) than any single model ensemble. There are numerous examples of how this multi-model ensemble (MME) approach yields superior forecasts compared to any single model. For example, Fig. 17.8, which is compiled from the NMME project data, compares the Ranked Probability Skill Score (RPSS) of the Climate Forecast System version 2 (CFSv2) (right panel; single NOAA operational model) to the grand NMME ensemble (left panel) for the DJF precipitation forecast for North America at six months lead. Details of how RPSS is calculated are omitted here; the important point to note is that larger values correspond to more useful information that can be translated into economic value. In terms of interpreting the results in Fig. 17.8, it is clear that the RPSS is larger for the MME compared to the single model. This improvement is particularly notable for Florida.

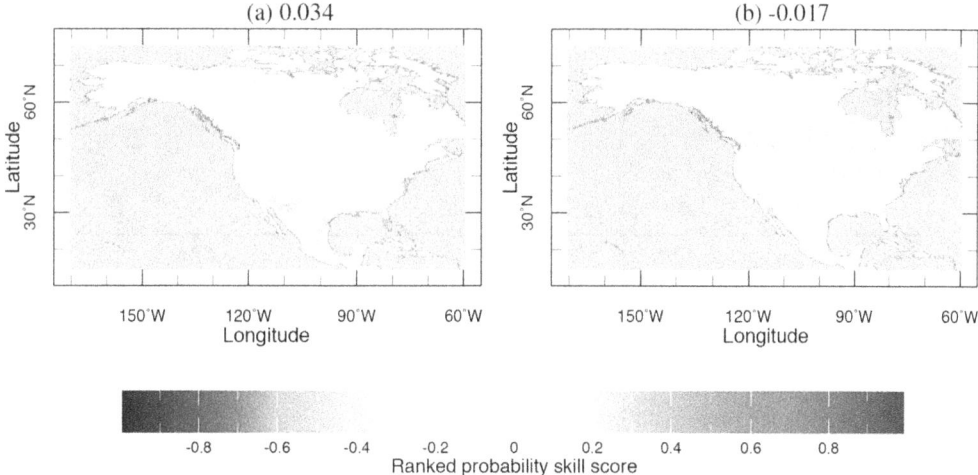

Figure 17.8. Precipitation forecast Rank Probability Skill Scores (RPSS) for the grand NMME (left panel) and for CFSv2 (right panel; single NOAA operational model). The skill is based on hindcasts initialized in July 1982-2010 and verifying the following DJF seasonal mean for tercile forecasts. Positive values indicate probabilistic skill that is better than climatology, and negative values indicate probabilistic skill that is worse than a climatological forecast. Area average RPSS is noted in the figure.

An example of a real-time operational North American rainfall anomaly forecast from the NMME project is presented in Fig. 17.9. The probability forecast indicates a 40-50% chance of below normal rainfall, allowing for the possibility (i.e. uncertainty) of near normal or even above normal rainfall throughout most of the southern tier of the US including Florida. The forecast also indicates a fairly strong probability of above normal rainfall in the northern tier of the US. Note that probability forecasts can be refined for different categories (e.g., quintiles) or even thresholds of exceedance. Finally, experimental CCSM4 forecasts with a regional southeast US focus are available at (http://benkirtman.weebly.com/climate-forecasts.html). CCSM4 is one of the many climate models included in the NMME suite.

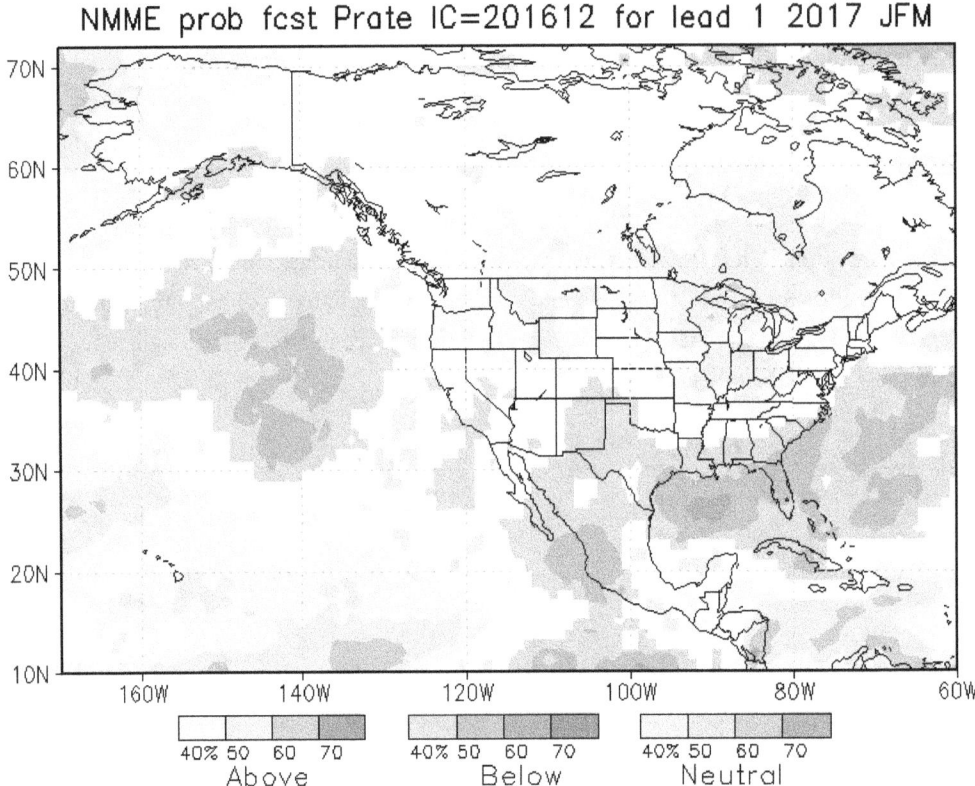

Figure 17.9. Example of NMME real-time operational rainfall forecast initialized 1 December 2017 and valid for January-February-March 2017 (JFM2017).

Regional Florida Prediction: Statistical Downscaling

The increased desire for localized climate prediction and projections makes the raw output from current global climate models, which are comparatively coarse in spatial resolution, inadequate (Giorgi et al. 2009). In order to overcome this limitation, various techniques for regionalization or downscaling of global model analyses or predictions have been adopted. These methods have been traditionally classified as either statistical or dynamical downscaling(Wilby and Wigley 1997). Although many different statistical downscaling tools exist (e.g. Wilby et al. 2004), they all essentially seek to relate the regional- tolocal-scale predictands with large-scale predictors (Hewitson and Crane 1996). Dynamic downscaling techniques use numerical climate models restricted to a regional domain of interest and forced at the lateral boundaries of the regional domain by the coarser global model or analysis (Warner et al. 1997). There is no clear consensus on one method being superior to the other (Wilby and Wigley 1997), although the statistical downscaling approach is far less computationally intensive than the dynamic downscaling approach. However, one of the two approaches may be more appropriate than the other for a particular region or end-use. For instance, the statistical downscaling approach may be less suitable in regions with unreliable historical local climate data or where the local climate data may have an insignificant relationship with large-scale climate variations. Similarly, dynamic downscaling may be inefficient where statistical approaches provide reliable predictions or projections, or where the intent is not to downscale the full 4-D (three dimensions of space and time) of the local climate.

Statistical Downscaling

For regional decision support, the resolution of the NMME global forecasts is far too coarse for direct use by, for example, water resource managers or to drive decision support models used for extreme sea states. Work adopted from Tian et al. (2014) illustrates an example of the spatial resolution challenge in downscaling the NMME data to the National Land Data Assimilation System phase 2 (NLDAS-2) grid (0.125x0.125; Fig. 17.10). Better tools are required in order to provide environmental predictions on spatial scales of a few kilometers. However, this need for better tools introduces another source of uncertainty that is inadequately evaluated. For instance, a common approach is to downscale global predictions by "forcing" a regional model with the large-scale information from the global models. This introduces the need to quantify the uncertainty in the large-scale forcing that comes from initial conditions and model formulation.

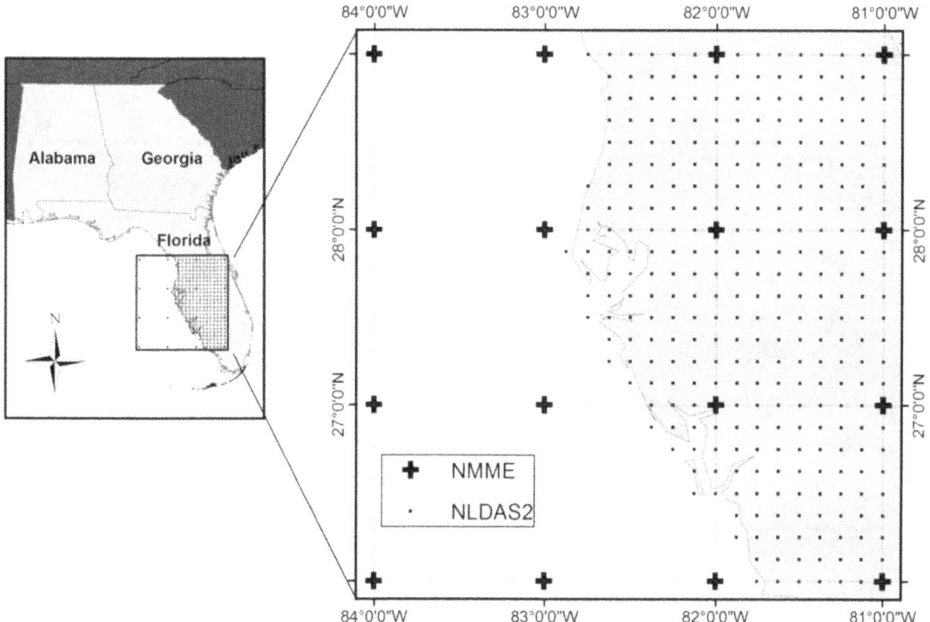

Figure 17.10. Example of large-scale NMME data to be downscaled to fine mesh for Regional Spectral Model (RSM) and other applications (Tian et al. 2014).

There are a number of statistical techniques for disaggregating or downscaling global-scale models such as those used in the NMME projects. Here, we briefly show results described in Tian et al. (2014); that is, (i) using quantile mapping on direct spatial disaggregation and bias correction of the NMME forecast, and (ii) the perfect prognosis approach using nonparametric locally weighted polynomial regression. Fig. 17.11 shows an example comparing skill scores for a one-month lead precipitation forecast for DJF using the two techniques from Tian et al. (2014). The skill scores used in Fig. 17.11 are the Mean Square Error Skill Score (MSESS) and the Brier skill score (BSS). Clearly, the disaggregation has more spatial heterogeneity than is possible when using the NMME forecast without the statistical downscaling.

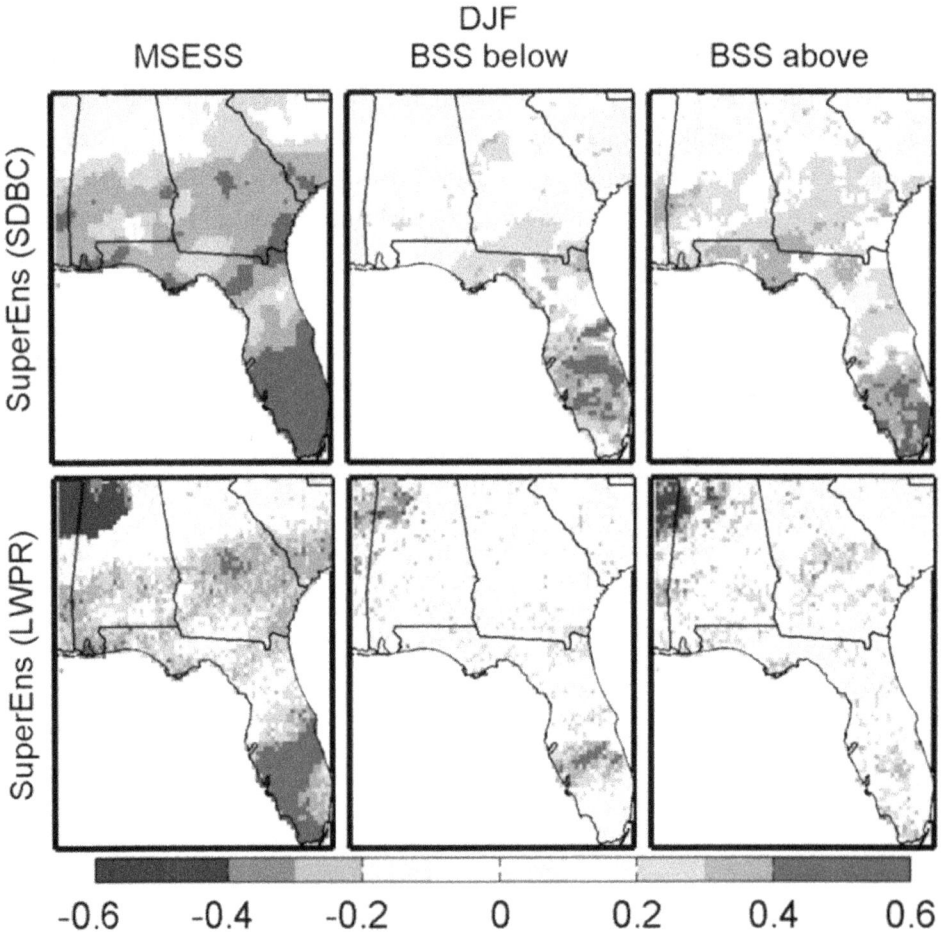

Figure 17.11. Precipitation forecast skill scores (MSESS, BSS) for one-month lead NMME seasonal forecasts downscaled using quantile mapping on direct spatial disaggregation and bias corrected forecast (top row) and nonparametric locally weighted polynomial regression (bottom row). Warm colors indicate skill above climatology whereas cold colors indicate skill worse than climatology. Figure adapted from Tian et al. (2014).

Dynamical Downscaling

The geography of Florida underscores the need for downscaling, as its peninsular structure is barely resolved in many of the current global climate models (Misra and Obeysekera 2011). Furthermore, the robust seasonal cycle of rainfall (Misra and DiNapoli 2013) and the significant contributions to Florida's hydroclimate from mesoscale events (ranging in spatial scales of 10-1000 km and temporal scales of one hour to a day) such as landfalling tropical cyclones (Knight and Davis 2009; Maxwell et al. 2012, 2013; Prat and Nelson 2013 a, b) and diurnal variations emanating from seabreeze thunderstorms (Misra et al. 2011b; Bastola and Misra 2013; Selman et al. 2013; Selman and Misra 2015) call for high resolution models to resolve these processes. Fig. 17.12 is a good illustration of dynamic downscaling, showing the accumulated rainfall from all landfalling tropical cyclones in the regional domain between 1948–2000 for El Niño (Figs. 17.12a-c) and La Niña (Figs. 17.12d-f) years. The dynamic downscaling was conducted at 10 km grid spacing from a coarser global reanalysis (Compo et al. 2011), which was at 2.5° (~300 km) grid spacing over a period of 104 years (1901–2004; Misra et al. 2012). The coarser analysis shows an unrealistically smooth distribution of rainfall from landfalling tropical cyclones in El Niño (Fig. 17.12a) and La Niña (Fig. 17.12d) years contrary to the corresponding observed rainfall distribution (Figs. 17.12c and f). The more detailed and inhomogeneous distribution of rainfall is described in the dynamic downscaling approach (Figs. 17.12b and e). There are, however, obvious differences between the observations (Figs. 17.12c and f) and the corresponding dynamic downscaling simulations (Figs. 17.12b and e) that relate to the limitations of the approach, including the use of an imperfect numerical model and the inherent chaotic nature of the regional climate system.

One of the major limitations of the dynamic downscaling approach is that it is significantly influenced by the quality of the coarser model forcing the lateral boundaries of the regional domain (Warner et al. 1997; Misra 2006). In the case of Florida, this becomes a significant issue when all current global climate models display a significant cold bias in the Gulf of Mexico and in the Caribbean Sea (Kozar and Misra 2012), which then translates to a dry bias over Florida (Selman et al. 2013). This cold bias in the global models is attributed to erroneous cloud simulations (Misra et al. 2009) and an erroneous Loop Current system in the Gulf of Mexico and the Caribbean Sea (Misra et al. 2017). The Loop Current system is a mesoscale ocean current system that is inadequately resolved in majority of the current global climate models (Liu et al. 2015).

More recently, Misra and Mishra (2017) used a regional coupled ocean–atmosphere model at 10 km grid spacing to downscale a coarser global ocean and atmospheric analysis over Florida in order to show that the SST variations from the variability of the Loop Current also influences the terrestrial summer season rainfall over peninsular Florida. For instance, they show that the systematic weakening of the Loop Current causes the Gulf of Mexico to become colder than normal. As a result, surface evaporation from the cold ocean surface reduces, which essentially results in less moisture for terrestrial convection over the comparatively warm peninsular Florida.

Studies such as Misra and Mishra (2017) show that the dynamic downscaling approach can be feasible for generating reliable predictions or projections for Florida despite an overwhelming display of systematic errors over the surrounding oceans by the current global models.

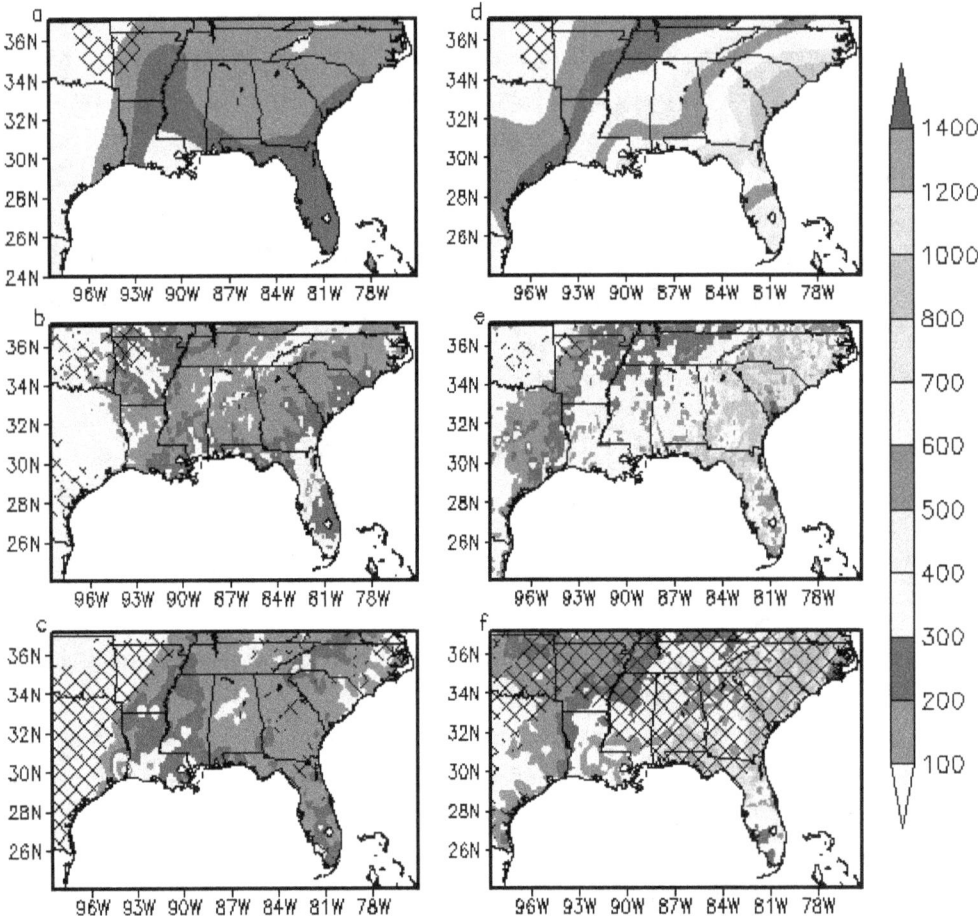

Figure 17.12. The composite rainfall from all landfalling hurricanes in the regional domain shown between 1948–2000 for El Niño years from a) global reanalysis (Compo et al. 2011), b) dynamic downscaling from global reanalysis (Misra et al. 2012), and c) observations (Higgins et al. 2000). Similarly, (d), (e) and (f) are the same as (a), (b), and (c) for La Niña years. The units are mm. The hashes represent statistical significance at 10% significance level from bootstrap method. Adapted from Misra et al. (2012).

Final Remarks

The demand continues to grow for prediction information on times scales of weeks to decades, as –many sectors of society are increasingly vulnerable to climate variability, and robust information saves lives and property. Indeed, there is also compelling evidence that as the climate continues to warm our vulnerabilities will continue to increase and that forecasts on response and adaptation timescales (i.e., weeks to decades) are essential for sustainable and resilient communities. It is also clear that there is much room for improvement in predicting regional climate variability and in using forecasts to inform decisions for societal benefit.

Clearly one of the most pressing challenges is improving regional forecasts. The current state-of-the-science is unclear about the best approach: do we invest in radical improvements to the resolution of global models? Or, do we focus on developing improved regional downscaling technology? Despite the lack of consensus on the best approach, there remains a pressing need to improve the global models and the regional downscaling techniques. This requires sustained efforts to enhance all of the building blocks of prediction systems (i.e., observations, data assimilation, and models).

Acknowledgments

The authors are grateful for feedback from the reviewers and the editors. The comments have greatly improved the chapter.

References

Anandhi, A. (2017). CISTA-A: Conceptual Model Using Indicators Selected by Systems Thinking for Adaptation Strategies in a Changing Climate: Case Study in Agro-Ecosystems. *Ecological Modelling, 345*, 41-55.

Alexander MA, Bladé I, Newman M, et al (2002) The Atmospheric Bridge: The Influence of ENSO Teleconnections on Air–Sea Interaction over the Global Oceans. J Clim 15:2205–2231. doi: 10.1175/1520-0442(2002)015<2205:TABTIO>2.0.CO;2

Barsugli JJ, Sardeshmukh PD (2002) Global atmospheric sensitivity to tropical SST anomalies throughout the Indo-Pacific basin. J Clim 15:3427–3442.
doi: 10.1175/1520-0442(2002)015<3427:GASTTS>2.0.CO;2

Cobb KM, Charles CD, Cheng H, Edwards RL (2003) El Nino/Southern Oscillation and tropical Pacific climate during the last millennium. Nature 424:271–276. doi: 10.1038/nature01779

Curtis S (2008) The Atlantic multidecadal oscillation and extreme daily precipitation over the US and Mexico during the hurricane season. Clim Dyn 30:343–351. doi: 10.1007/s00382-007-0295-0

Davis RE (1978) Predictability of Sea Level Pressure Anomalies Over the North Pacific Ocean. J Phys Oceanogr 8:233–246. doi: 10.1175/1520-0485(1978)008<0233:POSLPA>2.0.CO;2

Enfield DB, Mestas-Nuñez AM, Trimble PJ (2001) The Atlantic multidecadal oscillation and its relation to rainfall and river flows in the continental US. Geophys Res Lett 28:2077–2080.

Fuentes-Franco R, Giorgi F, Coppola E, Kucharski F (2016) The role of ENSO and PDO in variability of winter precipitation over North America from twenty first century CMIP5 projections. Clim Dyn 46:3259–3277. doi: 10.1007/s00382-015-2767-y

Gershunov A (1998) ENSO Influence on Intraseasonal Extreme Rainfall and Temperature Frequencies in the Contiguous United States: Implications for Long-Range Predictability. J Clim 11:3192–3203. doi: 10.1175/1520-0442(1998)011<3192:EIOIER>2.0.CO;2

Gershunov A, Barnett TP (1998a) ENSO Influence on Intraseasonal Extreme Rainfall and Temperature Frequencies in the Contiguous United States: Observations and Model Results. J Clim 11:1575–1586. doi: 10.1175/1520-0442(1998)011<1575:EIOIER>2.0.CO;2

Gershunov A, Barnett TP (1998b) Interdecadal Modulation of ENSO Teleconnections. Bull Am Meteorol Soc 79:2715–2725. doi: 10.1175/1520-0477(1998)079<2715:IMOET>2.0.CO;2

Goly A, Teegavarapu RSV (2014) Individual and coupled influences of AMO and ENSO on regional precipitation characteristics and extremes. Water Resour Res 50:4686–4709. doi: 10.1002/2013WR014540

Hu Q, Feng S, Oglesby RJ (2011) Variations in North American Summer Precipitation Driven by the Atlantic Multidecadal Oscillation. J Clim 24:5555–5570. doi: 10.1175/2011JCLI4060.1

Hudson D, Alves O, Hendon HH, Marshall AG (2011) Bridging the gap between weather and seasonal forecasting: intraseasonal forecasting for Australia. Q J R Meteorol Soc 137:673–689. doi: 10.1002/qj.769

Hewitson, B.C. and R.G. Crane, 1996: Climate downscaling: techniques and application. Climate Research, 7, 85-95.

Higgins RW, Shi W, Yarosh E, Joyce R (2000) Improved United States precipitation quality control system and analysis. NCEP/ CPC ATLAS No. 7.
 Also available at: http://www.cpc.ncep. noaa.gov/research_papers/ncep_cpc_atlas/7/index.html

Kirtman B, Anderson D, Brunet G, et al (2013) Prediction from Weeks to Decades. In: Asrar GR, Hurrell JW (eds) Climate Science for Serving Society. Springer Netherlands, pp 205–235

Kirtman BP, Min D, Infanti JM, et al (2014) The North American Multimodel Ensemble: Phase-1 seasonal-to-interannual prediction; Phase-2 toward developing intraseasonal prediction. Bull Am Meteorol Soc 95:585–601. doi: 10.1175/BAMS-D-12-00050.1

Liu, Y., Lee, S.-K., Enfield, D.B., Muhling, B.A., Lamkin, J.T., Muller-Karger, F., Roffer, M.A., 2015. Potential impact of climate change on the Intra-Americas Seas: part-1. A dynamic downscaling of the CMIP5 model projections. J. Mar. Syst. 148, 56–69, http://dx.doi.org/10.1016/j.jmarsys.2015.01.007.

Knight JR, Folland CK, Scaife AA (2006) Climate impacts of the Atlantic Multidecadal Oscillation. Geophys Res Lett 33:n/a-n/a. doi: 10.1029/2006GL026242

Kozar, M. and V. Misra, 2012: Evaluation of twentieth-century Atlantic Warm Pool simulations in historical CMIP5 runs Clim. Dyn., 41(9-10), 2375-2391, doi:10.1007/s00382-012-1604-9.

Leathers DJ, Yarnal B, Palecki MA (1991) The Pacific/North American teleconnection pattern and United States climate. Part I: Regional temperature and precipitation associations. J Clim 4:517–528.

Mantua NJ, Hare SR (2002) The Pacific decadal oscillation. J Oceanogr 58:35–44.

Maxwell JT, Soulé PT, Ortegren JT, Knapp PA (2012) Drought-busting tropical cyclones in the Southeastern Atlantic United States: 1950–2008. Ann Assoc Am Geogr 102(2):259–275.

Maxwell JT, Ortegren JT, Knapp PA, Soulé PT (2013) Tropical cyclones and drought amelioration in the gulf and Southeastern coastal United States. J Clim 26:8440–8452.

McCabe GJ, Palecki MA, Betancourt JL (2004) Pacific and Atlantic Ocean influences on multidecadal drought frequency in the United States. Proc Natl Acad Sci 101:4136–4141.

Misra, Vasubandhu, 2006: Addressing the Issue of Systematic Errors in a Regional Climate Model. J. Climate. 20, 801-818

Misra V, Carlson E, Craig R, Enfield D (2011) Climate scenarios: a Florida-Centric view.

Misra, V., L. Moeller, L. Stefanova, S. Chan, J. J. O'Brien, T. J. SmithIII, and N. Plant, 2011: The influence of Atlantic warm pool on panhandle Florida sea breeze. J. Geophys. Res., 116, doi: 10.1029/2010JD015367.

Misra, V. and J. Obeysekera, 2011: The inadequacies of IPCC AR4 models to project climate over Florida. In Climate Scenarios: A Florida-Centric View. Available from
 http://floridaclimate.org/climate_scenario_pdf.php

Misra, V. and S. DiNapoli, 2013: Understanding the wet season variations over Florida. Clim. Dyn., 40, 1361-1372. doi: 10.1007/s00382-012-1382-4.

Misra, V. and A. Mishra, 2016: The oceanic influence of the rainy season of Peninsular Florida. J. Geophys. Res., in press. doi: 10.1002/2016JD024824.

Misra, Vasubandhu, S. Chan, R. Wu, and E. Chassignet, 2009: Air-sea interaction over the Atlantic warm pool in the NCEP CFS.Geophys. Res. Lett., 36, L15702, doi: 10.1029/2009GL038525.

Misra, V., S. DiNapoli and S. Bastola, 2012: Dynamic downscaling of the 20th century reanalysis over the southeastern United States Regional Environmental Change, 13, S15-23, doi: 10.1007/s10113-012-0372-8.

Misra, V., A. Mishra, and H. Li, 2016: The sensitivity of the regional coupled ocean atmosphere simulations over the Intra-Americas Seas to the prescribed bathymetry. Dyn. Atm. and Ocean, http://dx.doi.org/10.1016/j.dynatmoce.2016.08.007.

Morss RE, Wilhelmi OV, Downton MW, Gruntfest E (2005) Flood Risk, Uncertainty, and Scientific Information for Decision Making: Lessons from an Interdisciplinary Project. Bull Am Meteorol Soc 86:1593–1601, 1527.

Moses CS, Anderson WT, Saunders C, Sklar F (2013) Regional climate gradients in precipitation and temperature in response to climate teleconnections in the Greater Everglades ecosystem of South Florida. J Paleolimnol 49:5–14. doi: 10.1007/s10933-012-9635-0

Nag, B. , V. Misra, and S. Bastola, 2014: Validating ENSO teleconnections on Southeastern United States Winter Hydrology Earth Interactions. DOI: EI-D-14-0007.1.

National Research Council (2010) Assessment of Intraseasonal to Interannual Climate Prediction and Predictability. The National Academies Press, Washington, DC

Obeysekera J, Irizarry M, Park J, et al (2011a) Climate change and its implications for water resources management in south Florida. Stoch Environ Res Risk Assess 25:495–516. doi: 10.1007/s00477-010-0418-8

Obeysekera J, Park J, Irizarry-Ortiz M, et al (2011b) Past and projected trends in climate and sea level for South Florida.

Philander SGH (1983) El Nino southern oscillation phenomena. Nature 302:295–301.

Prat OP, Nelson BR (2013a) Precipitation contribution of tropical cyclones in the Southeastern United States from 1998 to 2009 using TRMM satellite data. J Clim 26(3):1047–1062

Prat OP, Nelson BR (2013b) Mapping the world's tropical cyclone rainfall contribution over land using the TRMM multi-satellite precipitation analysis. Water Resour Res 49:7236–7254. doi:10.1002/wrcr.20527.

Rasmussen KL, Hill AJ, Toma VE, et al (2015) Multiscale analysis of three consecutive years of anomalous flooding in Pakistan. Q J R Meteorol Soc 141:1259–1276. doi: 10.1002/qj.2433

Rasmusson EM, Carpenter TH (1982) Variations in Tropical Sea Surface Temperature and Surface Wind Fields Associated with the Southern Oscillation/El Niño. Mon Weather Rev 110:354–384. doi: 10.1175/1520-0493(1982)110<0354:VITSST>2.0.CO;2

Rasmusson EM, Wallace JM (1983) Meteorological Aspects of the El Niño/Southern Oscillation. Science 222:1195–1202. doi: 10.1126/science.222.4629.1195

Ropelewski CF, Halpert MS (1986) North American precipitation and temperature patterns associated with the El Niño/Southern Oscillation (ENSO). Mon Weather Rev 114:2352–2362. doi: 10.1175/1520-0493(1986)114<2352:NAPATP>2.0.CO;2

Ruiz-Barradas A, Nigam S, Kavvada A (2013) The Atlantic Multidecadal Oscillation in twentieth century climate simulations: uneven progress from CMIP3 to CMIP5. Clim Dyn 41:3301–3315. doi: 10.1007/s00382-013-1810-0

Sardeshmukh PD, Hoskins BJ (1988) The generation of global rotational flow by steady idealized tropical divergence. J Atmospheric Sci 45:1228–1251.
doi: 10.1175/1520-0469(1988)045<1228:TGOGRF>2.0.CO;2

Schlesinger ME, Ramankutty N (1994) An oscillation in the global climate system of period 65-70 years. Nature 367:723–726.

Schneider N, Cornuelle BD (2005) The Forcing of the Pacific Decadal Oscillation. J Clim 18:4355–4373. doi: 10.1175/JCLI3527.1

Schubert S, Gutzler D, Wang H, et al (2009) A U.S. CLIVAR project to assess and compare the responses of global climate models to drought-related SST forcing patterns: overview and results. J Clim 22:5251–5272. doi: 10.1175/2009JCLI3060.1

Straus DM, Shukla J (2002) Does ENSO Force the PNA? J Clim 15:2340–2358. doi: 10.1175/1520-0442(2002)015<2340:DEFTP>2.0.CO;2

Sutton RT, Hodson DLR (2005) Atlantic Ocean Forcing of North American and European Summer Climate. Science 309:115–118. doi: 10.1126/science.1109496

Teegavarapu RS (2012) Floods in a changing climate: extreme precipitation. Cambridge University Press

Tian D, Martinez CJ, Graham WD, Hwang S (2014) Statistical Downscaling Multimodel Forecasts for Seasonal Precipitation and Surface Temperature over the Southeastern United States. J Clim 27:8384–8411. doi: 10.1175/JCLI-D-13-00481.1

Trenberth KE (1997) The definition of El Niño. Bull Am Meteorol Soc 78:2771–2777. doi: 10.1175/1520-0477(1997)078<2771:TDOENO>2.0.CO;2

Trenberth KE, Branstator GW, Karoly D, et al (1998) Progress during TOGA in understanding and modeling global teleconnections associated with tropical sea surface temperatures. J Geophys Res Oceans 103:14291–14324. doi: 10.1029/97JC01444

Trenberth KE, Shea DJ (2006) Atlantic hurricanes and natural variability in 2005. Geophys Res Lett 33:n/a–n/a. doi: 10.1029/2006GL026894

Warner, T. T., R. A. Peterson, and R. E. Treadon, 1997: A tutorial on lateral boundary conditions as a basic and potentially serious limitation to regional numerical weather prediction. Bull. Amer. Meteor. Soc., 78, 2599–2617.

Wilby, R.L., and T.M.L. Wigley, 1997: Downscaling general circulation model output: a review of methods and limitations. Progress in Physical Geography, 21, 530–548.

Wilby, R. L., et al., 2004: Guidelines for use of climate scenarios developed from statistical downscaling methods. IPCC task group on data and scenario support for impact and climate analysis (TGICA), http://ipcc.ddc.cru.uea.ac.uk/guidelines/StatDown_Guide.pdf

Zorn MR, Waylen PR (1997) Seasonal Response of Mean Monthly Streamflow to El Niño/Southern Oscillation in North Central Florida. Prof Geogr 49:51–62. doi: 10.1111/0033-0124.00055

CHAPTER 18

Future Climate Change Scenarios for Florida

Ben P. Kirtman[1], Vasubandhu Misra[2], Aavudai Anandhi[3], Diane Palko[1], and Johnna Infanti[4]

[1]*Rosenstiel School of Marine & Atmospheric Science, University of Miami, Miami, FL;* [2]*Florida Climate Institute/Center for Ocean-Atmospheric Prediction Studies/Department of Earth, Ocean and Atmospheric Science, Florida State University, Tallahassee, FL;* [3]*Biological and Agricultural Systems Engineering, Florida Agricultural and Mechanical University, Tallahassee, FL;* [4]*University Corporation for Atmospheric Research/Florida Atlantic University/University of Miami/United States Geological Survey*

This chapter describes both the nature of and anthropogenic mechanisms for climate change, as well as how scenarios and projections of future climate change are made. Specific emphasis is placed on understanding the changes over the near-term (i.e., adaption timescale) where the emission scenario has little impact vs. changes beyond the mid-century where the projections are conditional on the emission scenario. The various tools and models used to assess climate change are also summarized, and projections from global and regional models are presented. Finally, the new science of decadal prediction is presented as it has the potential to improve climate information in the near-term.

Key Messages

- The climate science community clearly understand that adaptation decision support needs robust regional information, and that the current generation of global models are not sufficient in this regard.
- Efforts to downscale the global models are promising but much remains to be done.

Keywords

Anthropogenically forced climate change; Decadal climate prediction; Climate projection; Climate scenario; Mitigation; Adaptation

Terminology and Definitions

The language of climate change and climate variability can often be confusing. In this section, we introduce terminology applicable to this chapter. The intent here is to clarify and simplify the discussion–we make no claim that this terminology list is complete, exhaustive, or universally excepted. Much of the discussion follows chapter 11 of the 2013 Intergovernmental Panel on Climate Change report (i.e., Kirtman et al. 2013). The important terms are first introduced in italic font.

Internally Generated vs. Externally Forced Climate Variability

The terms *climate change* and *climate variability* are typically used rather loosely. It is, perhaps, more well-defined to use the terms *externally forced climate variability* and *internally generated climate variability*. Externally forced climate variability (in the vernacular of climate change) describes how the climate system responds to changes in external forcing whether they be natural (e.g., changes in solar output, volcanoes, natural methane from permafrost melt, dust, continental drift) or anthropogenic, that is due to human activities (e.g., CO_2 concentrations from fossil fuel emissions, methane from natural gas production, land use and land cover change). Some confusion arises when it is unclear whether the externally forced climate variability is natural or anthropogenic. Throughout this chapter, we attempt to be clear about which type of variability we are referring to.

Internally generated climate variability (in the vernacular of climate variability) refers to the natural climate variability that would happen if all forcing (natural and anthropogenic) was fixed or unchanging. For example, the modes of climate variability discussed in Chapter 17 of this book —including the El Niño–Southern Oscillation (ENSO), the Pacific Decadal Oscillation (PDO), and the Atlantic Multidecadal Oscillation (AMO), among others—would occur without changes in the external forcing of the climate system. These modes are natural elements of the climate system that typically are due to interactions among the components of the climate systems (i.e., land–surface, sea–ice, ocean, and atmosphere). However, even though this internally generated climate variability exists without any changes in the external forcing, we cannot assume that changes in external forcing will not affect these natural modes. Indeed, the effect of increasing CO_2 levels on ENSO remains an active area of research, and remains very much an open science question.

Climate Prediction, Projection and Simulation, Scenario

There is also a distinction between a *climate projection* and a *climate prediction*. A climate projection is a statement about the future of the climate system that is conditional on the changes in the external forcing. For example, one might ask what is the state of the climate system 100 years from now if we assume CO_2 will increase by 1% per year or 2 % per year? The response would be very different if the *scenario* is a 1% vs. 2% per year increase. The science of climate projection, therefore, is highly dependent on the specific future scenario for the anthropogenic external forcing. In the Fifth Assessment Report of the Intergovernmental Panel on Climate Change (IPCC), these scenarios are referred to as Representative Pathway Concentrations (RCPs) and are typically formulated by economic assumptions since, for example, CO_2 emissions are well correlated with gross domestic production.

In contrast, a climate prediction is conditional on the external forcing and the initial condition (see Chapter 17 for a more detailed discussion of this). Simply put, a climate prediction attempts to capture the evolution of the natural modes of variability and, at the same time, the response to the changes in the external forcing. For the *seasonal predictions* discussed in Chapter 17, the initial condition is of paramount importance and the external forcing is relatively unimportant. For longer timescale prediction such as decadal both the initial condition and the evolving external forcing are important. When the timescales of interest are even longer (i.e., greater than say 20–30 years), then the initial condition is of much smaller importance and the external forcing is paramount. At very long timescales (i.e., beyond 30 years), for all practical purposes and assuming the same external forcing scenario, climate projection and climate prediction are indistinguishable. On the other hand, assuming the same external forcing scenario a ten-year prediction and projection may be very different.

We also need to make the distinction between a climate projection and a *climate simulation*. Much like a climate projection, a climate simulation is a computer model-based depiction of the evolution of a climate system conditional on the historical or past-observed external forcing. The projection is conditional on the assumed or projected external forcing into the future. Sometimes climate simulations are referred to as historical runs or historical simulations. Fig. 18.1, for example, shows an ensemble of climate simulations (gray curves). Each ensemble member or individual simulation was started with slightly different initial conditions and/or different models so that each simulation has different internally generated climate variability. However, all the ensemble members have the same externally prescribed forcing, so that the ensemble mean or average of all the ensemble members across all models is an estimate of the observed (black curve) externally forced climate over the past. The climate simulation can simply transition into a projection as the external forcing evolves into an assumed future evolution (various colored curves or the RCPs).

A *scenario* is a coherent and plausible description of a possible future state of the world. Scenarios are not projections or predictions, neither predicting nor forecasting future conditions. They differ from forecasts, which impose patterns extrapolated from the past onto the future. Since climate scenarios envisage assessment of future developments in complex systems, they are often inherently unpredictable, insufficiently assessed, and have high scientific uncertainties. The climate scenario differs from *climate projection* in that it refers to a description of the response of the climate system to a scenario of greenhouse gas and aerosol emissions, as simulated by a climate model. Climate projections alone rarely provide sufficient information to estimate future impacts of climate change because the model outputs commonly have to be manipulated and combined with observed climate data to be usable, for example, as inputs to impact models. Similarly, a *climate scenario* and a *climate change scenario* are also different, as the term climate change scenario refers to a representation of the difference of some plausible future climate from the current climate or a control climate, adapted from a climate model (IPCC 2001). A climate change scenario can be viewed as an interim step towards constructing a climate scenario because a climate scenario requires combining the climate change scenario with the observed current climate.

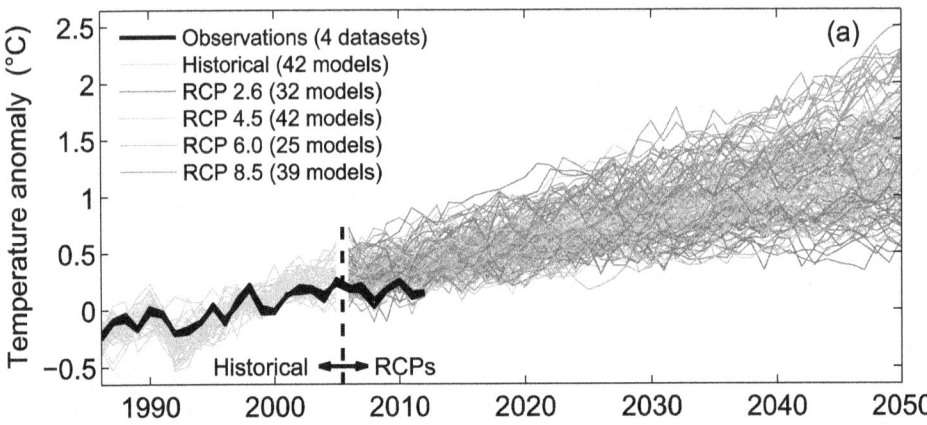

Figure 18.1. Climate simulations and projections of annual mean global mean surface temperature 1986–2050 (anomalies relative to 1986–2005). Projections under all RCPs from the Coupled Model Intercomparison Project Phase 5 (CMIP5) models (grey and colored lines, one ensemble member per model), with four observational estimates for the period 1986–2012 (black lines). Figure taken from Kirtman et al. (2013).

Near-Term vs. Long-Term Climate

In part driven by the distinction between climate projection and climate prediction, we also make the distinction between *near-term* climate and *long-term* climate. Near-term refers to the period from the present day to the mid-century and long-term refers to the period from the mid-century until 2100 and perhaps beyond. This distinction is useful from at least three specific perspectives. First, in the near-term the response to plausible differences in external forcing scenarios are relatively small. To be clear, the evolution of external forcing remains very important. We are simply acknowledging that any differences between plausible scenarios does not emerge until about the mid-century. Essentially, over the next 20–30 years or so, we have already committed to a certain amount of climate variability (i.e., warming) due to past anthropogenic external forcing. An example of this relative insensitivity to external forcing scenario and the increases in the global mean surface air temperature projections are shown in Fig. 18.2. Differences in the global mean surface temperature projections do not become significant until about the mid-century.

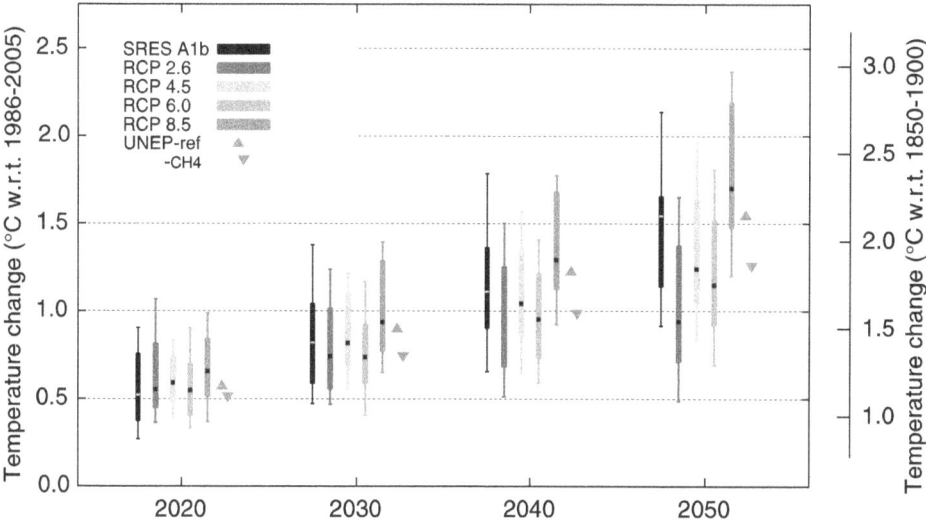

Figure 18.2. Near-term increase in global mean surface air temperatures (°C) across scenarios. Increases in 10-year mean (2016–2025, 2026–2035, 2036–2045 and 2046–2055) relative to the reference period (1986–2005) of the globally-averaged surface air temperatures. Results are shown for the CMIP5 model ensembles for RCP2.6 (dark blue), RCP4.5 (light blue), RCP6.0 (orange), and RCP8.5 (red) and the CMIP3 model ensemble (22 models) for SRES A1b (black). The multi-model median (square), 17 to 83% range (wide boxes), 5 to 95% range (whiskers) across all models are shown for each decade and scenario. Also shown are best estimates for a UNEP scenario (UNEP-ref, grey upward triangles) and one that implements technological controls on methane emissions (UNEP CH4, red downward-pointing triangles) (UNEP and WMO 2011; Shindell et al. 2012). Both UNEP scenarios are adjusted to reflect the 1986–2005 reference period. The right-hand floating axis shows increases in global mean surface air temperature relative to the early instrumental period (0.61 °C), defined from the difference between 1850–1900 and 1986–2005 in the Hadley Centre/Climate Research Unit gridded surface temperature data set 4 (HadCRUT4) global mean temperature analysis. Note that uncertainty remains on how to match the 1986–2005 reference period in observations with that in CMIP5 results. Figure from Kirtman et al. (2013).

Second, we want to make the distinction between *adaption* and *mitigation* in the context of near-term and long-term climate. This distinction is fairly straightforward since adaptation focuses on how ecosystems (including human activities) respond to both internally generated and externally forced climate variability. Adaptation is defined as 'adjustment in ecosystem management in response to actual or expected climatic stimuli or their effects, which moderates harm or exploits beneficial opportunities' (Anandhi 2017). There are three levels of adaptation, depending on the degree of change and the benefits of adaptation: 1) Incremental adaptation refers to changes in practices and technologies within an existing system. These are tactical choices requiring minimal financial investment, few cropping seasons for the mastery of associated managerial skills, and they can be reversed from one cropping season to another. 2) Systems adaptation are changes to an existing system, such as new crop types that are mapped against an increasing degree of change. 3) Transformational adaptation refers to the more radical end of a spectrum of change, such as a change inland use. Adaptations become systemic and then transformational in proportion to their irreversibility, capital requirements, life time, and impact

(Anandhi 2017). These are discussed in detail in Chapter 8. Adaptation is typically a near-term issue since we have already committed to a certain level of warming, and ecosystems will necessarily have to respond. Mitigation is about reducing or modifying external anthropogenic forcing either through reducing greenhouse gas emissions or through some sort of geoengineering solution to enhance to sinks of greenhouse gases (e.g., scrubbing CO_2 from the atmosphere). This is more of a long-term issue since the changes in external forcing or the emergence of a geoengineering solution will mostly affect long-term climate. Third, we noted above that decadal climate prediction is at the boundary where both initial condition and external forcing are important. Therefore, decadal climate prediction is primarily a near-term climate problem that is potentially useful for adaptation, whereas climate projections that reach 2100 are more aptly used for mitigation.

Criteria for Selection of Climate Scenario

Not all imaginable futures can be viable scenarios of future climate. The suitability of each type of scenario for use in policy-relevant impact assessments can be evaluated based on the following five criteria (Mearns et al. 2001; Anandhi 2017):

- Physical plausibility and realism: Changes in climate should be physically plausible, such that changes in different climatic variables are mutually consistent and credible.
- Consistency at regional level with global projections: Scenario changes in regional climate may lie outside the range of global mean changes but should be consistent with the theory and model-based results.
- Appropriateness of information for impact assessment: Scenarios should present climate changes at an appropriate temporal and spatial scale, for a sufficient number of variables, and over an adequate time horizon to facilitate impact assessment.
- Representativeness of regional climate: Scenarios should represent the potential range of future regional climate change.
- Accessibility: The information required for developing climate scenarios should be readily available and easily accessible.

Types of Scenarios

Four types of climate scenarios have been adapted in impact assessments (Mearns et al. 2001; Anandhi 2007), namely: incremental scenarios, analogue scenarios, a general category of "other scenarios," and scenarios based on the outputs from climate models. The most commonly used scenario type is based on outputs from climate models. The other three types have usually been applied with reference to or in conjunction with model-based scenarios.

- **Incremental scenarios** describe techniques where particular climatic (or related) elements are changed incrementally by arbitrary amounts (e.g., +1, +2, +3, +4°C change in temperature). These scenarios are also referred to as synthetic scenarios (IPCC 1994), as they

do not necessarily present a realistic set of changes that are physically plausible. They are usually adapted for exploring system sensitivity prior to the application of more credible, model-based scenarios (Anandhi et al. 2016).
- **Analogue scenarios** are constructed by identifying recorded climate regimes, which may resemble the future climate in a given region. Both spatial and temporal analogues have been used in constructing climate scenarios.
  - **Spatial analogues** are regions which currently have a climate analogous to that anticipated in the study region in the future. For example, using a region in Africa as a spatial analogue for the potential future climate over South Florida.
    - **Temporal analogues** make use of climatic information from the past as an analogue for possible future climate. They are of two types: palaeoclimatic analogues and instrumentally based analogues. Palaeoclimatic analogues are based on reconstruction of past climate periods from fossil evidence, such as plant or animal remains and sedimentary deposits. Examples of past periods are the mid-Holocene and the Last (Eemian) Interglacial. Periods of observed global scale warmth during the historical period have also been used as analogues of a greenhouse gas induced warmer world (instrumentally based analogues).
- **Scenarios Based on Outputs from Climate Models:** Climate models at different spatial scales and levels of complexity provide a major source of information for constructing scenarios. General circulation models (GCMs), regional climate models (RCMs), and a hierarchy of simple models produce information at the global scale.
  - **Scenarios from simple climate models:** As these models are seldom able to represent the non-linearities of some processes that can be captured by more complex models, the outputs from these models have been used mostly in conjunction with GCM information to develop scenarios using pattern-scaling techniques.
  - **Scenarios from GCMs:** From the early 1990s, GCM-based scenarios generally refer to outputs from coupled Atmosphere-Ocean GCMs (AOGCMs). AOGCM simulations start by modeling historical forcing by greenhouse gases and aerosols from the late 19th or early 20th century onwards. Climate scenarios based on these simulations are being increasingly adopted in impact studies along with scenarios based on ensemble simulations and scenarios accounting for multi-decadal natural climatic variability. There are several limitations that restrict the usefulness of these outputs for impact assessment: (1) their coarse spatial resolution compared to the scale of many impact assessments; (2) the difficulty of distinguishing an anthropogenic signal from the noise of inherent internal model variability; and (3) the difference in climate sensitivity between various models. In spite of these limitations, AOGCMs are widely used for developing climate scenarios for quantitative impact assessments.
    - **Downscaled scenarios:** The difficulty encountered in using the scenarios from GCMs has been the mismatch of spatial scales between GCMs and local impact assessments

(Anandhi et al. 2011). To overcome this mismatch, scenarios from GCMs at a global scale are translated to scenarios at regional or local scale using downscaling approaches. Two different downscaling approaches that are currently being pursued are dynamic downscaling and statistical downscaling. In the dynamic downscaling approach a RCM is embedded into GCM. There are two types of dynamic downscaling based on the types of nesting: one way nesting and two way nesting. Statistical downscaling involves developing quantitative relationships between large-scale atmospheric variables (predictors) and local surface variables (predictands). There are three types of statistical downscaling, namely weather types, weather generators, and transfer functions.

- **Other Types of Scenarios** Four additional types of climate scenarios have also been adopted in impact studies.
  - The first type involves extrapolating ongoing trends in climate that have been observed in some regions and that appear to be consistent with model-based projections of climate change. There are obvious dangers in relying on extrapolated trends, because if current trends in climate are pointing strongly in one direction, it may be difficult to defend the credibility of scenarios that posit a trend in the opposite direction, especially over a short projection period.
  - A second type of scenario uses empirical relationship between regional climate and global mean temperature from the instrumental record to extrapolate future regional climate on the basis of projected global or hemispheric mean temperature change. Again, this method relies on the assumption that past relationships between local- and broad-scale climates are applicable to the future conditions.
  - A third type of scenario is based on expert judgment, whereby estimates of future climate change are solicited from climate scientists. The results are sampled to obtain probability density functions of future change. The main criticism of expert judgment is its inherent subjectivity, including problems associated with the likely biases in questionnaire design and in comprehending information gathered from different scientists.
  - A fourth type of scenario is estimated from indicators. An indicator is defined as any variable that represents either the magnitude of an element (e.g., average annual precipitation), the variability of an element (e.g., coefficient of variation for annual precipitation) or the statistical relationship among elements (Anandhi 2017). Indicators are powerful tools to communicate climate change in relatively simple terms by portraying the interrelationships among climate and the ecosystems. They help reveal information on the impacts of climate change in the ecosystems, which can be useful in developing adaptation and mitigation strategies. For example, changes in first fall freeze or last spring freeze in Florida are useful in communicating some changes in climate for specific stakeholders and policy development. The scenarios developed from changes in

freeze are useful in portraying the interrelationships among climate and the citrus or strawberry growers for adapting/mitigating to the changes.

IPCC scenarios

In 1988, the Intergovernmental Panel on Climate Change (IPCC) was jointly established by the World Meteorological Organization (WMO) and the United Nations Environment Programme (UNEP) to assess the scientific, technical, and socio-economic information relevant to the understanding of climate change, its potential impacts, and options for adaptation and mitigation. Since its inception, reports by the IPCC have become the standard works of reference. They are widely used by policymakers, scientists, and other experts for assessing the causes of climate change, its potential impacts, and evolving response strategies. Further, the emission scenarios generated in them are widely used for driving AOGCMs to develop climate change scenarios, and the results are freely available for general use.

In 1992, the IPCC released a set of six global emissions scenarios (IS92a to f), called IS92 scenarios. These scenarios provide estimates of possible occurrences of greenhouse gases based on a wide array of assumptions. Out of the six scenarios, IS92a (also known as the "business as usual" scenario) has been widely adopted by the scientific community during the last decade. The IS92 scenarios were further updated in 2000 and the new set of emissions scenarios that were published in the Special Report on Emissions Scenarios (SRES) (Nakicenovic et al. 2000) are known as SRES scenarios. These SRES scenarios were constructed in a fundamentally different way, with a different range for each projection called a "storyline." There are four storylines (A1, A2, B1, and B2) that describe the way the world population, land use changes, new technologies, energy resources, economies, and political structure may evolve over the next few decades. Recently, four future scenarios' representative concentration pathways (RCPs) (Van Vuuren et al. 2011) have been used. The freely available, state-of-the-art multi-model dataset (multiple GCMs and RCPs) was designed to advance our knowledge of climate variability and climate change.

Near-Term and Long-Term Climate Projections

This section presents dynamical model-based near-term and long-term climate projections. We separate the results into those from the global models reported in the IPCC assessment (Stocker et al. 2013), North American Regional Climate Change Assessment Program (NARCCAP; Mearns et al. 2012), and the archive of statistically downscaled CMIP3 and CMIP5 Climate and Hydrology Projections (DHCP, Brekke et al. 2013).

Multi-Model Climate Projections from Global Models

The figures below (Fig. 18.3 and 18.4) show projections of surface temperature for eastern North America using global multi-model climate projections based on RCP4.5. The top panel shows an area average time series from 1900-2100 for the eastern third of the North America. The grey curves or the historical climate simulations and the colored curves are the results from the climate projections using the various RCPs as noted. The bottom rows show maps of the spatial distribution of the projected change over eastern North America for the near-term (2016–2035), the mid-century (2046–2065) and for the end of the century (2081–2100). The columns indicate, for each point on the map, the 25th, 50th, and 75th percentiles for the multi-model ensemble distribution. The hatching indicates regions where the differences of percentiles are less than the standard deviation of the model-estimated, internally generated present day climate variability. Simply put, the hatching indicates where the projections show little change relative to the present day.

Separate seasonal means for December through February (DJF) and June through July are shown in Figs 18.3 and 18.4. In terms of temperature, these are the extreme seasons and are often of the most interest. Typically, the temperature response is strongest when the background state is coldest; that is, in the higher latitudes and in the cold season (DJF). This is primarily because the land–atmosphere exchange through a comparatively stable atmospheric boundary layer is weaker than in the summer time. Usually in the summer season, the warming of the land surface often leads to increased atmospheric eddies allowing for a more robust exchange of heat and moisture fluxes between land and atmosphere, which moderates the response of land surface temperature to anomalous radiative forcing from increased greenhouse gas emissions. As expected, the temperature response is also largest in the long term. Florida is somewhat in contrast with the rest of eastern North America in that its largest temperature responses are in June through August season.

The rainfall response is presented in Figs. 18.5 and 18.6. In contrast to temperature, the hatched regions are more extensive indicating that the rainfall response does not exceed the internally generated climate variability of the present day. The exception to this is in the far southeast US, and in Florida in particular, where the enhanced dry season rainfall is relatively strong and positive across all timescales. For this emission scenario (RCP4.5), the signal during the wet season over Florida is relatively weak, but indicates small increases (<10%) in rainfall. The larger or stronger emissions scenario (RCP8.5; not shown) indicates a considerably stronger response over Florida in the long term. Interestingly, the multi-model mean in the June-August period at the end of the century with RCP8.5 indicates a 20–30% *reduction* in Florida relative to the present day, whereas the September-November period has a 10–20% increase in rainfall. This seasonal dependence in the differences and in scenario are particularly challenging for planning and responding.

FUTURE CLIMATE CHANGE SCENARIOS FOR FLORIDA • 543

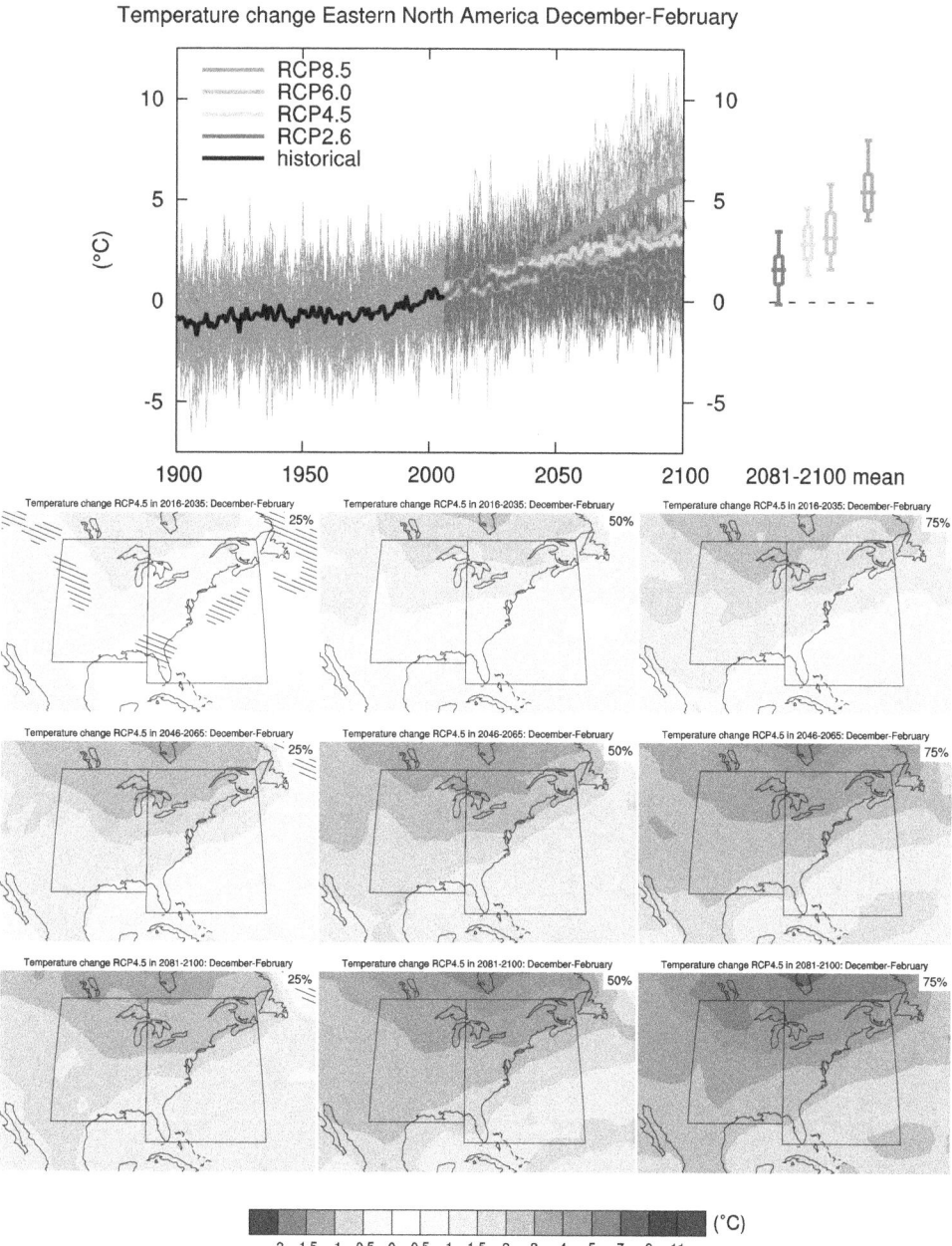

Figure 18.3. Time series of temperature change relative to 1986–2005, averaged over land grid points in eastern North America (25°N to 50°N, 85°W to 60°W) in December to February. Thin lines denote one ensemble member per model, thick lines the CMIP5 multi-model mean. On the right-hand side the 5th, 25th, 50th (median), 75th and 95th percentiles of the distribution of 20-year mean changes are given for 2081–2100 in the four RCP scenarios. (Below) Maps of temperature changes in 2016–2035, 2046–2065 and 2081–2100 with respect to 1986–2005 in the RCP4.5 scenario. For each point, the 25th, 50th and 75th percentiles of the distribution of the CMIP5 ensemble are shown; this includes both natural variability and inter-model spread. Hatching denotes areas where the 20-year mean differences of the percentiles are less than the standard deviation of model-estimated present-day natural variability of 20-year mean differences. Figure from IPCC 2013.

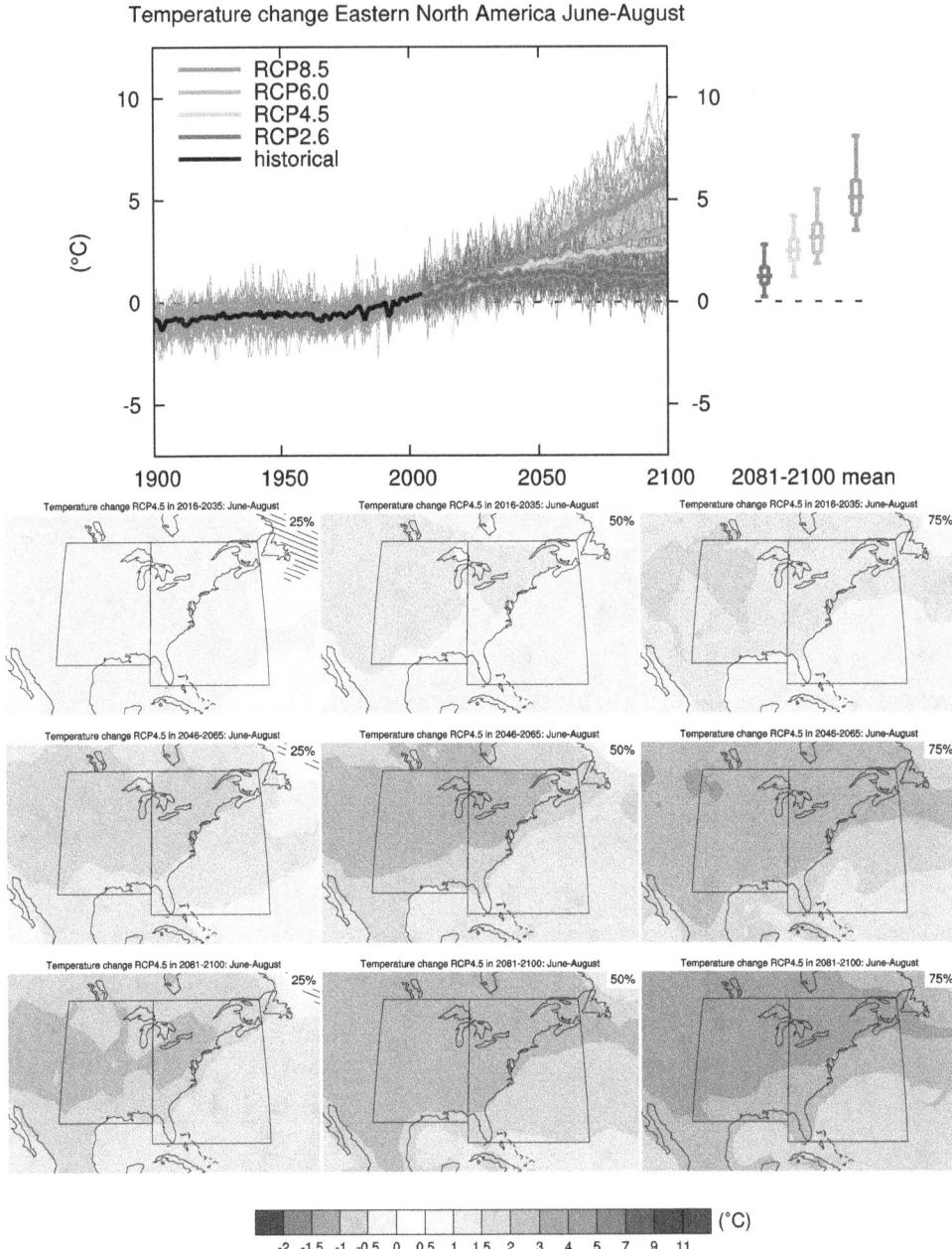

Figure 18.4. Time series of temperature change relative to 1986–2005 averaged over land grid points in eastern North America (25°N to 50°N, 85°W to 60°W) in June to August. Thin lines denote one ensemble member per model, thick lines the CMIP5 multi-model mean. On the right-hand side the 5th, 25th, 50th (median), 75th and 95th percentiles of the distribution of 20-year mean changes are given for 2081–2100 in the four RCP scenarios. (Below) Maps of temperature changes in 2016–2035, 2046–2065 and 2081–2100 with respect to 1986–2005 in the RCP4.5 scenario. For each point, the 25th, 50th and 75th percentiles of the distribution of the CMIP5 ensemble are shown; this includes both natural variability and inter-model spread. Hatching denotes areas where the 20-year mean differences of the percentiles are less than the standard deviation of model-estimated present-day natural variability of 20-year mean differences. Figure from IPCC 2013.

FUTURE CLIMATE CHANGE SCENARIOS FOR FLORIDA • 545

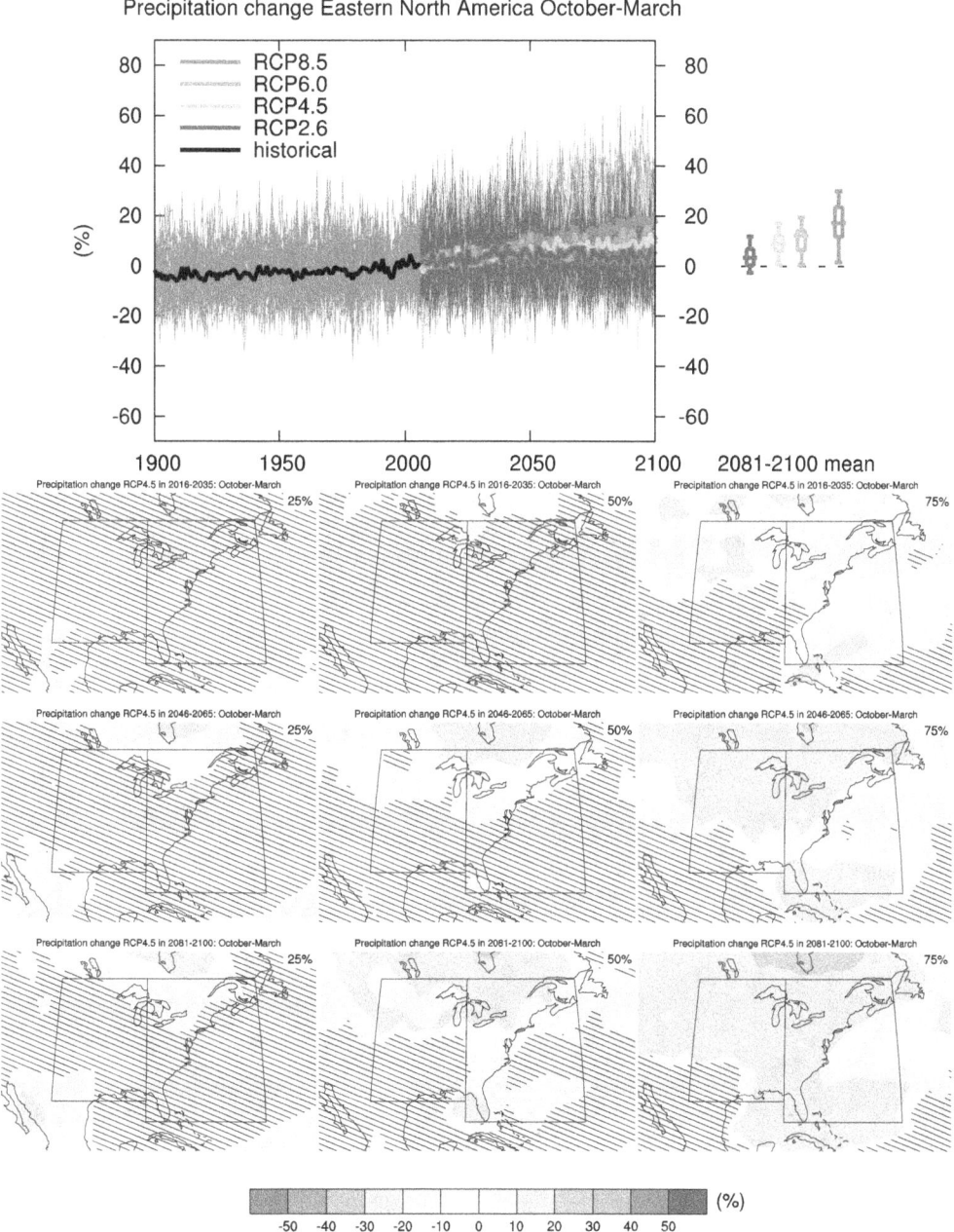

Figure 18.5. (Top) Time series of relative change with reference period 1986–2005 in precipitation averaged over land grid points in Eastern North America (25°N to 50°N, 85°W to 60°W) in October to March. Thin lines denote one ensemble member per model, thick lines the CMIP5 multi-model mean. On the right-hand side the 5th, 25th, 50th (median), 75th and 95th percentiles of the distribution of 20-year mean changes are given for 2081–2100 in the four RCP scenarios. (Bottom) Maps of precipitation changes in 2016–2035, 2046–2065 and 2081–2100 with respect to 1986–2005 in the RCP4.5 scenario. For each point, the 25th, 50th and 75th percentiles of the distribution of the CMIP5 ensemble are shown; this includes both natural variability and inter-model spread. Hatching denotes areas where the 20-year mean differences of the percentiles are less than the standard deviation of model-estimated present day natural variability of 20-year mean differences. Figure from IPCC 2013.

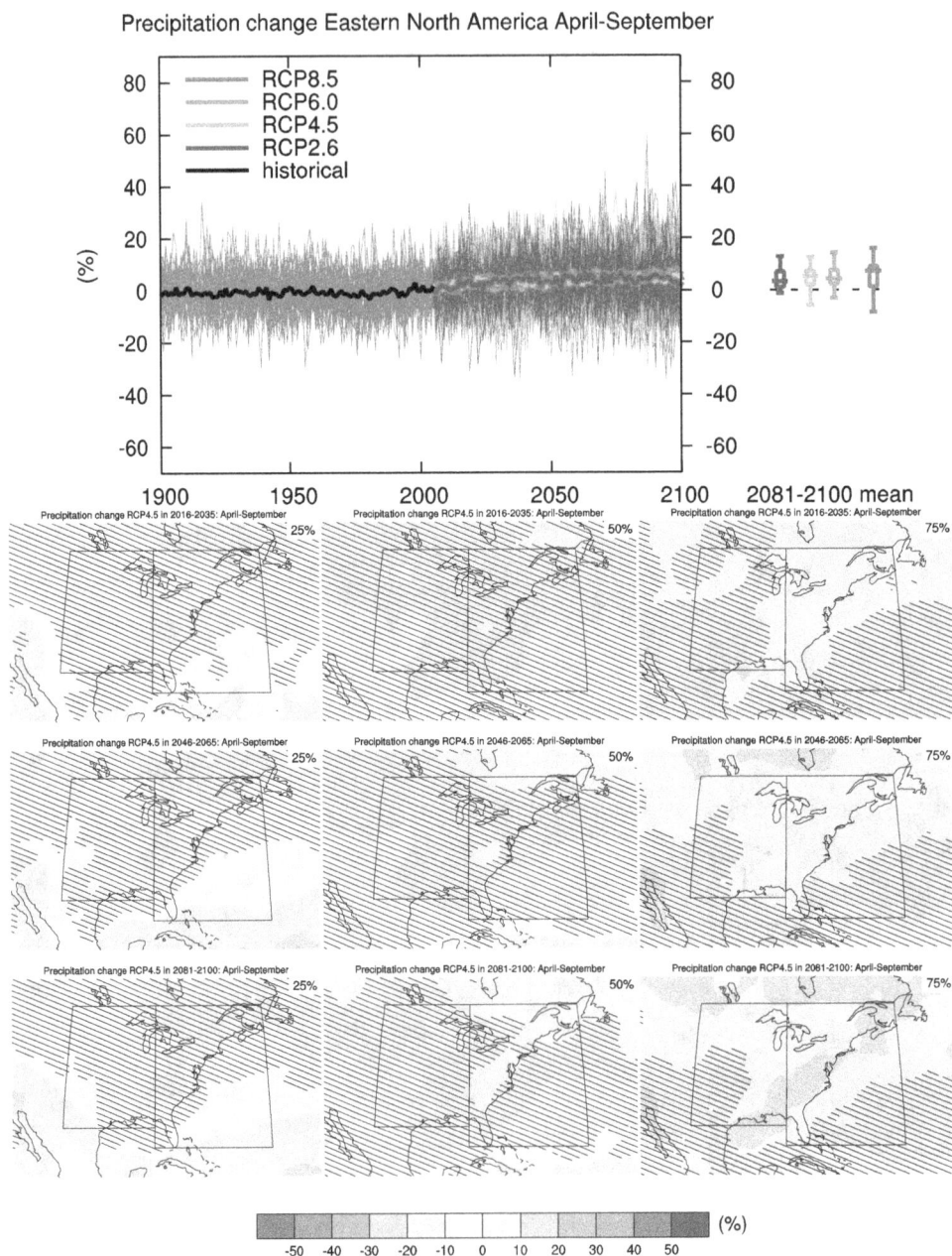

Figure 18.6. (Top) Time series of relative change relative to 1986–2005 in precipitation averaged over land grid points in Eastern North America (25°N to 50°N, 85°W to 60°W) in April to September. Thin lines denote one ensemble member per model, thick lines the CMIP5 multi-model mean. On the right-hand side the 5th, 25th, 50th (median), 75th and 95th percentiles of the distribution of 20-year mean changes are given for 2081–2100 in the four RCP scenarios. (Bottom) Maps of precipitation changes in 2016–2035, 2046–2065 and 2081–2100 with respect to 1986–2005 in the RCP4.5 scenario. For each point, the 25th, 50th and 75th percentiles of the distribution of the CMIP5 ensemble are shown; this includes both natural variability and inter-model spread. Hatching denotes areas where the 20-year mean differences of the percentiles are less than the standard deviation of model-estimated present-day natural variability of 20-year mean differences. Figure from IPCC 2013.

Regional Climate Projections

The global model results discussed above clearly show relatively little regional spatial resolution. This is particularly troublesome for decision makers at the regional level, as many users are unable to utilize GCM data that are on a coarse spatial grid (e.g. Obeysekera et al. 2011). There are a number of different statistical and dynamical techniques for downscaling the global scale models to the regional level (see also Chapter 17, which discusses downscaling of climate predictions for regional studies). In brief, dynamical downscaling translates large-scale GCM data to a finer grid using a regional climate model (Giorgi et al. 2001, 2009; Bastola and Misra 2014; among many others), and statistical downscaling uses assumptions of the relationships between large-scale fields and local climate (Wood et al. 2004; Maurer et al. 2007; among many others). Two popular datasets currently in use are the archive of statistically downscaled CMIP3 and CMIP5 Climate and Hydrology Projections (DHCP, Brekke et al. 2013) and the World Climate Research Programme (WCRP) Coordinated Regional climate Downscaling Experiment (CORDEX) (Giorgi et al. 2009), both of which are publically available. For CMIP3, the North American regional Climate Change Assessment Program (NARCCAP; Mearns et al. 2012) provides dynamically downscaled results. The interested reader is encouraged to visit the NARCCAP project (http://www.narccap.ucar.edu).

Statistical and dynamical downscaling of climate projections has often been used over the southeast and Florida. In studying the hydrological system of the Tampa Bay region, Hwang and Graham (2014) emphasized the importance of choosing the correct statistical downscaling that preserves the precipitation characteristics of the region in order to simulate the streamflow variations. Hwang et al. (2011) evaluated the fifth-generation Pennsylvania State University-National Center for Atmospheric Research Mesoscale Model (MM5) to dynamically downscale precipitation over the Tampa Bay region, and found the spatial patterns of precipitation to be realistic on daily, seasonal, and inter-annual timescales; they consider the data useful for multidecadal water resource planning in Tampa Bay. In another dynamical downscaling effort, Stefanova et al. (2012) studied seasonal, sub-seasonal, and diurnal variability of rainfall from the Center for Ocean-Atmospheric Prediction Studies (COAPS) Land-Atmosphere Regional Reanalysis for the Southeast at 10km resolution (CLARReS10), and found that that the downscaled reanalyses agreed with station and gridded observations for seasonal distribution and diurnal structure, but total precipitation was overestimated. CMIP3 climate projections were downscaled using this methodology, titled the COAPS Land-Atmosphere Regional Ensemble Climate Change Experiment for the Southeast United States at 10-km resolution (CLARREnCE10). Ning et al. (2011, 2012) use a Self-Organizing Map (SOM) strategy to statistically downscale CMIP5 precipitation over the mid-Atlantic region, determining that downscaling reduced the inter-GCM uncertainties for this region; the SOM strategy has been expanded to include Florida. For temperature, Keellings (2016) assessed the DHCP historical simulations and found that the mean and distribution of temperature matched well with observations, while extreme maximum daily temperatures were not well simulated.

Specific to Florida and based on CMIP3, Obeysekera et al. (2015) determined that for 2060, reasonable estimates for projected precipitation and temperature changes are +/- 10% and 1.5 degrees C, respectively. For CMIP5 (using DHCP), Dessalegne et al. (2016) found a wet bias in future precipitation, and percent changes in precipitation that range from -2.6 to 20.2% and changes in temperature ranging from 0.4 to 3.7 degrees C, depending on the RCP and time-period considered. However, Obeysekera et al. (2015) also pointed out the need for more information on seasonality of projected changes and extremes.

Fig. 18.7 shows the mid-century (2041 –2070) summer season (June to August) rainfall response to a relatively strong emission scenario from one global model (top) and three different regional NARCCAP models (remaining panels). All three of these regional models are forced by the particular global model at the regional boundaries of the North American sector. Focusing on Florida, it is clear that the projections from the regional models have considerably more spatial heterogeneity than the corresponding global model projections. Unfortunately, the regional models give remarkably different results on even relatively large scales. These differences are easily seen over the state of Florida, where one of the regional models has a reduction in rainfall, one is neutral, and one has a sizable increase in rainfall. This is precisely why regional climate projection remains a scientific challenge, and projections need to be presented in robust probabilistic format; however, we also note that these results are based on climate models included in CMIP3.

A more regional view of the precipitation change, though still on a coarse grid scale, is shown in Fig. 18.8 for the southern tier of the US and Caribbean in 2080–2099, with respect to 1986 – 2005 in June to September (left) and December to March (right) for the RCP4.5 scenario with 39 CMIP5 models. Because Florida's climate is modulated by many modes of natural variability (see Chapter 17), inter-decadal trends can be difficult to interpret, and there can be prolonged dry and wet periods related to decadal variability (Christensen et al. 2013; and references therein). CMIP5 models project an ensemble mean decrease in precipitation in precipitation over southern Florida and an increase in northern Florida in JJAS. In DJFM, there is an increase in precipitation. These results are robust over northern Florida (light hatching).

Though these results are more regionally focused, DHCP data has a spatial resolution more relevant to Florida, provided on a 0.125 degree x 0.125 degree grid as opposed to a 1.0 degree grid. Fig. 18.9 shows maps of the projected changed change in precipitation for 71 CMIP5 models and ensemble members. This Figure is intended to be analogous to Fig. 18.8, though it is shown as the percent change in precipitation and has not been normalized by the global mean surface temperature change as in Fig. 18.8. The downscaled results largely agree with the coarse-scale CMIP5 results, with decreasing precipitation in southern Florida and increasing precipitation in northern Florida in JJAS, and increasing precipitation overall in DJFM. We also note that results are robust across peninsular Florida in DJFM, whereas we only see robustness in the northern and extreme southernmost part of the domain in JJAS (stippling).

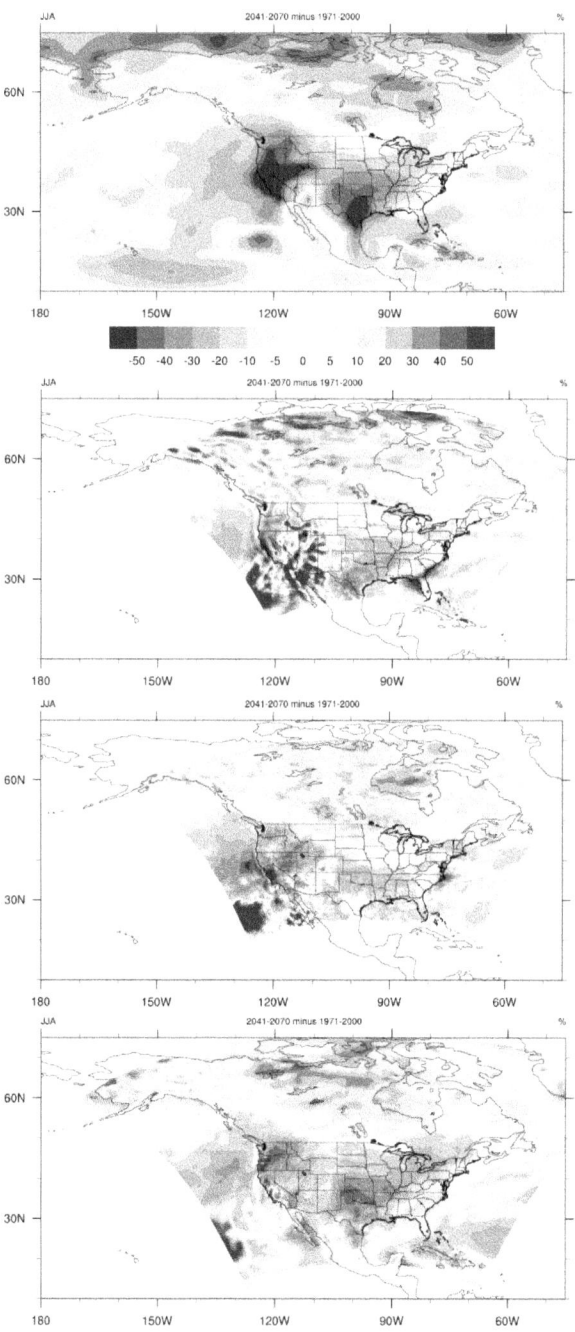

Figure 18.7. Panels show projections for North American June –August rainfall percent change during the mid-century (2041 –2070) based on a relatively high emission scenario for (top) a global model and for (remaining) three different regional models that are driving by the global model. Results are from the NARCCAP project.

550 • BEN P. KIRTMAN ET AL.

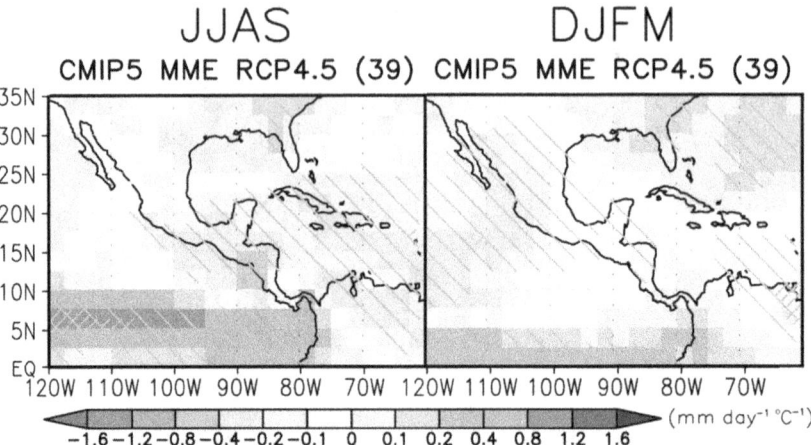

Figure 18.8. Maps of precipitation changes for southern North America and the Caribbean in 2080–2099 with respect to 1986–2005 in June through September (JJAS, left) and December through March (DJFM, right) in the RCP4.5 scenario with 39 CMIP5 models. Precipitation changes are normalized by the global annual mean surface air temperature changes in RCP4.5. Light hatching denotes where more than 66% of models have the same sign with the ensemble mean changes, while dense hatching denotes where more than 90% of models have the same sign with the ensemble mean change. Figure adapted from IPCC 2013 (Chapter 14).

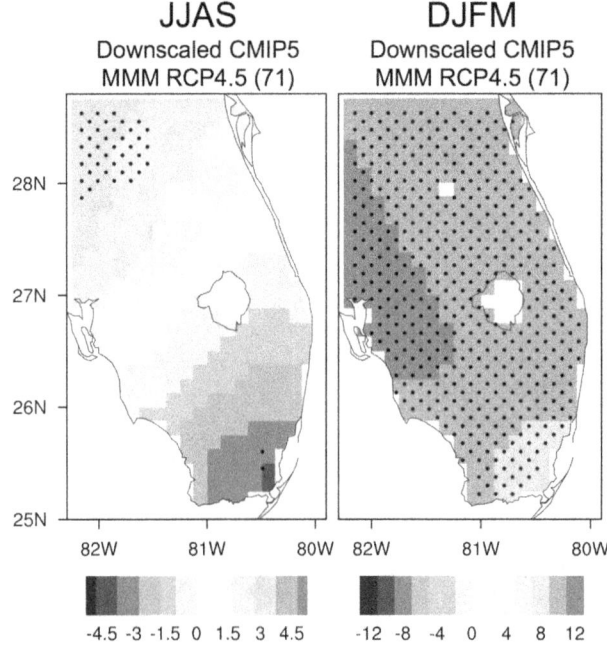

Figure 18.9. Maps of downscaled precipitation changes for peninsular Florida in 2080–2099 with respect to 1986–2005 in June through September (JJAS, left) and December through March (DJFM, right) in the RCP4.5 scenario with 71 CMIP5 models and ensemble members, from DHCP data. Precipitation changes are given as percent change in JJAS or DJFM from 1986–2005. Stippling denotes where more than 66% of models have the same sign with the ensemble mean changes.

For regions like Florida, whose terrestrial climate is dependent on the strong but mesoscale (of the order of ~10s of km) ocean currents (e.g. the Loop Current system), which transport warm waters from the tropics to the subtropical and higher latitude region, it becomes even more challenging to simulate or project the regional climate. For example, Misra et al. (2016) showed that the global models have significant errors in simulating the Loop Current. More recent publications point to conflicting estimates on the observed trends of the strength of the western boundary currents (Miller 2017). The readers are referred to Chapter 13 of this book for further discussion of the projected climate of the oceans around Florida.

Indicators as Tools for Developing Scenarios

Indicators estimated from simulations of global models, downscaled models, or observed climate variables can be used to develop scenarios at a local scale (e.g. a farm). Fig. 18.9 shows a decrease in precipitation during summer and an increase in precipitation during winter in South Florida. These changes can be translated using indicators to communicate technical data in relatively simple terms that portray the interrelationships among climate and other physical and biological elements of the ecosystem to help reveal evidence of the discernible impacts of climate change. For example, decreases in summer precipitation in South Florida is translated to trends (e.g. increases in drought, dry spells). Incremental scenarios that can be estimated from these indicator trends are +5%, +10%, +15%. These scenarios provide useful information for sustainable water resource planning and management in crop production and urban water supply. Similarly, increases in winter precipitation values in South Florida can be translated to trends (e.g. increases in flooding, wet spells indicators) etc. Incremental scenarios can be estimated from these indicator trends (e.g. +5%, +10%, 15%). These scenarios provide useful information for stormwater management and wetland management.

For example, a change in temperature (e.g. 0.5 °C) can translate to change in frost that translates to earlier spring and/or later fall seasons. Minimum temperature is the climate variable. Examples of indicators estimated from minimum temperature that portray the interrelationships among climate and the ecosystem can be frost day, last spring freeze, first fall freeze, and length of growing season. A frost day in this case is defined as a day with minimum temperature < 0 °C. Changes in the indicators are observed in Floridaand can provide important insights on the factors, processes, and structures in the ecosystem (e.g. deciding the planting day or variety of agricultural crops, the flowering of flora, the changes in the fauna life cycle, water requirement of flora and fauna). Changes in the near-term and long-term climate projections and decadal climate predictions, when translated to changes in indicators, promote developing adaptation and mitigation strategies that can protect and conserve Florida's unique ecosystems and natural resources.

Decadal Climate Prediction

Up to this point, we have only discussed results from projections. Typically, these results are shown in such a way as to minimize the internally generated climate variability. This minimization is often done by taking multi-year time averages. However, when examining the near-term there is a possibility that the internally generated near-term climate could be important and perhaps even predictable. Fig. 18.1 is suggestive in this regard. For instance, during the period 2005–2012, the projections are largely warmer than the observational estimates. The trace from the observational estimates lies in the lower tail of the climate projections. Is this because the models produce too much warming for a given level of external forcing (i.e., their so-called climate sensitivity is too large)? Alternatively, this could be because the projections make no attempt to capture the phasing of the internally generated climate variability. This is where the new science of decadal prediction comes in. As noted earlier, decadal predictions are dependent on both the initial state and the external forcing, and as such have the potential to predict the internally generated climate variability and capture the externally forced response (see Box 11.1 in Kirtman et al. 2013). Fig. 18.8 which follows a similar format as Fig. 18.1, shows the near-term projections and some early attempts at decadal predictions (black and red hatched regions). The decadal predictions suggest less warming than the projections in better agreement with the observational estimates (see Meehl and Tang 2012 and Smith et al. 2012 for details).

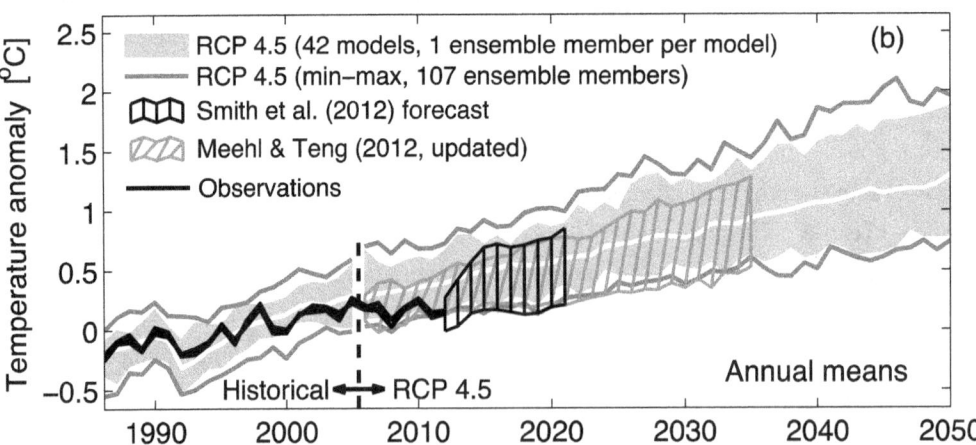

Figure 18.8. Projections of global mean, annual mean surface air temperature 1986–2050 (anomalies relative to 1986–2005), with four observational estimates as shown in Fig. 18.1. The shading illustrates the 5 to 95% range (grey and blue shades, with the multi-model median in white) of annual mean CMIP5 projections using one ensemble member per model from RCP4.5 scenario, and annual mean observational estimates (solid black line). The maximum and minimum values from CMIP5 are shown by the grey lines. Red hatching shows 5 to 95% range for predictions initialized in 2006 for 14 CMIP5 models applying the Meehl and Teng (2012) methodology. Black hatching shows the 5 to 95% range for predictions initialized in 2011 for eight models from Smith et al. (2013). Figure taken from Kirtman et al. (2013).

Since the science of decadal prediction is still relatively immature, it is not ready for regional decision support. However, there it holds considerable potential for providing near-term probabilistic information that takes both the external forcing and the internally generated variability into account (Meehl et al. 2009, 2013). As an example, we show (Fig. 18.9) the five-year forecast (2016-2020) for surface air temperature from a number of dynamic and statistical prediction systems (details in Smith et al. 2013).

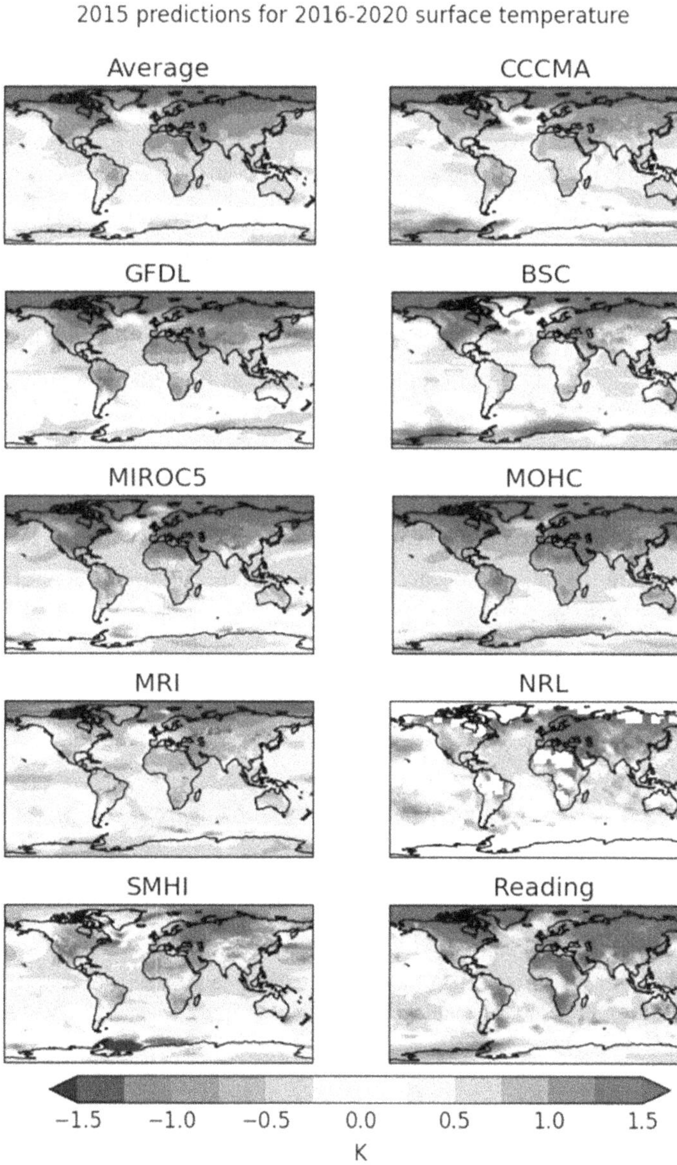

Figure 18.9. Decadal predictions for 2016 – 2020 from a number of different dynamic and statistical prediction systems (see Smith et al. 2013).

Final Remarks

The results described above (and in Chapter 17) clearly demonstrate that the science of regional externally forced and internally generated climate variability remains unresolved. Florida's climate, in particular, is especially difficult to project because of its narrow peninsula and the complex air – sea interactions associated with the surrounding oceans. Florida also sits at the boundaries between the tropics and the extra-tropics, and small shifts in how the global models represent the tropics and subtropics have profound impacts over Florida. Indeed, the global models have large uncertainties in the boundary between the tropics and extra-tropics, leading to large uncertainties in Florida projections from global models. The Fifth Assessment Report of the IPCC notes the need for extreme caution when using the global models for regional projections (see box 11.2 in Kirtman et al. 2013).

The climate science community clearly understands that adaption decisions need robust regional information, and that the current generation of global models are not sufficient in this regard. As such, there are a number of efforts to produce regional climate information using a variety of dynamical and statistical methodologies. All of these approaches show promise, but the science is relatively immature and robust projections and predictions will ultimately need to be tailored to the specific decision support requirements.

In terms of near-term climate, decadal prediction seems to also hold some promise. Part of the reason for this is that decadal predictions can be rigorously verified in terms of both the internally generated and externally forced variability, and they can be calibrated for robust probabilistic information. Moreover, decadal prediction can be performed at considerably higher spatial resolution than is possible with global projections; and there is compelling results indicating that this increased resolution will improve the fidelity of the predictions (see Siqueira and Kirtman 2016).

Acknowledgments

The authors are grateful for comments from reviewers and editors in improving this chapter presentation.

References

Anandhi, A. (2017). CISTA-A: Conceptual Model Using Indicators Selected by Systems Thinking for Adaptation Strategies in a Changing Climate: Case Study in Agro-Ecosystems. *Ecological Modelling, 345*, 41-55.

Anandhi, A., 2007. Impact Assessment of Climate Change on Hydrometeorology of Indian River Basin for IPCC SRES scenarios (Doctoral dissertation, PhD thesis, Indian Institute of Science, India).

Anandhi, A., Omani, N., Chaubey, I., Horton, R., Bader, D.A. and Nanjundiah, R.S., 2016. Synthetic Scenarios from CMIP5 Model Simulations for Climate Change Impact Assessments in Managed Ecosystems and Water Resources: Case Study in South Asian Countries.

Anandhi, A., 2010. Assessing impact of climate change on season length in Karnataka for IPCC SRES scenarios. Journal of Earth System Science, 119(4), pp.447-460.

Anandhi, A., Srinivas, V.V., Nanjundiah, R.S. and Nagesh Kumar, D., 2008. Downscaling precipitation to river basin in India for IPCC SRES scenarios using support vector machine. International Journal of Climatology, 28(3), pp.401-420.

Hwang, S., and W. Graham, Assessment of alternative methods for statistically downscaling daily GCM precipitation outputs to simulate regional streamflow, Journal of the American Water Resources Association, doi:10.1111/jawr.12154, 50(4), 1010-1032, 2014.

IPCC, 2013: Annex I: Atlas of Global and Regional Climate Projections [van Oldenborgh, G.J., M. Collins, J. Arblaster, J.H. Christensen, J. Marotzke, S.B. Power, M. Rummukainen and T. Zhou (eds.)]. In: *Climate Change 2013: The Physical Science Basis. Contribution of Working Group I to the Fifth Assessment Report of the Intergovernmental Panel on Climate Change* [Stocker, T.F., D. Qin, G.-K. Plattner, M. Tignor, S.K. Allen, J. Boschung, A. Nauels, Y. Xia, V. Bex and P.M. Midgley (eds.)]. Cambridge University Press, Cambridge, United Kingdom and New York, NY, USA.

Kirtman, B., S.B. Power, J.A. Adedoyin, G.J. Boer, R. Bojariu, I. Camilloni, F.J. Doblas-Reyes, A.M. Fiore, M. Kimoto, G.A. Meehl, M. Prather, A. Sarr, C. Schär, R. Sutton, G.J. van Oldenborgh, G. Vecchi and H.J. Wang, 2013: Near-term Climate Change: Projections and Predictability. In: *Climate Change 2013: The Physical Science Basis. Contribution of Working Group I to the Fifth Assessment Report of the Intergovernmental Panel on Climate Change* [Stocker, T.F., D. Qin, G.-K. Plattner, M. Tignor, S.K. Allen, J. Boschung, A. Nauels, Y. Xia, V. Bex and P.M. Midgley (eds.)]. Cambridge University Press, Cambridge, United Kingdom and New York, NY, USA.

Mearns, L.O., Hulme, M., Carter, T.R., Leemans, R., Lal, M., Whetton, P., Hay, L., Jones, R.N., Kittel, T., Smith, J. and Wilby, R., 2001. Climate scenario development. Advances in Geoecology, pp.739-768.

Mearns, L.O., et al., 2007, updated 2012. *The North American Regional Climate Change Assessment Program dataset*, National Center for Atmospheric Research Earth System Grid data portal, Boulder, CO. Data downloaded 2017-01-12. [doi:10.5065/D6RN35ST].

Meehl, G. A., and H. Y. Teng, 2012: Case studies for initialized decadal hindcasts and predictions for the Pacific region. *Geophys. Res. Lett.*, 39, L22705.

Meehl, G. A., et al., 2009: Decadal prediction: Can it be skillful? *Bull. Am. Meteorol. Soc.*, 90, 1467–1485.

Meehl, G. A., et al., 2013: Decadal climate prediction: An update from the trenches. *Bull. Am. Meteorol. Soc.*, doi:10.1175/BAMS-D-12-00241.1.

Miller, J. L., 2017: Ocean currents respond to climate change in unexpected ways. *Phys. Today*, 70, doi:10.1063/PT.3.3415.

Misra, V., A. Mishra, and H. Li, 2016: The sensitivity of the regional coupled ocean-atmosphere simulations over the Intra-Americas seas to the prescribed bathymetry. *Dyn. Atm. Ocn.*, 76, 29-51.

Shindell, D., et al., 2012a: Simultaneously mitigating near-term climate change and improving human health and food security. *Science*, 335, 183–189.

Siqueira, L. and B. Kirtman, 2016: Atlantic near-term climate variability and the role of a resolved Gulf Stream. *Geophys. Res. Lett.*, 10.1002/2016GL068694.

Smith, D.M., Scaife, A.A., Boer, G.J., Mihaela Caian, Francisco J. Doblas-Reyes, Virginie Guemas, Ed Hawkins, Wilco Hazeleger, Leon Hermanson, Chun Kit Ho, Masayoshi Ishii, Viatcheslav Kharin, Masahide Kimoto, Ben Kirtman, Judith Lean, Daniela Matei, William J. Merryfield, Wolfgang A. Müller, Holger Pohlmann, Anthony Rosati, Bert Wouters, Klaus Wyser *Clim Dyn* (2013) 41: 2875. doi:10.1007/s00382-012-1600-0.

Stocker, T.F., D. Qin, G.-K. Plattner, L.V. Alexander, S.K. Allen, N.L. Bindoff, F.-M. Bréon, J.A. Church, U. Cubasch, S. Emori, P. Forster, P. Friedlingstein, N. Gillett, J.M. Gregory, D.L. Hartmann, E. Jansen, B. Kirtman, R. Knutti, K. Krishna Kumar, P. Lemke, J. Marotzke, V. Masson-Delmotte, G.A. Meehl, I.I. Mokhov, S. Piao, V. Ramaswamy, D. Randall, M. Rhein, M. Rojas, C. Sabine, D. Shindell, L.D. Talley, D.G. Vaughan and S.-P. Xie, 2013: Technical Sum- mary. In: *Climate Change 2013: The Physical Science Basis. Contribution of Working Group I to the Fifth Assessment Report of the Intergovernmental Panel on Climate Change* [Stocker, T.F., D. Qin, G.-K. Plattner, M. Tignor, S.K. Allen, J. Boschung, A. Nauels, Y. Xia, V. Bex and P.M. Midgley (eds.)]. Cambridge University Press, Cambridge, United Kingdom and New York, NY, USA.

UNEP and WMO, 2011: Integrated Assessment of Black Carbon and Tropospheric Ozone. United Nations Environment Programme & World Meteorological Organization [Available at http://www.unep.org/dewa/Portals/67/pdf/BlackCarbon_SDM.pdf]

CHAPTER 19

Sea Level Rise

Gary Mitchum[1], Andrea Dutton[2], Don P. Chambers[1], and Shimon Wdowinski[3]

[1]*College of Marine Science, University of South Florida, St. Petersburg, FL;* [2]*Department of Geological Sciences, University of Florida, Gainesville, FL;* [3]*School of Environment, Arts and Society, Florida Atlantic University, Miami, FL*

Sea level rise is naturally a topic of concern to many Floridians. Our intention in this chapter is to give the reader enough information on this topic to inform decisions about future adaptation strategies. We begin by reviewing how we measure sea level and the reasons that sea level can change. At the global level, the problem is relatively simple in that globally averaged sea level can only increase if water is added to the ocean or the ocean warms. The situation is more complicated at the local level, where variations can occur (e.g., due to changes in wind and ocean current patterns, and differences in vertical land motion rates). We present summaries of global sea level change over several time scales, ranging from the modern day to the geological records. Although we have confidence in estimates of the rate of global mean sea level change, determining from observations whether the rate is increasing, or accelerating, is more challenging. Over the next century, sea level change in Florida is expected to follow the global trend reasonably closely, but on shorter time scales and in different localities some variations are inevitable. We end with a discussion of the future sea level rise projections for Florida that should form the basis for efforts to plan adaptation strategies.

Key Messages

- Unless greenhouse gas emissions are reduced, sea level will most likely increase by 1-2 meters over the next 50 to 100 years. The time scale is not certain, but the ultimate rise of sea level is. The only way to mitigate this risk is to reduce greenhouse gas emissions as soon as possible and to commit to lowered emissions in the future.
- The linkage between greenhouse emissions and sea level rise in incontrovertible. Sea level rise projections are often misinterpreted due a lack of understanding of this point. We cannot invoke any particular sea level rise projection without committing to the emission scenario associated with that sea level rise projection. The scatter seen in charts projecting sea level rise is due to the differing emission scenarios assumed, and is not due to uncertainty in the climate science that underlies the projections.
- On shorter time scales of a few years to a few decades, sea level rise fluctuations due to oceanic and atmospheric changes and vertical land motion can substantially increase the frequency of nuisance flooding events. Although these smaller sea level changes are likely ephemeral, these events can have large economic impacts.
- Sea level rise impacts in coming decades will be felt differently in different communities. Regional to local adaptations should be developed based on the best available science, and to support these efforts, scientists need to be involved at the local level. We do not discuss this point in our chapter, but would argue that an important outcome of this book is that local scientists, practitioners, and decision makers will have the information needed to inform at the local level.

Keywords

Sea level; Climate change; Vertical land motion; Last Glacial Maximum; Ice melt; Ocean warming; Tide gauges; Satellite altimetry

Introduction

Florida's vulnerability to sea level rise is obvious. Our population is predominantly in low-lying areas and our tourism-based economy depends heavily on the state's beaches. Most of you have heard of the impressive approach to planning for the sea level rise that has occurred in South Florida, but similar planning is now occurring in multiple regions around the state. As one example, the Tampa Bay region has formed the Climate Science Advisory Panel that is facilitated by Florida Sea Grant and the Tampa Bay Regional Planning Council, and other municipalities are taking similar independent actions.

It might seem that a state-wide approach would be better, but perhaps not. Different regions have unique problems and a one-size-fits-all approach is probably not best. Instead, different regions must plan to meet their own challenges and we should focus on providing local managers and political leaders with the best information and tools to help them. That is the aim of this chapter. We do not give a single sea level rise projection, but instead provide information that will give local decision makers the ability to best use the available tools and research.

This chapter is organized into four main sections. In the first, we review how sea level is measured on time scales ranging from millions to thousands of years to the instrumental record over the past 100 years, and we address why sea level changes at all. In the second section, we get to the actual data and see how sea level changes over the long time and space scales, where our information is the most reliable. In the third section, we examine regional changes in Florida and review our knowledge of what happens on shorter time scales, meaning season to season or year to year. This turns out to be very challenging. In the final section, we review projections of future sea level rise in Florida. Again, we show how the shorter our timeframe is for predictions and the more localized those predictions need to be, the more difficult the problem. On the other hand, long-term sea level changes can be projected with confidence.

How Do We Measure Sea Level and Why Does It Change?

Measuring Sea Level Changes

Tide gauge measurements of sea level extend back to the 18th century, and a fascinating history of the development of these measurements is given by David Cartwright (1999). The earliest tide measurements used a yard stick attached to a sea wall that was observed directly by a person to obtain measurements of the time and height of high and low tides. Later, a system was developed that consisted of a water surface following float inside a stilling well that served to dampen wave signals; when the motion of the float was coupled to a continuous pen recording on a strip chart, the modern tide gauge was born. This happened in Sheerness, England in 1827. These instruments are fully capable of observing climate change signals, but exist only along coastlines and on islands.

In the early 1990s, with the launch of the TOPEX/Poseidon satellite altimeter mission, global sea level measurements entered a new era. Satellite altimeters measure sea level globally on a roughly 10-day cycle by directly measuring the height of the sea level from space. While the precision at any point is not as good as a tide gauge, the key point is that the measurement is global. Sea level variations due to redistributing ocean volume from one point to another cancel out and the global average of the altimeter measurements is an excellent measurement of the changes in the volume of the ocean. We now have about 25 years of satellite altimetry measurements from multiple missions and we can determine the global mean sea level changes with unprecedented precision.

We can also infer past sea level changes via paleo methods. Kemp et al. (2015) have given an excellent review of these methods that explicitly separates sea level measurements according to time scale, which is a theme of this chapter. First, on the scale of millions of years, sea level is measured by inferring ice sheet volume from oxygen isotope data and assuming that ice lost or added means ocean volume has increased or decreased, or by dating the height of coastal geological features that are expected to stay near sea level. Second, since the last ice age, corals that grew near sea level are dated in tropical regions. Finally, over the past 2,000 years, sea level histories are obtained from salt marsh sediment cores, coral microatolls, and archaeological evidence.

Sea Level Changes Associated with Ocean Volume Change

The change in global mean sea level is a measure of the change in the ocean volume. Think for a moment about a bathtub containing still water. If you took a meter stick and measured the depth of the bathtub and multiplied times the area of the bathtub, you would get the volume of water in the tub. Now think about doing the same thing each day, month, or year. Why would the volume change with time? Suppose we turn on the water supply or open the drain. The water level would change because we have added or removed water from the tub. The only effective ways to do this in the ocean are to melt ice that is on the land and add the resulting water to the oceans, or to take water from the ocean via evaporation and turn it into ice on the land.

There is one other way to change the volume of the ocean, and that is to change to the average density of the ocean. In this case we have to think about the water in the bathtub being warmed. Why would that matter? When the water is warmed, it becomes less dense, expands, and takes up more space (more volume), so the water level increases. The reverse happens if the water is cooled.

How do we measure the amount of water added to the ocean and the average density of the ocean? Basically, the amount of water in the ocean is measured by satellite missions that measure the gravity field of the Earth, and the density of the water is determined by profiling floats that measure the density of the ocean. So how well can we monitor the change in the global mean sea level? Fig. 19.1 will be discussed fully in a later section; but for now, let us focus on the altimetry curve, which is the direct measurement of the global mean sea level from satellite altimetry and the sum of the global mean sea level change from the measurements of the ocean density and mass changes. The close agreement of these two independent estimates gives us confidence in the altimetry estimate of the global mean sea level change.

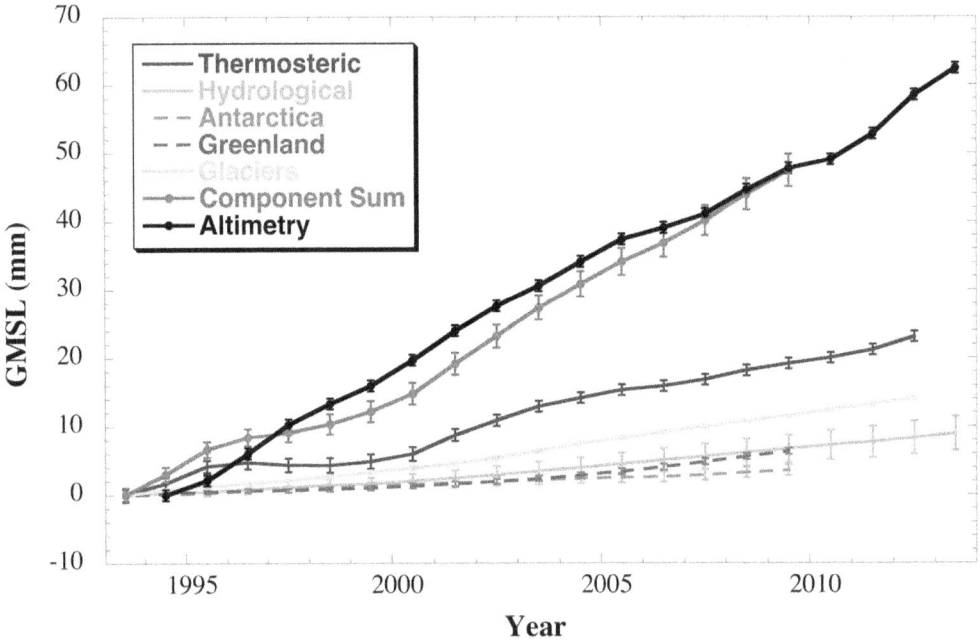

Figure 19.1 Three-year running means of global mean sea level rise (GMSL) from altimetry, its components, and the sum of the components from 1993.0-2014.0, as discussed in Chambers et al. (2016). Uncertainty bars are one standard error.

Regional Sea Level Changes Not Associated with Ocean Volume Change

The situation is much more complicated when we consider regional and local sea level changes due to contributions from oceanic, atmospheric, and geological processes. Spatial patterns of the regional changes are complex and time dependent. The global map of sea level change rates from satellite altimetry over the past 25 years provides an excellent example (Fig. 19.2), showing rates of change of both sign and magnitudes greater than 10 mm/yr, as compared to the globally averaged rate of about 3 mm/yr. The observed regional changes are mostly associated with changes in wind and ocean circulation patterns (e.g., Kohl and Stammer 2008; Levitus et al. 2005; Zhang and Church 2012; Timmerman et al. 2010; Qiu and Chen 2012). Closer to Florida, the most noticeable regional changes are observed in the northern Atlantic Ocean, where high rates of sea level rise have occurred along and north of the North Atlantic Current. High rates are also found along the southern edge of the North Atlantic gyre, in the subtropical Atlantic region.

Figure 19.2. Global map showing rates of sea surface height (SSH) obtained from satellite altimetry for the period 1993-2017. The rates are in mm/yr.

Melting ice sheets in polar regions affect regional variability of sea level changes due to two processes: changes in ocean circulation and gravitational attraction. Increases in freshwater forcing can affect ocean circulation, and are suggested as an explanation of regional sea level changes in the northern Atlantic region (e.g., Yin et al. 2009). Melting ice sheets also reduce the mass of water stored in polar regions and, consequently, change the Earth's gravity field. As a result, near a melting ice sheet we counterintuitively expect decreases in sea level height. The geographic pattern of sea level change that results is sometimes referred to as a sea level fingerprint (Mitrovica et al. 2001; 2009). Basically, sea level height reduces near the source of a fresh water supply to the ocean and increases further away from the melting ice sheet.

Another important process that affects regional and local relative sea level changes is vertical land movement. If the land is moving vertically, then the sea level will appear to be moving in the opposite direction. This not only complicates the interpretation of the tide gauge data, but it is also important for determining the local impacts of sea level change; i.e., if sea level is rising and the land is falling, then the impacts will be more severe. Land subsidence or uplift can reach rates of more than 20 mm/yr, as observed in New Orleans (Dixon et al. 2006), for example. Causes of vertical land motion vary from the continuing response to melting ice sheets in the Pleistocene to local sediment compaction. The regional-scale subsidence due to delayed mantle flow, which is termed Glacial Isostatic Adjustment, affects the coastlines of the United States (Sella et al. 2007; Tamisea and Mitrovica, 2011; Karegar et al. 2016). Local-scale land subsidence occurs in many locations along the coast, especially in sediment-rich areas, such as river deltas, and reclaimed land and wetlands. For example, the land subsidence in New Orleans has occurred mainly in new neighborhoods built on reclaimed wetlands (Dixon et al. 2006).

The Global, Long-Term Context

Global Changes during the Instrumental Period

Nearly continuous records of sea level extend back to the early 18th century for several locations in Northern Europe (Fig. 19.3). Although there are differences over short time periods, the rates of sea level change from 1800 to 2000 are very similar. By the late 1800s and early 1900s, more and more tide gauges were placed around the world, including around Australia, Asia, and North and South America. A weighted average of all these tide gauges (Fig. 19.3) has a similar rate to the three long tide gauges in Northern Europe after 1880, and the record from the global sea surface height measurements from altimetry agrees well with the tide gauge average. All of this indicates that the average of the sparse tide gauges is a reasonable estimate of global mean sea level change. Estimating acceleration is, however, more difficult.

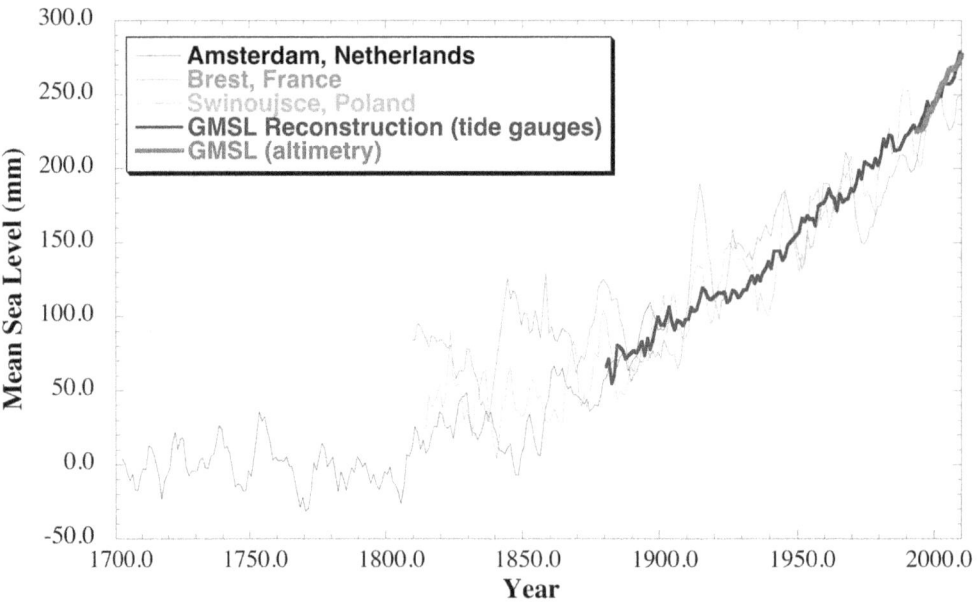

Figure 19.3. Five-year averages of sea level change recorded by tide gauges at three sites in Europe: The Netherlands, France, and Poland. The data have been corrected for vertical land movement predicted by a Post Glacial Rebound model (Peltier 2004). Data are from the Permanent Service for Mean Sea Level in Liverpool, UK. Also shown is yearly-averaged global mean sea level change from a weighted average of tide gauges (thick blue line) (Church and White 2011) and from satellite altimetry (thick red line) (Nerem et al. 2010).

The rate of global mean sea level rise since 1900 has been 1.7 ± 0.2 mm/year on average, while since 1993 the rate is higher at 3.2 ± 0.5 mm/year (Church et al. 2013; Rhein et al. 2013). Although this suggests an acceleration, it may reflect decade to decade fluctuations in sea level change rather than true long-term acceleration. Reconstructions of global mean sea level rise from tide gauges are unclear on this topic (e.g., Jevrejeva et al. 2008; Chambers et al. 2012; Rhein et al. 2013; Calafat et al. 2014; Natarov et al. 2016). To further complicate things, the change may not be a steady, continuous acceleration, (Woodworth et al. 2009). The net result is that in order to accurately detect accelerations with any confidence, one needs a very long record of fully global observations. This is only possible after 1993, with the advent of precision satellite altimetry.

Partitioning the observed global mean sea level into exact sources (i.e., density changes and addition of water to the ocean) is difficult for the time period before 1993, when observations of both ocean thermal expansion and the integrated mass loss of the Greenland and Antarctic ice sheets became available. After 1993, the partitioning is better known. Numerous studies have looked at the sea level budget over various time spans since 1993, and quantified how well one can partition the sources responsible for the observed trends (e.g., Church et al. 2013; Llovel et al. 2014; von Schuckmann et al. 2014; Dieng et al. 2015; Chambers et al. 2016; Reager et al. 2016). Here, we summarize the results of Chambers et al. (2016) for the period from 1993 through 2014 (Fig. 19.1).

The ocean density change due to warming is the largest single contributor to global mean sea level rise since 1993, accounting for about 40% of the trend. The upper ocean (0-700 m) explains about 28% of the trend, with approximately 8% coming from the middle layers (700-2000 m) and 4% from the deep ocean (> 2000 m depth). The contributors that combine to increase ocean mass (more water in the oceans) explain the remaining 60% of the trend. Glaciers and ice caps outside of Greenland and Antarctica caused 25% of sea level rise, hydrology (from pumping water from underground aquifers for irrigation) explained 15%, Greenland ice melt explained 12%, and Antarctic ice melt explained 7%. Note, however, the increasing separation of the contribution from Greenland relative to Antarctica (Fig. 9.1). This is due to an accelerated ice loss from Greenland that has been occurring since 2003 (Shepherd et al. 2012; Velicogna et al. 2014).

Sea Level Changes over the Past Few Millennia

Studies of sea level change over the last few millennia provide an important context for contemporary observations of sea level rise. Several high-resolution records of sea level change over the last few thousand years have been reconstructed using sediment cores from salt marshes (e.g., Gehrels et al. 2005; Kemp et al. 2011; Waller 2015). This technique relies on looking at the assemblages of foraminifera, which allows us to constrain the vertical position of sea level based on the presence and absence of the various foraminiferal taxa. This approach enables the position of sea level to be estimated with a high degree of certainty (< 0.10 m uncertainty). Similarly, a variety of dating techniques, including radiocarbon (carbon-14) dating, allow for precise age estimates. As a result, the salt marsh-derived sea level reconstructions for the past few millennia are considered to provide a very high-resolution and precise reconstruction of relative sea level through time.

Another approach to reconstructing sea level on this timescale relies on using coral microatolls as markers for the past position of sea level. Microatolls are corals that grow in very shallow water and therefore grow predominantly in a lateral direction, creating large, disc-shaped coral heads. Microatolls are renowned as very precise recorders of past sea level position and have been used to argue that there was very little change in global mean sea level during the last few millennia preceding the Industrial Era (e.g., Woodroffe et al. 2012).

Like the coral microatolls, data amassed from salt marshes also demonstrate very little change in local, or relative, sea levels over the last 2,800 years and prior to the Industrial Era (Kopp et al. 2016). These authors showed that global mean sea level varied by ±8 cm over the pre-Industrial Common Era, including a decline in sea level over a 400-year period that coincided with 0.2° C of global cooling. They also concluded that 20th century sea level rise was faster than during any of the 27 previous centuries (this includes the entire time window of their dataset). In this sense, the rapid warming that is associated with the Industrial Era and the associated increased emission of greenhouse gases (Rhein et al. 2013) appears to be coupled to a rapid rise in global mean sea level unlike any sea level rise that has occurred over the past 2,800 years.

Sea Level Changes on Geological Time Scales

On even longer timescales, fluctuations in temperature on geologic time scales have caused land-based ice sheets to grow and shrink in repeating cycles. Over the last million years, for example, large ice sheets have waxed and waned on 100,000-year cycles in the Northern Hemisphere, advancing over large tracts of North America, Europe, and Asia. Geologists have used the elevation of the Last Glacial Maximum (LGM) paleoshoreline, other markers of sea level position such as fossil corals that live near the sea surface, and models of glacial isostatic adjustment to estimate that sea level was 130-135 m lower then present (Yokoyama et al., 2001; Austermann et al. 2013; Lambeck et al. 2014; Dutton et al. 2015) (Fig. 19.4). As the Earth warms out of an ice age, rising temperatures cause land-based ice to melt and sea levels to rise. From the LGM to present, global mean sea level increased by about 130 m. This translates to an average of more than 80 cm of sea level increase per century. To put this in perspective, sea level rose by about 19 cm between 1901 and 2010, and is projected to increase by about 80 cm over the coming century (Church et al. 2013). The first lesson we can take from studying the geology is that increases in sea level similar to what are projected for the coming century are the norm in a warming climate, and it is the relatively slow rate of increase over the past few thousand years that is anomalous.

We are now in a warming period and it is natural to ask what the conditions were during the last warm period. Global mean temperatures were warmer than the pre-Industrial Era baseline by 1° C (similar to the temperature today), and atmospheric carbon dioxide concentrations were similar to the pre-Industrial Era value (280 ppm) during the Last Interglacial warm period that occurred about 125 thousand years ago. The best estimate for peak sea level during that time period is 6–9 m above present (Dutton et al. 2015). What are the implications of this? First, given physical limits on thermal expansion and melting of mountain glaciers, a significant amount of meltwater must have been derived from polar ice sheets in order to reach sea levels 6–9 m higher than present. Second, polar ice sheets have been very sensitive to past increases in global mean temperature of 1° C above the pre-Industrial Era level. And third, given that Greenland only partially melted during that time window, the high sea levels would have required approximately 5 m worth of sea level rise from melting of the Antarctic ice sheet (Dutton et al. 2015b). Relating these global numbers to the state of Florida, sea level rise during the last interglacial period inundated a significant fraction of the state, including most of South Florida (Fig. 19.4).

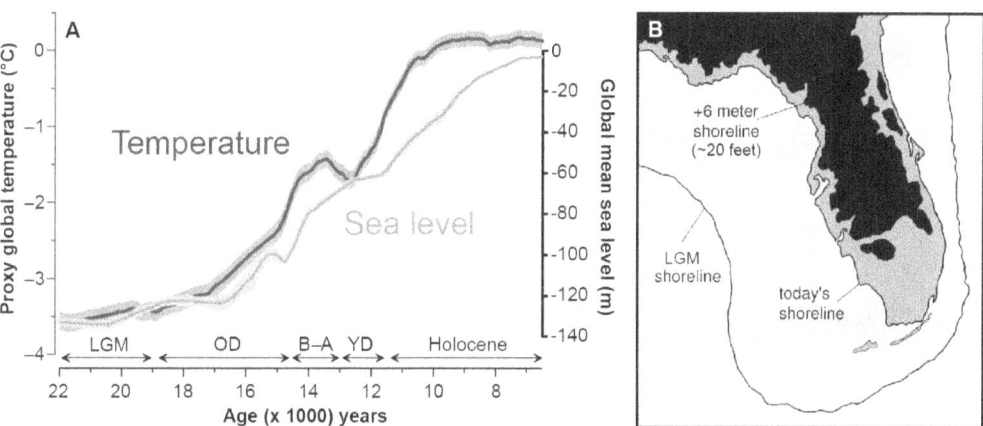

Figure 19.4. (A) Change in global mean temperature relative to the pre-Industrial Era (blue, Shakun et al. 2012), and in sea level (orange, Lambeck et al. 2014) from the LGM to 6,000 years ago. Abbreviations denoted on time axis represent climate intervals: Older Dryas (OD), Bolling-Allerod (B-A), Younger Dryas (YD). (B) Comparison of the shoreline during the LGM, today, and for a position representing ~ +6 meters (~ 20 feet) higher than present.

Sea Level Change in Florida

Difficulty of Determining Regional Rather Than Global Sea Level Changes

Long-term (> 60 years) sea level changes along the Florida coast, as determined from tide gauge measurements, are similar to the long-term global rates of 1.8-2.5 mm/yr (Church et al. 2013). Rates over shorter time spans, in particular after year 2000, are, however, higher and more variable, which illustrates the problem of estimating regional rather than global changes. The causes of these decadal scale changes are an active area of research, but there is a consensus that these changes are due to changes in wind and ocean current patterns (e.g., Yin et al. 2009; Sallenger et al. 2012; Ezer 2013; Ezer et al. 2013; Wdowinski et al. 2016; Rossby et al. 2014; Kopp 2013; Valle-Levinson et al. 2017). An unanswered question is whether the changes are ephemeral variations or sustained long-term changes. For example, decadal-scale changes in the rate of sea level rise occurring along the Florida coast is not unique to the post-2000 period, with long tide gauge records indicating that another accelerating period occurred during 1928–1948 (see the dashed black lines in Fig. 19.5). This does not mean that the recent changes are temporary, as accelerating change is forecast by climate models. Again, though, properly interpreting and projecting regional sea level change is still a challenging research problem.

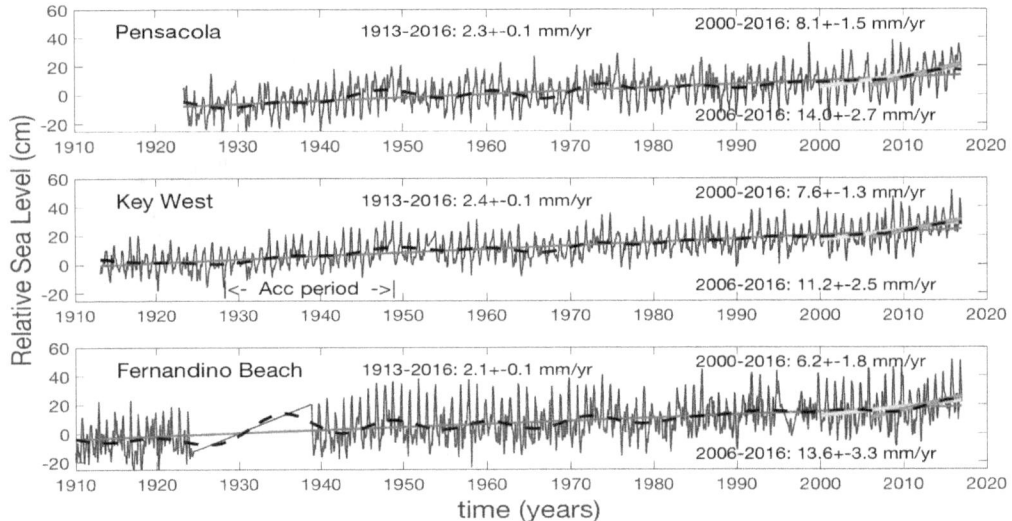

Figure 19.5. Sea level curves for three Florida sites with long tide gauge records (> 90 years) showing similar long-term rates of sea level rise (2.1–2.4 mm/yr). All three curves also show a decadal-scale acceleration in the rate of sea level rise between 1928–1948 and post-2000 (green line), which increased after 2006 (magenta). The data, yearly mean values, were obtained from the Permanent Service for Mean Sea Level (PSMSL, http://www.psmsl.org/). The rates of sea level rise were calculated using a least-square linear fit algorithm and are indicated by solid lines. Dashed black lines present low pass filters fit with a 10-year cutoff, which are indicative of decadal-scale sea level variability.

Some of the observed variability in the rate of relative sea level change along the Florida coast is related to local vertical land movements, induced mainly by land subsidence in sediment-rich areas, such as river deltas, reclaimed land, and wetlands. Tide gauges, being attached to the land, measure the relative motion between sea level and land subsidence, or uplift. A recent report by the National Oceanic and Atmospheric Administration (NOAA) (Zervas et al. 2013) based solely on the tide gauge data lists 12 tide gauge locations in Florida and estimates the vertical ground movements to be in the range 0.1-0.5 mm/yr, which suggests that vertical land motions are not a major contributor over most of Florida's coastline. A more robust method for estimating vertical land movements is based on global positioning system (GPS) measurements conducted at selected locations, which indicate that most Florida coastal land elevations are relatively stable, confirming the tide gauge-based results.

Interferometric Synthetic Aperture Radar (InSAR) is another geodetic technique widely used for detecting vertical land movements at the mm/yr level. This technique is particularly interesting because it can provide high spatial resolution maps (1-100 m pixel resolution) of land movements, unlike the point measurements obtained with GPS, and these give us information at the local scale. Although InSAR has been widely used to detect land subsidence (e.g., Dixon et al. 2006; Osmanaglu et al. 2011; Bock et al. 2012), it has rarely been used to measure land motions in Florida. Recently, though, Fiaschi and Wdowinski (2017) analyzed a synthetic aperture radar dataset acquired over southeastern Florida during the years 1994-2006 and detected 2-3 mm/yr of land subsidence in several localities in Miami Beach. The subsiding areas consist of houses that were built in 1920s and 1930s on reclaimed marshland, which was drained and filled with unconsolidated sediments. Although most of the land settlement occurs in the early years after reclaiming the land, these InSAR results indicate that the settlement process continues to affect these areas, which are located at low elevation and often subjected to coastal flooding. Extending InSAR analyses around Florida would be extremely valuable for advising decision makers at the city and county level.

Short-Term Variations in the Sea Level Change Rate

Although projections for future sea level rise typically depict a smooth rise over time, we know that there will be shorter-term variations (from seasonal- to decadal-scale) superimposed on this long-term pattern of sea level rise (Fig. 19.6). For example, along the coast of Florida there is a seasonal cycle of sea level variability that is primarily driven by meteorological and oceanographic processes. Wahl et al. (2014) have shown that tide gauge records along the Gulf of Mexico recorded a significant amplification of this seasonal sea level cycle from the 1990s onward. The net effect is that this change, coupled with a gradual rise in the base level of the sea surface, combined to double the risk of hurricane-induced flooding along the Florida Gulf Coast.

The seasonal cycle of sea level variability, with the highest sea levels occurring in the autumn months (Fig. 19.6), superimpose with twice-per-year maxima in the spring tide range. These peak tidal sea levels are a naturally occurring feature and are sometimes referred to as king tides. As sea level continues to rise, it will reach higher elevations during these times, and the duration and frequency of flooding will increase as the base level of the sea surface continues to rise. If certain areas already inundated with 1-2 ft of water during king tides, an anticipated sea level rise of 2 ft above present would translate to 3-4 ft of submergence during these events; the same is true for seasonal sea level extremes and storm surges caused by the winds. The point is that when a projection identifies a certain elevation for sea level at some time in the future, we can be certain that this elevation will be reached before then, and with increasing frequency, during the short-term extreme events.

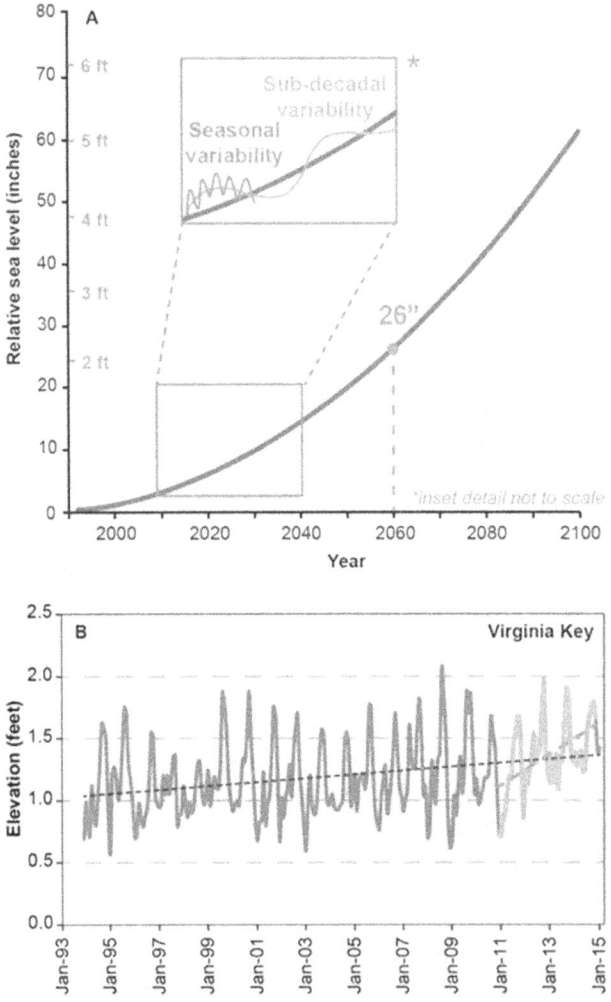

Figure 19.6. (A) Projection for sea level rise near Key West, with schematic variability in sea level on seasonal and sub-decadal time scales shown in inset. Projection for the year 2060 highlighted in orange. (B) Monthly mean sea level from Jan. 1993 to Feb. 2015 at the Virginia Key tide gauge (NOAA). The annual peaks are the seasonal king tides that occur in the autumn. The rate of sea level rise from Jan. 2011 to Dec. 2014 (gray dashed trendline) is higher than the longer-term average (black dashed line), demonstrating sub-decadal-scale variations in the rate of sea level rise.

Local precipitation can also magnify the effects of coastal flooding, particularly in coastal communities where the topographic gradient is low and there is no place for excess stormwater to go, especially if the ocean level is higher. It is not uncommon for multiple extreme high tide conditions to superimpose in time, which is another important observation that is not conveyed through simple sea level projection curves. For example, in October 2016, king tides, a storm swell associated with an offshore hurricane, and intense rainfall created a trifecta of conditions on one weekend that led to unusually high levels of coastal flooding in southeast Florida.

In addition to these short duration, intermittent, variations in sea level, there are also changes that persist for years to decades. As mentioned in the previous section, these changes are due to the sea level response to changes in the wind and ocean current patterns, which in turn respond to the large-scale changes in the Earth's temperature distribution. Basically, these changes provide a link from global warming to regional sea level change, and deciphering the relationships between the sea level changes and the changes in the winds and the ocean currents is an active research area. Understanding these changes is extremely important because the short-term events are on top of the background given by the larger-scale patterns, and can therefore amplify the impact of extreme events.

Sea Level Change in Florida Compared to Global Sea Level Changes

Clearly, understanding and projecting sea level change on time scales shorter than a few decades is challenging. Similar challenges exist when we attempt to project at specific locations or even regionally. Many people find it counterintuitive that we can more easily project global mean sea level for long lead times than we can determine what to expect in a particular harbor over the coming decade. And this is especially true when we try to project extreme sea level events. The reason for this is simply that there are many processes, as we have discussed, that affect regional or local sea level change on relatively short time scales; whereas for the global mean sea level, we only need to worry about the two processes that determine the ocean volume—adding water to the ocean and warming the ocean.

The result is that we have the most confidence in projections of global mean sea level on time scales longer than a few decades. However, we also know that regional and shorter time scale changes can complicate the application of these projections. So how useful are these global projections for Florida? We will give a possible answer to this question using the historical (20th century) tide gauge observations, but we need to be cautious about this. In essence, this approach uses past data to project future changes, and this implicitly assumes that the dynamics controlling future changes are the same as those observed during the past century. This assumption is worrisome. As many climate scientists say these days, the past is no longer a guide to the future.

The 20th century data from the tide gauges in and around Florida are shown in Fig. 19.7. For each tide gauge, we show the tide gauge record, the linear trend, and the global mean sea level reconstruction. Note that the gauges are plotted so that in the first column you move from Texas to the West Coast of Florida to the Keys; the second column goes from the Keys, along the East Coast of Florida and up to the Carolinas. We can see deviations between the gauges and the global reconstruction along the northern coast of the Gulf of Mexico and as we move north of Florida on the eastern coast. These differences can be attributed to vertical land motion. On the Florida coast, however, the sea level change during the 20th century is in good agreement with global sea level change. So, on the longer time scale, sea levels along Florida track the ocean volume changes (i.e., the global mean sea level) reasonably well. If we assume that this will remain true in the coming century, then we can conclude that the global sea level rise projections, which we will discuss in the next section, can be used as a zero order estimate of the sea level rise in Florida.

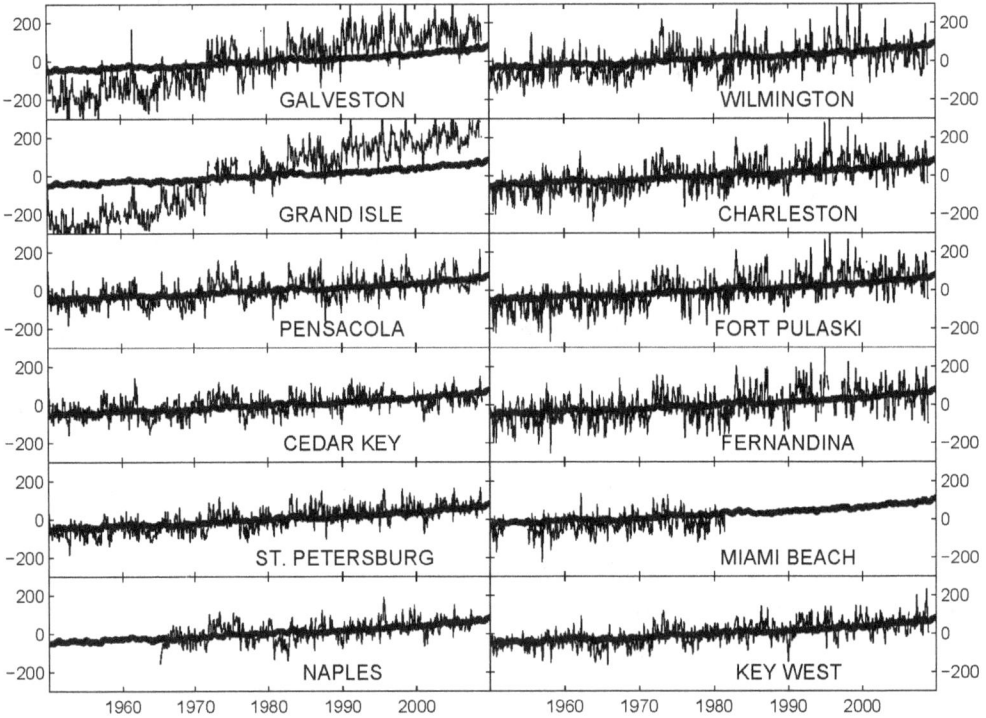

Figure 19.7. Sea level change around Florida as compared to the global reconstruction of Church and White (2011). On each panel, the solid curve is the long-term change observed at the tide gauge and the dashed curve is the global reconstruction. Units are millimeters. After Hine et al. (2016, Fig. 2.8).

Future Sea Level Rise

IPCC and National Climate Assessment Projections of Sea Level Change

This final section focuses on projections of future sea level rise in Florida. We will start with a brief summary of the global mean sea level projections based on the assessment by the Intergovernmental Panel on Climate Change (IPCC; Church et al. 2013) and the US National Climate Assessment (NCA; Parris et al. 2013). We have also included our own assessment of semi-empirical methods that were a contentious part of the IPCC assessment. The bulk of this section discusses projections for sea levels in Florida on several different time scales. We conclude with a short discussion of the uncertainties in the long-term, global projections and the prospects for reducing them.

The IPCC and NCA use somewhat different methods, but arrive at similar estimates for the sea level change by 2100. Note that when we refer to the NCA, we are referring the last assessment completed (Parris et al. 2013) and not the one that is presently under review. We decided to show the last official results, but we do acknowledge that if nothing changes substantially during the review, the sea level change by 2100 estimates will be somewhat higher than those given here. It is natural that our estimates will evolve as our observational series get longer and research continues, and this is not a cause for alarm. It only shows that our sea level projections and advice to the public needs to be regularly updated, as the NCA and IPCC assessments are updated about every five years. Concerning the methods, the NCA uses a panel of experts who review the literature and data and form a consensus projection. Interestingly, the older versions of the IPCC assessment used a similar method. The most recent, and presumably the next, IPCC assessment depends much more heavily on climate system models and requires only research that has been vetted in the peer-reviewed literature be used. There is, of course, a lag between research being conducted, papers being written, reviewed and read, and having the results implemented in the climate models, which slows the process somewhat.

An example projection using the NCA estimates is shown in Fig. 19.8. Again, note that using the IPCC assessment would give similar results. This particular assessment is from a report by the Tampa Bay Climate Science Advisory Panel mentioned at the beginning of this chapter. Bear in mind that it is tied to the Saint Petersburg tide gauge and thus tailored to the Tampa Bay region. The numbers will be discussed below, but we first want to stress the dependence of the sea level rise projection on the emission scenario chosen. To make this point we consider the IPCC's model-based method, but the point we will make also applies to the NCA approach. In order to do a sea level projection, we must first specify future greenhouse gas emissions into the atmosphere. This then allows models to compute the amount of surface warming, then the amount of ocean heating and ice melt, and finally the amount of global mean sea level increase. It is fallacious to look at a set of projections based on different emission scenarios and interpret the spread as uncertainty in the models. For a given scenario the spread between models is smaller than the difference between results for different scenarios.

Another point we have found confusing is that some people assume if we were to suddenly reduce emissions that the sea level would stop going up. We need to explain that if emissions are reduced to extremely small levels, it will still be a long time before the greenhouse gases already put into the atmosphere can be removed by natural processes. This idea and the concept of "committed" sea level rise has been nicely explained in an article by Strauss (2013).

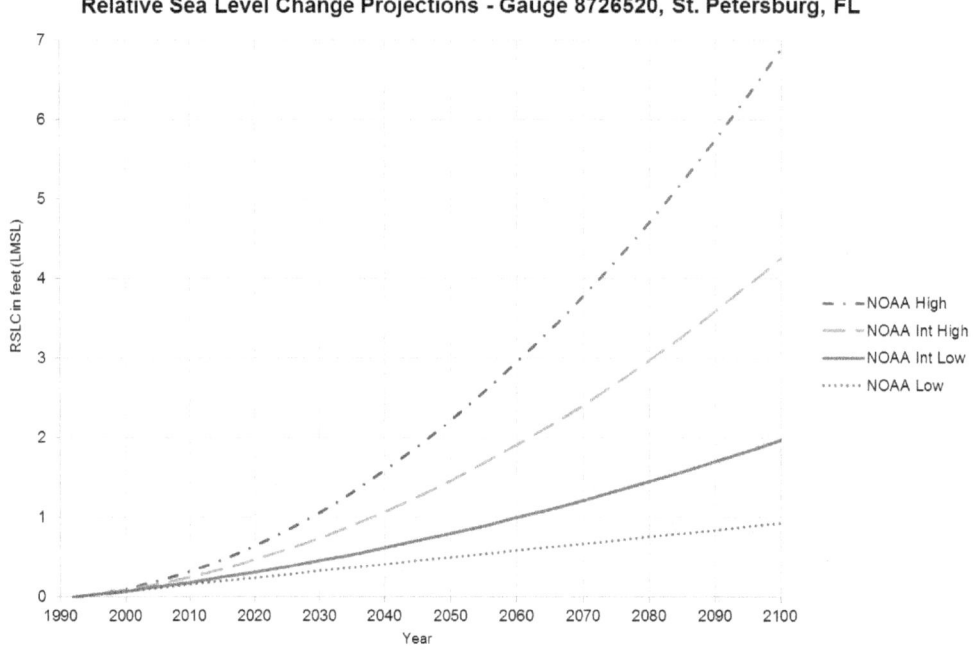

Figure 19.8 Sea level rise projections for Tampa Bay.

Empirical Projections of Sea Level Change

In addition to the IPCC and NCA methods, another method that has become popular for sea level projections is based on a semi-empirical scaling of globally averaged surface temperature to global sea level (Rahmstorf 2007; Grinsted et al. 2009). This approach is based on the observation that global averages of sea level and surface temperature tend to be highly correlated, and the physical understanding that as the Earth warms, sea level will increase from a combination of ocean warming and melting of land-based ice. After an empirical scaling is computed for historical surface temperature and global mean sea level measurements, the global average surface temperature taken from predictive climate models forced with different emissions scenarios can be scaled by the estimated parameter, and this will allow for prediction of a future sea level scenario.

The proponents of such models argue that surface temperature is better represented in ocean models than deep ocean warming, and that such a scaling can better represent the effects of ice dynamics than the process-based models used by the IPCC, which do not include ice dynamics (Church et al. 2013). However, these semi-empirical models also assume that the processes that drove sea level change in the past will be the same processes to drive it in the future. Considering the increasing contributions from the ice sheets, however, this assumption may not be entirely correct.

Some semi-empirical projections (Rahmstorf 2007; Grinsted et al. 2009) produce higher sea level predictions for 2100 than the process-based models for the same emission scenario, with upper ranges exceeding 1.5 m for the "business-as-usual" scenario (Church et al. 2013). However, the lowest probable range for the semi-empirical models does overlap with the highest probable range for the IPCC process-based models. These particular semi-empirical models may be biased, however, due to the time-period and sea level reconstructions used. For example, one group of semi-empirical models relied on temperature and sea level data mainly from the 20th century for calibration (Rahmstorf 2007). Another group used longer records of reconstructed global temperature but only a few regional sea level reconstructions (Grinsted et al. 2009). More recently, Kopp et al. (2016) used a sea level reconstruction for the past 3,000 years based on all globally available regional reconstructions and a statistical model that included global and regional patterns. When this global sea level reconstruction is combined with surface temperature to calibrate the scaling parameter, the resulting projections agree very well with the IPCC process-based model. Both means and ranges overlap. This suggests that the semi-empirical approach is quite sensitive to the data used, but after utilizing the most historical sea level data possible, the results are similar to the process-based model projections. This consistency increases our confidence in all of the approaches.

Sea Level Rise Projections for Florida on Various Timescales

As we have said, sea level rise projections are inextricably tied to projections of greenhouse emissions. Here, we present projections that assume emissions will continue to increase (i.e., the "business-as-usual" or "no action taken scenarios"). For planning purposes, we have to consider this to be the most likely scenario until we see evidence of aggressive mitigation actions within the United States and also globally. Under these scenarios, the last NCA and IPCC assessment suggest 2-3 ft of sea level rise on the 50 to 100-year time scale, but we expect that the next assessments will double these numbers. So on the 50 to 100-year time scale, we recommend planning for a 4-6 ft sea level increase. And, as stated previously, this global estimate is also what we expect to see in Florida over the next 50 to 100 years.

This projection for the next 50 to 100 years is where we have the most confidence. As explained, projecting over the next few years to 20 years or from 20 to 50 years is more difficult. That said, it is also very important for planning purposes. Cities and counties need to know what to plan for over the coming few years, developers need to plan for the coming decades, etc. Next, we will present our best estimates for the shorter time scales based on the processes discussed earlier in the chapter.

Future sea level increases will be mainly due to ice melting in Greenland and Antarctica. The spatial variations of sea level around the globe will depend on the location and rate of polar ice melt. But here in Florida, these contributions offset and we expect that the differences from the global rate will be less than 10%, meaning that if sea level increases globally by 100 cm, then Florida will experience increases within 90-110 cm.

On time scales of a few years to a few decades, the contribution of sea level changes due to changes in wind and ocean circulation patterns are potentially important but difficult to project, as discussed. This is because these wind and current pattern changes are expected to vary on multiple time scales, and because it is difficult to separate variability from trends. But for the next decade, we project that sea levels around Florida will be strongly affected by variability in rates due to atmosphere and ocean dynamics, translating to average sea level changes of about 10 cm, similar to the observed increase during 2006-2017. Given that these changes may be ephemeral rather than permanent, we could also see smaller changes; but in the most recent decade, sea level has been increasing sharply in parts of Florida and it is prudent to plan on the assumption that this type of short-term acceleration is possible in the future also.

On the 20 to 50-year time scale, we expect that sea level changes in Florida will be dominated by a combination of increasing global mean sea level, which will be seen in Florida as well, and high regional variability associated with wind and ocean current changes. Assuming that the present increased rates due to ocean and atmosphere changes continue, the projected sea level rise rates will be even higher than those observed since 2000 and sea level will rise by 30–45 cm, which is in the range of the NOAA-projected intermediate-low and intermediate-high curves. As stated earlier, however, we expect that the projected global rates will be higher than this, meaning that sea level increases of 50– 100 cm are definitely possible within the next 50 years.

Uncertainties in Projecting Global Mean Sea Level Change

We will conclude this chapter with a discussion of where the largest uncertainties in the long-term projections arise. Thermal expansion will likely be no more than 10– 30 cm over the next century, even at the highest "business-as-usual" emission scenario (Church et al. 2013). The contribution from glaciers not on the ice sheets will likely be of the same order, about 9–23 cm for the high emission scenario (Church et al. 2013). The ice sheets, on the other hand, hold huge amounts of ice—about 7 m of sea level equivalent for Greenland and 60 m for Antarctica. We are not suggesting that all of this ice will melt in the near future, but even a losing a small fraction (<2%) would raise sea level by a meter.

There are two major processes that act on the ice sheets and contribute to sea level rise. The first is the difference between summer surface melting and winter accumulation. This can be estimated from the atmospheric climate models, and is expected to be small over the next century (Church et al. 2013). The second process driving mass loss from the ice sheets is based on speeding up of the glaciers that move ice into the ocean, rapid thinning of the glaciers, or both. These dynamic processes can potentially lead to significant increases in sea level over decades. Our understanding of this process is still incomplete, but we have learned a great deal in the last decade from observations and models.

For instance, we now know that much of the bedrock under the Antarctic ice sheet actually sits below sea level, and that the bedrock slopes down from the coast. Most drainage glaciers in such regions, however, have a grounding line on a lip of bedrock above sea level (or just below) and a floating ice shelf. Both of these keep the glacier stable and limit the speed with which it can drain. However, in several regions the ice shelf has fractured, the glacier has sped up, and the leading edge of the glacier has retreated beyond the grounding line. Because of the slope of the bedrock, warmer ocean water can be forced under the ice sheet and melt the ice sheet from the bottom, making the glacier instable. This means the glacier will never be able to form another grounding line and will continue to discharge ice (and raise sea level) until it is gone. Evidence suggests this is already occurring in the Thwaites Glacier in the Amundsen Sea sector of Antarctica (Mouginot et al. 2014).

The problem related to predicting future sea level rise is modeling how long this will take. A recent model for all of Antarctica that includes ice dynamics found that under the "business-as-usual" emission scenario, the ice shelves begin to break apart and sea level starts to rise rapidly starting around 2050, going from no significant contribution before 2050 to an increase of 80 cm between 2050 to 2100 (DeConto and Pollard 2016). This is in contrast to the last IPCC report, which stated that these dynamic changes would contribute less than 23 cm of sea level rise by 2100. A better understanding of these dynamic ice processes is required in order to make more accurate projections.

References

Austermann J., J.X. Mitrovica, K. Latychev, G.A. Milne (2013). Barbados based estimate of ice volume at Last Glacial Maximum affected by subducted plate. Nat Geosci., 6, 553-7.

Bock, Y., S. Wdowinski, A. Ferretti, F. Novali, and A. Fumagalli 2012, Recent subsidence of the Venice Lagoon from continuous GPS and interferometric synthetic aperture radar, Geochemistry Geophysics Geosystems 13, doi:10.1029/2011gc003976.

Calafat, F. M., D. P. Chambers, and M. N. Tsimplis 2014, On the ability of global sea level reconstructions to determine trends and variability, *J. Geophys. Res. Oceans 119* 1572–1592, doi: 10.1002/2013JC009298.

Cartwright, D.E. 1999, Tides: A Scientific History. Cambridge University Press, Cambridge, United Kingdom, ISBN 0 521 62145 3.

Chambers, D. P., M. A. Merrifield, and R. S. Nerem (2012), Is there a 60-year oscillation in global mean sea level?, *Geophys. Res. Lett., 39*, L18607, doi: 10.1029/2012GL052885.

Chambers, D. P., A. Cazenave, N. Champollion, H. Dieng, W. Llovel, R. Forsberg, K. von Schuckmann, and Y. Wada (2016) Evaluation of the Global Mean Sea Level Budget between 1993 and 2014, *Surv. Geophys.*, doi: 10.1007/s10712-016-9381-3

Church, J. A., and N. J. White 2011, sea level rise from the late 19th to the early 21st century. *Surveys in Geophysics, 32*, 585-602.

Church, J. A., Clark, P. U., Cazenave, A., Gregory, J. M., Jevrejeva, S., Levermann, A., Merrifield, M. A., Milne, G. A., Nerem, R. S., Nunn, P. D., Payne, A. J., Pfeffer, W. T., Stammer, D., and Unnikrishnan, A. S. (2013) Sea level change, in: *Climate Change 2013: The Physical Science Basis. Contribution of Working Group I to the Fifth Assessment Report of the Intergovernmental Panel on Climate Change*, edited by: Stocker, T. F., Qin, D., Plattner, G.-K., Tignor, M., Allen, S. K., Boschung, J., Nauels, A., Xia, Y., Bex, V., and Midgley, P. M., Cambridge University Press, Cambridge, UK and New York, NY, USA.

DeConto, R. M., and D. Pollard (2016), Contribution of Antarctica to past and future sea level rise, *Nature, 531*, 591-597, doi: 10.1038/nature17145.

Dieng H., Cazenave A., von Shuckmann K., Ablain M. and Meyssignac B. (2015), Sea level budget over 2005-2013: missing contributions and data errors, *Ocean Science* 11, 789-802, doi:10.5194/os-11-789-2015.

Dixon, T., F. Amelung, A. Ferretti, F. Novali, F. Rocca, R. Dokka, G. Sella, S. Kim, S. Wdowinski, D. Whitman 2006. Subsidence and flooding in New Orleans, Nature, 441, 587-588.

Dutton, A., A.E. Carlson, A.J. Long, G.A. Milne, P.U. Clark, R. DeConto, B.P. Horton, S. Rahmstorf, and M.E. Raymo (2015). Sea level rise due to polar ice sheet mass loss during past warm periods. Science., 349, doi: 10.1126/science.aaa4019.

Dutton A., J.M. Webster, D. Zwartz, K. Lambeck, and B. Wohlfarth (2015b). Tropical tales of polar ice: Evidence of last interglacial polar ice sheet retreat recorded by fossil reefs of the granitic Seychelles islands. Quat. Sci. Rev., 107, 182–96.

Ezer, T. 2013. Sea level rise, spatially uneven and temporally unsteady: Why the US East Coast, the global tide gauge record, and the global altimeter data show different trends, *Geophysical Research Letters*, 40(20), 5439-5444, doi:10.1002/2013gl057952.

Ezer, T., L. P. Atkinson, W. B. Corlett, and J. L. Blanco 2013. Gulf Stream's induced sea level rise and variability along the U.S. mid-Atlantic coast, Journal of Geophysical Research-Oceans 118(2), 685-697, doi:10.1002/jgrc.20091.

Fiaschi, S. and S. Wdowinski (2017), The contribution of local land subsidence to coastal flooding hazard along the U.S. Atlantic coast, in preparation.

Gehrels, WR, Kirby JR, Prokoph A, Newnham RM, Achterberg EP, Evans H, et al. (2005). Onset of recent rapid sea-level rise in the western Atlantic ocean. Quat Sci Rev. 2005;24:2083–100.

Grinsted A, Moore JC, Jevrejeva S (2009) Reconstructing sea level from paleo and projected temperatures 200 to 2100 AD. *Clim Dyn 34*, 461–472.

Hine, A.C., D.P. Chambers, T.D. Clayton, M.R. Hafen and G.T. Mitchum (2016). Sea Level Rise in Florida: Science, Impacts, Options. University Press of Florida, Gainesville, Florida, ISBN 9780813062891.

Jevrejeva. S., J. C. Moore, A. Grinsted, and P. L. Woodworth (2008). Recent global sea level acceleration started over 200 years ago? *Geophysical Research Letters*, *35*, L08715, doi:10.1029/2008GL033611, 2008

Joughin, I., Smith, B. E., and Medley, B. (2014). Marine ice sheet collapse potentially under way for the Thwaites Glacier Basin, West Antarctica. *Science*, *344*(6185), 735-738, doi: 10.1126/science.1249055.

Karegar, M. A., Dixon, T. H., & Engelhart, S. E. (2016). Subsidence along the Atlantic Coast of North America: Insights from GPS and late Holocene relative sea level data. *Geophysical Research Letters*, *43*(7), 3126-3133.

Kemp AC, Horton B, Donnelly JP, Mann ME, Vermeer M, Rahmstorf S. (2011). Climate related sea-level variations over the past two millennia. Proc Natl Acad Sci., 11;108:11017–22.

Kemp, A.C., A. Dutton and M. Raymo (2015). Paleo Constraints on Future sea level Rise. *Curr. Clim. Change Rep.* 1, doi: 101007/s40461-015-0014-6.

Kohl, A., and D. Stammer (2008). Decadal sea level changes in the 50-year GECCO ocean synthesis. J. Clim. *21* 1876–1890.

Kopp, R. E. (2013). Does the mid-Atlantic United States sea level acceleration hot spot reflect ocean dynamic variability? *Geophysical Research Letters, 40(15)*, 3981-3985.

Kopp. R.E., A.C. Kemp, K. Bittermann, B.P. Horton, J.P. Donnelly, W.R. Gehrels, C.C. Hay, J.X. Mitrovica, E.D. Morrow, S. Rahmstorf (2016). Temperature-driven global sea level variability in the common era, *Proc. Nat. Acad. Sci. 113*, E1434-E1441, doi:10.1073/pnas.1517056113.

Lambeck, K., H. Rouby, A. Purcell, Y. Sun, and M. Sambridge M. (2014). Sea level and global ice volumes from the last glacial maximum to the Holocene. Proc. Natl. Acad. Sci., 111, 15296–303.

Levitus, S., J. Antonov, and T. Boyer (2005). Warming of the world ocean 1955–2003. Geophys. Res. Lett., **32**, L02604.

Llovel W., Willis J.K., Landerer F.W. and Fukumori I. (2014). Deep-ocean contribution to sea level and energy budget not detectable over the past decade, *Nature Climate Change*, online publication 5 October 2014, DOI: 10.1038/NCLIMATE2387.

Mitrovica, J. X., M. E. Tamisiea, J. L. Davis, and G. A. Milne (2001). Recent mass balance of polar ice sheets inferred from patterns of global sea level change. Nature, **409** 1026-1029.

Mitrovica, J. X., N. Gomez, and P. U. Clark (2009). The sea level fingerprint of West Antarctic collapse. Science, **323**, 753–753.

Mouginot, J., E. Rignot, and B. Scheuchl (2014), Sustained increase in ice discharge from the Amundsen Sea Embayment, West Antarctica, from 1973 to 2013, *Geophys. Res. Lett.*, 41 1576–1584, doi: 10.1002/2013GL059069.

Natarov, S. I., M. A. Merrifield, J. M. Becker, and P. R. Thompson (2016), Regional influences on reconstructed global mean sea level, *Geophys. Res. Lett.,* in press.

Nerem, R. S., D. P. Chambers, C. Choe, and G. T. Mitchum, Estimating mean sea level change TOPEX and from the Jason missions, *Marine Geodesy*, *33, Supplement 1*, 435-446, doi: 10.1080/01490419.2010.491031 2010.

Qiu, B., & Chen, S. (2012). Multidecadal sea level and gyre circulation variability in the northwestern tropical Pacific Ocean. *Journal of Physical Oceanography*, *42*(1) 193-206.

Osmanoglu, B., T.Dixon, S. Wdowinski, E. Cabral-Cano, and Y. Jiang, (2011), Mexico City subsidence observed with Persistent Scatterer InSAR, International Journal of Applied Earth Observation and Geoinformation, doi:10.1016/j.jag.2010.05.009.

Parris, A., P. Bromirski, V. Burkett, D. Cayan, M. Culver, J. Hall, R. Horton, K. Knuuti, R. Moss, J. Obeysekera, A. Sallenger, and J. Weiss. 2012. *Global Sea Level Rise Scenarios for the US National Climate Assessment*. NOAA Tech Memo OAR CPO-1. 37 pp.

Peltier, W.R (2004) Global Glacial Isostasy and the Surface of the Ice-Age Earth: The ICE-5G(VM2) model and GRACE, *Ann. Rev. Earth. Planet. Sci.*, *32*,111-149.

Peltier, W.R., and R.G. Fairbanks (2006). Global glacial ice volume and Last Glacial Maximum duration from an extended Barbados sea level record. *Quaternary Science Reviews*, *25*, 3322-3327, doi.org/10.1016/j.quascirev.2006.04.010.

Rahmstorf, S. (2007) A semi-empirical approach to projecting future sea level rise. *Science 315*, 368–370.

Reager J.T., A.S. Gardner, J.S. Famiglietti, D.N. Wiese, A. Eicker, M.-H. Lo (2016) A decade of sea level rise slowed by climate driven hydrology. *Science, 351*, 699–703. doi: 10.1126/science.aad8386.

Rhein, M., S. R. Rintoul, S. Aoki, E. Campos, D. Chambers, R. A. Feely, S. Gulev, G. C. Johnson, S. A. Josey, A. Kostianoy, C. Mauritzen, D. Roemmich, L. D. Talley and F. Wang 2013: Observations: Ocean. In: *Climate Change 2013: The Physical Science Basis. Contribution of Working Group I to the Fifth Assessment Report of the Intergovernmental Panel on Climate Change* (Stocker, T. F., D. Qin, G.-K. Plattner, M. Tignor, S. K. Allen, J. Boschung, A. Nauels, Y. Xia, V. Bex and P. M. Midgley (eds.)). Cambridge University Press, Cambridge, United Kingdom and New York, NY, USA.

Rossby, T., C. N. Flagg, K. Donohue, A. Sanchez-Franks, and J. Lillibridge (2014), On the long term stability of Gulf Stream transport based on 20 years of direct measurements, Geophys. Res. Lett., 41 114–120, doi: 10.1002/2013GL058636.

Sallenger, A. H., K. S. Doran, and P. A. Howd 2012. Hotspot of accelerated sea level rise on the Atlantic coast of North America, Nature Clim. Change 2(12), 884-888, doi:10.1038/nclimate1597.

Sella, G. F., Stein, S., Dixon, T. H., Craymer, M., James, T. S., Mazzotti, S., & Dokka, R. K. (2007). Observation of glacial isostatic adjustment in "stable" North America with GPS. Geophysical Research Letters, 34(2).

Shakun, J.D., Peter U. Clark, Feng He, Shaun A. Marcott, Alan C. Mix, Zhengyu Liu, Bette Otto-Bliesner, Andreas Schmittner, and Edouard Bard (2012). Global warming preceded by increasing carbon dioxide concentrations during the last deglaciation. *Nature*, 484, 49-54, doi: 10.1038/nature10915.

Shepherd, A., E.R. Ivins, A. Geruo, V.R. Barletta, M.J. Bentley, S. Bettadpur, and K.H. Briggs (2012). A reconciled estimate of ice-sheet mass balance. *Science, 338*, 1183-1189, doi: 10.1126/science.1228102.

Strauss, B.H. (2013). Rapid accumulation of committed sea level rise from global warming. Proceedings of the National Academy of Sciences, doi/10.1073/pnas.1312464110.

Tamisea, M.E. and Mitrovica, J.X. (2011). The moving boundaries of sea level change: Understanding the origins of geographic variability. Oceanography 24(2), 24-39.

Timmermann, A., S. McGregor, and F. F. Jin 2010: Wind effects on past and future regional sea level trends in the Southern Indo-Pacific. J. Clim. **23**, 4429–4437.

Valle-Levinson, A., A. Dutton, and J. B. Martin (2017). Spatial and temporal variability of sea level rise hot spots over the eastern United States, *Geophys. Res. Lett.*, *44*, 7876–7882, doi:10.1002/2017GL073926.

Velicogna I, Sutterley T.C., van den Broeke M.R. 2014 Regional acceleration in ice mass loss from Greenland and Antarctica using Grace time variable gravity data. *Geophys. Res. Lett.,* doi: 10.1002/2014GL061052.

Von Schuckmann K., Sallée J.B., Chambers D., Le Traon P.Y., Cabanes C., Gaillard C., Speich S., and Hamon M. (2014). Consistency of the current global ocean observing systems from an Argo perspective, *Ocean Sciences* 10, 547-557, doi:10.5194/os-10-547-2014.

Wahl, T., F. Calafat, and M. Luther (2014). Rapid changes in the seasonal sea level cycle along the US Gulf coast from the late 20th century, *Geophys. Res. Lett., 41,* 491-498, doi: 10.1002/2013GL058777.

Waller, M. (2015). Techniques and applications of plant macrofossil analysis in sea-level studies, In: Shennan, I., Long, A.J., Horton, B.P. (Eds.), Handbook of sea-level research. (pp. 183–190) Wiley-Blackwell.

Wdowinski, S., R. Bray, B. Kirtman, and Z. Wu (2016). Increasing flooding frequency and accelerating rates of sea level rise in Miami Beach, Florida, submitted, Ocean & Coastal Management, Volume 126, Pages 1-8, ISSN 0964-5691.

Woodroffe, C.D., H.V. McGregor, K. Lambeck, S.G. Smithers, and D. Fink; Mid-Pacific microatolls record sea-level stability over the past 5000 yr. *Geology* ; 40 (10): 951–954, doi: org/10.1130/G33344.1.

Woodworth, P. L., N. J. White, S. Jevrejeva, S. J. Holgate, J. A. Church, and W. R. Gehrels (2009). Evidence for the accelerations of sea level on multi-decade and century timescales, *Int. J. Climatol. 29,* 777–789, doi:10.1002/joc.1771.

Yin, J., M. E. Schlesinger, and R. J. Stouffer (2009). Model projections of rapid sea level rise on the northeast coast of the United States, Nature Geosci 2(4) 262-266.

Yokoyama, Y., De Deckker, P., Lambeck, K., Johnston, P., Fifield, L.K. (2001). Sea-level at the Last Glacial Maximum: evidence from northwester Australia to constrain ice volumes for oxygen isotope stage 2, Palaeogeography, Palaeoclimatology, Palaeoecology, 165, 281-297.

Zervas, C., S. Gill, and W. V. Sweet (2013). Estimating vertical land motion from long-term tide gauge records, in NOAA Tech. Rep. NOS CO-OPS 65 22 pp.

Zhang, X. B., and J. A. Church (2012). Sea level trends, interannual and decadal variability in the Pacific Ocean. Geophys. Res. Lett., **39**, L21701.

CHAPTER 20

Climate and Weather Extremes

Jennifer M. Collins[1], Charles H Paxton[2], Thomas Wahl[3], and Christopher T. Emrich[4]

[1]*School of Geosciences, University of South Florida, Tampa, FL;* [2]*Channelside Weather LLC, Tampa, FL;* [3]*Department of Civil, Environmental, and Construction Engineering, University of Central Florida, Orlando, FL;* [4]*National Center for Integrated Coastal Research, University of Central Florida, Orlando, FL*

This chapter examines Florida's extreme weather hazards: 1) why they happen, 2) their relation to interannual to multidecadal climate variability, and 3) the potential of each hazard and spatial variability across the state. The weather hazards indicated are under these broad categories: precipitation (rainfall, flooding, droughts), thunderstorms (lightning, hail, convective wind, tornadoes), tropical weather (tropical storms and hurricanes), and temperatures (extreme highs and lows). The conclusions section mainly addresses the challenge of attributing extreme events to human-induced climate change.

Key Messages

- The state of Florida is prone to various types of weather extremes, with tropical cyclones being the most dangerous in terms of impact potential.
- Most types of extreme events exhibit a seasonal cycle making it more likely for them to occur at certain times of the year. Many of them are also sensitive to large-scale climate variation resulting in strong interannual to multidecadal variability. The most important climate indicator for extreme weather in Florida is the El Niño Southern Oscillation (ENSO).
- Strong spatial variability of extreme events and related hazards exists across the state, with certain areas more susceptible to particular weather hazards than others. Maps of observed frequencies of the different types of events in the past can help identify hot spots.
- Attribution of climatic extremes is challenging because of insufficient observational records (with strong variability) and limitations in models with regards to resolution and complexity of the processes involved in the genesis of extreme events. This is also the reason for large uncertainties in future projections of most types of extreme events.
- There is, however, agreement that changes in the mean, which are better represented in climate models, will also affect the extremes (e.g., increasing mean sea level will affect storm surge heights and frequency).

Keywords

Weather extremes; Seasonality; Climate variability; Frequencies; Attribution

Introduction

Florida has a relatively dry cool season from October into May, a summer rainy season that lasts from June through September, and the overlapping hurricane season from June through November, which has varying impacts on the state from year to year. A narrower focus reveals that Florida has a distinct summer rainy thunderstorm season while the Florida Panhandle also has a distinct winter rainy season. The moisture-laden frontal systems bring winter rain to the panhandle but the rain typically dwindles as cold fronts move southward across the peninsula.

Florida has been divided into seven climate division zones (Fig. 20.1; NCDC 2016) from the Panhandle and North Florida (climatic zones 1 and 2) to the Florida Keys (zone 7) on the south end. Most of the state is subtropical with only the southern tip (the southern part of zones 5 and 6 and zone 7) in the tropical environment year-round. Obviously the northern part of the state (zones 1-2) is cooler during the winter and, because this area is adjacent to the continental U.S., the warming influence of the ocean on land surface temperatures is comparatively less than over the peninsula. The central portion of the Florida Peninsula (zones 3 and 4 and northern zones 5 and 6) experiences mild winters and nearly daily summer thunderstorms as east and west coast sea breezes merge.

Figure 20.1. Florida Climate Division Zones.

This chapter examines Florida's extreme weather hazards: 1) why they happen, 2) their relation to interannual to multidecadal climate variability, and 3) the potential of each hazard and spatial variability across the state. The weather hazards indicated are under these broad categories: precipitation (rainfall, flooding, droughts), thunderstorms (lightning, hail, convective wind, tornadoes), tropical weather (tropical storms and hurricanes), and temperatures (extreme highs and lows). The conclusions section of this chapter mainly addresses the challenge of attributing extreme events to human-induced climate change[1].

---

[1] The maps and descriptions of hazard potential areas that follow are based on either modeled deterministic probabilities, potentials derived from Geographic Information Systems (GIS) models or GIS-based historical frequency assessments. These findings were part of a Florida Department of Emergency Management funded mitigation project undertaken by scholars at the University of South Carolina (FDEM 2015). Many of the hazards discussed cannot be modeled in a probabilistic fashion. Rather, future event "probabilities" must be derived from historical frequencies. As such, we define "hazard potential areas" as those places across the state subjected to historical hazard events and likely at risk for future events of the same type and magnitude. In each case, except for flooding and wildfire, historical event geographies and associated frequencies were created and overlaid with a 12-mile hexagonal grid to create a state-level view of disaster potential. This grid system provides a user-friendly view of event data at a scale appropriate for state-level comparisons.

Rainfall and Related Hazards

Seasonal rainfall varies greatly from the Florida Panhandle to the Florida Keys. Fig. 20.2 indicates that the monthly average rainfall over the panhandle area of northwest Florida is greatest during July but the cool season months of January, February and March also have considerable rain. This pattern changes as we move southward over the peninsula to the Everglades region, which encompasses the southern part of the peninsula, with the warm season months of June-September having the most rainfall and with decreasing amounts southward during the cool season.

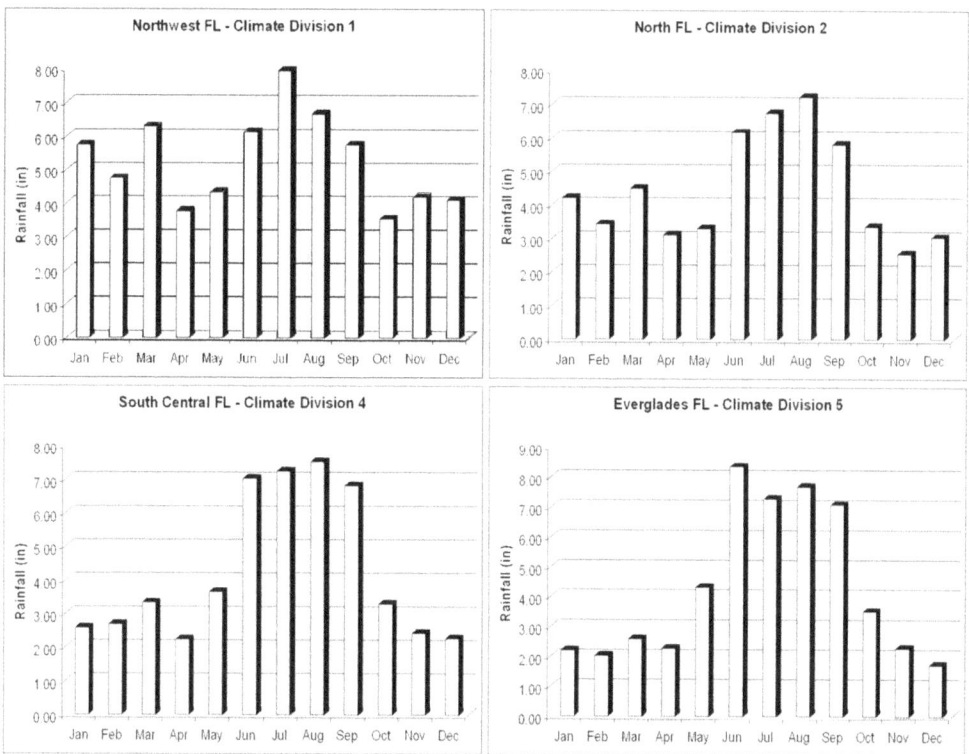

Figure 20.2. Average monthly precipitation for Florida Climate Divisions 1, 2, 4, and 5.

Floods

Flooding in Florida occurs in several different ways that include localized rainfall flooding, river flooding, and coastal flooding. Much of the flooding in Florida is associated with heavy rains that result from a stationary rainstorm, a series of repeat rainstorms, coastal convergence that produces a stationary area of rain, or a larger tropical weather system. Coastal flooding from storm surges occurs when strong winds from hurricanes or winter storms push shallow continental shelf waters onshore. In stronger storms, these inundating surges can be over 3.0 m

(10 ft). Often it is the combination of surge and rain that creates flooding in coastal areas (Wahl et al., 2015). In these scenarios, the surge covers drain pipes and the rainfall runoff backs up into the streets.

Flood Potential

Florida's low elevation above sea level and general lack of slope make flood hazards a threat for every county. Flood zones (Fig. 20.3), determined by the Federal Emergency Management Agency's (FEMA) National Flood Insurance Program, indicate areas at a 1% (100-year) annual probability of flooding. This "true risk" indication makes flood risk one of the easiest to quantify spatially. As such, flood risks and the mitigation of these risks are currently one of the only hazards accounted for in land-use planning, zoning, and building.

Flood hazard areas are fairly ubiquitous across much of the state but are highly conspicuous in South Florida, especially in the counties surrounding the Everglades, from Lake Okeechobee south to the Gulf of Mexico and east to the Atlantic Ocean. The entire Big Bend region of Florida, from Gulf County to Levy County and continuing into coastal Citrus and Hernando counties has a large amount of flood zone areas covering a majority of each of these counties.

The upland areas of St. Lucie and Martin counties have less land in flood zones. These, along with similar areas from inland Hernando County, north through Alachua and into Columbia and Suwanee counties, where pronounced ridges can be found, also have much more land area falling outside of federally-mandated flood zones.

Figure 20.3. Flood potential.

Flash Flood Potential

It is not possible to predict exactly where rain will fall across the state in the future. However, the real threat to life from extreme rainfall events is not the rain itself, but instead flash flooding in areas with large expanses of impervious surfaces, greater slopes, and soil less capable of rapid water infiltration. These characteristics of the landscape, when analyzed geospatially, produce a visual depiction of flash flood potential (Fig. 20.4)[2].

Flash flood potential varies greatly across Florida, but a few places are known to have greater potential for this type of flooding. Among the areas with higher hazard potential are the northern panhandle areas where higher terrain supports rapid downslope movement of water, major metropolitan areas where impervious surfaces impede water infiltration, and low-lying areas around major waterways. The county with the highest overall flash flood potential is Charlotte County. Here, a greater total land area exhibits characteristics associated with higher flood potential. Homestead, in Miami-Dade County, has the highest flood potential at a single location while Gadsden County in North Central Florida has the highest average potential across the state.

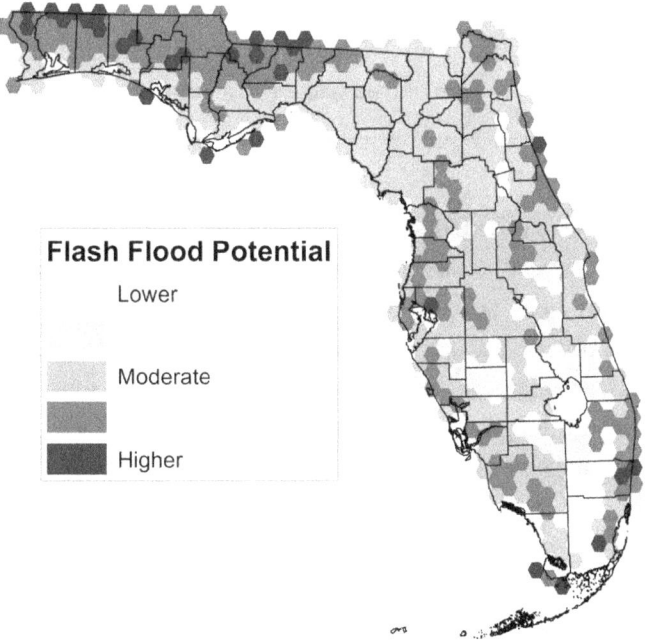

Figure 20.4. Flash flood potential index (FFPI) based on Zogg et al. (2013).

---

[2] Modeled from Zogg (2013), this flash flood potential index (FFPI) is not a risk per se since it is not tied to certain precipitation events and the associated occurrence probabilities, but rather the numerical representation of areas that have a higher flash flooding threat from extreme rainfall amounts.

Floods and Climate Variability

Extreme precipitation and discharge events[3] in Florida also show significant correlation with the Atlantic Multidecadal Oscillation (AMO), with increased values in AMO warm phases, in particular for precipitation events lasting >24 hours (associated with tropical cyclone landfalls) (Curtis 2008; Teegavarapu et al. 2013). There is, however, spatial variation in the AMO sensitivity. And in some areas in Southeast and Central Florida, as well as the Florida Panhandle, extreme precipitation events (and, in turn, extreme discharge events) appear to be less sensitive to AMO compared to the rest of the state. There is also a shift in the phasing: extreme precipitation events typically occur earlier in the year (June to August) during AMO cool periods, whereas they are observed later (August to October) during AMO warm periods.

Besides AMO, there are other large-scale climate indicators affecting precipitation patterns in Florida at interannual to decadal time scales, most notably the El Niño Southern Oscillation (ENSO): precipitation amounts during the Florida dry season (November to April) have been found to be larger in El Niño years compared to La Niña years (Teegavarapu et al. 2013). This is due to a general increase in storminess in the same season during El Niño years when the polar jet stream flows farther south allowing more frontal systems to reach Florida (e.g., Douglas and Englehart 1981). The Pacific Decadal Oscillation (PDO) has a similar influence on Florida climate as ENSO, but on longer decadal time scales (Misra et al. 2011).

Droughts

Florida is still vulnerable to periods of drought, even though the state has abundant precipitation by world standards. As might be expected, Florida droughts are most prevalent during the dry season when rain falls along fast-moving cold fronts and the amounts are minimal or non-existent. The weather patterns that bring drought to the state are linked to areas of stationary high pressure to the west extending deep into the atmosphere. Droughts are subtle, though. These high pressure systems usually bring a northerly flow that produces nice sunny and dry weather. But after a while without rain, plants begin to wilt, streams and lakes get much lower, and the ground hardens. Sunny days become an unwelcome prospect for residents during drought conditions. Droughts are a part of Florida's climatic history, however with the increasing population, water shortages become a serious problem. According to the Florida Climate Center (2016), since 1900 every decade has produced at least one severe and widespread drought somewhere within Florida, with the most severe droughts occurring in 1906, 1927, 1945, 1950, 1955, 1961, 1968, 1980, 1984, 1998, and 2006.

---

[3] Extreme hydrological events are typically defined as annual or monthly maxima or events that exceed some high threshold (usually expressed as percentile; e.g. the 99th percentile).

Drought Potential

Extended periods with lack of rain lead to hydrological (groundwater) and agricultural drought conditions. These types of droughts put pressure on water systems and can lead to conflict over the protection of water as a commodity. Using 16 years of data (2000-2015) from the U.S. Drought Monitor to calculate the average number of weeks in drought[4] per year provides a view of the possible frequency of future droughts given similar circumstances (Fig. 20.5). These periods of extended dry weather will likely increase in an uncertain climate future.

Three key areas across the state have seen higher than average weeks in drought during the period of observation. First, the entire northern border area, from Jackson County east to Madison County, has the highest historical number of drought weeks per year. Second, the area from Jackson County east to Nassau County and southeast to Volusia County has at least a moderate drought frequency. Finally, the area north of Lake Okeechobee, including Glades, Highlands, and Okeechobee counties, northward along the border between Osceola and Brevard/Indian River counties and southeast into western Martin and Palm Beach counties also has a moderate drought frequency.

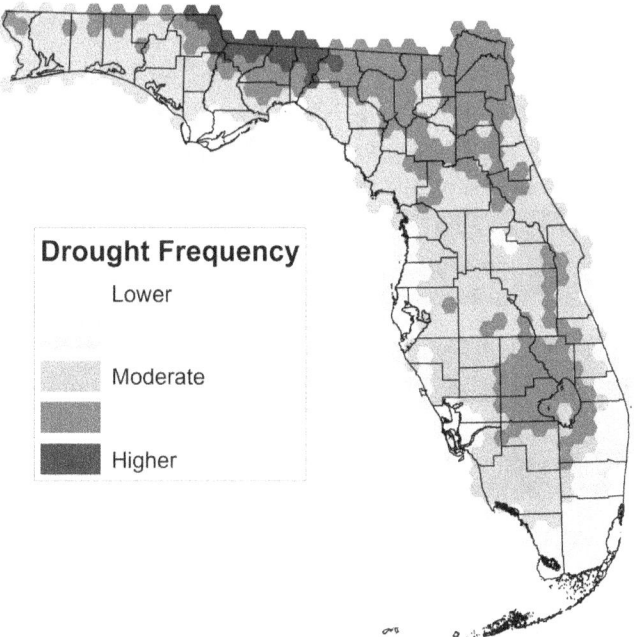

Figure 20.5. Drought frequency based on number of weeks in drought according to the U.S. Drought Monitor (2010-2015).[5]

---

[4] Drought weeks are calculated where Standardized Precipitation Index (SPI) values are less than the -2.0 (Extremely Dry) thresholds.

[5] http://drought.unl.edu/MonitoringTools/ClimateDivisionSPI.aspx

Droughts and Climate Variability

Similar to extreme precipitation events and as we see later with hurricanes, Florida droughts are affected by large-scale climate variability, but with opposing effects. Observations and model simulations show that dry winters are weakly associated with La Niña patterns (Seager et al. 2009, Schubert et al. 2009). It has been found that during cold AMO phases there is reduced streamflow over Lake Okeechobee (Enfield et al. 2001) and fewer drought events over Florida (McCabe 2004). On the other hand, less precipitation and warmer temperatures have been observed during La Niña years, increasing the drought risk throughout Florida (Brolley et al. 2007).

Florida Wildfires

Wildfires are directly related to lack of rainfall and resulting drought conditions. More than 3,300 Florida wildfires occur annually according to the Florida Forest Service (2016). Fire is dependent on three ingredients: 1) heat as a source of ignition, 2) fuel, and 3) oxygen. As might be expected, the wildfire season begins after the summer thunderstorm season ends, around the beginning of October when the relative humidity is low even though temperatures can be warm during the "cool" season. These drier conditions more readily evaporate the moisture from dead branches and leaves; brush and grasses dry out. As the dry season progresses, fires become more likely, with the height of the fire season occurring during the spring before the summer rainy season starts. Fires are most often set by humans' carelessly tossed cigarettes, campfires, burn piles, and vehicles, or by arson. However, a third of the wildfires are started by lightning. Spring lightning storms that bring little rain may ignite a fire.

Florida also has some unique vegetation containing combustible oils that burn even while the plant is alive. Flammable plants native to Florida include galberry, wiregrass, saw palmetto, cabbage palm, and pine trees. In fact, half of Florida surface fires are fed by grass fuels with a third of the fires burning in patches of palmetto and galberry. Crown or tree canopy fires are a rare occurrence. Some fires ignite drier areas of swamp containing peat and logs and, once started, these areas often smolder for weeks at a time while emitting abundant and thick water vapor-laden smoke.

Smoky fires near roadways, especially when combined with fog, have led to some horrific roadway crashes over the years. In these conditions, visibility can go to zero in a matter of meters. This happens when smoldering organic material adds fine particulates and heated water vapor to the air and mixes with fog. The fine particulates emitted from the burning provide a nuclei for water vapor to form on as it condenses into fog. The thick mixture of smoke and fog creates deadly mix drifts over roadways, reducing visibilities to zero instantaneously.

During the early morning hours of 9 January 2008, visibility dropped to zero as smoke from a prescribed burn the previous day combined with fog and drifted across I-4 in Polk County in Central Florida. Seventy cars and trucks collided killing five people and injuring 38 (Collins et al. 2009). A similar traffic accident occurred on I-75 in Alachua County near Gainesville a few years later, on 29 January 2012, resulting in 11 deaths and injuring 46 others. In the wake of these accidents, the Florida Highway Patrol has become much more proactive in closing portions of highways when the possibility of smoke and fog could limit visibilities for drivers.

Wildfire Potential

Historically, Florida was a peninsula perpetually on fire with little to no assistance in controlling blazes. These fires would "clean the slate" and allow keystone species such as Gopher Tortoises to flourish. Advances in early fire identification and better fire suppression techniques have steadily reduced the number of brush/wildfires and allowed undergrowth to build. While the state no longer burns from coast to coast, small fires that would have normally burned themselves out can turn into conflagrations requiring massive containment efforts. To assist state and local fire planning and mitigation efforts, the Southern Wildfire Risk Assessment produced the Wildfire Susceptibility Index for Florida in 2010 (SWRA 2016). Fig. 20.6 uses this data to build an understanding of where "an acre or more" will burn if ignited. In contrast to many of the other frequency data pieces, this assessment produces a real estimate of potential for burning.

Patterns of wildfire susceptibility vary across the state, but a majority of the south central counties are at an elevated wildfire susceptibility and a swath of highest susceptibility trends north and south between northern Osceola County and Okeechobee County. Additionally, areas from Charlotte County northeast through Highlands County, east into Glades County, and north to central Marion County are characterized by at least a moderate wildfire susceptibility.

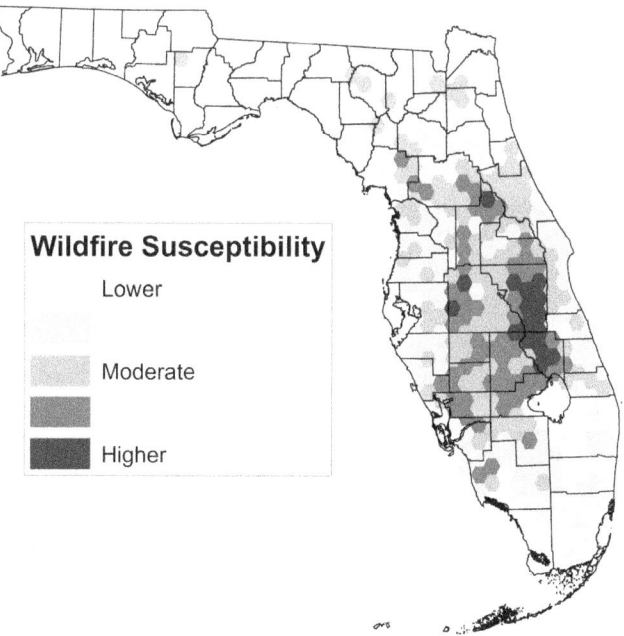

Figure 20.6. Wildfire susceptibility.

Wildfires and Climate Variability

Major wildfires often occur during drought conditions that are linked to La Niña. From 1981 to 2008, the most significant wildfires in Florida (Fig. 20.7; FreshfromFlorida.com 2017) burned 2,144,386 acres and most occurred either during La Niña or ENSO-neutral conditions. According to St. John's County Emergency Management (2017), the worst fires in Florida have occurred about every 20 years. The 1935 Big Scrub Fire in the Ocala National Forest, which occurred during a prolonged La Niña, was the fastest spreading fire in the history of the U.S., covering 35,000 acres in four hours. In 1956 during a strong La Niña, the Buckhead Fire burned 100,000 acres in Osceola National Forest in a single day. In the drought period of 1969 to 1976, during which a La Niña persisted except for several months when a brief El Niño episode brought more rain to South Florida during 1972-1973, fires in the Everglades gained national attention, with some fires reaching 50,000 acres. In 1985 during a La Niña, Florida had its first serious "wildland/urban interface" fire with the Palm Coast Fire, which burned 250 homes. This fire was important in introducing the state to the concept of the wildland/urban interface. In 1998 after a very strong El Niño that brought copious rainfall, which rapidly switched to a La Niña, fires ignited over much of the state and fire suppression organizations from 44 states responded. In July of 1998, Florida hosted the largest aerial suppression operation ever conducted in the United States. The 1998 fires received major media attention for almost two months, largely because of massive evacuations.

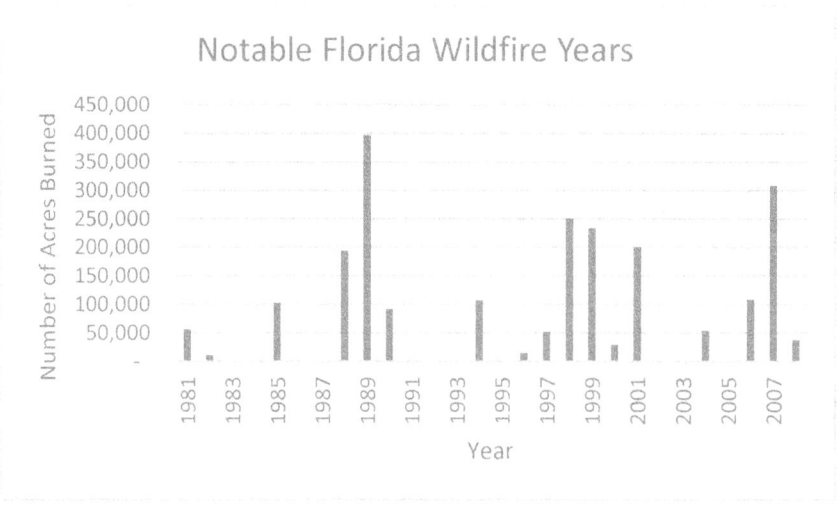

Figure 20.7. Notable Florida wildfire years and acreage burned.

Thunderstorms and Related Hazards

Thunderstorms develop from the abundant moisture in the form of water vapor over the state that is convectively lifted in an unstable atmosphere, then cools and condenses into growing cumulus clouds. As condensation occurs latent heat is released into the atmosphere, which adds to the instability and that increases rising motion in a towering cumulus cloud. The most notable convective cells develop within strong updrafts created by interactions with differential heating and wedged boundaries of slightly cooler air associated with sea breezes and outflows from existing convection. The stronger updrafts lift the water vapor in supercooled liquid form above the freezing level into an area of ice crystals. The mixed phase is an efficient process of cloud growth, as the supercooled vapor freezes onto the ice crystals and leads to a cumulonimbus (thunderstorm) cloud with the glaciated top. Severe weather typically develops in the colder area of the cloud.

Thunderstorms are steered by the average winds from the surface up to around 3 km. The position of the subtropical ridge, which is an extension of the Bermuda-Azores high pressure area, dictates the wind flow across the state as well as the timing and movement of convective clouds. When the ridge is north of Florida, the state experiences an easterly steering flow. When the ridge is south of Florida, the dominant steering flow is from the west. When the ridge is across the central peninsula, east winds flow over the south and west winds flow over the north. The speed of flow dictates the timing of sea breeze formation and movement inland. When the flow is from the east, sea breezes merge near the west coast. With westerly flow, sea breezes merge along the east coast. Under calm wind regimes, the east and west coast sea breezes merge in the central peninsula. Over the Florida Panhandle, the sea breeze propagates northward but is altered by east or west flow. Sea breeze convergence creates stronger updrafts and more vigorous thunderstorms.

The frequency of thunderstorms over certain parts of the state is also dictated by coastline shape. As sea breezes develop along the coasts and move inland, wind convergence can be focused by an area of land that juts into the water such as a cape. Apalachicola, Cape Canaveral, areas of Tampa Bay, and Fort Myers are areas of enhanced convergence. Landforms around Tampa Bay focus sea breeze convergence generating more frequent and persistent thunderstorms over 100 days per year, which is a greater number than any other part of the country. Other areas of Florida coastline that bend inland create the opposite effect with diverging wind inland from the area and less vigorous convection.

During the cool season, thunderstorms develop along frontal boundaries pushing into Florida where atmospheric moisture collects and is lifted in a more dynamic wind flow. We often see the jet stream near Florida, which creates stronger lift for thunderstorms. Winds that increase and flow from south to northwest with height can produce more persistent rotating thunderstorms. Additionally, during the cool season the mid-levels of the atmosphere are colder resulting in a more unstable atmosphere and more energetic thunderstorms when surface temperatures are warm.

Thunderstorms are responsible for much of the damaging weather that impacts Florida. As a peninsula surrounded by warm water in the sub-tropics, Florida has a unique geography that creates a tap for the deep layer atmospheric moisture, which is a prime ingredient for thunderstorms. Coupled with instability to lift the moisture, the intense vertical motions within thunderstorms produce twisting tornadoes, downbursts of rain-cooled air, large pounding hail, and deadly lightning.

During the warm season, from June through September, thunderstorms rumble around the state daily. Florida has many land interfaces of forest, farmland, urban and suburban areas that are dotted with more than 7,700 inland lakes over 40,000 m² (10 acres) including Lake Okeechobee, and all this sandwiched in between the Gulf of Mexico and Atlantic Ocean. The diverse land surfaces absorb shortwave solar energy at different rates, leading to differential heating that becomes obvious when cumulus clouds form over hotspots during the summer late mornings. Heating over the land creates lower pressure and a natural wind flow from the surrounding ocean water areas inland. These sea breeze boundaries create an area of focused rising motion where clouds are more likely to develop. At night, the pattern reverses with wind flowing offshore as land breezes.

Lightning

Ice crystals that form the tops of thunderstorms attract positively-charged particles, the liquid water droplets down below attract negatively-charged particles. This changes when a thunderstorm cloud passes over the ground, as positively-charged particles are drawn to the surface of the Earth. When the voltage difference becomes too great, negatively-charged stepped leaders, which are small segments of energy, travel downward as positively-charged stepped leaders move up from the ground. As the stepped leaders meet, the channel becomes a conduit for electrical current moving from the ground to the cloud, which creates lightning. Lightning can also occur in clouds, from cloud to cloud, and from cloud to air. The temperature of lightning is estimated at 27760° C (50,000° F), which creates a rapid expansion of air outward that creates a brief vacuum then the air rushes back. This is what creates the thunder. The speed of sound is much slower than the speed of light, so for roughly every five seconds after the flash, the sound has traveled a mile. Although very bright, the lightning channel is extremely narrow, about the size of a pencil.

Roughly 90% of all lightning occurs over land areas nearest to the Equator, and the most prolific lightning locations across the globe are across the central African continent and near Lake Maracaibo in Venezuela. Florida leads the nation in lightning strikes and is known as the lightning capital of the United States, experiencing about 100 days with thunderstorms each year across the central peninsula along the I-4 corridor. Florida also leads the nation in lightning deaths, averaging six per year.

Lightning Potential

Most of the lightning that strikes Florida occurs during the warm season across the central peninsula. During the warm season, tropical moisture and conditional instability lifted by diurnal sea breezes initiate convection that grows to 15 km (50 kft) until it encounters the stable stratosphere. Florida's coastal geography also shapes the sea breezes and areas with capes; areas of land extending seaward create coastal wind convergence, which amplifies the convection. It is these areas that have the most lightning. The synoptic scale wind flow determines where the lightning will be most prevalent from day to day. Over the panhandle, wind regimes have less impact on lightning than over the peninsula, where the east and west coast sea breezes collide and intensify the convection. Easterly winds promote sea breeze collisions and the strongest thunderstorms along the west coast of the peninsula. The converse is true with westerly winds. During the cool season, the most lightning occurs over the Florida Panhandle, where cold fronts carry more moisture and instability as they often lose it moving southward over the peninsula.

Mapping the average number of cloud-to-ground lightning flashes per year (1986-2012) from the National Central for Environmental Information provides a visual understanding of the historical frequency of this hazard (Fig. 20.8). Hillsborough County emerges as the leader in lightning frequency. This has to do with the shape of Tampa Bay, which forms the most prolific lightning storms with cloud-to-ground lightning strikes occurring about 25 times every 2.6 km^2 (one square mile) around Tampa and eastward along I-4 annually (Hodanish et al., 1997). Franklin County in the North Central Panhandle has the lowest lightning frequency. Seminole County, however, exhibits the highest average number of lightning flashes making it a location of concern for this type of hazard. Finally, Polk County, because of its size and location in relation to the "lightning belt" trending from Tampa to Orlando, has the highest overall number of strikes during the period of record. Other areas with higher historical lightning frequency include Pasco County north of Tampa.

Lightning and Climate Variability

A relationship between ENSO and global lightning activity has been found in both the intensity and position of lightning activity (Sátori et al., 2009). It has been observed that generally more lightning occurs in the tropical-extratropical land regions during warm El Niño episodes, especially in Southeast Asia. LaJoie and Laing (2008) specifically focused on the lightning activity along the U.S. Gulf Coast and noted that there is an ENSO influence, such that they observed the highest annual flash rates occurred in 1997 during the strongest El Niño event on record, while the lowest annual lightning flash rates occurred in 2000, which was more of a neutral/La Niña phase. This is consistent with what is observed, in general, globally.

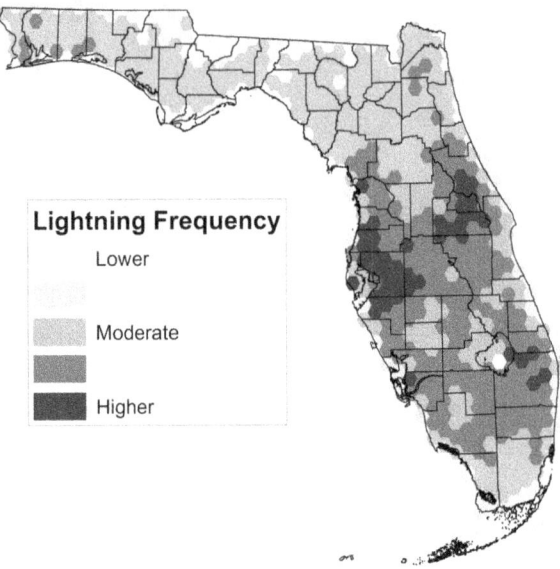

Figure 20.8. Historical cloud-to-ground lightning flash frequency (1986-2012).

Hail

Extreme updrafts produce supercooled liquid water vapor that freezes on frozen water droplets creating hail. Growth continues while the updraft is strong enough to suspend the hail. Hailstones sometimes aggregate into larger sizes or may be solely formed from frozen water layers. Hail falls when it moves out of the updraft, grows too large to be suspended, or when the updraft weakens. The largest hailstone recovered to date fell near Vivian South Dakota in 2010 and was over 200 mm (8 in) across with a mass of just under 1 kg (2.2 pounds).

According to the NOAA Storm Events database, 60% of the reports of Florida hail occur during the prolific summer thunderstorm season, with a peak in June; but, the largest hail falls during the cool season with a peak in March. During the cool season, the dynamics are stronger to produce supercell thunderstorms that produce large hail, and the atmosphere is colder and less melting occurs on the way down with lower freezing levels below 3 km. Supercell thunderstorms occur when atmospheric instability and wind shear are strong, thus organizing developing convection into a persistent rotating storm. During the warm season, wind shear is weak, the convective storms are short-lived, and the freezing levels are often above 4 km (13,000 feet). The typical maximum hail size reported during the warm season is around 50 mm (2.0 in). During the cool season the largest hail size reported has been twice that size, in the grapefruit category of 114 mm (4.5 in), and has only been reported in Florida three times in the past.

The number of Florida hail reports in the National Weather Service Storm Events database (Storm Events 2016) varies by decade and is biased by hail sizes reported (Table 20.1). The hail reports in the database begin during 1955 and averaged four reports per year during the 1950s.

During the 1960s, the reports averaged 14 per year because of increased sighting as population grew: up to 19 per year during the 1970s and 25 per year during the 1980s. During the 1990s, modernizations by the National Weather Service allowed for more robust inquiries into events and improved communication via cell phone and the Internet, which led to more than 130 reports per year in the 1990s and nearly 200 reports per year during the 2000s. The hail reports were biased primarily because of the popular and readily identified object, the golf ball (Table 20.1). When considering the reports of hail 32-51 mm (1.25-1.99 in) in size, golf ball-sized hail, 43-51 mm (1.7-1.99 in) was reported 629 times (69%) compared to 195 reports (31%) of smaller hail between 32-43 mm (1.25-1.69 in).

Table 20.1. Reported hail sizes in Florida 1955-2015.

Size (mm)	Size (in)	Number of Events	Percentage	
101-114	4.0-4.5	4	0.10	Softball to Grapefruit
76-101	3.0-3.99	9	0.21	Tea Cup
51-76	2.0-2.99	90	2.15	Hen Egg to Baseball
43-51	1.7-1.99	629	15.01	Golf Ball
38-43	1.5-1.69	110	2.63	Ping Pong Ball
32-38	1.25-1.49	85	2.03	Half Dollar
25-32	1.0-1.24	1003	23.94	Quarter
20-25	0.8-0.99	606	14.46	Nickel
19	0.75	1654	39.47	Penny

Hail Potential

Hail is another hazard that cannot be predicted well enough in advance to take any "real-time" precautionary measures against. Yet this hazard threat is capable of destroying roofs, windshields, and can cause major damage to crops across the state. Creating a geographic representation of historical hail events affords us a glimpse into where these events have interacted with human activity in the past. Using data from NOAA's Storm Prediction Center (1986-2015) to produce an average annual hail event surface (Fig. 20.9) provides us a "climatology" of events that can be utilized to identify areas of potential future human-hail interactions.

While human bias is evident in this depiction of hail events, with the major interstates and population centers popping out with higher frequency, we also see areas of higher threat in the less populated Polk County in Central Florida, in more rural western Duval County, and in areas along the northern I-4 corridor that have smaller populations. Orange County has the largest amount of land area historically impacted by hail events, followed by Hillsborough, Duval, and Polk counties. The highest historical frequency area is located from Sanford in Seminole County south to the northern edge of Orlando.

Hail and Climate Variability

Fig. 20.10 shows the number of reports of hail in Florida, one inch in diameter or greater, from 1995 (a time when reports were less likely to be biased due to National Weather Service staffing) to 2015. While ENSO is the most prominent indicator for Florida extreme weather, Fig. 20.10 does not provide evidence for a relation and hence we cannot identify robust relationships between hail and larger-scale climate variability, in general. For example, the lowest years (1995, 2000, 2002, 2010, 2013 and 2015) with under 60 reports across the state were varied with respect to the ENSO phase. The same is true for the years with the greatest number of reports (1998, 1999, and 2011), which had more than 100.

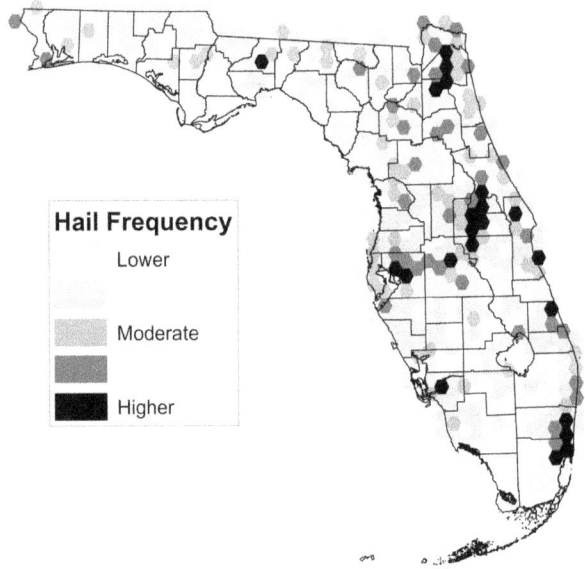

Figure 20.9. Historical hail event frequency (1986-2015).

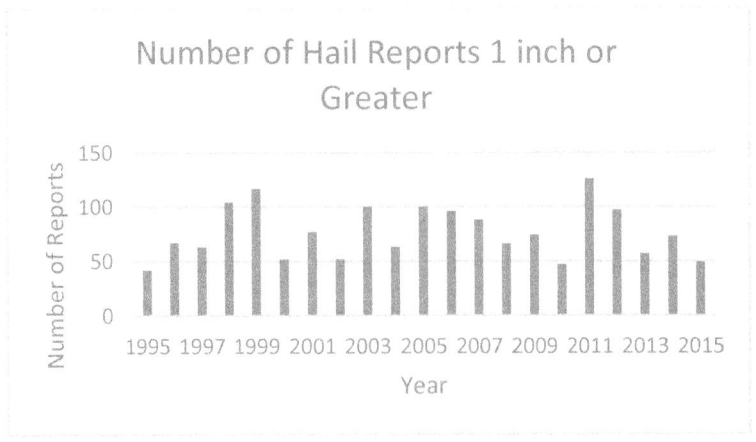

Figure 20.10. Number of hail reports one inch or greater.

Tornadoes

The strongest tornadoes that impact Florida develop along vigorous cold fronts during the cool season when instability, moisture, and wind shear is greatest. These strongest tornadoes usually develop from supercell convection organized by strong directional wind shear (changing wind direction with height). Under strong unidirectional wind shear, convective systems organized into lines near cold front boundaries will, at times form moderately strong tornadoes. The Fujita scale matches up tornado severity with numbers preceded by an "F". Since 2007, this scale was modified to the Enhanced Fujita scale preceded by an "EF". When the conditions are met for cool season tornadoes, they often occur in clusters over a several-hour time period. In the past, five tornado outbreaks with more than 10 fatalities have occurred in Florida and are described in Table 20.2. Grazulis (1993) defined a "tornado outbreak" as a group or family of six or more tornadoes spawned by the same weather system.

For the period 1991-2010, Florida leads the nation in the average annual number of EF0-EF5 tornadoes per 10,000 square miles, followed by Kansas and Maryland. State by state, Florida comes in third for the average total of tornadoes from 1991-2010, averaging 66 tornadoes per year; Texas, which has a large land area, averaged 155 and Kansas averaged 96. Florida comes in second for both the greatest number of tornadoes from 1980-1990, and the average number of days per year with tornadoes (28). That said, most of Florida's tornadoes are weak. When ranked by the strongest tornadoes (F3/EF3 and above) Florida ranks 25th with the frontrunners, which are all in areas with stronger instability and wind shear: Kansas at the top, followed by Arkansas, Texas, Tennessee, and Oklahoma.

Most of Florida's tornadoes are spawned during the warm season by single cell convection and are short-lived. These brief tornadoes typically cause damage to trees, aluminum awnings and carports, and only minor damage to well-constructed homes. Some of these tornadoes start as waterspouts and then move onshore causing damage along the shore.

Tropical cyclones typically produce fast 25 m s^{-1} (56 mph) northward-moving tornadoes on the east side of the storm, up to EF2 on the Enhanced Fujita scale. The most significant tropical tornado outbreak on record occurred during Hurricane Agnes in June 1972, creating the fifth-deadliest tornado outbreak in Florida history.

Tornado Potential

Historical tornado frequencies across the state derived by the NOAA Storm Prediction Center statistics (SPC 2015) provide the basis for identifying possible future threats from these powerful hazards. In Fig. 20.11, hex grids illuminate the areas where tornadoes have been reported across the state from 1955-2015. Areas of high frequency can be seen across the state. And although some of the deadliest tornadoes have been in the interior of the state, the majority of tornado reports come from the more densely populated coastal areas. The western three panhandle counties (Escambia, Santa Rosa, and Okaloosa) have the highest frequency of tornadoes over the period from 1955-2015. The Tampa Bay area is also characterized by higher numbers of tornadoes that often start as waterspouts and move onshore. The lower east coast area, from Miami north, also has a high frequency of reported tornadoes along with Orlando, the Space Coast area, and Jacksonville.

Tornadoes and Climate Variability

Interestingly, the number of tornadoes reported annually from 1995 to 2015 (Fig. 20.12) decreased over time. From 1995 through 2005, the average number of tornadoes reported each year in Florida was 76 with the greatest number of reports, over 100, during 1997, 1998, and 2004. From 2006 through 2015 the average annual number of tornadoes reported was 38 or about half that of the previous decade. The data at hand does not provide robust evidence for linking this decrease to climate variability.

Table 20.2. Florida's Deadliest Tornadoes.

Date	Event	Description
June 18-19 1972	Hurricane Agnes Tornadoes	Hurricane Agnes struck Florida with the fifth deadliest tornado outbreak in Florida history. Agnes travelled north, creating the costliest natural disaster in U.S. history at the time, with $3.5 billion (1972 dollars) in storm damage. Rainfall from Agnes created extreme flooding in Pennsylvania, New York, Maryland, Virginia, and Washington D.C., killing 122 people. Researchers Hagemeyer and Spratt (2002) examined the deadly tornadoes slamming Florida that spawned from Hurricane Agnes. The storm produced 28 tornadoes over the southern half of the peninsula, from near Daytona Beach to Key West. Two of the tornadoes produced F3 damage, nine were ranked as F2, 11 were F1, and the other six were F0. One tornado cut a path 100 yards wide through Okeechobee City, killing six people and destroying 50 mobile homes. Another tornado near La Belle in Hendry County killed a woman in a trailer. The tornadoes injured 140 people, destroyed 15 homes and more than 200 mobile homes.
February 2, 2007	2007 Groundhog Day Tornadoes	The tornadoes took 21 lives and injured another 76 people. Within minutes, a mesocyclone that meteorologists had been watching gained strength and a tornado warning was issued. Minutes later soon-to-be victims of the Villages community, who were not woken by the siren sound of the warning on their NOAA weather radios, awoke to the horrendous sound of their homes being destroyed. The tornadoes demolished 200 homes and damaged over 1,100 in Sumter County. Those were the lucky victims – all they lost was their home. The tornado continued moving eastward rapidly at 60 mph with over 160 mph winds and ravaged the Lady Lake area, killing eight people and destroying more than 100 homes in Lake County before it lifted. A second tornado developed minutes later and obliterated the Lake Mack area, killing 13 people and destroying over 500 homes. A third tornado touched down just east of the second tornado before lifting along the East Coast at New Smyrna Beach. This tornado outbreak was the second-deadliest on record for Florida, with damages of $218 million.

31 March 1962	March 1962 Tornado	A savage tornado ripped through Milton, Florida (about 30 miles northeast of Pensacola) claiming 17 lives and injuring 80. This was the highest death toll from a tornado in Florida up to that date. Along the seven-mile tornado path of destruction, three residential blocks were devastated. Homes were destroyed and trees left as stumps. The storm picked up a home with three residents inside and took them for the ride of their life. The tornado spun the house and gently dropped it a hundred yards away on the foundation of another home that was demolished by the twister leaving the residents unharmed. Relief workers, National Guardsmen, and sailors from nearby Whiting Field poured into the area to help the victims.
April 4, 1966	April 1966 Florida Tornado Outbreak	Eleven people died and more than 3,300 people were injured during the 1966 tornado outbreak that left a more than 100-mile (160-km) path of destruction across Central Florida. It occurred during the morning of April 4, 1966, beginning around 8:00 am. The strongest tornado, with an estimated intensity of F4, created a path of death and destruction -- from Clearwater on Central Florida's West Coast to Merritt Island near the Kennedy Space Center on the East Coast. At times, it was measured to be as wide as 300 yards. In addition to the deaths and injuries, more than 250 homes and businesses were destroyed. The tornado ripped through parts of the Forest Hills and Carrollwood areas, as well as the University of South Florida campus in Tampa. East of Tampa, in the town of Lakeland in Polk County, seven people lost their lives and a 55-foot radio tower was yanked from its concrete pilings and smashed to the ground. As the storm clouds continued moving eastward, citrus trees near Auburndale, were stripped bare and fruit was scattered on the ground. The second tornado moved ashore near the mouth of Tampa Bay and created destruction across over 100 miles to Cocoa Beach, where more than 20 frame homes, a shopping center, and 150 mobile homes were destroyed, and more than 100 were injured. This tornado outbreak is the fourth-deadliest documented in Florida.
23 February 1998	Central Florida Tornado Outbreak of February 1998	Below is an account from the National Weather Service in Melbourne, FL of the deadliest tornado outbreak in Florida, which occurred in February 1998. "During the late night and early morning hours of 22-23 February 1998 (Sunday - Monday), the most devastating tornado outbreak ever to occur in the state of Florida in terms of both loss of life and property damage, occurred from Kissimmee to Sanford to Daytona Beach. Forty-two people died as a result of the tornadoes and more than 260 others were injured. Over 3,000 structures were damaged, and more than 700 were completely destroyed. A total of seven confirmed tornadoes occurred that night. Four of the tornadoes were unusually long-lived and produced damage tracks of between 8 and 38 miles, resulting in the majority of damage and all of the fatalities. Uncommon for Florida tornadoes, the estimated wind speed for three of these twisters reached 200 mph which is on the high end of F3 intensity on the Fujita scale [and would be on the border between EF4 and EF5 intensity today]."

598 • JENNIFER M. COLLINS ET AL.

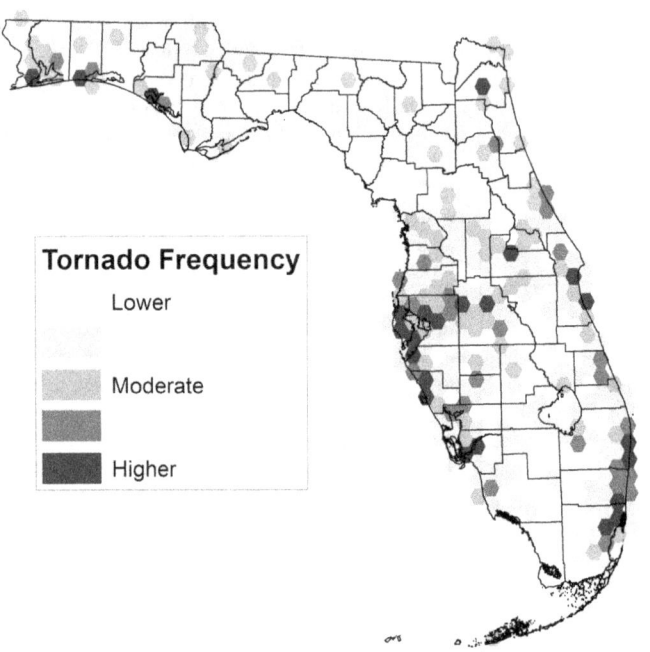

Figure 20.11. Historical tornado frequency (1955-2015).

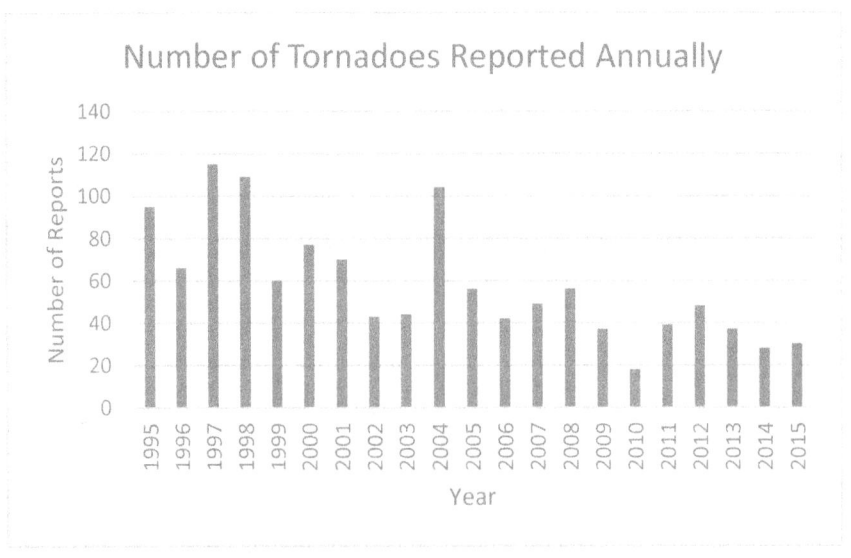

Figure 20.12. Number of tornadoes reported annually (1995-2015).

Damaging Thunderstorm Winds

Damaging thunderstorm winds are winds that cause damage to buildings or sturdy trees and are either estimated or measured at 50 knots or greater. Damaging thunderstorm winds across the state manifest in two ways. First, during the cool season when the atmospheric dynamics are stronger, thunderstorms develop near active cold fronts either in a linear or curved arrangement or as single strong super cells. These cool season thunderstorms are longer lasting than the summer thunderstorms that form within the very unstable tropical air mass and reach peak intensity and begin to dissipate within an hour.

Thunderstorm Wind Potential
Fig. 20.13 shows damaging thunderstorm wind frequencies over Florida derived from the NOAA Storm Prediction Center statistics (SPC 2015) from 1955-2015. Unlike tornado report frequencies, which have a coastal bias, the stronger thunderstorm winds are seen across interior areas of the state, as well. Some population bias is also indicated near major metropolitan areas.

Damaging Thunderstorm Winds and Climate Variability
Unlike tornado reports, which have decreased during the past decade, the reports of damaging thunderstorm winds have increased to more than 500 reports in 2011 and 2015 (Fig. 20.14). The decadal averages were 291 from 1995 through 2005 and 374 from 2006 through 2015. The data shows that the number of damaging wind reports appears to be linked to El Niño patterns with lesser annual reports linked to neutral or La Niña patterns.

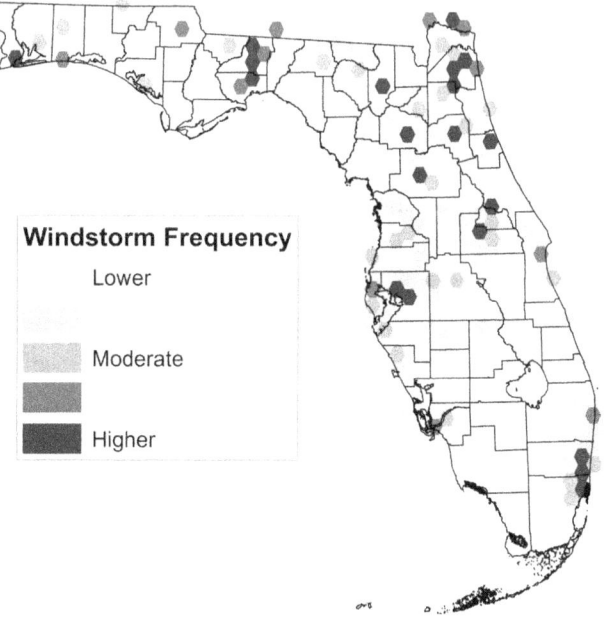

Figure 20.13. Historical damaging thunderstorm wind frequency (1955-2015).

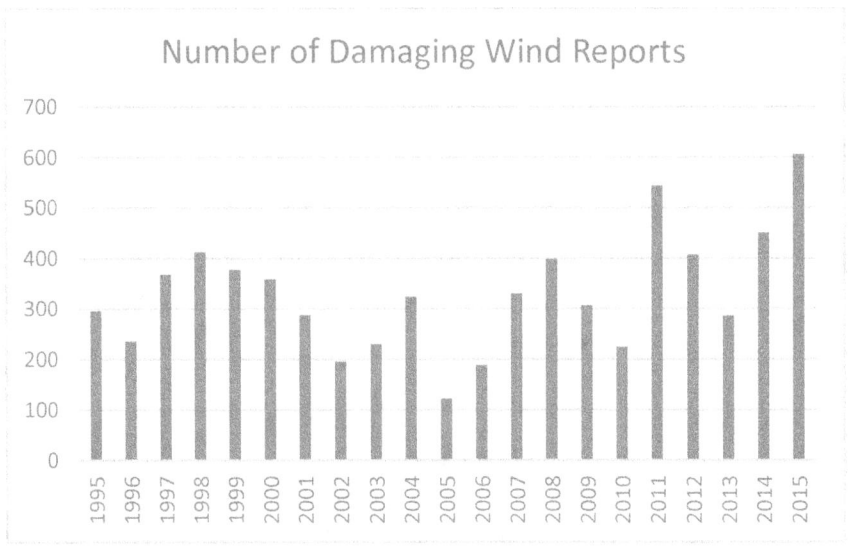

Figure 20.14. Number of damaging wind reports (1995 to 2015).

Tropical Weather

The term "hurricane" is used for North Atlantic Ocean and Northeast Pacific Ocean tropical cyclones. It comes from the Spanish word huracán, which was derived by Spanish explorers from references to hurakán (or 'god of the storm'), a word used by the Arawak people who inhabited the Greater Antilles and the Bahamas in the past. The Arawak may have taken it from the Mayan creator god, 'Hurakan', who blew the water.

Unlike mid-latitude low pressure systems with cold cores fueled by temperature contrasts between the poles and the tropics, hurricanes are warm core areas of low pressure fueled by warm ocean waters and tropical moisture that comes from the latent heat of vaporization being released by convection and condensation. Gray (1979) developed a set of six factors that he noted were necessary for tropical cyclone formation. Tropical cyclones are dependent on: 1) warm ocean waters above 27° C (80° F) over a deep layer (typically at least 60 m), 2) plentiful atmospheric moisture (i.e., high levels of mid-level relative humidity), 3) an unstable atmosphere that allows air to rise, 4) little or no vertical wind shear, which allows for strong vertical development, 5) sufficient Coriolis force to give the initial cyclonic spin, and 6) high relative vorticity.

Prior to the formation of a tropical cyclone, a disturbance is created through convergence of air. This convergence can form by more than one mechanism. For instance, it can be created along the Intertropical Convergence Zone where winds from the Southern Hemisphere cross the Equator and meet winds from the Northern Hemisphere. Another mechanism can be from African easterly waves that form off Africa. As air moves into these waves, convergence occurs. Regardless of the mechanism, the convergence causes air to rise and a disturbance is created. These disturbances become "tropical depressions" when they are well organized and have a cyclonic circulation. If a tropical depression intensifies with a sustained wind speed of 17 m s^{-1} (39 mph), then it becomes a "tropical storm" and is given a name. When a sustained wind speed of 34 m s^{-1} (74 mph) occurs the system becomes a "hurricane." Hurricanes are ranked by intensity from Category 1 to 5, using the Saffir-Simpson scale (Table 20.3), with the categories corresponding to anticipated wind damage to sturdy structures. Major or intense hurricanes are those designated as a Category 3 hurricane or higher.

The official hurricane season begins June 1 and stretches to the end of November. Most of the early season storms and those that occur in November are typically weak because the ingredients are lacking and the storms often form in the Caribbean Sea and/or Gulf of Mexico where waters are warmer. As the deep Atlantic Ocean waters warm during the summer, many seedling storms that become hurricanes move westward from the continent of Africa's coast (around 20°N) and grow into tropical depressions. These systems are often known as Cabo Verde (Cape Verde) storms because they transit near or across the archipelago off the African Coast. In the Atlantic Basin, hurricane activity increases greatly during August, peaking around September 10 and beginning to decrease into October. When just considering the storms impacting Florida, roughly 33% occur in September, 25% during October, and 20% during August. As sun angles get lower and water temperatures over the Atlantic Ocean cool, October storms are more likely to form in the warmer waters near Florida and are often stronger than June storms.

During the warm season when southward cold air transport to balance temperatures in the northern hemisphere is minimal, tropical cyclones carry excess heat energy from the tropics toward the poles. Hurricanes bring multiple impacts for Florida residents including a battering storm surge, damaging winds, flooding rains, and tornadoes. Storm surge, which is the most deadly of these hazards, is a rapid rise in coastal waters as the storm moves near shore. Surges can be up to 9 m (30 ft) in a Category 5 hurricane in the shallow coastal waters along the Gulf Coast.

Table 20.3. The Saffir-Simpson scale.

Category	Wind Speed (km h^{-1})	Wind Speed (mph)
1	119-153	74 – 95
2	154-177	96 – 110
3	178-208	111 – 129
4	209-251	130 – 156
5	≥252	≥157

Table 20.4. Significant hurricanes.

Date	Name	Description
September 1848	The 1848 Hurricane	A 15-foot storm surge within Tampa Bay reshaped much of the area when a major hurricane struck in late September of 1848. Damage and loss of life were minimal because only a few people, mostly military, lived in the Tampa Bay region at the time. However, the surge had a huge impact on the barrier islands. The high surge, rough surf, and strong currents cut several gaps or passes through barrier islands including: Stump Pass at Englewood, Casey's Pass at Venice, New Pass at Palm Island (which became Longboat and Lido keys), and John's Pass between Madeira Beach and Treasure Island.
September 1919	The Dry Tortugas Hurricane (also known as the Florida Keys Hurricane) of 1919	This was one of the few historically-severe hurricanes that was not a Cape Verde storm. Instead, it was believed to have formed near Guadeloupe in the Lesser Antilles. The storm struck the Florida Keys on September 10, 1919 as a Category 4 hurricane. As the storm progressed into the Gulf of Mexico, it attained one of its distinctive features – slow forward movement – before finally making another landfall in southern Texas on the evening of September 14. Because of its slow speed and wide circulation, this hurricane had tremendous storm surge impacts across most of the U.S. Gulf Coast, even in areas far from where it tracked. The storm caused approximately 800 fatalities from the Bahamas to Texas. Many of the victims were on ships; less than 100 deaths occurred in the then sparsely-populated Florida.
October 1921	The 1921 Hurricane (unnamed)	The last hurricane to impact the Tampa Bay area directly was the unnamed storm of October 1921. The estimated Category 3 storm made landfall just north of Tampa Bay, battered the city with winds exceeding 100 mph, and created a 10- to 12-foot storm surge in Tampa Bay. The storm caused $1-10 million dollars in damage and was responsible for six deaths (Ballingrud 2002).
September 1928	The 1928 Okeechobee Hurricane	The second-deadliest hurricane to strike U.S. soil, after the Galveston Hurricane of 1900, was the 1928 Okeechobee hurricane. This storm was blamed for more than 4,000 deaths, of which over 2,500 were Floridians and most occurring near its landfall area of West Palm Beach to Lake Okeechobee on September 17, 1928. The storm earned its "Okeechobee" moniker because its winds pushed a tremendous amount of Lake Okeechobee's water into the adjacent lowlands, drowning many people who lived on the north and south sides of the lake. The track of this ferocious hurricane then turned northward, taking it lengthwise up the Florida Peninsula, which only added to its toll on the Sunshine State.
September 1935	The 1935 Labor Day Hurricane	The deadly Category 5 Labor Day hurricane struck Florida's Upper Keys on September 2, 1935. It killed more than 400 people. At landfall, this storm may have had the lowest central pressure (892 mb) in U.S. hurricane history and its localized storm surge of up to 20 ft was devastating. After striking the Upper Keys and moving across southern Florida into the Gulf of Mexico, the storm continued moving north but weakened just offshore of Florida's West Coast before making landfall again near Cedar Key on September 4 (Map 7.5). This was the first of three Category 5 hurricanes to hit the U.S. during the 1900s.
August 1992	Hurricane Andrew (1992)	Hurricane Andrew, whose southern Florida landfall at Elliott Key and then in Homestead on August 24, 1992 as a Category 5,

		became the costliest natural disaster in Florida history, with over $25 billion in damages (1992 dollars). Andrew left the most damage of any hurricane in U.S. history and its death toll of 44 Floridians was substantial.
September/ October 1995	Hurricane Opal (1995)	Hurricane Opal was one of the most memorable storms of the 1995 hurricane season, causing more damage in Florida than elsewhere in the U.S. Opal was unusual in that it formed as a Cape Verde storm, but remained poorly organized until reaching just north of the Yucatan Peninsula where it was named a tropical storm on September 30, 1995. Opal then crossed the Yucatan from east to west, rapidly strengthening to Category 4 status over the Bay of Campeche before moving northeastward across the Gulf and slamming into the Florida Panhandle near Santa Rosa Island on October 4. Opal killed 50 people in Central America and 13 in the U.S. One of these deaths was in Florida during an F2 tornado that Opal spawned. Florida did bear a very large share of Opal's land and property damage, including major damage to Highway 98, which parallels the coast along Fort Walton Beach, Destin, and other resort areas.
August 2004	The 2004 Hurricanes: Charley, Frances, Ivan, and Jeanne	A decade-long period of relative quiet for Florida ended with a vengeance in 2004 when four historically powerful storms made landfall in or very near the state, and another weaker tropical storm (Bonnie) struck the Panhandle. According to the National Weather Service, the sixth-, eighth-, and tenth-costliest hurricanes (Ivan, Charley, and Frances, respectively) in U.S. history affected Florida directly during that 2004 season. Charley, a Cape Verde storm, is known for its small geographic extent and multiple landfalls across the Caribbean and southeastern U.S. and also for setting a record (with Tropical Storm Bonnie) on August 13 (a Friday the 13th, no less) as the only pair of named storms to hit a single state within a 24-hour period. Charley moved across the Dry Tortugas and then an unseasonably early atmospheric trough dipped southward across the Gulf, steering Charley northeastward. While being steered northeastward by the trough, Charley intensified rapidly before striking Cayo Costa along the southwest Florida coast with Category 4 winds. Charley continued inland over heavily-populated areas such as Orlando, Kissimmee, and New Smyrna Beach while still at hurricane strength. Charley re-emerged over the Atlantic Ocean making additional landfalls in South Carolina, and eventually merging with a frontal system as it moved over southern New England. Hurricane Frances, another Cape Verde storm that caused significant impacts, was a Category 2 at landfall near Hutchinson Island along Florida's southeast coast. Unlike Charley, Frances was a large, slowly-moving storm that caused more widespread damage but had less intense impacts. After wreaking havoc in the Bahamas as a Category 4 storm, Frances weakened prior to its Florida landfall and re-emerged in the Gulf of Mexico near Tampa before turning back to the northeast and striking land again near St. Marks. Hurricane Ivan made landfall west of Pensacola, on the Alabama coast. But Ivan's large extent, with an eye alone that extended for more than 40 miles, had its strongest impacts occur in northwest Florida on the right side of the storm's landfall. Ivan had been a Category 5 storm as it entered the Gulf of Mexico and

		had weakened substantially to Category 3 strength just prior to its September 16 landfall in Gulf Shores, Alabama. However by September 20, Ivan had made a huge clockwise loop, re-emerging in the Atlantic before crossing the southern Florida Peninsula and making landfall near Miami. After its second landfall, Ivan crossed the Florida Peninsula, re-emerging in the Gulf of Mexico, and making its third and final mainland U.S. landfall as a tropical depression near the Louisiana-Texas border. One of the most distinctive features of Ivan was its strength at low latitudes. For area residents, the most memorable damage was to the I-10 Escambia Bay Bridge, where huge segments of roadway were moved off the foundation by the storm surge, taking many weeks to be repaired. Storm surge varied widely along Ivan's path, along the different types of coastline configurations and water depths in the storm's path; areas near the Bay Bridge experienced the worst surges, at over 13 ft. A non-Cape Verde storm, Jeanne, followed Ivan into Florida by less than a week. Hurricane Jeanne's Florida landfall occurred on September 26 near Hutchinson Island at almost the same location as Hurricane Francis' landfall, but Jeanne quickly dissipated after moving inland and trekking north-northwest up the western side of the Florida Peninsula. Like Ivan, Jeanne emerged in the Atlantic after crossing Georgia, the Carolinas, Virginia, and Maryland. However, unlike Ivan, Jeanne had already made its loop before reaching Florida and did not reappear for a second landfall. Jeanne's impacts were magnified by its timing – the rain from Jeanne had fallen on ground already waterlogged by Ivan, exacerbating flooding and crop damage. In terrain like Florida's, large trees are easily destabilized by such soggy conditions, particularly because the roots tend to be shallow owing to the naturally-high water table.
August and October 2005	The 2005 Hurricanes: Katrina and Wilma	While Hurricane Katrina will always be remembered for its devastation in Louisiana, the storm also impacted Florida. In South Florida, Katrina caused 14 deaths, three of which were due to falling trees and another three due to drowning. Katrina dropped nearly 15 inches of rainfall imposed widespread damage and power outages, and a spawned tornado on Marathon Key on August 25, 2005. The pressure dropped to 985 mb and the intensity had increased abruptly just prior to that tornado's landfall. Hurricane Wilma holds the record low pressure in the Atlantic Ocean Basin, dropping almost 100 mb in one day to 882 mb. Wilma was one of the few intense hurricanes to have approached peninsular Florida from the west, after having clipped Mexico's Yucatan Peninsula. This northeastward turn occurred because Wilma's late-season lifespan made her more vulnerable than most storms to being steered by the eastern side of an upper-atmospheric trough. Wilma made landfall near Cape Romano, Florida, on October 24, 2005.

Tropical Storm and Hurricane Potential

Wind damage from tropical systems has ravaged every part of Florida at one time or another. Fortunately, since Hurricane Andrew (1993), we have not had a singular event cause catastrophic impacts. However, since that time many storms have produced moderate to severe damage. Understanding hurricane wind frequency in terms of return period or annual expected number of events enables planners and decision-makers to prepare for future threats. While hurricane paths are unpredictable, looking back over the past 25+ years provides valuable information pertaining to hurricane wind potential. Below, we calculate the average times per year an area can expect to experience tropical storm and hurricane force winds using Colorado State University's Extended Best Track dataset (Demuth et al. 2006).

Fig. 20.15 shows higher historical frequencies of tropical storm winds (≥ 17 m s^{-1} or ≥ 34 knots) reach much further inland, putting many more counties at a moderate to high potential. An area of elevated tropical storm wind frequency is present from Gainesville in Alachua County south to Key West and north again to Palm Beach. Coincidentally, the highest historical frequency of tropical storm winds occurs in Charlotte County. Here, more storm paths have crossed than any other place in the state during the period of observation.

Higher levels of hurricane wind frequency (≥ 33 m s^{-1} or ≥ 64 knots; Fig. 20.16) are shown over East Central Florida, from Osceola and Polk counties southeast to northern Palm Beach County, along with a swath trending northeast from western Monroe County through Broward County. Additionally, areas of southern Escambia and Santa Rosa as well as Brevard, Orange, Volusia, and Polk counties have at least a moderate potential for experiencing hurricane force winds. The pattern is slightly different when accounting for weaker but more frequent tropical storms, which are also capable of causing severe structural damage to homes and businesses.

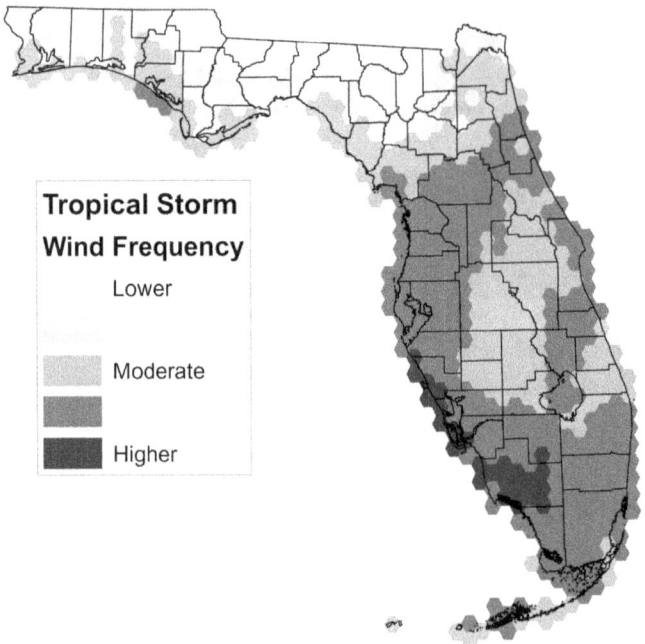

Figure 20.15. Historical tropical storm wind frequency (1988-2014).

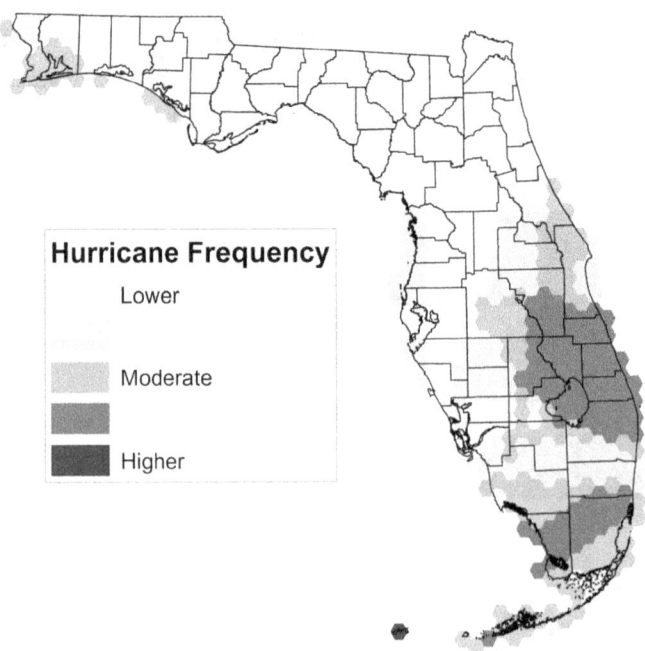

Figure 20.16. Historical hurricane wind frequency (1988-2014).

Hurricanes and Climate Variability

Hurricanes, and tropical cyclones in general, in the Gulf of Mexico and southeast of the U.S. are linked to larger-scale climate features and teleconnection patterns in multiple ways and at different time scales. Hurricanes can be responsible for multiple types of hazards such as extreme precipitation, tornadoes, or storm surges. The latter are the most dangerous and have the highest potential to cause catastrophic damage. Because of that, and due to the availability of long observational records from tide gauges, high coastal sea levels are often used as proxy to unravel the relationship between the variability in tropical cyclone activity at interannual to multidecadal time scales and climate variability/change. For Florida in particular, a strong relationship exists between extreme sea levels observed during the Atlantic tropical cyclone season and the Atlantic Multi-decadal Oscillation (AMO), which in turn is related to the size of the Atlantic Warm Pool that can favor the development of tropical cyclones; e.g., more landfalling cyclones were observed in AMO warm phases (Park et al., 2010a, 2010b). This leads to significant multi-decadal variations in extreme sea levels, as shown in Wahl et al. (2015, 2016), for the one in 100-year return water level (with a 1% probability of being exceeded in any given year) that is often used for design purposes by engineers or to define flood zones.

Another data source for this type of analysis is the "best-track" data (or HURDAT, which is short for HURricane DATabase) maintained by the National Hurricane Center. It contains information on a range of hurricane parameters (e.g., wind speed, central pressure, track), but it is considered more inhomogeneous due to rapid developments in observation techniques, for example, the use of aircrafts in the 1940s and satellites in the 1960s. The data show that ENSO also plays an important role in the variability of cyclone genesis at interannual time scales. Collins and Roache (2010) noted the presence of El Niño, when equatorial sea surface temperature is warm, inhibited tropical cyclone activity in the Atlantic for the 2009 season. On the other hand when equatorial sea surface temperature is cold (La Niña periods), hurricanes are more likely to affect the United States. The North Atlantic Oscillation (NAO), on the other hand, modulates the storm tracks and hence landfall locations; e.g., more hurricanes tend to affect the U.S. Gulf and southeast coast in years with below average NAO values (Elsner et al. 2001; Elsner 2003).

Florida's Record Maximum and Minimum Temperatures

Florida's all-time maximum temperature of 43° C (109° F) occurred on June 9, 1931 in Monticello (Jefferson County), just east of Tallahassee. Interestingly, the locations of Florida's all-time maximum and minimum temperatures are separated by only 64 km (40 mi). The all-time minimum temperature for Florida of -19° C (-2° F) occurred on February 13, 1899 in Tallahassee during an Arctic outbreak that impacted the entire eastern U.S. from February 10-13, 1899.

Historical Freezes in Florida

Several factors lead to freezes in Florida including: an air mass that becomes stationary in the frigid Arctic darkness, wind flow that then steers the air mass southeast through Canada, and preceding cold weather and snow cover along the path of the air mass over the central and eastern U.S. The following briefly summarizes Florida's historical freezes as noted by Citrus Mutual (2016).

Maximum and Minimum Temperature Potential

Identifying where temperature extremes occur enables us to understand where populations, crops, livestock, and infrastructure are most at risk. Utilizing 30 years (1985-2015) of daily temperature data from the National Climatic Data Center produces an average annual number of days where the daily low temperature has been below 32° (Fig. 20.17) or above 95° (Fig. 20.18). These visualizations enable planners and emergency managers to more completely identify where heat or cold protection might be needed or where crop losses may be expected.

Patterns of freezing temperatures follow a typical north-south trend where the panhandle has a higher frequency of freezing days compared to the southern part of the peninsula. This area abuts in the east to an area with a mixture of moderate to high freeze frequency: from Washington County in the panhandle, east to western Nassau, Duval and Clay counties, and south to northern Levy County.

Interestingly, the pattern of heat hazard days (above 95°) mimics cold days, for the most part, with a few notable exceptions (Fig. 20.18). The areas of highest heat hazard threat include Gadsden County south to the Gulf of Mexico, a small portion of Polk County in Central Florida, and a large portion of eastern Collier and northern Monroe counties. A large area extending from Jefferson County in the northeast to Baker and Clay counties, snaking south through Central Florida to Polk County, has at least a moderate frequency of heat hazard days.

CLIMATE AND WEATHER EXTREMES • 609

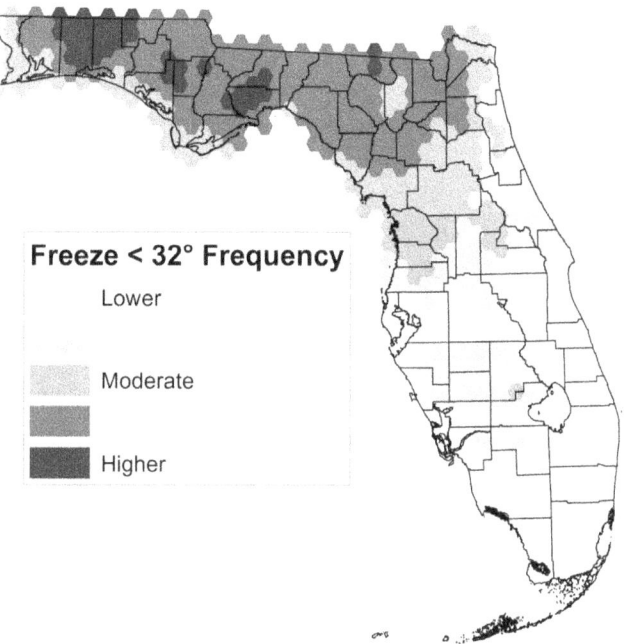

Figure 20.17. Frequency of temperatures below 32° F.

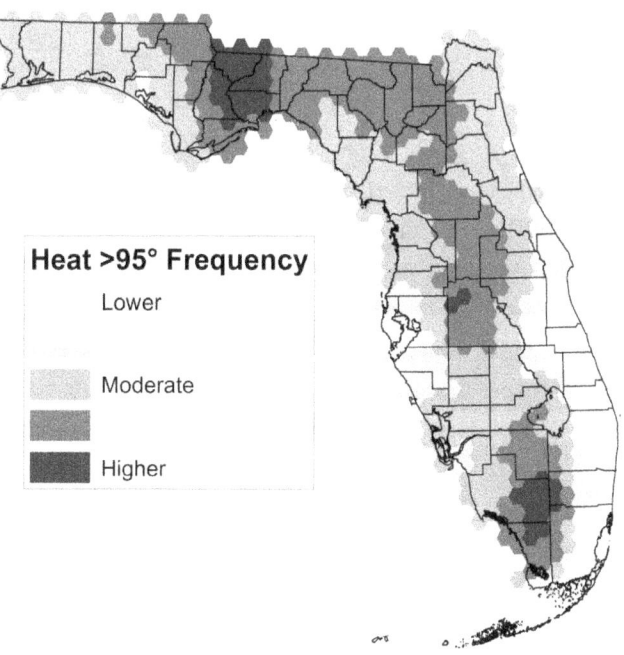

Figure 20.18. Frequency of temperatures above 95° F.

Temperatures and Climate Variability

Arctic outbreaks that create major freezes in Florida occur roughly every 10-20 years mainly during neutral ENSO or weak La Niña conditions, but they have also been observed in years with a positive ENSO index (Martsolf 2001; Gato-Maeda et al. 2008). A stronger connection exists between extreme freezes in Florida and the climate over the North Atlantic, as represented by the Arctic Oscillation (AO) or NAO (e.g. Wang et al. 2010). If those oscillations are negative, it means that there is higher than normal pressure over the North Atlantic and a lower than normal pressure in the equatorial Atlantic. This results in more storms affecting the northern part of Florida and increases the chances that polar air will be transported south. It has been shown, for example, that there is a 50% chance of a freeze when the AO index has extreme negative values, but close to a 0% chance in periods with extreme positive AO values (Hagemeyer 2007).

Heat waves and climate variability have been studied in Florida. Keellings and Waylen (2015) noted that the models indicate heat wave impacts are geographically opposing, resulting from warm or cool phases of ENSO. More intense heat wave events (in terms of increased magnitude, frequency, duration, and earlier timing) are brought to South Florida during the warm phase but at the same time there are diminishing events in the north. Alternatively, the cool phase ENSO amplifies heat wave events in North Florida while diminishing events in the south. Keellings and Waylen (2015) further noted that the warm phase of the AMO brings heat waves earlier in the summertime while also increasing their magnitude, frequency, and duration. This is consistent with other studies.

Outlooks into the Future

Attributing climate extremes and developing robust projections for the future remains a formidable challenge. And although models tend to agree (at least on the direction of expected changes) for near-term projections, the signals themselves are often small compared to the internal variability in the system. For longer time scales, different approaches and models may lead to more contrasting results, even in the direction of changes (IPCC 2012). One complication exists due to the fact that changes may be driven by competing factors. For example, while a warming of sea surface temperature favors the genesis of cyclones, an increase in the vertical wind shear in a warmer climate may have the opposite effect (although it may be minor) (Bruyère et al. 2012; Vecchi and Soden, 2007). However, a consensus has been reached among scientists, based on results derived with different methods, that the frequency of tropical cyclones will decrease in a warmer climate but that the intensity will increase (Bender et al. 2010; Knutsen et al. 2010), globally and in the Atlantic Basin. Uncertainties in regional projections of changes in tropical cyclone activity are even larger than those for global (average) projections (Christensen et al. 2013). For the southeastern U.S., Grinsted et al. (2013) used long tide gauge records as proxies for hurricane activity and applied a statistical downscaling approach that linked hurricane storm surge activity to changes in global sea surface temperature. They reported that a 1° C increase in global sea surface temperature could lead to a significant increase (two- to sevenfold) in the frequency of Katrina-like hurricane events. Although uncertainties in projections of tropical cyclone activity remain large, it is very likely and expected that continuous and potentially accelerating global (and regional) mean sea level rise will increase the storm surge potential along major coastline stretches, including Florida's, by shifting the base water level for storm surges upwards (Hunter 2012; Hunter et al. 2013).

Future changes in precipitation extremes during the summer season are closely linked to changes in tropical cyclones (Misra et al. 2011), as the latter contributes significantly to the number of extreme rainfall events observed throughout the state. For a high greenhouse gas emission scenario (Representative Concentration Pathway (RCP) 8.5) and the target year 2050, Gao et al. (2012) found an increase of ~20% (302 mm/yr to 365 mm/yr) in the annual total of daily extreme precipitation, but smaller changes (~5% increase) in the frequency of daily extreme events.

In addition to an increase in the mean global temperature, climate models also point to an increase in extreme heat wave events (combined high temperature and humidity) in the state of Florida, where 10 cities with the greatest expected increase in the number of dangerous heat days (i.e., heat index that combines temperature and humidity exceeds 104°F) by 2050 are located; as many as 130 dangerous heat days could occur by 2050, compared to 25 under the present climate (Climate Central 2017). Later in the century (2080 to 2100), extreme heat events that currently have a 5% probability of occurrence in any given year are projected to have a 50% to 100% chance of occurrence (Melilo et al. 2014).

Romps et al. (2014) suggested that the lightning flash rate is proportional to the convective available potential energy (CAPE) times the precipitation rate and, when applied to 11 climate models, CONUS lightning strikes were predicted to increase 12 ± 5% per degree Celsius of warming. Using 2003-2012 cloud-to-ground lighting data across the U.S. from the National Lightning Detection Network (NLDN), Koshak et al. (2015) found a 12.8% downward trend in total cloud-to-ground lightning count for the 10-year period, although the authors found a slow upward trend in the number of positive-polarity flashes. This was during a time when temperatures were warming but moisture trended downward.

Conclusion

This chapter describes various types of climate-sensitive extreme weather events relevant to the state of Florida. For each type of extreme event, the atmospheric/oceanographic conditions for its formation are summarized, the spatial variability of the potential for the different hazard types is depicted, and the links to large-scale climate variability are discussed.

Potential future changes in (selected) climate extremes in Florida are also briefly discussed, highlighting that large uncertainties continue to exist in the projections of such events. For most types of hazards, especially when focusing on individual events, it is challenging to attribute changes to anthropogenic influences given the strong variability in extremes in combination with strong seasonal to multi-decadal fluctuations compared to the small underlying low-frequency signal (IPCC 2012). In general, there is more confidence in the attribution of observed changes in the mean climate, which can in turn affect the risk imposed by certain extreme events (e.g. mean sea level rise shifting the base water level for storm surges upwards). Two distinct methods can be used for the attribution. The first one relies on observational records to determine potential changes in the frequency or magnitude of extremes. For this kind of analysis, the record timescales need to be long enough to account for a robust separation of natural variability and longer-term trends associated with anthropogenic influences. This is often not the case and hampers attribution. The second approach uses models that can be run over long time periods, with and without assuming a world that is affected by human-caused climate change. This is done by ignoring or including the changing concentration of greenhouse gases and aerosols in the atmosphere and comparing the results from unforced (control) runs and forced runs. In this approach, uncertainties can be large because some extremes are more challenging to model than others and may not be captured very well in the simulations, (e.g., due to insufficient spatial resolution). Often, a combination of both methods is required (i.e., validation of the models' ability to reproduce observations) and, when combined with sound physical principles, leads to the most robust results (NASEM 2016).

Extreme weather will continue to impact Florida in the future. It is the rapid changes in climate that can have a devastating effect on life on Earth, as humans contribute to a warming climate. For Florida, this could mean prolonged drought at one end of the state and flooding rains at the other end. It could also be an increase in devastating tropical cyclones. As sea levels rise, human habitations in low-lying coastal areas will become repeatedly inundated with these extreme weather events, eventually resulting in mass migrations to higher ground.

Acknowledgements

Hail, tornado, and wind data were collected from http://www.spc.noaa.gov/wcm/test.html. The authors would like to thank Maria Bower at the University of Central Florida for assistance in the literature review for some extremes and proof reading, and Amy Polen from University of South Florida for converting the references in the required style and proofreading.

References

Bender, Morris A., Thomas R. Knuston, Robert E. Tuleya, Joseph J. Sirutis, Gabriel A. Vecchi, Stephen T. Garner, and Isaac M. Held. 2010. "Modeled Impact of Anthropogenic Warming on the Frequency of Intense Atlantic Hurricanes." *Science* 327 (5964): 454–58.

Brolley, Justin, James O'Brien, Jutin Schoof, and David Zierden. 2007. "Experimental Drought Threat Forecast for Florida." *Agriculture and Forest Meteorology* 145 (1–2): 84–96.

Bruyère, Cindy L., Greg J. Holland, and Erin Towler. 2012. "Investigating the Use of a Genesis Potential Index for Tropical Cyclones in the North Atlantic Basin." *Journal of Climate* 25 (24): 8611–26.

Christensen, Jens H., Krishna Kumar Kanikicharla. 2013. "Climate Phenomena and Their Relevance for Future Regional Climate Change." *Climate Change 2013: The Physical Science Basis. Contribution of Working Group I to the Fifth Assessment Report of the Intergovernmental Panel on Climate Change.*

Citrus Mutual. 2016. "Timeline of Major Florida Freezes." http://flcitrusmutual.com/render.aspx?p=/industry-issues/weather/freeze_timeline.aspx.

Collins, Jennifer M. and David R. Roache. 2010. "The Inactive 2009 Hurricane Season in the North Atlantic Basin: An Analysis of Environmental Conditions. *Nat. Wea. Dig.* 34 (2): 117-128.

Collins, Jennifer M., Alicia N. Williams, Charles H. Paxton, Richard J. Davis, and Nicholas M. Petro. 2009. "Geographical, Meteorological, and Climatological Conditions Surrounding the 2008 Interstate-4 Disaster in Florida." *Paper of the Applied Geography Conferences* 32: 153–62.

Curtis, Scott. 2008. "The Atlantic Multidecadal Oscillation and Extreme Daily Precipitation over the US and Mexico during the Hurricane Season." *Climate Dyn* 30: 343.

Demuth, Julie L., Mark DeMaria, and John A. Knaff. 2006. "Improvement of Advanced Microwave Sounder Unit Tropical Cyclone Intensity and Size Estimation Algorithms." *J. Appl. Meteor.* 45: 1573–81.

Douglas, Arthur V., and Phillip J. Englehart. 1981. "On a Statistical Relationship between Autumn Rainfall in the Central Equatorial Pacific and Subsequent Winter Precipitation in Florida." *Mon. Wea. Rev.* 109: 2377–82.

Elsner, James B. 2003. "Tracking Hurricanes." *Bull. Am. Meteorol. Soc.* 84 (3): 353–56.

Elsner, James B., Brian H. Bossak, and Xu-Feng Niu. 2001. "Secular Changes to the ENSO-U.S. Hurricane Relationship." *Geophys. Res. Lett.* 28: 4123–26.

Emanuel, Kerry, Ragoth Sundararajan, and John Williams. 2008. "Hurricanes and Global Warming, Results from Downscaling IPCC AR4 Simulations." *Bull. Am. Meteorol. Soc.* 89 (3): 347–67. doi:10.1175/BAMS-89-3-347.

Enfield, David B., Alberto M. Mestas-Nunez, and Paul J. Trimble. 2001. "The Atlantic Multidecadal Oscillation and Its Relation to Rainfall and River Flow over the U.S." *Geophys. Res. Lett.* 28: 2077–80.

Gao, Yang, Joshua S. Fu, J. B. Drake, Yamg Liu, and Jean-Francois Lamarque. 2012. "Projected Changes of Extreme Weather Events in the Eastern United States Based on a High Resolution Climate Modeling System." *Environ Res Lett* 7 (4): 44025.

Goto-Maeda, Y, D. W. Shin, and J. O'Brien. 2008. "Freeze Probability of Florida in a Regional Climate Model and Climate Indices." *Geophys. Res. Lett.* 35 (11). doi:10.1029/2008GL033720.

Gray, Willaim M. 1979. "Hurricanes: Their Formation, Structure and Likely Role in the Tropical Circulation." *Royal Meteorological Society*, 155–218.

Grazulis, Thomas P. 1993. "Significant Tornadoes: 1680-1991." In *Environmental Films*, 1326pp. St. Johnsbury, VT.

Grinsted, Aslak, John C. Moore, and Svetlana Jevrejeva. 2013. "Projected Atlantic Hurricane Surge Threat from Rising Temperatures." *Proc. Natl. Acad. Sci. USA* 110: 5369–73. doi:10.1073/pnas.1209980110.

Hagemeyer, Bartlett C. 2007. "The Relationship between ENSO, PNA, and AO/NAO and Extreme Storminess, Rainfall, and Temperature Variability during the Florida Dry Season: Thoughts on Predictability and Attribution, Preprints, 19th Conference on Climate Variability and Change." *American Meteorological Society.*
http://www.weather.gov/media/mlb/research/19th_climate_JP2_16.pdf.

Hagemeyer, Bartlett C., and Scott M. Spratt. 2002. "Thirty Years after Hurricane Agnes—The Forgotten Florida Tornado Disaster." In *25th Conf. on Hurricanes and Tropical Meteorology*, 422–23. San Diego, CA: Amer. Meteor. Soc.

Hodanish, Stephen, David Sharp, Waylon Collins, Charles Paxton, and Richard E. Orville, 1997: "A 10-yr monthly lightning climatology of Florida": 1986–95. *Wea. Forecasting*, 12, 439–448.

Hunter, John. 2012. "A Simple Technique for Estimating an Allowance for Uncertain Sea-Level Rise." *Climate Change* 13: 239–52.

Hunter, J., J.A. Church, N.J. White, and Xuebin Zhang. 2013. "Towards a Global Regionally Varying Allowance for Sea-Level Rise." *Ocean Engineering* 71: 17–27.

IPCC. 2012. "Managing the Risks of Extreme Events and Disasters to Advance Climate Change Adaptation. A Special Report of Working Groups I and II of the Intergovernmental Panel on Climate Change."

Keellings, David and Peter Waylen. 2015. "Investigating Teleconnection Drivers of Bivariate Heat Waves in Florida Using Extreme Value Analysis." *Climate Dyn* 44 (11): 3383–91. doi:10.1007/s00382-014-2345-8.

Kennedy, Andrew J., Melissa L. Griffin, Steven L. Morey, Shawn R. Smith, and James O'Brien. 2007. "Effects of El Niño–Southern Oscillation on Sea Level Anomalies along the Gulf of Mexico Coast." *J. Geophys. Res.* 112 (C05). doi:10.1029/2006JC003904.

Knuston, Thomas R., John McBride, Johnny Chan, Kerry Emanuel, and Greg Holland. 2010. "Tropical Cyclones and Climate Change." *Nat. Geosci* 3 (3): 157–63. doi:10.1038/ngeo779.

Koshak, William J., Kenneth L. Cummins, Dennis E. Buechler, Brian Vant-Hull, Richard J. Blakeslee, Earle R. Williams, and Harold S. Peterson. 2015. "Variability of CONUS Lightning in 2003–12 and Associated Impacts." *Journal of Applied Meteorology and Climatology* 54: 15–41. doi:http://dx.doi.org/10.1175/JAMC-D-14-0072.1.

LaJoie, Mark, and Arlene Laing. 2008. "The Influence of the El Niño-Southern Oscillation on Cloud-to-Ground Lightning Activity along the Gulf Coast. Part I: Lightning Climatology." *Monthly Weather Review* 136: 2523–42.

Martsolf, J. David. 2001. "Relationship between Florida Freeze Incidence and Sea Surface Temperature in the Tropical Pacific." *Proc. Fla. State. Hortic. Soc.* 114: 19–21.

McCabe, Gregory J., Michael A. Palecki, and Julio L. Betancount. 2004. "Pacific and Atlantic Ocean Influences on Multi-Decadal Drought Frequency in the United States." *Proc. Natl. Acad. Sci. USA* 101: 4136–41.

Melilo, Jerry M., Terese C. Richmond, and Gary W. Yohe. 2014. "Climate Change Impacts in the United States: The Third National Climate Assessment." *U.S. Global Change Research Program*, 841 pp. doi:10.7930/J0Z31WJ2.

Mesonet. 2015. "Iowa State Mesonet System, Search for Storm Based Warnings." https://mesonet.agron.iastate.edu/vtec/search.php.

Misra, Vasubandhu, Elwood Carlson, Robin K. Craig, David Enfield, Benjamin Kirtman, William Landing, Sang-Ki. Lee, David Letson, Frank Marks, Jayantha Obeysekera, Mark Powell, and Sang-Ik Shin . 2011. "Climate Scenarios: A Florida-Centric View." *Florida Climate Change Task Force*.

National Academies of Sciences, Engineering and Medicine. 2016. "Attribution of Extreme Weather Events in the Context of Climate Change." *Washington, DC: The National Academies Press*. doi:10.17226/21852.

Park, Joseph C., Jayantha Obeysekera, and Jenifer Barnes. 2010. "Temporal Energy Partitions of Florida Extreme Sea Level Events as a Function of Atlantic Multidecadal Oscillation." *Ocean Science* 6: 587–93.

Park, Joseph C., Jayantha Obeysekera, Michelle Irizarry, Jenifer Barnes, and Winifred Park-Said. 2010. "Climate Links and Variability of Extreme Sea Level Events at Key West, Pensacola, and Mayport Florida." *J. Wtrwy, Port, Coast, Oc Engrg* 136: 350–56. doi:10.1061/(ASCE)WW.1943-5460.0000052.

Romps, David M., Joseph T. Seeley, David Vollaro, and John Molinari. 2014. "Projected Increase in Lightning Strikes in the United States due to Global Warming." *Science* 346 (6211): 851–54.

Sátoria, Gabriella, E. Williams, and I. Lempergera. 2009. "Variability of Global Lightning Activity on the ENSO Time Scale." *Atmospheric Research* 91 (2–4): 500–507. doi:http://dx.doi.org/10.1016/j.atmosres.2008.06.014.

Schubert, Siegfried, David Gutzler, Hailan Wang, Aiguo Dai, Tom Delworthe, Clara Deserd, Kirsten Findelle, Rong Fuf, Wayne Higginsg, Martin Hoerlingh, Ben Kirtmani, Randal Kostera, Arun Kumarg, David Leglerj, Dennis Lettenmaierk, Bradfield Lyonl, Victor Maganam, Kingtse Mog, Sumant Nigamn, Philip Pegiong, Adam Phillipsd, Roger Pulwartyo, David Rindp, Alfredo Ruiz-Barradasn, Jae Schemmg, Richard Seagerq, Ronald Stewartr, Max Suareza, Jozef Syktuss, Mingfang Tingq, Chunzai Wangt, Scott Weavera,c, and Ning Zengn. 2009. "A US CLIVAR Project to Assess and Compare the Responses of Global Climate Models to Drought-Related SST Forcing Patterns: Overview and Results." *J. Climate* 22: 5251–72.

Seager, Richard, Alexandria Tzanova, and Jennifer Nakamura. 2009. "Drought in the Southeastern United States: Causes, Variability over the Last Millennium, and the Potential for Future Hydroclimate Change." *Journal of Climate* 22: 5021–45. doi:http://dx.doi.org/10.1175/2009JCLI2683.1.

SWRA. 2016. "Southern Wildfire Risk Assessment." Southern Group of State Foresters Wildfire Risk Assessment Portal. Accessed February 12, 2017. https://www.southernwildfirerisk.com/

"States at Risk Project." 2017. *Climate Central*. Accessed February 1. statesatrisk.org.

"Storm Events Database." 2016. *NOAA*. https://www.ncdc.noaa.gov/stormevents/.
"Storm Prediction Center." 2015. http://www.spc.noaa.gov/wcm/test.html.
Teegavarapu, Ramesh S. V., Aneesh Goly, and Jayantha Obeysekera. 2013. "Influences of Atlantic Multidecadal Oscillation Phases on Spatial and Temporal Variability of Regional Precipitation Extremes." *J. Hydrol* 495: 74–93. doi:10.1016/j.jhydrol.2013.05.003.
Vecchi, Gabriel A., and Brian J. Soden. 2007. "Increased Tropical Atlantic Wind Shear in Model Projections of Global Warming." *Geophys. Res. Lett.* 34 (L08702). doi:10.1029/2006GL028905.
Wahl, Thomas, and Don P. Chambers. 2015. "Evidence for Multi-Decadal Variability in US Extreme Sea Level Records." *Journal of Geophysical Research Oceans* 120: 1527–44. doi:10.1002/2014JC010443.
Wahl, Thomas, and Don P. Chambers. 2016. "Climate Controls Multi-Decadal Variability in U.S. Extreme Sea Level Records." *Journal of Geophysical Research Oceans* 121. doi:10.1002/2015JC011057.
Wang, Chunzai, Hailong Liu, and Sang-Ki Lee. 2010. "The Record-Breaking Cold Temperatures during the Winter of 2009/2010 in the Northern Hemisphere." *Atmosph. Sci. Lett.* 11: 161–68. doi:10.1002/asl.278.
"Web-Based Risk Assessment Planning Project for the State of Florida." 2015. *FDEM*. http://artsandsciences.sc.edu/geog/hvri/web-based-risk-assessment-planning-project-state-florida#overlay-context=current-projects.
"Wildlife Risk Assessment Portal." 2016. *Florida Forest Service*. https://www.southernwildfirerisk.com/.
Zogg, Jeffrey, and Kevin Deitsch. 2013. "The Flash Flood Potential Index at WFO Des Moines."

INDEX

A

adaptation 2, 4-5, 36-39, 43, 112-113, 118, 125, 165, 204-205, 209, 213, 217-218, 227, 229, 233, 235-236, 239, 249, 252-253, 256-259, 270, 286, 297, 303-307, 311-312, 322-324, 326-327, 329, 334, 339-340, 349, 352, 355, 358-359, 367, 371-376, 427-428, 430-433, 438-441, 444-447, 529, 533, 537-538, 540-541, 551, 557
Aedes aegypti mosquito 140-142
aerosols 464, 501, 522, 535, 539, 612
agriculture 12, 35, 40, 51-52, 57, 70-72, 77, 106, 111, 126, 190, 219, 235-237, 239-241, 248-249, 252-253, 258, 260, 315, 317, 362, 493, 502
AgroClimate 257, 617
air conditioning 8, 35, 54, 161-163, 165-167, 181, 256
air quality 11, 132, 143, 162, 194, 212, 215, 235, 271
air-sea interaction 391, 419, 520
Apalachicola River 83, 86, 107, 109-110, 119-120, 396-397
aquaculture 77, 84, 240, 362, 366, 427-433, 445-447
aquifers 62, 84, 87, 94, 104, 106-107, 109, 112-114, 117, 119, 249-250, 279-280, 311-312, 314, 316-317, 319, 465, 476, 478, 563
Atlantic Meridional Overturning Circulation (AMOC) 391-394, 399-400, 404-405, 414, 416, 419, 500
Atlantic Multi-decadal Oscillation (AMO) 87-88, 119, 391, 398, 419, 485, 494, 498-501, 505, 511, 517-520, 522, 534, 584, 586, 607, 610
Atlantic Warm Pool (AWP) 485, 492, 495, 497-499, 505, 607

B

bacteria 114, 133, 135, 137, 139, 142-143
beaches 35, 38, 42, 63, 209, 215, 218, 221, 224-227, 229, 298, 301-303, 306, 319, 327, 342, 347-348, 355-356, 391, 404, 406, 410-413, 462, 472, 474, 558
big data 297, 306
biodiversity 51-54, 61-62, 71, 77-78, 109, 251, 258, 271, 339-340, 342-344, 349, 353, 359-361, 364-367, 369-376, 408-409, 411, 435
Building Resilience Against Climate Effects (BRACE) project 127, 129, 131-132, 135, 143

C

carbon dioxide (CO_2) 9-11, 236, 244, 280, 345, 457, 466-467, 480, 565
carbon emissions 2, 11, 13-14, 16, 284-285
carbon sequestration 11, 269, 271, 284-288
Caribbean climate 392, 399
catastrophes 179-182, 203
Citizens Property Insurance Corporation 186-187, 190, 196-198, 205, 210
climate models 83, 87-88, 96, 119, 130, 191, 271, 325, 391, 399-401, 404, 414-415, 517, 523-524, 527, 535, 538-539, 547-548, 566, 571-572, 574, 579, 611
climate predictions 278, 282, 511-512, 520-521, 524, 533-536, 538, 547, 551-552
climate projections 83, 93, 119, 271-273, 304, 391, 399, 401, 533-536, 538, 541-542, 547-548, 551-552
climate scenarios 90, 95, 140, 232, 249, 257, 276, 279, 400, 533, 535, 538-540
climate variability 93, 191, 239, 248, 252-253, 258, 260, 373, 375, 398, 401, 505, 511-512, 515-517, 520-522, 528-529, 533-537, 541-542, 552, 554, 579-580, 584, 586, 588, 591, 594, 596, 599, 607, 610-611
coastal areas 36, 66, 84, 90, 106, 119, 126, 180-181, 214-216, 231, 297, 302-303, 307, 311, 317, 327-328, 347-348, 360, 412, 582, 595, 612
communication 1-2, 5-7, 16, 112, 284, 307, 334
conservation 2, 12-15, 35, 37, 39, 43, 51-52, 62-65, 69, 72, 76, 153, 205-206, 211-212, 287-288, 318, 328, 351, 353, 367-376, 410, 431-432
conservation tillage 236, 252-253
coral reefs 7, 14, 212, 342, 345, 349, 354-355, 361, 391-392, 404, 406-411, 413, 415-416, 428, 433-434, 436-438, 440
cover crop 236, 252-253, 257-258
crop modeling 236, 243-244, 258-259
crops 35, 52, 54, 57, 74, 194, 235-238, 240-259, 284, 551, 593, 608
currents (ocean) 352, 391, 393-396, 398, 400, 404, 408-410, 413, 415-417, 419, 434-435, 471, 485, 551, 569, 602

D

Dania Beach 326-331
Davie (City of) 326, 331-332
decision support systems 236, 243-244, 252, 257, 370
development 1, 3-4, 8-12, 14-15, 37-39, 41-42, 51-57, 59-60, 62-63, 68, 72, 76-78, 112,

129, 133, 139, 145, 158, 184-185, 201, 206, 212-214, 216, 218, 220, 222, 224, 287, 303, 311-312, 314-315, 317, 322-323, 325, 328-329, 334, 344, 347, 360, 371, 396, 413, 434
disaster risk 179-180
disease 54, 126, 128-130, 132-134, 137, 139-143, 145-146, 235-237, 245, 257, 259, 270, 273, 277-278, 280, 286, 345, 351, 354, 361, 363-364, 366, 415
disturbance 270-271, 275-276, 284, 287, 344, 358, 366, 374, 600
diurnal variations 485-486, 488-490, 501, 505, 527, 547
downscaling 84, 88, 95, 110, 401, 415, 511, 524-529, 540, 547, 610
drought 36, 86, 94, 104, 106, 112, 114, 117, 125-126, 134, 143-144, 191, 216, 229, 233, 236, 238-239, 244, 251-257, 259, 275-276, 279-283, 303, 305, 314, 346-347, 366, 399, 473, 478, 499, 521, 551, 579-580, 584-586, 588, 612

E

ecology 65, 339, 343, 359
economics 2-5, 7-9, 16, 22, 26, 35-36, 38, 42, 59, 63, 83, 125, 145, 153, 155, 160-162, 164, 166-167, 174, 179-180, 182, 184-185, 200, 203-204, 212, 224, 230, 235-237, 239, 248, 252, 255, 259-260, 269-272, 275-276, 278, 281, 283, 286-288, 297-298, 302-305, 307, 311, 315, 317, 320, 322, 327-328, 336, 355, 362, 375, 407-408, 410, 413, 427-434, 438-439, 443-444, 447, 470-471, 480, 523, 534, 541, 557-558
ecosystems 12, 14, 36, 39, 51, 53, 61, 66-69, 71, 76-77, 85-87, 89-91, 94, 110, 113, 119-120, 125-126, 257-258, 269-271, 275-276, 280, 284-288, 302, 318, 328, 339-340, 342-347, 353-354, 359-361, 364-367, 370, 372-376, 400, 407, 410-414, 416-417, 419, 428, 430, 433, 435-436, 441, 443, 490, 537-538, 540, 551
ecosystem services 12, 14, 71, 76, 269-271, 276, 285-288, 407, 410-412, 414
education 3, 23, 25-26, 35, 38, 211, 306, 313, 334
El Niño-Southern Oscillation (ENSO) 87-88, 113, 119, 239, 257, 277, 391, 398, 415, 419, 473, 485, 491-495, 498-499, 505, 511-516, 520-522, 527-528, 534, 579, 584, 586, 588, 591, 594, 599, 607, 610
emergency management 16-17, 40, 43, 188, 219, 328, 580, 582, 588

employment 3, 8, 11, 16, 23, 25, 63, 214, 235, 237, 269-270, 297, 302, 307, 370, 374, 428-430
energy 1, 4, 9-14, 16, 21, 35, 37, 43, 96, 131, 153-156, 159-175, 203-204, 209-212, 214, 218-219, 245, 257, 259-260, 269, 284-285, 318, 327, 352, 373, 393, 395, 408, 411, 469, 471, 490, 501, 534, 539, 541, 564, 590, 601, 611
  coal 12, 14, 153-157, 167-168, 170, 174-175
  natural gas 12-14, 153-156, 158, 167, 170-171, 173-175, 534
  oil 7, 153-155, 157, 159, 165, 167-168, 174-175, 259
  renewable energy 13-14, 153-154, 159, 163-164, 167-172, 174-175, 204, 210
  solar 12-14, 16, 35, 70, 153-155, 164, 169-175, 237, 249, 396, 502, 590
energy efficiency 13, 153, 160-161, 163-164, 167, 170, 174-175, 210, 212
environmental hazards 22, 125-127, 130
Everglades 53-54, 56, 60-62, 71, 83, 85-87, 89-91, 93, 99, 118-120, 141, 217, 312, 314, 318, 331, 339, 342, 345-346, 354, 375, 412, 435, 471, 480, 581-582, 588
extreme events 21, 128, 130-131, 191, 204, 238-239, 251, 280, 297, 304, 320, 375, 567, 569, 579-580, 611-612
extreme heat 2, 125-129, 216, 611

F

fisheries 7, 366, 391, 407-410, 414, 416-417, 419, 427-445, 447-448
flooding 2, 17, 21, 35-36, 38-39, 42, 57, 60, 62, 83, 85, 90, 93, 103-105, 107-108, 114, 117-119, 126-128, 130-131, 134, 139-140, 144, 146, 154, 179, 181, 186-193, 195, 198, 200-201, 204, 209-211, 213-218, 220, 223, 230-234, 238-239, 245, 251, 255, 259, 280, 282, 303, 311-313, 316-320, 323-326, 328, 330-335, 348, 360-362, 365, 407, 441, 504, 551, 557, 567-568, 579-584, 596, 601, 604, 607, 612
Florida Hurricane Catastrophe Fund (FHCF) 186-187, 196-200, 205
Florida Keys 55, 59, 69, 103, 185, 311, 316, 339, 342, 347, 349, 354, 392, 396, 407, 438, 471-472, 480, 580-581, 602
Florida Panhandle 12, 55, 103, 109-110, 311, 317, 350, 473, 485, 491-492, 505, 579, 581, 584, 589, 591, 603
Florida platform 457, 460-465, 468, 470-472, 474, 479-481
forestry 9, 11-12, 14, 25, 52-53, 55, 57, 60-63, 66-71, 74-75, 93, 107, 109, 111, 168, 190,

269-273, 275-281, 283-288, 328, 334, 345-347, 364, 366, 411-412, 427, 463, 472-473, 586, 588, 590, 597
fossil fuels (see: energy)
freezes 53, 57-58, 67-68, 70, 239, 245, 247, 255, 258, 363, 540-541, 551, 589, 592, 608, 610
freshwater 53, 68-69, 71, 74-75, 84-85, 106, 112, 114, 118, 135, 142, 248-250, 316-317, 319, 339, 342, 344-346, 349-350, 352-354, 366, 396, 404, 412, 427-430, 432-433, 436, 441-446, 460, 472, 476, 486, 561
fruits 235-237, 240, 244-247, 256, 258-259, 361, 472, 597
Ft. Lauderdale 55, 217, 326, 517

G

General Circulation Model (GCM) 84, 87-88, 94-99, 101-102, 110-112, 539-541, 547
government 3-4, 9, 14-16, 21-22, 35, 37-43, 55, 59-60, 62-63, 160, 164, 181, 185, 187-188, 199, 205, 209-220, 222-224, 227-228, 231-233, 260, 297, 302-303, 316, 320, 327, 335-336
greenhouse gases 1, 3-4, 9-11, 13, 35, 95, 99, 153, 155, 157-158, 211-212, 218, 228-230, 284, 286, 302, 304, 326, 399-401, 412, 535, 538-539, 541-542, 557, 564, 571, 611-612
groundwater 7, 38, 69, 83-84, 91, 93-94, 98, 104, 106-107, 109, 112-114, 119, 237, 249-251, 254, 311-312, 316, 318-321, 323, 325, 327, 331, 346, 366, 432-433, 457, 476, 478, 481, 585
Gulf of Mexico 21, 66, 93-94, 103, 107, 110, 145, 158-159, 175, 301, 303, 334, 355, 392-394, 398, 400-404, 407-410, 412, 415, 417-419, 430, 432-433, 435-436, 439, 457, 460, 464-465, 472-474, 495, 499, 505, 527, 569, 582, 590, 601-604, 607-608

H

habitat 54, 56, 58, 61, 63, 68-69, 107, 140, 211, 215, 269, 271, 276, 285, 328, 339-340, 342-343, 346-356, 359-363, 366, 371-372, 376, 404, 407-409, 412-414, 417, 427-428, 431, 434, 436-444, 457, 476, 481
hail 579-580, 590, 592-594, 612
harmful algal blooms 35, 135, 365, 391, 414, 417, 419, 427, 432, 445-446
human health 11, 113, 125-131, 135, 137, 139, 141, 143-146, 320, 414, 417, 445
hurricanes (see: tropical cyclones (including hurricanes))

hydrology 60, 69, 279, 282, 325, 344, 346, 361, 376, 392, 431-432, 443, 541, 547, 563

I

ice melt 106, 467, 480, 557, 563, 571, 573
ice sheets 17, 157-159, 250, 457, 462, 466-469, 472-473, 480, 559, 561, 563-565, 572, 574-575
infrastructure 1, 3, 11, 16, 21-22, 25, 35-36, 38-39, 42-43, 54, 56, 59, 67, 76, 91, 94, 130, 145, 153, 188, 205, 209-210, 212-213, 215, 218-220, 222-224, 232, 303, 311-312, 314, 316, 318-320, 322-328, 331, 334-336, 360-361, 365, 370, 407, 412, 414, 428, 431-432, 438, 440, 448, 608
inland migration 7, 37, 39-40, 67
insects 133, 146, 255, 270, 273, 285, 357
insurance 22, 25, 39, 42, 103, 179-207, 210, 214, 227, 230, 328, 336, 582
  insurability 179, 195, 201, 203, 206
  reinsurance 179, 182-185, 190, 192, 196-198, 202-203, 206
Intergovernmental Panel on Climate Change (IPCC) 2, 36, 87, 95, 110, 114, 118, 126, 190-191, 195, 250, 254, 271, 303, 326, 356, 359, 371, 398-399, 418, 467, 533-535, 538, 541, 543-546, 550, 554, 570-573, 575, 610, 612
invasive species 52, 57-58, 61, 67, 142, 270, 286-287, 340, 346, 352, 354, 358-359, 361-364, 433, 441, 447
irrigation 84, 107, 235-237, 240, 245, 248-250, 252, 254-255, 259, 313-318, 501, 503-505, 563

J

Jacksonville 13, 36, 53, 55-56, 61, 103, 311, 498, 596
jobs (see: employment)

L

Lake Okeechobee 17, 57, 60, 62, 86, 89, 91, 312, 317, 342, 412, 471, 499, 505, 582, 585-586, 590, 602
land acquisition 15, 39, 63, 328
land use/land cover 9-12, 37-39, 41, 51-54, 57-62, 66, 68-78, 93, 106-107, 110-111, 113, 210-212, 218, 231, 259, 277, 285-286, 305, 328, 347, 351, 360-362, 366, 368-370, 431, 433, 441, 443, 457, 485, 491, 501, 505, 534, 541
La Niña (see: El Niño-Southern Oscillation)

Last Glacial Maximum 457, 468-469, 473-476, 478, 481, 557, 564
law 13, 40, 58, 194, 199, 209-210, 212-215, 223-224, 226-227, 372
lightning 125-127, 130-131, 464, 478, 579-580, 586, 590-592, 611
livestock 57, 77, 235-236, 240-241, 250-251, 256-260, 608

M

mangroves 14, 39, 53, 59, 67-69, 74-75, 327-328, 334, 346, 348-349, 353, 376, 407, 411-412, 434, 436
Miami 53, 55-56, 154, 156, 217, 444, 474, 517, 596, 604
Miami Beach 35, 59, 217, 303, 326, 567
Miami-Dade 21, 23, 29, 31-32, 35, 38, 42-43, 58, 145, 250, 320-321, 326, 583
mining 51-52, 55, 58, 77, 470
mitigation 1-2, 4-5, 13-15, 22, 39-41, 113, 125, 146, 161, 165, 175, 193, 201, 204-205, 214, 218, 258, 269, 271, 286, 288, 303, 305, 311, 322-328, 412-413, 418-419, 431-432, 499, 533, 537-538, 540-541, 551, 573, 582, 587

N

National Climate Assessment (NCA) 17, 220, 239, 570-573
National Flood Insurance Program (NFIP) 103, 186-187, 214, 227, 230-231, 328, 582
natural processes 51, 60, 228, 571
natural resources 14, 51-52, 55, 58-59, 62-63, 65, 112, 253, 270, 359, 371-374, 376, 551

O

ocean acidification 7, 280, 349, 354, 391, 416, 418, 427, 432, 436, 445-446
ocean climate 392, 398, 485
ocean warming 17, 391, 414-417, 419, 435, 557, 572
Orlando 24, 53-57, 237, 591, 593, 596, 603

P

Pacific Decadal Oscillation (PDO) 87-88, 485, 494, 505, 511, 515-516, 520-522, 534, 584
paleoclimate 457-458, 460-463, 465, 471, 475, 477, 481
Pensacola 53, 406-407, 412, 597, 603
phenology 112, 245, 278, 340, 352, 356-358, 435
planning 2, 5, 13-16, 37-41, 43, 51, 59-60, 77-78, 83, 85, 88, 90-93, 99, 112, 118, 154-155, 174, 180, 193, 200, 205, 209-219, 224, 228-229, 231, 233-234, 258, 281, 303, 305-307, 311, 313-314, 323, 325-328, 331, 334-336, 339-340, 351, 359, 367, 369-376, 441, 446-447, 542, 547, 551, 558, 573, 582, 587
policy 1, 4, 7, 11-16, 37-40, 164, 175, 184-185, 187, 189, 204-205, 209-212, 215-216, 218-221, 223-224, 227-230, 232-234, 252, 259, 269-271, 276, 285-288, 311, 327, 370, 372-373, 376, 445, 447, 540
population growth 8, 26, 51, 59, 70-71, 76-78, 83, 139, 302, 316, 361, 370, 433, 443-444
precipitation 37, 66, 68, 70, 83-84, 87-88, 90-94, 96, 98-108, 110-114, 117-119, 125-126, 130, 133-135, 139-140, 143, 145, 216, 220, 224, 233, 236-240, 245, 248-249, 251-254, 257-259, 269, 271-272, 274, 278-282, 288, 311, 313-318, 320, 323-325, 332, 335, 339, 345-347, 350-352, 356-357, 361-362, 364-366, 368, 398-399, 404, 431-432, 435-436, 441-442, 445, 447-448, 457-458, 460, 471, 475-476, 478-481, 486-490, 494-495, 497-501, 503, 505, 511-521, 523-524, 526-528, 540, 542, 545-551, 568, 579-586, 588, 596, 604, 607, 611
public health (see: human health)
Punta Gorda (City of) 334-335

R

rainfall (see: precipitation)
reefs (see: coral reefs)
resiliency 2, 15, 37, 68-69, 127, 175, 209, 212-213, 218-219, 230, 233-234, 249, 252, 258, 270, 275-277, 279, 284, 286-288, 305, 325-326, 339, 355, 361, 372-376, 416-417, 434, 529
risk assessment 40, 179, 189, 192-193, 203, 323-325, 522, 587
risk management 180-182, 185, 197, 204-205, 209, 230-231, 233
rivers 9, 53, 56, 58, 60, 63, 66, 71, 74, 83-86, 91, 93, 98, 101-105, 107, 109-110, 113, 115-117, 119-120, 130, 156, 158, 168, 174, 248, 279, 313, 316, 346, 350-351, 392, 395-398, 406-407, 409, 432, 435-436, 441, 443, 471, 480, 561, 566, 581

S

salinity 66-67, 69, 106, 111, 118, 236, 249-250, 259, 281, 319, 344-345, 348, 351-353, 365, 376, 396, 398, 400, 403-404, 407, 411, 414, 442, 445, 460
saltwater intrusion 2, 35, 38, 87, 90, 104, 107, 112, 216, 220, 236, 245, 249-250, 432, 442

satellite observations 157, 195, 227, 521, 557, 559-563, 607
scenario planning 339-340, 369-371
sea breeze 70, 464, 485-486, 490-491, 505, 527, 580, 589-591
sea level rise 2, 7, 16-19, 21, 38-39, 51, 66-69, 76-78, 84, 87, 90-93, 104, 108, 117, 125-127, 130, 145-146, 153-154, 157-159, 174, 193, 200-201, 204-205, 209-211, 213-220, 223, 225-229, 231, 233-236, 245, 249-250, 297, 303, 305, 311-312, 315-316, 318-323, 325-328, 330-331, 334-336, 339, 344-349, 351-355, 360, 364, 366, 368, 370-371, 375-376, 392, 404, 407, 410-414, 416, 419, 427-428, 431-432, 436, 438, 440, 442, 445, 448, 469, 504, 557-558, 560-568, 570-575, 611-612
seasonal cycle 359, 485-486, 497, 527, 567, 579
sea turtles 339, 342, 347-348, 351, 355-356, 363, 391, 407, 410, 413, 419
sod-based rotation 236, 252-254
Southeast Florida Regional Climate Change Compact 38, 90, 218-220, 326, 373
South Florida Water Management District 85, 118, 219, 314, 317, 331
St. Johns River region 86, 103-104, 107, 119-120
St. Petersburg 217
storm surge 21, 66-67, 104, 125-126, 145, 188-189, 200, 204, 213-214, 216, 297, 303, 311, 318, 334, 346-347, 349, 351, 354, 392, 407, 411, 567, 579, 581, 601-602, 604, 607, 611-612
streamflow 83, 95-96, 98-99, 101-102, 104, 109-110, 113, 279, 547, 586
Suwannee River 66, 83, 85-86, 107, 113, 115, 119

T

Tallahassee 58, 66, 130, 237, 607
Tampa Bay region 38, 54, 68, 83, 86, 93-96, 99, 119, 200-201, 219-220, 311, 317, 326, 547, 558, 571, 596, 602
temperature 7, 37, 66, 68, 70, 83-84, 87-88, 90-93, 96-99, 110, 113-114, 117-119, 125-126, 128-130, 133-135, 139-140, 142, 145, 162-163, 228, 235-239, 243, 245-253, 255-259, 272, 274, 278, 280-281, 288, 306, 311, 315, 320, 324-325, 339, 344-346, 348-350, 352-353, 355-357, 363-366, 368, 392, 395-398, 400, 402-404, 414-415, 418, 431-432, 437, 442, 445-447, 457-458, 460, 466-467, 478, 480, 486, 488-490, 493, 495, 499, 501, 503, 505, 511-513, 515-517, 520, 536-538, 540, 542-544, 547-548, 550-553, 564-565, 569, 572-573, 590, 600, 607-608, 610-611
thunderstorms 130, 457, 464, 480, 486, 490, 527, 579-580, 586, 589-592, 599
tide gauge 412, 557-559, 561-563, 565-571, 607, 610
toolbox of strategies 312, 325
tornadoes 191, 238, 579-580, 590, 595-599, 601, 603-604, 607, 612
tourism 3, 7, 35, 51-52, 54, 56, 126, 180, 184, 200, 209, 224, 226, 235, 297-298, 300-307, 315, 410, 414, 438, 486
transportation 1, 3, 9-12, 25, 35, 37, 51-52, 54-56, 153-155, 159-160, 167, 174-175, 219, 236, 297, 300-302, 304, 307, 311, 318-319, 325, 327, 336, 360-361
tropical cyclones (including hurricanes) 6, 21, 23, 35-36, 117, 125-126, 130-132, 135, 144-145, 179-180, 182, 184, 186-188, 190-191, 193, 197-199, 215, 221, 226, 233-234, 237-238, 273, 275-276, 280, 297, 303, 307, 312, 334, 347, 391, 398-400, 403, 413, 438, 442, 446, 485, 493-496, 499-500, 503-505, 517, 527-528, 568, 579-581, 584, 586, 595-596, 600-607, 610-612
Turkey Point Nuclear Generating Station 156-159, 173-174

U

uncertainties 6, 38, 83-84, 93, 112, 118-119, 125, 179, 189, 191-192, 233, 248, 276-277, 279, 281-282, 288, 297, 304, 306-307, 317, 320, 323, 326, 356, 368-373, 376, 401, 436, 501, 522-523, 525, 535, 537, 547, 554, 557, 560, 563, 570-571, 574, 579, 610-612
urban heat island 87, 501-502
urbanization 3, 8-11, 51, 59, 70, 84, 286, 486

V

vertical land motion 557, 561-562, 566-567, 569
vulnerability 2, 4, 14, 16-17, 22-34, 37-38, 41-43, 131, 195, 223, 229, 235, 252, 285, 297, 306, 311, 320-326, 328-329, 339-340, 347-348, 351-353, 359, 365, 367-369, 371-375, 413, 438-441, 505, 558
vulnerability assessment 37-38, 229, 323, 329, 340, 351, 353, 368-369, 371-375, 438, 440

W

water quality 68, 83-85, 104, 112, 114, 117-118, 120, 143, 223, 239, 271, 280-281, 285, 313, 317-318, 334, 352, 396, 407, 413, 447

water resources 14-15, 52, 62-63, 83-88, 90-91, 93, 99, 112-114, 119, 235-237, 239, 248-249, 259, 269-270, 279, 281, 286, 311, 313, 317, 366, 525, 547, 551
West Palm Beach 8, 217, 326, 602
wetlands 9, 11-12, 14, 51-52, 54, 60, 67, 70-72, 74-76, 84, 91, 93-94, 104, 106-107, 157, 216, 279-281, 312, 328, 339, 344-345, 349-350, 354, 367, 391, 411-414, 419, 433, 435-436, 473, 551, 561, 566
wildfire 36, 179, 216, 269-270, 276, 288, 347, 499, 580, 586-588
wildlife 52, 54, 56, 62-63, 67-68, 107, 269, 271, 276, 285, 287, 328, 342, 361, 365-366, 373, 442
wind 12, 131, 153, 164, 168, 170-172, 174-175, 185-186, 190, 192, 202-203, 205, 255, 283, 297, 303, 334, 344, 352, 391, 395, 399-400, 419, 435, 474, 491, 497, 499, 501, 503, 557, 560, 566, 569, 574, 579-580, 589-592, 595, 597, 599-601, 605-608, 610

Z

zoning 11, 39, 59, 201, 328, 582

www.ingramcontent.com/pod-product-compliance
Lightning Source LLC
Chambersburg PA
CBHW082318220526
45470CB00008B/2348